# Marine Conservation

Providing a comprehensive account of marine conservation, this book examines human use and abuse of the world's seas and oceans and their marine life and the various approaches to management and conservation. Healthy marine ecosystems – the goods and services that they provide – are of vital importance to human well-being. There is a pressing need for a global synthesis of marine conservation issues and approaches.

This book covers conservation issues pertinent to major groups of marine organisms, such as sharks, marine turtles, seabirds, and marine mammals; key habitats, from estuaries, wetlands, and Vcoral reefs to the deep sea; and local and regional to international initiatives in marine conservation.

An ideal resource for students, researchers, and conservation professionals, the book pays appropriate attention to the underlying marine biology and oceanography and how human activities impact marine ecosystems, enabling the reader to fully understand the context of conservation action and its rationale.

**P. Keith Probert**, now retired, was Associate Professor in the Department of Marine Science at the University of Otago, New Zealand, where he taught marine biology and ecology. His research has mainly concerned the ecology of sediment benthos, including the effects of human activities and the implications for marine conservation.

# Marine Conservation

P. KEITH PROBERT

Department of Marine Science
University of Otago

with an initial contribution from the late Norman A. Holme

**CAMBRIDGE**
**UNIVERSITY PRESS**

University Printing House, Cambridge CB2 8BS, United Kingdom

One Liberty Plaza, 20th Floor, New York, NY 10006, USA

477 Williamstown Road, Port Melbourne, VIC 3207, Australia

314-321, 3rd Floor, Plot 3, Splendor Forum, Jasola District Centre, New Delhi - 110025, India

79 Anson Road, #06-04/06, Singapore 079906

Cambridge University Press is part of the University of Cambridge.

It furthers the University's mission by disseminating knowledge in the pursuit of education, learning and research at the highest international levels of excellence.

www.cambridge.org
Information on this title: www.cambridge.org/9781108412629
DOI: 10.1017/9781139043588



First published 2017

*A catalogue record for this publication is available from the British Library*

ISBN 978-0-521-32685-8 Hardback
ISBN 978-1-108-41262-9 Paperback

For Margaret, Sarah, Philippa, and Matthew

and

Jan, Tom, and Will in memory of Norman Holme

Norman A. Holme ScD MA JP (1926–1989).
Photograph: David Nicholson.

Norman was for many years on the staff of the Marine Biological Association in Plymouth, UK, and is well known for his studies of seabed fauna, particularly of the English Channel, and for the Holme and McIntyre handbook on methods for the study of marine benthos. In later years, Norman became increasingly involved in marine conservation. He undertook surveys of British shores for the then Nature Conservancy Council, was very active in voluntary marine conservation organizations in South West England, and was the inspiration for the Wembury Marine Centre on the South Devon coast. Norman's plan to write a book on marine conservation was, however, sadly overtaken by his untimely death. This book is based largely on the outline that he had developed.

# CONTENTS

*Colour plates are to be found between pp. 276 and 277*

# PREFACE

This book owes its origin to marine biologist and conservationist Dr Norman Holme. Shortly before he died, Norman asked me if I would take on the book that he planned on marine conservation. As a result, he sent me his outline, notes, and materials that he had amassed for the book. I am very conscious of the long delay in bringing Norman's request to a completion due to other commitments that resulted in lapses and then the need to update and rewrite. Whilst much of the material that Norman intended to use has been superseded, I have largely retained the overall structure that he had developed. I only hope that at least in its approach and scope the book is not too dissimilar to what he had in mind.

The book is concerned with biological conservation in marine environments and covers an introduction to the marine environment, human impacts, and conservation; species-based approaches to conservation; major habitats, from coastal to deep sea; marine protected areas; and regional approaches to marine conservation. Considerable attention is given to the underlying biology, ecology, oceanography, and response of marine systems to human impacts. Such background is important in order to understand approaches to marine conservation and mitigatory measures that may be employed. The emphasis is on the issues, their context, and ways to address them but not specifically the practical implementation, which has been well covered elsewhere. Whilst the book moves from species to habitats, ecosystems, and regions, the chapters are relatively self-contained. Generic issues that crop up through the book are then examined in different contexts.

The book assumes a basic marine biological background and, it is hoped, should appeal to a reasonably wide audience. Primarily, it is seen as suitable for senior undergraduate and postgraduate courses but also of interest to practitioners in conservation and resource management.

I am very grateful to all those who provided feedback on draft chapters: Peter Batson, Katrin Berkenbusch, Michelle Boyle, Malcolm Clark, Mark Costello, John Darby, Simon Davy, Steve Dawson, Malcolm Francis, John Jillett, Chris Lalas, Will Probert, Dave Raffaelli, Ashley Rowden, Candida Savage, David Thompson, Steve Wing, and Anna Wood. I am also particularly indebted to Anna Wood for obtaining third-party permissions and for preparing figures and to Peter Batson for the many figures that he also prepared and edited.

I much appreciate the support I have received from the University of Otago – from colleagues in the Department of Marine Science, from staff in the University Library, and for periods of study leave. For leave spent at the University of Plymouth I thank Malcolm Jones, Colin Munn, and Ashley Rowden.

I am conscious that many other people have assisted in various ways, to all of whom I am most grateful and offer a collective and heartfelt thank-you.

Staff at Cambridge University Press, both past and present, have been enormously helpful. I thank in particular Alan Crowden, Dominic Lewis, Sarah Payne, and Megan Waddington.

The unstinting support of my wife Jan has been hugely important throughout this project. I can only apologize for all those evenings and weekends when I was otherwise occupied.

Lastly, I greatly appreciate the help, understanding, and patience of Margaret Holme in this endeavour to realize her late husband's idea.

# 1 Marine Environments

To early humans living on the coast, the oceans must have appeared as largely inhospitable expanses, difficult to access, seemingly limitless, and so virtually unexplored and unexploited. But even prehistoric cultures, especially once they became proficient seafarers, started to have substantial impacts on marine environments, such as on islands and near coastal settlements, in some cases causing local extinctions (Erlandson & Rick, 2010). Within historical times, particularly within the past century, the scale of exploitation of marine living resources and degradation of marine ecosystems have become global issues raising widespread concern. An estimated 40% of the world's oceans are now strongly affected by human impact with no area untouched (Halpern *et al.*, 2008).

Efforts at nature conservation have been, and indeed still are, directed largely at terrestrial environments, as is to be expected. In our use of living space and natural resources we have wrought enormous changes to terrestrial habitats and their biotas. And increasingly we are being disadvantaged by environmental consequences associated with our burgeoning population. Aside from any moral sensibility, we need to safeguard the functioning of healthy ecosystems on which we ultimately depend. On land, the need to afford protection to vulnerable species and habitats has long been recognized, and many practices of terrestrial conservation are well established. But widespread concern about the health of the world's oceans and use of their resources has surfaced only relatively recently, resulting in concerted moves towards marine conservation.

## Marine and Terrestrial Ecosystems

Ecological concepts have stemmed largely from our experience of the terrestrial environment, and programmes for nature conservation were established on land long before the emergence of marine conservation. Marine and terrestrial ecosystems, however, call for different approaches to conservation, and the attributes of marine ecosystems must be borne in mind in developing strategies for their conservation. It is worth considering, therefore, the distinguishing features of marine ecosystems and how they differ from those on land (e.g. Carr *et al.*, 2003).

The most obvious difference between terrestrial and marine systems is one of size. The oceans cover 362 million km$^2$, or 71% of the Earth's surface (Harris *et al.*, 2014). The difference, however, is far greater by volume. On land, the habitable zone is a comparatively thin veneer – generally some tens of metres in height – whereas the oceans have a mean depth of about 3.7 km, and this entire marine space is inhabited. As a result, the marine realm accounts for more than 99% of the habitable volume of the planet (Dawson, 2012).

For land organisms, biological tissue is far denser than the surrounding atmosphere. Water, on the other hand, some 800 times denser than air, is by comparison a very supportive medium, so many marine organisms can maintain near neutral or positive buoyancy and inhabit the water column at little energetic cost. The density of suspended particles also means that suspension feeding is of major importance in marine ecosystems, whereas the sparsity of suitable particles in air effectively rules out a comparable method of feeding for land animals.

The large thermal capacity of the ocean dampens short-term temperature variability. Most marine organisms need to contend with only small and gradual changes in temperature compared to the extremes often experienced by land organisms. So even at higher trophic levels, ectothermic, or 'cold-blooded,' animals predominate in the sea. However, the temporal variability in abiotic factors such as temperature differs radically between marine and terrestrial systems. Variance of temperature on land is relatively constant, at least over ecological timescales. In marine systems, on the other hand, whereas short-term variability is constrained, this variance increases over longer timescales. In other words, large, slow oceanographic processes account for more environmental variability than smaller, short-lived processes (Stergiou & Browman, 2005).

Inhabitants of terrestrial ecosystems are usually close to sources of primary production, whereas such proximity is rare in marine ecosystems. Most ocean space is deep sea, remote from the euphotic zone, and where the biota largely comprises microorganisms and animals dependent upon the flux of detritus from surface waters. The euphotic zone, where photosynthesis is possible, typically extends to depths of only a few tens of metres in coastal waters and usually to little more than 100 m in the open ocean. This is of similar dimension to the height of the photosynthetic layer on land as represented by the tallest trees. To counter the unsupportive environment of the atmosphere, land plants invest heavily in structural materials, such as cellulose and lignin, to elevate their photosynthetic tissues. In this way vegetation is usually a large and important component of terrestrial communities. Many terrestrial plant communities such as tropical forests are structurally complex and contain an immense diversity of plant species. This infrastructure in turn provides for a huge diversity of niches and a corresponding richness of associated animal species, of which more than two-thirds are insects. An analogous structural dimension is provided in coastal habitats by macroalgae, seagrasses, mangroves, and saltmarsh plants (see Chapter 12), although these are restricted to shallow waters. Phytoplankton, on the other hand, although they are the main primary producers in the marine environment, provide few habitat opportunities for associated species, except perhaps for microbes.

## Food Webs

Global net primary production has been estimated at about 105 x $10^9$ tonnes C $y^{-1}$, with roughly equal contributions from land and the oceans (Field *et al.*, 1998; Carr *et al.*, 2006), even though the biosphere is predominantly marine (Table 1.1). The nature of production, however, differs greatly between the two systems. Marine primary producers are mainly phytoplankton, which represent

**Table 1.1** Primary Production and Biomass for Marine and Terrestrial Biospheres

| | Oceans | Land |
|---|---|---|
| Total net primary production (x $10^9$ t C $y^{-1}$) | ~50 | ~60 |
| Total primary producer biomass (x $10^9$ t C) | 1 | 500 |
| Average turnover of biomass | 2–6 days | 19 years |

From Falkowski *et al.*, 1998; Field *et al.*, 1998; Westberry *et al.*, 2008; Huston & Wolverton, 2009.

only 0.2% of the global biomass of primary producers. Phytoplankton have an average turnover time of only 2–6 days and so can achieve high rates of production. By contrast, plant biomass on land is dominated by forests, and the turnover time of terrestrial primary producers averages 19 years (Field *et al.*, 1998). For many land ecosystems, the major primary producers are the largest and longest-lived organisms (notably trees), whereas in marine systems most primary producers are microscopic and short-lived.

Much terrestrial primary production is indigestible to herbivores and is broken down by decomposers and detritivores. This may explain why levels of herbivory in the sea can typically be 10–20 times those on land (May, 1994). The classical model of marine food chains was one in which a large proportion of phytoplankton production is grazed directly by zooplankton with few subsequent trophic steps. However, pelagic food webs incorporate far more components at low trophic levels than previously realized, and traditional distinctions between autotrophic phytoplankton and heterotrophic protists are often blurred. Particularly significant is the role played by a range of planktonic organisms in the smallest size classes, including cyanobacteria (of less than 2 μm) that may account for most of the pelagic primary production in warm ocean waters. There are also minute grazers (2–5 μm) and a range of protists (5–20 μm) that in turn are consumed by larger zooplankton (Longhurst, 2007).

Major food-web patterns are recognizable in the global ocean based on gross differences in nutrient supply, productivity, and trophic complexity. At one end of the spectrum are nutrient-poor open ocean waters with food webs of high trophic complexity culminating in low production of top consumers. This situation is typical of tropical and subtropical mid-ocean regions where a strong thermocline separates the warm surface layer from deeper cold water. The stability of this stratification keeps phytoplankton near the surface, meaning there is sufficient light to support their growth. However, stratification also inhibits mixing of the water column and thereby the input of deep, nutrient-rich water into the euphotic zone. In higher latitudes, as sea surface temperature falls in autumn, the difference in density between the surface and deep layers diminishes. This enables vertical mixing to occur during winter, which replenishes nutrients at the surface in readiness for a spring phytoplankton bloom. The productivity of such systems is dominated by this major seasonal event.

In some areas, notably coastal upwelling systems, the conditions favouring maximum production occur over significantly longer periods. Upwelling ecosystems occur mainly along the west coasts of continents at subtropical latitudes where there are prevailing offshore winds and strong eastern boundary currents. Here, surface waters are diverted offshore so that cold subsurface nutrient-rich water is drawn to the surface. Major coastal upwelling systems occur off Peru, Oregon and California, north-west and south-west Africa, and in the NW Indian Ocean (in this case driven by the seasonal monsoon). These nutrient-rich systems are characterized by few trophic levels and high productivities of fish, seabirds, and marine mammals. Whilst upwelling regions total only about 1% of the world ocean area, they account for about 20% of global fish landings (Mann, 2000).

Food webs of shelf waters are generally of intermediate complexity and productivity. Here nutrients are not lost into deep water and can be returned to the surface, particularly in mid- to high latitudes when the water column loses its thermal stability in winter. Also, estuaries can supply important nutrients to coastal waters. A significant proportion of primary production in coastal areas may be contributed by fringing macrophytes rather than by phytoplankton, such as by the highly productive kelp, saltmarsh, and mangrove systems.

Differences in marine and terrestrial food webs and the susceptibility of their respective trophic levels to harvesting make for contrasting patterns of human use. On land, where food production is based on agriculture, plants such as cereals and sugar cane are the primary harvest. Crops are increasingly being grown for herbivore production as our consumption of meat rises

(Tilman *et al.*, 2002). By contrast, our consumption of terrestrial carnivores is minuscule. An essentially inverse pattern pertains to the sea where the catch is predominantly of carnivores, in particular predatory fish. Some herbivores, such as clupeid fishes and bivalve molluscs, are also exploited, but primary producers (in this case macroalgae) have traditionally made up only a small proportion of the total marine biomass taken. With the catch coming largely from wild stocks, the trophic level of targeted marine species has not customarily been a concern. But with existing fisheries under intense pressure, there is interest in making greater use of organisms at lower trophic levels, such as herbivorous zooplankton. The difficulty is that such organisms tend to be uneconomic to harvest. In the case of agriculture, the trophic level of a harvest directly affects the energy efficiency and viability of the operation. The contribution of aquaculture to marine production is steadily increasing but will need to focus increasingly on organisms of low trophic status – algae, herbivores, and detritivores – and avoid methods that are environmentally harmful (see Chapter 5).

## Ocean Basins and Circulation

The world's land and sea areas are distributed very unequally. In the Southern Hemisphere, the marine area is four times greater than the land area, whilst in the Northern Hemisphere, it is only 1.5 times larger. In fact, the latitudinal distributions of sea and land in the two hemispheres are almost mirror images (Fig. 1.1). The temperate zone of the Southern Hemisphere is almost entirely maritime, but in the Northern Hemisphere this is where the landmasses are concentrated. This disparity

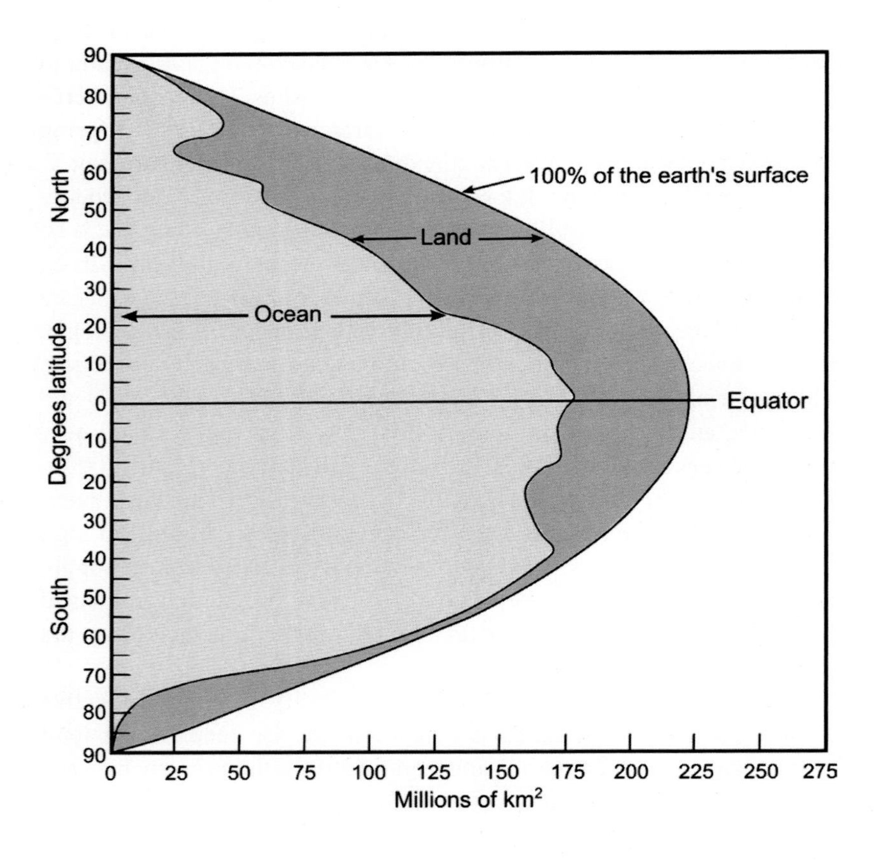


**Fig. 1.1** Distribution of ocean and land by latitude. Areas are based on 5° latitude intervals. The areal extent of ocean differs markedly between the two hemispheres, particularly at mid-latitudes. From Duxbury, A.C. & Duxbury, A.B. (1994). *An Introduction to the World's Oceans*, 4th edn. Dubuque, IA: Wm. C. Brown. Copyright McGraw-Hill Education.

has major implications for the global distribution of marine environments. Continental shelf, estuarine and brackish systems are, for instance, better represented in the Northern Hemisphere. A corresponding distribution of human population and associated scientific endeavour has meant too that marine environments of the Northern Hemisphere have in general been more intensively researched than their southern counterparts.

The global marine environment can be defined in terms of the major oceans and associated seas, their circulation, and bathymetric zonation. The Pacific Ocean accounts for roughly 47% of the total ocean area, the Atlantic Ocean plus Arctic Ocean and Mediterranean Sea 28%, and the Indian Ocean 20% (Harris *et al.*, 2014). The southern sectors of these oceans comprise the Southern Ocean at about 6%, which, although less well defined geographically, has distinctive hydrological and biological characteristics. The northern limit of the Southern Ocean is taken as the Antarctic Polar Front (at 50–60° S), a major circumglobal zone where surface Antarctic and subantarctic surface water masses meet (see Chapter 17).

Oceanic circulation is driven chiefly by the prevailing pattern of global winds and differences in the salinity and temperature (i.e. the density) of water masses. The pattern of surface currents is dominated by major gyres, huge circular flows within each of the ocean basins that move clockwise in the Northern Hemisphere and anticlockwise in the Southern Hemisphere (Fig. 1.2). An important exception is the Southern Ocean where, uninterrupted by landmasses, there is a circumglobal eastward flow. The Earth's rotation tends to compress gyres on the western side of ocean basins to produce intense western boundary currents, such as the Gulf Stream (NW Atlantic), Kuroshio (NW Pacific), and Agulhas (SW Indian Ocean). The swift, warm flow of the Gulf Stream contributes to the North Atlantic Current, which flows across the Atlantic and has a strong moderating influence on the climate of NW Europe.

Large-scale zonal differences in the relative importance of evaporation and precipitation produce slight density differences of surface waters, and at the fronts where different water masses meet, the denser water sinks back into the interior of the ocean. Such fronts, marked by relatively abrupt changes in physical and chemical characteristics between water masses, can function as important biogeographical boundaries, at least for many pelagic species, such as the Antarctic Polar Front already mentioned and the Kuroshio Front separating subtropical and subpolar water masses (Clayton *et al.*, 2014).

The large-scale pattern of currents and frontal systems incorporates, however, a high degree of ecological complexity. Conspicuous features, at scales of up to a few hundred kilometres, include large eddies pinched off from currents. These are temporary islands of water with distinctive physical and biological characteristics. Eddies carry with them populations of organisms entrained from the parent water body and may persist for a year or so before eventually decaying and becoming indistinguishable from the surrounding water. In the vicinity of fronts, the convergence of water masses and increased nutrient availability can result in enhanced productivity and higher concentrations of pelagic organisms. As zones of increased food supply, fronts can be important feeding grounds. Apex predators such as seabirds and marine mammals are often associated with fronts (Bost *et al.*, 2009).

The pattern of deep-water circulation is driven largely by differences in density between water masses. In particular, the sinking in polar regions of dense (colder and more saline) water sustains a convective flow, known as the global thermohaline circulation, that links the major ocean basins (Fig. 1.3). Many deep-water species have distributions consistent with features of deep oceanic circulation. The mid-slope demersal fish fauna of temperate Australia and New Zealand, for example, has obvious links with that of the temperate North Atlantic, a pattern reflecting the circulation of intermediate water masses between ocean basins (Koslow *et al.*, 1994).

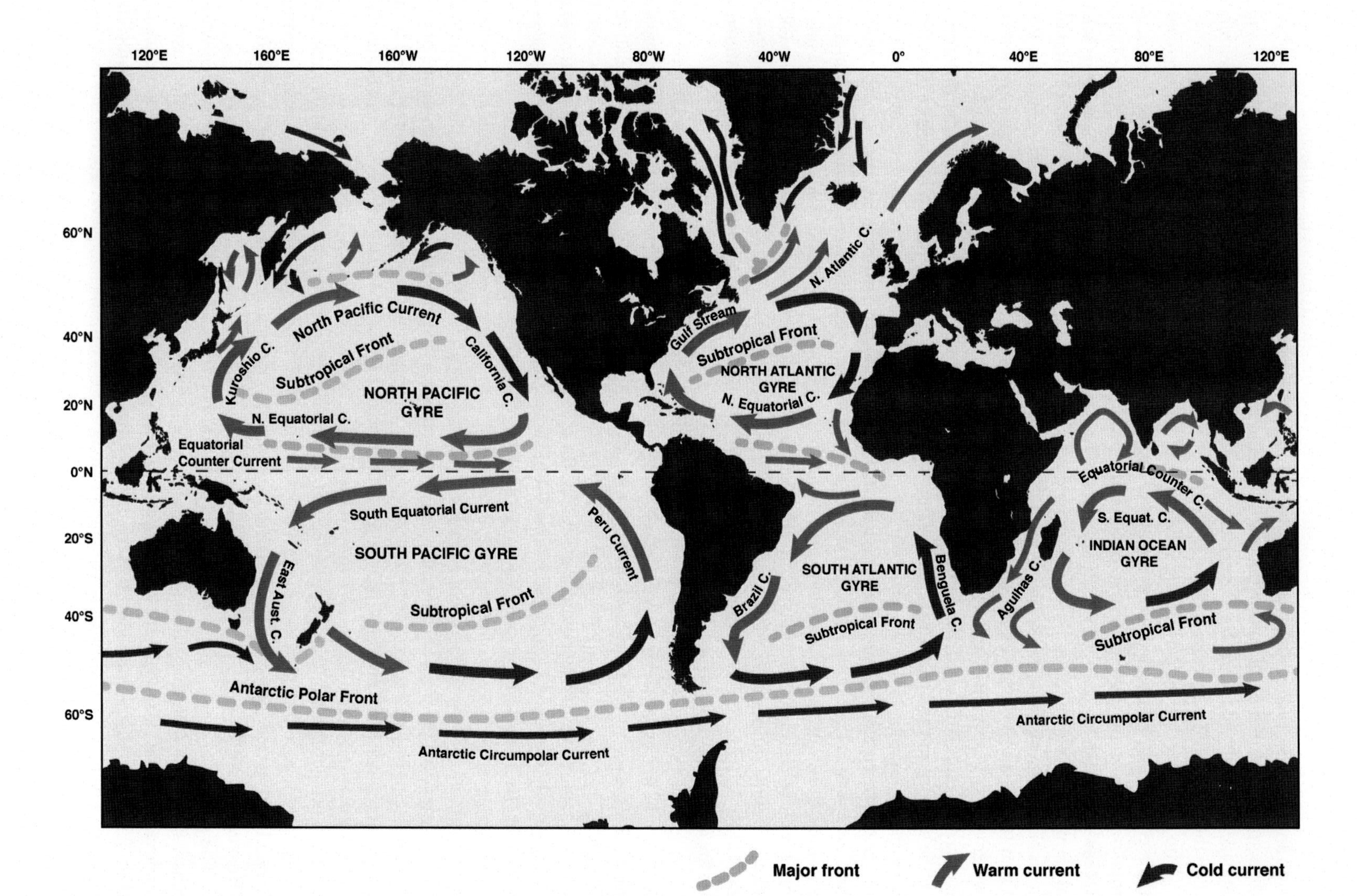


**Fig. 1.2** The major surface currents and fronts of the oceans. The pattern of circulation is dominated by the subtropical gyres, apart from the circumglobal eastward flow in the Southern Ocean. (A black and white version of this figure will appear in some formats. For the colour version, please refer to the plate section.)

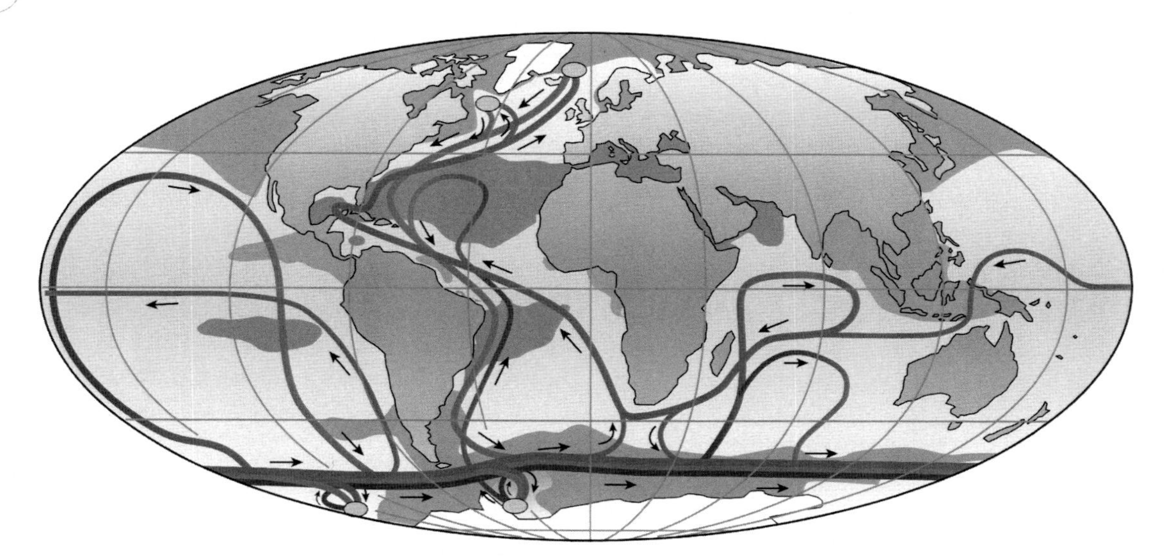

**Fig. 1.3** A simplified diagram of the global thermohaline circulation. Near-surface waters (red lines) flow towards the main regions of deep-water formation (yellow ovals) – in the northern North Atlantic, the Ross Sea, and the Weddell Sea – and recirculate at depth as deep currents (blue lines) and bottom currents (purple lines). Green shading, salinity above 36; blue shading, salinity below 34. Reprinted by permission from Macmillan Publishers Ltd: *Nature*. Rahmstorf, S. (2002). Ocean circulation and climate during the past 120,000 years. *Nature*, 419, 207–14, copyright 2002. (A black and white version of this figure will appear in some formats. For the colour version, please refer to the plate section.)

## Biogeography and Bathymetric Zones

A key requirement for conservation is a sound biogeographic framework (Lourie & Vincent, 2004). Ideally, we need to understand not just the distribution of species and habitats but also regional differences in ecosystem functioning and the environmental drivers underlying these patterns. Such information is essential, for example, as a basis by which to establish networks of marine protected areas that are adequately representative (see Chapter 15) and to develop strategies for managing exploited and vulnerable species. We are still some way from a detailed understanding of the biogeography of the seas, but at least for surface waters there is broad agreement between global schemes for categorizing the marine environment for biogeographic purposes.

Broad divisions of the world's surface ocean can be identified, related to the major climatic zones and ocean basins, and defined by physico-chemical parameters, notably temperature and salinity, wind-streams, and surface currents. On this basis various biogeographic schemes have been proposed for defining biologically meaningful areas at a range of spatial scales. A synthesis by Spalding *et al.* (2007) provides a global biogeographic classification of the world's coastal and shelf areas (out to the 200-m isobath) based on data from benthic and pelagic biotas (Fig. 1.4). This nested system comprises 12 realms, within which are 62 provinces and then 232 ecoregions. These realms and provinces represent the relative importance of various biotic and abiotic factors, such as taxonomic coherence and degree of endemism, and geomorphological, hydrological, and geochemical characteristics (Box 1.1). So, for example, the Temperate Northern Atlantic realm contains six provinces, each of which contains up to several ecoregions (Table 1.2). Delimiting marine areas by their characteristic ecology and living resources has been applied to coastal waters with the development of the concept of large marine ecosystems (see Chapter 17).

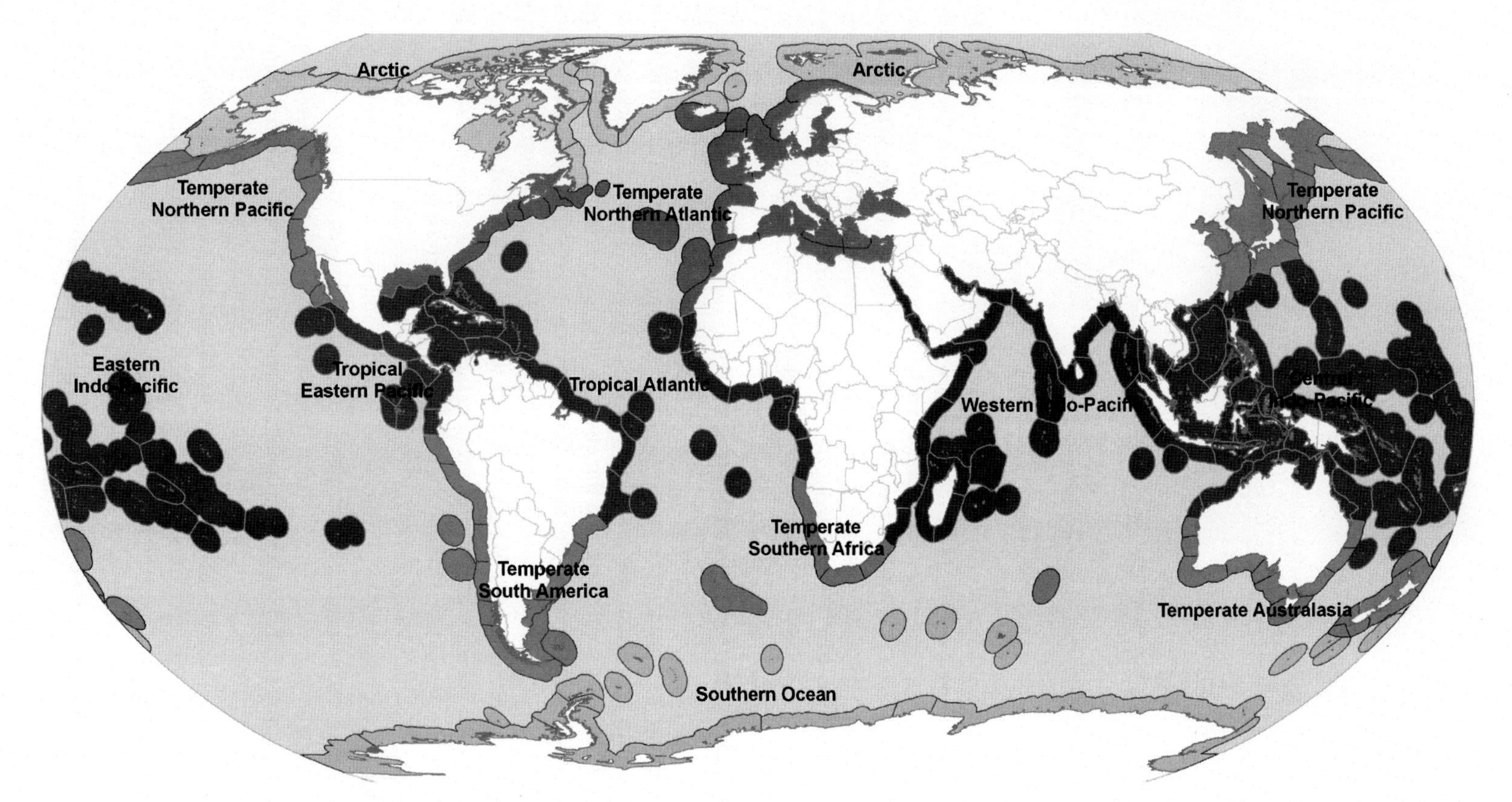


**Fig. 1.4** Biogeographic realms of coastal and shelf areas, with ecoregion boundaries outlined. From Spalding, M.D., Fox, H.E., Allen, G.R., *et al.*, Marine ecoregions of the world: a bioregionalization of coastal and shelf areas, *BioScience*, 2007, Vol. 57 (7), pages 573–83. By permission of American Institute of Biological Sciences. (A black and white version of this figure will appear in some formats. For the colour version, please refer to the plate section.)

**Table 1.2** Provinces and Ecoregions of the Temperate Northern Atlantic Realm

| Province | Ecoregion |
|---|---|
| Northern European Seas | South and West Iceland<br>Faroe Plateau<br>Southern Norway<br>Northern Norway and Finnmark<br>Baltic Sea<br>North Sea<br>Celtic Seas |
| Lusitanian | South European Atlantic Shelf<br>Saharan Upwelling<br>Azores Canaries Madeira |
| Mediterranean Sea | Adriatic Sea<br>Aegean Sea<br>Levantine Sea<br>Tunisian Plateau/Gulf of Sidra<br>Ionian Sea<br>Western Mediterranean<br>Alboran Sea |
| Cold Temperate Northwest Atlantic | Gulf of St. Lawrence – Eastern<br>Scotian Shelf<br>Southern Grand Banks – South<br>Newfoundland<br>Scotian Shelf<br>Gulf of Maine/Bay of Fundy<br>Virginian |
| Warm Temperate Northwest Atlantic | Carolinian<br>Northern Gulf of Mexico |
| Black Sea | Black Sea |

From Spalding, M.D., Fox, H.E., Allen, G.R., *et al*., Marine ecoregions of the world: a bioregionalization of coastal and shelf areas, *BioScience*, 2007, Vol. 57 (7), pages 573–83. By permission of American Institute of Biological Sciences.

Spalding *et al.* (2012) present a parallel biogeographic classification of oceanic epipelagic waters and semi-enclosed areas to 200 m water depth, similarly based on the distribution of taxa and major underlying oceanographic drivers, notably water movements (e.g. gyres, currents, and upwellings), nutrients, and temperature (Fig. 1.5, Box 1.1). They describe 37 pelagic provinces, large areas each with a coherent suite of oceanographic factors and distinct species assemblages. And these provinces are nested into four realms (Northern Coldwater, Indo-Pacific Warmwater, Atlantic Warmwater, and Southern Coldwater). The realms are much larger scale regions that are still distinguishable at higher taxonomic levels. The provinces can also be grouped into major biomes – systems with distinct oceanographic processes (polar, gyre, eastern boundary currents,

## Box 1.1 Coastal and Off-Shelf Biogeographic Areas

### Coastal

**Realms** – Very large regions of coastal, benthic, or pelagic ocean across which biotas are internally coherent at higher taxonomic levels as a result of a shared and unique evolutionary history. Realms have high levels of endemism, including unique taxa at generic and family levels in some groups. Driving factors behind the development of such unique biotas include water temperature, historical and broadscale isolation, and the proximity of the benthos.

**Provinces** – Large areas defined by the presence of distinct biotas that have at least some cohesion over evolutionary time frames. Provinces will hold some level of endemism, principally at the level of species. Although historical isolation will play a role, many of these distinct biotas have arisen as a result of distinctive abiotic features that circumscribe their boundaries. These may include geomorphological features (isolated island and shelf systems, semi-enclosed seas); hydrographic features (currents, upwellings, ice dynamics); or geochemical influences (broadest-scale elements of nutrient supply and salinity).

**Ecoregions** – Areas of relatively homogeneous species composition, clearly distinct from adjacent systems. The species composition is likely to be determined by the predominance of a small number of ecosystems and/or a distinct suite of oceanographic or topographic features. The dominant biogeographic forcing agents defining the ecoregions vary from location to location but may include isolation, upwelling, nutrient inputs, freshwater influx, temperature regimes, ice regimes, exposure, sediments, currents, and bathymetric or coastal complexity.

### Off-Shelf

**Realms** – Very large regions across which biotas are internally coherent at higher taxonomic levels as a result of a shared and unique evolutionary history. High levels of endemism, including unique taxa at generic and family levels in some groups. Distribution of individual species often does not encompass all of a realm, but coherence is often present at generic or family levels.

**Provinces** – Large areas of epipelagic ocean that can be defined by large-scale, spatially, and temporally stable (or seasonally recurrent) oceanographic drivers. Host distinct species assemblages that share a common history of co-evolution. Oceanographic drivers may include major ocean gyres, equatorial upwellings, upwelling zones at basin edges, semi-enclosed pelagic basins, and large-scale transitional elements. Taxonomic refinement will typically be driven by isolation at the scale of ocean basin and hemisphere.

**Biomes** – Groupings of provinces with common oceanographic processes (boundary current systems, mid-oceanic gyres, etc.). These may be separated by large physical distances and have very different evolutionary histories. Therefore expected to host ecosystems with comparable structural and functional properties but not the same species.

From Spalding, M.D., Agostini, V.N., Rice, J. & Grant, S.M. (2012). Pelagic provinces of the world: a biogeographic classification of the world's surface pelagic waters. *Ocean & Coastal Management*, **60**, 19–30; Spalding, M.D., Fox, H.E., Allen, G.R., *et al.* (2007). Marine ecoregions of the world: a bioregionalization of coastal and shelf areas. *BioScience*, **57**, 573–83.

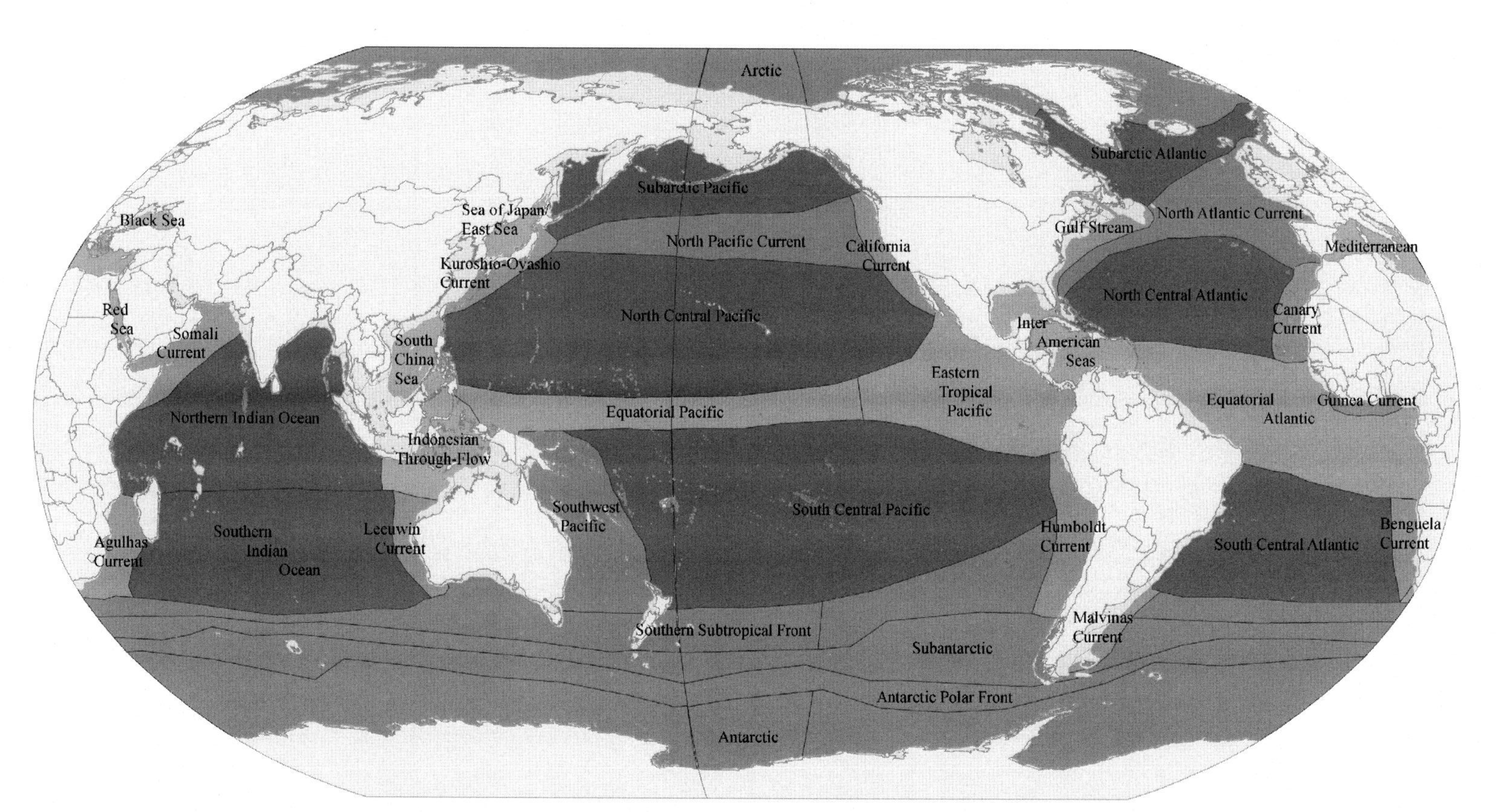


**Fig. 1.5** Biogeographic provinces of surface pelagic waters. The colours represent the different biomes, for example: polar (Arctic), gyre (Subarctic Pacific), transitional (North Pacific Current), eastern boundary current (California Current), western boundary current (Kuroshio-Oyashio Current), equatorial (Equatorial Pacific), and semi-enclosed seas (South China Sea). Reprinted from *Ocean & Coastal Management*, Vol. 60, Spalding, M.D., Agostini, V.N., Rice, J. & Grant, S.M, Pelagic provinces of the world: a biogeographic classification of the world's surface pelagic waters, pages 19–30, copyright 2012, with permission from Elsevier. (A black and white version of this figure will appear in some formats. For the colour version, please refer to the plate section.)

western boundary currents, equatorial, transitional, and semi-enclosed seas). The various biomes support communities that are functionally similar but not necessarily closely related taxonomically given their often wide geographic separation. Biogeographic boundaries in such schemes are rarely sharply defined, nor are they static – there may be considerable spatial variation over various temporal scales.

At the regional level, a number of classifications of marine habitats have been developed (Costello, 2009). One of the most comprehensive for coastal benthic habitats is the marine habitat classification for Britain and Ireland, intended particularly for management and conservation (Connor *et al.*, 2004). It also contributes to the marine component of the European Union Nature Information System (EUNIS) habitat classification developed by the European Environment Agency. Such classifications comprise hierarchies of levels from large-scale divisions based on differences in oceanographic and geomorphic characteristics down to individual communities or biotopes (see Chapter 10).

Marine biogeographic classifications have been developed mainly for near-surface waters. Environmental conditions in the ocean differ far more dramatically vertically than they do horizontally, particularly for such factors as temperature, light, and food supply. In the tropics, the difference in water temperature at the surface and at 1 km depth could be 20°C, whereas at the surface such a temperature difference would occur over thousands of kilometres.

Major bathymetric divisions of the ocean are recognized for the pelagic and benthic realms (Fig. 1.6). The upper water column to a depth of about 200 m, known as the epipelagic zone, includes the zone that receives sufficient light for photosynthesis, the euphotic zone. A water depth of 100–200 m marks the edge of the continental shelf in most parts of the world. Shelf waters are referred to as 'coastal' or 'neritic'. From about 200–1000 m, corresponding to upper continental slope depths, lies the mesopelagic zone, a region where there is still enough light for vision but not for photosynthesis. The regions below are permanently sunless. The bathypelagic zone extends from about 1000–3000 m water depth and the abyssopelagic from 3000 to 6000 m. In terms of the benthic environment, these two deeper zones correspond roughly to lower continental slope and abyssal plain depths. The deepest regions occur in the ocean trenches, the hadal zone, at some 6000–11 000 m (see Chapter 14).

As we have seen, in biogeographic classifications a number of biotic provinces are recognized for the epipelagic zone. Similar biotic distributions are recognizable in the mesopelagic zone; indeed, many species undergo diurnal vertical migration between the two zones. On the other

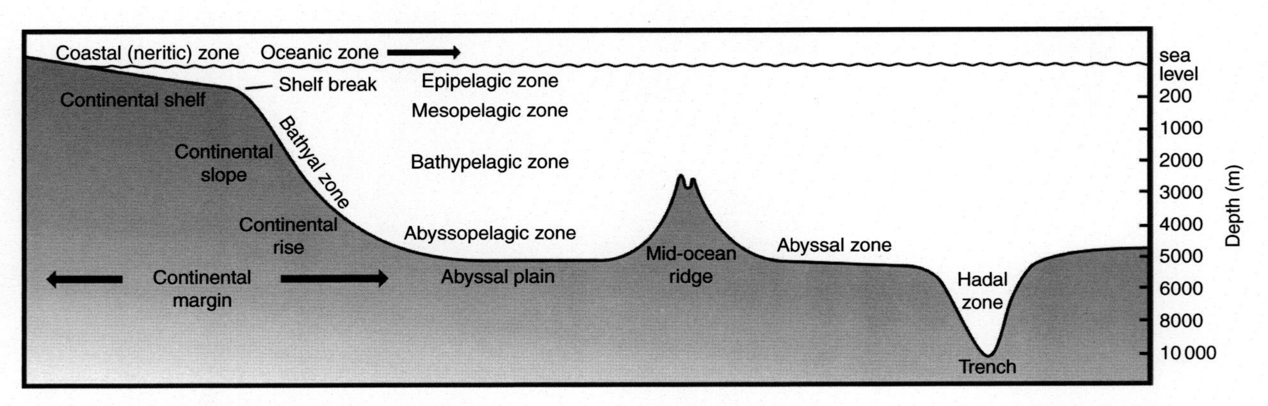


**Fig. 1.6** Section of an ocean basin to show major topographic features and bathymetric zones for benthic and pelagic environments.

hand, many bathy- and abyssopelagic species appear to have very broad distributions through the Atlantic, Pacific, and Indian Oceans. For bathyal (slope), abyssal, and hadal benthos, there is evidence of identifiable faunas related, for example, to ocean basins and regimes of surface production. Biogeographic provinces for deep-sea benthic systems have been proposed but still need further data on species' distributions to test them (see Chapter 14).

There is potential for marine species to have larger geographical ranges than species on land. Apart from continental land barriers, the oceans are contiguous and lack the obvious physical boundaries that can occur between ecosystems on land. Although many marine species are sedentary, their planktonic larvae may be relatively long-lived and able to disperse widely. For example, larval durations of about a year, or in extreme examples up to a few years, have been reported for a range of taxa, providing the potential for larvae to be transported thousands of kilometres (Strathmann & Strathmann, 2007). Even so, currents, fronts, eddies, or other oceanographic features may impede larval dispersal, and a long planktonic stage does not necessarily mean long-distance dispersal and high connectivity between populations. In fact, Weersing and Toonen (2009) found larval duration to be poorly correlated with genetic differentiation across a broad range of marine taxa. In some habitats, species with restricted ranges appear to be common. The Cape Verde archipelago off western Africa is home to 56 species of cone snail, 43 of them each restricted to a single island (Peters *et al.*, 2013). Even the dispersal of highly mobile pelagic species may be restricted by oceanographic barriers. The harbour porpoise is widely distributed in cold coastal waters of the North Atlantic and North Pacific. In the eastern North Atlantic, there is evidence of strong barriers to gene flow between porpoises from Iberian waters and those to the north. This separation coincides with marked oceanographic changes that for this species are likely to affect food availability (Fontaine *et al.*, 2007). Questions of dispersal and connectivity are important in conservation, for example in considering the function and effectiveness of marine protected areas (see Chapter 15).

## Marine Biodiversity

The Convention on Biological Diversity (CBD) defines biological diversity, or biodiversity, as 'the variability among living organisms from all sources, including, *inter alia*, terrestrial, marine and other aquatic ecosystems and the ecological complexes of which they are part; this includes diversity within species, between species and ecosystems' (Article 2). Biological diversity can thus be considered at various levels of organization. In terms of conservation effort, ecological diversity, as outlined above, may be the most effective to shape long-term strategies and goals. By conserving the diversity of healthily functioning marine ecosystems, the maintenance of other levels of diversity will follow. Taxonomic diversity is, however, easier to categorize and quantify, and it is concern over loss of species that often drives biodiversity conservation initiatives. Biodiversity is thus most often discussed at the species level.

The total number of extant species so far described is about 1.5 million (Costello *et al.*, 2012). How many remain undescribed is difficult to assess as many groups are still poorly known. Most of the genetic diversity of the oceans may reside with microbial organisms (Sogin *et al.*, 2006), yet their biodiversity has barely been explored. Conservation has so far concerned itself almost entirely with macroscopic organisms or at least eukaryote species (those with cells that have a distinct nucleus and organelles). Estimates of the number of living eukaryote species vary considerably, but recent analyses indicate a total of about 5 to 8.7 million (Mora *et al.*, 2011; Costello *et al.*, 2013). So far, some 0.23 million eukaryotic marine species have been described, but the total number may be of the order of 0.3–1.0 million (Appeltans *et al.*, 2012) to 2.2 million (Mora *et al.*, 2011). There is considerable uncertainty around such figures. Fisher *et al.* (2015), for instance, estimated that coral reefs alone could support 0.8 million multicellular species. A major advance in documenting the biodiversity of the oceans has been the Census of Marine Life, an international research programme from 2000 to 2010 on marine life across the world's seas and oceans (www.coml.org). It set up the Ocean Biogeographic Information System (www.iobis.org), which publishes millions of locations for more than 100 000 marine species online.

The above estimates indicate that up to about 20% of extant eukaryote species are marine. But a very different pattern of biodiversity emerges at higher taxonomic levels. Of the 33 or so animal phyla, 28 (85%) occur in marine habitats, and about half of these are exclusively marine, such as the ctenophores (comb jellies), brachiopods (lamp shells), and echinoderms (sea stars, sea urchins, and allies) (Box 1.2). By contrast, 16 animal phyla occur in the terrestrial-freshwater realm, only one of which is exclusive (Ray & Grassle, 1991). Animal diversity on land is overwhelmingly dominated by a few major taxa, in particular insects. Marine species, on the other hand, are not only spread among considerably more phyla, they are also far more equitably distributed across those phyla. Several reasons have been proposed to explain these contrasts in biodiversity between land and sea. The basic differentiation of animal phyla occurred in the seas of the Late Precambrian, Cambrian, and Ordovician before the invasion of freshwater and land in the Silurian and Devonian, and seemingly many marine groups never colonized freshwater or land for physiological or anatomical reasons. The far greater species richness on land is largely attributable to the huge diversification of insects and flowering plants.

Much of the primary production, herbivory, and predation in the sea involves smaller organisms than on land, and it has been suggested that globally there are fewer species in smaller size classes because such organisms typically have wider geographical distributions (May, 1994). It is argued that free-living microbial organisms are typically ubiquitous, being so abundant, unrestricted in their dispersal, and with low extinction probabilities, such that only larger organisms have biogeographies as such (Finlay, 2002). However, whilst marine micro-eukaryote species may be globally distributed, they may also be genetically very diverse (Šlapeta *et al.*, 2006). Small phytoplankton cannot provide physical support for metazoans comparable to the role played by land plants, which may also limit marine biodiversity. Nevertheless, as the smallest size classes of planktonic organisms become better known, it is likely that a far higher microbial diversity than is currently known will be revealed.

The diversity of photosynthetic species in the sea also appears to be low compared to that on land. Of the vascular plants alone, about 308 000 living species are known, the vast majority (about 95%) being flowering plants (Christenhusz & Byng, 2016). No more than a few hundred of these can be considered truly marine, although they play key roles in coastal wetlands (see Chapter 12). There are an estimated 5000 species of marine phytoplankton (Tett & Barton, 1995) and about 9300 species of seaweeds (www.algaebase.org). In addition, benthic microalgae can be important (or the major) primary producers in sheltered coastal habitats, such as on estuarine sediments, though their biodiversity is still poorly known.

It is convenient to recognize two major categories of marine organisms depending on their predominant adult lifestyle: pelagic organisms that inhabit the water column and benthic organisms that inhabit the sea floor. The distinction is not always clear-cut for species that live close to the sea floor, and many species have both pelagic and benthic phases in their life cycle. Even so, the great

### Box 1.2 Major Groups of Eukaryotic Marine Organisms

Some group names (e.g. protists, macroalgae) are used for convenience rather than as accepted taxa.

Seaweeds (macroalgae): brown algae (Phaeophyceae), green algae (Chlorophyta), red algae (Rhodophyta)
Tracheophyta (vascular plants): mangroves, seagrasses, saltmarsh plants
Protists (unicellular eukaryotes): ciliates, foraminiferans, coccolithophores, diatoms, dinoflagellates
Porifera: calcareous sponges (Calcarea), horny sponges (Demospongiae), glass sponges (Hexactinellida)
Ctenophora: comb jellies
Cnidaria: sea anemones (Actiniaria), black corals (Antipatharia), stony corals (Scleractinia), gorgonians, sea pens (Octocorallia), hydroids, stylasterid corals, siphonophores (Hydrozoa), jellyfish (Scyphozoa)
Platyhelminthes: flatworms
Nemertea: ribbon worms
Polychaeta: bristleworms
Mollusca: chitons (Polyplacophora), bivalves, clams (Bivalvia), snails (Gastropoda), octopuses, squids (Cephalopoda)
Brachiopoda: lamp shells
Bryozoa: moss animals
Chaetognatha: arrow worms
Nematoda: round worms
Chelicerata: sea spiders (Pycnogonida), mites (Acari)
Crustacea: barnacles (Cirripedia), copepods (Copepoda), ostracods (Ostracoda), mantis shrimps (Stomatopoda), amphipods, isopods, mysids, tanaidaceans (Peracarida), krill (Euphausiacea), caridean shrimps, prawns, crabs, lobsters (Decapoda)
Echinodermata: sea stars (Asteroidea), sea urchins (Echinoidea), brittlestars (Ophiuroidea), sea lilies, feather stars (Crinoidea), sea cucumbers (Holothuroidea)
Tunicata: ascidians (sea squirts), salps
Chondrichthyes (cartilaginous fishes): sharks, rays, chimaeras (see Chapter 6)
Actinopteri: sturgeons and teleosts (bony fishes) (see Chapter 6)
Reptilia: marine turtles, sea snakes (see Chapter 7)
Aves: seabirds (see Chapter 8), waders (see Chapter 11)
Mammalia: otters, pinnipeds (sea lions, seals) (Carnivora), dugong, manatees (Sirenia), whales, dolphins, porpoises (Cetacea) (see Chapter 9)

majority of marine species are predominantly benthic as adults, the sea floor being a more multifaceted environment than the overlying water mass. It is estimated that more than 90% of known marine animal species are benthic rather than pelagic (May, 1994).

Genetic diversity – the 'diversity within species' in the CBD definition – concerns the frequency and diversity of genes and/or genomes. Gene flow might be expected to be high in the marine environment, given the relative absence of barriers and high dispersal capability of many species. For example, although total genetic diversity of marine and freshwater fishes is similar, genetic differentiation of subpopulations is significantly lower in marine fishes. This implies that, in general, more migrations occur among subpopulations of marine fish than among those of freshwater fish (Ward *et al.*, 1994). Various mechanisms can, however, lead to genetic differences accumulating in high-dispersal marine species (Palumbi, 1994), and molecular techniques are now revealing hitherto unsuspected levels of intraspecific genetic variability in marine organisms (Bucklin *et al.*, 2011).

Many taxa that in the past have been classified as single species are now known – thanks largely to DNA sequencing – to comprise two or more cryptic species: species that are more or less impossible to distinguish morphologically yet are genetically distinct. For instance, mitochondrial and nuclear DNA markers show that *Ciona intestinalis*, regarded as a common shallow-water ascidian of temperate to boreal regions, in fact comprises two species with largely disjoint distributions, one occurring in the Mediterranean Sea, neighbouring NE Atlantic, and Pacific and the other in the North Atlantic. Crossing experiments show the two species to be reproductively isolated (Caputi *et al.*, 2007). Cryptic marine species are proving far more common than previously realized and may number some tens of thousands (Appeltans *et al.*, 2012).

The recognition of cryptic species can have important implications for environmental assessment and conservation. Considered the most abundant and wide-ranging coral of the tropical western Atlantic, *Orbicella annularis* has been much used in studies of environmental degradation and global climate change. However, the taxon comprises at least three species, two of which show significant differences in growth rate and oxygen isotopic ratios, parameters that are used to assess past climatic conditions (Knowlton *et al.*, 1992). Similarly, morphologically similar capitellid polychaetes, originally believed to be one species, *Capitella capitata*, and considered a key indicator of organically polluted sediments, in fact comprise a complex of species that display wide differences in life-history features, including both benthic and planktonic larvae (Méndez *et al.*, 2000). Cryptic species may raise significant conservation issues. A threatened species could turn out to be more than one cryptic species, each even more vulnerable and requiring different conservation measures (Bickford *et al.*, 2007).

## Global Patterns of Biodiversity

Biotas can differ markedly in their species richness, with some habitats and geographic regions being far more diverse than others. Being able to characterize patterns of biodiversity and understand the structuring factors is important for informing management and conservation. A range of factors might be expected to influence large-scale patterns of species richness, such as tectonic and evolutionary history, habitat area and availability, temperature, primary productivity, oxygen concentration, and environmental stability. In terms of global patterns, the most obvious trend is latitudinal, where species richness tends to be highest in equatorial regions and decline with increasing latitude north and south, a pattern very evident in the terrestrial biosphere. A similar pattern related to water temperature is seen for the marine biosphere, with species richness highest at low latitudes, at least to water depths of around 2000 m, and with peaks in the tropical Indo-west Pacific and western Atlantic regions. But for water depths greater than 2000 m, maximum richness is seen in temperate latitudes (30–50°) and regions near continental margins, corresponding to areas of high flux of particulate organic carbon (Tittensor *et al.*, 2010; Woolley *et al.*, 2016).

Diversity and availability of habitat strongly influence benthic species richness. Certain tropical habitats, such as coral reef and mangrove systems, provide for a particularly high degree of structural complexity and support a correspondingly high diversity of benthic epifauna. Witman *et al.* (2004) found the species richness of benthic epifauna of shallow (10–15 m) rock-wall habitats to peak around equatorial regions and fall away towards higher latitudes in both hemispheres. Sediment infauna, on the other hand, shows only a weak latitudinal gradient in species richness (Hillebrand, 2004). Some groups of organisms are obvious exceptions. For seaweeds, the regions with highest genus richness are in temperate latitudes, probably because of availability of large areas of suitable habitat and the role of major ocean currents (Kerswell, 2006). Importantly for conservation, hotspots of species richness are often regions with medium or high human impacts (Halpern *et al.*, 2008; Tittensor *et al.*, 2010).

## Variability in Marine Systems

Variability is a natural characteristic of ecological systems and occurs at a wide range of spatial and temporal scales. Population variability may depend, for example, on the responses of organisms to environmental factors, biological interactions such as predation, and a species' life-history characteristics. At one extreme are large-scale movements and migrations of animals leading to seasonal and longer-term changes in patterns of abundance; at the other extreme are gradual shifts in the geographical distribution of species in response to changing ocean temperature patterns. Physical factors are particularly important in determining the distribution of planktonic organisms, from large-scale distributional patterns due to transport by currents to smaller-scale patchiness generated by entrainment in eddies. Organisms with short generation times, such as many planktonic species, may be able to respond rapidly to take advantage of favourable environmental conditions.

A primary environmental factor influencing benthic populations is the nature of the seabed. Even apparently subtle changes in the physico-chemical and biotic characteristics of the substratum can result in considerable differences in the distribution and abundance of benthic populations. Again this occurs across a range of scales. Many benthic invertebrates, particularly those of the more physically variable coastal and estuarine habitats, respond rapidly to environmental factors and display wide fluctuations in abundance. Benthic environments are affected by a wide variety of natural disturbances at a range of scales (Fig. 1.7). Large areas, often extending to many square kilometres, can be affected by storms, unusually low winter temperatures, salinity reductions, and oxygen depletion. Typically at the other end of the scale (at $m^2$ or $cm^2$) are biological disturbances, such as excavations produced by bottom-feeding fish and marine mammals and sediment reworking by infauna (Kaiser *et al.*, 2005).

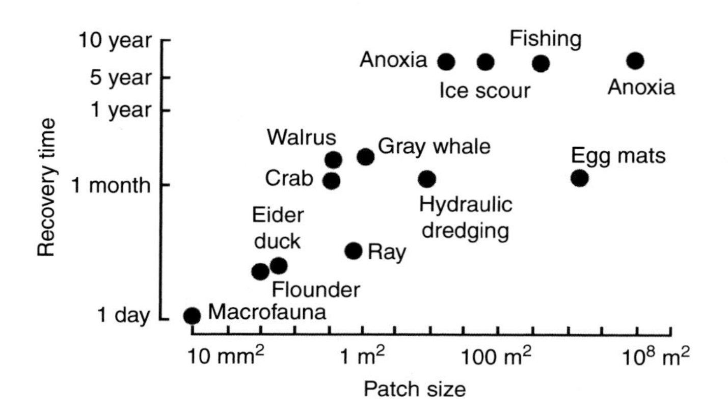


**Fig. 1.7** Relationship between different scales of disturbance, both natural and anthropogenic, and their approximate recovery time. From Hall, S.J., Raffaelli, D. & Thrush, S.F. (1994). Patchiness and disturbance in shallow water benthic assemblages. In *Aquatic Ecology: Scale, Pattern and Process*, ed. P.S. Giller, A.G. Hildrew & D.G. Raffaelli, pp. 333–75. Oxford, UK: Blackwell Scientific Publications. © 1994 by the British Ecological Society.

Natural disturbances play a vital role in structuring marine communities. By creating patches, disturbances provide opportunities for other species. Rather than being homogeneous assemblages, communities are more akin to mosaics, with individual pieces at different stages of recovery after the last perturbation. The frequency, extent, and intensity of disturbances within a particular habitat are key factors affecting the composition, abundance, and diversity of its biota. Thus the benthic environment of an estuary has a regime of natural disturbance and recovery very different from that occurring in the deep sea.

Knowledge of the natural variability of wild populations is needed to assess impacts of human activities and make informed decisions about management and conservation. Natural disturbances operate over a wide range of spatial scales, and superimposed on this background may be various human-induced disturbances also operating at a range of scales (Kaiser *et al.*, 2005). The problem is distinguishing between the effects of human activities and natural background fluctuations in space and time, and it may not be feasible to establish an unequivocal link between cause and effect for many human activities. Some events, like disease outbreaks, harmful algal blooms, or unusually harsh winter temperatures, can remove more than 90% of a population. Such mass mortalities may often be natural phenomena but may also result from natural and anthropogenic factors occurring in concert (Fey *et al.*, 2015).

## North Atlantic Oscillation

Temporal variability inherent in atmosphere-ocean interactions has major implications for marine ecosystems. Changes can occur at the decade-to-secular scale, such as the North Atlantic Oscillation (NAO), a periodic shift in the relative gradient between the subpolar low-pressure and subtropical high-pressure regions that drives winter westerly winds across the North Atlantic (Longhurst, 2007). The NE Atlantic experiences milder winter conditions during positive NAO phases, when westerlies are stronger, and harsher conditions when the gradient weakens, allowing the inflow of cold Siberian air masses to intensify (Fig. 1.8).

Changes in the NAO have been linked to major changes in marine ecosystems of the North Atlantic, including shifts in the abundance of different plankton species and impacts on fish

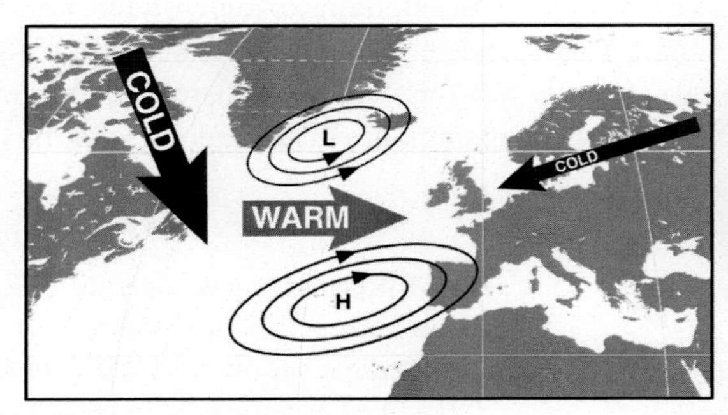


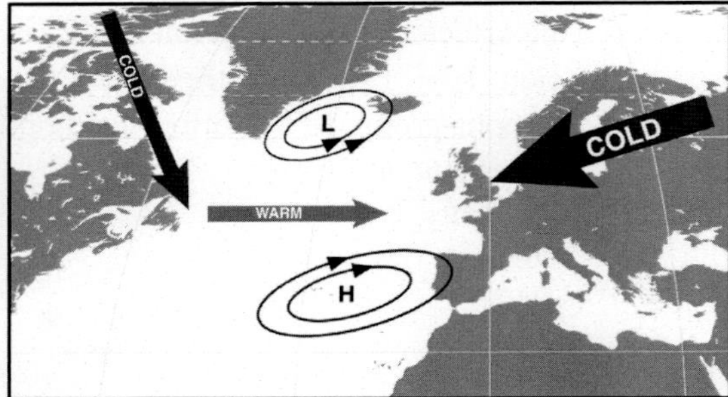


**Fig. 1.8** The atmospheric circulation over the North Atlantic indicating the positive and negative modes of the North Atlantic Oscillation (NAO). The modes relate to the relative difference in air pressure between the subpolar low (L) and the subtropical high (H) and cold and warm air masses. Positive mode (left), with a large pressure difference, strong westerlies, reduced inflow of Siberian air masses, and a mild European winter. Negative mode (right), with a small pressure difference, a weaker band of westerlies, a stronger inflow of Siberian air masses, and a severe European winter. From Alheit, J. & Hagen, E. (1997). Long-term climate forcing of European herring and sardine populations. *Fisheries Oceanography*, 6, 130–9. © 1997 Blackwell Science Ltd.

productivity (Parsons & Lear, 2001). A striking example is the correspondence between the prevailing wind direction (NAO index) and alternating periods of herring and sardine abundance in European waters. Herring, a predominantly arctic-boreal species, is favoured during periods of reduced westerly winds (i.e. negative NAO index), whereas the warmer water sardine or pilchard is associated with periods of intensified westerlies (positive NAO index) (Alheit & Hagen, 1997; Coombs *et al.*, 2010).

Shifts in distribution are likely to be most pronounced in populations near the edge of a species' geographic range, in this case where the distributions of the two fish species overlap but where shifts have also been intensified by overfishing and subsequent recruitment failure (Southward *et al.*, 1988). Such data derived from sampling in the western English Channel date back more than a century and illustrate the importance of long time-series observations for our understanding of the dynamics and functioning of marine ecosystems and in forecasting effects of global climate change (Harris, 2010).

The NAO is occasionally punctuated by marked ocean climate anomalies. Two exceptional periods reported for the North Sea in the late 1970s and the late 1980s were associated with unusual oceanic incursions of different origin. The 1970s anomaly was associated with the influx of cold, low salinity water into the North Sea and reduced inflow of warm Atlantic water, significantly decreasing the flux of nutrients from oceanic sources. This coincided with sudden changes in the abundance of macrobenthos, fish, and birds in the southern North Sea and the appearance of more cold-water species. By contrast, in the late 1980s a high NAO index and increased flow of relatively warm Atlantic water into the North Sea was accompanied by exceptionally high phytoplankton biomass, an unprecedented influx of oceanic species, and a sudden increase in macrobenthic biomass in the southern North Sea (Edwards *et al.*, 2002).

Inter-decadal variability is also well documented for the monsoon zone. Off the south-west coast of India there are long-term shifts in the strength of the monsoon, upwelling intensity, diatom stocks, and landings of oil sardine (Longhurst, 2007). A dozen or so low-frequency ocean-atmosphere interactions are in fact recognized from around the world, and there are probably connections between them.

## El Niño-Southern Oscillation

Oceanographic variability also occurs over shorter timescales, with periodicities ranging from a few years to seasonal cycles. Particularly striking are El Niño events, the appearance from time to time of warm surface water in the eastern equatorial Pacific, notably off Ecuador and Peru. Such anomalous warmings are, however, part of a larger-scale perturbation of an ocean-atmosphere interaction centred on the tropical Pacific Ocean and known as the El Niño-Southern Oscillation (ENSO), which typically has a return interval of 2–7 years (Glantz, 2001; Longhurst, 2007). Normally the difference between the low atmospheric pressure over Indonesia and the high in the south-eastern Pacific maintains the South-East Trades. These persistent winds mean that a thick layer of warm surface water piles up in the western equatorial Pacific, whereas in the east the thermocline is shallow, allowing cold nutrient-rich water to upwell along the coast. Every few years, however, there is a change: differences in atmospheric pressure across the Pacific diminish, the trade winds weaken, and the warm water pooled in the west surges across into the eastern Pacific. The deepening of the thermocline off Ecuador and Peru during an El Niño means that the deeper, nutrient-rich water is less available for upwelling (Fig. 1.9). Upwelling systems are noted for their high biological productivity, and the sudden decline of productivity off South America during an El Niño has dramatic impacts on the food chain, notably on the abundance of the Peruvian anchovy, seabirds, and marine mammals.

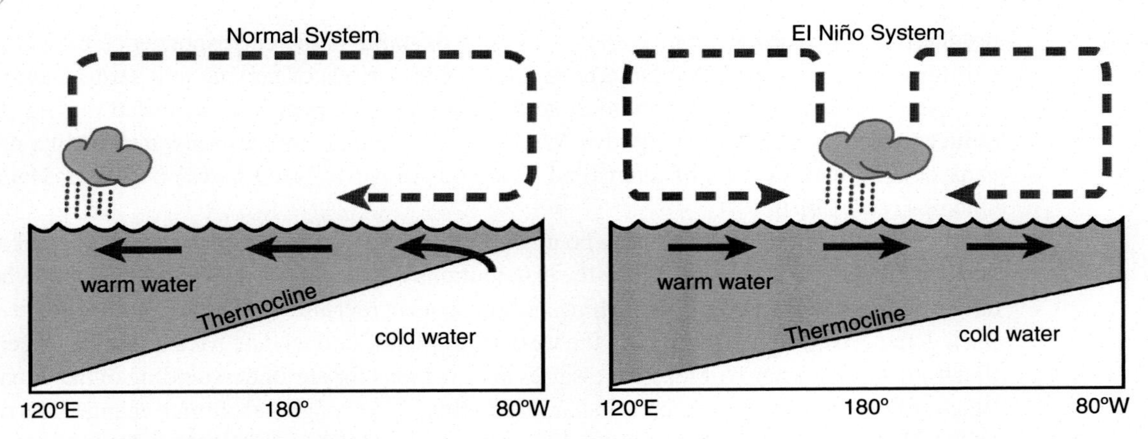


**Fig. 1.9** The development of El Niño. Under normal conditions, surface water in the equatorial Pacific Ocean is pushed westward by strong trade winds, and with a shallow thermocline off the coast of Peru, upwelling occurs that supports high productivity. El Niño conditions develop when the trade winds weaken and warm surface water flows eastward, deepening the thermocline and inhibiting the Peruvian upwelling. Lamont-Doherty Earth Observatory.

Depending on their intensity, ENSO events can have wide-ranging implications for marine ecosystems given their potential impact on sea surface temperatures, salinity and sea state, nutrient availability and productivity, algal blooms, and changes in the relative abundances of competing species (Glynn, 1988). The number of green turtles nesting at rookeries in the western Pacific has been correlated with an index of the Southern Oscillation, probably via a nutritional pathway, with turtle numbers peaking after El Niño events (Poloczanska *et al.*, 2009). There is a lag in the relationship such that it may be possible to forecast the size of a nesting population up to 2 years in advance, which may be of considerable value to marine turtle management. El Niño activity in the equatorial Pacific is probably linked to climate anomalies at higher latitudes, such as the eastward propagation of sea surface temperature anomalies around the Southern Ocean (White & Peterson, 1996). These have a periodicity of 4–5 years, taking 8–10 years to encircle the pole, and link with inter-annual variability in sea-ice cover, a factor of critical importance to populations of Antarctic krill and their dependent predators in the Southern Ocean (see Chapter 17).

The ENSO event of 1997–8, one of the strongest on record, was responsible for major disturbances to marine ecosystems, both locally and globally, including widespread bleaching (loss of symbiotic algae) and mortality of corals (see Chapter 13). Severe ENSO events are probably among the greatest natural perturbations known on our planet in terms of areas affected and biological consequences.

At a smaller scale are seasonal cycles, and here too we may witness occasional anomalies. An exceptional heat wave in Europe in summer 2003 resulted in unusually high sea surface temperatures. In the NW Mediterranean region, Garrabou *et al.* (2009) reported the mass mortality of at least 25 benthic species (mainly sponges and gorgonian corals) inhabiting shallow-water (up to 40 m water depth) rocky areas. Such events are likely to occur more often given the climate warming projections, and indeed there is evidence of this for the Mediterranean Sea as a whole (Rivetti *et al.*, 2014). Natural catastrophes can also occur over timescales of hours, such as tropical cyclones that cause localized devastation of coral reefs (see Chapter 13).

## Spatial and Temporal Scales

Broadly speaking, because of differences in density and viscosity, physical processes of the ocean and atmosphere operate over different spatial and temporal scales. At short timescales, the ocean is less variable than the atmosphere, and for the pelagic environment, physical and biological processes can be closely coupled (Fig. 1.10). Natural large-scale changes in open sea systems typically occur over periods of years to decades, and pelagic populations have evolved reproductive strategies that can respond accordingly. As a result, marine systems can naturally undergo major shifts in species composition. By contrast, terrestrial systems can operate over much longer timescales, of the order of centuries in the case of forest growth, and land populations have adapted to cope with atmospheric variability as short-term noise. Marine systems may thus be able to respond more readily than terrestrial systems to global environmental changes imposed by human activity since these are occurring rapidly, at timescales comparable to those of natural large shifts in open sea systems (Steele, 1991).

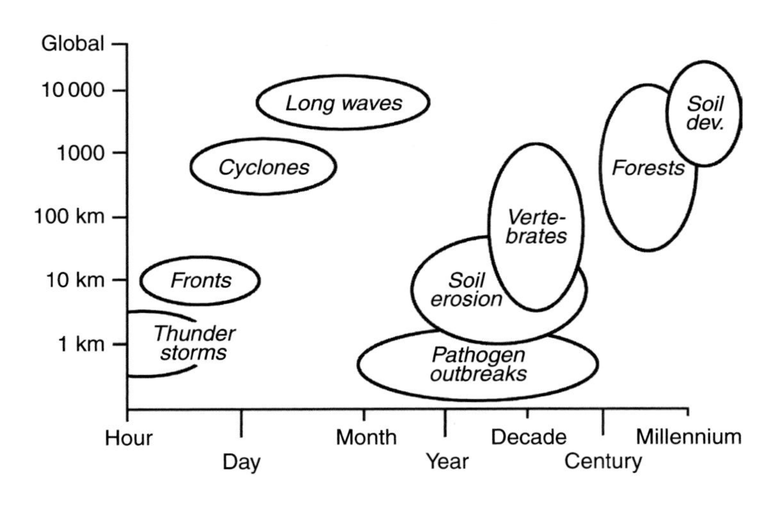


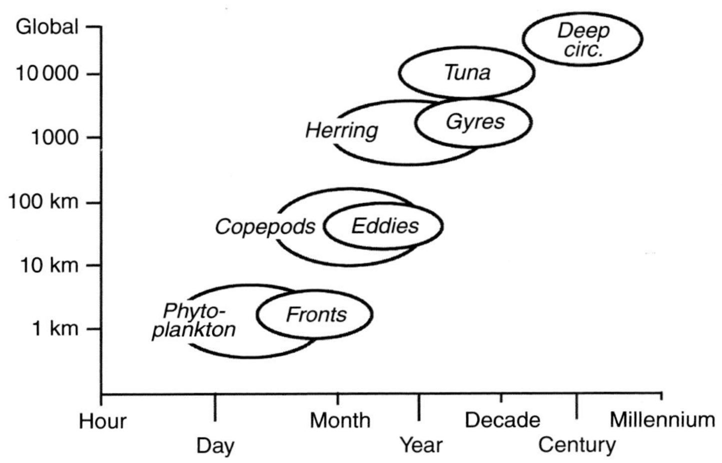


**Fig. 1.10** Relationship between spatial and temporal scales for atmospheric processes and terrestrial groups showing their marked separation in time and for oceanographic processes and pelagic groups showing their marked coupling. From Steele, J.H. & Henderson, E.W. (1994). Coupling between physical and biological scales. *Philosophical Transactions of the Royal Society of London B*, 343, 5–9. By permission of the Royal Society.

## Reproductive Strategies

Many marine species have reproductive strategies characterized by a highly variable rate of recruitment of larvae or juveniles to adulthood. A typical strategy is to ensure that the maximum number of planktonic larvae are committed to the water column, a corollary of which is that parental investment towards the survival of individual offspring is correspondingly minimized. By contrast, many land organisms produce relatively small numbers of young to which they need to devote a degree of care before they can afford to relinquish them. There are marine species that produce small numbers of young that are brooded or otherwise protected for an extended period and others whose life-history strategy lies between these extremes of the spectrum. But many marine species produce huge numbers of eggs, commonly millions in the case of bony fish that produce pelagic eggs. A cost of this tactic is extremely high egg and larval mortality with only a few percent surviving through the juvenile period (Palumbi & Hedgecock, 2005). The loss between the time of reproduction and recruitment of the new cohort to the population may be due to various physico-chemical and biotic factors, the combined effects of which are usually of far greater significance than the variability contributed by the size of the parent population. For most marine species, recruitment variability results directly from variation in mortality during early life-history stages and, within limits, is often seemingly independent of the adult population size, a point of particular importance in fisheries management (see Chapter 5). Populations of marine species are frequently, though unpredictably, dominated by a single year class when favourable circumstances lead to a recruitment peak. On occasions, larval survival may be exceptionally high and result in a population outbreak, such as those of the crown-of-thorns starfish on coral reefs (see Chapter 13). The fact that many marine species have small, pelagic dispersal stages that show high spatial and temporal variability adds to the difficulty of understanding population dynamics and forecasting recruitment events, which has important implications for marine conservation. Life-history characteristics have a major bearing on the vulnerability of species to overexploitation (see Chapter 4).

# A Sea Ethic

In this chapter we have looked at major features of the marine environment and its biological diversity. These have various implications for the unique challenges of marine conservation. In that respect we also need to consider how these aspects intersect societal attitudes. In general, public regard for the marine environment is often poorly developed, and a 'sea ethic' that recognizes intrinsic values beyond the purely utilitarian has yet to gain wide acceptance (Dallmeyer, 2005). There are various reasons for this. That most of the biosphere is unseen and undiscovered has not fostered an environmental awareness of the marine realm; in fact, often the opposite, as our impacts resulting from exploitation and waste disposal have to a degree been hidden or not even considered given the obvious immensity of the oceans. Also, few marine species engage public concern compared to the many, often emblematic animals and plants on land. In the sea, virtually all organisms are seen as exploitable. There has existed, at least in Western societies, a widespread view that those who exploit marine resources should not be hampered by notions of boundary and property that we take for granted on land, a view that has exacerbated human impacts on the sea and its resources – as we shall see in subsequent chapters.

## REFERENCES


Alheit, J. & Hagen, E. (1997). Long-term climate forcing of European herring and sardine populations. *Fisheries Oceanography*, **6**, 130–9.

Appeltans, W., Ahyong, S.T., Anderson, G., *et al.* (2012). The magnitude of global marine species diversity. *Current Biology*, **22**, 2189–202.

Bickford, D., Lohman, D.J., Sodhi, N.S., *et al.* (2007). Cryptic species as a window on diversity and conservation. *Trends in Ecology and Evolution*, **22**, 148–55.

Bost, C.A., Cotté, C., Bailleul, F., *et al.* (2009). The importance of oceanographic fronts to marine birds and mammals of the southern oceans. *Journal of Marine Systems*, **78**, 363–76.

Bucklin, A., Steinke, D. & Blanco-Bercial, L. (2011). DNA barcoding of marine metazoa. *Annual Review of Marine Science*, **3**, 471–508.

Caputi, L., Andreakis, N., Mastrototaro, F., *et al.* (2007). Cryptic speciation in a model invertebrate chordate. *Proceedings of the National Academy of Sciences*, **104**, 9364–69.

Carr, M.-E., Friedrichs, M.A.M., Schmeltz, M., *et al.* (2006). A comparison of global estimates of marine primary production from ocean color. *Deep-Sea Research II*, **53**, 741–70.

Carr, M.H., Neigel, J.E., Estes, J.A., *et al.* (2003). Comparing marine and terrestrial ecosystems: implications for the design of coastal marine reserves. *Ecological Applications*, **13** (Suppl.), S90–107.

Christenhusz, M.J.M. & Byng, J.W. (2016). The number of known plants species in the world and its annual increase. *Phytotaxa*, **261**, 201–17.

Clayton, S., Nagai, T. & Follows, M.J. (2014). Fine scale phytoplankton community structure across the Kuroshio Front. *Journal of Plankton Research*, **36**, 1017–30.

Connor, D.W., Allen, J.H., Golding, N., *et al.* (2004) *The marine habitat classification for Britain and Ireland. Version 04.05.* Joint Nature Conservation Committee, Peterborough, UK. Available at http://jncc.defra .gov.uk/marinehabitatclassification.

Coombs, S.H., Halliday, N.C., Conway, D.V.P. & Smyth, T.J. (2010). Sardine (*Sardina pilchardus*) egg abundance at station L4, Western English Channel, 1988–2008. *Journal of Plankton Research*, **32**, 693–7.

Costello, M.J. (2009). Distinguishing marine habitat classification concepts for ecological data management. *Marine Ecology Progress Series*, **397**, 253–68.

Costello, M.J., May, R.M. & Stork, N.E. (2013). Can we name Earth's species before they go extinct? *Science*, **339**, 413–16.

Costello, M.J., Wilson, S. & Houlding, B. (2012). Predicting total global species richness using rates of species description and estimates of taxonomic effort. *Systematic Biology*, **61**, 871–83.

Dallmeyer, D.G. (2005). Toward a sea ethic. In *Marine Conservation Biology: the Science of Maintaining the Sea's Biodiversity*, ed. E.A. Norse & L.B. Crowder, pp. 410–21. Washington, DC: Island Press.

Dawson, M.N. (2012). Species richness, habitable volume, and species densities in freshwater, the sea, and on land. *Frontiers of Biogeography*, **4**, 105–16.

Duxbury, A.C. & Duxbury, A.B. (1994). *An Introduction to the World's Oceans*, 4th edn. Dubuque, IA: Wm. C. Brown.

Edwards, M., Beaugrand, G., Reid, P.C., Rowden, A.A. & Jones, M.B. (2002). Ocean climate anomalies and the ecology of the North Sea. *Marine Ecology Progress Series*, **239**, 1–10.

Erlandson, J.M. & Rick, T.C. (2010). Archaeology meets marine ecology: the antiquity of maritime cultures and human impacts on marine fisheries and ecosystems. *Annual Review of Marine Science*, **2**, 231–51.

Falkowski, P.G., Barber, R.T. & Smetacek, V. (1998). Biogeochemical controls and feedbacks on ocean primary production. *Science*, **281**, 200–6.

Fey, S.B., Siepielski, A.M., Nusslé, S., *et al.* (2015). Recent shifts in the occurrence, cause, and magnitude of animal mass mortality events. *Proceedings of the National Academy of Sciences*, **112**, 1083–8.

Field, C.B., Behrenfeld, M.J., Randerson, J.T. & Falkowski, P. (1998). Primary production of the biosphere: integrating terrestrial and oceanic components. *Science*, **281**, 237–40.

Finlay, B.J. (2002). Global dispersal of free-living microbial eukaryote species. *Science*, **296**, 1061–3.

Fisher, R., O'Leary, R.A., Low-Choy, S., *et al.* (2015). Species richness on coral reefs and the pursuit of convergent global estimates. *Current Biology*, **25**, 500–5.

Fontaine, M.C., Baird, S.J.E., Piry, S., *et al.* (2007). Rise of oceanographic barriers in continuous populations of a cetacean: the genetic structure of harbour porpoises in Old World waters. *BMC Biology*, **5**, article 30.

Garrabou, J., Coma, R., Bensoussan, N., *et al.* (2009). Mass mortality in Northwestern Mediterranean rocky benthic communities: effects of the 2003 heat wave. *Global Change Biology*, **15**, 1090–103.

Glantz, M.H. (2001). *Currents of Change: Impacts of El Niño and La Niña on Climate and Society*, 2nd edn. Cambridge, UK: Cambridge University Press.

Glynn, P.W. (1988). El Niño-Southern Oscillation 1982–1983: nearshore population, community, and ecosystem responses. *Annual Review of Ecology and Systematics*, **19**, 309–45.

Hall, S.J., Raffaelli, D. & Thrush, S.F. (1994). Patchiness and disturbance in shallow water benthic assemblages. In *Aquatic Ecology: Scale, Pattern and Process*, ed. P.S. Giller, A.G. Hildrew & D.G. Raffaelli, pp. 333–75. Oxford, UK: Blackwell.

Halpern, B.S., Walbridge, S., Selkoe, K.A., *et al.* (2008). A global map of human impact on marine ecosystems. *Science*, **319**, 948–52.

Harris, P.T., Macmillan-Lawler, M., Rupp, J. & Baker, E.K. (2014). Geomorphology of the oceans. *Marine Geology*, **352**, 4–24.

Harris, R. (2010). The L4 time-series: the first 20 years. *Journal of Plankton Research*, **32**, 577–83.

Hillebrand, H. (2004). Strength, slope and variability of marine latitudinal gradients. *Marine Ecology Progress Series*, **273**, 251–67.

Huston, M.A. & Wolverton, S. (2009). The global distribution of net primary production: resolving the paradox. *Ecological Monographs*, **79**, 343–77.

Kaiser, M.J., Hall, S.J. & Thomas, D.N. (2005). Habitat modification. In *The Sea, Volume 13, The Global Coastal Ocean: Multiscale Interdisciplinary Processes*, ed. A.R. Robinson & K.H. Brink, pp. 927–70. Cambridge, MA: Harvard University Press.

Kerswell, A.P. (2006). Global biodiversity patterns of benthic marine algae. *Ecology*, **87**, 2479–88.

Knowlton, N., Weil, E., Weigt, L.A. & Guzman, H.M. (1992). Sibling species in *Montastraea annularis*, coral bleaching, and the coral climate record. *Science*, **255**, 330–3.

Koslow, J.A., Bulman, C.M. & Lyle. J.M. (1994). The mid-slope demersal fish community off southeastern Australia. *Deep-Sea Research I*, **41**, 113–41.

Longhurst, A. (2007). *Ecological Geography of the Sea*, 2nd edn. Burlington, MA: Academic Press.

Lourie, S.A. & Vincent, A.C.J. (2004). Using biogeography to help set priorities in marine conservation. *Conservation Biology*, **18**, 1004–20.

Mann, K.H. (2000). *Ecology of Coastal Waters: with Implications for Management*, 2nd edn. Malden, MA: Blackwell Science.

May, R.M. (1994). Biological diversity: differences between land and sea. *Philosophical Transactions of the Royal Society of London B*, **343**, 105–11.

Méndez, N., Linke-Gamenick, I. & Forbes, V.E. (2000). Variability in reproductive mode and larval development within the *Capitella capitata* species complex. *Invertebrate Reproduction and Development*, **38**, 131–42.

Mora, C., Tittensor, D.P., Adl, S., Simpson, A.G.B. & Worm, B. (2011). How many species are there on Earth and in the ocean? *PLoS Biology*, **9**, e1001127.

Palumbi, S.R. (1994). Genetic divergence, reproductive isolation, and marine speciation. *Annual Review of Ecology and Systematics*, **25**, 547–72.

Palumbi, S.R. & Hedgecock, D. (2005). The life of the sea: implications of marine population biology to conservation policy. In *Marine Conservation Biology: the Science of Maintaining the Sea's Biodiversity*, ed. E.A. Norse & L.B. Crowder, pp. 33–46. Washington, DC: Island Press.

Parsons, L.S. & Lear, W.H. (2001). Climate variability and marine ecosystem impacts: a North Atlantic perspective. *Progress in Oceanography*, **49**, 167–88.

Peters, H., O'Leary, B.C., Hawkins, J.P., Carpenter, K.E. & Roberts, C.M. (2013). *Conus*: first comprehensive conservation Red List assessment of a marine gastropod mollusc genus. *PLoS ONE*, **8**, e83353.

Poloczanska, E.S., Limpus, C.J. & Hays, G.C. 2009. Vulnerability of marine turtles to climate change. *Advances in Marine Biology*, **56**, 151–211.

Rahmstorf, S. (2002). Ocean circulation and climate during the past 120,000 years. *Nature*, **419**, 207–14.

Ray, G.C. & Grassle, J.F. (1991). Marine biological diversity. *BioScience*, **41**, 453–7.

Rivetti, I., Fraschetti, S., Lionello, P., Zambianchi, E. & Boero, F. (2014). Global warming and mass mortalities of benthic invertebrates in the Mediterranean Sea. *PLoS ONE*, **9**, e115655.

Šlapeta, J., López-García P. & Moreira, D. (2006). Global dispersal and ancient cryptic species in the smallest marine eukaryotes. *Molecular Biology and Evolution*, **23**, 23–9.

Sogin, M.L., Morrison, H.G., Huber, J.A., *et al.* (2006). Microbial diversity in the deep sea and the underexplored 'rare biosphere'. *Proceedings of the National Academy of Sciences*, **103**, 12115–20.

Southward, A.J., Boalch, G.T. & Maddock, L. (1988). Fluctuations in the herring and pilchard fisheries of Devon and Cornwall linked to change in climate since the 16th century. *Journal of the Marine Biological Association of the United Kingdom*, **68**, 423–45.

Spalding, M.D., Agostini, V.N., Rice, J. & Grant, S.M. (2012). Pelagic provinces of the world: a biogeographic classification of the world's surface pelagic waters. *Ocean & Coastal Management*, **60**, 19–30.

Spalding, M.D., Fox, H.E., Allen, G.R., *et al.* (2007). Marine ecoregions of the world: a bioregionalization of coastal and shelf areas. *BioScience*, **57**, 573–83.

Steele, J.H. (1991). Can ecological theory cross the land-sea boundary? *Journal of Theoretical Biology*, **153**, 425–36.

Steele, J.H. & Henderson, E.W. (1994). Coupling between physical and biological scales. *Philosophical Transactions of the Royal Society of London B*, **343**, 5–9.

Stergiou, K.I. & Browman, H.I. (eds.). (2005). Bridging the gap between aquatic and terrestrial ecology. *Marine Ecology Progress Series*, **304**, 271–307.

Strathmann, M.F. & Strathmann, R.R. (2007). An extraordinarily long larval duration of 4.5 years from hatching to metamorphosis for teleplanic veligers of *Fusitriton oregonensis*. *Biological Bulletin*, **213**, 152–9.

Tett, P. & Barton, E.D. (1995). Why are there about 5000 species of phytoplankton in the sea? *Journal of Plankton Research*, **17**, 1693–1704.

Tilman, D., Cassman, K.G., Matson, P.A., Naylor, R. & Polasky, S. (2002). Agricultural sustainability and intensive production practices. *Nature*, **418**, 671–7.

Tittensor, D.P., Mora, C., Jetz, W., *et al.* (2010). Global patterns and predictors of marine biodiversity across taxa. *Nature*, **466**, 1098–101.

Ward, R.D., Woodwark, M. & Skibinski, D.O.F. (1994). A comparison of genetic diversity levels in marine, freshwater, and anadromous fishes. *Journal of Fish Biology*, **44**, 213–32.

Weersing, K. & Toonen, R.J. (2009). Population genetics, larval dispersal, and connectivity in marine systems. *Marine Ecology Progress Series*, **393**, 1–12.

Westberry, T., Behrenfeld, M.J., Siegel, D.A. & Boss, E. (2008). Carbon-based primary productivity modeling with vertically resolved photoacclimation. *Global Biogeochemical Cycles*, **22**, GB2024.

White, W.B. & Peterson, R.G. (1996). An Antarctic circumpolar wave in surface pressure, wind, temperature and sea-ice extent. *Nature*, **380**, 699–702.

Witman, J.D., Etter, R.J. & Smith, F. (2004). The relationship between regional and local species diversity in marine benthic communities: a global perspective. *Proceedings of the National Academy of Sciences*, **101**, 15664–9.

Woolley, S.N.C., Tittensor, D.P., Dunstan, P.K., *et al.* (2016). Deep-sea diversity patterns are shaped by energy availability. *Nature*, **533**, 393–6.

# 2 Human Impacts

Human activities affect the marine environment in a multitude of ways, but particularly through the exploitation of its living and non-living resources and as a result of pollution and other forms of environmental modification. This chapter provides an overview of these human impacts and their implications for marine conservation. Specific aspects are examined in more detail in later chapters, especially those relating to exploitation of living marine resources and modification of marine habitats.

Globally, the magnitude of pressures on the natural environment is inseparable from the exponential growth of the world's human population and patterns of economic development. The world population eventually reached 1 billion in about 1800 but now exceeds 7 billion, with an adult human biomass of about 290 million tonnes (Walpole *et al.*, 2012). Projections indicate world populations of about 9.7 billion by 2050 and 11.2 billion by 2100 (UN, Department of Economic and Social Affairs, Population Division, 2015) (Fig. 2.1). Such is our influence on the planet that the period since human impacts have become comparable to natural processes is being proposed as a new geological epoch, the Anthropocene, with a possible start date sometime in the industrial era (Corlett, 2015).

Human environmental impact is driven not just by population size but also by affluence (gross domestic product per capita) (Dietz *et al.*, 2007). The extremes of human deprivation and affluence

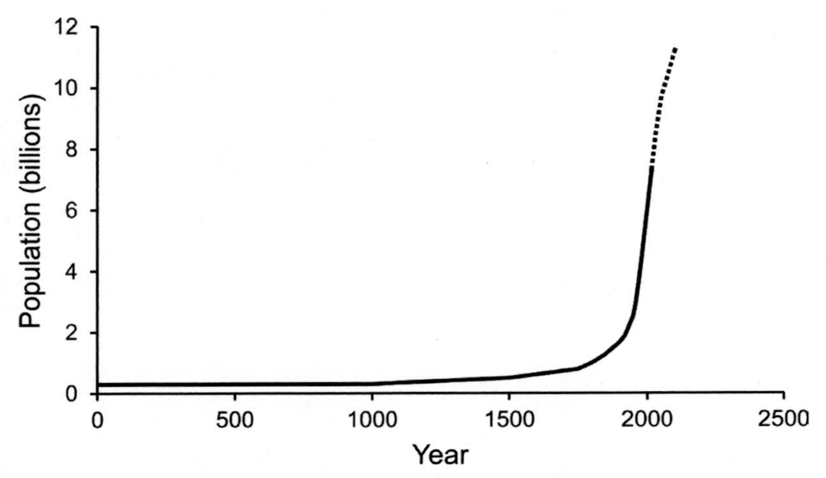


**Fig. 2.1** World human population to 2015 and projected population (dashed line) to 2100. Based on data from United Nations, Department of Economic and Social Affairs, Population Division (1999; 2015).

both contribute to global environmental problems, and achieving more sustainable use of the world's resources depends ultimately on population control and moving towards an economic order founded on more equitable apportioning of the world's resources (Kitzes *et al.*, 2008).

Although human activities can disturb marine ecosystems in many ways, five broad types of impact can be recognized: exploitation of resources, pollution, physical alteration of habitat, introduction of alien species, and climate change (Norse, 1993; Thorne-Miller, 1999) (Box 2.1). This order reflects the typical historical sequence of human disturbance to marine ecosystems (Fig. 2.2) and the current ranking of threat to marine biodiversity worldwide (Costello *et al.*, 2010). Overfishing usually precedes other disturbances, at least for coastal systems, and the modifications it causes may predispose an ecosystem to other stresses, such as eutrophication and non-native species (Jackson *et al.*, 2001). In the longer term, however, human-induced climate change may represent the greatest threat to marine ecosystems, especially in combination with other disturbances.

## Box 2.1 Major Types of Human Impacts on Marine Ecosystems

### Extraction of Living Resources (Direct and Indirect Impacts)

Fish, invertebrates, marine mammals, sea turtles, seabirds, aquaculture

### Pollution

Organic wastes and nutrients
Petroleum hydrocarbons
Persistent organic pollutants
Persistent debris (notably plastics)
Metals
Radionuclides

### Physical Alteration

Bottom fishing
Seabed mining (minerals, building aggregates)
Dredging and spoil disposal
Reclamation
Coastal structures
Altered riverine discharge (freshwater abstraction, dams, sediment input)

### Biological Invasions

Fouling, ballast water, fisheries, aquaculture, canals

### Global Climate Change

Global warming, ocean acidification, sea-level rise, increased UV radiation

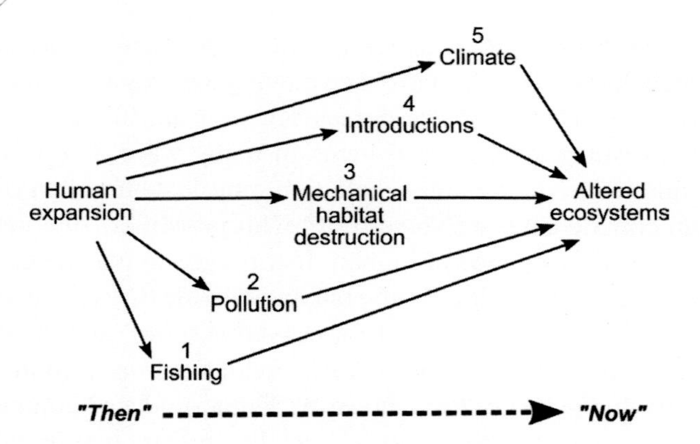


**Fig. 2.2** Typical historical sequence of human disturbances affecting coastal ecosystems. From Jackson, J.B.C., Kirby, M.X., Berger, W.H., *et al.* (2001). Historical overfishing and the recent collapse of coastal ecosystems. *Science*, 293, 629–38. Reprinted with permission from AAAS.

## Global Climate Change

Since about the mid-eighteenth century, as a result of increasing industrialization and changes in land use, humans have been gradually modifying the Earth's atmosphere and thereby its climate. Atmospheric and oceanic systems are closely coupled, and for marine life these climate changes have enormous ramifications.

### Global Warming

Certain gases in the atmosphere trap infrared radiation, making the Earth's surface considerably warmer than it would otherwise be – the greenhouse effect. Water vapour (clouds) and carbon dioxide are the most important of these greenhouse gases, with lesser contributions from methane, ozone, and nitrous oxide. There is a scientific consensus that since the Industrial Revolution humans have enhanced the greenhouse effect, causing a gradual global warming that will have environmental reverberations for centuries. This enhancement is mainly due to carbon dioxide ($CO_2$). In pre-industrial times the concentration of carbon dioxide in the atmosphere was about 280 ppm, but now as a result of the burning of fossil fuels and deforestation it has exceeded 400 ppm. Also acting as greenhouse gases and contributing significantly to global warming are methane and nitrous oxide, the concentrations of which have also increased through human activities. Water vapour will increase as a result of global warming, further enhancing the effect. Since 2010, total greenhouse gas emissions (in gigatonnes of $CO_2$-equivalent per year) have been increasing at more than 2% per year and are now at concentrations not seen in at least the past 800 000 years (IPCC, 2014). Numerous models have been developed to examine future climate change scenarios, though the complexity of the system and incomplete understanding mean that there remain many uncertainties about the magnitude and rate of temperature change from greenhouse gas emissions, the likely environmental consequences, and the degree of regional variation.

The Intergovernmental Panel on Climate Change (IPCC) was established by the UN in 1988 to evaluate the scientific information relating to climate change, the environmental and socio-economic impacts, and possible response strategies. Since 1990 it has published a series of assessment reports with projections of anthropogenic warming, sea-level rise, and other environmental impacts. In its Fifth Assessment Report, the IPCC examined different greenhouse gas concentration trajectories based on such factors as population size, economic activity, lifestyle, energy use, land-use patterns, technology, and climate policy (IPCC, 2014). Projections include a stringent

**Table 2.1** IPCC Projections for Global Increases (Relative to the 1986–2005 Period) in Air Surface Temperature, Sea Level, and Ocean Acidification to 2100 for Different Greenhouse Gas Emission Scenarios

Representative concentration pathways (RCPs) are projections for greenhouse gas concentrations based on socio-economic development and climate policy.

| Emissions scenario | Peak in greenhouse gas emissions | Surface temperature: mean (and range) in °C | Sea-level rise: mean (and range) in metres | Ocean acidification: pH decrease and percent increase in acidity |
|---|---|---|---|---|
| Stringent scenario (RCP 2.6) | ~2020 | 1.0 (0.3–1.7) | 0.40 (0.26–0.55) | 0.06–0.07 15–17% |
| Intermediate scenario (RCP 4.5) | ~2040 | 1.8 (1.1–2.6) | 0.47 (0.32–0.63 | 0.14–0.15 38–41% |
| Intermediate scenario (RCP 6.0) | ~2080 | 2.2 (1.4–3.1) | 0.48 (0.33–0.63) | 0.20–0.21 58–62% |
| Very high emissions scenario (RCP 8.5) | Continues to rise throughout twenty-first century | 3.7 (2.6–4.8) | 0.63 (0.45–0.82) | 0.30–0.32 100–109% |

Data from IPCC (2014). *Climate Change 2014: Synthesis Report*. Contribution of Working Groups I, II, and III to the Fifth Assessment Report of the Intergovernmental Panel on Climate Change (Core Writing Team, R.K. Pachauri & L.A. Meyer (eds.)). Geneva, Switzerland: IPCC.

mitigation scenario, two intermediate scenarios, and one scenario with very high greenhouse gas emissions (Table 2.1). Projections for intermediate scenarios indicate a global mean increase in surface air temperature of about 2°C by the end of the century. Given the fundamental influence of temperature on physical, chemical, and biological processes, such warming can be expected to impact marine ecosystems globally, both directly and indirectly (Drinkwater *et al.*, 2010; Hoegh-Guldberg & Bruno, 2010; Doney *et al.*, 2012). Temperature change, by its effect on metabolism, has major implications for the growth and activity of organisms, their reproduction, timing of spawning and migration, interactions, and distribution. A warming climate could also make many marine populations more susceptible to disease. In recent decades there has been a worldwide increase in reports of diseases affecting marine organisms (Harvell *et al.*, 1999). Warmer surface waters and increased stratification of the water column will also tend to reduce vertical mixing and thereby the supply of nutrients and oxygen to the euphotic zone, with effects on primary production and food webs. Decreases in primary production affect not just pelagic organisms but also the benthos through changes in the flux of organic matter to the seabed, with potentially fundamental effects on ocean ecosystems.

Global warming causes a rise in mean sea level, particularly as a result of thermal expansion of seawater and the melting of ice sheets and glaciers. Recent analysis indicates a rate of global mean sea-level rise of 1.2 mm per year for the twentieth century but accelerating since the 1990s to the current rate of about 3.0 mm per year (Hay *et al.*, 2015). Substantial geographic variability in sea-level change can, however, be expected, associated with regional changes in climate and ocean circulation. Coastal areas vulnerable to sea-level rise will be more susceptible to coastal erosion, loss of intertidal and wetland habitat, and flooding of other low-lying coastal habitats and atolls.

Sea surface temperature data show climate change impacts to vary considerably with latitude, with warming being particularly marked in mid- to high latitudes of the Northern Hemisphere

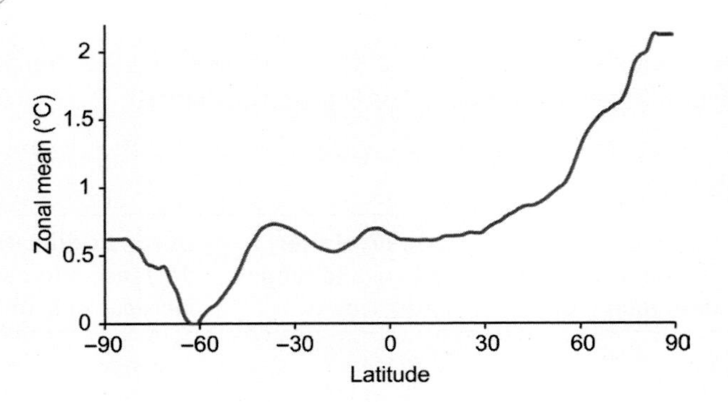


**Fig. 2.3** Mean sea surface temperature anomaly for 2010–15 relative to the mean for 1951–80 as a function of latitude. From GISTEMP Team (2016). GISS Surface Temperature Analysis (GISTEMP). NASA Goddard Institute for Space Studies. Dataset accessed 28 July 2016 at http://data.giss.nasa.gov/gistemp/. Hansen, J., Ruedy, R., Sato, M. & Lo, K. (2010). Global surface temperature change. *Reviews of Geophysics*, 48, article RG4004.

(Fig. 2.3). An increased influx of icesheet meltwater together with differences in surface warming could modify patterns of ocean circulation. In particular, by lowering the density of high latitude surface water, such effects could weaken the input of deep water into the global thermohaline circulation (Hansen *et al.*, 2016). If ultimately this were, for instance, to seriously diminish the overturning circulation in the North Atlantic, it would have enormous consequences for the marine ecosystems of the region, including its important seabird populations, not to mention the huge social and economic upheaval. Also, reductions in sea ice have major implications for polar marine ecosystems including all those organisms, from ice algae to marine mammals, that depend on this habitat (see Chapter 17).

Among the clearest evidence of marine organisms responding to anthropogenic global warming are the many examples of species across multiple groups showing poleward shifts in their distribution. Globally, average shifts of about 30 km per decade have been reported (Poloczanska *et al.*, 2013). In Australia, the black sea urchin *Centrostephanus rodgersii* has extended its range southward along the east coast from New South Wales to Tasmania. This may have important consequences for community structure in its extended range. Like other grazing urchins this species strongly modifies its habitat, converting macroalgal-dominated reefs to 'barrens' that have a much impoverished fauna (Ling, 2008). Some communities are likely to be more vulnerable to global warming than others. Stuart-Smith *et al.* (2015) examined the geographic and thermal distribution of 3920 reef fish and invertebrate species and found that most of them have ranges centred in either temperate or tropical zones, implying that species in subtropical communities are already living close to the upper limits of their temperature distribution.

Effects of climatic warming on the oceans may be exacerbated by low-frequency variability. Over the past 40 years there appears to have been a substantial change in the biogeography of copepod assemblages in the eastern North Atlantic: a northward extension by more than 10° latitude of warm-water species and a decrease in the number of colder-water species, shifts that Beaugrand *et al.* (2002) suggest are consistent with the climatic warming of the Northern Hemisphere combined with the North Atlantic Oscillation (see Chapter 1).

As this illustrates, marine ecosystems are contending not just with anthropogenic climate change. They may also be subject to natural environmental variability, from short-term perturbations to low-frequency atmosphere-ocean interactions, as well as to other human impacts, from overfishing to various types of pollution. And these factors may interact with or reinforce each other. Strong El Niño events on top of climatic warming are resulting in the mass bleaching of warm-water corals, a reaction to stress whereby the corals lose their symbiotic algae (see Chapter 13).

Human-induced climate change has fundamental implications for the structure and functioning of marine ecosystems. We are starting to tease out some of the consequences for the marine biosphere of this enormous alteration we have set in motion, but it is still far from clear how this will unfold. The 1992 UN Framework Convention on Climate Change (and its Kyoto Protocol) aims to at least stabilize greenhouse gas concentrations in the atmosphere to avert dangerous environmental repercussions. In 2015 the Conference of the Parties to the Convention adopted the Paris Agreement, which aims to hold 'the increase in the global average temperature to well below 2°C above pre-industrial levels and to pursue efforts to limit the temperature increase to 1.5°C above pre-industrial levels, recognizing that this would significantly reduce the risks and impacts of climate change' (Article 2) (UNFCCC Secretariat, 2015). To realize such targets, the world's nations need to greatly reduce greenhouse gas emissions as rapidly as possible and become increasingly ambitious over time (Rogelj *et al.*, 2016). Even so, limiting global warming to 2°C may not avert major climate change impacts on marine and terrestrial ecosystems (Hansen *et al.*, 2016), and because of the lag in response time, actions to cut emissions will take a long time to show any effect. The consequences of climate change due to human activities likely represent the most serious environmental problem we face.

## Ocean Acidification

There is a further major aspect to increasing atmospheric carbon dioxide – ocean acidification. As carbon dioxide is absorbed by seawater, it produces a weak acid, resulting in a lowering of the pH and a decrease in the concentration of carbonate ions in the water. Seawater is normally slightly alkaline with an average pH in pre-industrial times of about 8.1. But since then the average pH of surface waters has reduced by about 0.1 units, and a further reduction of up to 0.2–0.3 units is projected by 2100 (Table 2.1). Whilst these sound like small decreases, the pH scale is logarithmic and the Earth's surface waters have probably not experienced such a change of pH for millions of years (Feely *et al.*, 2004).

A lowering of the carbonate ion concentration in seawater makes it energetically more costly for calcifying organisms to produce their various shells and skeletons. Such organisms are very numerous and include coralline algae, corals (both shallow tropical and deep cold-water reef builders), and molluscs, crustaceans, bryozoans, and echinoderms. Larval stages of benthic invertebrates may be even more susceptible. Important calcifiers of the permanent plankton are foraminiferans, coccolithophores, and pteropods, which in turn are important components of marine food webs. Calcification is not, however, the only process sensitive to changes of pH. Other physiological processes are affected as well as aspects of seawater chemistry such as nutrient dynamics (Fabry *et al.*, 2008; Guinotte & Fabry, 2008, Pörtner, 2008). A meta-analysis of more than 200 studies shows that overall, across a wide range of marine organisms, ocean acidification results in decreased survival, calcification, growth, development, and abundance. Taxa vary in their responses, but in general heavily calcified organisms, including calcified algae, corals, and molluscs, and the larval stages of echinoderms are the most adversely affected (Kroeker *et al.*, 2013).

The solubility of calcium carbonate increases as temperature decreases and pressure increases, so there is a water depth – the saturation horizon – below which calcium carbonate will tend to dissolve. Calcifying organisms use two forms of calcium carbonate: calcite and aragonite. The saturation horizons for these two forms are gradually becoming shallower in the world's oceans under ocean acidification, slowly reducing the habitable space where calcification remains a going concern. Aragonite is about 50% more soluble in seawater than calcite, so calcifiers that use mainly aragonite, such as scleractinian corals and pteropods, are more vulnerable than those that mainly use calcite. Since colder waters can take up more carbon dioxide, acidification is expected to impact on marine calcifiers in high latitudes more severely.

Some ocean regions, such as upwelling systems, have a naturally lower pH as a result of raised $CO_2$ from increased organic matter remineralization, making them particularly susceptible to ocean acidification. For instance, by 2050 much of the California Current System is forecast to become under-saturated in aragonite throughout the year, with potentially major effects on this diverse ecosystem (Gruber *et al.*, 2012).

Already there is evidence of large-scale impacts on coral reefs, indicating that increasing temperature and declining aragonite saturation are making it harder for corals to maintain skeletal carbonate (see Chapter 13). Projections indicate that unless atmospheric carbon dioxide concentrations can be kept below about 450 ppm, then coral reefs are likely to progressively erode and become dominated by algae (Hoegh-Guldberg, 2011).

## Ozone Depletion

Marine life is also affected by human-induced atmospheric changes that increase the amount of ultraviolet (UV) radiation reaching the Earth's surface. Within the lower stratosphere, about 20–25 km above the Earth's surface, is a layer where ozone attains relatively high concentrations (about 10 ppm). This layer shields the Earth from the sun's UV radiation. In the 1970s, concern was raised that some industrial chemicals can accelerate the breakdown of ozone in the stratosphere. Among the most important of these are chlorofluorocarbons (CFCs), synthetic volatile compounds used mainly as refrigerants and aerosol propellants (and which are also powerful greenhouse gases). CFCs are unreactive in the lower atmosphere, but when eventually they reach the stratosphere they are broken down by UV radiation, releasing chlorine atoms and causing an ozone-destroying chain reaction. In the 1980s there was dramatic confirmation that CFCs were depleting the stratospheric ozone layer with the discovery of a seasonal ozone 'hole' over Antarctica where ozone levels were greatly reduced (see Chapter 17). The hole develops during winter (the very low temperatures and resulting ice particles in the stratosphere catalyse the reactions) and reaches its greatest extent by early spring; it is presently up to at least 25 million $km^2$. Ozone depletion, although generally weaker, is also seen in the warmer Arctic (Solomon *et al.*, 2014).

UV radiation has many, mostly harmful, effects on living organisms, damaging proteins and nucleic acids and negatively affecting growth, development, and reproduction of marine invertebrates and fish (Dahms & Lee, 2010). The wave-bands of UV radiation that reach the biosphere are UV-A (315–400 nanometre (nm)) and UV-B (280–320 nm). UV-B can penetrate clear ocean water to depths of 20–30 m and UV-A to 40–60 m. However, UV-B appears the more harmful as it can directly damage DNA. Early developmental stages are generally the most susceptible. The various lethal and sub-lethal effects of UV radiation on marine organisms, particularly in combination with other climatic stressors, may substantially impair populations of individual species, but the consequences for marine ecosystems are at this stage less clear (Häder *et al.*, 2015).

Concerns about the effects of increased UV radiation on human health (e.g. increased risk of skin cancer and cataracts) and on the natural environment led to the Montreal Protocol, an international agreement to reduce and phase out the production of the principal ozone-depleting substances. The protocol entered into force in 1989 and has been ratified by all UN countries with high overall compliance. Ozone-depleting substances in the stratosphere are now decreasing, and there is evidence of the onset of healing of the ozone layer, but full recovery is not expected before mid-century (Solomon *et al.*, 2016). The Montreal Protocol illustrates what can be achieved by concerted international action to mitigate effects of anthropogenic climate change.

# Exploitation of Marine Resources

## Living Resources

The greatest impact of humans on the marine biosphere has been through the exploitation of its living resources. Since prehistoric times humans have exploited fish, shellfish, and other marine organisms for food and other uses (Erlandson & Rick, 2010). Large sought-after marine animals have in particular suffered from intense exploitation. An analysis by Lotze and Worm (2008) of more than 250 records of large fishes, sea turtles, coastal birds, and marine mammals shows today's populations to have declined by on average 84% from historical abundance levels (Fig. 2.4). And analyses by Edgar *et al.* (2014) comparing effective marine protected areas with fished coasts indicate that as a result of fishing, total fish biomass on temperate and tropical reefs has declined 63% from historical levels and shark biomass by 93%. Even artisanal fishing at relatively low levels can markedly impact seafood species that have vulnerable life-history characteristics (Pinnegar & Englehard, 2008). Some societies have developed practices that deliver sustainable yields. In the Pacific basin, for example, traditional marine conservation methods include reef and lagoon tenure systems where the right to fish a particular area is controlled by the local social unit so as to constrain overexploitation (see Chapter 5). But more familiar has been a history of expansion and discovery by societies seeing no need to question the seas' apparent inexhaustibility, thereby developing little in the way of a conservation tradition. This perspective has become untenable under the demands of a rapidly growing human population and an increasing capability to extract living resources and as our understanding of the sea and its sustainable yields has developed. Gross overexploitation of highly vulnerable species dates back centuries, and nowadays there are few marine populations of commercial significance (stocks) that are not overstressed. Capture fisheries remove a biomass of more than 100 million tonnes annually (Pauly & Zeller, 2016), and only about 10% of assessed stocks are not fully or overfished (FAO, 2016).

By their nature, large-scale Western-style fisheries were historically predisposed to overexploitation. There have been few restrictions on fisheries beyond narrow territorial waters before states began extending their jurisdiction in the form of exclusive economic and fishing zones. The

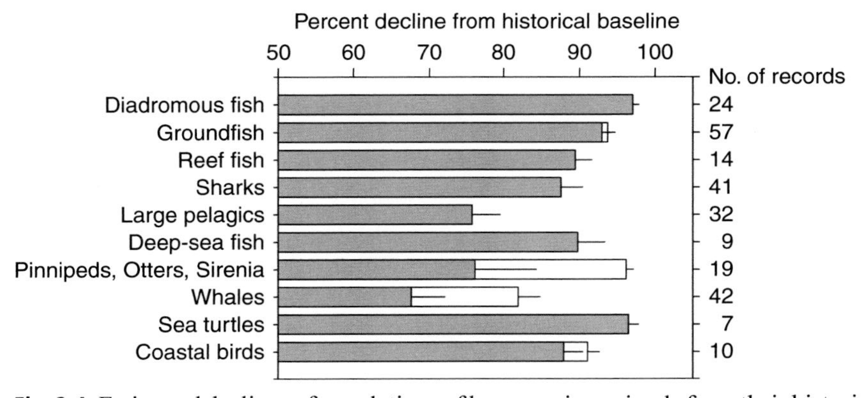


**Fig. 2.4** Estimated declines of populations of large marine animals from their historical baselines, based on 256 records from 95 studies. The average decline and standard error for each group are shown to the low point in abundance (white bar) and, if different, to the most recent abundance estimate (grey bar). From Lotze, H.K. & Worm, B. (2008). Historical baselines for large marine animals. *Trends in Ecology and Evolution*, 24, 254–62. With permission from Elsevier.

problem of overexploitation resulting from uncontrolled access to common property resources by users pursuing private gain was described by Gordon (1954). The theme was further expounded by Hardin (1968) as the 'tragedy of the commons', by analogy to the overexploitation of common land through unrestricted grazing. In this scenario each person who grazes animals on the common can maximize their personal gain by continually adding to their herd. The personal benefit of doing this outweighs any negative consequences to individuals of overgrazing because these are shared by the community as a whole, at least until the carrying capacity of the commons is reached. Many fishes and invertebrates undergo wide natural fluctuations of abundance, and this can accelerate overexploitation. A run of good years encourages additional investment, but during poor years there is strong economic and political pressure not to disinvest (Ludwig *et al.*, 1993). The term 'tragedy of the commons' is often applied to fisheries, although there are successful community-based systems of fisheries in industrialized countries (Acheson, 2005). But common property as an institution for resource management is now rarely considered viable, at least in developed countries with their market economies dominated by the concept of private property rights. The principal options advocated for addressing the tragedy of the commons are by government control of the resource and regulation of the users and by privatizing the resource, such as by assigning individual quotas, or spatial rights (see Chapter 5).

Environmental consequences of the large-scale removal of fish, whales, and other marine organisms can extend far beyond effects on the targeted species (Dayton *et al.*, 1995; Hall, 1999; Jennings *et al.*, 2001). Removing key predators from marine communities may result in significant shifts in community structure, including the replacement of exploited species by others at a similar trophic position or an increase in production at lower trophic levels. There is evidence for such control by high-level predators from most oceanic ecosystems (Baum & Worm, 2009). There are also many examples from shallow-water systems, such as kelp forests, rocky subtidal and intertidal habitats, and coral reefs, where removal of a predator can result in a 'trophic cascade' involving a series of interactions propagating down the food web (e.g. carnivore-herbivore-algae) (Pinnegar *et al.*, 2000).

Some fisheries are damaging mainly because of the large quantity of unintentional catch, or bycatch, that is also caught. Differences in life-history characteristics can mean that bycatch species are more vulnerable to fishing pressure than the target species, as in the incidental catch of some sharks and rays, turtles, seals, and small cetaceans. Fishing may also entail significant physical disturbance or destruction of habitat, as in certain types of trawling and reef fishing (see Chapter 5).

There are also innumerable marine species that are exploited not for food but for raw materials, jewellery and curios, the aquarium trade, medicinal compounds, and scientific and educational purposes. Many of the active compounds in pharmaceutical drugs are derived from terrestrial plants and microorganisms. Research over recent decades has shown that marine organisms are also a potentially rich source of compounds with biomedical potential. Many thousands of novel compounds spanning a wide range of chemical structures have been identified from marine organisms, and a number are already approved drugs or in clinical development (Mayer *et al.*, 2010). Sponges, cnidarians, ascidians, and other sessile organisms, particularly from reef habitats, have been the main target as many of them employ potent metabolites to deter competitors and predators. Concerns have been raised about overcollecting, especially given the low natural concentrations of such metabolites in organisms and if the target organism is rare or restricted in its distribution. However, advances in screening technology mean that smaller amounts can now be used for assessment, and usually once the compound is in clinical development it has been synthesized (Hunt & Vincent, 2006).

A number of marine laboratories and scientific suppliers trade in specimens of marine organisms for research, museums, and school and university classes. In most cases the effects of such collecting on wild populations would be negligible. Some potentially vulnerable species are, however, taken in large numbers. In the USA alone it is estimated that some 100 000 spiny dogfish are used each year for school dissections. The species is already heavily impacted by commercial fisheries and, like many chondrichthyans, is a long-lived species of low fecundity. Balcombe (2000) recommends that biological supply companies should be required to conduct environmental impact assessments before collecting from wild populations and that dissection of species whose populations are known to be overexploited and/or in decline should be discontinued.

### Non-living Resources

The oceans are also important sources of non-living resources, in particular seabed minerals and petroleum hydrocarbons. The offshore oil industry has expanded rapidly since the 1960s and 1970s following advances in technology and a rise in the value of crude oil and gas. The main oil production regions are the Middle East, Europe and Eurasia, and North America (BP, 2015). Oil exploration and production can result in various impacts on marine environments depending on the location and nature of the development. Among potential effects are those due to the chronic discharge of contaminants, especially of cuttings, drilling muds, and production water; the physical disturbance from the placement of structures and pipelines; and the risk of a blowout and the effect of a major oil spill on sensitive coastal habitats and on aggregations of birds, mammals, and turtles (Boesch *et al.*, 1987).

Important hard mineral deposits that are mined from the seabed are sands and gravels dredged from shelf areas to supply the building industry. Other seabed minerals of beaches and coastal waters include placer deposits in which metals such as tin, gold, and platinum have become concentrated. Important at certain deep-sea sites – but yet to be mined at large commercial scales – are sulphide deposits containing copper, zinc, lead, cadmium, gold, and silver; phosphorite nodules; and nodules rich in manganese, copper, nickel, and cobalt (see Chapter 14).

## Physical Alteration of Habitat

Human activities commonly result in marine habitats being physically modified, often radically. This is especially pertinent for coastal areas, which are home to an increasing number of people. In many parts of the world, such as in the Mediterranean and in tropical countries, this pressure is exacerbated by the growth of tourism. Tourism is a growing sector of the global economy, much of it impacting coastal regions (Orams, 1999; Davenport & Davenport, 2006). As well as the direct loss and fragmentation of habitat, increasing urbanization of coastal areas results in degradation and pollution of the coastal environment. A major cause of irreversible loss of habitat is the drainage of coastal wetlands and infilling of estuaries (see Chapters 11 and 12). Poor land management practices on adjacent catchment areas can also result in important physical impacts on coastal ecosystems. Large-scale deforestation and changes in land use can leave soils susceptible to erosion and result in massive inputs of terrestrial sediment to coastal areas (Thrush *et al.* 2003).

Modifying the natural flow of a river can impact on marine systems in a variety of ways. Dam construction tends to decrease the discharge of surplus water to the sea, reducing the nutrient concentrations and biological productivity of adjacent waters (see Chapter 10). A reduction in the delivery of alluvium can lead to coastal retreat, whilst the dam itself can hinder or obstruct spawning migrations (see Chapter 6). Large-scale diversion of rivers for irrigation can be devastating for enclosed systems, such as the inland brackish seas of central Asia.

Beyond the shore, widespread physical alterations to seabed habitats are caused by trawling, dredging and spoil disposal, and extraction of aggregates and other seabed minerals (see Chapter 10)

## Biological Invasions

As a result of human activities, many marine species have been carried to areas outside their native range, where they have then established viable populations. There are numerous examples where such species have spread and significantly impacted native habitats and communities. The process has presumably been going on since humans first embarked on maritime exploration and has impacted marine habitats worldwide, particularly estuarine and coastal areas (Molnar *et al.*, 2008). Many nearshore marine species, possibly more than 1000, thought to be naturally cosmopolitan in distribution may in fact be the result of pre-nineteenth-century invasions (Carlton, 1999).

Such movement of species has been both intentional and accidental. Many species have been purposely introduced, usually in attempts to develop new fisheries and aquaculture industries. Important in this regard, at least since the 1870s, has been the worldwide transportation of commercial oysters. The movement of oysters and other shellfish around the world has also, however, resulted in the unintentional introduction of numerous accompanying species. There are also many species that have been accidentally introduced over the centuries, particularly as a result of shipping, notably as fouling and boring organisms and those that have been carried in ships' ballast water. Other major human-mediated pathways for exotic species are aquaculture and canals (Fig. 2.5). The Suez Canal has enabled hundreds of Red Sea species to become established in the

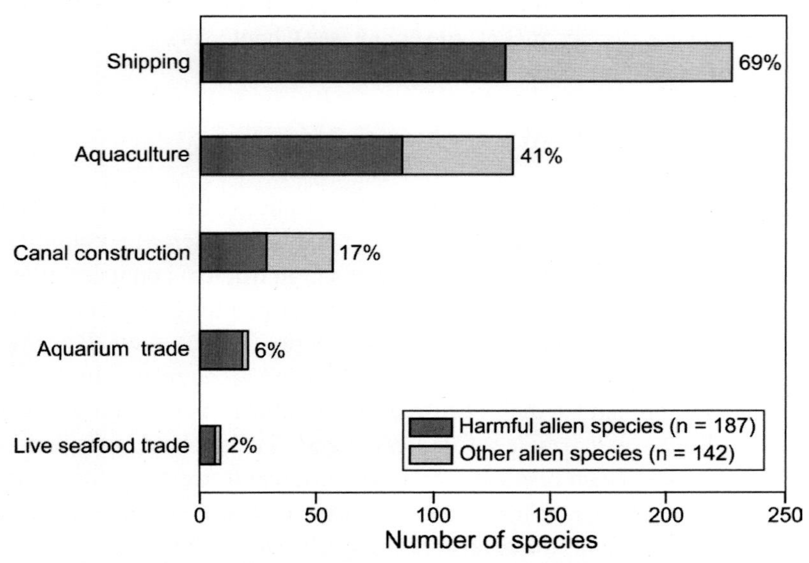


**Fig. 2.5** The main human-assisted pathways known or likely to be taken by alien marine species. The darker shading shows the proportion rated as having a high ecological impact. The percent of the total number of species in the assessment ($n = 329$) is indicated. Republished with permission of the Ecological Society of America from Assessing the global threat of invasive species to marine biodiversity, Molnar, J.L., Gamboa, R.L., Revenga, Carmen, R. & Spalding, M.D., *Frontiers in Ecology and the Environment*, 6 (9), copyright 2008; permission conveyed through Copyright Clearance Center, Inc.

eastern Mediterranean (see Chapter 16). Proposals to build a sea-level canal across the Panama isthmus would have enormous implications for the exchange of west Atlantic and east Pacific species. Just how many introductions in total humans might be responsible for is unknown – many will go unrecognized – but they likely amount to some thousands (Carlton & Ruiz, 2005).

Invasions by marine species have not, as far as we know, resulted in any global extinctions (Briggs, 2012), but their impact can still be dramatic. Some species have traits that make them especially successful as invaders, and some habitats appear to be more readily invaded than others. Systems with low species diversity and those subject to stressors appear to be among the most susceptible. Enclosed waters, estuaries, ports, and harbours are habitats that typically are most heavily invaded (see Chapter 11), although they also tend to be areas where, because of human activities, the opportunities for invasion are greatest (Carlton & Ruiz, 2005). There is evidence that increased species richness decreases invasion success, probably because available space is used more completely and efficiently in species-rich communities. Loss of biodiversity may thus facilitate invasion (Stachowicz *et al.*, 1999).

Examples abound where alien species have become prominent components of coastal communities. There are well-known examples of non-indigenous species in European waters that have been introduced by various means (Raffaelli & Hawkins, 1996). The barnacle *Austrominius modestus*, endemic to southern Australia and New Zealand, was discovered on the south-east coast of England in 1945, presumably having arrived as a hull-fouling species, and is now a major component of sheltered and estuarine rocky shore faunas of NW Europe. And as a result of the transport of non-native oysters, the slipper limpet *Crepidula fornicata* was introduced from the east coast of North America to western Europe, where it competes with oysters for space and food. Similarly, an important predator of oysters, the American oyster drill *Urosalpinx cinerea*, was found in oyster beds in SE England in the late 1920s.

One of the more conspicuous introduced species of tidal flats of NW Europe is the Pacific oyster *Crassostrea gigas*, a NW Pacific native (Troost, 2010). The species was introduced to Dutch waters in the mid-1960s – following the demise of European oyster stocks – and subsequently to other European countries. It has spread rapidly in NW European estuaries, significantly impacting these systems. In particular, the Pacific oyster forms extensive reef-like structures, substantially altering the benthic habitat in invaded estuaries, although the increase in structural complexity may in fact promote biodiversity (Markert *et al.*, 2010). Also, on account of the high filter-feeding rate of *C. gigas*, the oyster reefs exert a strong control on the estuarine phytoplankton, which in turn likely affects higher trophic levels. The Pacific oyster is a good example of a species that possesses the traits of a successful invader (Table 2.2).

Marine invasions are no doubt greatly underestimated, partly because of uncertainties over identification. Given the ubiquity of cryptic species that are morphologically similar, it is likely that many introduced species have gone undetected (Geller *et al.*, 2010). European crabs of the genus *Carcinus* consist of a species pair: *C. maenas* (Fig. 2.6) and *C. aestuarii* on Atlantic and Mediterranean shores, respectively. Mitochondrial DNA analysis shows that both species invaded South Africa and Japan, whereas previously it was thought only *C. maenas* was in South Africa and only *C. aestuarii* in Japan (Geller *et al.*, 1997).

The exchange of huge volumes of ships' ballast water between marine regions has the potential to promote the introduction of marine organisms on a massive scale. David (2015) estimated that some 3 billion tonnes of ballast water was discharged from vessels engaged in international seaborne trade in 2013. And every day, of the order of 3000 to 7000 marine species (excluding microorganisms) are likely to be in transit in the ballast water of the world's shipping (Gollasch *et al.*, 2015). These are mostly small, free-living, often larval, stages with crustaceans, molluscs, polychaetes, and algae usually well represented. Some organisms can survive in sediment in the bottom of ballast tanks for months, especially if they develop resting stages. There is strong evidence that toxic

**Table 2.2** Traits of Successful Invaders

Different characteristics may be needed for different stages of invasion (i.e. initial colonization, establishment in the receiving habitat, and subsequent range expansion by natural means).

| Stage | Trait |
|---|---|
| Colonization | *r*-selected life history strategy:<br>Rapid growth<br>Rapid sexual maturation<br>High fecundity<br>Generalists:<br>Ability to colonize wide range of habitat types<br>Broad diet<br>Tolerant of wide range of environmental conditions<br>Gregarious behaviour<br>Genetic variability and phenotypic plasticity<br>Ability to recolonize after population crash |
| Establishment | Lack of natural enemies<br>Ecosystem engineer<br>Association with humans<br>Repeated introductions<br>Genetic variability and phenotypic plasticity<br>Competitiveness |
| Natural range expansion | Traits of successful colonists (see above)<br>Dispersability |

Reprinted from *Journal of Sea Research*, Vol. 64, Troost, K., Causes and effects of a highly successful marine invasion: case-study of the introduced Pacific oyster *Crassostrea gigas* in continental NW European estuaries, pages 145–65, copyright 2010, with permission from Elsevier.

**Fig. 2.6** The green shore-crab (*Carcinus maenas*) is native to NW Europe, where it occurs on a variety of open coast to estuarine shores, but has been introduced to many other temperate locations around the world. It is a highly successful invader, tolerant of a wide range of habitats, and an important predator. Photograph: Vital Signs Program at the Gulf of Maine Research Institute. (A black and white version of this figure will appear in some formats. For the colour version, please refer to the plate section.)

dinoflagellates, which can also survive as resting cysts, have spread by this means, resulting in the potential for widespread occurrence of toxic outbreaks (Bolch & de Salas, 2007). It is now apparent that many bioinvasions may be attributable to ballast water transport. Hayes and Sliwa (2003) identified more than 850 marine species from around the world that appear to have been successfully introduced into new regions by ships.

Modelling and field data indicate that invasion risk is greatest where there is high shipping intensity between tropical and subtropical ports (those with generally high species richness) of similar temperature and salinity and where the journeys are of intermediate distance (about 8000–10 000 km). At shorter distances non-native species are less likely to be transported, whereas on longer journeys there is less chance of survival. Focusing on mitigative measures at selected high-risk ports, particularly in SE Asia and the Middle East, would considerably reduce the risk of ballast water introductions globally (Seebens *et al.*, 2013).

To reduce the risk of introductions, the International Maritime Organization (IMO) adopted the Ballast Water Management Convention in 2004, with entry into force in September 2017 (www.imo.org.). Under the convention, ships should wherever possible exchange 95% of their ballast water at least 200 nm from the nearest land and in water at least 200 m deep. Performance standards are measured in terms of maximum number of viable organisms allowed to be discharged. This type of ballast water exchange is based on the assumption that coastal species will be unlikely to survive offshore environmental conditions, and vice versa (but see McCollin *et al.*, 2008). In addition, many onboard treatment methods have been approved by IMO, involving, for example, mechanical separation, electrolysis/electrochlorination, and UV radiation (David & Gollasch, 2015).

The primary focus on controlling bioinvasions must always be on preventive measures that address all identified vector routes (Bax *et al.*, 2001). Eradication may be achievable if a response programme can be organized quickly, so public awareness programmes to increase the chances of early detection and developing action plans for responding to new invasions are important. Infestations of the invasive green alga *Caulerpa taxifolia* were discovered in 2000 at two locations in southern California, but a rapidly mounted eradication effort was deemed to have succeeded after 2 years. In the Mediterranean Sea, by contrast, a delay of several years following discovery of the same alga enabled it to spread across thousands of hectares of coastal habitat (Simberloff *et al.*, 2013; see Chapter 16). Unless the invading species is discovered early on and its distribution still localized, then subsequent eradication is virtually impossible.

## Pollution

Humans have always used the sea as a repository for wastes. For thousands of years the effects of such inputs were likely to have been very localized. But with the rapid rise in human population and with urban and industrial growth, it has become evident that the seas are not sufficiently vast to accommodate unlimited inputs of wastes from human activities without suffering widespread adverse effects. Nevertheless, much of the world's waste still ends up in the sea. This also represents a tragedy of the commons: there is an incentive for individuals and industries to dispose of wastes in the cheapest way possible – usually the most polluting – so that the eventual costs in terms of environmental degradation are borne by society as a whole.

An organization that has done much to increase our understanding of marine pollution is the Joint Group of Experts on the Scientific Aspects of Marine Environmental Protection (GESAMP). Sponsored by several UN bodies, GESAMP is a multidisciplinary group of independent experts that provides advice on scientific aspects of marine pollution and environmental protection (Wells *et al.*, 2002). Its first meetings were in 1969, and until the early 1990s, its focus was marine pollution. It has since broadened its mandate to encompass marine environmental protection and management more generally. Its various working groups have published numerous studies, particularly on the evaluation of potentially harmful substances, assessment of effects of marine pollutants, the monitoring and control of pollution, and the state of the global marine environment.

A commonly adopted definition of marine pollution is that given by GESAMP as 'the introduction by man, directly or indirectly, of substances or energy into the marine environment (including estuaries) resulting in such deleterious effects as harm to living resources, hazards to human health, hindrance to marine activities including fishing, impairment of quality for use of sea water and reduction of amenities' (GESAMP, 1990a). A similar definition was adopted by the United Nations Convention on the Law of the Sea (Article 1) (UN, 1983).

Pollution by the above definition is linked to human activities, although that is not necessarily easily established. Many substances that enter the sea, such as organochlorine pesticides and plastics, are synthetic and thus attributable only to human activity. But some inputs due to humans are similar to natural inputs, such as the seepage of oil from the seabed, areas where the geology furnishes high levels of metals or radioactivity, or an upwelling region that results in an unusually high deposition of organic material and oxygen depletion. Also, by the GESAMP definition, the effects of pollution are deemed harmful. A distinction is sometimes made between pollution and contamination, the latter being the presence of a substance at a concentration that is above natural background levels but without producing significant adverse effects. Making this distinction can be questionable, though (Taylor, 1993). Views differ as to what constitutes 'harm', and anthropogenic effects may often be indistinguishable against the typically high background variability of biological systems. However, in practice it is often necessary to formally recognize an acceptable level of harm, as for example under the terms of a discharge consent.

An enormous variety of substances enter the marine environment as a result of human activities. Impacts are many and varied, and it is possible here only to highlight major issues. There are dangers in attempting to generalize about marine pollution as regional differences and the vulnerability of habitats and taxa can vary greatly. Specific pollution problems are, however, dealt with in more detail in later chapters. The literature on marine pollution has grown enormously in recent decades. Clark (2001) provides a useful introductory treatment.

## Sources and Types of Pollutants

Rough estimates of the relative contribution of pollutants entering the sea indicate that runoff and land-based discharges contribute some 44%, the atmosphere 33%, maritime transportation 12%, dumping 10%, and offshore production 1% (GESAMP, 1990a). Thus by far the greatest amounts come from land-based and atmospheric sources. Pollutants entering from the land include those from point sources, such as discharges of waste through pipelines carrying sewage and industrial effluents, as well as diffuse sources such as runoff from urban and agricultural areas and groundwater seeps. Among the major causes of concern in the marine environment on a global basis are sewage, nutrient inputs, plastic litter, and synthetic organic compounds (Box 2.2).

### Organic Wastes and Nutrients

Most of the waste discharged to coastal and estuarine waters is organic matter. Sewage is by far the most important contributor, but other wastes such as those from food-processing plants and pulp mills can also be significant (Fig. 2.7). It is common practice in many parts of the world for sewage to be discharged untreated into the sea. Apart from aesthetic considerations, such effluents are a major concern as human pathogens in sewage pose a threat to public health (e.g. hepatitis, cholera, and gastrointestinal diseases). Levels of faecal bacteria in sewage, notably

## Box 2.2 Major Categories of Marine Pollutants

Organic matter
- Sewage and pulp-mill, abattoir, and food industry waste

Nutrients (nitrates, phosphates)
- Organic wastes and agricultural and urban runoff

Petroleum hydrocarbons
- Paraffins, naphthenes, and polycyclic aromatic hydrocarbons

Persistent organic pollutants
- Organochlorine pesticides, polychlorinated biphenyls

Debris
- Plastics (macro- to microplastic particles)

Metals
- Heavy metals (mercury, cadmium, copper, lead)
- Organometals (tributyltin)

Thermal and radioactive discharges
- Power station cooling water

faecal coliforms or enterococci, can be used to designate water quality standards, for instance, in classifying water as suitable for bathing or for shellfish gathering. However, these broad-scale indicators have limited ability to link elevated bacterial counts to a particular source of pollution. Recent advances in DNA-based microbial source tracking techniques can help pinpoint pollutant sources by distinguishing between human and livestock sewage contamination. Sewage also generally contains other contaminants such as nutrients, heavy metals, and hydrocarbons.

Sewage and other organic wastes are subject to bacterial degradation. The organic loading creates a high demand for dissolved oxygen by aerobic bacteria, and when this demand exceeds the rate at which oxygen is taken up by diffusion, then oxygen depletion occurs. Organic matter can still be broken down by anaerobic bacteria, but the process is slower, so that waste is liable to accumulate, and the end products include toxic compounds such as methane and hydrogen sulphide. To what extent problems arise will depend therefore on the rate of input of the effluent and its oxygen demand and on the hydrography of the receiving environment: whether the prevailing conditions promote the dilution and dispersion of effluent and re-oxygenation of the water.

Sewage and other domestic wastes also have elevated concentrations of nutrients, most importantly forms of nitrogen and phosphorus. Nutrients also enter the marine environment from industrial discharges and agricultural runoff and via emissions to the atmosphere (nitrogen oxide from fuel combustion). Since the Industrial Revolution, human activity has hugely increased the supply of reactive, biologically available forms of nitrogen to the environment, particularly since the discovery of the Haber-Bosch process – the manufacture of ammonia from combining nitrogen and hydrogen – enabling the production of synthetic nitrogen fertilizer and expansion of agriculture. The use of synthetic nitrogen fertilizer has increased dramatically in recent decades – exacerbated by an increased per capita consumption of meat – with more than half the fertilizer ever used being

**Fig. 2.7** Waste outfall from an abattoir. Pollutants enter the marine environment primarily from land-based sources, with organic wastes and nutrients being the main contributors. Anthropogenic nutrient enrichment affects coastal waters in many parts of the world, resulting in degradation of ecosystem structure and function. Photograph: P.K. Probert. (A black and white version of this figure will appear in some formats. For the colour version, please refer to the plate section.)

produced since 1985 (Howarth, 2008). Inputs of nutrients to the marine environment through runoff and other routes have increased greatly as a result, although the relative importance of these sources can vary greatly from region to region.

The input of nutrients to the sea stimulates the growth of phytoplankton and macrophytes. In moderation, this might enable larger populations of consumers to be sustained, an effect that could be seen as beneficial where, for instance, it leads to a more productive fishery. But problems arise when the input of nutrients exceeds the normal assimilative capacity of the system. Then the growth of algae begins to outstrip the capacity of the system to use this increased food source through its normal channels of energy flow. This phenomenon, known as eutrophication, can substantially alter the structure of communities. Andersen *et al.* (2006) define eutrophication as 'the enrichment of water by nutrients, especially nitrogen and/or phosphorous and organic matter, causing an increased growth of algae and higher forms of plant life to produce an unacceptable deviation in structure, function and stability of organisms present in the water and to the quality of water concerned compared to reference conditions'.

An increase in the rate of nutrient supply and ensuing eutrophication can have various impacts on marine ecosystems (see e.g. Gray, 1992; Rabalais, 2005). The primary effect is the increase in primary production, particularly by phytoplankton but also by macroalgae. There is also typically a change in the composition of the algal communities, and this may entail a major shift in the dominant group. In phytoplankton communities, for example, this may be a shift from diatoms to dinoflagellates and in some cases result in blooms of toxic algal species. Reduced light penetration as a result of algal blooms may adversely affect submerged vegetation through shading and competition. Eutrophic conditions also favour certain types of macroalgae, notably opportunistic green seaweeds, which can form extensive mats in shallow bays and inlets. One of the initial signs of eutrophication in macrophyte communities and on coral reefs is a proliferation of epiphytic algae.

Since under eutrophic conditions more algal material is produced than can be consumed, organic material settles to the seabed. This produces a characteristic response among the sediment benthos, encouraging dense populations of relatively few species that thrive in organic-rich sediments. The bacterial degradation of accumulated organic matter can deplete the oxygen concentration of the water, producing hypoxia (oxygen concentrations below 2 mg $L^{-1}$) or even anoxia ($< 1$ mg $L^{-1}$). As the oxygen content declines, there will be some behavioural changes of the fauna. Initially fish will leave the area and burrowing species will move up to the sediment surface. But if the oxygen concentration drops below about 1 mg $L^{-1}$, macrofauna will be largely eliminated. The overproduction of organic material can lead to extensive smothering of the seabed, the development of anoxic conditions with the production of hydrogen sulphide, and mass mortalities of organisms. There are numerous nutrient-enriched coastal zones of the world, some up to thousands of $km^2$ in area, that experience severe oxygen depletion and loss of fauna, at least periodically (see Chapter 10).

There is no simple direct relationship between the amount of nutrient added to a given system and the growth of algal populations. Particularly important are the flushing characteristics of the system and thus the flux of nutrients in relation to the normal assimilative capacity of the ecosystem (GESAMP, 1990b). Some coastal environments and ecosystems are thus more sensitive than others to the effects of eutrophication. Among the more vulnerable are estuaries, lagoons, inlets, and enclosed bays, since they are usually relatively shallow areas with limited water exchange. Being adapted to low nutrient conditions, coral reef systems are especially sensitive to effects of eutrophication (see Chapter 13).

Sewage can be treated in various ways depending on the quality of the effluent that can be safely discharged (Lester, 1990; Clark, 2001). Options range from screening, or simply removing the larger solids (primary treatment), to secondary treatments in which the oxygen demand of the resultant liquid is further reduced by promoting aerobic degradation, such as by trickling the effluent through a bed of porous stones that provides a large surface area for the microbial community. More advanced treatments, such as tertiary treatment, can be employed to remove finely suspended solids and to reduce the levels of nutrients and pathogens.

Whilst it is feasible to treat point-source discharges to reduce nutrient loads, non-point sources present a far more difficult problem. Diffuse nutrient inputs from agricultural runoff are important in many coastal areas. Often, however, the application of fertilizers appears to be excessive and wasteful, and there may be the potential to improve farming practices so as to make better use of fertilizers without jeopardizing food production (Gabric & Bell, 1993).

Eutrophication is one of the most damaging of human impacts on coastal ecosystems worldwide. Frequently, however, such coastal areas are also subject to other human impacts, and it may not be apparent which source of disturbance is primarily responsible for an observed habitat degradation and alteration of community structure. As mentioned earlier, removal of top predators through overfishing can have significant flow-on effects. This is known as top-down control, where a predator influences the trophic levels below it. By contrast, nutrient supply is bottom-up control as it determines the amount of energy that can enter the food web. Overfishing, by relaxing predation pressure on grazers, could permit increased growth of algae and so have an outcome similar to that of nutrient enrichment. Increases in plant biomass in shallow benthic habitats may often be driven more by consumers (top-down control) than by nutrients (bottom-up control) (Heck & Valentine, 2007).

## Petroleum Hydrocarbons

Humans have for centuries used petroleum deposits for a variety of purposes, but it was not until the development of oil fields in the USA in the mid-nineteenth century followed by the invention of the internal-combustion engine that the modern oil industry mushroomed. World consumption of oil now exceeds 4 billion tonnes per year (BP, 2015).

Through human activities, the input of petroleum hydrocarbons to the sea has greatly increased, although estimates vary widely. The total amount of oil now entering the marine environment from all sources has been estimated at 1.3 million tonnes per year (National Research Council, 2003). Nearly half the total input is estimated to be from natural seepage, where oil-bearing strata outcrop at the seabed, and the other half from human activities. Inputs resulting from human activities are due largely to operational discharges from vessels (discharge of fuel oil sludge and cargo washings), land-based runoff (such as from urban runoff, municipal wastes, and refineries), and shipping and offshore platform accidents. The relative importance of different sources can vary greatly from region to region depending on patterns of population and industrialization, shipping, and offshore activities. Inputs can also fluctuate considerably from year to year due, for instance, to the significance of tanker or blowout accidents. In recent decades, with the development of supertankers, there is the potential for large quantities of oil to be released rapidly from ships, with locally disastrous consequences. Tanker spills and other accidental releases in fact account for less than 10% of the total inputs of oil (National Research Council, 2003). It is, however, this form of oil pollution that attracts the most attention, although the number of large spills has steadily declined in recent decades (Fig. 2.8). On the other hand, there are other significant sources on the horizon. Sunken ships contain a huge reservoir of oil that will eventually be released to the sea. Michel *et al.* (2005) estimated that worldwide there are more than 8500 wrecks, nearly 1600 of them tankers, that pose a significant oil pollution risk. The volume of oil they contain is of the order of 2.5 to 20 million tonnes. More than three-quarters of these wrecks date from World War II and are reaching a state of corrosion that their releases of oil might be expected to peak over the next few decades.

Crude oil is a complex and highly variable mixture containing thousands of different compounds. The hydrocarbon component (usually > 95%) comprises alkanes or paraffins (aliphatic hydrocarbons), naphthenes (alicyclic hydrocarbons), and aromatic compounds. Other components include organic compounds containing nitrogen, sulphur, and oxygen and trace metals such as nickel and vanadium coupled with organic compounds.

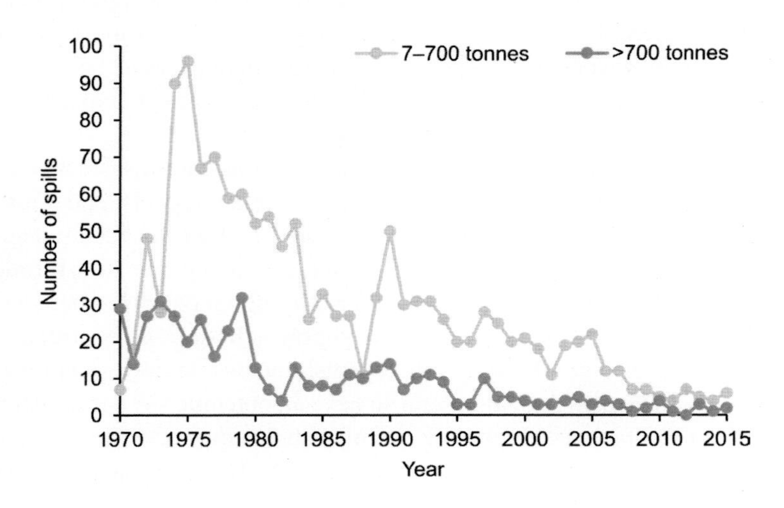


**Fig. 2.8** The number of medium (7–700 tonnes) and large (> 700 tonnes) oil spills from 1970 to 2015. From ITOPF (2016). Oil Tanker Spill Statistics 2015. The International Tanker Owners Pollution Federation Ltd.

Oil spilled at sea immediately starts to undergo physical and chemical changes, though the nature of this weathering depends on a host of factors relating to the type of oil and the prevailing environmental conditions. Spilled oil rapidly spreads over the sea surface as a thin film or slick to be moved by winds and currents. A significant amount of the more volatile (low molecular weight) fractions may quickly evaporate. It is also these components that are the most soluble. Turbulence will promote the formation of emulsions. In rough weather, many crude oils can in a few hours form a stable and viscid water-in-oil emulsion, so-called 'chocolate mousse', which being resistant to degradation is difficult to treat if it comes ashore. Photochemical degradation can be important at the slick surface, leading to the formation of more soluble products. Microbial activity is important too in the decomposition of oil but again is highly variable depending upon such factors as temperature and availability of nutrients and oxygen. Balls of tar are often the final product of weathering of oil in water and, being relatively inert, can be transported long distances before they eventually degrade. If the source of a spill is not obvious, there may be a problem in identifying the culprit from analysis of the oil, given its chemical complexity and the weathering it has undergone. Analytical techniques are being refined to assist in fingerprinting the source of oil spills (e.g. Wang *et al.*, 2006).

It is difficult to generalize about the marine environmental impact of oil spills. Each spill is unique owing to the many variables that relate to the type and state of the oil, the receiving environment, and susceptibility of the affected organisms. Biological effects of oil pollution are due mainly to direct toxic and physical effects and, indirectly, to clean-up efforts. Oils differ in their toxicity. Acute toxic effects are due mainly to the lower molecular weight compounds, as these are the more soluble fractions. However, since these are also the components lost most rapidly through evaporation, large-scale mortalities due primarily to toxicity tend to be infrequent. The main physical effects arise through surface coating and smothering, which can prevent or impede activities such as movement and breathing. Organisms most at risk and among the most conspicuous casualties of oil pollution are those obliged to cross the air-sea interface, such as diving and surface-dwelling seabirds, fur-covered marine mammals, and intertidal organisms. Among habitats particularly sensitive to oil spill damage are sheltered coastal habitats, polar regions, and coral reefs (Samiullah, 1985).

In order to mitigate damage, a first concern must be placed on measures to reduce the amount of oil released into the sea. Several international agreements dealing with marine oil pollution have been adopted since 1954. These address such issues as navigational standards and shipping safety, design and operation of tankers, contingency plans and pollution response, designation of sensitive geographic areas, and compensation. A shortcoming has been the limited progress on measures to combat land-based inputs, a major source of petroleum hydrocarbons into the sea.

Changes to oil tanker design and operation have been introduced with the aim of reducing oil pollution. After unloading, an oil tanker needs to take on seawater as ballast to maintain stability. Since the normal practice has been to use the same tanks for cargo and ballast, the pumping out of ballast can result in the discharge of residual oil. One way of reducing this problem (the load-on-top system) is to retain the dirty ballast until the oil has floated to the surface. The underlying water, whilst not completely clean, can then be pumped out. The oil that is left is pumped to a slops tank, and fresh oil is loaded on top. Alternatively, maritime authorities may require tankers to have separate compartments for ballast and oil (segregated ballast). Also, single-hull oil tankers are being phased out in favour of double-hull vessels.

Mounting an effective response to an oil spill requires a high degree of forward planning and preparedness. The scale of the operation clearly depends on the size of the spill. This may call for a tiered response, ranging from a local contingency plan appropriate to a small spill within a particular port to more serious emergencies that require a national plan or even an integrated regional response involving neighbouring countries. Whatever the level, a number of issues need to be addressed in a contingency plan. These include defining the responsibilities of authorities and agencies involved; assessing the risk of spills (frequency and magnitude), the probable fate of oil slicks, the resources at risk and priorities for protection, and the appropriate clean-up strategies and their implementation (location and mobilization of equipment and personnel); and ensuring effective coordination and communication.

There are several regional centres around the world concerned with oil pollution contingency planning and response to oil spill emergencies. A wide variety of gear has been developed to treat oil spills by either recovering or containing oil or hastening its degradation. Oil spill response equipment includes remote sensing instruments; booms that can be deployed to concentrate spilled oil so as to facilitate its recovery or to protect certain parts of a coastline; recovery devices, especially mechanical skimmers, for removing oil from the sea surface; sorbents for soaking up oil; and chemical dispersants that enhance the break-up of oil into fine droplets. Early dispersants tended to be toxic (considerable biological damage from the *Torrey Canyon* spill in 1967 was due to the liberal use of toxic dispersants), but formulations of low toxicity are now available. Nevertheless, breaking up oil makes it easier for it to enter food chains. Physically removing oil whilst it is still at sea is the least damaging environmentally, and there needs to be greater emphasis on developing more efficient mechanical methods (MacKenzie, 2011). Whilst there is public pressure on authorities to act when a spill occurs, inaction may in some circumstances be more appropriate. In the case of coastal wetlands, for example, once oil has come ashore, clean-up efforts may be as damaging as the spill itself (see Chapter 12).

Under international conventions for oil pollution liability, financial claims can be made against tanker owners in the case of damage from oil spills (Mason, 2003). The International Oil Pollution Compensation Funds provide compensation for damage from oil pollution that results from spills of persistent oil from tankers. Compensation can be granted if a claimant (who has a legal right to claim under national law) has suffered quantifiable economic loss. But in the case of unexploited natural resources that have no owner and no direct market value, a claim for damage may be inadmissible. The regime extends out to the EEZ of contracting states; liability for high-seas oil pollution remains very restricted.

## Persistent Organic Pollutants

Persistent organic pollutants (POPs) are organic compounds, mostly synthetic and halogenated, that are in general highly resistant to chemical and bacterial breakdown and can potentially persist in the environment for many years. Also, being semi-volatile they can be transported long distances in the atmosphere, and whilst very insoluble in water they are lipid-soluble and so liable to accumulate in fatty tissues to concentrations far higher than ambient. Given these properties, POPs can pose a serious environmental risk, including to marine organisms (Box 2.3).

A major group of POPs are the organochlorine pesticides. The best known is DDT, introduced as an insecticide in 1939. It was used extensively through the 1950s and 1960s, proving extremely successful in the control of agricultural pests and of insects carrying human diseases. It is still used heavily in some parts of the world, mainly in the tropics in the fight against malaria, although since the 1970s most developed countries have adopted less persistent insecticides. Owing to its refractory nature and heavy and widespread use and aided by aerial dispersal, DDT has become globally distributed. Its principal breakdown product (DDE) can be equally as toxic to marine organisms.

Also of particular concern are the polychlorinated biphenyls (PCBs), which have been used in a wide variety of industrial applications since the 1930s (e.g. as insulating fluids in electrical plant and as additives to paints, adhesives, and hydraulic oils). Global production of PCBs has declined

## Box 2.3 Persistent Organic Pollutants

### Organochlorine (or Chlorinated Hydrocarbon) Pesticides

- DDT (dichlorodiphenyltrichloroethane)
- Cyclodiene insecticides (aldrin, dieldrin, lindane, chlordane, hepatachlor)
- Toxaphene (a mixture of chlorinated camphenes)

### Polychlorinated Biphenyls

- More than 200 isomers

### Dioxins

- Polychlorinated dibenzodioxins
- Polychlorinated dibenzofurans

### Brominated Flame Retardants

- Polybrominated diphenyl ethers
- Perfluorinated compounds

considerably since the 1970s – commercial production in the USA ceased in 1977 – but, because of persistence and difficulties associated with disposal, environmental problems due to PCBs may be long-lasting.

POPs tend to be passed up the food chain so that top predators, such as fish-eating birds and mammals, are usually the organisms most at risk of accumulating harmful levels. DDT and PCBs have been linked to reproductive failure in seabirds and marine mammals (see Chapters 8 and 9). High levels of POPs can accumulate in coastal areas, particularly in semi-enclosed waters downstream of heavily developed industrial and agricultural areas, such as in the Bohai Sea, NE China, where surveys of sediments and molluscs show high levels of DDT and other POPs, indicating their recent (and illegal) use (Zhang *et al.*, 2009). Organochlorines continue to be dispersed globally, particularly via the atmosphere from developing countries in the tropics, and there appears to have been no decline of organochlorine concentrations in the open ocean (Loganathan & Kannan, 1994).

Dioxins comprise another group of anthropogenic POPs that are persistent and fat-soluble and that bioaccumulate. Some are highly toxic. The main sources of dioxins are from combustion of organic materials and from the manufacture of wood-processing chemicals and herbicides, but there has been a significant reduction in industrial sources in recent years (Clark, 2001).

Other POPs of concern include the brominated flame retardants. These have a wide range of industrial applications and have also been detected in marine wildlife worldwide. Their known toxic effects, or structural similarity to other problematic compounds, are cause for concern (Allsopp *et al.*, 2009).

There are difficulties in disposing of organochlorines safely. High-temperature incineration breaks down the compounds to carbon dioxide and hydrochloric acid, but toxic byproducts, notably dioxins, may be released.

The Stockholm Convention on Persistent Organic Pollutants of 2001 is a global treaty, now with more than 150 signatories, that aims to protect human health and the environment from POPs. The convention so far identifies some 20 chemicals that are to be eliminated or restricted in their use (e.g. DDT for disease vector control) or where unintentional production is to be curtailed (e.g. dioxins released from manufacturing processes and incineration). Also, various countries have adopted or are developing programmes to regulate the use of chemical substances and thereby help address environmental impacts, such as the European Union's Registration, Evaluation, Authorisation and Restriction of Chemicals (REACH) programme, which came into force in 2007.

## Plastics

In recent decades, persistent debris has become a ubiquitous and serious pollutant of the world's oceans. It is believed that about 80% of debris in the marine environment comes from sources on land and the other 20% from ocean-based sources. Major sources of marine debris are litter from coastal populations, sewage and storm drains, fishing, and wastes from ships and boats (Allsopp *et al.*, 2006). Plastic items are by far the most abundant type of persistent marine debris, typically making up 50–80% of the debris that is stranded on beaches, floating on the ocean, and on the seabed (Derraik, 2002; Barnes *et al.*, 2009). Major types of plastic debris entering the seas are fishing gear, particularly nets and synthetic rope; packing materials and litter, such as strapping bands, bags, cups, bottles, six-pack holders, and sheeting; and raw plastics in the form of small pellets (Pruter, 1987; GESAMP, 1990a; Allsopp *et al.*, 2006).

Plastics are synthetic organic polymers. Those most commonly produced (as a proportion of the global plastics production in 2007) are low- and high-density polyethylenes (38%), polypropylene (24%), and polyvinylchloride (19%) (Andrady, 2011). Most plastics are not readily degraded, many are buoyant, and they can accumulate in large amounts at sea, particularly in the vicinity of heavily populated coasts, major shipping routes, fishing grounds, semi-enclosed seas, and the subtropical ocean gyres, where eddies and convergences naturally concentrate drifting materials (Barnes *et al.*, 2009; Cózar *et al.*, 2015).

Estimates indicate that about 200 000 t of large (> 200 mm) and up to 240 000 t of small (< 200 mm) plastic particles are floating in the global ocean (Eriksen *et al.*, 2014; van Sebille *et al.*, 2015). The small plastic particles include the pellets and granules, up to a few mm in diameter, that are the raw materials for the plastics industry. These have become globally distributed in surface waters and on beaches as a result of spillages on wharves, during handling and transport, and at processing plants from where pellets may be transported to coastal waters by rivers and stormwater drains (Redford *et al.*, 1997). Granule densities of >1000 per metre have been recorded on shores of remote non-industrialized islands in the SW Pacific (Gregory, 1999).

Of growing concern are even smaller particles, from micron to millimetre size, that are being found in the marine environment in increasing amounts. These come from products that contain plastic microbeads as an abrasive, such as cosmetic exfoliants and industrial cleaners, and from the fragmentation of larger plastic particles (Andrady, 2011; Cole *et al.*, 2011; Wright *et al.*, 2013). Vast numbers of microbeads are likely to be entering the marine environment. For the USA alone, Rochman *et al.* (2015) give a conservative estimate of 8 trillion microbeads released into aquatic habitats per day. Microplastics can be ingested by filter and deposit feeders, with largely unknown effects. With a high surface-to-volume ratio they may be effective carriers of potentially harmful POPs and other compounds added during manufacture or subsequently adsorbed from seawater. Setälä *et al.* (2014) showed that various zooplankton taxa will ingest plastic microspheres and that particles can be transferred from one trophic level to another.

Jambeck *et al.* (2015) calculated that in 2010 the world's coastal countries generated 275 million tonnes of plastic waste, of which 4.8–12.7 million tonnes entered the ocean. Estimates of the total mass of plastic particles afloat in the ocean (< 0.5 million tonnes) are, however, an order of magnitude less than this. Ingestion by organisms and degradation are likely to be important mechanisms responsible for removing microplastics from the sea surface (Eriksen *et al.*, 2014). Also, deep-sea sediments appear to be a major sink for plastics. Large, dense plastic particles may accumulate on the seabed. But also, in a global study, Woodall *et al.* (2014) found plastic microfibres (2–3 mm in length and less than 0.1 mm in diameter) of a wide variety of polymer types to be up to four orders of magnitude more abundant per unit volume in deep-sea sediments than in heavily contaminated surface water gyres.

Various factors affect interactions between marine animals and plastic debris, including abundance, distribution and nature of the debris, and response of organisms. Convergences and drift lines, which serve to concentrate debris, are also often important feeding areas for seabirds and turtles. Fish and other marine life are often actively attracted to floating debris, as they are to fish-aggregating devices. Predators may be drawn to prey on animals already entangled, whilst some marine mammals, notably juvenile seals, may be attracted to debris as objects to play with.

Assessing the effects of plastic debris on marine wildlife can be difficult, principally because the interactions are occurring over vast areas and few of the animals affected are likely to be recovered. Nevertheless, available data indicate that plastics pose major threats to marine animals, especially from entanglement and ingestion, and may cause widespread mortality, principally among seals, small cetaceans, seabirds, turtles, and fish. Entanglement or ingestion has been reported for 395 marine species, including all species of sea turtle and more than half of all known species of marine mammal and seabird (Table 2.3). Of the total number of species, 53 (13%) are categorized as threatened on the IUCN Red List (Gall & Thompson, 2015).

Entanglement can result in drowning, impaired ability to feed or escape from predators, or debilitating wounds (Laist, 1987). A particular problem is caused by packing bands and fishing nets that are lost or discarded, which can then entangle marine organisms over extended periods, so-called ghost fishing. Entanglement of seals is a worldwide problem (see Chapter 9). Ingested plastic may block or damage the gut or suppress feeding (Laist, 1987). Ingestion is a particular threat to surface-feeding seabirds (see Chapter 8). Plankton-feeding species can mistake raw pellets and small plastic fragments as prey, and even larger pieces of floating debris may be taken by albatrosses. Adult birds may be able to regurgitate ingested plastics but thereby feed plastic items to their chicks.

**Table 2.3** Marine Debris Entanglement and Ingestion Records for Sea Turtles, Seabirds, and Marine Mammals
The figures in brackets show the percentage of species in the group affected.

| Group | Number of known species | Number of species with entanglement records | Number of species with ingestion records | Total number of species with either entanglement or ingestion records |
|---|---|---|---|---|
| Sea turtles | 7 | 7 (100%) | 6 (86%) | 7 (100%) |
| Seabirds | 312 | 79 (25%) | 122 (39%) | 174 (56%) |
| Marine mammals | 115 | 52 (45%) | 30 (26%) | 62 (54%) |

Reprinted from *Marine Pollution Bulletin*, Vol. 92, Gall, S.C. & Thompson, R.C., The impact of debris on marine life, pages 170–9, copyright 2015, with permission from Elsevier.

Marine debris can provide a substratum for a wide range of attached organisms, such as algae, foraminiferans, bryozoans, barnacles, serpulid polychaetes, hydroids, and molluscs, and may be a significant vector for dispersal of alien species (Winston *et al.*, 1997; Gregory, 2009). It is estimated that, compared to natural debris, drifting plastic has more than doubled the dispersal opportunities for marine organisms (Barnes, 2002). The increased abundance of the bryozoan *Electra tenella* along the Atlantic coast of Florida appears to be due chiefly to the large quantities of drift plastic on which the organism encrusts (Winston, 1982).

The rates at which plastics degrade in the marine environment are not well determined. Decomposition may range from a year to a few decades for plastic bags and foam cups to a few centuries for bottles and fishing line (Ocean Conservancy, 2010). Breakdown is strongly influenced by physical factors. Photodegradation, embrittlement, and abrasion are especially important in the disintegration of beached plastics (Cooper & Corcoran, 2010). Under sunny conditions on low- to middle-latitude beaches, domestic plastic containers may collapse in less than 18 months, whereas at higher latitudes they may survive for up to a few years. However, most common plastics appear to endure even longer if they remain at sea (Gregory, 1991). Eventually, such items contribute to the increasing pool of microscopic plastic particles that is destined to remain indefinitely in the environment, given that the great majority of plastics in use are non-biodegradable thermoplastics.

The main instrument of international legislation to address plastics pollution is Annex V of the MARPOL Convention (see below), which entered into force in 1988. MARPOL addresses vessel-source pollution (and thus complements provisions of the London Convention that are concerned with the dumping at sea of land-generated wastes). Annex V prohibits the disposal of plastics at sea from ships and restricts the discharge of other garbage into coastal waters, with stricter requirements for designated special areas. It also requires ports to have garbage reception facilities for ships. The effectiveness of Annex V is, however, unclear. There is evidence that it is widely ignored but also of reductions in plastic litter at sea (Derraik, 2002). Concern over marine debris has focused mainly on material discharged from vessels, but as we have seen most marine debris derives from activities on land. Land-based discharges of marine debris have yet to be adequately addressed by international controls, although marine litter is one of the target categories of the UN Global Programme of Action for the Protection of the Marine Environment from Land-based Activities (UNEP, 1995). But overall, there appears to be no evidence that inputs of plastics to the marine environment are reducing (Moore, 2008), and in fact the cumulative quantity of plastic waste that is available to enter the ocean from land is forecast to increase by an order of magnitude by 2025 (Jambeck *et al.*, 2015).

Fundamental to this issue are efforts to minimize waste, restrict excessive use of plastics, develop viable recycling strategies, and promote less damaging solutions where impacts occur. There is increasing support for banning plastic microbeads in cosmetics, especially as natural, harmless alternatives are available (Rochman *et al.* 2015). Biodegradable plastics are being developed, particularly starch-based polymers, that decompose into less environmentally harmful end products (Hammer *et al.*, 2012). Some countries have legislation requiring that plastics have a certain level of degradability. It is, however, important that such plastics are truly biodegradable, decomposing into organic and inorganic components and not just breaking down into microplastics (Bilkovic *et al.*, 2012). Biodegradable plastics might still persist long enough to be harmful, and by being viewed as environmentally more acceptable than conventional plastics they may in fact engender more wasteful practices (Allsopp *et al.*, 2006).

## Metals

As a result of natural biogeochemical cycles, enormous quantities of metals enter the sea from river runoff and via the atmosphere. But for some metals these inputs have been greatly increased by anthropogenic activities such as mining, burning fossil fuels, and various manufacturing processes. Levels of contamination are generally higher in coastal waters than in the open ocean, reflecting the importance of land-based sources (Davis, 1993). With some metals there may also be considerable contamination via atmospheric inputs. Certain metals are essential in trace amounts for normal metabolic processes (e.g. as constituents of respiratory pigments or enzymes), but all metals tend to be harmful at high concentrations. The higher atomic weight metals are generally the ones most toxic to marine life – mercury, lead, and cadmium in particular (Fowler, 1990; UNEP/GPA, 2006). Numerous factors, however, affect the bioavailability and toxicity of a particular metal, including the form of the metal, its interactions with other metals, environmental factors (e.g. salinity, temperature, redox, and metal-binding components of sediments), and the organism's condition, life-cycle stage, and capacity to detoxify the metal. Many laboratory studies have demonstrated harmful effects of metals on organisms, although few studies have shown clear impacts of metals on marine invertebrate assemblages (Mayer-Pinto *et al.*, 2010).

Of obvious concern are organometals, notably tributyltin (TBT), an anti-fouling compound considered perhaps the most toxic substance ever deliberately introduced into the marine environment (Mee & Fowler, 1991). TBT has been linked to detrimental effects in many marine species and is responsible in particular for declines of gastropod populations by disrupting their reproduction. TBT concentrations in coastal waters have generally declined in recent years following bans on the use of TBT anti-fouling paints, but the retention of TBT in sediments of ports and harbours is a problem that will persist for decades (Antizar-Ladislao, 2008). Organometals are most likely to pose a threat in estuarine or enclosed waters (see Chapter 11).

TBT is an example of an endocrine-disrupting chemical, a compound that mimics natural hormones or otherwise interferes with endocrine systems. Other endocrine disrupters include various PCBs and organochlorine pesticides, plasticizers, surfactants, and pharmaceuticals. Many are oestrogenic in action, affecting reproduction, and bioaccumulate in lipids. Of concern, therefore, is their impact on top predators, especially those in highly contaminated enclosed waters (Porte *et al.*, 2006; Burkhardt-Holm, 2010).

## Radioactivity

Radioactivity of the oceans is due largely to naturally occurring radionuclides (predominantly potassium-40) (Clark, 2001). But since the late 1940s the oceans have also received radioactive inputs from human activities, chiefly from the testing of nuclear weapons, nuclear power station accidents, and discharges from reprocessing plants (IAEA, 2005). Nuclear fallout from weapons testing in the atmosphere has been the major anthropogenic source. In terms of caesium-137, as one of the most representative anthropogenic radionuclides, input to the world ocean from global fallout has been some 600 PBq $^{137}$Cs (where 1 becquerel (Bq) = 1 nuclear decay per second and P (peta) = $10^{15}$). Caesium-137 has a half-life of 30 years, and nuclear fallout has diminished since international treaties were adopted during the 1960s that limit the testing of nuclear weapons. Atmospheric tests have been continued by some non-signatory nations, such as France, which has carried out tests on atolls in French Polynesia (Van Dyke, 1991). But since 1974 all such testing appears to have been underground and without atmospheric fallout. Among the most contaminated sites are Enewetak Atoll in the Marshall Islands (central Pacific Ocean), used as a US nuclear testing facility, and Chernaya Bay, a fjord on southern Novaya Zemlya (Arctic Ocean), a nuclear test site of the former Soviet Union (Yablokov, 2005).

Accidents at nuclear power stations have resulted in significant inputs of anthropogenic radionuclides to the marine environment. A large release of radioactive material into the environment occurred in 1986 as a result of an explosion at the Chernobyl nuclear power station in Ukraine, notably about 100 PBq $^{137}Cs$, of which 10–20% fell into European seas, particularly the Baltic Sea (Aarkrog, 2003). Major accidental releases have also occurred at the Fukushima Dai-ichi nuclear power plant on the east coast of Honshu after an earthquake and tsunami in 2011. Although the total release was considerably smaller than for Chernobyl, the accident was more important as a source of ocean radioactivity as, in addition to atmospheric releases, there were also direct ocean discharges in the range of 4–40 PBq $^{137}Cs$ (Buesseler, 2014). Caesium levels in fish off Fukushima remained elevated more than a year after the accident, particularly in demersal fish, indicating that the seabed may be acting as a chronic source of contamination (Buesseler, 2012).

Radionuclides are released during the reprocessing of spent fuel (to recover uranium and plutonium for reuse in reactors). The most significant of such discharges is that from the Sellafield plant in NW England, which has piped radioactive wastes into the Irish Sea since 1957, with a cumulative input of some 39 PBq $^{137}Cs$ (Buesseler, 2014), but the reactor is now being decommissioned. Radionuclides such as $^{137}Cs$ can be traced from Sellafield into the North Atlantic and Arctic, and the discharge is implicated in radio-caesium contamination of marine mammals in UK waters (Allsopp *et al.*, 2009).

Packaged low- and intermediate-level nuclear wastes have been dumped at numerous deep-sea sites from the late 1940s until 1983 when the practice was suspended under a moratorium of the London Convention. A ban on the dumping of all radioactive wastes under the convention entered into force in 1994. The picture, however, is likely incomplete, particularly given the secrecy surrounding military accidents and disposal (Yablokov, 2005).

Given the possible effects of radiation, notably genetic damage and various forms of cancer, radioactive contamination is understandably a sensitive issue in terms of human health. In the marine environment, anthropogenic radionuclides can bioaccumulate, for some isotopes up to many thousands of times, and at maximum dose rates some genetic aberration may occur in individual marine organisms. There is, however, a view that for observed levels of radioactive contamination there appears little evidence to expect deleterious effects on marine ecosystems (Fowler & Fisher, 2004).

## Marine Pollution Legislation

Whilst every effort needs to be made to reduce waste, improve recycling, and prevent contaminants being released to the marine environment, some wastes are unavoidable. Ideally, waste disposal policies should address all options for disposal and the full internal and external costs they are likely to entail. External costs, such as environmental degradation and effects on human health, may be very difficult to assess. Indeed, a major concern about marine waste disposal is the often high level of uncertainty of the environmental risk. Also, there is a widely held view that by impacting on resources that are more obviously a common heritage, marine waste disposal indirectly affects everyone, whereas the effects of land disposal are likely to be more localized (Kite-Powell *et al.*, 1998). Inevitably, foreign materials will enter the sea, and in some situations marine disposal may be considered the least damaging option. What needs to be ensured is that inputs are managed so as not to undermine the long-term viability of marine ecosystems. Unfortunately we are a long way from this goal, and huge efforts are needed at all levels to check marine pollution. Efforts to control marine pollution by international legislation date from the 1950s and initially were directed mainly at curbing the discharge of oil from vessels at sea. A principle of the 1972 Stockholm Declaration – regarded as the basis of modern approaches to environmental management – is that states 'shall take all possible steps to prevent pollution of the seas ... ' (Principle 7). Since then the international regime has been augmented to address pollution sources at regional and global levels (Box 2.4).

## Box 2.4 Principal International Agreements Concerning Marine Pollution and Entry into Force (EIF)

### The United Nations Convention on the Law of the Sea

Articles 207–12 concern 'international rules and national legislation to prevent, reduce and control pollution of the marine environment' (EIF 1994; see Chapter 3).

### Convention on the Prevention of Marine Pollution by Dumping of Wastes and Other Matter (the 'London Convention')

Parties shall 'promote the effective control of all sources of pollution of the marine environment, and … especially to take all practicable steps to prevent the pollution of the sea by the dumping of waste and other matter that is liable … to harm living resources and marine life' (EIF 1996; 1996 Protocol EIF 2006).

### International Convention for the Prevention of Pollution from Ships (MARPOL)

Annex I Regulations for the Prevention of Pollution by Oil (EIF 1983)
Annex II Regulations for the Control of Pollution by Noxious Liquid Substances in Bulk (EIF 1983)
Annex III Prevention of Pollution by Harmful Substances Carried by Sea in Packaged Form (EIF 1992)
Annex IV Prevention of Pollution by Sewage from Ships (EIF 2003)
Annex V Prevention of Pollution by Garbage from Ships (EIF 1988)
Annex VI Prevention of Air Pollution from Ships (EIF 2005)
(Convention (1973) and Protocol (1978) EIF 1983)

### International Convention on the Control of Harmful Anti-fouling Systems on Ships

'To reduce or eliminate adverse effects on the marine environment and human health caused by anti-fouling systems', notably organotin compounds (EIF 2008).

### International Convention on Oil Pollution Preparedness, Response and Co-operation

'Parties … to take all appropriate measures … to prepare for and respond to an oil pollution incident.' In particular, ships and operators of offshore units are required to have oil pollution emergency plans for responding promptly and effectively to oil pollution incidents. Parties shall also have appropriate oil spill combating equipment, and detailed plans for responding to oil pollution incidents (EIF 1995).

### International Convention for the Control and Management of Ships' Ballast Water and Sediments (Ballast Water Management Convention)

'To prevent, minimize and ultimately eliminate the transfer of harmful aquatic organisms and pathogens through the control and management of ships' ballast water and sediments.' Ships in international traffic are required to manage ballast water and sediments according to specific management standards (EIF 2017).

One of the first global instruments aimed at protecting the marine environment from pollution was the 1972 London Convention. It entered into force in 1975, and currently some 87 states are party to the convention. Essentially it concerns the deliberate dumping at sea of land-generated waste. The present convention prohibits the dumping of hazardous substances (Annex I, the 'black list', e.g. organohalogens, mercury, cadmium, plastics, oil, and high-level radioactive wastes), whilst certain somewhat less harmful substances (Annex II, the 'grey list') can be dumped at sea under special permit. Other substances can be dumped under a permit issued from the relevant national authority, subject to provisions (Annex III). The 1972 Convention is, however, being superseded by a 1996 protocol that entered into force in 2006. This adopts instead a precautionary approach so that parties are prohibited from dumping any waste that is not in the protocol's Annex I (the so-called 'reverse list', or materials that may be considered suitable for dumping, such as dredged material; sewage sludge; fish waste; vessels and offshore platforms or other structures at sea; inert, inorganic geological material; and organic material of natural origin). The protocol also prohibits the practice of incineration at sea.

A major agreement concerning pollution from vessels at sea is the 1973 International Convention for the Prevention of Pollution from Ships, as amended by a protocol of 1978 and known as MARPOL. MARPOL has a number of annexes, the first of which entered into force in 1983. Some 140 countries are now parties to the convention, representing virtually all the world's merchant shipping. MARPOL regulates marine pollution from ships but not the disposal of land-generated waste by dumping from ships and the discharge of substances directly arising from the exploration or exploitation of seabed minerals. The convention includes six annexes. Signatories must accept Annexes I and II, but compliance with the others is voluntary. Annex I deals with oil and, in particular, extends earlier legislation on the control of ballast discharges and improving tanker operations and construction, including the requirement for double hulls; and Annex II concerns noxious liquid substances carried in bulk and establishes categories of substances according to their threat to the marine environment. The voluntary annexes concern harmful substances carried in packaged form (Annex III); sewage (IV); garbage, including plastics (V); and air pollution (VI).

MARPOL provides for countries to identify special areas where the need to control pollution is much greater than elsewhere. Designated areas include the North, Baltic, Black, Mediterranean, and Red Seas; the Persian, Oman, and Aden Gulf areas; the Wider Caribbean; and the Antarctic area. For an area to be accorded special area status certain criteria must be satisfied. These mainly concern oceanographic and ecological conditions, vessel traffic characteristics, and port reception facilities. Important oceanographic conditions concern the extent to which harmful substances would be retained in the waters or sediments. Ecological considerations include the need to protect certain species and habitats from harmful substances (depleted, threatened, or endangered species; areas of high natural productivity; spawning, breeding, and nursery areas for important species and migratory routes for seabirds and mammals; rare or fragile ecosystems (e.g. coral reefs, mangroves, seagrass beds, and wetlands); and critical habitats for marine resources).

It is difficult to assess the effectiveness of international conventions in reducing pollution. GESAMP (1990a) reported a gradual decrease in the amounts of industrial wastes and sewage sludge dumped since the London Convention entered into force. Estimates also indicate that globally the input of oil from anthropogenic sources into the sea has decreased in recent decades and that this is largely attributable to the implementation of international agreements, in particular entry into force of MARPOL Annex I (GESAMP, 1993). Aerial surveillance data from 1991 to 2010 of the southern North Sea indicate, for instance, that MARPOL has contributed to a decline in ship-source oil pollution (Lagring *et al.*, 2012). A key question relating to the efficacy of Annex I will be the effect of the conversion from single-hull to double-hull tankers, the results of which will not become apparent for some years (Mattson, 2006). With respect to Annex V, studies in the Gulf of Alaska indicate that the passage of MARPOL has contributed to a decline in the deposition of trawl netting on beaches and to an observed decline in the entanglement rate of fur seals (Johnson, 1994).

There are also various regional conventions that concern the prevention and control of marine pollution from waste disposal at sea. For example, the Oslo Convention, which came into force 1974, is similar to the London Convention in its provisions but is restricted to the NE Atlantic and North Sea. Other areas covered by regional conventions include the Baltic, Mediterranean, South Pacific, wider Caribbean, and Red Seas and Gulf of Aden (Table 16.1).

As pointed out earlier, marine pollution derives very largely from land-based sources. No international binding legislation has been adopted to regulate land-based pollution, nor does such an instrument look likely (Dyoulgerov, 2000). Major stumbling blocks include the wide range of pollutants that are entering coastal waters from countless point and non-point sources, the implications of such measures for industry as well as for urban authorities and agriculture, and the relative economic development in different parts of the world. There has, however, been some progress with non-mandatory instruments, notably the Global Programme of Action for the Protection of the Marine Environment from Land-based Activities (GPA) of 1995. The GPA is under the implementation of UNEP and has been adopted by more than 100 governments and the European Commission. Under the GPA there is a commitment to address nine main areas, namely sewage, persistent organic pollutants, radioactive substances, heavy metals, oils, nutrients, sediment mobilization, marine litter, and the physical alteration and destruction of habitats. Overall, progress has been made with respect to persistent organic pollutants, radioactive substances, and oils, but with the other categories it appears the situation has not improved overall or indeed has worsened (UNEP/GPA, 2006). Various regional agreements include regulations for the control of land-based sources (confined mainly to point sources and sewage treatment), and pursuing a regional approach has advantages (Dyoulgerov, 2000).

For atmospheric sources of pollution there has been little action at the global level, and there are few provisions in regional agreements, again because of economic pressures (GESAMP, 1990a).

The UN Convention on the Law of the Sea (Part XII) deals with protection and preservation of the marine environment, including measures to prevent, reduce, and control all sources of marine pollution. It requires states to cooperate on a global basis and, as appropriate, on a regional basis, directly or through international organizations, in formulating and elaborating international rules, standards, and recommended practices and procedures for the protection and preservation of the marine environment. States have an obligation to adopt national and international legislation, as appropriate, to minimize marine pollution from land-based sources, from seabed activities subject to national jurisdiction, from activities in the international seabed area, and from dumping, vessels, and atmospheric sources. The provisions are intended to be consistent with existing treaties on marine pollution. Although very desirable, it is debatable whether in reality a truly international approach to arresting marine pollution can be implemented. From a practical point of view, an effective legal framework is probably more likely to result from regional measures but nevertheless based on accepted international criteria. It still remains necessary for nations to implement measures by enacting appropriate domestic legislation.

An agency of the UN with particular responsibility for maritime activities is the International Maritime Organization (IMO), established in 1958. One of its major responsibilities concerns international action to prevent marine pollution, and it administers international conventions of navigation, safety, and pollution, including the London Convention and MARPOL (see www.imo.org).

## Assessing Environmental Effects

There are widely recognized procedures for evaluating the environmental effects of developments and projects. These include environmental impact assessment and monitoring, although each is open to a variety of definitions. An environmental impact assessment (EIA) is usually understood to mean a formal process designed to critically examine the likely environmental consequences of a development. It can be defined as 'the process of identifying, predicting, evaluating and mitigating the biophysical, social, and other relevant effects of development proposals prior to major decisions being taken and commitments made' (IAIA, 2010). Environmental monitoring is intended to provide feedback on the adequacy of an EIA to check that any changes do not exceed those forecast.

The process of EIA was developed in the 1960s and first became mandatory in the USA following the National Environmental Policy Act, which took effect in 1970. This requires federal agencies to prepare a detailed statement on the environmental impact of proposed actions that significantly affect the quality of the environment and of alternatives to those actions. Since then most countries have enacted legislation with requirements for environmental impact assessment, and it has been incorporated into international initiatives such as the Convention on Biological Diversity.

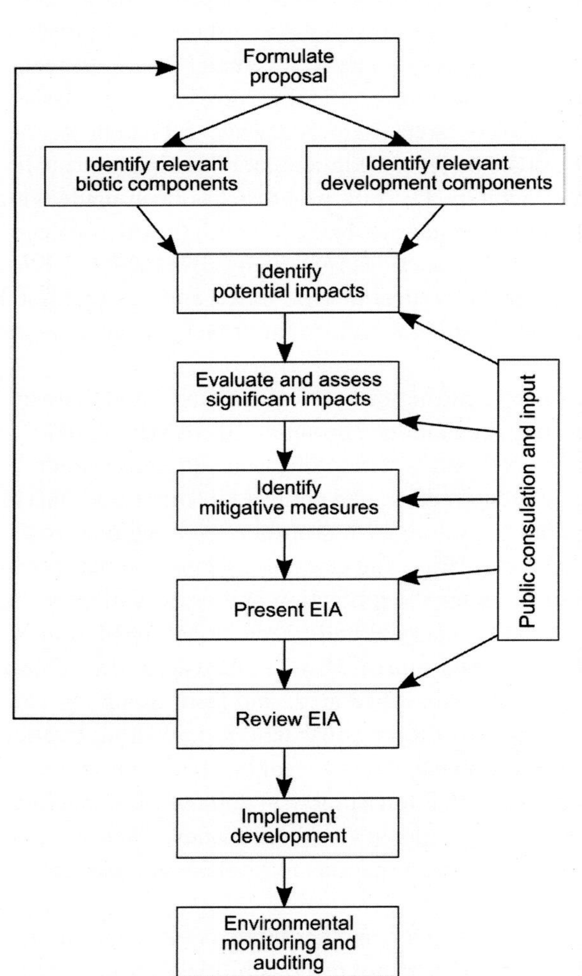


**Fig. 2.9** A flowchart of the main components of an environmental impact assessment.

A number of EIA methodologies have been proposed, and the steps taken will depend on the scale and type of development and on the regulatory requirements and legislation of the country concerned. The procedure should, nevertheless, fulfil certain criteria, in particular by being objective, comprehensive and quantitative. Key components of an EIA are a detailed specification of the project in terms of its main elements and stages (e.g. development and operational phases) as well as alternatives considered, information on all components of the environment that may be affected, analysis of significant environmental effects, and an examination of ways to mitigate harmful effects (Fig. 2.9). Whilst potential impacts may be identified, evaluating the significance of interactions is rarely straightforward. Various points need to be addressed, such as the duration of impacts, whether they are reversible or irreversible, and to what extent they may have secondary or tertiary consequences. There is a wide range of techniques for identifying impacts, from simple checklists and matrices to the use of composite indices and overlay map methods (Glasson *et al.*, 2005). Modelling may often be used, for instance, to forecast the dispersion of dredge spoil at a dump site or a pollutant from an outfall.

To carry out an EIA effectively good biological data are clearly required, not just in terms of the species present but also information on their life-history patterns, migratory behaviour, and respective tolerance to particular sources of disturbance. Often an important consideration is the status of any especially vulnerable or threatened species in the area. Frequently, however – and this is a problem with many marine EIAs – the local biota is poorly known. A baseline survey may be needed, and the lead-in time this requires, particularly when information on temporal variability is necessary, can be a source of contention between environmental and development interests. An essential aspect of any EIA is identifying particularly deleterious impacts and examining measures for their mitigation. There may be viable and less damaging alternatives in terms of the scale or location of the development or in its operational procedures.

## Baseline Surveys

Whilst baseline surveys are commonly undertaken to monitor possible changes in marine systems, the true baseline condition will often be unknown. Succeeding generations of biologists working in an area that is gradually being altered by, for example, nutrient input or fishing will establish slightly different baselines. Each may consider the ecosystem they are familiar with as being relatively healthy. But given the cumulative changes over time, the most recent baseline would represent a significantly disturbed system to the earlier biologists. There is good evidence that baselines do shift in this way (Pauly, 1995; Sheppard, 1995; Pinnegar & Engelhard, 2008). It needs to be borne in mind too that marine systems are responsive to natural shifts at time scales on the order of generations to confound the detection of human impacts over extended periods (see Chapter 1). Also, apart from the problem of distinguishing between natural and human factors, it may be difficult to separate the effects of different human factors. Overexploitation of herbivorous fish in coral reef communities makes it more likely, for instance, that organic pollution will lead to a shift from coral to algae (see Chapter 13).

## Monitoring

If a development proceeds, monitoring programmes will often need to be established. Where a contaminant is being released, compliance monitoring will be needed to ensure that concentrations in the receiving environment do not exceed agreed-upon or legal levels. Contamination is measured in terms of physico-chemical parameters, but monitoring of effects is assessed in terms of biology. Monitoring should provide a check that the environmental impacts are within the range of those forecast and be able to trigger changes to operational procedures in the event of unanticipated harmful effects attributable to the development or discharge. The perennial problem in such situations is distinguishing between an anthropogenic effect and natural variability – the issue of separating a 'signal' from background 'noise'. A commonly used sampling design that is suitable for detecting whether a disturbance is associated with a change in a biological community, such as a significant change in the mean abundance of a species, is a Before-After-Control-Impact (BACI) design. To take account of the natural spatial and temporal variability such a design needs to incorporate replicate control sites and replicate sampling occasions (Fig. 2.10).

Effects of stress or pollution may be detectable at different levels of biological organization, from the biochemical up to the population, community, and ecosystem levels (Fig. 2.11). Biochemical methods for detecting contaminants include measuring the detoxifying activity in the cell. Most animals produce detoxifying proteins when exposed to certain heavy metals and organic contaminants. Whilst these may provide an early warning of contamination, such responses are very specific and not necessarily of significance at, say, the population level. Effects of pollutants are less well understood at higher levels of organization, where it becomes increasingly difficult to link cause and effect. However, it is the effects at these more complex levels that are generally more relevant to environmental impact and monitoring studies.

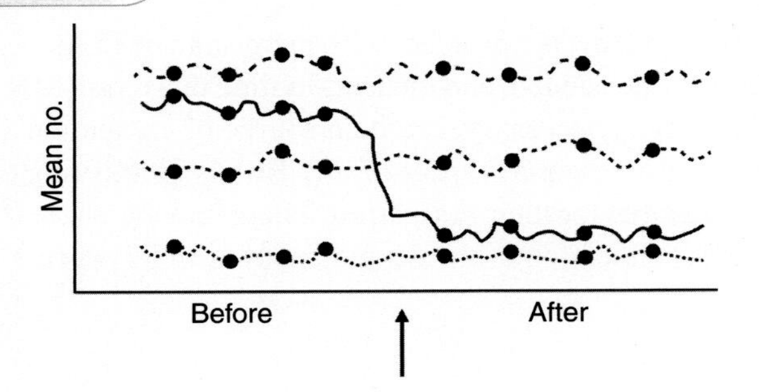


**Fig. 2.10** Before-After-Control-Impact (BACI) sampling programme with three control sites (broken lines) and one impacted site (solid line). The arrow indicates the time of the impact. There are four sampling occasions before and four after the impact. Reprinted from *Journal of Experimental Marine Biology and Ecology*, Vol. 161, Underwood, A.J., Beyond BACI: the detection of environmental impacts on populations in the real, but variable, world, pages 145–78, copyright 1992, with permission from Elsevier.

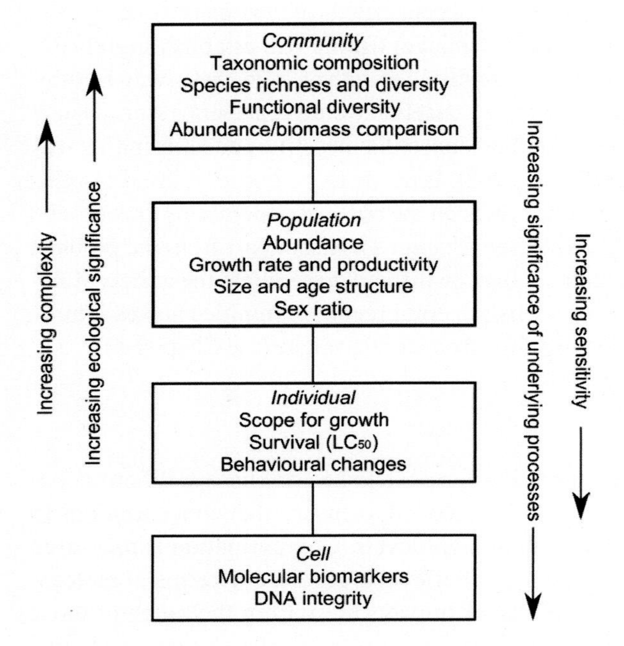


**Fig. 2.11** Effects of pollution can be examined at different levels of biological organization, for example, from the cellular to the community level. Species richness is the number of species in the community, whereas species diversity also quantifies the relative abundance of species (evenness). Functional diversity quantifies the range of functions performed by species (e.g. feeding types). The abundance/biomass comparison is a graphical technique (species ranked by cumulative percentage dominance of abundance and biomass) that indicates the relative importance of *r*- and *K*-selected species (see Chapter 4): *r*-selected species tend to be abundant in stressed communities but being small-bodied do not amount to a large biomass, whereas in an equilibrium community, large-bodied *K*-selected species have low abundance but large biomass. Scope for growth is an index of physiological stress: a measure of the energy available for growth. $LC_{50}$ is a measure of toxicity: the lethal concentration of a substance that will kill 50% of individuals within a specified time. Molecular biomarkers are proteins (e.g. metallothioneins, mixed function oxygenases) that are synthesized as detoxicants.
From various sources, in particular Wilson (1988) and Clark (2001).

Most marine environmental impact studies focus on responses at the community level. Often only one component of the biota is examined on the assumption that its response is representative as well as for practical considerations. Different biotic components offer different advantages and disadvantages in terms of practical use and appropriateness (Warwick, 1993). Plankton samples taken over relatively large areas will be valuable for monitoring regional changes but of limited use in assessing local impacts since the water masses they inhabit are continually moving. Similarly, on account of their mobility, fish are often more useful in studies of widespread rather than local impact, except for species that show strong site fidelity such as many reef fish. Fish are sometimes used to examine the incidence of morphological and pathological effects of pollutants (e.g. ulcers or fin erosion). Birds, particularly waders, can be among the most conspicuous predators of coastal and estuarine areas, often feeding in large numbers on intertidal flats. Various bird census methods are in use, but the often large seasonal and daily movements can pose problems. Local or national (as well as international) organizations regularly undertake and coordinate bird counts (Prater & Lloyd, 1987).

## Use of Benthos in Impact and Monitoring Studies

Marine impact and monitoring studies often focus on benthic communities as indicators of the general health of marine ecosystems. Benthic communities have obvious advantages over pelagic communities for such studies. Many benthic organisms are relatively sedentary or immobile, and their habitat is more two-dimensional than that occupied by the often highly mobile pelagic organisms. They thus tend to be responsive to local (as well as far-reaching) impacts and are easier to sample quantitatively, particularly in the case of sediment benthos. Also, the responses of sediment macrofauna to anthropogenic stress and pollution have been examined in more detail than for any other component of the marine biota. Responses of benthic communities to perturbation may be detectable at a higher taxonomic level than that of species, which can greatly reduce the time and cost needed for analysing samples (Warwick, 1988).

## Community Responses to Disturbance

Typically there is a gradational pattern in community structure in response to a disturbance. Such patterns have been described from numerous studies of the benthic impact of effluent discharges and disposal sites. In broad terms a severe impact might produce a more or less azoic zone in the immediate vicinity of the disturbance, beyond that a zone characterized by a few opportunistic species, then an intermediate zone occupied by opportunistic and equilibrium species (i.e. those representative of the mature unimpacted community), and finally a 'normal' zone supporting an equilibrium community (Fig. 11.4). Various parameters can be used to describe changes in community structure along disturbance gradients and thereby to delimit and quantify an impact. A wide variety of techniques have been developed for analysing spatial or temporal changes in community structure, such as univariate methods in which relative abundances of species are reduced to a single index (e.g. a species diversity index), graphical/distributional methods in which relative abundances are plotted for comparison, and multivariate methods of classification and ordination. The latter are widely employed, particularly given software packages such as Plymouth Routines in Multivariate Ecological Research (PRIMER) that enable biotic patterns and the environmental factors, including measures of disturbance, that best explain them, to be examined.

## Regional and Global Monitoring

In assessing the environmental impact of a particular local development it may be possible to identify specific parameters, from the biochemical up to the community level, that are appropriate to monitor. However, where the aim is to monitor the general health of a marine area that may be subject to multiple stresses, there is a need to develop a more integrated approach that measures attributes of whole ecosystems. Such a programme would need to incorporate a suite of indicators, such as primary productivity, nutrient cycling, species diversity, disease prevalence, size spectrum, and contaminant concentrations across trophic levels (Harding, 1992).

A number of regional monitoring plans have been developed, notably within UNEP's Regional Seas Programme (see Chapter 16). These have been based on common methodologies and include inter-calibration exercises to ensure comparability of data. A major programme undertaken by the National Oceanic and Atmospheric Administration has been the monitoring of contaminant levels in biota and sediments at numerous estuarine and coastal sites throughout the USA since 1984 (see Chapter 11). UNEP has also been active in the development of global programmes for monitoring marine pollution (Gerges, 1994).

There is, however, an unfortunate dearth of long-term monitoring programmes that target ecological data. In fact, for only two regions do we have data dating back to at least the mid-twentieth century, namely for the North Atlantic, North Sea, and western English Channel (programmes undertaken by the marine laboratories in Plymouth, UK, and the Continuous Plankton Recorder Survey) and for the NE Pacific off the coast of California (the California Cooperative Oceanic Fisheries Investigations). These programmes collect oceanographic, plankton, and fisheries data, time-series information that is invaluable in understanding long-term variability in ocean ecosystems and the likely effects of human impacts, in particular climate change. There is a pressing need for such marine ecological monitoring to be to undertaken at a series of sites so as to achieve global coverage (Koslow & Couture, 2013).

An aspect of regional and international monitoring has been the use of organisms to provide an integrated measure of contaminants over large spatial and temporal scales. There is no universally suitable organism – species differ among other things in their capacity to accumulate and regulate contaminants – but the most extensively used animal has been the blue mussel *Mytilus edulis*, which was adopted in 1976 for the US Mussel Watch Monitoring Program. Several aspects of the species appeared to make it a suitable candidate for such a programme. It is an extensively studied species with a wide geographic distribution in sheltered coastal areas (often those most exposed to contamination) and able to accumulate large amounts of many contaminants. It was later discovered that the taxon in fact comprises three sibling species and that at least two differ in growth rate, which has implications for comparability of data on contaminant uptake (Lobel *et al.*, 1990; McDonald *et al.*, 1991; but see also Blackmore & Wang, 2003). Biomonitoring programmes using mussel species and/or other bivalves have been implemented in other parts of the world, including Europe and SE Asia.

Regional approaches to marine environmental protection can be effective and workable options, but ultimately also essential is greater international cooperation in developing and implementing protection measures as well as in the exchange of information and in the provision of technical and financial assistance. Integral too is an approach that is more holistic in perspective so that mitigative measures do not simply result in the transfer of damage or hazards to other environments. Existing international instruments for marine environmental protection represent a considerable achievement, but it has long been recognized that they nevertheless provide for a somewhat piecemeal approach (GESAMP, 1991). However, we do at least now understand the main elements and principles for responsible governance of the marine environment – notably as derived from the UN Convention of the Law of the Sea and from subsequent international forums – elements that are needed for a more integrated and comprehensive strategy consistent with the concept of sustainable development. In the next chapter we look at the emergence of these key principles for management and conservation of the marine environment.

## REFERENCES

Aarkrog, A. (2003). Input of anthropogenic radionuclides into the World Ocean. *Deep-Sea Research II*, **50**, 2597–606.

Acheson, J.M. 2005. Developing rules to manage fisheries: a cross-cultural perspective. In *Marine Conservation Biology: The Science of Maintaining the Sea's Biodiversity*, ed. E.A. Norse & L.B. Crowder, pp. 351–61. Washington, DC: Island Press.

Allsopp, M., Page, R., Johnston, P. & Santillo, D. (2009). *State of the World's Oceans*. London: Springer.

Allsopp, M., Walters, A., Santillo, D. & Johnston, P. (2006). *Plastic Debris in the World's Oceans*. Amsterdam, Netherlands: Greenpeace International.

Andersen, J.H., Schlüter, L. & Ærtebjerg, G. (2006). Coastal eutrophication: recent developments in definitions and implications for monitoring strategies. *Journal of Plankton Research*, **28**, 621–8.

Andrady, A.L. (2011). Microplastics in the marine environment. *Marine Pollution Bulletin*, **62**, 1596–605.

Antizar-Ladislao, B. (2008). Environmental levels, toxicity and human exposure to tributyltin (TBT)-contaminated marine environment: a review. *Environment International*, **34**, 292–308.

Balcombe, J. (2000). *The Use of Animals in Higher Education: Problems, Alternatives, & Recommendations*. Washington, DC: The Humane Society Press.

Barnes, D.K.A. (2002). Invasions by marine life on plastic debris. *Nature*, **416**, 808–9.

Barnes, D.K.A., Galgani, F., Thompson, R.C. & Barlaz, M. (2009). Accumulation and fragmentation of plastic debris in global environments. *Philosophical Transactions of the Royal Society B*, **364**, 1985–98.

Baum, J.K. & Worm, B. (2009). Cascading top-down effects of changing oceanic predator abundances. *Journal of Animal Ecology*, **78**, 699–714.

Bax, N., Carlton, J.T., Mathews-Amos, A., *et al.* (2001). The control of biological invasions in the world's oceans. *Conservation Biology*, **15**, 1234–46.

Beaugrand, G., Reid, P.C., Ibañez, F., Lindley J.A. & Edwards, M. (2002). Reorganization of North Atlantic marine copepod biodiversity and climate. *Science*, **296**, 1692–4.

Bilkovic, D.M., Havens, K.J., Stanhope, D.M. & Angstadt, K.T. (2012). Use of fully biodegradable panels to reduce derelict pot threats to marine fauna. *Conservation Biology*, **26**, 957–66.

Blackmore, G. & Wang, W.-X. (2003). Comparison of metal accumulation in mussels at different local and global scales. *Environmental Toxicology and Chemistry*, **22**, 388–95.

Boesch, D.F., Butler, J.N., Cacchione, D.A., *et al.* (1987). An assessment of the long-term environmental effects of U.S. offshore oil and gas development activities: future research needs. In *Long-Term Environmental Effects of Offshore Oil and Gas Development*, ed. D.F.Boesch & N.N.Rabalais, pp. 1–53. London: Elsevier Applied Science Publishers Ltd.

Bolch, C.J.S. & de Salas, M.F. (2007). A review of the molecular evidence for ballast water introduction of the toxic dinoflagellates *Gymnodinium catenatum* and the *Alexandrium* '*tamarensis* complex' to Australasia. *Harmful Algae*, **6**, 465–85.

BP (2015). *Statistical Review of World Energy 2015*. London: BP. Available at www.bp.com/statisticalreview (accessed 18 January 2016).

Briggs, J.C. (2012). Marine species invasions in estuaries and harbors. *Marine Ecology Progress Series*, **449**, 297–302.

Buesseler, K.O. (2012). Fishing for answers off Fukushima. *Science*, **338**, 480–2.

Buesseler, K O. (2014). Fukushima and ocean radioactivity. *Oceanography*, **27**, 92–105.

Burkhardt-Holm, P. (2010). Endocrine disruptors and water quality: a state-of-the-art review. *International Journal of Water Resources Development*, **26**, 477–93.

Carlton, J.T. (1999). The scale and ecological consequences of biological invasions in the world's oceans. In *Invasive Species and Biodiversity Management*, ed. O.T.Sandlund, P.J.Schei & A.Viken, pp. 195–212. Dordrecht, Netherlands: Kluwer.

Carlton, J.T. & Ruiz, G.M. (2005). The magnitude and consequences of bioinvasions in marine ecosystems: implications for conservation biology. In *Marine Conservation Biology: The Science of Maintaining the Sea's Biodiversity*, ed. E.A.Norse & L.B.Crowder, pp. 123–48. Washington, DC: Island Press.

Clark, R.B. (2001). *Marine Pollution*, 5th edn. Oxford, UK: Clarendon Press.

Cole, M., Lindeque, P., Halsband, C. & Galloway, T.S. (2011). Microplastics as contaminants in the marine environment: a review. *Marine Pollution Bulletin*, **62**, 2588–97.

Cooper, D.A. & Corcoran, P.L. (2010). Effects of mechanical and chemical processes on the degradation of plastic debris on the island of Kauai, Hawaii. *Marine Pollution Bulletin*, **60**, 650–4.

Corlett, R.T. (2015). The Anthropocene concept in ecology and conservation. *Trends in Ecology & Evolution*, **30**, 36–41.

Costello, M.J., Coll, M., Danovaro, R., *et al.* (2010). A census of marine biodiversity knowledge, resources, and future challenges. *PLoS ONE*, **5**, e12110.

Cózar, A., Sanz-Martín, M., Martí, E., *et al.* (2015). Plastic accumulation in the Mediterranean Sea. *PLoS ONE*, **10**, e0121762.

Dahms, H.-U. & Lee, J.-S. (2010). UV radiation in marine ectotherms: molecular effects and responses. *Aquatic Toxicology*, **97**, 3–14.

Davenport, J. & Davenport, J.L. (2006). The impact of tourism and personal leisure transport on coastal environments: a review. *Estuarine, Coastal and Shelf Science*, **67**, 280–92.

David, M. (2015). Vessels and ballast water. In *Global Maritime Transport and Ballast Water: Management Issues and Solutions*, ed. M.David & S.Gollasch, pp. 13–34. Dordrecht, Netherlands: Springer.

David, M. & Gollasch, S. (2015). Ballast water management systems for vessels. In *Global Maritime Transport and Ballast Water: Management Issues and Solutions*, ed. M.David & S.Gollasch, pp. 109–32. Dordrecht, Netherlands: Springer.

Davis, W.J. (1993). Contamination of coastal versus open ocean surface waters: a brief meta-analysis. *Marine Pollution Bulletin*, **26**, 128–34.

Dayton, P.K., Thrush, S.F., Agardy, M.T. & Hofman, R.J. (1995). Environmental effects of marine fishing. *Aquatic Conservation: Marine and Freshwater Ecosystems*, **5**, 205–32.

Derraik, J.G.B. (2002). The pollution of the marine environment by plastic debris: a review. *Marine Pollution Bulletin*, **44**, 842–52.

Dietz, T., Rosa, E.A. & York, R. (2007). Driving the human ecological footprint. *Frontiers in Ecology and the Environment*, **5**, 13–18.

Doney, S.C., Ruckelshaus, M., Duffy, J.E., *et al.* (2012). Climate change impacts on marine ecosystems. *Annual Review of Marine Science*, **4**, 11–37.

Drinkwater, K.F., Beaugrand, G., Kaeriyama, M., *et al.* (2010). On the processes linking climate to ecosystem changes. *Journal of Marine Systems*, **79**, 374–88.

Dyoulgerov, M.F. (2000). Global legal instruments on the marine environment at the year 2000. In *Seas at the Millennium: an Environmental Evaluation*, ed. C.Sheppard, pp. 331–48. Oxford, UK: Elsevier Science Ltd.

Edgar, G.J., Stuart-Smith, R.D., Willis, T.J., *et al.* (2014). Global conservation outcomes depend on marine protected areas with five key features. *Nature*, **506**, 216–20.

Eriksen, M., Lebreton, L.C.M., Carson, H.S., *et al.* (2014). Plastic pollution in the world's oceans: more than 5 trillion plastic pieces weighing over 250,000 tons afloat at sea. *PLoS ONE*, **9**, e111913.

Erlandson, J.M. & Rick, T.C. (2010). Archaeology meets marine ecology: the antiquity of maritime cultures and human impacts on marine fisheries and ecosystems. *Annual Review of Marine Science*, **2**, 231–51.

Fabry, V.J., Seibel, B.A., Feely, R.A. & Orr, J.C. (2008). Impacts of ocean acidification on marine fauna and ecosystem processes. *ICES Journal of Marine Science*, **65**, 414–32.

FAO (2016). *The State of World Fisheries and Aquaculture 2016: Contributing to Food Security and Nutrition for All.* Rome: FAO.

Feely, R.A., Sabine, C.L., Lee, K., *et al.* (2004). Impact of anthropogenic $CO_2$ on the $CaCO_3$ system in the oceans. *Science*, **305**, 362–6.

Fowler, S.W. (1990). Critical review of selected heavy metal and chlorinated hydrocarbon concentrations in the marine environment. *Marine Environmental Research*, **29**, 1–64.

Fowler, S.W. & Fisher, N.S. (2004). Radionuclides in the biosphere. In *Marine Radioactivity*, ed. H.D.Livingston, pp. 167–203. Amsterdam: Elsevier.

Gabric, A.J. & Bell, P.R.F. (1993). Review of the effects of non-point nutrient loading on coastal ecosystems. *Australian Journal of Marine and Freshwater Research*, **44**, 261–83.

Gall, S.C. & Thompson, R.C. (2015). The impact of debris on marine life. *Marine Pollution Bulletin*, **92**, 170–9.

Geller, J.B., Darling, J.A. & Carlton, J.T. (2010). Genetic perspectives on marine biological invasions. *Annual Review of Marine Science*, **2**, 367–93.

Geller, J.B., Walton, E.D., Grosholz, E.D. & Ruiz, G.M. (1997). Cryptic invasions of the crab *Carcinus* detected by molecular phylogeography. *Molecular Ecology*, **6**, 901–6.

Gerges, M.A. (1994). Marine pollution monitoring, assessment and control: UNEP's approach and strategy. *Marine Pollution Bulletin*, **28**, 199–210.

GESAMP (IMO/FAO/UNESCO/WMO/WHO/IAEA/UN/UNEP Joint Group of Experts on the Scientific Aspects of Marine Pollution) (1990a). *The State of the Marine Environment. GESAMP Reports and Studies*, **39**, 111 pp.

GESAMP (IMO/FAO/UNESCO/WMO/WHO/IAEA/UN/UNEP Joint Group of Experts on the Scientific Aspects of Marine Pollution) (1990b). *Review of Potentially Harmful Substances. Nutrients. GESAMP Reports and Studies*, **34**, 40 pp.

GESAMP (IMO/FAO/UNESCO/WMO/WHO/IAEA/UN/UNEP Joint Group of Experts on the Scientific Aspects of Marine Pollution) (1991). *Global Strategies for Marine Environmental Protection. GESAMP Reports and Studies*, **45**, 36 pp.

GESAMP (IMO/FAO/UNESCO/WMO/WHO/IAEA/UN/UNEP Joint Group of Experts on the Scientific Aspects of Marine Pollution) (1993). *Impact of Oil and Related Chemicals and Wastes on the Marine Environment. GESAMP Reports and Studies*, **50**, 180 pp.

Glasson, J., Therivel, R. & Chadwick, A., eds. (2005). *Introduction to Environmental Impact Assessment*, 3rd edn. London: Routledge.

Gollasch, S., Minchin, D. & David, M. (2015). The transfer of harmful aquatic organisms and pathogens with ballast water and their impacts. In *Global Maritime Transport and Ballast Water: Management Issues and Solutions*, ed. M.David & S.Gollasch, pp. 35–58. Dordrecht, Netherlands: Springer.

Gordon, H.S. (1954). The economic theory of a common-property resource: the fishery. *Journal of Political Economy*, **62**, 124–42.

Gray, J.S. (1992). Eutrophication in the sea. In *Marine Eutrophication and Population Dynamics*, ed. G.Colombo, I.Ferrari, V.U.Cecherelli & R.Rossi, pp. 3–15. Fredensborg, Denmark: Olsen & Olsen.

Gregory, M.R. (1991). The hazards of persistent marine pollution: drift plastics and conservation islands. *Journal of the Royal Society of New Zealand*, **21**, 83–100.

Gregory, M.R. (1999). Plastics and South Pacific island shores: environmental implications. *Ocean & Coastal Management*, **42**, 603–15.

Gregory, M.R. (2009). Environmental implications of plastic debris in marine settings–entanglement, ingestion, smothering, hangers-on, hitch-hiking and alien invasions. *Philosophical Transactions of the Royal Society B*, **364**, 2013–25.

Gruber, N., Hauri, C., Lachkar, Z., *et al.* (2012). Rapid progression of ocean acidification in the California Current System. *Science*, **337**, 220–3.

Guinotte, J.M. & Fabry, V.J. (2008). Ocean acidification and its potential effects on marine ecosystems. *Annals of the New York Academy of Sciences*, **1134**, 320–42.

Häder, D.-P., Williamson, C.E., Wängberg, S.-Å., *et al.* (2015). Effects of UV radiation on aquatic ecosystems and interactions with other environmental factors. *Photochemical & Photobiological Sciences*, **14**, 108–26.

Hall, S.J. (1999). *The Effects of Fishing on Marine Ecosystems and Communities*. Oxford, UK: Blackwell Science.

Hammer, J., Kraak, M.H.S. & Parsons, J.R. (2012). Plastics in the marine environment: the dark side of a modern gift. *Reviews of Environmental Contamination and Toxicology*, **220**, 1–44.

Hansen, J., Ruedy, R., Sato, M. & Lo, K. (2010). Global surface temperature change. *Reviews of Geophysics*, **48**, article RG4004.

Hansen, J., Sato, M., Hearty, P., *et al.* (2016). Ice melt, sea level rise and superstorms: evidence from paleoclimate data, climate modeling, and modern observations that 2 °C global warming could be dangerous. *Atmospheric Chemistry and Physics*, **16**, 3761–812.

Hardin, G. (1968). The tragedy of the commons. *Science*, **162**, 1243–8.

Harding, L.E. (1992). Measures of marine environmental quality. *Marine Pollution Bulletin*, **25**, 23–7.

Harvell, C.D., Kim, K., Burkholder, J.M., *et al.* (1999). Emerging marine diseases – climate links and anthropogenic factors. *Science*, **285**, 1505–10.

Hay, C.C., Morrow, E., Kopp, R.E. & Mitrovica, J.X. (2015). Probabilistic reanalysis of twentieth-century sea-level rise. *Nature*, **517**, 481–4.

Hayes, K.R. & Sliwa, C. (2003). Identifying potential marine pests – a deductive approach applied to Australia. *Marine Pollution Bulletin*, **46**, 91–8.
Heck, K.L. Jr. & Valentine, J.F. (2007). The primacy of top-down effects in shallow benthic ecosystems. *Estuaries and Coasts*, **30**, 371–81.
Hoegh-Guldberg, O. (2011). The impact of climate change on coral reef ecosystems. In *Coral Reefs: An Ecosystem in Transition*, ed. Z.Dubinsky & N.Stambler, pp. 391–403. New York: Springer-Verlag.
Hoegh-Guldberg, O. & Bruno, J.F. (2010). The impact of climate change on the world's marine ecosystems. *Science*, **328**, 1523–8.
Howarth, R.W. (2008). Coastal nitrogen pollution: a review of sources and trends globally and regionally. *Harmful Algae*, **8**, 14–20.
Hunt, B. & Vincent, A.C.J. (2006). Scale and sustainability of marine bioprospecting for pharmaceuticals. *Ambio*, **35**, 57–64.
IAEA (2005). *Worldwide Marine Radioactivity Studies (WOMARS): Radionuclide Levels in Oceans and Seas*. IAEA-TECDOC-1429, 187 pp. Vienna: International Atomic Energy Agency.
IAIA (2010). Principles of environmental impact assessment best practice. International Association for Impact Assessment (www.iaia.org).
IPCC (2014). *Climate Change 2014: Synthesis Report*. Contribution of Working Groups I, II and III to the Fifth Assessment Report of the Intergovernmental Panel on Climate Change (Core Writing Team, R.K.Pachauri & L.A.Meyer (eds.)). Geneva, Switzerland: IPCC.
Jackson, J.B.C., Kirby, M.X., Berger, W.H., *et al.* (2001). Historical overfishing and the recent collapse of coastal ecosystems. *Science*, **293**, 629–38.
Jambeck, J.R., Geyer, R. & Wilcox, C., *et al.* (2015). Plastic waste inputs from land into the ocean. *Science*, **347**, 768–71.
Jennings, S., Kaiser, M.J. & Reynolds, J.D. (2001). *Marine Fisheries Ecology*. Oxford, UK: Blackwell Science.
Johnson, S.W. (1994). Deposition of trawl web on an Alaska beach after implementation of MARPOL Annex V legislation. *Marine Pollution Bulletin*, **28**, 477–81.
Kite-Powell, H.L., Hoagland, P. & Jin, D. (1998). Policy, law, and public opposition: the prospects for abyssal ocean waste disposal in the United States. *Journal of Marine Systems*, **14**, 377–96.
Kitzes, J., Wackernagel, M., Loh, J., *et al.* (2008). Shrink and share: humanity's present and future Ecological Footprint. *Philosophical Transactions of the Royal Society B*, **363**, 467–75.
Koslow, J.A. & Couture, J. (2013). Follow the fish. *Nature*, **502**, 163–4.
Kroeker, K.J., Kordas, R.L., Crim, R., *et al.* (2013). Impacts of ocean acidification on marine organisms: quantifying sensitivities and interaction with warming. *Global Change Biology*, **19**, 1884–96.
Lagring, R., Degraer, S., de Montpellier, G., *et al.* (2012). Twenty years of Belgian North Sea aerial surveillance: a quantitative analysis of results confirms effectiveness of international oil pollution legislation. *Marine Pollution Bulletin*, **64**, 644–52.
Laist, D.W. (1987). Overview of the biological effects of lost and discarded plastic debris in the marine environment. *Marine Pollution Bulletin*, **18**, 319–26.
Lester, J.N. (1990). Sewage and sewage sludge treatment. In *Pollution: Causes, Effects, and Control*, 2nd edn, ed. R.M.Harrison, pp. 33–62. Cambridge, UK: The Royal Society of Chemistry.
Ling, S.D. (2008). Range expansion of a habitat-modifying species leads to loss of taxonomic diversity: a new and impoverished reef state. *Oecologia*, **156**, 883–94.
Lobel, P.B., Belkhode, S.P., Jackson, S.E. & Longerich, H.P. (1990). Recent taxonomic discoveries concerning the mussel *Mytilus*: implications for biomonitoring. *Archives of Environmental Contamination and Toxicology*, **19**, 508–12.
Loganathan, B.G. & Kannan, K. (1994). Global organochlorine contamination trends: an overview. *Ambio*, **23**, 187–91.
Lotze, H.K. & Worm, B. (2008). Historical baselines for large marine animals. *Trends in Ecology and Evolution*, **24**, 254–62.
Ludwig, D., Hilborn, R. & Walters, C. (1993). Uncertainty, resource exploitation, and conservation: lessons from history. *Science*, **260**, 17 & 36.
MacKenzie, D. (2011). Operation clean-up. *New Scientist*, **212** (2836), 46–9.

Markert, A., Wehrmann, A. & Kröncke, I. (2010). Recently established *Crassostrea*-reefs versus native *Mytilus*-beds: differences in ecosystem engineering affects the macrofaunal communities (Wadden Sea of Lower Saxony, southern German Bight). *Biological Invasions*, **12**, 15–32.

Mason, M. (2003). Civil liability for oil pollution damage: examining the evolving scope for environmental compensation in the international regime. *Marine Policy*, **27**, 1–12.

Mattson, G. (2006). MARPOL 73/78 and Annex I: an assessment of its effectiveness. *Journal of International Wildlife Law and Policy*, **9**, 175–94.

Mayer, A.M.S., Glaser, K.B., Cuevas, C., *et al.* (2010). The odyssey of marine pharmaceuticals: a current pipeline perspective. *Trends in Pharmacological Science*, **31**, 255–65.

Mayer-Pinto, M., Underwood, A.J., Tolhurst, T. & Coleman, R.A. (2010). Effects of metals on aquatic assemblages: what do we really know? *Journal of Experimental Marine Biology and Ecology*, **391**, 1–9.

McCollin, T., Shanks, A.M. & Dunn, J. (2008). Changes in zooplankton abundance and diversity after ballast water exchange in regional seas. *Marine Pollution Bulletin*, **56**, 834–44.

McDonald, J.H., Seed, R. & Koehn, R.K. (1991). Allozymes and morphometric characters of three species of *Mytilus* in the Northern and Southern Hemisphere. *Marine Biology*, **111**, 323–33.

Mee, L.D. & Fowler, S.W. (1991). Organotin biocides in the marine environment: a managed transient? *Marine Environmental Research*, **32**, 1–5.

Michel, J., Etkin, D.S., Gilbert, T., Urban, R., Waldron, J. & Blocksidge, C.T. (2005). Potentially polluting wrecks in marine waters. An issue paper prepared for the 2005 International Oil Spill Conference, Miami, Florida. 40 pp.

Molnar, J.L., Gamboa, R.L., Revenga, C. & Spalding, M.D. (2008). Assessing the global threat of invasive species to marine biodiversity. *Frontiers in Ecology and the Environment*, **6**, 485–92.

Moore, C.J. (2008). Synthetic polymers in the marine environment: a rapidly increasing, long-term threat. *Environmental Research*, **108**, 131–9.

National Research Council (2003). *Oil in the Sea III: Inputs, Fates, and Effects*. Washington, DC: National Academies Press.

Norse, E.A. (ed.) (1993). *Global Marine Biological Diversity: a Strategy for Building Conservation into Decision Making*. Washington, DC: Island Press.

Ocean Conservancy (2010). *International Coastal Cleanup 2010 Report*. Washington, DC: Ocean Conservancy.

Orams, M. (1999). *Marine Tourism: Development, Impacts and Management*. London: Routledge.

Pauly, D. (1995). Anecdotes and the shifting baseline syndrome of fisheries. *Trends in Ecology and Evolution*, **10**, 430.

Pauly, D. & Zeller, D. (2016). Catch reconstructions reveal that global marine fisheries catches are higher than reported and declining. *Nature Communications*, **7**, 10244.

Pinnegar, J.K. & Engelhard, G.H. (2008). The 'shifting baseline' phenomenon: a global perspective. *Reviews in Fish Biology and Fisheries*, **18**, 1–16.

Pinnegar, J.K., Polunin, N.V.C., Francour, P., *et al.* (2000). Trophic cascades in benthic marine ecosystems: lessons for fisheries and protected-area management. *Environmental Conservation*, **27**, 179–200.

Poloczanska, E.S., Brown, C.J., Sydeman, W.J., *et al.* (2013). Global imprint of climate change on marine life. *Nature Climate Change*, **3**, 919–25.

Porte, C., Janer, G., Lorusso, L.C., *et al.* (2006). Endocrine disruptors in marine organisms: approaches and perspectives. *Comparative Biochemistry and Physiology C*, **143**, 303–15.

Pörtner, H.-O. (2008). Ecosystem effects of ocean acidification in times of ocean warming: a physiologist's view. *Marine Ecology Progress Series*, **373**, 203–17.

Prater, A.J. & Lloyd, C.S. (1987). Birds. In *Biological Surveys of Estuaries and Coasts*, ed. J.M.Baker & W.J.Wolff, pp. 374–403. Cambridge, UK: Cambridge University Press.

Pruter, A.T. (1987). Sources, quantities, and distribution of persistent plastics in the marine environment. *Marine Pollution Bulletin*, **18**, 305–10.

Rabalais, N.N. (2005). The potential for nutrient overenrichment to diminish marine biodiversity. In *Marine Conservation Biology: The Science of Maintaining the Sea's Biodiversity*, ed. E.A.Norse & L.B.Crowder, pp. 109–22. Washington, DC: Island Press.

Raffaelli, D. & Hawkins, S. (1996). *Intertidal Ecology*. London: Chapman & Hall.

Redford, D.P., Trulli, H.K. & Trulli, W.R. (1997). Sources of plastic pellets in the aquatic environment. In *Marine Debris: Sources, Impacts, and Solutions*, ed. J.M.Coe & D.B.Rogers, pp. 335–43. New York: Springer-Verlag.

Rochman, C.M., Kross, S.M., Armstrong, J.B., *et al.* (2015). Scientific evidence supports a ban on microbeads. *Environmental Science & Technology*, **49**, 10759–61.

Rogelj, J., den Elzen, M., Höhne, N., et al. (2016). Paris Agreement climate proposals need a boost to keep warming well below 2 °C. *Nature*, **534**, 631–9.

Samiullah, Y. (1985). Biological effects of marine oil pollution. *Oil & Petrochemical Pollution*, **2**, 235–64.

Seebens, H., Gastner, M.T. & Blasius, B. (2013). The risk of marine bioinvasion caused by global shipping. *Ecology Letters*, **16**, 782–90.

Setälä, O., Fleming-Lehtinen, V. & Lehtiniemi, M. (2014). Ingestion and transfer of microplastics in the planktonic food web. *Environmental Pollution*, **185**, 77–83.

Sheppard, C. (1995). The shifting baseline syndrome. *Marine Pollution Bulletin*, **12**, 766–7.

Simberloff, D., Martin, J.-L., Genovesi, P., *et al.* (2013). Impacts of biological invasions: what's what and the way forward. *Trends in Ecology & Evolution*, **28**, 58–66.

Solomon, S., Haskins, J., Ivy, D.J. & Min, F. (2014). Fundamental differences between Arctic and Antarctic ozone depletion. *Proceedings of the National Academy of Sciences*, **111**, 6220–5.

Solomon, S., Ivy, D.J., Kinnison, D., *et al.* (2016). Emergence of healing in the Antarctic ozone layer. *Science*, **353**, 269–74.

Stachowicz, J.J., Whitlatch, R.B. & Osman, R.W. (1999). Species diversity and invasion resistance in a marine ecosystem. *Science*, **286**, 1577–9.

Stuart-Smith, R.D., Edgar, G.J., Barrett, N.S., Kininmonth, S.J. & Bates, A.E. (2015). Thermal biases and vulnerability to warming in the world's marine fauna. *Nature*, **528**, 88–92.

Taylor, P. (1993). The state of the marine environment: a critique of the work and role of the Joint Group of Experts on Scientific Aspects of Marine Pollution (GESAMP). *Marine Pollution Bulletin*, **26**, 120–7.

Thorne-Miller, B. (1999). *The Living Ocean: Understanding and Protecting Marine Biodiversity*, 2nd edn. Washington, DC: Island Press.

Thrush, S.F., Hewitt, J.E., Norkko, A., Cummings, V.J. & Funnell, G.A. (2003). Macrobenthic recovery processes following catastrophic sedimentation on estuarine sandflats. *Ecological Applications*, **13**, 1433–55.

Troost, K. (2010). Causes and effects of a highly successful marine invasion: case-study of the introduced Pacific oyster *Crassostrea gigas* in continental NW European estuaries. *Journal of Sea Research*, **64**, 145–65.

UN (1983). *The Law of the Sea: United Nations Convention on the Law of the Sea with Index and Final Act of the Third United Nations Conference on the Law of the Sea*. New York: United Nations.

UN, Department of Economic and Social Affairs, Population Division (1999). *The World at Six Billion*. ESA/P/WP.154. New York: United Nations.

UN, Department of Economic and Social Affairs, Population Division (2015). *World Population Prospects: The 2015 Revision, Key Findings and Advance Tables*. Working Paper No. ESA/P/WP.241. New York: United Nations.

Underwood, A.J. (1992). Beyond BACI: the detection of environmental impacts on populations in the real, but variable, world. *Journal of Experimental Marine Biology and Ecology*, **161**, 145–78.

UNEP (1995). *Global Programme of Action for the Protection of the Marine Environment from Land-based Activities*. Washington, DC: United Nations Environment Programme. 60 pp.

UNEP/GPA (2006). *The State of the Marine Environment: Trends and Processes*. The Hague: Coordination Office of the Global Programme of Action for the Protection of the Marine Environment from Land-based Activities of the United Nations Environment Programme.

UNFCCC Secretariat (2015). *Adoption of the Paris Agreement*. United Nations Framework Convention on Climate Change. FCCC/CP/2015/L.9/Rev.1.

Van Dyke, J.M. (1991). Protected marine areas and low-lying atolls. *Ocean & Shoreline Management*, **16**, 87–160.

van Sebille, E., Wilcox, C., Lebreton, L., *et al.* (2015). A global inventory of small floating plastic debris. *Environmental Research Letters*, **10**, 124006.

Walpole, S.C., Prieto-Merino, D., Edwards. P., *et al.* (2012). The weight of nations: an estimation of adult human biomass. *BMC Public Health*, **12**, article 439.
Wang, Z., Stout, S.A. & Fingas, M. (2006). Forensic fingerprinting of biomarkers for oil spill characterization and source identification. *Environmental Forensics*, **7**, 105–46.
Warwick, R.M. (1988). The level of taxonomic discrimination required to detect pollution effects on marine benthic communities. *Marine Pollution Bulletin*, **19**, 259–68.
Warwick, R.M. (1993). Environmental impact studies on marine communities: pragmatical considerations. *Australian Journal of Ecology*, **18**, 63–80.
Wells, P.G., Duce, R.A. & Huber, M.E. (2002). Caring for the sea – accomplishments, activities and future of the United Nations GESAMP (the Joint Group of Experts on the Scientific Aspects of Marine Environmental Protection). *Ocean & Coastal Management*, **45**, 77–89.
Wilson, J.G. (1988). *The Biology of Estuarine Management*. London: Croom Helm.
Winston, J.E. (1982). Drift plastic – an expanding niche for a marine invertebrate? *Marine Pollution Bulletin*, **13**, 348–51.
Winston, J.E., Gregory, M.R. & Stevens, L.M. (1997). Encrusters, epibionts, and other biota associated with pelagic plastics: a review of biogeographical, environmental, and conservation issues. In *Marine Debris: Sources, Impacts, and Solutions*, ed. J.M.Coe & D.B.Rogers, pp. 81–97. New York: Springer-Verlag.
Woodall, L.C., Sanchez-Vidal, A., Canals, M., *et al.* (2014). The deep sea is a major sink for microplastic debris. *Royal Society Open Science*, **1**, 140317.
Wright, S.L., Thompson, R.C. & Galloway, T.S. (2013). The physical impacts of microplastics on marine organisms: a review. *Environmental Pollution*, **178**, 483–92.
Yablokov, A.V. (2005). Meta-analysis of the radioactive pollution of the ocean. In *Strategic Management of Marine Ecosystems*, ed. E.Levner, I.Linkov & J.-M.Proth, pp. 11–27. Dordrecht, Netherlands: Springer.
Zhang, P., Song, J. & Yuan, H. (2009). Persistent organic pollutant residues in the sediments and mollusks from the Bohai Sea coastal areas, North China: an overview. *Environment International*, **35**, 632–46.

# 3 Conservation

How we perceive the natural environment is fundamental to our perspective on conservation. Our views on the natural world are neither static nor absolute – and often ambivalent – and cultural differences affect the way we interact with and interpret our natural environment. Within cultures, the values placed on natural features and biota – not just exploitable resources but also non-material aspects – have changed and will continue to do so. Inevitably, one's approach to nature conservation has a cultural element.

Belief systems have strongly influenced the way we see nature. In pagan animism, plants, animals, and natural objects were believed to be endowed with spirits, and the need to placate them may have afforded some protection from human interference. Subsequently, it has been argued that Judeo-Christian beliefs have exerted a major influence on Western attitudes to nature, in particular a view that the natural world was created for humans to subjugate and exploit (White, 1967). Christian belief in the perfection of God's design nevertheless taught that 'even the most apparently noxious species served some indispensable human purpose ... that all created species had a necessary part to play in the economy of nature' (Thomas, 1983). Followers of other beliefs, such as Hindus and Buddhists, hold that all living things are to be respected and that nature and human society are closely interrelated, a view that has gained currency in Western culture as deep ecology.

A familiarity with nature does not necessarily result in conservation. Whilst our ancestors may have been well attuned to their natural environment, the notion that they limited their use of resources may be modern wishful thinking (Grayson, 2001). When it comes to open-access resources, they were, at least on land and along the coastal fringe, as adept at overexploitation as their descendants, and any sustainability may have been more by default, the result of limited technology and low population density. Governed by self-interest, humans appear to have no innate tendency to exercise restraint in this regard (Ridley, 1996).

## Meaning of Conservation

What do we mean today by 'management' and 'conservation'? Neither term admits to any rigorous definition, each being used in somewhat different ways. Holdgate (1991) takes management to mean 'human intervention in the dynamic processes which determine the composition of plant and animal communities, so as to maintain a particular desired pattern or series of processes'. Conservation, from the Latin 'to keep entire', implies a particular type of natural resource management so as to maintain as far as possible the natural structure and function of ecosystems and avoid unsustainable human impacts. The word is often used as essentially synonymous with the concept of sustainable development. The World Conservation Strategy (IUCN/UNEP/WWF, 1980), for instance, defines conservation as 'the management of human use of the biosphere so that it may yield the greatest sustainable benefit to present generations while maintaining its potential to meet the needs and aspirations of future generations'.

As a form of management, the practice of conservation may entail human intervention in the control of natural populations, including the taking of animal life. There is, on the other hand, a protectionist philosophy that wildlife populations should not be subject to human interference and that killing animals is unethical. Both conservationist and protectionist viewpoints have been incorporated into marine wildlife policy, notably in relation to marine mammals (Manning, 1989). Both viewpoints agree on the need to protect endangered populations. Protectionist concerns, however, extend to the protection of individuals, irrespective of whether the population in question is threatened. In terrestrial systems this situation commonly applies to large vertebrates, but apart from marine mammals in particular, vertebrates in marine systems often engender little public sympathy.

## Rationale for Conservation

Various reasons can be advanced in support of conservation, but certain main lines of argument can be recognized: ethical, aesthetic, scientific, economic, and ecological. The ethical basis affirms that it is immoral to exploit living organisms, but how this is expressed may vary considerably among cultures. Moral arguments tend to be advanced most strongly in the case of mammals, reflecting public sentiment. There is also a less anthropocentric viewpoint that all species have a right to exist, irrespective of human values or material needs. It is often explicit, however, that the goal of conservation concerns the human use of resources rather than the maintenance of ecosystem integrity. This dichotomy, harking back to the theological arguments earlier, remains a fundamental question relevant to modern approaches to resource management and conservation (Stanley, 1995).

Aesthetic arguments for conservation emphasize the uniqueness and natural beauty of living things. Conspicuous marine vertebrates and spectacular shallow-water habitats, such as reefs, would be among the more obvious candidates to invite aesthetic considerations. This is reflected by the popularity of marine natural history topics in the media as well as by pursuits such as diving, whale watching, and other forms of ecotourism. But aesthetic considerations may carry little overall potential to drive conservation efforts in the sea where, for the public at large, most habitats and organisms are inaccessible or have limited appeal.

One aspect of the scientific argument for conservation stresses the scientific insights that may be lost with the extinction of a species. Compared to the terrestrial realm, few marine species have been driven to global extinction as a result of human activities, although some extinctions in the sea have probably gone undetected (Carlton *et al.*, 1999). Also relevant as a scientific argument is the potential value of scientific information that is generated by the implementation of measures to minimize human impact. One of the values advocated for marine protected areas is their use in the study of biological and ecological processes in areas less confounded by human disturbance.

Conservation has long been promoted for economic reasons, at least for species of direct commercial value, and increasingly the potential economic value of species is advanced as a reason to protect marine biodiversity because it likely harbours a wide range of natural products and genetic materials that could prove useful to us (Arrieta *et al.*, 2010). Cone snails, for instance, are predators that paralyse their prey using venom injected from their proboscis. The several hundred species of cone snails produce an unparalleled diversity of venom peptides (more than 100 000) that are being used in the development of analgesics and other drugs (Han *et al.*, 2008).

There may be pitfalls in assigning a purely economic value to biodiversity. On solely financial grounds it can be argued that it makes more sense to hunt attractive high-biomass resources to extinction to maximize present value as this would provide a better return on investment than waiting for a population to recover to the point where it could sustain an annual catch. Species targeted by whaling and deep-sea fisheries, for example, have such low population growth rates as to be effectively non-renewable resources on human timescales (Clark, 1973; Norse *et al*., 2012).

A major reason why the natural world is in crisis is that traditional economic models fail to take account of ecological constraints and of ecosystem services – the benefits humans derive, directly or indirectly, from ecosystems, such as maintaining water quality, nutrient cycling, and coastal protection (Costanza *et al*., 1997, 2014) (Table 3.1). The value of these services can be estimated in monetary terms (Table 3.2). Ecosystems provide a range of irreplaceable processes and functions, and maintaining these life-support systems is essential for our welfare. This is the most powerful argument for conservation, although in terms of popular appreciation and support its rationale may seem less than tangible. The importance, for example, of coastal wetlands in nutrient cycling or of coral reefs in coastal protection may not be immediately obvious, and

**Table 3.1** Marine Ecosystem Services

| Ecosystem service | Explanation/examples |
|---|---|
| **Provisioning services** | |
| Seafood | Fish, shellfish, seaweed |
| Seawater | Seawater used in shipping, industrial cooling, desalination |
| Raw materials | Algae (non-food), building aggregate, salt |
| Genetic resources | Genetic material from marine organisms for non-medicinal uses |
| Medicinal resources | Marine-derived pharmaceuticals |
| Ornamental resources | Shells, aquarium animals, pearls, corals |
| **Regulating services** | |
| Air purification | Removal of dust, sulphur dioxide, carbon dioxide |
| Climate regulation | Contribution of a marine ecosystem to maintenance of a favourable climate through impacts on the hydrological cycle, temperature regulation, and the contribution to climate-influencing substances in the atmosphere |
| Disturbance, prevention, or moderation | Contribution of marine ecosystem structures (e.g. coastal wetlands) to dampening the intensity of environmental disturbances such as storm floods, tsunamis, and hurricanes |
| Regulation of water flows | Contribution of marine ecosystems to the maintenance of localized coastal current structures (e.g. effect of macroalgae on localized current intensity) |
| Waste treatment | Removal of pollutants by marine ecosystems (e.g. breakdown by microorganisms; filtering of water by shellfish) |

**Table 3.1** (cont.)

| Ecosystem service | Explanation/examples |
|---|---|
| Coastal erosion prevention | Maintenance of dunes by coastal vegetation; macroalgal beds reducing nearshore scouring |
| Biological control | Contribution of marine ecosystems to maintenance of healthy population dynamics to support ecosystem resilience through maintaining food-web structure and flows |
| **Habitat services** | |
| Migratory and nursery habitat | Contribution of a particular habitat to migratory species' populations through provision of critical habitat for feeding, reproduction, and juvenile maturation |
| Gene pool protection | Contribution of marine habitats to maintenance of viable gene pools through natural selection/evolutionary processes |
| **Cultural services** | |
| Recreation and leisure | Bird watching, whale watching, beachcombing, sailing, recreational fishing, scuba diving |
| Aesthetic experience | Contribution that a marine ecosystem makes to the existence of a surface or subsurface landscape that generates a noticeable emotional response within the observer |
| Inspiration for culture, art and design | Contribution that a marine ecosystem makes to the existence of environmental features that inspire elements of culture, art, and/or design |
| Spiritual experience | Contribution that a marine ecosystem makes to formal and informal religious experiences |
| Information for cognitive development | Contribution that a marine ecosystem makes to education and research |
| Cultural heritage and identity | Importance of coastal and marine environments in cultural traditions and folklore |

Reprinted from *Journal of Environmental Management*, Vol. 130, Böhnke-Henrichs, A., Baulcomb, C., Koss, R., de Groot, R.S. & Hussain, S., Typology and indicators of ecosystem services for marine spatial planning and management, pages 135–45, copyright 2013, with permission from Elsevier.

small organisms that support essential services of ecosystems are overlooked or worse. 'It is a failing of our species that we ignore and even despise the creatures whose lives sustain our own' (Wilson, 1992).

There is, nevertheless, a growing awareness of the interrelatedness of the natural world and that even at small, local levels the relevance of conservation may be recognized. We still have much to understand about ecological processes, and there is a strong argument for taking an overall precautionary approach in our interactions with the natural environment. Opposition to conservation can, however, be widespread and deep-seated, usually where there are conflicts with immediate socio-economic needs.

**Table 3.2** Economic Value of Marine Ecosystem Services
Value estimates are in 2007 $US.

| Biome | Value in 2011($ ha$^{-1}$ y$^{-1}$) |
|---|---|
| Marine | 1 368 |
| Open ocean | 660 |
| Coastal | 8 944 |
| Estuaries | 28 916 |
| Seagrass/algal beds | 28 916 |
| Tidal marsh/mangroves | 193 843 |
| Coral reefs | 352 249 |
| Shelf | 2 222 |

Reprinted from *Global Environmental Change*, Vol. 26, Costanza, R., de Groot, R., Sutton, P., *et al.*, Changes in the global value of ecosystem services, pages 152–8, copyright 2014, with permission from Elsevier.

## Origins of a Conservation Conscience

We tend to think of conservation as a relatively modern concept. Yet laws to regulate hunting and the establishment of wilderness reserves, at least on land, are known from ancient Egypt, some 3000 years ago. The conservation of natural resources is a topic that has in fact occupied many civilizations (Alison, 1981).

There exists a well-recorded history of conservation law in England dating from about 700 AD, enacted by numerous assizes and charters – including the Magna Carta of 1215, which contained environmental sections – and later by means of statutes or acts of Parliament. An early example of fisheries regulations, dating from 1278, concerned salmon fishing in northern England, where regulations included closures and restrictions on the use of nets. In the 1420s, statutes were introduced to protect fish by prohibiting the setting of nets and to address the deterioration of coastal marshes and sea banks (Graham, 1956; Alison, 1981). The term 'conservation' appears to date from about this time with the 'conservators' of statutes who were responsible for the upkeep of the River Thames and entrusted with its 'conservacie' (Thomas, 1983). Later evidence of concern over fisheries comes from an act of 1558 relating to the preservation of fish 'which heretofore hath been much destroyed in rivers and streams, salt and fresh, within this Realm' as well as acts in the eighteenth century stipulating a minimum mesh size for nets and size limits for several marine fisheries and acts concerned with maintaining oyster fisheries (Graham, 1956).

The late nineteenth century saw a spate of conservation statutes introduced, but like much of the earlier legislation the primary concern was management of edible species. Wider issues relating to non-game animals, non-marketable plants, and habitat destruction were not considered, even though from the early 1800s there was increasing evidence of environmental contamination and degradation caused by industrialization (Alison, 1981). The devastating impact of the colonial exploitation of tropical islands had in fact become apparent from the early eighteenth century, giving rise to early examples of Western-style conservation strategies and legislation. The early history of conservation tends to demonstrate that countries try to curb environmental damage only when their economic interests are in danger (Grove, 1992).

From the late nineteenth century we start to see a rethinking of our view of nature and of our place in the natural world. With Darwinism and the emergence of ecology, humans started to recognize themselves as an integral part of the biota and began to value more explicitly the intrinsic properties of nature, not just its material 'resources', a term still with echoes of dominion. The evolution of environmental ethics is well illustrated in North America through the philosophies of Emerson, Thoreau, Muir, Pinchot, and Leopold (Callicott, 1992). A major legislative development in the USA was the passage of the Yellowstone Park Act of 1872, which stimulated an interest in conservation worldwide and helped lay the foundations for modern conservation policy (Jones, 1987). Although the act provided for the establishment of wilderness areas for public enjoyment rather than specifically for biological conservation, it was influential in the development of an approach to conservation in which economic consideration was not the main or only criterion. Economic arguments have continued to prevail, often when ecological factors indicate a need for restraint, but more recent approaches recognize that the pursuit of conservation cannot be isolated from economic and social goals; the three are ultimately interdependent, as emphasized, for instance, by Mangel *et al.* (1996) in their principles for conserving wild living resources.

The late 1960s to early 1970s witnessed an important shift in attitudes to conservation and a new urgency. The interrelatedness of components of the biosphere was more fully appreciated, as were its implications; humans faced an environmental dilemma that could no longer be ignored. The word 'ecology', coined a century earlier, entered the popular vocabulary as the ecological basis for pressing global problems became only too clear. At the heart of this new perception was a growing anxiety about the impact of a soaring human population on the global environment, notably through the depletion of non-renewable resources, the rates of habitat destruction, and species' extinction together with evidence of widespread environmental degradation and pollution. Public awareness was raised by the media, popular books on environmental issues, a new climate of environmental activism, and the first major international meetings on global environmental issues and resource management (Box 3.1).

## Major Environmental Forums

A landmark event of this period was the United Nations Conference on the Human Environment held in Stockholm in 1972 as it represents the emergence of international environmental law at the global level. The Stockholm Declaration defined general principles of environmental protection, and these have formed the basis of modern approaches to the management of the environment and natural resources. Also that year, as a result of the Stockholm conference, the UN General Assembly established the United Nations Environment Programme (UNEP) to 'serve as a focal point for environmental action and co-ordination within the United Nations system'. UNEP aims to foster environmental action worldwide and has a largely catalytic role in encouraging and supporting governments in the development and coordination of strategies of environmental management. It works closely with other UN bodies as well as with governments and national and international organizations but lacks any executive authority. From its inception, UNEP identified the health of the oceans as one of its priorities and has pursued this, mainly by means of a regional approach, most notably through its Regional Seas Programme (see Chapter 16).

### Box 3.1 Major International Forums Concerning Global Management and Conservation of the Natural Environment

**UN Conferences on the Law of the Sea (1958, 1960, 1973)**
UN Convention on the Law of the Sea (adopted 1982, entry into force 1994).
Rules for delimiting jurisdictional zones by coastal states, legal framework for major uses of the oceans and management of marine resources.

**UN Conference on the Human Environment (1972)**
Emergence of international environmental law and modern approaches to environmental management. United Nations Environment Programme (UNEP) and UNEP Regional Seas Programme established.

**World Conservation Strategy (1980)**
Develops the concept of sustainable development and aims 'to help advance the achievement of sustainable development through the conservation of living resources'.

**World Commission on Environment and Development (1983–7)**
The Brundtland Commission's elaboration of a new perspective on sustainable development, published as *Our Common Future* (Brundtland, 1987).

**UN Conference on Environment and Development (1992)**
Agenda 21 for sustainable development, Convention on Biological Diversity, Framework Convention on Climate Change, and promotion of the precautionary principle.

**World Summit on Sustainable Development (2002)**
The Johannesburg Declaration and plan of action.

**UN Conference on Sustainable Development (2012)**
A third global summit on sustainable development.

Other significant international developments that have occurred since then include the World Conservation Strategy (WCS) (IUCN/UNEP/WWF, 1980), intended to stimulate a more focused approach to the management of living resources and to provide guidance on how this might be achieved. The WCS identified coastal systems, including estuaries, wetlands, and coral reefs, as among the most threatened ecosystems whose ecological processes are essential for food production, health, and other aspects of human survival. It also highlighted marine ecosystems of exceptional diversity, problems of overfishing and associated incidental mortality, the need to strengthen regional approaches to marine conservation strategies, and the conservation of living resources beyond national jurisdictions. The WCS promoted the concept of 'sustainable development' to help advance an integration of conservation and development. This recognized the reality of resource limitation and the carrying capacities of ecosystems but the need to reconcile this with the human quest for economic and social betterment.

## Sustainable Development

The need to address ecological and societal needs was elaborated in particular in *Our Common Future*, a seminal report released in 1987 by the Brundtland Commission, the UN-sponsored World Commission on Environment and Development. Sustainable development is here defined as 'development that meets the needs of the present without compromising the ability of future generations to meet their own needs'. It contains two key concepts: 'the concept of "needs", in particular the essential needs of the world's poor, to which overriding priority should be given; and the idea of limitations imposed by the state of technology and social organization on the environment's ability to meet present and future needs' (Brundtland, 1987).

The concept of sustainable development has generated much debate, especially its practical implementation, but has been interpreted differently by advocates and detractors. Some argue that it overlooks the fundamental relationship – or contradiction – between human population growth, resource consumption, and environmental degradation. Nevertheless, the key issues of sustainable development relating to ecological integrity, economic security, and social equity are still just as relevant today, if not more so. Significant, for instance, in the post-Brundtland political landscape have been the rise of neoliberal ideology and campaigns to discredit environmental concerns, notably climate change, and concerted attacks on sustainable development and its subsequent elaboration in Agenda 21 (see below). *Our Common Future* very successfully brought the concept of sustainable development to deliberations on environmental matters – from those at the international to local levels – and saw it incorporated as an underlying principle in many environmental initiatives. But it has resulted in few enactments and failed to direct ongoing international discourse, such as for the Rio+20 Conference (see below) (Borowy, 2014).

The concept of sustainable development has tended to polarize debate on conservation strategy on whether such initiatives should have the safeguarding of biological diversity as a strict priority, notably through the use of protected areas, or should also incorporate human welfare needs such as alleviating poverty. This 'parks versus people' debate has tended to highlight the two ends of an ideological spectrum, whereas exploring pragmatic solutions to integrate biodiversity protection and human well-being using a variety of tools might be more productive (Miller *et al.*, 2011).

Twenty years after the Stockholm conference, the United Nations Conference on Environment and Development (UNCED), the 'Earth Summit', was held in Rio de Janeiro in 1992. Among the main outcomes of UNCED were the Rio Declaration promoting the principle of sustainable development and Agenda 21, an action plan for the twenty-first century intended as a blueprint for sustainable development and addressing a wide range of environmental concerns (UNCED, 1992). The principal UNCED prescriptions concerning the world's coasts and oceans, Chapter 17 of Agenda 21, cover several themes, including integrated coastal and ocean management, marine environmental protection, and sustainable fisheries. The only hard laws to result from Rio were the Convention on Biological Diversity and the Framework Convention on Climate Change, the latter giving rise to the Kyoto Protocol with its requirements to reduce greenhouse gas emissions (see Chapter 2).

A third international conference on sustainable development held in Rio de Janeiro in 2012 (the so-called Rio+20 Conference, coming 20 years after the Rio Earth Summit) essentially reaffirmed past efforts aimed at advancing sustainable development, including those to promote the conservation and sustainable use of the oceans and seas, and failed to deliver a binding agreement. To many observers, Rio+20 was especially disappointing in not significantly progressing the implementation of sustainable development and produced considerable disenchantment with major international forums.

## Precautionary Approach

The Rio Declaration was significant in its advocacy of another key concept in environmental protection, the 'precautionary approach'. This states that 'In order to protect the environment, the precautionary approach shall be widely applied by States according to their capabilities. Where there are threats of serious irreversible damage, lack of full scientific certainty shall not be used as a reason for postponing cost-effective measures to prevent environmental degradation' (Principle 15). Precautionary instruments are increasingly being adopted in environmental agreements, recognizing that data are often inadequate to accurately forecast the outcome of impacts and that there is typically a high degree of uncertainty when it comes to environmental consequences. So in the case of a new fishery, exploitation should proceed only at a rate consistent with the acquisition of data needed to determine sustainable catch levels. The usual scenario in our use of marine resources has been to forge ahead with exploitation with little understanding of the full implications of our actions and to institute controls only when obvious damage has occurred. But the measures then taken are usually as much to protect the economic investment that has been built up (Norse, 1993). By contrast, under a precautionary approach, environmental protection management becomes proactive rather than reactive and serves to promote responsibility to future generations rather than just allocation of resources among present users, as characterizes traditional approaches to resource use (Garcia, 1994).

Taking the prevention of environmental damage as a central premise necessitates a fundamental shift in our view of where responsibility should reside in providing relevant environmental evidence and criteria. For economic and political reasons this burden of proof has fallen to the public sector in the form of regulatory authorities rather than to those profiting from the exploitation, who are not required to argue that their actions will not be environmentally damaging, a stance that contrasts strongly to our expectations elsewhere, such as where human health and safety are of direct concern (Dayton, 1998).

## Biodiversity Targets

Another major outcome of Rio was the Convention on Biological Diversity, which came into force in 1993. Its main objectives are 'the conservation of biological diversity, the sustainable use of its components and the fair and equitable sharing of the benefits arising out of the utilization of genetic resources' (Article 1). The convention emphasizes *in situ* conservation, particularly conservation of ecosystems rather than species *per se*. And to this end, among other obligations, parties are required to develop national strategies for the conservation and sustainable use of biodiversity, as far as possible to establish a system of protected areas to conserve biodiversity, to prepare environmental impact assessments of proposed projects that are likely to have significant adverse effects on biodiversity, and to minimize adverse impacts.

The World Summit on Sustainable Development, held in Johannesburg in 2002, recognized the critical importance of the marine environment for global food security and well-being of many nations. Its plan of implementation urges nations to implement the various existing instruments to promote sustainable fisheries, marine environmental protection, and conservation of marine life. The plan also sets targets for marine conservation, including the application of an ecosystem approach to sustainable development of the oceans, maintaining or restoring fisheries stocks to levels that can produce the maximum sustainable yield, establishing marine protected area networks, and a significant reduction in the current rate of loss of biological diversity (WSSD, 2002).

The World Summit endorsed targets set by the Convention on Biological Diversity, a major goal being to significantly reduce rates of biodiversity loss by 2010. Most indicators, however, showed that targets had not been met and biodiversity continued to decline. This included marine indicators showing negative trends, such as shorebird populations worldwide, the extent of mangroves and seagrass beds, and the condition of coral reefs (Butchart *et al.*, 2010). The CBD subsequently adopted a new strategic plan for the period 2011–20 to stem the loss of biodiversity and revised targets (Aichi Biodiversity Targets). These include eliminating subsidies harmful to biodiversity, sustainably managing all fish and invertebrate stocks, minimizing anthropogenic pressure on coral reefs, and including at least 10% of the total marine area in protected areas (Normile, 2010). But whether such targets can be policed and adequately measured is debatable. Rands *et al.* (2010) also identify fundamental changes needed so that biodiversity issues are appropriately incorporated into socio-economic decision making. They argue that biodiversity should be managed as a public good with its economic value integral to and properly accounted for in management decisions. This needs to be supported by a policy and regulatory framework that provides suitable incentives and penalties but also, in a wider context, institutions and governance to enable such market models to operate.

## Key Principles of Conservation

We now understand the basic precepts to guide the conservation and sustainable management of living resources. Mangel *et al.* (1996), for example, present broad principles for the conservation of wild living resources (Box 3.2) and potential mechanisms for their implementation, and

### Box 3.2 Principles for the Conservation of Wild Living Resources

**Principle I** – Maintenance of healthy populations of wild living resources in perpetuity is inconsistent with unlimited growth of human consumption of and demand for those resources.

**Principle II** – The goal of conservation should be to secure present and future options by maintaining biological diversity at genetic, species, population, and ecosystem levels; as a general rule neither the resource nor other components of the ecosystem should be perturbed beyond natural boundaries of variation.

**Principle III** – Assessment of the possible ecological and sociological effects of resource use should precede both proposed use and proposed restriction or expansion of ongoing use of a resource.

**Principle IV** – Regulation of the use of living resources must be based on understanding the structure and dynamics of the ecosystem of which the resource is a part and must take into account the ecological and sociological influences that directly and indirectly affect resource use.

**Principle V** – The full range of knowledge and skills from the natural and social sciences must be brought to bear on conservation problems.

**Principle VI** – Effective conservation requires understanding and taking account of the motives, interests, and values of all users and stakeholders, but not by simply averaging their positions.

**Principle VII** – Effective conservation requires communication that is interactive, reciprocal, and continuous.

From Mangel, M., Talbot, L.E., Meffe, G.K., *et al.* (1996). Principles for the conservation of wild living resources. *Ecological Applications*, **6**, 338–62.

Costanza *et al.* (1998) propose core principles to guide the sustainable governance of the oceans, the so-called Lisbon Principles (Box 3.3). These sets of principles recognize a number of key aspects common to conservation problems (and which will arise in subsequent chapters as we examine management and conservation of marine resources). They emphasize, for instance, that conservation issues have scientific, social, and economic aspects and that underlying the need for conservation is the human per capita demand for resources. Also, using a n environmental resource should be seen not as a right but as a privilege that carries responsibilities and takes into account the full costs of using it. The precautionary principle is included, with the burden of proof resting on the user to demonstrate that activities will not have harmful environmental effects.

## Box 3.3 Lisbon Principles for Sustainable Governance of the Oceans

**Principle 1: Responsibility** – Access to environmental resources carries attendant responsibilities to use them in an ecologically sustainable, economically efficient, and socially fair manner. Individual and corporate responsibilities and incentives should be aligned with each other and with broad social and ecological goals.

**Principle 2: Scale-Matching** – Ecological problems are rarely confined to a single scale. Decision-making on environmental resources should (i) be assigned to institutional levels that maximize ecological input, (ii) ensure the flow of ecological information between institutional levels, (iii) take ownership and actors into account, and (iv) internalize costs and benefits. Appropriate scales of governance will be those that have the most relevant information, can respond quickly and efficiently, and are able to integrate across scale boundaries.

**Principle 3: Precaution** – In the face of uncertainty about potentially irreversible environmental impacts, decisions concerning their use should err on the side of caution. The burden of proof should shift to those whose activities potentially damage the environment.

**Principle 4: Adaptive Management** – Given that some level of uncertainty always exists in environmental resource management, decision-makers should continuously gather and integrate appropriate ecological, social, and economic information with the goal of adaptive improvement.

**Principle 5: Full cost allocation** – All of the internal and external costs and benefits, including social and ecological, of alternative decisions concerning the use of environmental resources should be identified and allocated. When appropriate, markets should be adjusted to reflect full costs.

**Principle 6: Participation** – All stakeholders should be engaged in the formulation and implementation of decisions concerning environmental resources. Full stakeholder awareness and participation contributes to credible, accepted rules that identify and assign the corresponding responsibilities appropriately.

From Costanza, R., Andrade, F., Antunes, P., *et al.* (1998). Principles for sustainable governance of the oceans. *Science*, **281**, 198–9.

## International Legislation

We need to consider further the international legislative framework in which marine conservation operates. But first we need to look at some general aspects of international legislation.

Public international law includes treaty law, in which the obligations of states are set forth in treaties and conventions, and customary law, which, although excluded from international treaties, states generally feel obliged to observe (Bowman *et al.*, 2010). In contrast to such 'hard law', with its binding legal obligations, international environmental protection and conservation includes an increasing element of 'soft law', more aspirational in nature and consisting of recommendations or declarations resulting from international conferences. Whilst not enforceable in the manner of hard law, such initiatives may in time become accepted as norms and recognized as customary international law.

The formulation of conventional international law frequently involves protracted negotiations, and after the signing of a treaty there may be a considerable period before the requisite number of parties have served ratification and the treaty enters into force. International law also differs fundamentally from domestic law in that adherence by parties is based on consent; there is no legislature to provide enforcement. States are thus not compelled to comply, and they agree only if they wish to be party to the treaty in question. The rule of unanimous consent is a particular difficulty in international negotiations and tends to result in agreements of limited power. Furthermore, international treaties often provide for parties to exempt themselves from aspects of a treaty by making 'reservations'. Whilst reservations facilitate wider participation, their widespread use dilutes the effectiveness of legislation. An exception is the UN Convention on the Law of the Sea, which parties have to adopt without reservations – as a 'package deal'. In order for an international treaty to become fully effective, states that ratify or accede to a treaty need to enact their own national legislation for the specific purpose of implementing the treaty. Whilst national law is easier to enforce than international law, appropriate national legislation is often not adopted.

International legal instruments dealing with the marine environment cover two principal areas: protection of the marine environment and protection of marine organisms. The former mainly concerns marine pollution, which we looked at in Chapter 2, and we shall come across examples of the latter in later chapters. The body of marine environmental legislation comprises some 30 global plus countless regional and bilateral agreements (Dyoulgerov, 2000). The overall framework, however, is provided by the UN Convention on the Law of the Sea.

## United Nations Convention on the Law of the Sea

The most significant achievement in international marine environmental law is the Third United Nations Convention on the Law of the Sea (UN, 1983). However, we first need to look at some basic concepts of maritime jurisdiction. The use of the seas has for centuries been governed by two basic concepts: *res communis*, the idea that the seas are common to all, and *res nullius*, the idea that the seas initially belong to no one and are thus available for claimants to establish sovereign rights. *Res nullius* has been applied universally by coastal states to waters immediately adjacent to coastlines, thus giving rise to the internationally accepted concept of a territorial sea, a region over which a coastal state can assert its exclusive control in the interests of national security and to restrict foreign fisheries. Beyond the territorial waters were the high seas or international waters. Application of *res communis* to the high seas has prevailed since the early seventeenth century, largely following its advocacy by the Dutch jurist Hugo Grotius in his *Mare Liberum* of 1609. It became customary for coastal states to claim territorial waters to a distance of 3 nm (5.5 km), this being the range of a good cannon. From the point of view of fisheries, attempts to extend exclusive rights to take in presumably inexhaustible stocks beyond the territorial sea would previously have seemed pointless. But by the late nineteenth century, coastal states were starting to question whether a 3-nm territorial sea was indeed adequate.

The question of defining maritime zones was addressed at international forums during the mid-twentieth century, notably at the First (1958) and Second (1960) United Nations Conference on the Law of the Sea. But these failed to reach agreement on limits, and the whole question of maritime jurisdiction became increasingly in need of resolution. During the 1960s many countries had made unilateral claims to a 12-nm territorial sea and to exclusive fishing zones beyond. Also, with advances in technology, the oil and gas industry was moving into deeper water. Development of the current international regime for ocean space dates from 1967 when the UN General Assembly discussed the concept of the international seabed and its resources as the 'common heritage of mankind' to be used exclusively for peaceful purposes. Eventually, in 1973, the Third UN Conference on the Law of the Sea was convened and the convention finally adopted in 1982. The UN Convention on the Law of the Sea (UNCLOS) entered into force in 1994 and has now been ratified by nearly 170 parties. Many of the provisions of UNCLOS had, however, entered into international customary law long before the convention came into force.

The 1982 convention defines ocean space in terms of national jurisdiction and provides a legal framework for all major uses of the oceans, including the protection and preservation of the marine environment and the management and conservation of marine resources, and provides mechanisms for the settlement of disputes.

UNCLOS established rules for delimiting jurisdictional zones, including methods for determining the baseline from which they are to be measured (normally the low-water line along the coast) (Fig. 3.1). Internal waters, landward of the baseline, such as within bays and estuaries, are sovereign territory. The convention gives coastal states the right to extend their jurisdiction to a territorial sea up to 12 nm (22 km) from the coastal baseline, which most states now claim. The territorial sea is also sovereign territory. Coastal states may also declare a contiguous zone beyond their territorial sea of up to 24 nm from the baseline to give them certain additional jurisdiction, and about 40% of maritime countries have done so. Most significantly, the convention allows a state to claim an exclusive economic zone (EEZ) beyond and adjacent to the territorial sea that extends up to 200 nm (370 km) from the baseline. Within its EEZ a state has 'sovereign rights for the purpose of exploring and exploiting, conserving and managing the natural resources, whether living or non-living' of the seabed and overlying waters, but shall have due regard to the rights and duties of other states (UN, 1983).

With the establishment of EEZs, responsibility for the management of the contained marine resources is largely transferred from international to national jurisdiction. This has major implications for conservation but, as we shall see in later chapters, has not averted overexploitation. UNCLOS requires a coastal state to determine the allowable catch of the living resources in its EEZ and to ensure through proper conservation and management that the maintenance of the living resources is not endangered by overexploitation. These measures are also designed to maintain or restore populations of exploited species at levels that produce the maximum sustainable yield and take into consideration the effects on species associated with or dependent upon exploited species. Where the state does not have the capacity to take the total allowable catch of its EEZ it is obliged to give other states access to the surplus, subject to conditions. States have a responsibility to cooperate in the conservation of highly migratory species (specified in Annex I), marine mammals, and diadromous fish both within and beyond their EEZs.

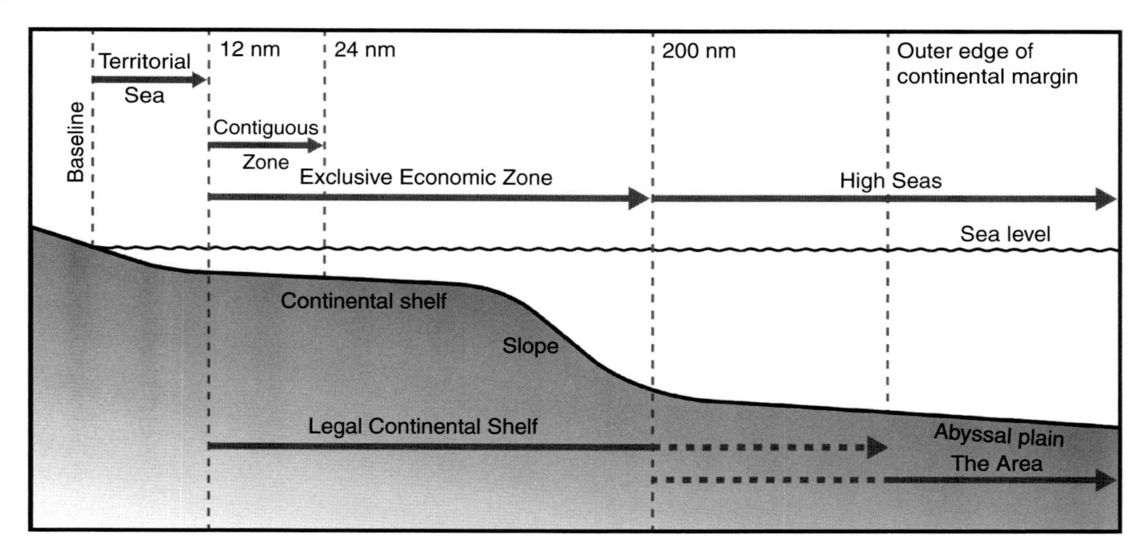


**Fig. 3.1** Maritime zones as defined by the UN Convention on the Law of the Sea. A legal continental shelf is potentially claimable to the outer edge of the continental margin, hence the broken lines for the outer limit of the shelf and the inner limit of the 'Area'. 1 nautical mile (nm) = 1.85 km.

Whilst in most cases the EEZ will be wider than the state's physical continental shelf, the convention allows coastal states to claim a legal continental shelf up to 200 nm from the coastal baseline. Where the physical shelf extends beyond 200 nm, the convention provides formulas for determining the legal outer edge. A state can elect the outer limit to be up to 350 nm from the coastal baseline (or in some cases beyond this for geomorphological reasons) or up to 100 nm beyond the 2500-m isobath. Legally, all states can claim a continental shelf that is at least 200 nm wide and extends to the outer edge of the continental margin. The coastal state has the right to exploit certain natural resources of its legal continental shelf, namely the mineral and other non-living resources of the seabed and sedentary living organisms. Where the legal shelf lies within the EEZ, then the state can also exploit the overlying waters, but where the shelf extends beyond the EEZ, the state does not have rights to the overlying waters, which remain the high seas.

The high seas are that area not included in EEZs, territorial seas, internal waters, or archipelagic waters of a state. With the proclamation of 200-nm EEZs, some 40% of the world's oceans is now under the jurisdiction of coastal states. In the Pacific Ocean in particular, large areas that were once high seas are now under national jurisdiction as a result of the claims of archipelagic states. An archipelagic state may, within certain limits, draw its baseline around its outermost islands to enclose archipelagic waters. From this baseline it is then entitled to delimit its territorial sea and EEZ. The high seas are open to all states as no state may claim sovereignty of any part of the high seas. All states have the right to fish on the high seas, but states also have a duty to cooperate with each other in the conservation and management of the living resources of the high seas, and this specifically includes marine mammals. Activities on the sea surface and in the water column are governed by the provisions relating to the high seas. The seabed beyond the limits of national jurisdiction (i.e. 200 nm or the legal continental shelf, whichever is farthest) is known as the 'Area', and its mineral resources are declared the 'common heritage of mankind'. To oversee and regulate the exploitation of such resources, the convention established an independent international organization, the International Seabed Authority (see Chapter 14).

## International Organizations

The United Nations has been and continues to be a major force in promoting and formulating marine environmental protection and the conservation and management of marine resources. This work is undertaken by the UN itself as well as by certain of its constituent bodies and specialized agencies, such as UNEP and the Food and Agriculture Organization of the United Nations (FAO).

A number of other international organizations play vital roles in initiating and promoting marine conservation and goading national governments into action. Established in 1948 as the first and now largest global environmental organization, the International Union for Conservation of Nature (IUCN) is a membership of governments, non-governmental organizations, research institutions, and conservation agencies active in some 140 countries working to secure the conservation of nature and to ensure that human use of natural resources is appropriate, sustainable, and equitable. One of IUCN's major functions is the gathering, processing, and publication of conservation information. This includes the Red List of Threatened Species, an active inventory of the conservation status of species (see Chapter 4). IUCN also acts as the secretariat for several major international conservation treaties, including the Convention on International Trade in Endangered Species of Wild Fauna and Flora (CITES; see Chapter 4) and the Ramsar Convention (see Chapter 11). An integral part of the IUCN is its commissions, specialist groups that provide technical expertise in key areas. These include the Species Survival Commission, which has numerous taxa-oriented groups (including those for marine turtles, cetaceans, and sharks) that focus on endangered and vulnerable organisms and recommend and promote measures for their conservation; and the World Commission on Protected Areas, which works to develop ways and means for the selection, establishment, and management of protected areas, including marine protected areas, and has a task force specifically concerned with high-seas protected areas.

On IUCN's initiative, the World Wide Fund for Nature (WWF) was founded in 1961 as an international organization to raise funds and create awareness of and support for conservation programmes worldwide. It is the largest private non-profit organization dedicated to nature conservation and best known for its campaigns to save endangered species.

Founded in Canada in 1971, Greenpeace is an international environmental group with some three million members worldwide that has also achieved prominence, particularly through active campaigns on such issues as marine pollution, global climate change, whaling, sealing, drift netting, nuclear testing, and the use of Antarctica. Many countries also have marine conservation organizations operating at the national level.

To conservation organizations such as Greenpeace, multilateral development banks have generally been seen as contributing to environmental problems by funding projects that barely, if at all, consider ecological sustainability, including those impacting on marine systems (Norse, 1993). But more recently these institutions have been addressing the environmental implications of projects. Notably, the Global Environment Facility (GEF) was established in 1991 to help promote environmentally sustainable programmes worldwide. Its key players are the World Bank, the UN Environment Programme, and the UN Development Programme, and it now has some 180 member states. Its focus is on developing countries and on projects related to biodiversity, climate change, international waters, land degradation, the ozone layer, and persistent organic pollutants. It also serves as the financial mechanism for certain environmental conventions, including the Convention on Biological Diversity. As the largest public environmental fund it has the potential to play a significant role in international marine conservation.

## Emergence of Marine Conservation

Widespread interest in marine conservation has developed only relatively recently. The overexploitation of marine living resources and pollution and degradation of coastal habitats have ceased to be solely localized problems and have emerged as global issues, and there is now a wider understanding that marine, atmospheric, and terrestrial systems are closely coupled. Awareness is increasing of the urgent need for the wise management of marine and coastal environments for long-term ecological and economic benefit, although to many societies the health of the world's seas and oceans would still appear to be of little practical relevance; terrestrial surroundings will always be our immediate concern. A factor that has also contributed to a hitherto-relaxed attitude to marine conservation is that few marine species are known to have become extinct in historical times compared with the countless land and freshwater species that have already been eradicated as a result of human activity (see Chapter 4).

Marine conservation has been impeded by attitudes and unwarranted assumptions about the marine environment (Ballantine, 1995). Constrained by our terrestrial viewpoint, we have not accorded the sea the scientific effort proportional to its size and importance. We hardly question whether marine space should be a common property solely to exploit and have tended to allocate resources only to those who in some way disturb marine environments. We are only now starting to assign value to undisturbed areas and to recognize that the public at large, and not just exploitive users, has 'rights'. As with environmental issues in general, we put most of our energy into attending to specific problems rather than seeking general solutions and being precautionary in our approach. There is, however, little incentive for individuals to be precautionary in their use of the sea if the effects of selfish actions are dissipated among all the users of a common property and where negative feedback is unlikely to be a constraint in the short term. But this has now caught up with us. What has particularly drawn attention to marine conservation has been the realization that the oceans' biological resources are not inexhaustible, especially in the face of modern fishing technology. Not only the food resources of the oceans but a wide variety of marine natural products are the focus of exploitation, raising questions of their sustainability. Enormous mineral resources exist on the seabed. Marine living and non-living resources are becoming increasingly valuable. On land, coastal zones are coming under intense pressure with the littoralization of the world population – the increasing number of people living near a coast. Marine tourism and recreation are growth industries, exacerbating these stresses. Among the most vulnerable ecosystems overall are coastal wetlands such as saltmarsh and mangrove areas, lagoons and estuaries, coral reefs, and sandy beaches.

This chapter has focused on the emergence of marine conservation mainly in the context of international environmental developments and the legislative framework. But at the same time, vitally important to marine conservation have been advances in our understanding of the underpinning science, of the functioning and variability of coastal and ocean ecosystems, of marine biodiversity, dispersal, population biology, and life-history traits of marine organisms, topics that will repeatedly crop up.

As conservation biology has developed, one of the trends has been an emergence of more holistic approaches. Early initiatives tended to focus on particular species or sectors, but subsequent approaches have attempted to address the management and conservation of species and habitats in broader contexts, considering ecosystems as a whole and, correspondingly, by integrating the various strands of management, as for example in coastal zone management (see Chapter 10). However, conservation efforts are often driven largely by concern for protecting particular species, and we need to look first at some of these approaches in the next chapters.

## REFERENCES

Alison, R.M. (1981). The earliest traces of a conservation conscience. *Natural History*, **9** (5), 72–7.

Arrieta, J.M., Arnaud-Haond, S. & Duarte, C.M. (2010). What lies underneath: conserving the oceans' genetic resources. *Proceedings of the National Academy of Sciences*, **107**, 18318–24.

Ballantine, W.J. (1995). Networks of 'no-take' marine reserves are practical and necessary. In *Marine Protected Areas and Sustainable Fisheries*, ed. N.L. Shackell & J.H.M. Willison, pp. 13–20. Wolfville, Nova Scotia: Science and Management of Protected Areas Association.

Böhnke-Henrichs, A., Baulcomb, C., Koss, R., de Groot, R.S. & Hussain, S. (2013). Typology and indicators of ecosystem services for marine spatial planning and management. *Journal of Environmental Management*, **130**, 135–45.

Borowy, I. (2014). *Defining Sustainable Development for Our Common Future. A History of the World Commission on Environment and Development (Brundtland Commission)*. Abingdon, UK: Routledge,

Bowman, M., Davies, P. & Redgwell, C. (2010). *Lyster's International Wildlife Law*, 2nd edn. Cambridge, UK: Cambridge University Press.

Brundtland, G.H. (1987). *Our Common Future*. Oxford, UK: Oxford University Press.

Butchart, S.H.M., Walpole, M., Collen, B., *et al.* (2010). Global biodiversity: indicators of recent declines. *Science*, **328**, 1164–8.

Callicott, J.B. (1992). Principal traditions in American environmental ethics: a survey of moral values for framing an American ocean policy. *Ocean & Coastal Management*, **17**, 299–308.

Carlton, J.T., Geller, J.B., Reaka-Kudla, M.L. & Norse, E.A. (1999). Historical extinctions in the sea. *Annual Review of Ecology and Systematics*, **30**, 515–38.

Clark, C.W. (1973). Profit maximization and the extinction of animal species. *Journal of Political Economy*, **81**, 950–61.

Costanza, R., Andrade, F., Antunes, P., *et al.* (1998). Principles for sustainable governance of the oceans. *Science*, **281**, 198–9.

Costanza, R., d'Arge, R., de Groot, R., *et al.* (1997). The value of the world's ecosystem services and natural capital. *Nature*, **387**, 253–60.

Costanza, R., de Groot, R., Sutton, P., *et al.* (2014). Changes in the global value of ecosystem services. *Global Environmental Change*, **26**, 152–8.

Dayton, P.K. (1998). Reversal of the burden of proof in fisheries management. *Science*, **279**, 821–2.

Dyoulgerov, M.F. (2000). Global legal instruments on the marine environment at the year 2000. In *Seas at the Millennium: An Environmental Evaluation*, ed. C. Sheppard, pp. 331–48. Oxford, UK: Elsevier Science Ltd.

Garcia, S.M. (1994). The Precautionary Principle: its implications in capture fisheries management. *Ocean & Coastal Management*, **22**, 99–125.

Graham, M. (1956). *Concepts of conservation*. Paper presented at the International Technical Conference on the Conservation of the Living Resources of the Sea, Rome, 18 April–10 May 1955. New York: United Nations. Document A/CONF.10/L.2, 13 pp.

Grayson, D.K. (2001). The archaeological record of human impacts on animal populations. *Journal of World Prehistory*, **15**, 1–68.

Grove, R.H. (1992). Origins of western environmentalism. *Scientific American*, **267** (1), 22–7.

Han, T.S., Teichert, R.W., Olivera, B.M. & Bulaj, G. (2008). *Conus* venoms – a rich source of peptide-based therapeutics. *Current Pharmaceutical Design*, **14**, 2462–79.

Holdgate, M.W. (1991). Conservation in a world context. In *The Scientific Management of Temperate Communities for Conservation*, ed. I.F. Spellerberg, F.B. Goldsmith & M.G. Morris, pp. 1–26. Oxford, UK: Blackwell.

IUCN/UNEP/WWF (1980). *World Conservation Strategy: Living Resource Conservation for Sustainable Development*. Gland, Switzerland: IUCN.

Jones, G.E. (1987). *The Conservation of Ecosystems and Species*. Beckenham, UK: Croom Helm.

Mangel, M., Talbot, L.E., Meffe, G.K., *et al.* (1996). Principles for the conservation of wild living resources. *Ecological Applications*, **6**, 338–62.

Manning, L.L. (1989). Marine mammals and fisheries conflicts: a philosophical dispute. *Ocean & Shoreline Management*, **12**, 217–32.

Miller, T.R., Minteer, B.A. & Malan, L.-C. (2011). The new conservation debate: the view from practical ethics. *Biological Conservation*, **144**, 948–57.
Normile, D. (2010). U.N. biodiversity summit yields welcome and unexpected progress. *Science*, **330**, 742–3.
Norse, E.A. (ed.) (1993). *Global Marine Biological Diversity: A Strategy for Building Conservation into Decision Making.* Washington, DC: Island Press.
Norse, E.A., Brooke, S., Cheung, W.W.L., *et al.* (2012). Sustainability of deep-sea fisheries. *Marine Policy*, **36**, 307–20.
Rands, M.R.W., Adams, W.M., Bennum, L., *et al.* (2010). Biodiversity conservation: challenges beyond 2010. *Science*, **329**, 1298–303.
Ridley, M. (1996). *The Origins of Virtue*. London: Viking.
Stanley, T.R. Jr (1995). Ecosystem management and the arrogance of humanism. *Conservation Biology*, **9**, 255–62.
Thomas, K. (1983). *Man and the Natural World: Changing Attitudes in England 1500–1800*. London: Allen Lane.
UN (1983). *The Law of the Sea: United Nations Convention on the Law of the Sea with Index and Final Act of the Third United Nations Conference on the Law of the Sea*. New York: United Nations.
UNCED (1992). *Agenda 21 – Chapter 17: Protection of the Oceans, All Kinds of Seas, Including Enclosed and Semi-Enclosed Seas, and Coastal Areas and the Protection, Rational Use and Development of their Living Resources*. United Nations Conference on Environment & Development. Rio de Janeiro, Brazil, 3 to 14 June 1992.
White, L. Jr (1967). The historical roots of our ecologic crisis. *Science*, **155**, 1203–7.
Wilson, E.O. (1992). *The Diversity of Life*. Cambridge, MA: Belknap Press.
WSSD (2002). *Plan of Implementation*. World Summit on Sustainable Development. New York: United Nations.

# 4 Species Conservation

Conservation efforts often target particular species. A focus of early conservation was protecting game animals or species of economic or utilitarian value. In Britain, the first measure taken to protect wild birds, the 1869 Act for the Preservation of Sea-Birds, prohibited the killing of seabirds during the breeding season, as 'sea-birds are useful in destroying grubs and worms, in acting as scavengers in the harbours, in warning vessels off the rocks during fogs by their cries and in hovering over and pointing out to fishermen the locality of the shoals of fish' (Barclay-Smith, 1959). The first multilateral treaty aimed at protecting a marine resource was concerned with valuable commercial species, a 1911 convention among the USA, Great Britain, Japan, and Russia 'for the preservation and protection of the fur seals and sea otters which frequent the waters of the North Pacific Ocean'. Many agreements, action plans, and other initiatives have since come into force aimed at protecting particular marine species, especially in the case of marine vertebrates where species-specific conservation measures commonly apply.

## Extinctions

Reducing the risk of species' extinction is a priority of many conservation strategies, as for instance in the World Conservation Strategy and subsequent major accords (see Chapter 3). We risk impairing the effective functioning of ecosystems with the loss of species and their genetic potential, and aside from any moral obligation, we cannot predict what species may become useful to us as sources of food and natural products. The great majority of species that have existed have come and gone over the aeons of evolutionary history. The fossil record shows several periods of upheaval when mass extinctions have occurred. The current accelerated rate of species loss, attributable directly or indirectly to human activities, is projected to rival the five big mass extinctions of the past 540 million years (Ceballos *et al.*, 2015). It is difficult to estimate how many species are extinct or imperilled as a result of human activities; for the great majority of species there is simply no information on their status. An extrapolation by Régnier *et al.* (2015) suggests that already 7% of described, present-day species are extinct. An indication of the likely impact just of climate change on global biodiversity comes from an analysis of more than 300 estimates of taxon-specific responses (both observed and predicted) by Maclean and Wilson (2011). Expressing population decline in terms of IUCN Red List criteria (see below), they projected that on average 11% of species are at risk of extinction from climate change by 2100 across all major taxonomic groups and ecoregions. Their analysis indicated that marine taxa are particularly threatened.

The main causes to date of human-induced extinction – often occurring in combination – are the alteration, fragmentation, and destruction of habitat and overexploitation (Vié *et al.*, 2009). On land, where humans are often competing directly with other species for living space, appropriation

**Table 4.1** The Main Threats Impacting Animal and Macrophyte Species on the IUCN Red List That Are Categorized as Occurring in the Marine Biome and Threatened
i.e. critically endangered, endangered, and vulnerable, $n = 1093$ species

| Threat | Species impacted (%) |
|---|---|
| Biological resource use (mainly fishing and harvesting marine resources) | 74 |
| Climate change and severe weather | 45 |
| Pollution | 45 |
| Invasive and other problematic species | 40 |
| Residential, commercial, and tourism/recreation development | 40 |
| Human intrusions and disturbance (mainly recreational activities) | 32 |
| Transportation and service corridors (principally shipping lanes) | 27 |

of habitat is most often the underlying cause of extinction. But most marine extinctions in historical times are attributable to exploitation (55%) and then to habitat loss and degradation (37%) (Dulvy *et al.*, 2003). Overexploitation is the primary risk for three-quarters of threatened marine species on the IUCN Red List (Table 4.1). The impact of fishing, from both direct and indirect effects, represents a significant extinction risk for many marine species, including for target species with high fecundity, especially when individuals enter the fishery before maturity and when the oldest, most fecund age classes are lost (Dulvy *et al.*, 2003; Myers & Ottensmeyer, 2005). Many fisheries also pose a major threat to non-target species, including sharks and rays, turtles, seabirds, and small cetaceans, whose life-history characteristics render them vulnerable to even low fishing pressure. Furthermore, fishing methods such as bottom trawling can radically alter and destroy seabed habitat and may eliminate, at least locally, long-lived, sessile benthic species that provide habitat complexity and shelter for many associated species. Loss and degradation of critical habitat also occurs through coastal development, sedimentation, nutrient enrichment, and other forms of pollution. There are many examples where deliberate or accidental introductions of species have had devastating effects on native terrestrial and freshwater biotas. Such impacts appear to be most damaging where native species occur as small insular populations and where their defences against competitors and predators are inferior to those of the invader (Vermeij, 1986). Humans have been responsible for the successful introduction of many marine species, especially to estuarine, brackish, and coastal habitats worldwide (Carlton, 1989). But whilst alien marine species may have dramatic impacts on recipient biotas, they do not appear to present an important extinction threat for native species (Dulvy *et al.*, 2003; Briggs, 2007).

Extinctions can be considered at different scales, from local extinction when a population or populations are lost from a small area of a species' range to regional extinction where the loss occurs over a substantial part of the range to global or complete extinction (Carlton *et al.*, 1999). Some 20 marine species are known to have become globally extinct in historical times (Table 4.2) and more than 100 to have suffered regional or local extinction of their populations (Dulvy *et al.*, 2003). The best-known examples of global extinctions come from mammals and seabirds, such as the sea

**Table 4.2** Global Extinctions of Marine Species in Historical Times

| Species | Location | Extinction cause | Last sighted |
|---|---|---|---|
| Steller's sea cow *Hydrodamalis gigas* | NW Pacific | Exploitation | 1768 |
| Sea mink *Neovison macrodon* | NW Atlantic | Exploitation | 1880 |
| Caribbean monk seal *Monachus tropicalis* | Gulf of Mexico, Caribbean | Exploitation | 1952 |
| Great auk *Pinguinus impennis* | N Atlantic | Exploitation | 1844 |
| Labrador duck *Camptorhynchus labradorius* | NW Atlantic | Exploitation | 1875 |
| Auckland Islands merganser *Mergus australis* | SW Pacific | Exploitation | 1902 |
| Canary Islands oystercatcher *Haematopus meadewaldoi* | NE Atlantic | Invasion | 1913 |
| Spectacled cormorant *Phalacrocorax perspicillatus* | NW Pacific | Exploitation | 1850 |
| Green wrasse *Anampses viridis* | Mauritius | Unknown | 1839 |
| Galápagos damselfish *Azurina eupalama* | Galápagos Islands | Habitat loss | 1982 |
| Rocky shore limpet *Lottia edmitchelli* | NE Pacific | Habitat loss | 1861 |
| White abalone *Haliotis sorenseni* | NE Pacific | Exploitation | ? |
| Periwinkle *Littoraria flammea* | China | Habitat loss | 1840 |
| Atlantic eelgrass limpet *Lottia alveus* | NW Atlantic | Habitat loss | 1929 |
| Ivell's sea anemone *Edwardsia ivelli* | Southern UK | Habitat loss | 1983 |
| Fire coral *Millepora boschmai* | E Pacific | Mass bleaching | 1998 |

**Table 4.2** (cont.)

| Species | Location | Extinction cause | Last sighted |
|---|---|---|---|
| Starlet coral *Siderastrea glynni* | E Pacific | Mass bleaching | 1998 |
| Turkish towel alga *Gigartina australis* | Sydney Harbour, Australia | Habitat loss, pollution | 1892 |
| Bennett's seaweed *Vanvoorstia bennettiana* | Sydney Harbour, Australia | Habitat loss, pollution | 1916 |

Reprinted with permission from Dulvy, N.K., Sadovy, Y. & Reynolds, J.D. (2003). Extinction vulnerability in marine populations. *Fish and Fisheries*, 4, 25–64. © 2003 Blackwell Publishing Ltd.

mink, Caribbean monk seal, Steller's sea cow, and great auk. Thorough research of highly impacted coastal areas can reveal substantial numbers of local extinctions, such as for the Wadden Sea, which has lost more than 50 species in historical times (Lotze *et al.*, 2005).

Extinctions may be thought less likely to occur in marine than in terrestrial systems. If marine species tend to have larger, more widely distributed populations and access to more effective refuges than terrestrial species, then bigger disturbances are needed to effect their demise. However, many marine species do not have large ranges, and widespread species can show high levels of genetic differentiation or even comprise multiple cryptic species. Also, a large range does not necessarily insure against extinction (Vermeij, 1993), particularly given modern fishing techniques (see Chapter 6). Extinction risk has yet to be assessed for many marine taxa, which may partly explain the lower rates of extinction recorded for marine than for terrestrial systems. For marine and non-marine groups that are taxonomically well known, the proportion of species threatened with extinction is in fact similar, at about 20–25% (Webb & Mindell, 2015).

The wider repercussions of an extinction depend on the role that a species plays in an ecosystem. For those that occupy a pivotal role, extinctions are likely to result in significant adjustments in the abundance and population structure of surviving species. If a key predator or grazer is heading for extinction, there will come a point where it no longer interacts significantly with other species, so-called ecological extinction (Estes *et al.*, 1989), which may trigger a trophic cascade (see Chapter 2). Sea otters are an important predator of benthic invertebrates, including herbivorous urchins, and their extinction in the NE Pacific strongly influenced the structure of its kelp forest ecosystem (see Chapter 9).

Another term used is commercial extinction, for when a stock has been reduced to such a level that exploiting it is no longer profitable. In practice, however, this may not often occur. Most fisheries are multi-species, and it may be that a rare species continues to be taken because it is still economic to fish other target species. Also, the value of some species simply rises with rarity, making them more desirable and exaggerating extinction pressure (Courchamp *et al.*, 2006), such as sturgeon species that provide the most sought-after caviar (see Chapter 6).

Species differ in their vulnerability to anthropogenic extinction. Potential vulnerability is a function of several factors, including economic value, distribution, abundance, population structure, life-history characteristics, behaviour, and species' interactions. Where such features are sufficiently well known it may be possible to identify susceptible species. Often the best predictors of vulnerability to fishing pressure are large body size and late maturity (Reynolds *et al.*, 2005).

The capacity of a population to recover from severe overexploitation or some other catastrophic event depends on life-history traits of the species. Some species are naturally resilient; they are tolerant of stress and can rapidly increase in numbers once conditions are favourable. Others recover only if they have a protracted period free from further disturbance. One way of viewing this is in relation to the concept of *r*- and *K*-selection, based on the parameters of population growth models, where *r* is the intrinsic rate of population increase and *K* is the maximum or asymptotic population size. The concept differentiates between contrasting life-history strategies: a 'quantitative' versus a 'qualitative' approach. In essence, under *r*-selected traits, maximum resources are devoted to reproduction but with the smallest practicable amount allocated to individual offspring so as to produce as many offspring as possible. With *K*-selected traits, on the other hand, the investment is towards promoting increased survivorship, so the strategy is to direct resources into the production of a few extremely fit offspring. Typically, therefore, *r*- and *K*-selected species exhibit certain characteristics (Table 4.3). It is important to bear in mind that *r*- and *K*-selected species represent the ends of a continuum; many species fall between these extremes or cannot be easily categorized. The theory of *r*- and *K*-selection has been superseded by age-specific demographic models that better explain the evolution of life-history traits (Reznick *et al.*, 2002), but it continues to be useful as a way of describing contrasting life-history strategies. In the next few chapters we examine groups in which *K*-selected species typically prevail, including the chondrichthyans (Chapter 6), sea turtles (Chapter 7), seabirds (Chapter 8), and marine mammals (Chapter 9) – in other words animals with a reduced capacity to recover from human disturbances and a higher risk of extinction.

**Table 4.3** Typical Characteristics of *r*-Selected (Opportunistic) and *K*-Selected (Equilibrium) Species

| Characteristic | *r*-selected species | *K*-selected species |
|---|---|---|
| Adult body size | Small | Large |
| Life span | Short | Long |
| Age at maturity | Early | Delayed |
| Growth rate | High | Low |
| Natural mortality | High | Low |
| Mobility | Low | High |
| Reproductive periods per year | Repeated | Single/few |
| Fecundity | High | Low |
| Offspring | Many small | Few large |
| Colonizing time | Early | Late |
| Niche | Broad | Narrow |
| Habitat variability | Variable/unpredictable | Constant/predictable |

From various sources.

## Assessing Population Viability

Appraising the risk of extinction requires an understanding of the potential significance of major factors on population viability. Extinction can be due to systematic factors, when an essential requirement of a population is removed (e.g. loss of habitat), or when something lethal is introduced (e.g. a pollutant), and clearly a priority in any conservation programme must be to alleviate such pressures. There are also, however, various sources of stochasticity, or uncertainty, bearing on populations that become increasingly important as population size diminishes. Principal among these are genetic uncertainty (in particular the decline of heterozygosity in small populations limiting their ability to cope with environmental change), demographic stochasticity (e.g. natural variations in sex ratio, fecundity, and survival), environmental variation (e.g. fluctuations of habitat parameters and the influence of competitors, predators, parasites, and diseases), and – as an extreme type of environmental uncertainty – natural catastrophes (e.g. floods, hurricanes, and extreme ENSO events) (Shaffer, 1981). These factors may reinforce one another, thereby potentially accelerating extinction. For most population sizes, environmental uncertainty and natural catastrophes tend to be the most significant.

A modelling tool often used in conservation biology to assess extinction risk of small populations is population viability analysis (PVA), where the probability of extinction is computed based on the size of the population and the various stochastic processes impacting on it as well as the deterministic forces known to be reducing population size, such as overexploitation and habitat loss. PVA programs generally use multiple random sampling to simulate the influence of factors and their interaction (e.g. Lacy, 2000). To endure the sources of uncertainty likely to occur over a specified time, a population will need to exceed a certain minimum size. PVA is frequently used to compute a minimum viable population (MVP), which is the smallest isolated population having a particular probability of remaining extant (e.g. at a 90% level) over a stipulated time (e.g. 100 years) despite the foreseeable effects of stochastic variation. Viability analysis can be used for indicating the relative vulnerability of small populations and hence the urgency of implementing conservation measures and the likely outcome of management options.

There is a danger an MVP might be interpreted as a level to which a population can decline, whereas species will vary considerably in this regard, and estimates need to be based on levels of probability that ensure a high margin of safety. Detailed demographic and ecological data are required for such projections to have reasonable predictive power, information that is often not available for threatened species. These simulation programs have been designed primarily for *K*-selected species. For most species a minimum population size of 1000–5000 individuals is likely needed to ensure its evolutionary potential or at least 100–500 individuals to avoid inbreeding depression and loss of reproductive fitness (Traill *et al.*, 2010; Frankham *et al.*, 2014).

Assessing population viability highlights a central problem in marine conservation, namely the degree of uncertainty inherent in assessing critical population size and how to account for this in guiding management. Uncertainty, particularly in fisheries management, can result in mitigative measures being deferred and overexploitation thereby continuing. There are some general tactics to help reduce the role of uncertainty, in particular adopting a precautionary approach to setting catch limits. Also important are regulatory mechanisms to ensure that corrective measures are implemented when needed as well as management tools that are not so vulnerable to uncertainty – such as minimum size limits so that recruits reproduce before they enter the fishery – and greater use of marine protected areas (Botsford & Parma, 2005).

Slooten *et al.* (2000) describe a situation where management decisions were being delayed because of uncertainty attached to the estimated risk posed by gillnet mortality to Hector's dolphin, a threatened species endemic to New Zealand's coastal waters (Fig. 4.1). Using a model that

**Fig. 4.1** Hector's dolphin, a small (up to 1.5 m in length) dolphin endemic to New Zealand's coastal waters that is threatened with extinction. Photograph: S. Dawson. (A black and white version of this figure will appear in some formats. For the colour version, please refer to the plate section.)

specifically incorporates uncertainty and testing the effect of assumptions using sensitivity analysis, they showed that bycatch mortality indeed represented a high risk to the population and that delaying gillnet restrictions was unwarranted. New protection measures were introduced in 2008 – mainly further restrictions on gillnetting – but modelling results indicated these were unlikely to halt the overall decline of Hector's dolphin. On the other hand, with complete protection and no fishing mortality, the total population is projected to roughly double by 2050 to around 15 000 individuals (Slooten & Dawson, 2010).

For populations below a minimum viable level there may be interventions that can be implemented to avert an otherwise inexorable extinction. For the highly endangered Mediterranean monk seal, Durant and Harwood (1992) modelled the effects on small isolated populations of demographic and environmental stochasticity from discrete catastrophic events. They used the models to identify demographic characteristics whose occurrence could be used to predict when small seal populations were most at risk of extinction. To be useful, a predictor of extinction should be easily measured, have high predictive power, give a reasonable warning of extinction, and occur during the life history of most populations. No one characteristic, however, has all these desirable properties. Certain characteristics have high predictive power but give insufficient time for preventative action (e.g. complete failure to produce or rear any pups in 2 successive years). Other indicators, whilst lacking particularly high predictive power, have the advantage of providing some warning (at least 10 years) and occur in almost all simulations (e.g. complete failure to produce pups in 1 year). But by monitoring a suite of demographic characteristics (e.g. in this case population size, pup production, and sex ratios), it may be feasible to forecast when the

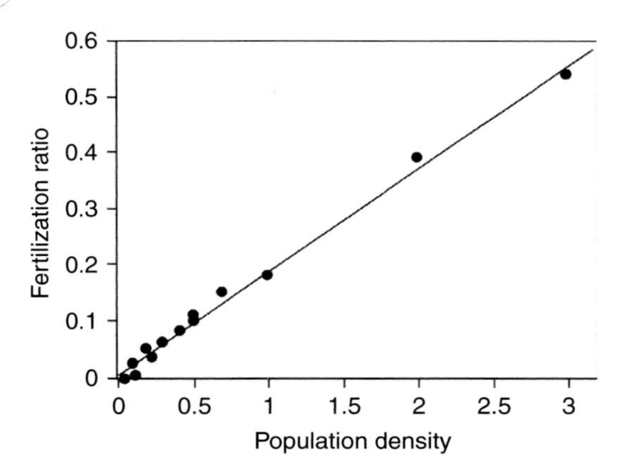


**Fig. 4.2** Results of a simulation model to examine the effect of population density (in a randomly dispersed population) on the fertilization ratio (the proportion of spawned eggs that are fertilized) in a free-spawning invertebrate. Reprinted from Claereboudt, M. (1999). Fertilization success in spatially distributed populations of benthic free-spawners: a simulation model. *Ecological Modelling*, 121, 221–33. With permission from Elsevier.

remaining populations are most at risk of extinction and, depending upon the degree of risk, what conservation efforts are appropriate.

The concept of minimum viable populations has been most often applied to mobile animals with internal fertilization, in particular for mammalian populations. In marine systems, where many species rely on broadcast spawning and are also sedentary, the minimum viable population density may be a more useful measure to assess the probability of extinction (Hockey & Branch, 1994). At low population sizes or densities, many species show a decrease in per capita population growth rate. This phenomenon, known as the Allee effect, increases the likelihood of local and global extinction (Courchamp *et al.*, 1999). Various factors may be involved, although most attention has focused on the reduction in social interactions, in particular the lower probability of receptive mate encounters at low population densities, and lower sperm concentrations in the water column. This in turn might result in a lower rate of recruitment, further exacerbating the shortage of mate encounters and creating an extinction vortex. There is evidence that Allee effects may be of significance for broadcast spawners (Gascoigne & Lipcius, 2004). A modelling study by Claereboudt (1999) indicates that for a free-spawning benthic species, the proportion of spawned eggs that are fertilized increases linearly with population density (Fig. 4.2). At a low population density of 0.5 individuals $m^{-2}$ in a randomly dispersed population, fertilization success was projected to be less than 10%. The main consequence of the Allee effect is that there will be a critical density below which the group concerned, be it a population, colony, or species, will become extinct even though it contains reproductive individuals.

## Assessment of Extinction Risk

An important step in developing and prioritizing conservation strategies is assessing the status of species and the threats they may face. Various agencies have devised systems to define and categorize degrees of threat, but the most widely used system is the IUCN Red List of Threatened Species, now essentially a standard for categorizing species in terms of extinction risk. The number of species assessed for the Red List is currently about 80 000, with the most thorough data relating to vertebrates and flowering plants. Nearly 12 000 species are currently listed for the marine biome, of which almost three-quarters are fish and other vertebrates. The main groups of marine invertebrates listed so far are corals, cone snails, squids and cuttlefishes, lobsters, and sea cucumbers. Seagrass and mangrove species are listed, but to date relatively few seaweeds are (Table 4.4).

**Table 4.4** Organisms on the IUCN Red List That Occur in the Marine Biome, Showing the Number (and Percentage) of Species in Each Group, and for Each Group the Number (and Percentage) of Species Listed as Threatened (Critically Endangered, Endangered, and Vulnerable) and Data Deficient

The vascular plants are mainly seagrass and mangrove species, the corals mainly stony corals, the gastropods mainly cone snails, the reptiles mainly sea snakes and turtles, and the birds mainly waders and seabirds.

| Group | No. of species | Threatened | Data deficient |
|---|---|---|---|
| Seaweeds | 75 (0.6) | 15 (20.0) | 55 (73.3) |
| Vascular plants | 151 (1.3) | 21 (13.9) | 14 (9.3) |
| Corals | 882 (7.6) | 239 (27.1) | 166 (18.8) |
| Gastropods | 814 (7.0) | 50 (6.1) | 179 (22.0) |
| Squids and cuttlefishes | 494 (4.2) | 5 (1.0) | 291 (58.9) |
| Lobsters | 254 (2.2) | 3 (1.2) | 87 (34.3) |
| Sea cucumbers | 371 (3.2) | 16 (4.3) | 244 (65.8) |
| Hagfishes | 76 (0.7) | 9 (11.8) | 30 (39.5) |
| Sharks and rays | 1063 (9.1) | 176 (16.6) | 463 (43.6) |
| Bony fishes | 6310 (54.3) | 331 (5.2) | 1017 (16.1) |
| Reptiles | 90 (0.8) | 13 (14.4) | 25 (27.8) |
| Birds | 861 (7.4) | 171 (19.9) | 5 (0.6) |
| Marine mammals | 136 (1.2) | 33 (24.3) | 46 (33.8) |
| Other | 54 (0.5) | 11 (20.4) | 13 (24.1) |
| Total | 11 631 | 1093 (9.4) | 2635 (22.7) |

Data from www.iucnredlist.org (accessed 26 March 2016).

There are eight Red List categories, applicable to taxa at or below the species level (IUCN, 2001) (Fig. 4.3):

- Extinct: No reasonable doubt that the last individual has died.
- Extinct in the wild: Known only to survive in cultivation, in captivity, or as a naturalized population (or populations) well outside the past range.
- Critically endangered: Facing an extremely high risk of extinction in the wild.
- Endangered: Facing a very high risk of extinction in the wild.
- Vulnerable: Facing a high risk of extinction in the wild.

- Near threatened: Close to qualifying for or is likely to qualify for a threatened category in the near future.
- Data deficient: Inadequate information to make a direct or indirect assessment of its risk of extinction based on its distribution and/or population status.
- Not evaluated: Has not yet been evaluated against the criteria.

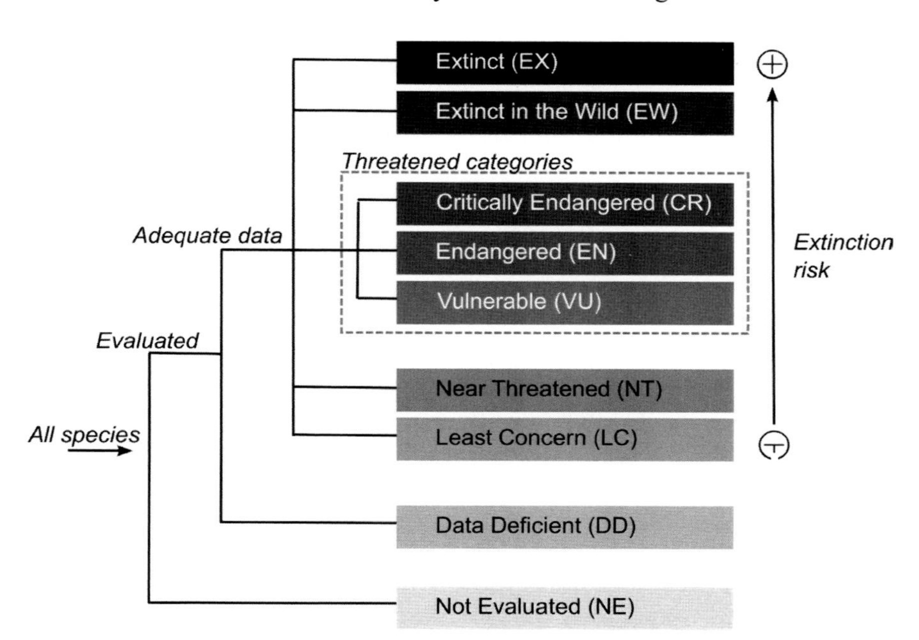


**Fig. 4.3** The IUCN Red List categories for classifying species according to their extinction risk. From IUCN (2001). *IUCN Red List Categories and Criteria: Version 3.1.* Gland, Switzerland, and Cambridge, UK: IUCN.

Species are assigned to categories based on criteria related to population trend, size and structure, and geographic range. For example, for threatened species – those in the categories critically endangered, endangered, and vulnerable – one of the criteria is based on quantitative analysis of extinction (though not necessarily a PVA). For species assessed as critically endangered this is an extinction probability of at least 50% within 10 years or three generations, for endangered at least 20% within 20 years or five generations, and for vulnerable at least 10% within 100 years. Overall, 9% of listed marine species are categorized as threatened (Table 4.4), but 12% if only those species having adequate information to be assessed are considered (i.e. excluding data-deficient species). (To complement its species Red List, the IUCN is developing a Red List for ecosystems based on quantitative criteria designed to assess risk of ecosystem collapse (Keith *et al.*, 2013).)

Detailed information is needed to evaluate species for the Red List, and most marine species are poorly known or undescribed. Overall, nearly a quarter of the marine species listed are categorized as data deficient, but for some groups, such as cephalopods and sea cucumbers, inadequate information means that most of the species cannot be assessed for extinction risk (Table 4.4). There is also a particular difficulty with marine organisms of determining when a species is extinct. The definition of extinct may be especially difficult to apply to marine invertebrates where possibly hundreds of 'species' have not been recorded since the eighteenth and nineteenth centuries (Carlton, 1993). By the time a species is listed its numbers may already be critically low and the prognosis for recovery very poor. A species may be locally rather than globally extinct, and in the case of a species that plays a key role in a community, the natural functioning of that ecosystem will already be significantly altered. One particularly contentious area is the inclusion of commercially exploited fishes as threatened species in the Red List (see Chapter 6).

## Box 4.1 US Endangered Species Act (ESA)

Under the ESA, the federal government has responsibility to protect:

- Endangered species: any species that is in danger of extinction throughout all or a significant portion of its range
- Threatened species: any species that is likely to become endangered in the foreseeable future
- Critical habitat for threatened or endangered species

Candidate species are evaluated according to any of the following factors:

- Present or threatened habitat loss or modification
- Overexploitation
- Disease or predation
- Inadequacy of existing protection
- Other natural or human factors affecting its continued existence

Assessments are made on the basis of 'the best scientific and commercial data available'.

The IUCN assessments have no international legal standing, but a number of countries have enacted legislation (often based on IUCN criteria) identifying threatened species within their jurisdiction and the need for protection. An example of domestic legislation concerned with threatened species is the US Endangered Species Act of 1973 (Box 4.1). This is the primary instrument of federal efforts to protect endangered species but not the sole one (in the marine area there is also the Marine Mammal Protection Act). Primary purposes of the act are to provide for the conservation of endangered and threatened species and the ecosystems upon which they depend. For marine species, the National Marine Fisheries Service (Department of Commerce) is charged with identifying and listing endangered and threatened species. Few marine species are listed, mainly marine mammals and turtles. 'Species' under the act includes subspecies and genetically distinct vertebrate populations (e.g. some Pacific salmon populations are listed). Once a species is listed, a recovery plan is to be prepared. Few species have yet recovered to the point where they can be taken off the list, the gray whale being a notable exception. There has been much debate about the effectiveness of the act. An analysis of population trends of more than 1000 species listed as threatened and endangered under the act indicates that the longer species are listed, the more likely they are to be recovering (Taylor *et al.*, 2005). Critics, on the other hand, argue that remedial action is not triggered until populations are already dangerously low and that to maximize support for conserving biodiversity it would be more effective to devote this amount of effort to protecting ecosystems (e.g. Miller, 1996).

Lists of threatened species provide an assessment of extinction risk to help raise awareness of these species and promote their conservation. However, they have some limitations, and Possingham *et al.* (2002) caution against threatened species lists being used for purposes for which they were not intended. For instance, there is invariably very uneven taxonomic treatment in such lists, and threatened species are not necessarily good indicators of habitat value or of the best way to allocate resources. A contentious issue in this context is the concept of conservation triage (Traill *et al.*, 2010). Nevertheless, given the scale of conservation problems and the realities of available funding, it is inevitable that decisions for prioritizing effort will need to become more explicit (Bottrill *et al.*, 2008).

## International Trade in Endangered Species

World trade in wildlife amounts to billions of dollars annually and poses a major threat to many species. The trade involves live animals and plants as well as a huge range of natural products such as food and clothing items, jewellery, medicines, and tourist curios. Modern efforts to control such trade date from the late nineteenth century, though initially mainly to protect birds taken for the feather trade. Provisions to regulate international trade in wildlife were included in the Western Hemisphere Convention of 1940, but whilst ahead of its time for its broad objectives to protect biodiversity, the convention has essentially remained dormant (Bowman *et al.*, 2010). It was not until 1973 that a comprehensive international convention on wildlife trade was agreed upon, namely the Convention on International Trade in Endangered Species of Wild Fauna and Flora (CITES). CITES entered into force in 1975 and now has some 180 parties.

Appendices to CITES list the species to be protected. Those in Appendix I are considered to be threatened with extinction, and international commercial trade in them is prohibited. Marine species listed here include the coelacanths, sawfishes, sea turtles, dugong, sea otter, monk seals, and several whale and dolphin species. Appendix II includes species that, whilst not currently threatened with extinction, may become so unless trading in them is strictly regulated. Exporting such species requires a permit, one of the conditions being that the export will not be detrimental to the species' survival. Among the taxa listed here are a number of coral groups including all stony corals (Scleractinia), giant clams, queen conch, seahorses, several shark species, southern elephant seal, and the remaining cetaceans. The convention requires that trading in Appendix II species 'should be limited in order to maintain that species throughout its range at a level consistent with its role in the ecosystems in which it occurs and well above the level at which the species might become eligible for inclusion in Appendix I' (Article IV 3) – in other words, the need to safeguard against ecological extinction. The criteria for listing species in these appendices relate to the size of the wild population, the area of distribution, and whether there is a marked decline in the population size in the wild (Box 4.2). As to be expected, CITES appendices include in particular species that are vulnerable to overexploitation on account of their life-history characteristics (*K*-selected) and/or have a very restricted distribution. Nevertheless, there can be difficulties obtaining appropriate data on population size and geographic range for marine species so as to meet criteria for their inclusion.

In common with other international treaties there is, however, provision under CITES for parties to claim exemption from conditions they find unacceptable. Parties can enter a reservation on particular species (usually if they already have important trade in the species). Reservations on marine species include those by whaling nations on baleen whales. But whilst it may be difficult for parties to ignore their obligations under CITES, not all parties are strictly complying with the conditions, and illegal trade continues to be a problem. Import controls are often ineffective, and enforcement agencies do not necessarily have the resources to enable products to be correctly identified. Whilst a large number of countries are signatories to CITES, only a small number have so far adopted domestic legislation for the specific purpose of implementing it.

Article XIV states that CITES shall not affect the provisions of other international treaties or regional trade agreements. Many commercially important mammals, fishes, and invertebrates are regulated by international and regional agreements. An important issue in the management of marine living resources is the conservation of species that occur outside national jurisdictions. Under CITES, if a state transports into its territory specimens that were taken on the high seas, it must be satisfied that the introduction will not be detrimental to the survival of that species and, in the case of Appendix I species, that the specimen is not to be used primarily for commercial purposes.

## Box 4.2 Criteria for Listing Species in the CITES Appendices

The criteria are set out in annexes as summarized below.

### Annex 1: Biological Criteria for Appendix I

A species is considered to be threatened with extinction if it meets or is likely to meet at least one of the following criteria: the wild population is small, it has a restricted area of distribution, or it shows a marked decline in abundance.

### Annex 2: Criteria for the Inclusion of Species in Appendix II

A species should be included if it meets at least one of the following criteria: trade must be regulated to avoid the species becoming eligible for inclusion in Appendix I in the near future or to ensure that the harvest of specimens from the wild is not reducing the wild population to a level at which its survival might be threatened by continued harvesting or other influences.

### Annex 3: Special Cases

Listing different populations of a species in more than one appendix should be avoided as it creates enforcement problems.

### Annex 4: Precautionary Measures

Proposals to amend Appendix I or II should take a precautionary approach. Where there is uncertainty about the status of a species or the impact of trade on the conservation of a species, parties shall act in the best interest of the conservation of the species.

### Annex 5: Definitions, Explanations, and Guidelines

Terms used in the annexes are defined and explained, including area of distribution, decline, and small wild population.

### Annex 6: Format for Proposals to Amend the Appendices

Information and instructions for the submission of a proposal to amend the appendices.

For the full text of the annexes, see www.cites.org/eng/res/09/09-24R16.php.

International agreements specifically for wide-ranging migratory species of the high seas mainly concern fish stocks (see Chapter 5) and whales (see Chapter 9). There is also the Convention on the Conservation of Migratory Species of Wild Animals, the Bonn Convention of 1979, which entered into force in 1983. Its main objectives are to provide strict protection for species in danger of extinction throughout all or part of their range (Appendix I species) and to persuade states to conclude agreements for the conservation and management of migratory species that require international agreements for their conservation or would benefit from international cooperation (Appendix II species). The convention is potentially of value to a number of marine species not adequately protected by existing treaties but has been slow to attract signatories (now more than 120 parties) (Bowman *et al.*, 2010). It includes memoranda of understanding (not legally binding) relevant to migratory sharks, sea turtles, and marine mammals.

## Protected Areas, Reintroductions, and Captive Breeding

Marine protected areas and enforcement of fishery restrictions have vital roles to play in conservation of particular marine species, and a number of reserves and sanctuaries have been established primarily for marine vertebrates. We shall look at examples of these in later chapters.

Reintroducing species that have become locally extinct has often been attempted, especially with terrestrial species. But the process is rarely straightforward and requires a high management effort, and the failure rate is high (Slobodkin, 1986). Research on salmonids illustrates some of the difficulties with captive breeding programmes. Whilst it may be possible to maintain genetic diversity within captive populations over several generations, there may be rapid loss of fitness, and it is not in fact clear whether captive-bred lines would successfully establish themselves in the wild. Importantly, such programmes do not address the underlying reasons for the decline of the wild populations (Fraser, 2008).

Aquaculture can have a role for certain species, such as for giant clams to help restock overexploited reefs (see Chapter 5) and for some fishes to ease the pressure on overexploited wild populations (see Chapter 6). Laboratory culture and cryopreservation techniques are well established for micro- and macroalgae and can be used to maintain species now rare in the wild (Brodie *et al.*, 2009). In the case of certain invertebrates and algae there may be strong economic incentives to culture or cryogenically preserve species that are rare in the wild and that have been found to be sources of valuable biomedical compounds. But it is questionable whether *ex situ* methods could play a substantial role in conservation of marine species, except as a last resort. Marine aquariums working cooperatively have a role to play in captive breeding and reintroduction programmes for threatened species, but perhaps more significantly they have major opportunities to engage visitors about issues in marine conservation, such as the trade in ornamental fish (Tlusty *et al.*, 2013; see Chapter 13). The total number of substantial public aquariums in the world is estimated at well over 315, and together they may attract as many as 450 million visitors each year (Penning *et al.*, 2009).

## Focal Species

Effective conservation must take account of the communities and habitats to which a species belongs. So can species-based approaches be useful in achieving wider conservation objectives? Does focusing on species of particular significance for ecological or social reasons help in the wider conservation and management of communities, habitats, or ecosystems? The possible value of so-called focal species has generated considerable debate in recent years, and various types of focal species have been proposed, such as indicator, keystone, umbrella, and flagship species (Zacharias & Roff, 2001). Focal species are meant, in different ways, to be indicative of the larger community to which they belong and are advocated for the management and conservation of natural environments. A distinction between the different types is not always clear, however.

An indicator species is one whose presence denotes the composition or condition of a particular habitat, community, or ecosystem. The most obvious types of indicator species are those that provide a major structural component, such as dominant mangrove species or algae that form kelp forests. These systems contain numerous associated species, many of which may be difficult to observe, so it would be helpful if community composition could be largely predicted from a conspicuous indicator species. It may not, however, be appropriate to designate a single indicator species, especially where trophic pathways are complex.

Keystone species are those considered to play a disproportionately large role in community structure. The concept was first applied to certain predatory species, in particular to sea stars on rocky shores that prevent mussels from dominating the community, but has since been extended to species that may have overriding influences in other ways, and numerous marine species have been proposed as keystone species (Roff & Zacharias, 2011). The keystone role of a species may, however, change over its geographical range and life history, and identifying keystone species (e.g. by predator exclusion or other manipulative experiments) may be difficult.

Potentially of particular relevance to conservation are keystone species that create or modify habitat, so-called ecosystem engineers (Braeckman *et al.*, 2014) (though some might equally be considered indicator species). Some species are important habitat formers by way of the physical structure they provide – most obviously corals – and thereby support many associated species. Other ecosystem engineers exert a large influence on community structure because their activity significantly modifies the habitat, notable examples being animals that disturb or rework bottom sediments, such as the walrus or ghost shrimps.

The idea of umbrella species is that their conservation will also conserve other species in that area, and to that extent they have similarities to indicator species, whilst flagship species are ones that can be used to help raise awareness, public support, and funding for conservation programmes. In the marine sphere, marine mammals are especially employed in this way.

## Invertebrates

Most of the species-based effort in marine conservation concerns vertebrates, notably certain fishes, marine turtles, seabirds, and marine mammals, particularly on account of their high profile and the conspicuous impact of exploitation on their populations. These groups and the particular conservation issues they raise are examined in the next few chapters. The conservation of other major groups of marine organisms has attracted far less attention. Protection of marine prokaryotes appears so far to have been barely considered, yet some are known only from particular habitats and may be at a higher risk of extinction than generally assumed (Arrieta *et al.*, 2010).

Gaining support for invertebrate conservation presents a major challenge and may even be derided. Marine invertebrates play many crucial roles in ecosystems, providing services and goods (see Chapter 3), but ecological importance has yet to align with conservation effort (Collier *et al.*, 2016). A survey by Kellert (1993) showed that in general the public dislike or are indifferent to invertebrates and unlikely to support significant conservation efforts for invertebrates. Results also indicated that marine invertebrates are among groups where the general public has the least basic knowledge, although there is evidence that educational programmes may greatly assist in raising public awareness of the crucial ecological importance of invertebrates. Certain invertebrates that attract public interest could be used as flagship species to raise the profile of marine invertebrate conservation (Guerra *et al.*, 2011). Policy makers, however, may often be unaware of invertebrate conservation problems, and the dearth of basic information on invertebrates needed to inform conservation – on their taxonomy, distribution and abundance, biology and ecology, and responses to habitat changes – needs addressing by appropriate funding and support (Cardoso *et al.*, 2011).

Marine invertebrates receive only minor attention in international conventions. The Ramsar Convention is of potential importance to invertebrates of coastal wetlands (see Chapter 11), whilst the Convention on the Conservation of Antarctic Marine Living Resources recognizes the key role of Antarctic krill in the Southern Ocean marine ecosystem (see Chapter 17; Fig. 17.3). Some countries have legislation protecting particular marine invertebrates from overcollection. Marine protected areas may benefit marine invertebrates subject to excessive exploitation, but few have been created specifically for them.

We still have a limited understanding of patterns of invertebrate biodiversity in the sea. Some major biomes, such as the deep-sea benthos, are poorly studied, as are certain major faunal components. The meiofauna, for instance, includes a rich fauna of minute invertebrates, the true diversity of which has barely been assessed (Fig. 4.4). Some marine invertebrate species are at risk, in particular those of restricted and vulnerable habitats, such as certain wetland and lagoonal species (see Chapter 11) and commensal and parasitic invertebrates associated with threatened marine vertebrates. The small deep-water limpet *Tentaoculus balantiophaga* has been found only within the spent egg cases of skates (Marshall, 1996), a group vulnerable to overfishing.

Relatively few marine invertebrate species, such as fisheries species, are likely to be the focus of conservation programmes. Even allowing for the fact that most marine invertebrates are unknown to science, many are effectively excluded from consideration on account of our lack of knowledge of their biology and ecology, practical difficulties of identification, small size, occurrence in inaccessible habitats not significantly affected by human activities, public and political prejudice, and the assumption that they will in any case be covered by other conservation programmes (New, 1993). Indeed, rather than tackling invertebrate conservation as a separate problem, the best option may be to incorporate it into broader, regional approaches to biodiversity protection (Strayer, 2006).

The first documented case of historical extinction of a marine invertebrate was that of the Atlantic eelgrass limpet *Lottia alveus* (Carlton *et al.*, 1991). Known from Labrador to Long Island Sound, this limpet lived only on the blades of eelgrass, the seagrass *Zostera marina*. North Atlantic eelgrass beds declined massively in the early 1930s due to 'wasting disease' (see Chapter 12). Eelgrass populations survived in low salinity areas, but these were outside the physiological range of *L. alveus*, no verified specimens of which have been collected since 1929. In this case, the extinction does not appear to be linked to human activity, but there are other possible extinctions of gastropods where vulnerable and restricted coastal habitats have apparently been destroyed by human activity (Carlton, 1993). Several other marine invertebrates have possibly become extinct in historical time (Carlton *et al.*, 1999). There are many taxa that have not been reported since the eighteenth

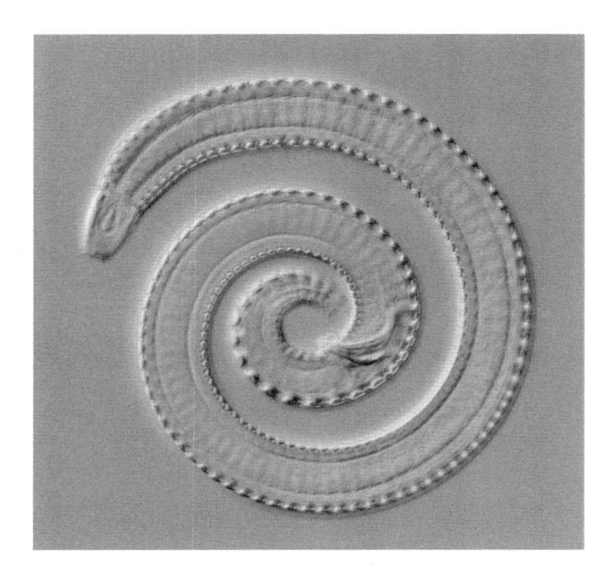

**Fig. 4.4** A deep-sea nematode (*Metadasynemella* sp.). The meiofauna inhabiting seabed sediments includes a huge diversity of minute invertebrates, typically dominated by nematodes. An estimated 50 000 marine nematode species are as yet undiscovered. Photograph: D. Leduc (NIWA).

and nineteenth centuries, and since most marine invertebrates have yet to be described, historical extinctions may well have been overlooked (Carlton, 1993).

Macrofaunal invertebrates often have high fecundity and disperse as planktonic larvae. It is thus generally believed that they tend to be more widely distributed than terrestrial species and less likely to suffer human-induced extinction. The potential for larval dispersal may, however, be overestimated. Not all species have planktonic larvae, and of those that do, larval duration may be no more than a few hours. But how far larvae can travel is in most cases unknown. Most exploited invertebrates are benthic and tend to be relatively immobile and have patchy distributions, so population refugia may often exist (Jamieson, 1993). On the other hand, patchiness also means that invertebrate fisheries can be difficult to regulate because fishing effort is often dispersed over large geographic areas and involves many small fishers. The main regulatory tool used is a minimum legal size (MLS). This should be set so that individuals have at least one spawning season before reaching exploitable size, although for older fisheries MLS limits may have no biological basis. For recreational fishers there may be a daily catch limit of MLS animals, but commonly for commercial invertebrate fisheries there is no quota, and virtually all legal-sized animals are taken from an area during the season.

The white abalone (*Haliotis sorenseni*), the first marine invertebrate to be listed as endangered under the US Endangered Species Act, occurred historically from southern California to Baja California, but overexploitation has brought it to the verge of extinction. Its very low population densities now mean little if any recruitment through Allee effects and indicate the need for active enhancement programmes, such as release of captive-bred individuals if the species is to survive (Stierhoff *et al.*, 2012). Many other marine invertebrate populations have been severely depleted locally or have disappeared. Marine invertebrate species most at risk from overfishing are likely to have certain characteristics: have a relatively large size, long life span, and low annual recruitment; occur in intertidal or shallow subtidal habitats (facilitating their capture); have a restricted geographic range near human settlements (so minimizing the existence of isolated, refuge populations); and be highly sought after (to warrant exploitation at very low densities and encourage illegal fishing) (Jamieson, 1993).

## Seaweeds

Interest in conservation of marine macrophytes has largely focused on systems dominated by vascular plants – seagrass, mangrove, and saltmarsh species – wetland systems that are examined later (see Chapter 12). Conservation of seaweeds by contrast has attracted far less attention despite their key role in the structure and function of coastal habitats. This neglect is in part due to seaweeds being in general difficult to identify and a scarcity of algal taxonomists, so seaweed floras for many parts of the world are poorly known.

The number of marine macroalgae currently recognized is estimated to be in the region of 6000 to 8500 species, of which about two-thirds are red seaweeds, but since so many species are likely to be undescribed the total may be two to three times higher (World Conservation Monitoring Centre, 1992; Norton *et al.*, 1996). Seaweed diversity is generally higher in temperate latitudes than in the tropics and lowest at the poles. Hotspots of seaweed diversity include southern Australia, Japan, Philippines, Mediterranean, Pacific North America, NE Atlantic, and the Caribbean (Phillips, 2001; Keith *et al.*, 2014).

Significant threats to marine macroalgae include coastal development, eutrophication, siltation, reduced light penetration, and the introduction of non-indigenous species (Walker & Kendrick, 1998; Mineur *et al.*, 2015). Introduced seaweeds (mainly by way of hull fouling and aquaculture) are of concern given the role they may assume in restructuring nearshore communities. Particularly invasive seaweeds competing directly with native species include *Codium fragile* in the NW Atlantic and *Caulerpa taxifolia* and *Caulerpa racemosa* in the Mediterranean (see Chapter 16). Other conspicuous algal introductions include the spread of the Asian kelp *Undaria pinnatifida* in southern Australia, New Zealand, and NW Europe and the Japanese brown seaweed *Sargassum muticum,* which has spread along coasts of Pacific North America and NW Europe and altered the structure of native communities (Schaffelke *et al.*, 2006; Williams & Smith, 2007).

Kelps and other seaweeds that dominate many rocky shores and shallow reefs may be considered keystone species for their major structural influence on benthic habitats. Similarly important are unattached nodular or branching coralline algae (of several genera) that form calcareous gravelly deposits known as rhodolith or maerl beds (Nelson, 2009). The largest known bed – about 21 000 $km^2$, comparable to the area of coral reef on the Great Barrier Reef – occurs on the Abrolhos shelf off Brazil, with other important beds off southern Japan and western Australia, in the Gulf of California and Mediterranean Sea, and at a number of locations in NW Europe (Amado-Filho *et al.*, 2012) (Fig. 4.5). The highly structured nature of these beds provides for a high diversity of associated species, with some reported only from rhodolith beds. However, rhodoliths have a variety of industrial uses, particularly as an agricultural fertilizer, and some beds are mined. The main commercial beds in Europe are in Brittany where annual production is about 500 000 tonnes. But given that rhodolith beds accumulate very slowly (about 1 mm $yr^{-1}$), they can hardly be considered renewable resources (Bosence & Wilson, 2003; BIOMAERL Team, 2003). As unique, vulnerable habitats of restricted distribution, rhodolith beds are highly significant from a conservation viewpoint, and there is an urgent need to develop suitable management plans for pristine beds, such as the designation of marine protected areas (BIOMAERL Team, 2003).

**Fig. 4.5** A European maerl bed. The unattached coralline algae provide a complex habitat supporting a diverse community. Photograph: Getty Images/Mark Webster. (A black and white version of this figure will appear in some formats. For the colour version, please refer to the plate section.)

Assessing the conservation status of seaweed species is currently often impossible given the generally limited knowledge of marine floras. Nevertheless, there is evidence that some seaweeds may now be extinct or in need of protection (Brodie *et al.*, 2009). The red seaweed *Vanvoorstia bennettiana*, described last century from Port Jackson, New South Wales, is now categorized as extinct on the IUCN Red List. The only other seaweeds so far listed are from the Galápagos Islands (74 species, 15 of them threatened). As a conservation strategy, Phillips (1998) advocates identifying species likely to be vulnerable, such as those that are habitat specialists, have restricted geographical ranges, occur in heavily populated areas, or are subject to human exploitation. Importantly, threatened species need to be given legal protection (Brodie *et al.*, 2009). In terms of their conservation, algae still face similar problems to most groups of invertebrates, where ecological significance seldom translates into public empathy and political resolve.

## REFERENCES

Amado-Filho, G.M., Moura, R.L., Bastos, A.C., *et al.* (2012) Rhodolith beds are major $CaCO_3$ bio-factories in the tropical South West Atlantic. *PLoS ONE*, **7**, e35171.

Arrieta, J.M., Arnaud-Haond, S. & Duarte, C.M. (2010). What lies underneath: conserving the oceans' genetic resources. *Proceedings of the National Academy of Sciences*, **107**, 18318–24.

Barclay-Smith, P. (1959). The British contribution to bird protection. *Ibis*, **101**, 115–22.

BIOMAERL Team (2003). Conservation and management of northeast Atlantic and Mediterranean maerl beds. *Aquatic Conservation: Marine and Freshwater Ecosystems*, **13**, S65–76.

Bosence, D. & Wilson, J. (2003). Maerl growth, carbonate production rates and accumulation rates in the northeast Atlantic. *Aquatic Conservation: Marine and Freshwater Ecosystems*, **13**, S21–31.

Botsford, L.W. & Parma, A.M. (2005). Uncertainty in marine management. In *Marine Conservation Biology: The Science of Maintaining the Sea's Biodiversity*, ed. E.A. Norse & L.B. Crowder, pp. 375–92. Washington, DC: Island Press.

Bottrill, M.C., Joseph, L.N., Carwardine, J., *et al.* (2008). Is conservation triage just smart decision making? *Trends in Ecology and Evolution*, **23**, 649–54.

Bowman, M., Davies, P. & Redgwell, C. (2010). *Lyster's International Wildlife Law*, 2nd edn. Cambridge, UK: Cambridge University Press.

Braeckman, U., Rabaut, M., Vanaverbeke, J., Degraer, S. & Vincx, M. (2014). Protecting the commons: the use of subtidal ecosystem engineers in marine management. *Aquatic Conservation: Marine and Freshwater Ecosystems*, **24**, 275–86.

Briggs, J.C. (2007). Marine biogeography and ecology: invasions and introductions. *Journal of Biogeography*, **34**, 193–8.

Brodie, J., Andersen, R.A., Kawachi, M. & Millar, A.J.K. (2009). Endangered algal species and how to protect them. *Phycologia*, **48**, 423–38.

Cardoso, P., Erwin, T.L., Borges, P.A.V. & New, T.R. (2011). The seven impediments in invertebrate conservation and how to overcome them. *Biological Conservation*, **144**, 2647–55.

Carlton, J.T. (1989). Man's role in changing the face of the ocean: biological invasions and implications for conservation of near-shore environments. *Conservation Biology*, **3**, 265–73.

Carlton, J.T (1993). Neoextinctions of marine invertebrates. *American Zoologist*, **33**, 499–509.

Carlton, J.T., Geller, J.B., Reaka-Kudla, M.L. & Norse, E.A. (1999). Historical extinctions in the sea. *Annual Review of Ecology and Systematics*, **30**, 515–38.

Carlton, J.T., Vermeij, G.J., Lindberg, D.R., Carlton, D.A. & Dudley, E.C. (1991). The first historical extinction of a marine invertebrate in an ocean basin: the demise of the eelgrass limpet *Lottia alveus*. *Biological Bulletin*, **180**, 72–80.

Ceballos, G., Ehrlich, P.R. & Barnosky, A.D., *et al.* (2015). Accelerated modern human-induced species losses: entering the sixth mass extinction. *Sciences Advances*, **1**, e1400253.

Claereboudt, M. (1999). Fertilization success in spatially distributed populations of benthic free-spawners: a simulation model. *Ecological Modelling*, **121**, 221–33.

Collier, K.J., Probert, P.K. & Jeffries, M. (2016). Conservation of aquatic invertebrates: concerns, challenges and conundrums. *Aquatic Conservation: Marine and Freshwater Ecosystems*, **26**, 817–37.

Courchamp, F., Angulo, E., Rivalan, P., *et al.* (2006). Rarity value and species extinction: the anthropogenic Allee effect. *PLoS Biology*, **4**, e415.

Courchamp, F., Clutton-Brock, T. & Grenfell, B. (1999). Inverse density dependence and the Allee effect. *Trends in Ecology & Evolution*, **14**, 405–10.

Dulvy, N.K., Sadovy, Y. & Reynolds, J.D. (2003). Extinction vulnerability in marine populations. *Fish and Fisheries*, **4**, 25–64.

Durant, S.M. & Harwood, J. (1992). Assessment of monitoring and management strategies for local populations of the Mediterranean monk seal *Monachus monachus*. *Biological Conservation*, **61**, 81–92.

Estes, J.A., Duggins, D.O. & Rathbun, G.B. (1989). The ecology of extinctions in kelp forest communities. *Conservation Biology*, **3**, 252–64.

Frankham, R., Bradshaw, C.J.A. & Brook, B.W. (2014). Genetics in conservation management: revised recommendations for the 50/500 rules, Red List criteria and population viability analyses. *Biological Conservation*, **170**, 56–63.

Fraser, D.J. (2008). How well can captive breeding programs conserve biodiversity? A review of salmonids. *Evolutionary Applications*, **1**, 535–86.

Gascoigne, J. & Lipcius, R.N. (2004). Allee effects in marine systems. *Marine Ecology Progress Series*, **269**, 49–59.

Guerra, Á., González, Á.F., Pascual, S. & Dawe, E.G. (2011). The giant squid *Architeuthis*: an emblematic invertebrate that can represent concern for the conservation of marine biodiversity. *Biological Conservation*, **144**, 1989–97.

Hockey, P.A.R. & Branch, G.M. (1994). Conserving marine biodiversity on the African coast: implications of a terrestrial perspective. *Aquatic Conservation: Marine and Freshwater Ecosystems*, **4**, 345–62.

IUCN (2001). *IUCN Red List Categories and Criteria: Version 3.1.* Gland, Switzerland, and Cambridge, UK: IUCN.

Jamieson, G.S. (1993). Marine invertebrate conservation: evaluation of fisheries over-exploitation concerns. *American Zoologist*, **33**, 551–67.

Keith, D.A., Rodríguez, J.P., Rodríguez-Clark, K.M., *et al.* (2013). Scientific foundations for an IUCN Red List of Ecosystems. *PLoS ONE*, **8**, e62111.

Keith, S.A., Kerswell, A.P. & Connolly, S.R. (2014). Global diversity of marine macroalgae: environmental conditions explain less variation in the tropics. *Global Ecology and Biogeography*, **23**, 517–29.

Kellert, S.R. (1993). Values and perceptions of invertebrates. *Conservation Biology*, **7**, 845–55.

Lacy, R.C. (2000). Structure of the VORTEX simulation model for population viability analysis. *Ecological Bulletins*, **48**, 191–203.

Lotze, H.K., Reise, K., Worm, B., *et al.* (2005). Human transformations of the Wadden Sea ecosystem through time: a synthesis. *Helgoland Marine Research*, **59**, 84–95.

Maclean, I.M.D. & Wilson, R.J. (2011). Recent ecological responses to climate change support predictions of high extinction risk. *Proceedings of the National Academy of Sciences*, **108**, 12337–42.

Marshall, B.A. (1996). A new subfamily of the Addisoniidae associated with cephalopod beaks from the tropical southwest Pacific, and a new pseudococculinid associated with chondrichthyan egg cases from New Zealand (Mollusca: Lepetelloidea). *Veliger* **39**, 250–9.

Miller, G. (1996). Ecosystem management: improving the Endangered Species Act. *Ecological Applications*, **6**, 715–17.

Mineur, F., Arenas, F., Assis, J., *et al.* (2015). European seaweeds under pressure: consequences for communities and ecosystem functioning. *Journal of Sea Research*, **98**, 91–108.

Myers, R.A. & Ottensmeyer, C. A. (2005). Extinction risk in marine species. In *Marine Conservation Biology: The Science of Maintaining the Sea's Biodiversity*, ed. E.A. Norse & L.B. Crowder, pp. 58–79. Washington, DC: Island Press.

Nelson, W.A. (2009). Calcified macroalgae – critical to coastal ecosystems and vulnerable to change: a review. *Marine and Freshwater Research*, **60**, 787–801.

New, T.R. (1993). Angels on a pin: dimensions of the crisis in invertebrate conservation. *American Zoologist*, **33**, 623–30.

Norton, T.A., Melkonian, M. & Andersen, R.A. (1996). Algal biodiversity. *Phycologia*, **35**, 308–26.

Penning, M., Reid, G. McG., Koldewey, H., *et al.* (eds.). (2009). *Turning the Tide: A Global Aquarium Strategy for Conservation and Sustainability.* Bern, Switzerland: World Association of Zoos and Aquariums.

Phillips, J.A. (1998). Marine conservation initiatives in Australia: their relevance to the conservation of macroalgae. *Botanica Marina*, **41**, 95–103.

Phillips, J.A. (2001). Marine macroalgal biodiversity hotspots: why is there high species richness and endemism in southern Australian marine benthic flora? *Biodiversity and Conservation*, **10**, 1555–77.

Possingham, H.P., Andelman, S.J., Burgman, M.A., *et al.* (2002). Limits to the use of threatened species lists. *Trends in Ecology & Evolution*, **17**, 503–7.

Régnier, C., Achaz, G., Lambert, A., *et al.* (2015). Mass extinction in poorly known taxa. *Proceedings of the National Academy of Sciences*, **112**, 7761–6.

Reynolds, J.D., Dulvy, N.K., Goodwin, N.B. & Hutchings, J.A. (2005). Biology of extinction risk in marine fishes. *Proceedings of the Royal Society B*, **272**, 2337–44.

Reznick, D., Bryant, M.J. & Bashey, F. (2002). *r*- and *K*-selection revisited: the role of population regulation in life-history evolution. *Ecology*, **83**, 1509–20.

Roff, J. & Zacharias, M. (2011). *Marine Conservation Ecology*. London: Earthscan.

Schaffelke, B., Smith, J.E. & Hewitt, C.L. (2006). Introduced macroalgae – a growing concern. *Journal of Applied Phycology*, **18**, 529–41.

Shaffer, M.L. (1981). Minimum population sizes for species conservation. *BioScience*, **31**, 131–4.

Slobodkin, L.B. (1986). On the susceptibility of different species to extinction: elementary instructions for owners of a world. In *The Preservation of Species: The Value of Biological Diversity*, ed. B.G. Norton, pp. 226–42. Princeton, NJ: Princeton University Press.

Slooten, E. & Dawson, S.M. (2010). Assessing the effectiveness of conservation management decisions: likely effects of new protection measures for Hector's dolphin (*Cephalorhynchus hectori*). *Aquatic Conservation: Marine and Freshwater Ecosystems*, **20**, 334–47.

Slooten, E., Fletcher, D. & Taylor, B.L. (2000). Accounting for uncertainty in risk assessment: case study of Hector's dolphin mortality due to gillnet entanglement. *Conservation Biology*, **14**, 1264–70.

Stierhoff, K.L., Neuman, M. & Butler, J.L. (2012). On the road to extinction? Population declines of the endangered white abalone, *Haliotis sorenseni*. *Biological Conservation*, **152**, 46–52.

Strayer, D.L. (2006). Challenges for freshwater invertebrate conservation. *Journal of the North American Benthological Society*, **25**, 271–87.

Taylor, M.F.J., Suckling, K.F. & Rachlinski, J.J. (2005). The effectiveness of the Endangered Species Act: a quantitative analysis. *BioScience*, **55**, 360–7.

Tlusty, M.F., Rhyne, A.L., Kaufman, L., *et al.* (2013). Opportunities for public aquariums to increase the sustainability of the aquatic animal trade. *Zoo Biology*, **32**, 1–12.

Traill, L.W., Brook, B.W., Frankham, R.R. & Bradshaw, C.J.A. (2010). Pragmatic population viability targets in a rapidly changing world. *Biological Conservation*, **143**, 28–34.

Vermeij, G.J. (1986). The biology of human-caused extinctions. In *The Preservation of Species: The Value of Biological Diversity*, ed. B.G. Norton, pp. 28–49. Princeton, NJ: Princeton University Press.

Vermeij, G.J. (1993). Biogeography of recently extinct marine species: implications for conservation. *Conservation Biology*, **7**, 391–7.

Vié, J.-C., Hilton-Taylor, C. & Stuart, S.N. (eds.) (2009). *Wildlife in a Changing World – an Analysis of the 2008 IUCN Red List of Threatened Species*. Gland, Switzerland: IUCN.

Walker, D.I. & Kendrick, G.A. (1998). Threats to macroalgal diversity: marine habitat destruction and fragmentation, pollution and introduced species. *Botanica Marina*, **41**, 105–12.

Webb, T.J. & Mindell, B.L. (2015). Global patterns of extinction risk in marine and non-marine systems. *Current Biology*, **25**, 506–11.

Williams, S.L. & Smith, J.E. (2007). A global review of the distribution, taxonomy, and impacts of introduced seaweeds. *Annual Review of Ecology, Evolution, and Systematics*, **38**, 327–59.

World Conservation Monitoring Centre (1992). *Global Biodiversity: Status of the Earth's Living Resources*. London: Chapman & Hall.

Zacharias, M.A. & Roff, J.C. (2001). Use of focal species in marine conservation and management: a review and critique. *Aquatic Conservation: Marine and Freshwater Ecosystems*, **11**, 59–76.

# 5 Fishing

Cod fisheries have for centuries shaped the social and economic development of maritime communities of the North Atlantic (Kurlansky, 1998). Yet the cod fishery off Newfoundland, one of the world's richest, was brought to commercial extinction in 1992. Seemingly, cod had rarely been fished at sustainable levels over the previous 30 years. Stock size and recruitment had been overestimated, which had fuelled increased investment (Hutchings & Myers, 1994; Schrank, 2005). This is a stark example of the plight of many fisheries and of our ability to decimate the ocean's living resources when adequate management and conservation measures are lacking.

Our greatest impact on the marine environment has been through exploiting its living resources. How fisheries (in their broadest sense) are managed – or often mismanaged – therefore has crucial implications for the conservation of marine biodiversity. Fisheries and their management are huge topics, and only certain aspects can be addressed here. In this chapter we look in particular at methods of fishing and the problem of overfishing and examine the effects of fishing, including impacts on target species and on the associated communities and habitats. We also look at aspects of management to address problems of overfishing and environmentally damaging practices, including customary and ecosystem-based approaches and the use of marine protected areas and artificial reefs. Finally we examine aquaculture and fisheries introductions and the conservation issues they raise.

## Methods of Fishing

Humans presumably made use of seafood at the earliest opportunities. Remains from archaeological sites along the coast of South Africa dating from 250 000 to 50 000 years ago show, for example, that these people collected shellfish from the shore, particularly limpets, mussels, and turban shells (Steele & Klein, 2008). The earliest evidence for fishhook manufacture, from East Timor, dates from 23 000 to 16 000 years ago (O'Connor *et al.*, 2011). Archaeological and other evidence from many parts of the world indicate that fishing by prehistoric societies sometimes had substantial impacts on coastal ecosystems (Jackson *et al.*, 2001). Classic signs of overfishing are evident from faunal remains at archaeological sites on Caribbean islands, where a comparison between sites of early (1850–1280 years Before Present) and late (1415–560 years BP) occupation indicates that heavy exploitation of reef fishes resulted in declines in the size and trophic level of species taken and also the proportion of reef fish biomass in the faunal record (Wing & Wing, 2001). Prototypes of the more important kinds of fishing gear were already in common use in Ancient Egypt, including various types of longlines, nets, traps, seine, and beam trawl (Aleem, 1972).

There exists a huge diversity of fishing gear, although there are only a few principal methods of fishing (e.g. King, 1995; Sainsbury, 1996). Traps include various types of baited pots and devices for catching carnivorous crustaceans, molluscs, and fish and barrier and fence traps to divert coastal fish into a retaining area. Hook and line gear ranges from simple hand-held lines to longlines kilometres in length with thousands of hooks set on short sidelines for surface or bottom deployment. There are also various types of hooked lures including those that are towed (trolled) to catch pelagic fish and jigs that are jerked vertically to snag squid. Gillnets, now usually made of monofilament nylon, are deployed vertically to catch various pelagic or demersal species. Setnets are gillnets anchored to the seafloor, whereas driftnets are gillnets that drift with the currents. Driftnets can be very long – nets more than 50 km in length have been used in open ocean pelagic fisheries. By contrast, trawls and dredges are actively towed, such as mid-water trawls used for schooling fish. In the demersal otter trawl, the net is held open horizontally by heavy otter boards that act as paravanes. The upper edge of the mouth of the net is buoyed by floats, whilst the bottom edge of the mouth normally drags over the seabed. For rough bottoms this groundrope is usually rigged with heavy steel bobbins or rollers. Dredges, usually steel framed, are used mainly for shellfish, such as scallops, oysters, and clams. Encircling nets are used to surround a school of fish, as with a beach seine used from the shore. In purse seining, a vessel tows an encircling net up to 2 km long around a school of pelagic fish, and then the purse line running around the lower edge of the net is hauled in to close off the bottom of the net. Purse seining and trawling are the most important commercial methods.

Commercial fishing has become far more efficient over the past century with technological advances in fishing gear, vessel design and operation, and processing of catch. The advent of steam-driven vessels towards the end of the nineteenth century meant that fish could be pursued and gear actively towed and hauled on powered capstans, and this enabled the otter trawl to come into use. Other major innovations have been the introduction of freezing, mechanized filleting, factory vessels, synthetic fibres, sonar, and positioning systems. There are still many small inshore fisheries employing traditional methods, but most fish are now taken on the open sea by large-scale commercial operations using technologically advanced methods.

## Overfishing

The term 'overfishing' implies a level of extraction that should not be exceeded, and there are various biological and economic ways to quantify this. Overfishing has traditionally been defined in terms of metrics used in fisheries management, such as fishing beyond the level of effort that produces maximum sustainable yield (see below). Such definitions are derived from single-species population models. Definitions that incorporate ecosystem attributes to take into account the wider impacts of fishing would be a useful development (Murawski, 2000). But in the meantime we can take overfishing simply to mean exploitation that exceeds a species' capacity to recover.

Until the late nineteenth century, a common view was that most fish stocks were inexhaustible and that open sea stocks at least could not be significantly depleted (given the fishing technology of the day). However, a review of the state of demersal fisheries of England and Wales at the turn of the century came to a rather different view: 'all these various independent sources of information display a melancholy unanimity …. We have … to face the established fact that the bottom fisheries are not only exhaustible, but in rapid and continuous process of exhaustion; that the rate at which sea fishes multiply and grow, even in favourable seasons, is exceeded by the rate of capture' (Garstang, 1900). The International Council for the Exploration of the Sea (ICES) was established in 1902 to initiate a cooperative international programme of research into the hydrography and fisheries of the North Atlantic and, significantly, formed an Overfishing Committee the same year.

Concerns about overexploitation were thus being expressed more than a century ago. We have no estimates of the world's marine catch then, but it was probably only a few million tonnes. The global fisheries catch has since increased dramatically. Whilst large-scale industrial fisheries dominate, a substantial biomass is also taken by artisanal (small-scale commercial) fisheries, subsistence fisheries (needed for personal/family consumption), bycatch (unintentional catch), and illegal and otherwise unreported catch. Taking all these fisheries sectors into account, the global marine catch peaked in 1996 at an estimated 130 million tonnes. It has since declined and in 2010 was estimated at about 110 million t (comprising industrial 67%, artisanal 20%, discarded bycatch 9%, subsistence 3%, and recreational 1%) (Pauly & Zeller, 2016) (Fig. 5.1).

Fisheries vary greatly in their size and status and from country to country (Worm *et al.*, 2009). Nevertheless, global fisheries landings are principally from large stocks that are fished by developed nations. Marine capture fisheries are dominated by relatively few species. Those that contribute most to the global catch (each averaging more than 1 million t annually for 2003–12) comprise small pelagic species (Peruvian anchovy, sardinellas, Atlantic herring, chub mackerel, scads, Japanese anchovy, and European pilchard), tunas (skipjack and yellowfin), cods (Alaska pollock, Atlantic cod, and blue whiting), and largehead hairtail. Also, only a few fishing areas account for most of the global catch, with more than 70% of production in recent years coming from the north-west, western central, and south-east Pacific; NE Atlantic; and eastern Indian Ocean (FAO, 2016).

Fishing pressure has resulted in a steady decline in the proportion of assessed stocks that are being fished within biologically sustainable levels – from 90% in 1974 to 69% in 2013. Nearly 60% of stocks are now fully fished and so have no room for further growth, and more than 30% are being fished at a biologically unsustainable level – their abundance is below the level that can produce the maximum sustainable yield (FAO, 2016).

The decline of primary target stocks typically results in a shift towards other, generally less desirable species, often ones that would formerly have been discarded. High-value species are mostly the larger, slow-growing, high-trophic-level fishes. An assessment by Christensen *et al.* (2014) based on 200 ecosystem models indicates that over the past 100 years the biomass of predatory fish in the world's oceans has declined by 66%, with 54% of the decline having occurred in just the past 40 years. And, probably as a result of predator release, the biomass of prey fish has increased.

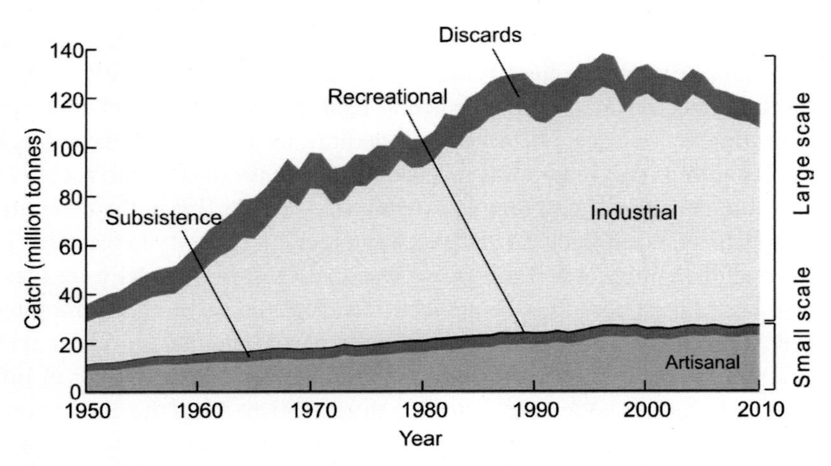


**Fig. 5.1** Estimates of global marine fisheries catches from 1950 to 2010 reconstructed for the EEZs of all maritime countries and the high seas for all fisheries sectors: large-scale (industrial) and small-scale sectors (artisanal, subsistence, recreational), with discards (overwhelmingly from industrial fisheries). From Pauly, D. & Zeller, D. (2016). Catch reconstructions reveal that global marine fisheries catches are higher than reported and declining. *Nature Communications*, 7, article 10244. Licensed under a Creative Commons Attribution 4.0 International License: http://creativecommons.org/licenses/by/4.0/.

**Fig. 5.2** The concept of fishing down marine food webs. As fisheries (blue arrow) deplete large, high-trophic-level species, they then target smaller, lower-trophic-level species. The accompanying decline in large epibenthic invertebrates represents the impact of bottom trawling. From Pauly, D. (2007). The *Sea Around Us* Project: documenting and communicating global fisheries impacts on marine ecosystems. *Ambio*, 36, 290–5. © Royal Swedish Academy of Sciences 2007. (A black and white version of this figure will appear in some formats. For the colour version, please refer to the plate section.)

Large-scale removal of predators provides for more opportunistic replacements, generally smaller, short-lived, lower-trophic-level planktivorous pelagic fish and invertebrates (Pauly *et al.*, 1998) (Fig. 5.2). Serial depletion of fish populations by overfishing thus progresses from species with life-history traits that render them highly vulnerable to overexploitation to those with less vulnerable life histories (Cheung *et al.*, 2007). With a reduction in the number of year classes in targeted populations and a shift towards more *r*-selected species, greater inter-annual variability of abundance is to be expected, thereby compounding management problems.

## Assessment of Catch

The vulnerability of fish stocks depends on various factors involving the biology and ecology of the species, environmental conditions, and socio-economic pressure. The dynamics of a fish population depend on the balance between additions via growth and recruitment and losses via natural and fishing mortality. Overfishing can be described in terms of its effect on recruitment and growth. Recruitment overfishing (typified by fisheries for small pelagic species and species of low reproductive capacity) is a level of fishing at which the adult stock is reduced to the extent that recruits produced are insufficient to maintain the population, whereas growth overfishing (as seen in many demersal species) is a level at which recruits entering the fishery are caught before they can grow to an optimum marketable size (King, 1995). An unexploited population is considered to have approached the carrying capacity of the environment, at which point further growth becomes constrained by density-dependent factors such as availability of food or space. But reducing the population size by fishing produces a compensatory response and a surplus production as the basis for a fishery. This response may involve an increase in recruitment or growth rate or a decrease in natural mortality rate. The relationship between fishing effort and yield can be described by a parabolic curve, with yield at first increasing to a maximum, known as the maximum sustainable yield (MSY), but falling as exploitation rate increases still further (King, 1995) (Fig. 5.3). The concept of MSY has traditionally been used to guide fisheries, although in practice management for MSY has been the exception rather than the rule. MSY was incorporated in the UN Convention on the Law of the Sea (UNCLOS) under provisions concerning the conservation of the living resources of EEZs and the high seas (Articles 61 and 119). It is recognized that MSY has shortcomings and may be more appropriate as an upper limit of exploitation rather than as a goal for management (Worm *et al.*, 2009). Other biological limits or reference points are now increasingly being proposed

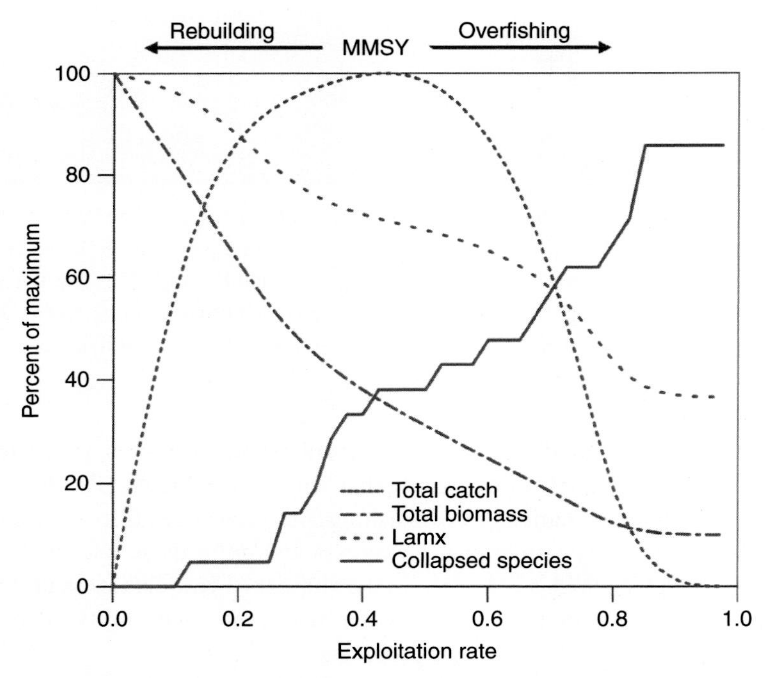


**Fig. 5.3** Effects of increasing exploitation rate on a fish community. Exploitation rate is the proportion of fish biomass caught each year, MMSY is the multispecies maximum sustainable yield, collapsed species are those that have been reduced to less than 10% of their unfished biomass, and Lmax is the average maximum length that species can reach. In this example an exploitation rate of ~0.45 gives the MMSY. Reducing exploitation rate to ~0.25 would still yield ~90% of maximum catch but greatly reduce the proportion of collapsed species and allow rebuilding. The model is based on a Georges Bank fish community. From Worm, B., Hilborn, R., Baum, J.K., *et al.* (2009). Rebuilding global fisheries. *Science*, 325, 578–85. Reprinted with permission from AAAS.

to regulate fishing mortality. Froese *et al.* (2008) show that instead of aiming for MSY, fishing has far less impact on stocks if fish are caught after they have reached the size – so-called optimum length – where unfished cohorts reach maximum biomass. This can deliver the same catch as under MSY, but by maintaining an age, size, and biomass structure similar to that of an unexploited stock, it also helps to conserve overall ecosystem stability.

Year-to-year recruitment can vary by orders of magnitude, depending on various biotic and abiotic factors, though these are difficult to determine. It has commonly been assumed that recruitment is unaffected by the size of the spawning stock. Such a relationship is often not evident because most fish are so fecund that a small difference in the survival rate of very young stages will have a more significant effect on subsequent recruitment than the number of mature adults. However, with increasing depletion of a population, there comes a point where the small number of adult fish results in recruitment failure (Pauly, 1994).

Recruitment variability is a major source of uncertainty in fisheries management, as too are imprecise estimates of stock abundance and stock-recruitment parameters. Uncertainty should dictate more precautionary approaches to management, including the use of more conservative reference points in preference to MSY. Precautionary models have been adopted by the International Whaling Commission and the Convention on the Conservation of Antarctic Marine Living Resources (see Chapters 9 and 17). Traditionally, such reference points and models have been applied solely to target species. But with increasing concern about the wider effects of fishing activities, additional parameters may be needed that relate to attributes of non-target species and assemblages.

## Impacts of Fishing

Fishing can have major direct and indirect impacts on marine ecosystems, affecting not just populations of target species but also associated communities and habitats (see e.g. Jennings & Kaiser (1998), Hall (1999), and Kaiser & de Groot (2000) for detailed treatments). Establishing causal relationships can, nevertheless, be difficult given, in particular, the complexity of fisheries-ecosystems interactions, the confounding effects of natural variability, and the fact that many ecosystems have long been impacted by fishing, leaving few suitable controls (Goñi, 2000).

Biological and economic effects of overfishing of stocks are summarized in Box 5.1. As fishing is generally size-selective it tends to modify the size and age structure of a population, so as fishing mortality increases, average body size and age of the population decrease. Such populations appear to be inherently less stable than unexploited stocks and more susceptible to the effects of environmental variability (Anderson *et al.*, 2008; Shelton & Mangel, 2011). Fishing may also skew the sex ratio when the sexes differ in body size or where (as in many reef fish) mature individuals change sex. The removal of large old females may disproportionately impact the number of eggs produced as maternal body mass in fish tends to be a good predictor of fecundity (Rideout & Morgan, 2010). The larvae from older fish may also have higher survival rates. Older female black rockfish provision their larvae with significantly larger oil globules than younger females, which appears to strongly influence the growth and survival of the larvae (Berkeley *et al.*, 2004). Removing large old females by fishing may thus compromise the stability and sustainability of the population.

Size selection by fishing can induce evolutionary changes in exploited populations, including maturation at younger age and smaller body size, and decreased genetic diversity. Such changes may in turn have implications for the affected communities and ecosystems and the sustainability of those fisheries (Belgrano & Fowler, 2013). Hauser *et al.* (2002), for instance, demonstrated a significant decline of genetic diversity in a population of snapper over its history of exploitation. Although

### Box 5.1 Biological and Economic Effects of Overfishing on Single- and Multi-species Stocks

#### Single-Species Fisheries

(i) Reduction of size (length and weight) of the animals caught, hence
- usually a reduction in value per unit weight

(ii) Reduction in biomass of stock, hence
- reduction of catch per unit effort

(iii) Reduction of total catch (at high levels of effort), hence
- lowered overall food supply
- increased prices
- need to import substitutes and hence increased nutritional deficiency among poorer sector of the population

(iv) Increased fluctuation of stock due to fewer age groups and to reduced buffering of recruitment fluctuations, hence
- more frequent occurrence of periods with extremely low catches
- increasing risk of occasional recruitment failure, inclusive of total collapse of stock and fishery

(v) Lowered income among fishers, hence
- social ills including conflict between pauperized small-scale fishers and their large-scale industrial competitors

#### Multi-species Fisheries

(i) Same as (i) above, plus:

(ii) Massive changes in species composition of catch, i.e.
- disappearance of previously important high-value species
- increase of unmarketable species, and hence
- reduction in average value of species mix
- loss of biodiversity

From Springer *On the Sex of Fish and the Gender of Scientists: Collected Essays in Fisheries Science*, 1994, page 94, Pauly, D., Chapman & Hall, with permission of Springer.

the number of individuals in an exploited population may number millions, those that determine the genetic properties of a population may be orders of magnitude less, exposing the stock to loss of genetic variability and, thereby, adaptability.

## Alteration of Community Structure

Fishing can fundamentally alter the species composition of a community. Typically, this means that as the larger, more valuable, more vulnerable species are depleted, the smaller, more productive, but less valuable and usually less vulnerable species become more abundant. This tends to simplify food-web structure, with shorter-lived species dominating and making the fishery more susceptible to environmental factors (Pauly *et al.*, 2002). The collapse in the early 1990s of the NW Atlantic

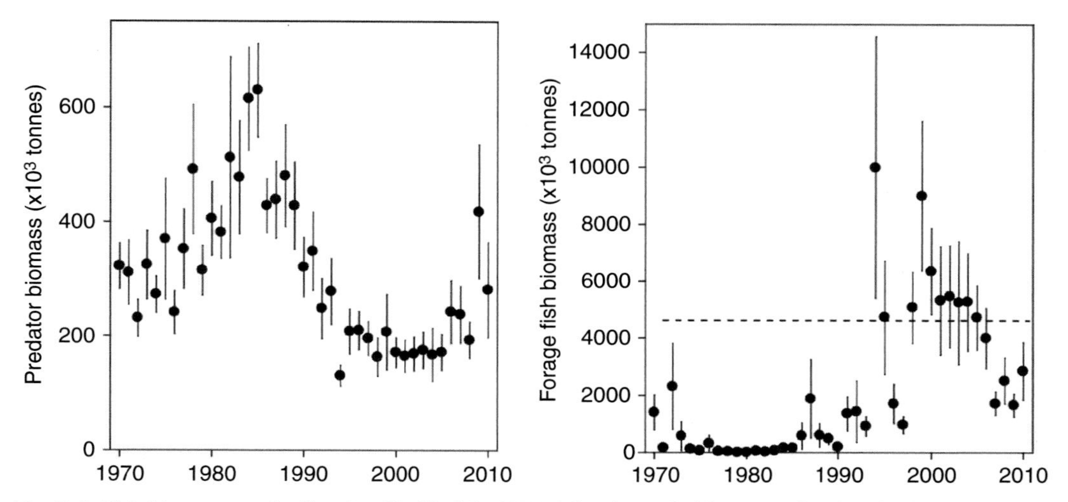


**Fig. 5.4** Fish biomass on the Scotian Shelf of the NW Atlantic. Left: biomass of cod and other large-bodied demersal species, showing the collapse of the fishery in the early 1990s. Right: biomass of small-bodied planktivorous 'forage' species and the estimated carrying capacity of the ecosystem (dashed line). Reprinted by permission from Macmillan Publishers Ltd: *Nature*. Frank, K.T., Petrie, B., Fisher, J.A.D. & Leggett, W.C. (2011). Transient dynamics of an altered large marine ecosystem. *Nature*, 477, 86–9, copyright 2011.

cod fishery precipitated a restructuring of the food web. With the demise of cod and other large demersal fishes, the food web became dominated by opportunistic forage species, notably northern sand lance, capelin, and Atlantic herring, and macroinvertebrates (Fig. 5.4). These planktivorous fish appear to have competed with and preyed on the early life-history stages of the collapsed demersal species, thereby hindering their recovery. However, the populations of the forage species subsequently crashed, having exceeded the zooplankton food available, and since 2005–6 there is evidence that the demersal species are recovering (Frank *et al.*, 2011).

Such shifts may have repercussions up the food chain, impacting on seabird and marine mammal populations, and natural fluctuations may exacerbate the problem. Hamre (1994) describes the consequences of overexploitation on the dynamics of herring, capelin, and cod in the Norwegian-Barents Sea ecosystem. Herring is here the key prey species, cod the main predator, and capelin an opportunistic species whose abundance is largely influenced by that of young herring and cod. Herring fishing increased dramatically in the mid-1960s, and when after a few years this stock was exhausted, fishing effort shifted to capelin. But with the demise of herring, capelin were now also increasingly important as prey for cod. Warm climate conditions in the early 1980s favoured the recruitment of cod, but with a lack of juvenile herring and reduced stock of capelin, the rapidly growing cod stock grazed down all other available food fishes, resulting from 1986 in an abrupt food shortage, not only for cod but also for seals and seabirds such as common guillemot, thousands of which washed ashore apparently having starved to death. For the eastern North Atlantic, however, the increased abundance of sandeels may have benefited seabird populations (see Chapter 8).

A consumer may strongly influence food-web structure so that its removal markedly affects the biomass of the trophic level it exploits. Such top-down control is well known in coastal benthic systems, such as on rocky shores and in kelp forest habitats. There is also increasing evidence of such top-down control in open sea systems where overfishing of high trophic-level predators, such as marine mammals, sharks, and large teleost fish, results in increases in the abundance of prey species, such as fish mesopredators (mid-trophic level predators) and invertebrates (Baum & Worm, 2009). In some cases these effects propagate through two or more trophic levels, though the circumstances under which such cascades occur are still unclear.

Ecosystems may exist in different configurations and switch from one state to another relatively suddenly in a so-called regime shift. Drivers of such shifts include climate change, overfishing, destruction of structural habitat, and introduction of alien species (de Young *et al.*, 2008). Once a switch has occurred it may be difficult for the system to revert to its previous state. The Black Sea provides a dramatic example where overfishing has triggered trophic cascades that have resulted in regime shifts. Two such shifts have occurred, the first in the 1970s following the depletion of top fish predators (dolphins, Atlantic bonito, bluefish, Atlantic bluefin tuna, swordfish) and the second in the 1990s related to a massive reduction in planktivorous fish (sprat, anchovy, Mediterranean horse mackerel) and a population explosion of an introduced ctenophore (*Mnemiopsis leidyi*) taking advantage of the surplus zooplankton. In turn, the decline in zooplankton saw a major increase in phytoplankton biomass (Daskalov, 2002; Daskalov *et al.*, 2007). It appears that blooms of jellyfish and ctenophores are becoming increasingly common in coastal waters worldwide, especially in East Asia, possibly related to overfishing and other anthropogenic disturbances (Purcell *et al.*, 2007).

## Impact of Bottom Fishing

The most extensive anthropogenic impact on benthic habitats worldwide results from the use of fishing gear towed across the sea floor, notably from trawling and dredging. It is estimated that bottom fishing has modified about 50 million km$^2$ of seabed habitat globally (Halpern *et al.*, 2008). The benthic impact of bottom fishing varies considerably depending in particular on the nature of the seabed habitat and its biota, the type of gear used and how often it impacts a particular area of seabed, and the type and frequency of natural background disturbances. On nearshore sandy bottoms where benthic communities are typically characterized by species adapted to marked natural disturbance, such as from storms and tidal currents, trawling impact may be relatively limited. But on the outer continental shelf and upper slope, where frequency and intensity of natural disturbance are much lower and where structurally complex benthic habitat can develop, the impact of mobile fishing gear will generally be far more severe (see Chapter 14).

An obvious effect of bottom trawling is mortality of the benthic organisms, killed either directly by being in the path of the trawl or indirectly as a result of the disturbance or from predation. In the south-eastern North Sea, Bergman and van Santbrink (2000) observed direct mortalities from the passage of a trawl of about 5–40% of initial densities for various polychaetes, gastropods, crustaceans, and sea stars and 20–65% for bivalve species. Epifauna are especially vulnerable to trawling, and their loss is of particular concern where they contribute importantly to habitat structure. In the central Mediterranean, Mangano *et al.* (2013) found that heavily trawled areas of the shelf and upper slope lacked *K*-selected structuring species such as crinoids but had an abundance of *r*-selected scavenging species.

Benthic habitats differ in their vulnerability to impact from bottom fishing gear. In general, hard substrata (cobble, boulder) are more vulnerable to trawling and dredging impacts than soft-sediment substrata (sand, mud), as they recover far more slowly (Grabowski *et al.*, 2014). Auster (1998) proposed a conceptual model of fishing gear impact across a gradient of habitat types from simple sedimentary bedforms with little vertical structure to complex boulder habitats with high vertical relief, abundant attached epifauna, and a diversity of interstitial space (Fig. 5.5). It may be possible to assess the relative sensitivity of benthic communities to fishing impact based on substratum type and the biological traits of the component species, at least for regions where the biology and ecology

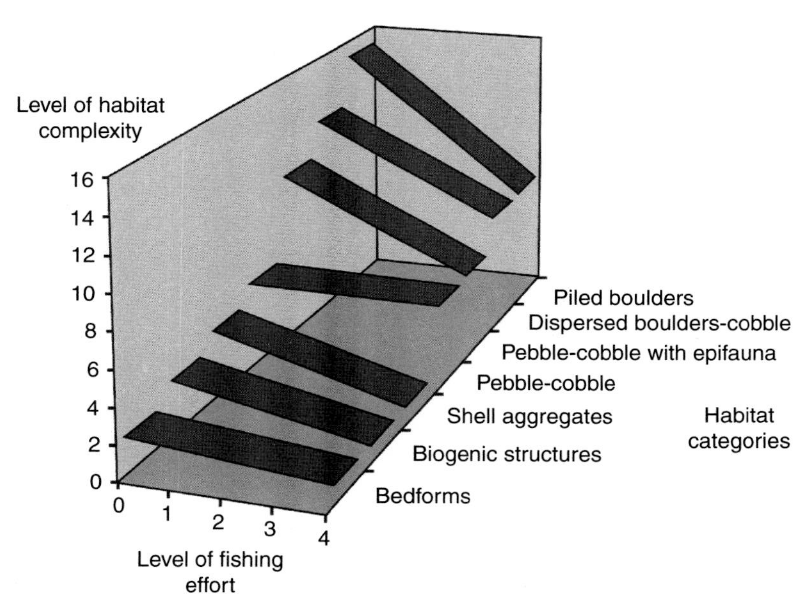


**Fig. 5.5** A conceptual model of the response of seafloor habitat types to increases in fishing effort. Fishing effort increases from left to right along the x axis, from 0 indicating no anthropogenic impact to 4 indicating the maximum effort required to produce the greatest change in habitat complexity. The y axis is a comparative index of habitat complexity. The habitat categories represent common types found across the north-eastern continental shelf of the USA. From Auster, P.J. (1998). A conceptual model of the impacts of fishing gear on the integrity of fish habitats. *Conservation Biology*, 12, 1198–203. © Society for Conservation Biology.

of those species are well known (Tyler-Walters *et al*., 2009). Such information can then be used to guide fisheries management.

Altering the benthic landscape may adversely affect fish or shellfish that make use of sedimentary or biogenic features. Many demersal fishes depend on the benthos for food and refugia. Predation on juvenile fish may increase in less complex habitats where there are fewer opportunities for shelter and subsequently lead to decreases in recruitment. Habitat-forming bryozoans – those with large heavily calcified colonies – are particularly well represented in New Zealand shelf waters in places contributing significantly to benthic habitat structure and supporting a diverse associated fauna. They are, however, readily damaged by bottom fishing (Wood *et al*., 2012). More than a century of oyster dredging off southernmost New Zealand has, for instance, reduced the extent and complexity of bryozoan habitat, and the modified seabed appears to be less suitable as habitat for blue cod (Carbines *et al*., 2004).

Over time, bottom fishing tends to reduce the heterogeneity of benthic habitats, with consequent loss of biodiversity, and change the relative importance of demographic and functional components. Impacted communities tend, for example, to be dominated by juvenile stages, smaller opportunistic species, and mobile scavengers (Thrush & Dayton, 2002; de Juan *et al*., 2007). Other potentially significant effects of bottom fishing include increased resuspension of bottom sediment, subsequent smothering effects – likely to impact suspension-feeding animals in particular – and changes in nutrient fluxes across the sediment-water interface.

Some countries have now restricted or banned bottom trawling, for instance, in the Asia-Pacific region (Loh & Jaafar, 2015), in recognition of the damaging and unsustainable nature of this type of fishing. Other fishing practices that can severely degrade benthic habitat are blast fishing and muro-ami carried out on coral reefs (see Chapter 13).

## Impact of Pelagic Fisheries

Driftnet and pelagic longline fisheries have also attracted attention because of concerns over their impact on ocean ecosystems and their sustainability. These fisheries can remove large numbers of predatory fish and cause heavy incidental mortality of other species vulnerable to overexploitation, including marine mammals, seabirds, turtles, and sharks. Longline fisheries have, for instance, been implicated in global declines of tuna and billfish diversity since the early 1950s (Worm *et al.*, 2005) and are responsible for heavy mortality of albatrosses (see Chapter 8).

Driftnetting increased rapidly from the mid-twentieth century as synthetic nets of smaller mesh size were introduced, and large-scale driftnet fisheries soon developed, particularly in international waters (Caddell, 2010). The North Pacific has been especially targeted by high-seas driftnet fisheries, notably those for Pacific salmon, flying squid, and tunas and other scombroids (Northridge, 1991). The largest of these fisheries, the North Pacific squid driftnet fishery, started in 1978 and has mainly involved Japan, Taiwan, and Korea. By the late 1980s, about 800 vessels were taking part seasonally, each deploying some 40 km or more of netting. It is estimated that in 1989, marine mammals entangled by this fishery would have included at least 2200 northern fur seals, 19 000 northern right whale dolphins, 6000 Dall's porpoises, and 11 000 Pacific white-sided dolphins. And Dall's porpoise is the major non-target casualty of the Japanese North Pacific salmon fishery. Seabirds have also drowned in large numbers in the North Pacific squid and salmon fisheries, particularly shearwaters, puffins, and albatrosses (see Chapter 8). Whilst attention has focused on industrial driftnet fisheries on the high seas, coastal artisanal fisheries can be responsible for comparable amounts of netting. Coastal driftnet fisheries in Sri Lanka, for example, may have taken several tens of thousands of small cetaceans, mainly spinner and striped dolphins (Northridge, 1991).

Widespread censure of the scale and perceived impact of driftnet fishing led to various restrictions being introduced at national and regional levels and eventually to a resolution adopted by the UN General Assembly (UNGA Resolution 46/215) calling for a global moratorium on all large-scale pelagic driftnet fishing on the high seas (large-scale being interpreted as driftnets longer than 2.5 km) to be implemented in 1992. Although UNGA resolutions are not legally binding, a number of states, regional fisheries management organizations, and other bodies enacted legislation to control the use of driftnets (Caddell, 2010). The global moratorium has generally been observed, but enforcement of fisheries regulations on the high seas is problematic, and illegal fleets continue to operate.

## Ghost Fishing

Fishing gear that is lost or abandoned at sea may continue to function in the water and result in animal mortalities, a phenomenon known as 'ghost fishing' (Matsuoka *et al.*, 2005). Ghost fishing is mostly likely to occur with gear that operates passively, such as gillnets, pots, and traps, and may be a serious problem when large volumes of gear are deployed and when the gear is made of durable synthetic materials that do not readily degrade. There is the potential for caught animals to act as bait and attract new animals, which in turn become caught. Lost gillnets may remain in a configuration where they continue to fish effectively for months or even years before they eventually deteriorate or collapse on themselves. In a simulated ghost fishing experiment carried out over nine months, Kaiser *et al.* (1996) observed a sequence of captures in setnets, mainly of crabs and other scavengers. The scale of ghost fishing mortality is difficult to quantify, but for various US crab and lobster fisheries, it is estimated that of the millions of pots deployed each year, the proportion lost or abandoned is typically of the order of 10–50%, potentially accounting for substantial mortality (Bilkovic *et al.*, 2012). The most effective preventative measure is clearly to minimize gear loss, but ensuring adequate marking of gear and prohibiting the discarding of gear at sea would also help. It may also be feasible to incorporate biodegradable components that in time render the gear ineffective, such as escape panels in crab pots (Bilkovic *et al.*, 2012).

## Bycatch and Gear Modification

Fishing methods can be very unselective and result in large quantities of bycatch being caught and usually killed. Bycatch mortality of *K*-selected species may represent a greater threat than the mortality experienced by the target species. Examples of this include bycatch of chondrichthyans, sea turtles, seabirds, and marine mammals (examined in later chapters).

Certain measures can help address the issue of unwanted catch, such as the use of more selective gear and excluder devices (Kennelly & Broadhurst, 2002), and implementing no-discard policies as the norm so that any discarding at sea has to be adequately justified. Trawl gear can be modified to make it more selective (Sacchi, 2008). This can be achieved most simply by increasing the codend mesh size or by the inclusion of square-mesh panels in the codend as, unlike diamond-shaped mesh, square mesh does not close up as the net fills. Other modifications include the use of sorting grids, escape panels, and separator panels. By exploiting behavioural differences between species, separator trawls with upper and lower compartments have been used to retain target species whilst allowing non-target species to escape. Excluder devices can be incorporated in trawls to reduce the incidental catch of turtles and pinnipeds (see Chapters 7 and 9).

Semi-pelagic trawls, in which the whole trawl, apart from the otter boards, is raised off the seabed by means of floats on the headline and weights on the footrope, can lessen damage to benthos and the amount of bycatch whilst maintaining catches of target species. Brewer *et al.* (1996) assessed three differently rigged versions of a demersal trawl at sites in northern Australia. One was rigged to fish in a standard demersal configuration with its footrope on the seabed, whilst the other two fished semi-pelagically with their footropes raised (0.4 or 0.8 m) above the seabed. Catches of the main target species (lutjanid snappers) by the three trawl types were not significantly different, but the semi-pelagic trawls took significantly lower numbers of bycatch species, including sponges and other epibenthic invertebrates. More selective methods, such as traps or longlines, may be feasible for some fisheries and be less damaging to the benthos, although in the case of longlines, methods used also need to minimize incidental mortality of seabirds (see Chapter 8). Assessing the environmental impact of trawling is, however, severely constrained by the scarcity of unfished sites – a benefit of no-take marine protected areas.

## Fisheries Management

To address the problems of overfishing and environmentally damaging practices so as to maintain fisheries that are sustainable in terms of target species and associated biodiversity requires concerted national, regional, and international action on a number of fronts.

UNCLOS and subsequent international instruments provide the framework for fisheries management at a global level. The partitioning of ocean space into waters under national jurisdiction and the high seas governs a nation's rights to exploit fishery resources and its management and conservation obligations. Around 90% of the world's fish catch is estimated to come from EEZs (FAO, 2012). Management here is subject largely to the domestic legislation of the country in question, though ideally according to internationally agreed guidelines, such as the FAO Code of Conduct for Responsible Fisheries (see below). Whilst a relatively small amount of the global fish catch is from the high seas, the proportion has grown significantly since the mid-twentieth century. Managed under international and regional instruments, these fisheries also present major conservation issues.

Under the 1982 UNCLOS, the high seas are declared a global commons where all states, whether coastal or land-locked, have the right to fish. And to that end the convention includes provisions that call on parties to cooperate in the conservation and management of the living resources of the high seas (Articles 116–20). It became evident, however, that these general provisions were not sufficient as a basis for global fisheries management, and since the 1990s, building on the UNCLOS framework, further instruments have been elaborated, including the Compliance Agreement, the Fish Stocks Agreement, and the Code of Conduct for Responsible Fisheries (Box 5.2).

## Box 5.2 Global Fisheries Management Instruments and Entry into Force (EIF)

### Compliance Agreement (1993, EIF 2003)

Promotes compliance with conservation and management measures by fishing vessels (more than 24 m in length) on the high seas. Key requirements are that each party:

- ensures that fishing vessels entitled to fly its flag do not engage in any activity that undermines international conservation and management measures
- prohibits any vessel entitled to fly its flag to fish on the high seas unless it has been authorized, and such a vessel must observe the authorization conditions
- maintains a record of vessels entitled to fly its flag and authorized to fish on the high seas and submits a list of authorized vessels to FAO

### Fish Stocks Agreement (1995, EIF 2001)

Sets out principles for the conservation and management of straddling and highly migratory fish stocks. Covers a range of general principles, including:

- adopt measures, based on the best scientific evidence, to ensure long-term sustainability of fish stocks
- take measures to prevent or eliminate overfishing and excess fishing capacity
- apply the precautionary approach
- assess the impacts of fishing on target stocks and on associated or dependent species
- adopt conservation and management measures for species belonging to the same ecosystem
- protect marine biodiversity
- minimize waste, discards, and bycatch
- collect and share complete and accurate fisheries data, including on bycatch
- promote and conduct scientific research
- implement and enforce conservation and management measures

## Code of Conduct for Responsible Fisheries (1995)

Sets out principles and international standards of behaviour for responsible practices to ensure the effective conservation, management, and development of living aquatic resources with due respect for the ecosystem and biodiversity. Addresses in particular:

- general principles of management to promote sustainability of resources
- fisheries management measures
- fishing operations
- aquaculture development
- the integration of fisheries into coastal area management
- post-harvest practices and trade
- fisheries research

International Plans of Action

- conservation and management of sharks
- incidental catch of seabirds in longline fisheries
- management of excess fishing capacity
- illegal, unreported, and unregulated (IUU) fishing

## Port State Measures (2009, EIF 2016)

Aims to prevent, deter, and eliminate illegal, unreported, and unregulated fishing.

- applies to fishing vessels seeking entry into a designated port of a state which is not their flag state
- vessels engaged in IUU fishing are prohibited from using these ports and landing their catches
- blocks the products of IUU fishing from reaching markets

The living resources of the high seas relate to those beyond national jurisdiction. But many species migrate between EEZs and international waters, so efforts by a coastal state to conserve such a stock can be thwarted. The management of these trans-boundary fishery resources was inadequately addressed by UNCLOS. Given that they include some of the most commercially valuable species, the question of overfishing of high-seas stocks had become a source of contention between coastal states and distant-water fishing nations. To address this issue, the UN in 1995 concluded the Fish Stocks Agreement. The agreement focuses on stocks that straddle national and international waters and highly migratory stocks that migrate large distances over several national boundaries and the high seas. The distinction between straddling and highly migratory stocks is not always clear, but straddlers include various stocks of cod, mackerel, halibut, and pollock, whilst highly migratory species include most tunas, swordfish, sauries, oceanic sharks, marlins, and other billfish. The Fish Stocks Agreement also developed and strengthened the role of regional bodies as the means to effect the management of high-seas fisheries resources. There are now 18 regional fisheries management organizations (RFMOs), which together oversee management of most of the high seas. About half the RFMOs are concerned only with certain species, such as tunas (see Chapter 6), whereas the others have broader responsibilities for the stocks in their region.

The Compliance Agreement requires vessels fishing on the high seas to comply with international conservation and management measures. In particular, the agreement addresses the problem of fishing vessels operating under a flag of convenience (FOC). This means that the vessel is registered in a country that is not the vessel's country of ownership and is a way to flout environmental (and other) obligations. UNCLOS requires there to be 'a genuine link' between the nationality of a vessel and the flag it flies (Article 91), with the country that registers the vessel being responsible for it. But a number of countries allow vessels to operate under their flag and overlook violations of environmental and other laws, including RFMO and other fisheries management agreements. Valuable high-seas stocks, such as tunas and Patagonian toothfish, although included in international agreements, are nevertheless being illegally plundered by FOC vessels.

The Code of Conduct for Responsible Fisheries (FAO, 1995) is a voluntary code to address overfishing and environmentally damaging practices in all fisheries and has been influential in the subsequent development of a number of national and international initiatives concerned with sustainable fisheries. The code complements in particular the Fish Stocks Agreement and covers several main areas (Box 5.2). It takes a similar ecosystem-based approach in its general principles and fisheries management measures but is wider in scope and applies to waters both within and beyond national jurisdiction. Additional to the code are international plans of action to address particular issues, and there are currently four of these (Box 5.2).

Whilst these instruments provide an important basis for managing sustainable fisheries, they have failed to stem the decline of many stocks. Many nations have yet to ratify the Fish Stocks Agreement, the number of ratifications (now more than 80) being half that for UNCLOS. And RFMOs have in general proved ineffective in conserving fish stocks; two-thirds of stocks fished on the high seas and under RFMO management are either depleted or overfished (Cullis-Suzuki & Pauly, 2010a). Among the major problems are RFMOs setting catch quotas that are too high and the scale of illegal fishing. The Port State Measures that entered into force in 2016 will help reduce illegal fishing (Box 5.2).

Overall compliance with the Code of Conduct is also low. Although countries often endorse the code in their official policies on fisheries, there is a poor record of actual implementation. There appear to be various reasons for this, notably because of socio-economic implications and administrative and political inaction (Hosch *et al.*, 2011). An evaluation by Pitcher *et al.* (2009) of the 53 countries landing 96% of the global marine catch shows an average compliance score for the code barely above a fail threshold. Countries fared worst on such critical issues as introducing ecosystem-based management, controlling illegal fishing, reducing excess fishing capacity, and minimizing bycatch and destructive fishing practices. Using ecological indicators to quantify the ecosystem effects of fishing indicates that countries with higher levels of compliance with the code demonstrate significantly improved fisheries sustainability (Coll *et al.*, 2013). Implementing the Code of Conduct is voluntary, but such results emphasize the importance to countries, regardless of socio-economic situation, of actively putting into practice appropriate fisheries management measures to ensure sustainability of stocks and integrity of ecosystems.

It is unrealistic to expect a single management regime to suit all fisheries; there are often widely different biological and socio-economic factors that need to be taken into account for a particular location. But among the management tools that are most often used, at least in developed countries, are gear restrictions and modifications, closed areas, a reduction of fishing capacity, reductions in total allowable catch, and catch shares. A combination of tools is likely to be appropriate depending on the local context (Worm *et al.*, 2009) (Box 5.3).

## Box 5.3 Fisheries Management Tools

Fisheries management actions mainly comprise measures to limit the weight of catch that can be taken, measures to control the catching power of fishers, and technical measures that control the catch that can be made for a given effort (Jennings *et al.*, 2001). Among the more important tools are:

- Total allowable catch (TAC) that can be taken from a stock
- Individual quotas so as to allocate a TAC
- Licences to restrict the number of boats or fishers
- Gear restrictions (e.g. mesh size, ban on gear that is too damaging)
- Vessel restrictions (e.g. vessel size)
- Restrictions on the size of fish that can be caught or landed
- Time closures (e.g. to protect particular life-history stages)
- Area closures (e.g. to protect spawning areas, vulnerable habitat)

In reviewing a range of fisheries from around the world, Dankel *et al.* (2008) identified aspects of management that are generally associated with successful and unsuccessful management. Important qualities of successful management include:

- Robust management. Management that will execute decisions that may be unpalatable but also show flexibility where, for instance, there is a high degree of uncertainty.
- Biological limits. Reference points to regulate fishing mortality that are based on accurate assessments and that are clear and quantifiable.
- Implementation. Management decisions based on stock assessments and conservation strategies are actually implemented.
- Consensus. A management plan based on consensus between stakeholders and with clear objectives defined from the outset.

Single-nation, single-stock fisheries were often the more successful, whereas those with management problems tend to be characterized by fleet overcapacity, unclear objectives, and illegal activity. There must be substantial reductions in the world's fishing fleet capacity and level of illegal fishing, and a major shift in fisheries management towards more precautionary and ecosystem-oriented approaches.

## Reduction of Effort

Stock assessment methods have weaknesses, and there can be a lack of scientific consensus over safe levels of exploitation (Ludwig *et al.*, 1993). But even when adequate and unambiguous data are available, poor management often results from policy makers and politicians failing to act on the results of scientific assessments and allowing overfishing and overdevelopment of the industry. Using data for the period 1987 to 2011, O'Leary *et al.* (2011) examined the extent to which European politicians adopted the scientifically recommended total allowable catch. They analysed the decisions for 11 quota-managed species across nine management zones and found that in 68% of cases politicians set the TACs above those recommended from scientific stock assessments and on average a third higher than scientifically recommended levels. And in no cases when a moratorium was advised was that adhered to. Simulation modelling indicated that the politically adjusted TACs markedly increased the probability of a stock collapse.

There are typically strong political and economic pressures not to reduce catches. Fishing needs to continue for loans to be repaid, but this requires further investment in technology as stocks dwindle. Such pressures act like a ratchet, leading inexorably to overcapacity and overfishing (Ludwig *et al.*, 1993; Pitcher, 2001). A huge imbalance exists between the cost of some wild capture fisheries and the revenues generated, maintained through direct and indirect subsidies. Many governments subsidize their fisheries. An assessment by Mora *et al.* (2009) indicates that 91% of the world's EEZs have fisheries sectors that rely to some extent on subsidies. Global fisheries subsidies have been estimated (for 2003) at $25–29 billion per year (Sumaila *et al.*, 2010). Most subsidies promote overcapacity and permit overexploited, uneconomic fisheries to continue. But since world fisheries provide income to some 200 million people directly or indirectly, there is socio-economic pressure on governments to maintain unsustainable practices. There is, however, no room for major increases in world catches of traditional species, and a drastic reduction in effort will have to occur. The world fishing fleet is perhaps two to three times larger than that needed for sustainable levels (Pauly *et al.*, 2002). Some countries have adopted policies to curb overcapacity of their fishing fleets, although the success of such schemes appears to vary considerably. EU fishing capacity has declined in terms of numbers of vessels and gross tonnage, whereas although China has reduced the number of its vessels, the fleet's combined engine power has in fact increased (FAO, 2014).

## Use of Catch Shares

The common property nature of fishery resources is often identified as a fundamental reason for the demise of fisheries in general, in providing little incentive for individual fishers to conserve stocks, at least where economic factors prevail. Stock depletion and the failure of international fisheries commissions to provide effective management were important in effecting UNCLOS. EEZs greatly extended the responsibility and opportunity of nations to manage coastal marine resources, but marine enclosure has not averted the 'tragedy of the commons' situation; in many cases it has simply meant that open access was no longer global but national.

There are various ways to try to regulate fishing effort, such as by limiting entry or the number of licences, by areal and/or seasonal closures, and by restrictions on gear. But by and large, such measures have failed to control overfishing. Another approach is to manage fisheries using incentives rather than input controls, in particular by providing fishers with exclusive rights to a certain fraction of a total allowable catch or to fishing a certain area. It is argued that this should promote a greater sense of ownership and less incentive to overfish, and by giving greater certainty about catch, fishers can better plan their fishing season. The main quota share system is the use of individual transferable quotas (ITQs), where fishers have individual harvesting rights that can be traded (Grafton, 1996). There are now about 250 fisheries, mostly large-scale ones, that use some form of ITQ management (Thébaud *et al.*, 2012; van Putten *et al.*, 2014). An analysis of fisheries worldwide by Costello *et al.* (2016) indicates that rights-based fishery management has a key role to play in the recovery of global fisheries. ITQs are primarily instruments for promoting economic efficiency rather than conservation (Hanneson, 1996) and as such still need to be used in conjunction with tools to protect ecosystem structure and function.

## Customary Marine Tenure

ITQ systems were promoted in response to the assumption that the institution of common property inevitably leads to overfishing. Ruddle *et al.* (1992) maintain that this reflects the dominating influence of Western fisheries management and that whilst such a premise may work for offshore fisheries, it is not necessarily applicable to small-scale inshore fisheries. They discuss systems of customary marine tenure, of which there are many examples still in use in the Pacific basin, whereby local communities, within the context of often complex socio-cultural arrangements sustainably manage fishery resources. There is a wide range of such systems, but typically they share a number of general features. The local community, in effect the sole owner, controls the marine habitats in question and the exploitive uses. Control is vested in traditional authorities with the sea rights of individuals dependent on social factors, exploitation is governed by use rights applicable to defined territories, and regulations are enforced. The community would typically have an intimate knowledge of the area and its natural history and an awareness of the potential to deplete its natural resources. Whilst such systems are based on a community's own body of customary law, they are not inflexible to changing circumstances. There has been a resurgence of interest in customary management systems and how they might be integrated with practices of Western fisheries to produce hybrid systems for managing small-scale inshore fisheries (Johannes, 2002; Aswani & Ruddle, 2013). The role of customary marine tenure in coral reef fisheries is discussed later (see Chapter 13).

It is important in this regard to bear in mind that the great majority of the world's fisheries are in fact small-scale operations and occur mainly in the developing world. (FAO has no agreed definition, however, of what constitutes a 'small-scale fishery'.) These stocks are unlikely to be assessed and managed (as these are costly procedures), and in general they appear to be in poor condition. Reforming small-scale fisheries is urgently needed to rebuild fisheries that are important to so many local communities (Costello *et al.*, 2016).

## Ecosystem-Based Fishery Management

It is essential that fisheries become integrated into wider ecological approaches to management and conservation. Fisheries management has traditionally taken a single-species approach, seeking to maximize catch of target species with little or no consideration of the broader implications for that community, habitat, and ecosystem. Ecosystem-based fishery management (EBFM), also known as the ecosystem approach to fisheries (EAF), aims to be more holistic by taking account of indirect impacts whilst still needing to optimize economic value. EBFM seeks to maintain the key attributes of the ecosystem in question in terms of its structure and function and thereby reduce the likelihood of irreversible changes. The approach should thus minimize ecosystem impact, such as from excessive incidental capture and use of fishing practices that are damaging to the habitat. This is likely to require increased effort in researching the implications of particular fishing practices and adopting precautionary practices where such understanding is not yet adequate (Pikitch *et al.*, 2004). We are still, however, some way from EBFM as such. Implementing single-species management is proving challenging enough, and achieving MSY or other metric of fishing level for numerous species in a system may be impracticable (Hilborn & Ovando, 2014). New tools need to be developed for overall ecosystem management, such as suitable indicators for defining overfishing in an EBFM context. The FAO has adopted an ecosystem approach to fisheries to help implement its Code of Conduct for Responsible Fisheries and has developed a suite of tools to assist in implementing this approach. The tools, targeted mainly at fisheries agencies within developing countries, cover four main steps:

- Step 1: planning initiation and scope
- Step 2: identification of assets, issues, and priorities
- Step 3: development of the EAF management system
- Step 4: implementation, monitoring, and performance review

For each step there is a list of tools available in the EAF toolbox that span a range of complexity. A set of tools can then be selected appropriate to local circumstances (Fletcher & Bianchi, 2014). More comprehensive, integrated approaches to marine resource management are consistent with the broad goals of conservation and are already being used and developed, notably under coastal zone management and the large marine ecosystem concept. The latter, whilst developed from fisheries management, has potential to guide conservation of large, ecologically coherent areas (see Chapter 17).

Pitcher (2005) proposes a process of restoration ecology of marine ecosystems to rebuild fisheries, a so-called 'back-to-the-future' (BTF) approach (Fig. 5.6). This uses simulation models of past ecosystems for chosen points in time (based on major historical changes in the fisheries) that are reconstructed from a variety of sources (e.g. archaeology, historical records, local and traditional environmental knowledge). The various candidate models are then evaluated in terms of ecological, economic, and social benefits that each might deliver and tested for the implications for local extinctions, climate change, and other sources of uncertainty. Next, the sustainable fishing options are determined for the potential, at least partially restored, ecosystems. Various criteria are used to ensure that any fishery is sustainable and undertaken responsibly, for example, to minimize bycatch and damage to habitat, include aboriginal fisheries and traditional target species, minimize risk to charismatic species, exclude fisheries based on juveniles, maximize community support, demonstrate sustainability (e.g. 100-year simulations), and incorporate adaptive management. Candidate fisheries could be assessed against the FAO Code of Conduct for Responsible Fisheries. The instruments needed to achieve a policy goal are determined, which may include use of marine protected areas, restrictions on time and places of fishing, effort controls, and quotas. In terms of restoration targets, BTF adopts the concept of an 'optimum restorable biomass', which recognizes that as some depletion will result from fishing, a biomass restoration target is more appropriately that of a historical rather than a restored past ecosystem. Also central to the BTF approach is strong community and stakeholder involvement, without which management is likely to fail. By way of example, Pitcher (2005) discusses simulation models developed to approximate the marine ecosystem for northern British Columbia at distinct periods in fisheries development (in this case 1750, 1900, 1950, and 2000 AD) and to evaluate the trade-offs between the economic benefits and biodiversity impacts for different restoration goals.

## Certification Schemes

Consumer awareness of fishing impacts and a demand for seafood that is appropriately sourced are likely to play an increasingly significant role in marine conservation. In particular, products can be certified to show that they come from sustainable fisheries that follow precautionary management practices, as specified in the Code of Conduct and other instruments. An analysis by Stokstad (2011) indicates that certified fisheries do show improved management, although the question remains

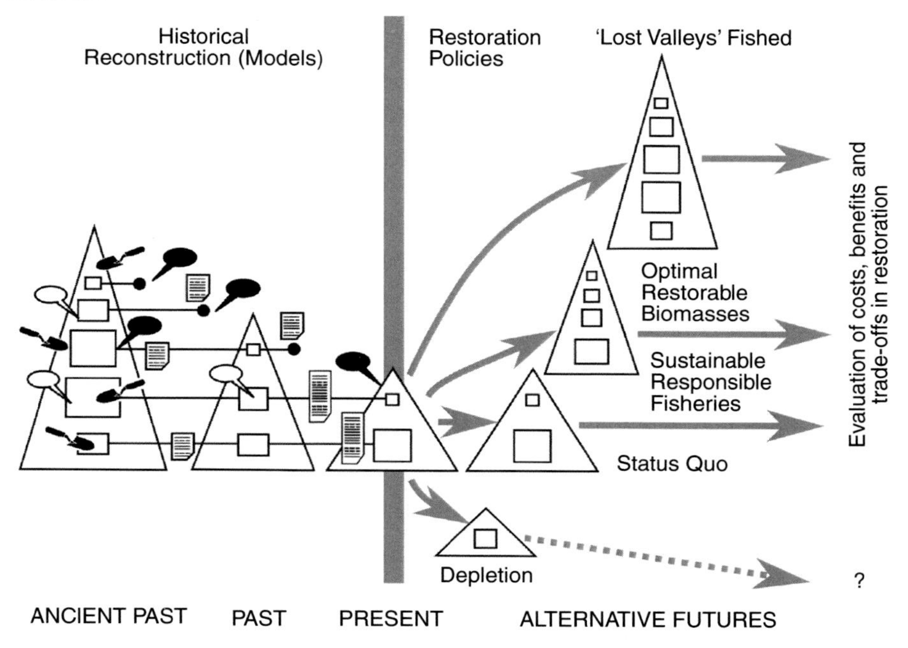


**Fig. 5.6** Diagram illustrating the 'back-to-the-future' (BTF) concept for the adoption of past ecosystems as future policy goals. Triangles at the left represent a time series of historical ecosystem models, constructed at appropriate past times before the present (thick grey vertical line), where the vertex angle is inversely related and the height directly related to biodiversity and internal connectance. Time lines of some representative species in the ecosystems are indicated; size of the boxes represents relative abundance, and solid circles represent local extinctions (= extirpations). Sources of information for constructing and tuning the ecosystem models are illustrated by symbols for historical documents (*paper sheet symbol*), data archives (*tall data table symbol*), archaeological data (*trowel*), the traditional environmental knowledge of indigenous peoples (*open balloons*), and local environmental knowledge of coastal communities (*solid balloons*). At right are alternative future ecosystems, representing further depletion, the *status quo*, or restoration to 'lost valleys' that may be used as alternative policy goals. Restored 'lost valleys' may be fished with sustainable, responsible fisheries designed according to specified criteria and aiming at optimal restorable biomasses determined using objective quantitative policy searches. Final choice of BTF policy goals is made by comparing trade-offs, costs, and benefits among possible futures using socio-economic and ecological objectives agreed upon among industrial and small-scale fishers, governments, conservationists, coastal communities, and other stakeholders in order to maximize compliance. Diagram does not show evaluation of risks from climate fluctuations and model parameter uncertainty. Springer *Hydrobiologia*, The sea ahead: challenges to marine biology from seafood sustainability, Volume 606, 2008, page 173, Pitcher, T.J. © Springer Science+Business Media B.V. 2008. With permission of Springer.

to what extent this is a direct cause. One of the main eco-labelling schemes is that of the Marine Stewardship Council, founded in 1997, which has now certified about 200 fisheries. Some certifications have proved controversial. Currently only a relatively small proportion of the global catch is certified, with a focus on particular species and markets. Interest in eco-labelling is strongest in wealthy, environmentally conscious markets of Europe and North America, whereas developing countries are at present poorly represented in such schemes, although demand for eco-labelled products is growing rapidly (FAO, 2014). Some eco-labelling schemes are concerned with single issues, such as the protection of dolphins in tuna fisheries (Box 9.2).

## Marine Protected Areas

Fishery closures have been used for centuries in Western societies to restrict fishing temporarily in certain locations so as to protect, for instance, spawning stock, nursery grounds, or other essential fish habitat (Fogarty *et al*., 2000). There is now increasing support for the view that, as an integral part of EBFM, selected areas should be closed to fishing on a long-term basis to provide refugia for impacted species and habitat. This is one of the potential functions of marine protected areas (MPAs). The broader use of MPAs as a means to conserve biodiversity in general is discussed later (see Chapter 15). However, as fishery management adopts more ecosystem-oriented approaches, there is increasing commonality between the goals of fisheries management and those of wider marine conservation (Jennings, 2009). The need to take a wider view of fisheries management and conserve not only target stocks but also their habitat is now being incorporated into legislation. Maintaining benthic habitat complexity needs to be incorporated as an aim of fisheries management, such as through the use of protected areas or restrictions on or modifications of gear. For example, the key legislation governing fisheries management in US waters, the Magnuson-Stevens Fishery Conservation and Management Act, requires that fishery management plans describe and identify 'essential fish habitat', defined as those waters and substrata necessary for spawning, breeding, feeding, or growth to maturity, and 'minimize to the extent practicable adverse effects on such habitat caused by fishing' (under the Sustainable Fisheries Act of 1996, an amendment to the Magnuson-Stevens Act).

No-take MPAs – areas permanently closed to fishing – may help conserve exploited stocks and their habitats (Gell & Roberts, 2003; Allsopp *et al*., 2009). Heavily exploited species – particularly relatively sedentary species – can attain higher population densities and larger average body size within no-take MPAs than in fished areas (e.g. Lester *et al*., 2009). MPAs may then benefit fisheries as a result of adults moving into adjacent unprotected areas, provided this spillover more than compensates for the loss of fishing ground in the MPA. Also, MPA populations may export eggs and larvae to adjacent fishing grounds, especially given that the protected females will tend to be larger and thus disproportionately more fecund.

Kerwath *et al*. (2013) found from catch per unit effort (CPUE) data that following the establishment in 1990 of the Goukamma MPA on the South African coast, the adjacent fishery for Roman seabream (an endemic species) recovered. CPUE began to increase one year after the MPA was established and after 10 years was twice the pre-reserve rate. The pattern of recovery is best explained by the spillover of adult fish followed by larval export from the MPA (Fig. 5.7). There was no evidence that establishing the 40-$km^2$ MPA disadvantaged fishers by resulting in an overall reduction in catch or increased travel distances. Similarly, Goñi *et al*. (2010) found that spillover of adult spiny lobster from a western Mediterranean marine reserve provided a net benefit to the local fishery, offsetting the loss of fishing grounds in the MPA. The spillover resulted in gains of >10% in mean annual catch, mainly because the emigrant lobsters were larger. A fisheries model analysis by Buxton *et al*. (2014) indicates, however, that any net benefit from spillover depends on the quality of fisheries management: no-take MPAs are unlikely to deliver a spillover benefit if recipient areas are already effectively managed but may be beneficial if stocks are severely depleted.

No-take MPAs have yet to be widely accepted and implemented in fisheries management; indeed, they are generally opposed by fishers. Fishers need to know what proportion of habitat may be designated as an MPA and the level of exploitation that can be sustained outside. MPAs that incorporate different levels of protection may be an option if an entirely no-take MPA is unachievable. Although not as effective as total exclusion, partial protection can still be advantageous in terms of greater density and biomass of fish (Sciberras *et al*., 2015).

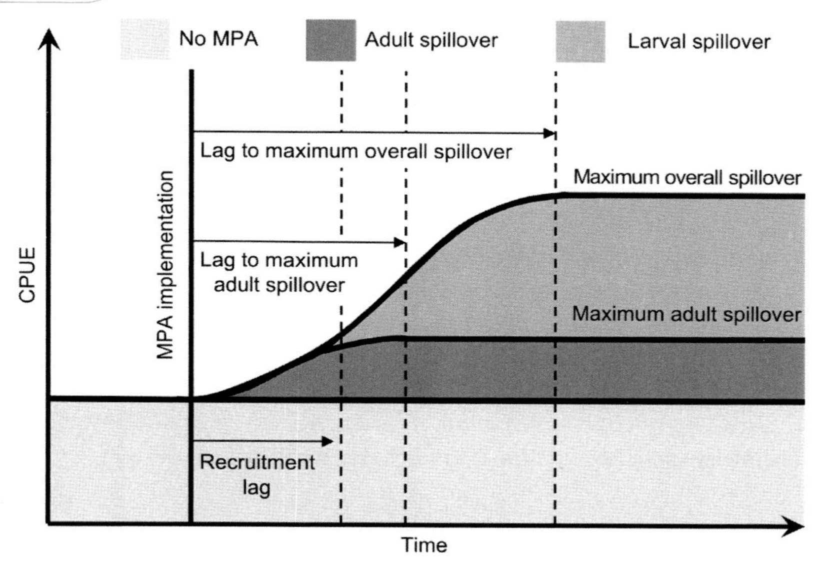


**Fig. 5.7** Possible mechanisms responsible for the increase in fisheries catch per unit effort (CPUE) around MPAs. Spillover of adult fish begins once there is a density gradient across the MPA boundary, resulting in an increase in CPUE in the fished area. CPUE reaches an asymptote when the biomass of adult fish inside the MPA attains its maximum. By contrast, larval export can begin only after the larvae spawned by the increasing population inside the MPA have settled and grown to recruitment size. The combined effect reaches its limit when maximum spawning capacity inside the MPA has been restored. Reprinted by permission from Macmillan Publishers Ltd: *Nature Communications*. Kerwath, S.E., Winker, H., Götz, A. & Attwood, C.G. (2013). Marine protected area improves yield without disadvantaging fishers. *Nature Communications*, 4, article 2347, copyright 2013.

Protected areas may be easier to apply than other fisheries management measures. Nevertheless, the establishment of MPAs can raise a number of ecological, socio-economic, management, and regulatory issues (Hilborn *et al.*, 2004). In particular, fishing effort displaced by MPAs may result in damaging impacts on target and non-target species and habitats in areas outside if such effects have not been taken into account and mitigating measures put in place. Increased pressure on non-MPA areas will be exacerbated if issues of fishing overcapacity have not been addressed. And the extent to which vulnerable habitat types are captured within MPAs will have an important bearing on the overall environmental impact that results from any such reallocation of effort. Reallocation of fishing effort may especially affect small-scale fishers, although not necessarily compromising their socio-economic well-being (Stevenson *et al.*, 2013). But the incentive to poach in MPAs may be considerable.

Fisheries, as we have seen, are subject to a high degree of uncertainty, notably from recruitment variability and the difficulties of estimating safe levels of fishing mortality. Even for the most data-rich fisheries, stock assessments may prove inadequate, given the level of variability, and unforeseen collapses can occur. Fisheries managers are increasingly exploring more precautionary strategies that can accommodate inherently high levels of uncertainty. In this regard MPAs offer a relatively straightforward method of insurance, as a way to buffer a proportion of the population from possible collapse.

However, MPAs for fisheries management still need to be used in conjunction with other tools, such as restrictions on vessels, gear, areas and times for fishing, and catch quotas or other rights. One strategy for small-scale coastal fisheries may be to link no-take MPAs with adjacent, designated areas where fishers have access rights to enhanced biomass resulting from spillover and larval export (Afflerbach *et al.*, 2014; Barner *et al.*, 2015).

Individual MPAs for fisheries management can be small (< 5 km across) to provide for effective spillover of sedentary target species and export of larvae to fished areas (Sale *et al.*, 2005; Green *et al.*, 2014). But overall, for sustainable fisheries, MPAs may need to occupy a considerable proportion of fisheries habitat. Where possible, networks of replicated MPAs are needed based on the habitat requirements of the targeted species to provide insurance against catastrophic events and stock collapse but also to enable the effectiveness of MPA management strategies to be tested. Studies indicate that for protection to deliver maximum benefits, around 20–40% of fishing ground area needs to be included in MPAs, depending on fishing pressure and the effectiveness of fisheries management outside the reserve (Gell & Roberts, 2003; Green *et al.*, 2014). Presently less than 1% of the world ocean is within no-take MPAs (Thomas *et al.*, 2014). Whilst the annual cost of maintaining 20–30% of the world's oceans as MPAs would be substantial, it is estimated to be similar to the total amount currently spent on subsidies to fisheries (Cullis-Suzuki & Pauly, 2010b).

## Artificial Reefs

Managing sustainable fish stocks depends on the conservation of the necessary habitat for all life-history stages. Major arguments advanced for the conservation of, for example, estuaries, coastal wetlands, temperate and tropical reefs, and reef-like structures at shelf and slope depths derive from the known use of such habitats by commercially important fish species. In some cases, such as with seagrass and wetland sites, there may be opportunities to restore degraded habitat (see Chapter 12) and possibly to reintroduce species that have become locally extinct. A further option to enhance biodiversity, including fish populations, is to provide alternative habitat using either natural or artificial materials, an approach that has been used particularly to make reef or reef-like habitat. It has long been known that submerged objects such as wrecks can be good fishing sites. There may be various reasons for this phenomenon: fish seeking shelter from currents or predators, using the structures for orientation, or exploiting the colonizing epibiota as food.

Although the building of floating and seabed structures to attract fish and shellfish dates back centuries, interest in such techniques has grown dramatically in recent decades (Clark, 1996). Fish aggregating devices, floating structures that are moored or drifting, are widely used to attract pelagic fish for capture, particularly in tropical and subtropical waters. Nearly half of the world's tropical tuna catch is now caught by purse seiners fishing on drifting fish aggregating devices (Fonteneau *et al.*, 2013) (see Chapter 6). Here, however, we focus on artificial reefs: submerged structures deliberately placed on the seabed to mimic characteristics of a natural reef (OSPAR, 1999). Scrap materials, such as old tyres, building rubble, cars, and vessels, have in the past commonly been used to construct artificial reefs, and there is interest in turning disused oil production platforms into artificial reefs. Often, however, such waste materials have been used with little regard for their suitability. Care is also needed that they are not a source of contaminants and that their use is not essentially an opportunity to dispose of unwanted items under the guise of conservation. More recently, the trend has been towards the construction of artificial reefs made from purpose-built modules, usually made from concrete and designed to provide optimum habitat characteristics.

Artificial reefs are widely promoted as a means of enhancing fisheries – at least at a local level – with some countries funding major programmes. Japan has made substantial investment since the mid-1970s to develop new fishing grounds based on artificial reefs. And in the Gulf of Mexico, decommissioned oil and gas structures are the basis of a major artificial reef programme for fisheries enhancement with more than 120 decommissioned platforms used in the largest of these rigs-to-reef programmes (Kaiser, 2006).

Whilst artificial reef technology has become a major industry, important questions about the effect of artificial reefs on fish populations have yet to be fully resolved. A critical issue is whether artificial reefs simply attract fish from natural reef areas or actually enhance production (Pickering & Whitmarsh, 1997). If biomass is simply redistributed from natural habitat to artificial reefs, then this may increase the proportion of stock that can be exploited unless fishing at artificial reef sites is not suitably managed. However, an artificial reef may attract fish but also enhance production by providing greater access to food and shelter. That artificial reefs can enhance local production is indicated by a study of the largest artificial reef system in the Mediterranean, in the Bay of Marseille. Using stable isotopes as chemical tracers, Cresson *et al.* (2014) showed the importance of artificial reefs as a source of food (primarily invertebrate filter feeders) for fishes.

Artificial reefs have been developed primarily for artisanal and recreational fisheries, but they have a number of potential functions besides fisheries enhancement (Seaman, 2007). By acting as an obstruction, artificial reefs can protect sensitive areas from trawling, such as seagrass beds or nursery grounds for juvenile fish. Most of the artificial reefs in the Mediterranean help protect seagrass beds from trawl damage (Jensen, 2002). Artificial reefs can be used in conjunction with MPAs or as no-take reserves for conservation of coastal ecosystems. Some artificial reef systems are designed for aquaculture to provide habitat for commercially important seaweeds and shellfish, such as kelps, mussels, abalone, lobsters, and urchins. They may also serve to promote recreational diving and provide acceptable alternatives to natural areas of high conservation value, for biodiversity conservation, and as mitigation for destroyed habitats. However, an assessment of artificial reef performance indicated that only half the case studies met their objectives (Baine, 2001). An artificial reef may not be the most appropriate solution, but if it is, clear objectives, good planning, and ongoing management will be essential for its success. Among the main reasons for artificial reefs failing to meet objectives are size – most being probably far too small to be effective – siting, inadequate monitoring, and illegal fishing.

## Aquaculture

Aquaculture – the farming of aquatic organisms – is increasingly advocated as a way to help bridge the gap between the yield from wild capture fisheries, most of which are fully or overexploited, and the rising demand for fish. Marine aquaculture production has risen dramatically in recent years and now amounts to some 27 million tonnes annually of invertebrates and fish and a similar amount of seaweeds (FAO, 2016). Seaweed aquaculture is undertaken almost entirely in Asia with the main species being *Kappaphycus alvarezii, Eucheuma* spp., and *Saccharina japonica* (together nearly 70% of production in 2014). Farmed seaweeds are grown mainly as a source of carrageenans (polysaccharides widely used in food and other industries) and for direct consumption. Numerous countries practise animal mariculture. Molluscs account for most of the global production (60% in 2014), followed by finfish (24%) and crustaceans (16%). Bivalves (*Ruditapes, Crassostrea*), salmon (*Salmo, Oncorhynchus*), and shrimps (*Penaeus*) are among the chief taxa that are farmed.

Aquaculture is not without its environmental costs. Some forms of aquaculture in fact exacerbate the global fishery crisis because of the high levels of fishmeal and oil used in the feeds. Particularly significant is China's rapidly growing aquaculture sector, which is increasingly moving from traditional culture systems that do not need formulated feeds to the culture of high-value species that require feeds, such as penaeid shrimps. As the largest importer of fishmeal, China accounts for about one-third of the world trade in fishmeal (Cao *et al.*, 2015). In the case of carnivorous fish, such as salmon, their aquaculture requires up to five times as much fish biomass as feed as is produced (Naylor *et al.*, 2000). The demand for fishmeal and fish oil for aquaculture feeds is likely to have a major influence on the future state of capture fisheries and underlines the importance of finding alternative sources for feed ingredients. Possibilities here include recovering ingredients from fish-processing wastes and using oils from terrestrial plants and microorganisms that are rich in omega-3 fatty acids, providing that the agricultural practices themselves are sustainable (Allsopp *et al.*, 2009; Naylor *et al.*, 2009; Cao *et al.*, 2015).

Destruction and degradation of coastal habitats as a result of mariculture developments are of particular concern. Sea-cage rearing of salmonids (especially salmon) has become a major industry in a number of cool temperate regions, notably in Norway and Chile (FAO, 2016). Inlets and fjords are among the most suitable sites for this form of mariculture, which can present significant implications for coastal management and conservation (Fig. 5.8). Intensive fish farming can generate large amounts of nitrogen-rich organic waste from faeces and uneaten food, which then accumulate on the seabed. The degree of enrichment and benthic impact depend on the farm and its operation (stocking density, farm size, food conversion ratio) and characteristics of the site (water depth, current strength, and sediment mud content). Typically, impacts occur up to 40–70 m around farms. In the most polluted zone directly under a farm, the sediments can have high sulphide concentrations and support only an impoverished benthic fauna of opportunistic species (Giles, 2008). And increased levels of dissolved nutrients from excretion and released from reducing sediments may enhance primary production in enclosed waters. Limiting the impact of fish farming on the benthos and on water quality may be possible through husbandry practices and careful site selection. Similar issues arise from the intensive culture of molluscs, notably mussels and oysters, grown on hanging long lines or racks or from rafts. Deposition of faeces and pseudofaeces (rejected particles) can similarly result in organic enrichment of the sediment and attendant impacts (Cranford *et al.*, 2009).

In tropical coastal areas, the building of aquaculture ponds is a major cause of mangrove destruction (see Chapter 12). Mangroves provide shelter and nursery areas for the juveniles of many coastal and

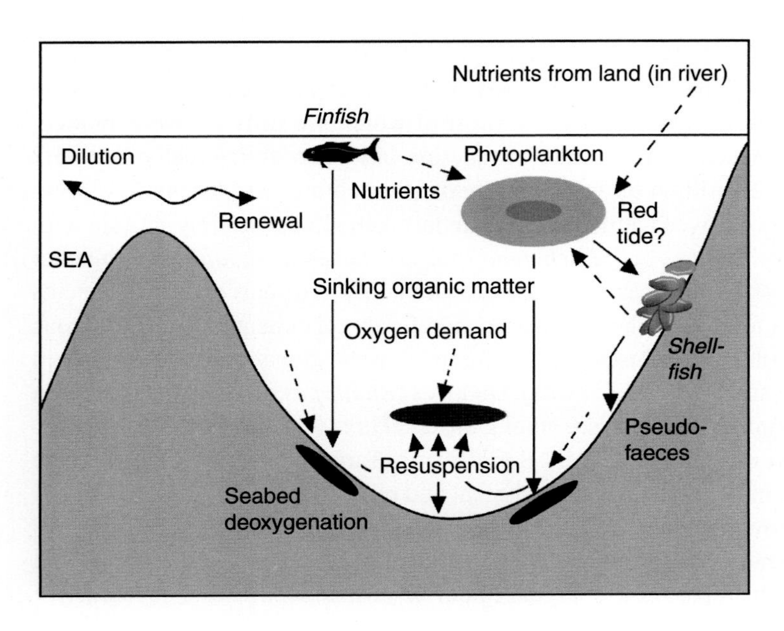


**Fig. 5.8** Effects of aquaculture in a fjord. Springer *Aquaculture in the Ecosystem,* Fish farm wastes in the ecosystem, 2008, page 7, Tett, P. © 2008 Springer Science + Business Media B.V. With permission of Springer.

offshore finfish and shellfish species, and their loss can substantially reduce the landings from these wild-caught fisheries. For Thai shrimp ponds developed in mangroves, Naylor *et al.* (2000) estimated that for every 1 kg of shrimp farmed, 0.4 kg of fish and shrimp have been lost from the capture fisheries.

Other potential environmental impacts of mariculture relate to the use of bioactive compounds (e.g. antibiotics, hormones, pesticides, and antifoulants), the spread of disease, and the genetic impact of cultured fish on natural fish populations, given the increasing numbers that are intentionally or accidentally released. The parental brood stock used in hatcheries is often of mixed origin so the genetic makeup of resulting offspring may depart significantly from that of the indigenous population, and fitness of the native population may thus be diminished.

As a rapidly growing food sector there is a pressing need for aquaculture to tackle questions of sustainability and environmental impact. The FAO Code of Conduct for Responsible Fisheries (Article 9) outlines the principal issues that states should address in promoting responsible development and management of aquaculture, including questions of sustainability, ecosystem integrity, genetic diversity, introduction of non-native species, discharge of effluents, and use of drugs and chemicals. Eco-certification schemes for aquaculture are increasingly being developed, comparable to those for capture fisheries, which may help reduce negative environmental impacts and promote sustainable practices. At present, however, only a small share of aquaculture production is certified. Including small-scale producers, such as in Asia – where most aquaculture takes place – is likely to prove difficult for economic and technical reasons (Jonell *et al.*, 2013).

More sustainable are aquaculture systems that mimic natural ecosystem processes. In integrated multi-trophic aquaculture (IMTA), species from different levels of the food web are co-cultured to enable wastes to be recycled. For instance, seaweeds can be cultivated to make use of nutrients from shrimp or fish culture (Chopin *et al.*, 2007) or deposit-feeding sea cucumbers used to exploit biodeposits from oyster farms (Paltzat *et al.*, 2008).

Aquaculture has a role to play in conservation. In the case of threatened species that are amenable to cultivation there may be opportunities to use hatchery-reared individuals for restoration programmes, including re-establishing populations in marine reserves. Such programmes include those for abalone (*Haliotis* spp.), giant clams (Tridacninae), trochus (*Tectus niloticus*), queen conch (*Lobatus gigas*), and North Pacific salmon (*Oncorhynchus* spp.).

Giant clams, for example, are found in coral reefs of the Indo-West Pacific, where they contribute importantly to the structural complexity of reefs and provide habitat for associated organisms (Neo *et al.*, 2015). They have also for centuries been an essential part of the diet in many Indo-Pacific cultures. But more recently, giant clams have been subject to large-scale exploitation and poaching given the demand for clam meat from restaurants and live animals for the aquarium trade and the use of the shell for ornaments and jewellery. Giant clams are very vulnerable to overexploitation. Most of the 13 species are protected under Appendix II of CITES, and four of the species are listed as threatened on the IUCN Red List. The two largest species, *Tridacna gigas* and *T. derasa*, now appear to be extinct over large parts of their range. Methods have been developed for the production of juvenile giant clams in hatcheries for the restocking of reefs (Okuzawa *et al.*, 2008).

## Introductions

Many fish and shellfish species have been transported from one part of the world to another with the aim of developing aquaculture programmes or establishing wild populations. Attempts to improve fisheries by intentional introductions have been made for more than a century, particularly in the former Soviet Union, in North America, and on oceanic islands, although most such endeavours have failed. Attempts to introduce anadromous salmonids have, in particular, been the subject of enormous effort in many parts of the world (see Chapter 6).

Successful introductions are virtually irreversible and can have serious environmental repercussions, in particular impacts on native species and habitats, genetic effects from the interbreeding of wild and farmed stock (which may have been genetically modified), and the introduction of associated parasites and pathogens. It is, therefore, imperative that countries adopt strict measures to evaluate the likely risks of a proposed introduction. Also, since introduced species may well move into waters of adjacent countries, any legislation needs to apply to the whole continental area in question. Reducing the risk of harmful effects arising from introductions thus needs international cooperation in the application of methods for assessing candidate species and their likely environmental impact. A code of practice adopted by ICES sets forth procedures to reduce the risk of harmful effects from the introduction and transfer of marine and brackish water organisms ('introductions' being non-indigenous species intentionally or accidentally transported and released into an environment outside their present range and 'transfers' being releases within areas of established populations) (ICES, 2005). The code includes a series of recommendations that range from the initial planning stages to proceeding with an introduction and including steps to take before releasing genetically modified or polyploid organisms. Although the code was developed for ICES member countries (i.e. North Atlantic and adjacent seas), it has worldwide applicability.

In this chapter we have seen a number of the major problems confronting fisheries, such as the common property nature of resources, overcapacity in the industry, variability of fish populations, and the impacts of fishing on associated species and habitats, and the challenges these pose for effective management and conservation. The effect of human exploitation on living marine resources varies greatly depending on the scale and method of capture and the biology and ecology of the species in question. In the following four chapters we look at particular groups of marine vertebrates that have been heavily exploited or otherwise impacted by humans and the conservation issues they raise. And later we examine conservation issues relating to fisheries in the deep sea (Chapter 14) and in international waters (Chapter 17).

## REFERENCES


Afflerbach, J.C., Lester, S.E., Dougherty, D.T. & Poon, S.E. (2014). A global survey of 'TURF-reserves', Territorial Use Rights for Fisheries coupled with marine reserves. *Global Ecology and Conservation*, **2**, 97–106.

Aleem, A.A. (1972). Fishing industry in Ancient Egypt. *Proceedings of the Royal Society of Edinburgh (B)*, **73**, 333–43.

Allsopp, M., Page, R., Johnston, P. & Santillo, D. (2009). *State of the World's Oceans*. London: Springer.

Anderson, C.N.K., Hsieh, C.-h., Sandin, S.A., *et al.* (2008). Why fishing magnifies fluctuations in fish abundance. *Nature*, **452**, 835–9.

Aswani, S. & Ruddle, K. (2013). Design of realistic hybrid marine resource management programs in Oceania. *Pacific Science*, **67**, 461–76.

Auster, P.J. (1998). A conceptual model of the impacts of fishing gear on the integrity of fish habitats. *Conservation Biology*, **12**, 1198–203.

Baine, M. (2001). Artificial reefs: a review of their design, application, management and performance. *Ocean & Coastal Management*, **44**, 241–59.

Barner, A.K., Lubchenco, J., Costello, C., *et al.* (2015). Solutions for recovering and sustaining the bounty of the ocean: combining fishery reforms, rights-based fisheries management, and marine reserves. *Oceanography*, **28**, 252–63.

Baum, J.K. & Worm, B. (2009). Cascading top-down effects of changing oceanic predator abundances. *Journal of Animal Ecology*, **78**, 699–714.

Belgrano, A. & Fowler, C.W. (2013). How fisheries affect evolution. *Science*, **342**, 1176–7.

Bergman, M.J.N. & van Santbrink, J.W. (2000). Mortality in megafaunal benthic populations caused by trawl fisheries on the Dutch continental shelf in the North Sea in 1994. *ICES Journal of Marine Science*, **57**, 1321–31.

Berkeley, S.A., Chapman, C. & Sogard, S.M. (2004). Maternal age as a determinant of larval growth and survival in a marine fish, *Sebastes melanops*. *Ecology*, **85**, 1258–64.

Bilkovic, D.M., Havens, K.J., Stanhope, D.M. & Angstadt, K.T. (2012). Use of fully biodegradable panels to reduce derelict pot threats to marine fauna. *Conservation Biology*, **26**, 957–66.

Brewer, D., Eayrs, S., Mounsey, R. & Wang, Y.-G. (1996). Assessment of an environmentally friendly, semi-pelagic fish trawl. *Fisheries Research*, **26**, 225–37.

Buxton, C.D., Hartmann, K., Kearney, R. & Gardner, C. (2014). When is spillover from marine reserves likely to benefit fisheries? *PLoS ONE*, **9**, e107032.

Caddell, R. (2010). Caught in the net: driftnet fishing restrictions and the European Court of Justice. *Journal of Environmental Law*, **22**, 301–14.

Cao, L., Naylor, R., Henriksson, P., *et al.* (2015). China's aquaculture and the world's wild fisheries. *Science*, **347**, 133–5.

Carbines, G., Jiang, W.M. & Beentjes, M.P. (2004). The impact of oyster dredging on the growth of blue cod, *Parapercis colias*, in Foveaux Strait, New Zealand. *Aquatic Conservation: Marine and Freshwater Ecosystems*, **14**, 491–504.

Cheung, W.W.L., Watson, R., Morato, T., Pitcher, T.J. & Pauly, D. (2007). Intrinsic vulnerability in the global fish catch. *Marine Ecology Progress Series*, **333**, 1–12.

Chopin. T., Yarish, C. & Sharp, G. (2007). Beyond the monospecific approach to animal aquaculture – the light of integrated multi-trophic aquaculture. In *Ecological and Genetic Implications of Aquaculture Activities*, ed. T.M. Bert, pp. 447–58. Dordrecht, Netherlands: Springer.

Christensen, V., Coll, M., Piroddi, C., *et al.* (2014). A century of fish biomass decline in the ocean. *Marine Ecology Progress Series*, **512**, 155–66.

Clark, J.R. (1996). *Coastal Zone Management Handbook*. Boca Raton, FL: CRC Press.

Coll, M., Libralato, S., Pitcher, T.J., Solidoro, C. & Tudela, S. (2013). Sustainability implications of honouring the Code of Conduct for Responsible Fisheries. *Global Environmental Change*, **23**, 157–66.

Costello, C., Ovando, D., Clavelle, T., *et al.* (2016). Global fishery prospects under contrasting management regimes. *Proceedings of the National Academy of Sciences*. **113**, 5125–9.

Cranford, P.J., Hargrave, B.T. & Doucette, L.I. (2009). Benthic organic enrichment from suspended mussel (*Mytilus edulis*) culture in Prince Edward Island, Canada. *Aquaculture*, **292**, 189–96.

Cresson, P., Ruitton, S. & Harmelin-Vivien, M. (2014). Artificial reefs do increase secondary biomass production: mechanisms evidenced by stable isotopes. *Marine Ecology Progress Series*, **509**, 15–26.

Cullis-Suzuki, S. & Pauly, D. (2010a). Failing the high seas: a global evaluation of regional fisheries management organizations. *Marine Policy*, **34**, 1036–42.

Cullis-Suzuki, S. & Pauly, D. (2010b). Marine protected area costs as 'beneficial' fisheries subsidies: a global evaluation. *Coastal Management*, **38**, 113–21.

Dankel, D.J., Skagen, D.W. & Ultang, Ø. (2008). Fisheries management in practice: review of 13 commercially important fish stocks. *Reviews in Fish Biology and Fisheries*, **18**, 201–33.

Daskalov, G.M. (2002). Overfishing drives a trophic cascade in the Black Sea. *Marine Ecology Progress Series*, **225**, 53–63.

Daskalov, G.M., Grishin, A.N., Rodionov, S. & Mihneva, V. (2007). Trophic cascades triggered by overfishing reveal possible mechanisms of ecosystem regime shifts. *Proceedings of the National Academy of Sciences*, **104**, 10518–23.

de Juan, S., Thrush, S.F. & Demestre, M. (2007). Functional changes as indicators of trawling disturbance on a benthic community located in a fishing ground (NW Mediterranean Sea). *Marine Ecology Progress Series*, **334**, 117–29.

de Young, B., Barange, M., Beaugrand, G., *et al.* (2008). Regime shifts in marine ecosystems: detection, prediction and management. *Trends in Ecology and Evolution*, **23**, 402–9.

FAO (1995). *Code of Conduct for Responsible Fisheries*. Rome: FAO. www.fao.org/docrep/005/v9878e/v9878e00.HTM.

FAO (2012). *The State of World Fisheries and Aquaculture 2012*. Rome: FAO.

FAO (2014). *The State of World Fisheries and Aquaculture: Opportunities and Challenges*. Rome: FAO.

FAO (2016). *The State of World Fisheries and Aquaculture 2016. Contributing to food security and nutrition for all*. Rome: FAO.

Fletcher, W.J. & Bianchi, G. (2014). The FAO – EAF toolbox: making the ecosystem approach accessible to all fisheries. *Ocean & Coastal Management*, **90**, 20–6.

Fogarty, M.J., Bohnsack, J.A. & Dayton, P.K. (2000). Marine reserves and resource management. In *Seas at the Millennium: an Environmental Evaluation*, ed. C. Sheppard, pp. 375–92. Amsterdam; New York, NY: Pergamon.

Fonteneau, A., Chassot, E. & Bodin, N. (2013). Global spatio-temporal patterns in tropical tuna purse seine fisheries on drifting fish aggregating devices (DFADs): taking a historical perspective to inform current challenges. *Aquatic Living Resources*, **26**, 37–48.

Frank, K.T., Petrie, B., Fisher, J.A.D. & Leggett, W.C. (2011). Transient dynamics of an altered large marine ecosystem. *Nature*, **477**, 86–9.

Froese, R., Stern-Pirlot, A., Winker, H. & Gascuel, D. (2008). Size matters: how single-species management can contribute to ecosystem-based fisheries management. *Fisheries Research*, **92**, 231–41.

Garstang, W. (1900). The impoverishment of the sea: a critical summary of the experimental and statistical evidence bearing upon the alleged depletion of the trawling grounds. *Journal of the Marine Biological Association of the United Kingdom*, **6**, 1–69.

Gell, F.R. & Roberts, C.M. (2003). Benefits beyond boundaries: the fishery effects of marine reserves. *Trends in Ecology and Evolution* **18**, 448–55.

Giles, H. (2008). Using Bayesian networks to examine consistent trends in fish farm benthic impact studies. *Aquaculture*, **274**, 181–95.

Goñi, R. (2000). Fisheries effects on ecosystems. In *Seas at the Millennium: An Environmental Evaluation*, ed. C. Sheppard, pp. 117–33. Oxford, UK: Elsevier Science Ltd.

Goñi, R., Hilborn, R., Díaz, D., Mallol, S. & Adlerstein, S. (2010). Net contribution of spillover from a marine reserve to fishery catches. *Marine Ecology Progress Series*, **400**, 233–43.

Grabowski, J.H., Bachman, M., Demarest, C., *et al.* (2014). Assessing the vulnerability of marine benthos to fishing gear impacts. *Reviews in Fisheries Science & Aquaculture*, **22**, 142–55.

Grafton, R.Q. (1996). Individual transferable quotas: theory and practice. *Reviews in Fish Biology and Fisheries*, **6**, 5–20.

Green, A.L., Fernandes, L., Almany, G., *et al.* (2014). Designing marine reserves for fisheries management, biodiversity conservation, and climate change adaptation. *Coastal Management*, **42**, 143–59.

Hall, S.J. (1999). *The Effects of Fishing on Marine Ecosystems and Communities*. Oxford, UK: Blackwell Science.

Halpern, B.S., Walbridge, S., Selkoe, K.A., *et al.* (2008). A global map of human impact on marine ecosystems. *Science*, **319**, 948–52.

Hamre, J. (1994). Biodiversity and exploitation of the main fish stocks in the Norwegian – Barents Sea ecosystem. *Biodiversity and Conservation*, **3**, 473–92.

Hanneson, R. (1996). On ITQs: an essay for the special issue of Reviews in Fish Biology and Fisheries. *Reviews in Fish Biology and Fisheries*, **6**, 91–6.

Hauser, L., Adcock, G.J., Smith, P.J., Bernal Ramirez, J.H. & Carvalho, G.R. (2002). Loss of microsatellite diversity and low effective population size in an overexploited population of New Zealand snapper (*Pagrus auratus*). *Proceedings of the National Academy of Sciences*, **99**, 11742–7.

Hilborn, R. & Ovando, D. (2014). Reflections on the success of traditional fisheries management. *ICES Journal of Marine Science*, **71**, 1040–6.

Hilborn, R., Stokes, K., Maguire, J.-J., *et al.* (2004). When can marine reserves improve fisheries management? *Ocean & Coastal Management*, **47**, 197–205.

Hosch, G., Ferraro, G. & Failler, P. (2011). The 1995 FAO Code of Conduct for Responsible Fisheries: adopting, implementing or scoring results? *Marine Policy*, **35**, 189–200.

Hutchings, J.A. & Myers, R.A. (1994). What can be learned from the collapse of a renewable resource? Atlantic cod, *Gadus morhua*, of Newfoundland and Labrador. *Canadian Journal of Fisheries and Aquatic Sciences*, **51**, 2126–46.

ICES (2005). *ICES Code of Practice on the Introductions and Transfers of Marine Organisms 2005*. Copenhagen, Denmark: International Council for the Exploration of the Sea.

Jackson, J.B.C., Kirby, M.X., Berger, W.H., *et al.* (2001). Historical overfishing and the recent collapse of coastal ecosystems. *Science*, **293**, 629–38.

Jennings, S. (2009). The role of marine protected areas in environmental management. *ICES Journal of Marine Science*, **66**, 16–21.

Jennings, S. & Kaiser, M.J. (1998). The effects of fishing on marine ecosystems. *Advances in Marine Biology*, **34**, 201–352.

Jennings, S., Kaiser, M.J. & Reynolds, J.D. (2001). *Marine Fisheries Ecology*. Oxford, UK: Blackwell Science Ltd.

Jensen, A. (2002). Artificial reefs of Europe: perspective and future. *ICES Journal of Marine Science*, **59**, S3–13.

Johannes, R.E. (2002). The renaissance of community-based marine resource management in Oceania. *Annual Review of Ecology and Systematics*, **33**, 317–40.

Jonell, M., Phillips, M., Rönnbäck, P. & Troell, M. (2013). Eco-certification of farmed seafood: will it make a difference? *Ambio*, **42**, 659–74.

Kaiser, M.J. (2006). The Louisiana artificial reef program. *Marine Policy*, **30**, 605–23.

Kaiser, M.J., Bullimore, B., Newman, P., Lock, K. & Gilbert, S. (1996). Catches in 'ghost fishing' set nets. *Marine Ecology Progress Series*, **145**, 11–16.

Kaiser, M.J. & de Groot, S.J. (eds.) (2000). *Effects of Fishing on Non-Target Species and Habitats: Biological, Conservation and Socio-Economic Issues*. Oxford, UK: Blackwell Science.

Kennelly, S.J. & Broadhurst, M.K. (2002). By-catch begone: changes in the philosophy of fishing technology. *Fish and Fisheries*, **3**, 340–55.

Kerwath, S.E., Winker, H., Götz, A. & Attwood, C.G. (2013). Marine protected area improves yield without disadvantaging fishers. *Nature Communications*, **4**, article 2347.

King, M. (1995). *Fisheries Biology, Assessment and Management*. Oxford, UK: Fishing News Books.

Kurlansky, M. (1998). *Cod: A Biography of the Fish That Changed the World*. London: Jonathan Cape.

Lester, S.E., Halpern, B.S., Grorud-Colvert, K., *et al.* (2009). Biological effects within no-take marine reserves: a global synthesis. *Marine Ecology Progress Series*, **384**, 33–46.

Loh, T.-L & Jaafar, Z. (2015). Turning the tide on bottom trawling. *Aquatic Conservation: Marine and Freshwater Ecosystems*, **25**, 581–3.

Ludwig, D., Hilborn, R. & Walters, C. (1993). Uncertainty, resource exploitation, and conservation: lessons from history. *Science*, **260**, 17, 36.

Mangano, M.C., Kaiser, M.J., Porporato, E.M.D. & Spanò, N. (2013). Evidence of trawl disturbance on mega-epibenthic communities in the Southern Tyrrhenian Sea. *Marine Ecology Progress Series*, **475**, 101–17.

Matsuoka, T., Nakashima, T. & Nagasawa, N. (2005). A review of ghost fishing: scientific approaches to evaluation and solutions. *Fisheries Science*, **71**, 691–702.

Mora, C., Myers, R.A., Coll, M., *et al.* (2009). Management effectiveness of the world's marine fisheries. *PLoS Biology*, **7**, e1000131.

Murawski, S. A. (2000). Definitions of overfishing from an ecosystem perspective. *ICES Journal of Marine Science*, **57**, 649–58.

Naylor, R.L., Hardy, R.W., Bureau, D.P., *et al.* (2009). Feeding aquaculture in an era of finite resources. *Proceedings of the National Academy of Sciences*, **106**, 15103–10.

Naylor, R.L., Goldburg, R.J., Primavera, J.H., *et al.* (2000). Effect of aquaculture on world fish supplies. *Nature*, **405**, 1017–24.

Neo, M.L., Eckman, W., Vicentuan, K., Teo, S. L.-M. & Todd, P.A. (2015). The ecological significance of giant clams in coral reef ecosystems. *Biological Conservation*, **181**, 111–23.

Northridge, S.P. (1991). Driftnet fisheries and their impacts on non-target species: a worldwide review. *FAO Fisheries Technical Paper*, **320**, 115 p.

O'Connor, S., Ono, R. & Clarkson, C. (2011). Pelagic fishing at 42,000 years before the present and the maritime skills of modern humans. *Science*, **334**, 1117–21.

Okuzawa, K., Maliao, R.J., Quinitio, E.T., *et al.* (2008). Stock enhancement of threatened species in Southeast Asia. *Reviews in Fisheries Science*, **16**, 394–402.

O'Leary, B.C., Smart, J.C.R., Neale, F.C., *et al.* (2011). Fisheries mismanagement. *Marine Pollution Bulletin*, **62**, 2642–8.

OSPAR (1999). *OSPAR Guidelines on Artificial Reefs in Relation to Living Marine Resources*. OSPAR 99/15/1-E. London, UK: OSPAR Secretariat. 5 pp.

Paltzat, D.L., Pearce, C.M., Barnes, P.A. & McKinley, R.S. (2008). Growth and production of California sea cucumbers (*Parastichopus californicus* Stimpson) co-cultured with suspended Pacific oysters (*Crassostrea gigas* Thunberg). *Aquaculture*, **275**, 124–37.

Pauly, D. (1994). *On the Sex of Fish and the Gender of Scientists: Collected Essays in Fisheries Science*. London: Chapman & Hall.

Pauly, D. (2007). The *Sea Around Us* Project: documenting and communicating global fisheries impacts on marine ecosystems. *Ambio*, **36**, 290–5.

Pauly, D., Christensen, V., Dalsgaard, J., Froese, R. & Torres, F. Jr (1998). Fishing down marine food webs. *Science*, **279**, 860–3.

Pauly, D., Christensen,V., Guénette, S., *et al.* (2002). Towards sustainability in world fisheries. *Nature*, **418**, 689–95.

Pauly, D. & Zeller, D. (2016). Catch reconstructions reveal that global marine fisheries catches are higher than reported and declining. *Nature Communications*, **7**, article 10244.

Pickering, H. & Whitmarsh, D. (1997). Artificial reefs and fisheries exploitation: a review of the 'attraction versus production' debate, the influence of design and its significance for policy. *Fisheries Research*, **31**, 39–59.

Pikitch, E.K., Santora, C., Babcock, E.A., *et al.* (2004). Ecosystem-based fishery management. *Science*, **305**, 346–7.

Pitcher, T.J. (2001). Fisheries managed to rebuild ecosystems? Reconstructing the past to salvage the future. *Ecological Applications*, **11**, 601–17.

Pitcher, T.J. (2005). Back-to-the-future: a fresh policy initiative for fisheries and a restoration ecology for ocean ecosystems. *Philosophical Transactions of the Royal Society B*, **360**, 107–21.

Pitcher, T.J. (2008). The sea ahead: challenges to marine biology from seafood sustainability. *Hydrobiologia*, **606**, 161–85.

Pitcher, T., Kalikoski, D., Pramod, G. & Short, K. (2009). Not honouring the code. *Nature*, **457**, 658–9.

Purcell, J.E., Uye, S. & Lo, W.-T. (2007). Anthropogenic causes of jellyfish blooms and their direct consequences for humans: a review. *Marine Ecology Progress Series*, **350**, 153–74.

Rideout, R.M. & Morgan, M.J. (2010). Relationships between maternal body size, condition and potential fecundity of four north-west Atlantic demersal fishes. *Journal of Fish Biology*, **76**, 1379–95.
Ruddle, K., Hviding, E. & Johannes, R.E. (1992). Marine resources management in the context of customary tenure. *Marine Resource Economics*, **7**, 249–73.
Sacchi, J. (2008). The use of trawling nets in the Mediterranean. Problems and selectivity options. *Options Méditerranéennes: Série B. Etudes et Recherches*, **62**, 87–96.
Sainsbury, J.C. (1996). *Commercial Fishing Methods: An Introduction to Vessels and Gears*, 3rd edn. Oxford, UK: Fishing News Books.
Sale, P.F., Cowen, R.K., Danilowicz, B.S., *et al.* (2005). Critical science gaps impede use of no-take fishery reserves. *Trends in Ecology and Evolution*, **20**, 74–80.
Schrank, W.E. (2005). The Newfoundland fishery: ten years after the moratorium. *Marine Policy*, **29**, 407–20.
Sciberras, M., Jenkins, S.R., Mant, R., *et al.* (2015). Evaluating the relative conservation value of fully and partially protected marine areas. *Fish and Fisheries*, **16**, 58–77.
Seaman, W. (2007). Artificial habitats and the restoration of degraded marine ecosystems and fisheries. *Hydrobiologia*, **580**, 143–55.
Shelton, A.O. & Mangel, M. (2011). Fluctuations of fish populations and the magnifying effects of fishing. *Proceedings of the National Academy of Sciences*, **108**, 7075–80.
Steele, T.E. & Klein, R.G. (2008). Intertidal shellfish use during the Middle and Later Stone Age of South Africa. *Archaeofauna*, **17**, 63–76.
Stevenson, T.C., Tissot, B.N. & Walsh, W.J. (2013). Socioeconomic consequences of fishing displacement from marine protected areas in Hawaii. *Biological Conservation*, **160**, 50–8.
Stokstad, E. (2011). Seafood eco-label grapples with challenge of proving its impact. *Science*, **334**, 746.
Sumaila, U.R., Khan, A.S., Dyck, A.J., *et al.* (2010). A bottom-up re-estimation of global fisheries subsidies. *Journal of Bioeconomics*, **12**, 201–25.
Tett, P. (2008). Fish farm wastes in the ecosystem. In *Aquaculture in the Ecosystem*, ed. M. Holmer, K. Black, C.M. Duarte, N. Marbà & I. Karakassis, pp. 1–46. Dordrecht, Netherlands: Springer.
Thébaud, O., Innes, J. & Ellis, N. (2012). From anecdotes to scientific evidence? A review of recent literature on catch share systems in marine fisheries. *Frontiers in Ecology and the Environment*, **10**, 433–7.
Thomas, H.L., MacSharry, B., Morgan, L., *et al.* (2014). Evaluating official marine protected area coverage for Aichi Target 11: appraising the data and methods that define our progress. *Aquatic Conservation: Marine and Freshwater Ecosystems*, **24** (Suppl. 2), 8–23.
Thrush, S.F. & Dayton, P.K. (2002). Disturbance to marine benthic habitats by trawling and dredging: implications for marine biodiversity. *Annual Review of Ecology and Systematics*, **33**, 449–73.
Tyler-Walters, H., Rogers, S.I., Marshall, C.E. & Hiscock, K. (2009). A method to assess the sensitivity of sedimentary communities to fishing activities. *Aquatic Conservation: Marine and Freshwater Ecosystems*, **19**, 285–300.
van Putten, I., Boschetti, F., Fulton, E.A., Smith, A.D.M. & Thébaud, O. (2014). Individual transferable quota contribution to environmental stewardship: a theory in need of validation. *Ecology and Society*, **19** (2), article 35.
Wing, S.R. & Wing, E.S. (2001). Prehistoric fisheries in the Caribbean. *Coral Reefs*, **20**, 1–8.
Wood, A.C.L., Probert, P.K., Rowden, A.A. & Smith, A.M. (2012). Complex habitat generated by marine bryozoans: a review of its distribution, structure, diversity, threats and conservation. *Aquatic Conservation: Marine and Freshwater Ecosystems*, **22**, 547–63.
Worm, B., Hilborn, R., Baum, J.K., *et al.* (2009). Rebuilding global fisheries. *Science*, **325**, 578–85.
Worm, B., Sandow, M., Oschlies, A., Lotze, H.K. & Myers, R.A. (2005). Global patterns of predator diversity in the open oceans. *Science*, **309**, 1365–9.

# 6 Fishes

The term 'fish' is used to encompass three main groups of vertebrates: the jawless vertebrates, such as the hagfishes; the cartilaginous fishes, mainly the rays and sharks; and the teleosts and other bony fishes. The bony fishes account for about 90% of fish species and include the best-known fisheries species such as the anchovies, mackerels, cods, and tunas. About 17 000 marine fish species are currently known, which is roughly half the total number of fish species, but several thousand more are probably yet to be discovered. The great majority of marine species are from coastal habitats. Fishes show a marked latitudinal gradient of species richness increasing from polar to tropical regions. Among the richest areas are the Indo-West Pacific, especially the Indo-Malay-Philippines archipelago, and the Caribbean (Nelson, 2006; Eschmeyer *et al*., 2010).

Our knowledge of fish biodiversity and imperilment is still poor compared to that for other vertebrates, but it appears that very few marine fish species have become globally extinct in historical time (Dulvy *et al*., 2003; Helfman, 2007). Even for heavily exploited fish species, the risk of biological extinction has seldom been considered an issue as it has generally been argued that fisheries are likely to become uneconomic at stock densities well above those at which risk of extinction becomes significant. But there are exceptions, notably in the case of species that are particularly valuable or highly aggregated or have low fecundity. Life-history characteristics have a major bearing on vulnerability, with the larger, slower-growing, and later-maturing species being especially at risk (Reynolds *et al*., 2005; Cheung *et al*., 2007; Field *et al*., 2009). The most vulnerable species may not in fact be those that are targeted but those taken as bycatch (as discussed later). Loss and degradation of habitat can also be important for fish species of restricted distribution, whether exploited or not, particularly those of enclosed basins.

The IUCN Red List currently includes some 500 marine fishes that are threatened (i.e. categorized as critically endangered, endangered, or vulnerable). Well represented are sharks, sawfishes, sturgeons, seahorses, groupers and other coral reef species, and species from Indo-China and the Caribbean (reflecting the abundance of coral reef species). To an extent, however, this reflects groups that have been the focus of IUCN assessment. A number of species were added following a revision of the extinction probability criteria for threatened species (see Chapter 4). As a result, important commercial species such as Atlantic cod (*Gadus morhua*), haddock (*Melanogrammus aeglefinus*), southern bluefin tuna (*Thunnus maccoyii*), bigeye tuna (*T. obesus*), and Atlantic halibut (*Hippoglossus hippoglossus*) became categorized as threatened. This has led to disagreement between fisheries managers and conservation biologists over the appropriateness of categorizing as threatened species that appear to have high reproductive potential and exhibit marked natural population fluctuations from year to year (Mace & Hudson, 1999). Fisheries managers argue that such quantitative criteria overestimate extinction risk for marine fishes and that to achieve an appropriate surplus production requires a major reduction of virgin biomass. And even if overexploited, at least those species that are widely distributed and *r*-selected are most unlikely to become extinct. Others maintain that there appears to be little evidence that marine fishes are in general less vulnerable than non-marine fishes and that to exempt them from the population-decline criteria used to assign extinction risk would be inconsistent with a precautionary approach to management (Hutchings, 2001).

Very few marine and diadromous fishes are included in CITES. Listed in Appendix I are the sawfishes, two species of sturgeon, the totoaba (a large sciaenid endemic to the Gulf of California), and the coelacanths; and Appendix II includes eight species of shark (including the basking, great white, and whale shark), the two species of manta ray, the sturgeons (apart from those in Appendix I), the humphead wrasse, and all seahorses. The inclusion of other commercial fish species in CITES could be an important tool in fisheries management, given that the convention was specifically established to protect species vulnerable to the forces of international trade (Vincent *et al.*, 2014).

## Conservation Issues

Fisheries can be broadly classified according to general biology of the species and fishery characteristics and the extent to which these raise conservation and management problems (Beverton *et al.*, 1984). Major categories of information to be considered include the following (though the factors may be relevant to other exploited groups):

- Stability of habitat. Does the species live in a physically unstable environment, such as an upwelling area or at an ocean front?
- General ecology. Are the adults pelagic or demersal? Is the habitat near the centre or the edges of the general geographical range of the species? Where is the species located in the food chain? Does the species have specialized habitat requirements, and are those habitats particularly vulnerable to anthropogenic disturbance?
- Life-history characteristics. Does the species have a long premature phase during which it is potentially exploitable? Is it short- or long-lived? Does it have a high or low reproductive capacity? Where does the species lie on the *r*-*K* spectrum?
- Catchability. Is it a shoaling species, or does it form spawning aggregations? Is it readily detectable (e.g. by sonar or from the air)? Does the species have access to habitat refuges (i.e. a part of its range which is not accessible to fishing)?
- Commercial value. Is the species particularly sought after so that exploitation is still economic at low population density?

In this chapter we look at certain fish groups vulnerable to overexploitation and that illustrate a number of these factors. Groups particularly susceptible to recruitment overfishing, and where there is a danger of at least local extinction, are those of large size and low fertility. Many cartilaginous fishes are considered especially vulnerable in this regard and are examined in some detail. Also, fisheries for deep-sea demersal species have developed in recent decades and illustrate dramatically how life-history characteristics should severely constrain exploitation. In the case of tunas, their highly migratory nature and high commercial value are major factors relating to their conservation. Heavy exploitation is also an important pressure on certain anadromous species, often compounded by degradation of the habitat used in their migration between the sea and freshwater. We also look at the trade in some species that are taken not for food but for medicinal and display purposes and the conservation issues that they present.

# Chondrichthyans

Cartilaginous fishes, the Chondrichthyes, currently number some 1040 species with many more still to be described. There are two main groups, the sharks and the batoids (sawfishes, guitarfishes, skates, and rays) with about 465 and 540 species, respectively. There are also some 40 chimaeras (ratfishes and ghost sharks) (Dulvy *et al.*, 2014). Chondrichthyans are predators or scavengers and almost entirely marine, occurring from tropical to polar waters and from shallow to deep-sea habitats and reaching their highest diversity in the Indo-West Pacific region. They range in size from pygmy sharks, with a maximum length of only 20–25 cm, to the world's largest fish, the zooplankton-feeding whale shark (*Rhincodon typus*), which can attain a maximum length of 20 m.

Chondrichthyans differ fundamentally in many ways from the bony fishes. In addition to their cartilaginous skeleton, they have dermal denticles and multiple gill openings (except in the chimaeras), and the jaws, paired fins, and tail fin are usually well developed. Males possess copulatory organs (claspers) for internal fertilization. In oviparous species, each fertilized egg is encased in a horny capsule and left on the seabed to hatch. In the many ovoviviparous species, however, the fertilized eggs develop inside the female, and the young are born at an advanced stage of development. As we shall see, the reproductive biology of cartilaginous fishes has major consequences for their conservation.

## Chondrichthyan Fisheries

Chondrichthyans are fished for a variety of products in addition to their meat. Shark fins, prized as a base for soup, are the most valuable product. Certain species are taken primarily for their liver as a source of vitamin A or for squalene oil, used in pharmaceutical and various other industries. Other species provide hides for leather and as an abrasive, teeth and fin spines are used for jewellery, and certain large sharks are sought for their jaws. There is also considerable use of chondrichthyans for teaching and biomedical research.

World commercial landings of chondrichthyans as reported by the FAO increased steadily over the latter half of the twentieth century from some 270 000 t in 1950 to nearly 900 000 t in the early 2000s, but catches have since declined 10–15% as a result of overfishing (Fig. 6.1). Official statistics,

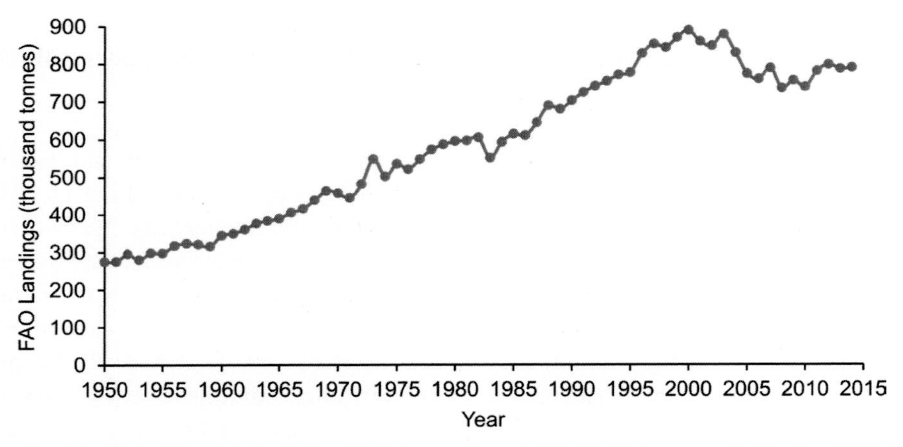


**Fig. 6.1** Global landings of chondrichthyans from 1950 to 2014 as reported by the FAO. Data extracted from FAO Global Capture Production 1950–2014. www.fao.org/fishery/statistics/global-capture-production/query/en (accessed 15 March 2016).

however, undoubtedly underestimate total catches and often do not distinguish between species or even between major groups of chondrichthyans. Many nations have not regularly supplied catch statistics, and cartilaginous fishes are often an important but unrecorded bycatch. Recreational catches are also significant in many places but remain unreported. Given the scale of unreporting, total catch of chondrichthyans may be three to four times greater than that reported (Dulvy *et al.*, 2014). Nevertheless, compared to landings of bony fish, relatively few countries have major chondrichthyan fisheries. Cartilaginous fishes have not in general been esteemed for their flesh, although in recent years some shark species have become sought after. The largest catches are from tropical regions, particularly the Indian Ocean, western central Pacific, and eastern central Atlantic (FAO, 2014). Over recent years, Indonesia, India, Spain, and Taiwan have been the countries mainly involved in fisheries, each averaging annual chondrichthyan catches of more than 40 000 t. Species that have been heavily fished in recent years include the blue shark (*Prionace glauca*), spiny dogfish (*Squalus acanthias*), shortfin mako (*Isurus oxyrinchus*), small-spotted catshark (*Scyliorhinus canicula*), and tope or school shark (*Galeorhinus galeus*) (Lack & Sant, 2009).

A factor now impacting heavily on global shark populations is the demand for shark fins to supply Asian markets. Shark fin soup has for centuries been important in Chinese cuisine, traditionally served on special occasions. Its consumption was discouraged during the Cultural Revolution, but since the 1980s demand has soared as the Chinese economy has grown and as the centre for the shark fin trade has shifted from Hong Kong to mainland China (Clarke *et al.*, 2007; Dell'Apa *et al.*, 2014). Fins of the blue shark dominate the market (Fig. 6.2), but many other taxa are taken, including shortfin mako, silky (*Carcharhinus falciformis*), dusky (*C. obscurus*), bull (*C. leucas*),

**Fig. 6.2** The blue shark (*Prionace glauca*), heavily impacted by shark finning and as bycatch. Photograph: M. Francis. (A black and white version of this figure will appear in some formats. For the colour version, please refer to the plate section.)

hammerhead (*Sphyrna* spp.), and thresher (*Alopias* spp.) sharks (Clarke *et al.*, 2006a). Often the fins are sliced off at sea and the less valuable carcass is thrown back. The discarded shark may still be alive but unable to survive for long. Clarke *et al.* (2006b) estimated that over the period 1996–2000, 26–73 million sharks were taken annually for their fins. This represents an average shark biomass of 1.7 million t, which is about twice the FAO estimate of the global chondrichthyan catch for that period (Fig. 6.1). Several regional fisheries management organizations (RFMOs) have brought in regulations that fins should not total more than 5% of the weight of the sharks onboard. Many countries have similar domestic regulations or require that sharks are landed with their fins attached, as in European Union countries (Council Regulation (EC) No 605/2013) and in the USA (Shark Conservation Act of 2010). Nevertheless, unregulated and illegal shark finning continue to be major problems. The demand for shark fins is driven primarily by Chinese cultural beliefs, indicating that educational campaigns to influence public attitude to shark fin consumption have an important role to play (Dell'Apa *et al.*, 2014).

In some parts of the world the recreational catch of chondrichthyans can be comparable to or even greater than that taken commercially (Fowler *et al.*, 2005). Among the important game species are threshers, mackerel sharks (Lamnidae), houndsharks (Triakidae), and requiem sharks (Carcharhinidae). The impact of recreational fishing on populations of grey nurse (sand tiger) shark (*Carcharias taurus*) in Australian waters has resulted in protective legislation. There is a growing trend for recreational anglers to release sharks, although post-release mortality may be significant, even when fishers care about the survival of released sharks and recognize the importance of appropriate handling (Lynch *et al.*, 2010; Press *et al.*, 2016).

## Shark Meshing Programmes

Large-mesh gillnets and/or baited lines are deployed off some swimming beaches around the world with the aim of reducing the risk of shark attacks. The main species implicated are the great white shark (*Carcharadon carcharias*), tiger shark (*Galeocerdo cuvier*), and bull shark, with the most extensive programmes being those in Australia (New South Wales (since 1937) and Queensland (1962)) and South Africa (KwaZulu-Natal (1952)). Such programmes typically report a consistent decline in the number of sharks caught, probably reflecting an initial impact on resident individuals and then a catch mainly of incoming migrants, against a background of declining populations due to fishing pressure (Reid *et al.*, 2011). The annual catch averages about 1500 sharks in the Australian programme and 1200 in South Africa (Fowler *et al.*, 2005). Although catches are small they may significantly deplete endemic populations (Field *et al.*, 2009). Most sharks caught are, however, harmless species, and netting programmes also catch other animals, including rays, large teleosts, turtles, and marine mammals (Krogh & Reid, 1996). There may be ways of reducing unnecessary bycatch in shark meshing programmes by attending to the siting of nets, the depth at which they are set, and careful timetabling of net deployment (Paterson, 1979). Baited lines (drumlines) are less indiscriminate and can be as effective as nets in catching white and tiger sharks, but a combination of lines and nets may be recommended (Cliff & Dudley, 2011).

## Problems of Chondrichthyan Fisheries

Life-history characteristics of chondrichthyans tend to be very different from those of most teleosts (Anderson, 1990; Hoenig & Gruber, 1990). Typically, chondrichthyans reach relatively large adult size with a rate of growth often slower than in teleosts. They also generally live far longer, many having a maximum age of at least 20–30 years, and there is evidence that great white sharks can reach at least 70 years (Hamady *et al.*, 2014). Related to longevity is a natural mortality rate lower than in most teleosts. Sexual maturity is often later in chondrichthyans, commonly in the range of 2–8 years for many sharks but as high as 25 years for spiny dogfish. Chondrichthyans are also characterized by low fecundity. Skates are among the more prolific with some species laying as many as 100 eggs per female per year, but this is still orders of magnitude lower than the number of eggs spawned annually by many teleosts. In some sharks there is a gestation period of 2 years and a reproductive cycle of 3 years. Thus chondrichthyans typically display life-history traits of *K*-selected species, characteristics that render them especially vulnerable to overexploitation (see Chapter 4). Worm *et al.* (2013) estimated that chondrichthyan populations were on average being exploited at a rate of about 6–8% of biomass per year, whereas populations rebound at an average rate of only 5% per year.

There are many examples of chondrichthyan fisheries that have expanded rapidly and then collapsed during a relatively short period of overexploitation, in some cases within a few years (Anderson, 1990; Stevens *et al.*, 2000). Among these are the Californian fishery of 1938–44 for tope or school shark, which were exploited to meet a wartime demand for vitamin A (from the liver); a similar fishery in south-eastern Australian waters that collapsed in the 1950s; and the Norwegian fishery for porbeagle (*Lamna nasus*) in the NW Atlantic in the 1960s. Similarly, heavy exploitation of spiny dogfish led to the decline of the British Columbian fishery in the late 1940s and the Scottish-Norwegian fishery from the late 1950s.

Management and conservation of chondrichthyan stocks tend to be hampered by the often very limited information available on catch, abundance, and biological parameters of species. There has been the difficulty of obtaining age and growth estimates for chondrichthyans as they lack calcareous otoliths, whereas bony fish can usually be readily aged from growth rings on their otoliths. Various methods have been used to try to overcome this problem, in particular basing age estimates on growth bands in other hard structures, notably vertebral centra or dorsal fin spines (Goldman *et al.*, 2012). But such methods have their limitations. For example, the deposition of vertebral bands may stop in old sharks when somatic growth ceases. Francis *et al.* (2007) found that this led to older porbeagle sharks being under-aged by as much as 50%. Providing appropriate data are available, certain stock assessment models developed for teleost stocks can be applied to chondrichthyan fisheries despite the differences in biology (Bonfil, 2005).

A major obstacle to managing chondrichthyan exploitation is that much of the catch comes not from directed fisheries but as incidental and unreported bycatch of fisheries for more valuable teleost species. This bycatch may be a relatively small component of the total catch yet impact chondrichthyan populations far more significantly than the target species. The vulnerability of chondrichthyans in heavily exploited mixed fisheries and as bycatch is illustrated by the skates. Although skates are among the more prolific chondrichthyans, their shape and large size at hatching mean that the young can be equally as vulnerable as mature fish to a variety of fishing methods. The flapper skate (*Dipturus batis*) is the world's largest skate – females can reach almost 3 m in length. It was once abundant at shelf and upper slope depths of north-western Europe, but as a result of heavy fishing over several decades – targeted fishing and as bycatch – it is now absent over much of its former range (Iglésias *et al.*, 2010). And other North Atlantic skates now appear to be missing from substantial parts of their range (Dulvy & Reynolds, 2002). These species illustrate that body size – which tends to correlate with vulnerable life-history parameters – is a reasonable indicator of local extinction vulnerability in skates (see also Field *et al.*, 2009). They also illustrate the difficulty of achieving realistic conservation measures, since the most effective action would be to ban most fishing over areas extensive enough to provide for self-sustaining populations.

Similar problems are encountered in tropical multi-species fisheries. About 150 species contribute significantly to the Gulf of Thailand demersal trawl fishery, but massive changes in species composition have been recorded since the 1960s, including the virtual disappearance of rays and sawfish and the decline of other chondrichthyans (Pauly, 1988). This raises a recurring issue in fisheries management: the extent to which it is acceptable to imperil species that are vulnerable on account of their life-history characteristics in order to realize the maximum yield from the community as a whole.

The growth of high-seas driftnet and longline fisheries heightened concern over the bycatch of chondrichthyans. At least 40 species are potentially at risk from these pelagic fisheries, with some species being taken in prodigious numbers, especially blue sharks (Bonfil, 1994) (Fig. 6.2). For most oceanic regions, blue sharks constitute at least half the shark bycatch in longline fisheries (Oliver *et al.*, 2015). Worm *et al.* (2013) estimated the global catch of sharks in longline fisheries for the year 2000 to be about 23.6 million individuals, or 852 000 t. And despite the moratorium on large-scale driftnet fisheries on the high seas (see Chapter 5), large numbers of sharks and other bycatch are still caught. During the period 1990–2000, the driftnet fishery for albacore in the eastern North Atlantic took at least 780 000 blue sharks, of which some 90% were discarded dead, likely contributing to a regional decline of the species (Rogan & Mackey, 2007).

Large numbers of chondrichthyans are also taken in deep-water trawl fisheries. Deep-water chondrichthyans are even more vulnerable to exploitation than shelf and pelagic species. They have lower growth rates, mature later, have smaller litters, and live longer (Rigby & Simpfendorfer, 2015).

The catch of sharks in particular might be expected to have far-reaching effects on marine ecosystems, given that millions of these top predators are being removed from the oceans each year. Modelling the large-scale removal of sharks indicates, however, that the effects can be unpredictable and species that respond most dramatically are not necessarily important as prey (Stevens *et al.*, 2000). Nevertheless, there is evidence that large sharks can exert strong top-down effects in marine ecosystems and that major declines in their populations over recent decades have resulted in structural changes to marine communities (Ferretti *et al.*, 2010) (Fig. 6.3). Reduced predation pressure can favour increased abundance of small chondrichthyans, marine mammals, and sea turtles, although

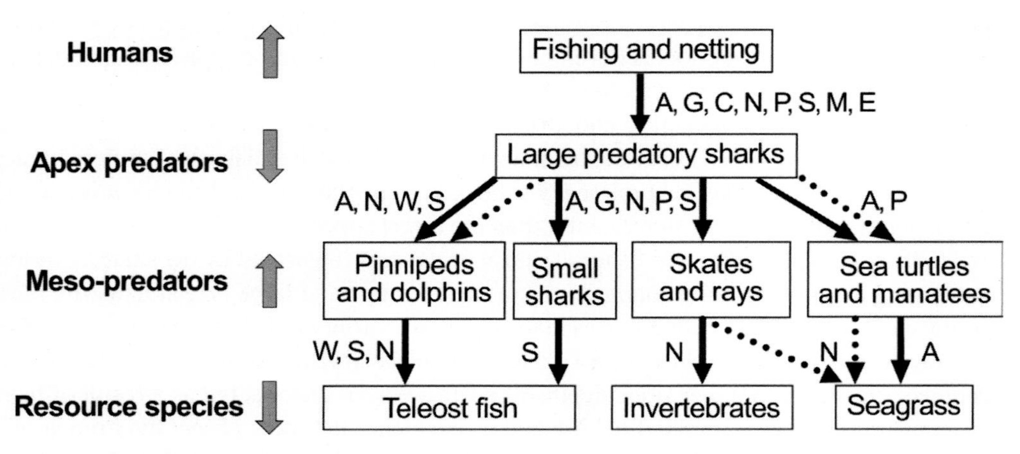


**Fig. 6.3** Ecosystem effects of fishing large sharks. Trophic interactions (solid arrows) and behavioural interactions (dotted arrows) are shown between humans, large and mesopredator elasmobranchs, and their prey species. Arrows on the left indicate overall population trends of the functional groups. Regions in which interactions have been documented are indicated by letters: A: Australia; C: Caribbean; E: Europe; G: Gulf of Mexico; M: Mediterranean Sea; N: North American East Coast; P: Central Pacific; S: South Africa; W: North American West Coast. Reprinted with permission from Ferretti, F., Worm, B., Britten, G.L., Heithaus, M.R. & Lotze, H.K. (2010). Patterns and ecosystem consequences of shark declines in the ocean. *Ecology Letters*, 13, 1055–71. © John Wiley & Sons Ltd/ CNRS.

mesopredator release – an increase in middle-level predators – may be masked by fishing pressure. Evidence of a trophic cascade comes from fisheries data for 1970–2005 for the US Atlantic coast where a major reduction in great sharks has allowed mesopredator prey species to increase, including bottom feeders, which have in turn depleted bivalve populations. These data show a consistent decline in abundance from overfishing for 11 species of apex predatory sharks (a decline of at least 80% for some species) and a corresponding increase in the abundance of 12 species of their chondrichthyan prey (small sharks, rays, and skates). Increased abundance of one of these mesopredators, the cownose ray (*Rhinoptera bonasus*), has been implicated in the demise of local scallop fisheries (Myers *et al.*, 2007).

## Conservation of Chondrichthyans

Many chondrichthyan populations are being severely overexploited due to targeted fisheries and by being taken as bycatch. Dulvy *et al.* (2014) estimate that on the basis of IUCN Red List criteria, one-quarter of chondrichthyan fishes are threatened with extinction, mainly because of overfishing, a conservation status that is among the worst for major vertebrate groups. Biodiversity hotspots where chondrichthyans are particularly at risk include the Indo-Pacific, Red Sea, and the Mediterranean Sea. Few countries have developed management plans for chondrichthyan stocks. Those that have are developed countries, whereas developing countries are responsible for most of the landings (Barker & Schluessel, 2005). Some small sharks, such as the spiny dogfish and gummy shark (*Mustelus antarcticus*), have been the basis of long-running sustainable fisheries. But in general, chondrichthyans offer limited opportunities for exploitation.

A number of aspects need to be addressed in management plans for chondrichthyan fisheries (see e.g. Camhi *et al.*, 1998; Barker & Schluessel, 2005; Dulvy *et al.*, 2008). A fundamental requirement is an adequate understanding of the biology of the species in question, particularly from studies of age, growth, and reproduction. A major obstacle to sustainable chondrichthyan fisheries is that few species have been researched in detail; compared to bony fishes, chondrichthyans have been largely neglected in this regard because of their generally low economic value. From what is known of the biology of chondrichthyans, it is clear that management of these fisheries generally demands a highly precautionary approach. Effective management measures might include limited entry early in the development of the fishery, catch quotas (including bycatch) for commercial as well as for recreational and artisanal fisheries, size limits, closed seasons, protection of critical habitat (e.g. nursery and pupping grounds), use of more selective fishing gear and bycatch reduction devices, protection of threatened species, and prohibition of finning. Also, with the growth in shark fisheries, in part due to an expansion of trade in shark fins and other products, management needs improved documentation and monitoring of the international trade in shark products, controls on trade in rare chondrichthyans threatened by overfishing, and evaluation of further species for possible listing on CITES appendices. Improved international collaboration, through ratification and adherence to fisheries treaties is particularly important given that many sharks are highly migratory.

There exists, however, in Western society widespread antipathy to sharks and other chondrichthyans. The perceived danger of sharks has a major impact on attitudes, even though the relative risk of a shark attack is minute; on average about 70 attacks per year have been recorded worldwide in recent years (Burgess, 2011). Sharks have been projected positively in some cultures, featuring in legends and rituals as gods, guardian spirits, and incarnated ancestors (Magnuson, 1987; Dell'Apa *et al.*, 2014). But, in general, shark conservation has yet to gain widespread public acceptance. There are positive signs of change as reflected, for instance, in the number of organizations actively promoting the conservation of chondrichthyans (Fowler, 1999). There is also a rapidly expanding interest worldwide in shark tourism, providing an opportunity to observe large species in the wild, such as great white, basking (*Cetorhinus maximus*), and whale sharks (e.g. Pierce *et al.*, 2010). Well-managed shark watching operations are likely to have limited environmental impact, can heighten public awareness, and make economic sense. Shark diving in Palau is estimated to generate $18 million per year, whereas the maximum fishery value of the 100 or so reef sharks that interact with the tourism industry would be less than 0.1% of this (Vianna *et al.*, 2012). Cisneros-Montemayor *et al.* (2013) forecast that in 20 years' time the direct visitor expenditure generated by shark ecotourism could more than double to at least $780 million. By comparison, the landed value of global shark fisheries is currently about $630 million and declining.

Concern about the decline of certain shark species has motivated some governments to introduce protective legislation. South Africa introduced legal protection for the great white shark in 1991 (Compagno, 1991), to be followed by several other countries. Other chondrichthyan species included in protective legislation of various countries include the whale shark and basking shark. Such legislation tends to be applied only to largest, more charismatic species, whereas others, including some skates and rays, need similar if not greater protection (Camhi *et al.*, 1998).

For many chondrichthyans, such as oceanic sharks, marine protected areas (MPAs) need to be very large to be effective, which can then have the attendant problems of enforcing adequate conservation measures in high-seas areas (see Chapter 15). Nevertheless, RFMOs could make a valuable contribution to management of such populations (Dulvy *et al.*, 2008). Smaller MPAs that protect critical habitat may be more likely to succeed. This requires good information on the biology and ecology of species and establishing which life-history stages are most important for promoting population recovery – and chondrichthyans have a wide diversity of life-history strategies. A number of species have distinct nursery grounds, such as rig (*Mustelus lenticulatus*), a small shark endemic to New Zealand that uses estuaries as nursery areas. Francis (2013) tagged juvenile rig with acoustic transmitters and tracked them to estimate the size of estuarine MPA needed to protect juveniles. Neonate and young juveniles are not necessarily the most critical life stages, and focussing instead on protecting large juveniles nearing maturity might be more advantageous in some species (Kinney & Simpfendorfer, 2009).

Currently, few management regimes specifically target the conservation of sharks. Within its framework of the Code of Conduct for Responsible Fisheries, the FAO has developed an International Plan of Action for the Conservation and Management of Sharks ('shark' being used to include all chondrichthyans). The plan is voluntary, but each participating state is to adopt a national plan to ensure the conservation and management of sharks and their long-term sustainable use. To date, however, there has been limited implementation of the plan of action (Lack, 2014). Certain existing international agreements are potentially very relevant to chondrichthyan conservation, including the UN Fish Stocks Agreement concerning straddling and highly migratory fish stocks (see Chapter 5), which non-participating states should be requested to join. Also, the Convention on the Conservation of Migratory Species of Wild Animals (which has a specific Memorandum of Understanding on the Conservation of Migratory Sharks), the Convention on Biological Diversity, and CITES provide opportunities for advancing chondrichthyan conservation that parties should explore and enact where possible. A further development in chondrichthyan conservation was the establishment in 1991 of the Shark Specialist Group as part of the IUCN's Species Survival Commission. It aims 'to determine the status and conservation needs of the taxa; promote the implementation of necessary research and management programmes; press for the wise management and sustainable use of chondrichthyan species; and to ensure their conservation through the development of conservation strategies and the promotion of specific projects to be carried out by appropriate organizations and governments' (Gruber & Fowler, 1991).

Dulvy *et al.* (2014) recommend a series of management actions that are needed to help rebuild threatened chondrichthyan populations and properly manage the fisheries (Box 6.1). In particular,

## Box 6.1 Recommended Management Actions That Are Needed to Help Rebuild Threatened Chondrichthyan Populations and Properly Manage Associated Fisheries

### Fishing Nations and Regional Fisheries Management Organizations (RFMOs) Are Urged to:

- Implement, as a matter of priority, scientific advice for protecting habitat and/or preventing overfishing of chondrichthyan populations
- Draft and implement plans of action pursuant to the International Plan of Action (IPOA – Sharks), which include, wherever possible, binding, science-based management measures for chondrichthyans and their essential habitats
- Significantly increase observer coverage, monitoring, and enforcement in fisheries taking chondrichthyans
- Require the collection and accessibility of species-specific chondrichthyan fisheries data, including discards, and penalize non-compliance
- Conduct population assessments for chondrichthyans
- Implement and enforce chondrichthyan fishing limits in accordance with scientific advice; when sustainable catch levels are uncertain, set limits based on the precautionary approach
- Strictly protect chondrichthyans deemed exceptionally vulnerable through ecological risk assessments and those classified by IUCN as critically endangered or endangered
- Prohibit the removal of shark fins whilst onboard fishing vessels and thereby require the landing of sharks with fins naturally attached
- Promote research on gear modifications, fishing methods, and habitat identification aimed at mitigating chondrichthyan bycatch and discard mortality

### National Governments Are Urged to:

- Propose and work to secure RFMO management measures based on scientific advice and the precautionary approach
- Promptly and accurately report species-specific chondrichthyan landings to relevant national and international authorities
- Take unilateral action to implement domestic management for fisheries taking chondrichthyans, including precautionary limits and/or protective status where necessary, particularly for species classified by IUCN as vulnerable, endangered, or critically endangered, and encourage similar actions by other range states
- Adopt bilateral fishery management agreements for shared chondrichthyan populations
- Ensure active membership of the Convention on International Trade in Endangered Species (CITES), the Convention on the Conservation of Migratory Species of Wild Animals (CMS), RFMOs, and other relevant regional and international agreements
- Fully implement and enforce CITES chondrichthyan listings based on solid non-detriment findings, if trade in listed species is allowed
- Propose and support the listing of additional threatened chondrichthyan species under CITES and CMS and other relevant wildlife conventions

- Collaborate on regional agreements and the CMS migratory shark Memorandum of Understanding, with a focus on securing concrete conservation actions
- Strictly enforce chondrichthyan fishing and protection measures and impose meaningful penalties for violations

From Dulvy, N.K., Fowler, S.L., Musick, J.A., *et al.* (2014). Extinction risk and conservation of the world's sharks and rays. *eLife*, 00590.

fishing nations and RFMOs are urged to conduct population assessments of chondrichthyans, implement scientific advice for preventing overfishing, and ensure fisheries data are collected and accessible. They also need to significantly increase fisheries monitoring and enforcement, implement the FAO International Plan of Action, protect highly vulnerable chondrichthyans (notably critically endangered or endangered species on the IUCN Red List), and prohibit the removal of shark fins at sea. And national governments are urged to secure RFMO management measures that are based on scientific advice and the precautionary approach, accurately report species-specific chondrichthyan landings, implement domestic management of fisheries that is suitably precautionary and/or protective, adopt and enforce relevant regional and international agreements (e.g. RFMOs, CITES, Convention for the Conservation of Migratory Species), and strictly enforce chondrichthyan fishing and protection measures. There is a pressing need for concerted and effective international action to conserve sharks and cartilaginous fishes in general. Herndon *et al.* (2010) argue for the formation of an international commission for the conservation and management of chondrichthyan fishes using experience gained from the operation of the International Whaling Commission, given the similarities between life-history characteristics of sharks and cetaceans.

## Deep-Sea Fishes

Some deep-sea species off oceanic islands have for centuries been the basis of sustainable small-scale artisanal fisheries, such as the deep handline fisheries for oilfish (*Ruvettus pretiosus*) in the South Pacific and Indian Ocean and the fishery for black scabbardfish (*Aphanopus carbo*) off Madeira. More recently, large-scale commercial fisheries for deep-sea species have developed. As traditional inshore stocks became depleted, so attention turned to possible deeper water stocks beyond the continental shelf. At first, the target tended to be the deeper range of species already overexploited in shallower waters. So, for instance, once inshore stocks of Atlantic halibut (*Hippoglossus hippoglossus*) had been depleted off Massachusetts, the fishery moved into deeper waters, so that by the later nineteenth century the fishery was conducted almost entirely at upper slope depths. By the turn of the century these stocks too had been exhausted, and by the 1940s the species was commercially extinct off the USA (Moore, 1999).

Deep-water demersal fisheries target species living at depths of about 200–1500 m on the continental slope or on seamounts, most being taken by trawling or longlining. Important commercial species include the roundnose grenadier (*Coryphaenoides rupestris*), roughhead grenadier (*Macrourus berglax*), Greenland halibut (*Reinhardtius hippoglossoides*), and witch flounder (*Glyptocephalus cynoglossus*), fished mainly in the North Atlantic; sablefish (*Anoplopoma fimbria*) in the North Pacific; hoki (*Macruronus novaezelandiae*) and orange roughy (*Hoplostethus atlanticus*) mainly in the SW Pacific; and Patagonian toothfish (*Dissostichus eleginoides*) in the Southern Ocean (Merrett & Haedrich, 1997).

**Fig. 6.4** Orange roughy, a species of upper continental slope depths that like many deep-sea fishes is highly vulnerable to overexploitation on account of its life-history characteristics. Photograph: Neil Bagley, NIWA.

Orange roughy, for example, is a large (typically up to 50 cm long) demersal species that occurs in most temperate regions at depths of 500–1500 m (Fig. 6.4). A major fishery for it began in New Zealand waters in the late 1970s, peaking a decade later with annual catches of more than 50 000 t (Clark, 2001). Although catches have since been greatly reduced, there is concern over the long-term sustainability of stocks. The Challenger Plateau stock off the west coast of New Zealand was reduced to 20% of its estimated virgin biomass in 10 years, whereas long-term sustainable yield of this stock is estimated at only 1–2% of virgin biomass (Clark & Tracey, 1994). Stocks were being overfished more rapidly than essential management data were being gathered. Life-history characteristics make orange roughy extremely vulnerable to even moderate fishing pressure. It is a slow-growing, long-lived species, attaining a maximum age of about 150 years, maturing at about 30 years, with only one short spawning period per year and comparatively low fecundity (Clark *et al.*, 1994; Andrews *et al.*, 2009; Norse *et al.*, 2012).

Orange roughy fisheries illustrate problems typical of large-scale fisheries for deep-water species. These fisheries have generally been pursued with little prior knowledge of the biology of the target species – its stocks, age structure, growth, maturation, fecundity, and mortality – let alone any understanding of the species' interactions and role in the community; in other words, with little regard for precautionary principles. In general, demersal fishes decline in productivity with increasing depth of occurrence, shown by increasing age at 50% maturity, and a decline in the growth coefficient, maximum fecundity, and potential rate of population increase (Drazen & Haedrich, 2012). Deep-water demersal fish very often do not reach maturity until 10–20 years and may live to 50–100 years (Moore & Mace, 1999; Norse *et al.*, 2012). Many of these species are *K*-selected, making them vulnerable to overexploitation and requiring decades to recover from overfishing. There have been numerous deep-water demersal fisheries where a valuable new stock has been discovered but over a few years has been overfished and crashed. Since deep-water fishing is far more costly than inshore fishing, catches need to be more valuable. This creates great economic pressure to overexploit, particularly once the fishery has become overcapitalized. The decline in relative abundance of some deep-sea fishes means that they meet the IUCN criteria for critically endangered species (Devine *et al.*, 2006). Life-history characteristics indicate that fisheries for deep-water species need to be developed cautiously and only so far as knowledge of population structure and reproductive biology as well as the wider ecological impacts permits. In reality, deep-sea fisheries are rarely sustainable, and without government subsidies most would be uneconomic (Merrett & Haedrich, 1997; Roberts, 2002; Norse *et al.*, 2012). Clarke *et al.* (2015) suggest a maximum depth limit for deep-sea fisheries based on a depth at which environmental costs outweigh commercial benefit. Their analysis of trawl survey data over 35 years indicates that for the NE Atlantic a depth limit of 600 m would be appropriate and consistent with EU fisheries legislation. Between 600 and 800 m, the commercial value of the catch decreases significantly, whereas the biodiversity of the demersal fish community, and hence the proportion of discarded non-target species, increases. Also at this depth range, the ratio of discarded to commercial biomass increases, including the ratio of sharks and rays – as taxa especially vulnerable to overexploitation.

Orange roughy is one of a number of exploited species that form spawning or feeding aggregations on seamounts. Other species that have been targeted by deep-sea trawl fisheries on seamounts include alfonsino (*Beryx splendens*), black cardinalfish (*Epigonus telescopus*), southern boarfish (*Pseudopentaceros richardsoni*), rattails (grenadiers), oreos, and toothfish (*Dissostichus* spp.) (Clark, 2009). Life-history characteristics of seamount-aggregating fishes render these species exceptionally susceptible to overfishing, and similar in their vulnerability to those listed as threatened in the IUCN Red List (Cheung *et al.*, 2007) (Fig. 6.5). Furthermore, seamounts support benthic communities that are highly vulnerable to damage from trawling (see Chapter 14).

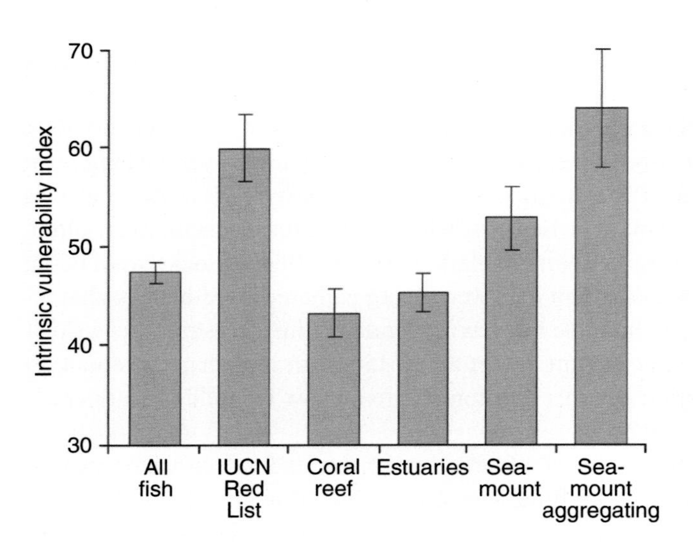


**Fig. 6.5** Vulnerability of marine fishes to exploitation, for the categories: all marine fish species, species listed as threatened in the IUCN Red List, species associated with coral reefs, species associated with estuaries, species recorded on seamounts, and species that aggregate on seamounts. The intrinsic vulnerability index is based on life-history traits (including maximum length, age at first maturity, longevity, von Bertalanffy growth parameter, natural mortality, fecundity, spatial behaviour, and geographic range). Mean and 95% confidence limits. Redrawn but not modified from Cheung, W.W.L., Watson, R., Morato, T., Pitcher, T.J. & Pauly, D. (2007). Intrinsic vulnerability in the global fish catch. *Marine Ecology Progress Series*, 333, 1–12. With permission from Inter-Research.

## Coelacanths

Coelacanths are large (< 2 m) deep-sea fishes of intense scientific interest, highly susceptible to overexploitation and at risk of extinction. They are the sole surviving members of a group of lobe-finned fishes close to the origin of the tetrapods that first colonized land (Amemiya *et al.*, 2013). Only two living

**Fig. 6.6** A coelacanth at a lava cave at 150–200 m water depth in the Comoro islands. The fish occupy caves during the day and emerge at night to feed. Photograph: JAGO-Team, GEOMAR.

species are known: *Latimeria chalumnae* from the western Indian Ocean (off the coast of Tanzania, the Comoro islands, and Mozambique Channel to South Africa) and *L. menadoensis* from off Sulawesi, Indonesia. Both were known to local fishermen, but *L. chalumnae* only came to the attention of scientists in 1938, and *L. menadoensis* not until 1997. *L. chalumnae* is thus the better known. Most have been caught at water depths of 100–400 m but occur to depths of several hundred metres. They inhabit steep rocky slopes where they have relatively small home ranges, frequenting submarine caves and overhangs during the day and emerging at night to forage on fish and squid (Fig. 6.6). They are long-lived (possibly exceeding 100 years), slow-growing fish with a low reproductive rate (Bruton & Stobbs, 1991; Bruton, 1995; Erdmann, 2006; Fricke *et al.*, 2011). In the Comoros, *L. chalumnae* is caught on handlines in a traditional artisanal fishery but not as a target species; it is taken as bycatch in the fishery for oilfish. Population monitoring since the early 1990s indicates that the total population off the island of Grande Comoro now probably numbers no more than 200–300 individuals (Hissmann *et al.*, 1998).

Coelacanths are categorized as threatened in the IUCN Red List and on Appendix I of CITES and accorded protected status in South Africa, Comoros, and Indonesia. Although capture and trading of the species are banned, this does not prevent incidental catch, and local artisanal fisheries appear to be the main threat to coelacanths. Comorans depend on fishing, so measures to relieve fishing pressure and find acceptable alternatives to oilfish are important. One option being used is the nearshore deployment of fish aggregating devices that can be reached by the artisanal fishers and, by attracting pelagic species, can help relieve the pressure on coelacanths. Also useful would be the establishment of marine protected areas, such as the area off the south-western part of Grande Comore, where there is a high coelacanth density (Bruton & Stobbs, 1991). A coelacanth MPA has been proposed for the northern coastal region of Tanzania. The coelacanth population here appears to be genetically distinct from *L. chalumnae* to the south (Nikaido *et al.*, 2011). The abundance of the Indonesian coelacanth has yet to be estimated. However, given the significance of its discovery, *L. menadoensis* has from the outset been the subject of a proactive conservation campaign in Indonesia (Erdmann, 2006). Coelacanth conservation depends to a large extent on public awareness and support and involvement of the local communities. Also, as a prime example of 'living fossils', coelacanths have generated huge interest internationally, appearing in language, literature, and art forms (Fricke, 1997). As such, and as an iconic symbol, they have a flagship role to play in promoting public marine education and conservation (Balon *et al.*, 1988; Fricke, 2001; Erdmann, 2006).

## Tunas

Tunas are among the most spectacular and highly prized of fish. They are mostly large, migratory, epipelagic species of tropical to temperate zones and exemplify a number of issues in marine conservation.

The global catch of tuna is in excess of 4 million t, with the main commercial species being skipjack (*Katsuwonus pelamis*) (60% of catch), yellowfin (*Thunnus albacares*) (25%), bigeye (*T. obesus*) (9%), albacore (*T. alalunga*) (5%), and bluefin (Atlantic bluefin *T. thynnus*, Pacific bluefin *T. orientalis*, and southern bluefin *T. maccoyii*) (1%) (ISSF, 2011). Tuna are taken mainly using purse seine, pelagic longline, and pole-and-line. Tuna (and other pelagic species) are attracted to floating objects, and artificial fish aggregating devices (FADs) that can be free floating or anchored are now extensively used to concentrate tuna for purse seining.

Tuna vary considerably in their life-history characteristics and vulnerability to fishing mortality, but two broad patterns are evident. The smaller, more tropical species, such as skipjack and yellowfin, reach maturity in 1–2 years, spawn year round, and live for about 10 years, reaching maximum lengths of 1–2 m. The commercial catch is dominated by these two species. Far less fecund are the larger bluefin tunas, which can also exploit temperate waters. They take several years to reach maturity, have restricted breeding seasons and locations, live for a maximum age of 15–20 years, and can reach at least twice the length of their more tropical counterparts (Safina, 2001, Schaefer, 2001). Bluefin tunas are also exceedingly valuable for sashimi (raw tuna), a Japanese delicacy, and can sell for thousands of dollars per kilogram at Tokyo's fish market. Southern bluefin is categorized as critically endangered on the IUCN Red List and Atlantic bluefin as endangered.

Management and conservation of tuna are the responsibility of five RFMOs that specifically cover tuna stocks of the major ocean regions (Box 6.2). Generally, however, stocks have not been exploited sustainably and have declined sharply (Cullis-Suzuki & Pauly, 2010). For example, the International Commission for the Conservation of Atlantic Tunas (ICCAT), with currently 51 contracting parties, oversees the stocks of tuna species that occur in the Atlantic and its adjacent seas, but in most cases stock biomasses have fallen dramatically since the RFMO came into force in 1969. Of particular note has been the collapse of Atlantic bluefin tuna in the western Atlantic as a result of consistent overfishing. Catches peaked in the mid-1960s at nearly 19 000 t, but in recent years the total allowable catch has been around 1500–2000 t (ICCAT, 2011). Despite evidence of the severe decline in stocks of west Atlantic bluefin over a number of years, the commission has consistently set quotas far higher than its scientific advisers have recommended (Safina & Klinger, 2008).

Bluefin tuna are also under pressure from the practice of tuna ranching. Wild juveniles are caught and held for months in cages where they are fattened to marketable size. They are fed mainly wild-caught pelagic fish such as sardines, anchovies, and mackerel, and given the high metabolic rates of tuna – they maintain a body temperature warmer than ambient – they require high inputs of food,

**Box 6.2 Regional Fisheries Management Organizations with Particular Responsibility for the Management and Conservation of Tunas**

### Inter-American Tropical Tuna Commission (IATTC)

Entered into force: 1950
Species under management: tuna and other fish taken by tuna fishing vessels in the eastern Pacific Ocean
The IATTC provides the secretariat for the Agreement on the International Dolphin Conservation Program, which aims to reduce incidental dolphin mortalities in the tuna purse seine fishery in the agreement area.

### International Commission for the Conservation of Atlantic Tunas (ICCAT)

Entered into force: 1969
Species under management: tunas and tuna-like species in the Atlantic Ocean and its adjacent seas

### Commission for the Conservation of Southern Bluefin Tuna (CCSBT)

Entered into force: 1994
Species under management: southern bluefin tuna throughout its distribution

### Indian Ocean Tuna Commission (IOTC)

Entered into force: 1996
Species under management: tunas and tuna-like species in the Indian Ocean

### Western and Central Pacific Fisheries Commission (WCPFC)

Entered into force: 2004
Species under management: tunas and other highly migratory fish stocks (see Annex I of the UN Convention on the Law of the Sea) in the Western and Central Pacific Ocean

of up to 20 kg of baitfish for each 1 kg of ranched tuna (Naylor & Burke, 2005). Tuna ranching now accounts for most of the catch of Atlantic bluefin tuna in the Mediterranean. Ranches are estimated to have a capacity far exceeding ICCAT's total allowable catch for the Mediterranean, suggesting a high level of illegal fishing (Allsopp *et al.*, 2009). Tuna ranching is also a lucrative industry in other parts of the world, such as South Australia (for southern bluefin) and Mexico (Pacific bluefin). Ultimately, the industry will need to breed tuna in captivity, although such aquaculture will still be unsustainable if it depends on high inputs of feed fish.

Tuna illustrate some important issues in the management and conservation of marine resources. Their high value – particularly of bluefin tunas – has fuelled rampant overfishing (both legal and illegal), with strong lobbying and political interference that has kept catch quotas above sustainable levels recommended by scientific advisors. Safina (2001) considers that management of bluefin has probably become more complicated than that of any other species as regulations are continually revised whilst ignoring the fundamental issue of overfishing. Typically a fishery becomes uneconomic long before there is a danger of the target species becoming extinct, but in the case of bluefin tuna, commercial extinction may be close to its biological extinction (see Chapter 4). In addition, tunas are highly migratory, and fishing occurs mainly on the high seas. As such, the management of stocks falls to the tuna RFMOs, which – despite the evidence of declining populations – have largely failed to prevent overexploitation.

Much of the concern about tuna fisheries – especially in the public perception – has focused not on the plight of the tuna stocks themselves but on the issue of bycatch. Pelagic longline and purse seine fisheries – the predominant methods of catching tuna – are a major source of mortality of some populations of seabirds, sea turtles, marine mammals, and sharks (Gilman, 2011). What particularly ignited the issue was the high bycatch of dolphins in the east Pacific purse seine fishery for yellowfin tuna, although with changes in fishing practice dolphin mortality was able to be greatly reduced (see Chapter 9).

## Diadromous Fishes

Only about 230–250 fish species are known to be diadromous – to migrate between the sea and freshwater to complete their life cycle – but a number of these are important fisheries species that are heavily exploited. Also, diadromous fish are potentially vulnerable through their use of a range of habitats, including estuaries and brackish waters, which are often among the most heavily impacted of coastal environments (McDowall, 1999). The most critical conservation problems concern anadromous species – those that ascend rivers to spawn – notably sturgeons and salmonids.

### Sturgeons

Sturgeons are of great biological and commercial interest but represent one of the crises in fish conservation. Nearly all the 25 species are threatened. They belong to an ancient group that retains many primitive characteristics, and as the source of black caviar (their roe), they are highly sought after. Their commercial value, life-history characteristics, and habitat requirements render them especially vulnerable to overfishing and environmental disturbance.

Sturgeons (Acipenseridae) occur in temperate latitudes of the Northern Hemisphere. All spawn in freshwater, and most migrate between freshwater and brackish or fully marine coastal waters to mature. They tend to be slow growing, maturing at 5–30 years, spawning only every few years, and attaining large size and considerable age. There are historical records of sturgeon reaching a maximum age of at least a century and a tonne in weight, but the largest individuals nowadays are unlikely to exceed 100 kg (Billard & Lecointre, 2001).

Fifteen sturgeon species are under commercial pressure, primarily because of their roe (Pikitch *et al.*, 2005). Caviar is one of the most expensive foods, with top-quality products selling for more than $10 000 per kg (Bronzi *et al.*, 2011). Consumers exaggerate the value of caviar from rare species, further fuelling overexploitation (Gault *et al.*, 2008). As we saw also for bluefin tuna, fishing for rare species on the brink of extinction can still be profitable in the case of markets for luxury goods.

Some 80–90% of the world's caviar comes from the Caspian Sea, where the main commercial species are beluga (*Huso huso*), Russian (*Acipenser gueldenstaedtii*), stellate (*A. stellatus*), and Persian sturgeon (*A. persicus*). Official statistics show large fluctuations in total sturgeon landings from the Caspian Sea, with peaks at 20 000–30 000 t in the early years of the twentieth century and mid-1970s to mid-1980s, but since then annual catches have plummeted to less than 1000 t. A sharp decline occurred in the early 1990s following the dissolution of the Soviet Union, which had exerted overall regulation of the fishery. Fishing rights devolved to the new independent states, but the black market in caviar increased dramatically. Indeed, the illegal catch is estimated to be several times the legal catch (Pourkazemi, 2006; Helfman, 2007).

Sturgeon populations of the Caspian Sea, as elsewhere, have also been severely impacted by loss and degradation of habitat. Construction of dams, water abstraction for irrigation, and pollution from intensive agriculture and industrial developments of surrounding catchments have drastically reduced available spawning grounds and water quality. As well as fishing regulations, efforts to combat these problems – though with mixed success – include introduction of fishways and elevators to overcome dams, creation of artificial spawning grounds downstream of large dams, and breeding and restocking programmes (Billard & Lecointre, 2001).

Also, in an effort to lessen dependence on wild stocks, there has been a rapid increase in sturgeon aquaculture. This has occurred particularly since the 1990s, mainly in Eurasia and the USA, and most caviar now comes from farmed sturgeon (Fig. 6.7). The most commonly used species for aquaculture are the Siberian (*A. baerii*), sterlet (*A. ruthenus*), and stellate sturgeons (Bronzi *et al.*, 2011). Accurate methods of identification and monitoring of aquaculture fish are needed to prevent wild-caught sturgeon being laundered as farmed product (Raymakers, 2006).

The inclusion of sturgeon species in CITES has helped regulate international trade in caviar. Shortnose sturgeon (*A. brevirostrum*) of eastern North America and the Atlantic sturgeon (*A. sturio*) are included in Appendix I, so no trading of these species is permitted, whilst all other sturgeon species are in Appendix II, meaning that any trading must not jeopardize the survival of those species in the wild. But illegal trade remains a major problem.

Rosenthal *et al.* (2006) summarize the actions that urgently need to be taken to protect and rehabilitate sturgeon populations, notably in the areas of stock assessment and fisheries management,

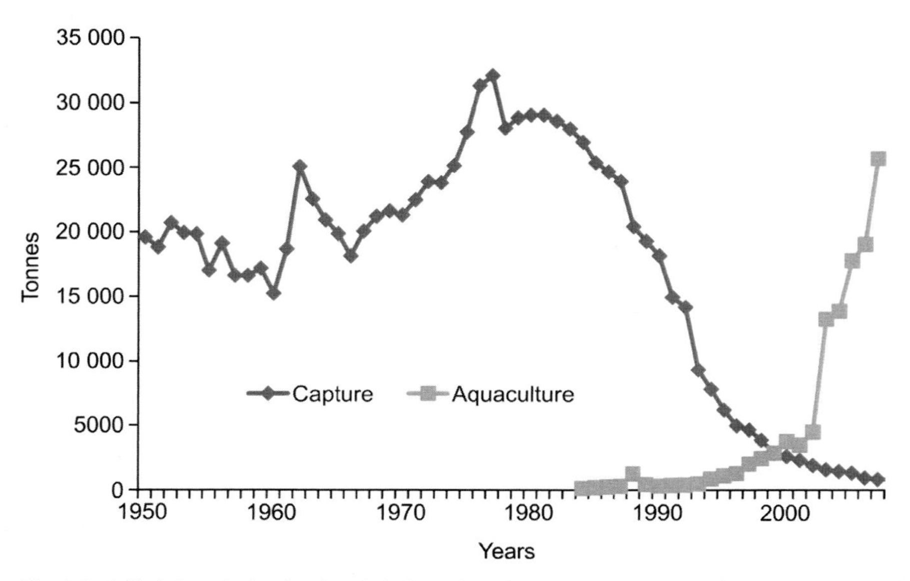


**Fig. 6.7** Official statistics for the global catch and aquaculture production of sturgeon (all species) since 1950. Reprinted with permission from Bronzi, P., Rosenthal, H. & Gessner, J. (2011). Global sturgeon aquaculture production: an overview. *Journal of Applied Ichthyology*, 27, 169–75. © 2011 Blackwell Verlag, Berlin. Based on FAO (Fisheries and Aquaculture Information and Statistics Service). (2009). Aquaculture production: quantities 1950–2005. FISHSTAT Plus – Food and Agriculture Organization of the United Nations. Universal software for fishery statistical time series (online or CDROM; accessed February 2010).

habitat protection and restoration, stock enhancement and rehabilitation, pollution and abatement measures, aquaculture development, public awareness, and the development of adequate national and international regulatory instruments.

## Salmon

All species of salmon are native to Northern Hemisphere waters, although natural distributions have been much altered by successful introductions to non-native waters, including to the Southern Hemisphere. Several anadromous salmon are of conservation concern, notably the Atlantic salmon (*Salmo salar*) of the northern Atlantic and the seven species of Pacific salmon (*Oncorhynchus* spp.) that spawn in river systems from California round to northern Japan. Wild salmon populations have generally been declining over the past century, major reasons being excessive exploitation, degradation and loss of freshwater habitat from pollution, water abstraction and dam construction, and interbreeding with hatchery-raised fish (Leidy & Moyle, 1998).

Adult salmon typically return to their natal streams to spawn, so each species comprises a number of genetically distinct populations. This raises problems for management and in evaluating extinction risk when there are many local populations as well as hatchery fish that pose a genetic risk (Wainwright & Kope, 1999). The concept of evolutionary significant units (ESUs) has been used in such situations, meaning a population (or group of populations) that is (1) 'substantially reproductively isolated from other conspecific populations units' and (2) 'an important component in the evolutionary legacy of the species' (Waples, 1991). The US Endangered Species Act allows such distinct populations to be listed, with the attention on the progeny of naturally spawning fish. But how distinct should ESUs be to warrant conservation action, and how are they to be incorporated into management plans? Dodson *et al.* (1998) proposed operational conservation units based on identifying ESUs and the socio-economic issues with which they interact. An operational unit would typically encompass several ESUs (Fig. 6.8).

Recovery programmes for salmon need to adopt an ecosystem approach around salmon life history, including restoration of natural ecosystem processes and functions, restoration of more natural habitat for all life-history stages (including migrations), reduction of mortality sources (including harvesters), and planning of hydro-power mitigation measures. Salmon need to be managed for population and life-history diversity, and reserves could be used to protect remaining core populations and intact habitats (Williams *et al.*, 1999). Effective conservation of salmon clearly requires coordinated programmes that may involve a wide range of agencies, given that salmon migrate between freshwater, estuarine, and marine systems and across geopolitical boundaries from the local and regional to national and international level.

Attempts to enhance salmon fisheries through hatchery programmes began more than a century ago. Salmon aquaculture – predominantly of Atlantic salmon – has increased dramatically in recent decades, and now only a small proportion of salmon entering the market is wild caught. In farming operations the fry are raised from eggs in freshwater hatcheries to the juvenile stage, and these smolt are then transferred to sea cages to reach market size, which takes 1–2 years. Intensive aquaculture has eased the pressure on wild salmon but introduced its own problems. Salmon regularly escape from sea cages and then interact with wild populations. It has been estimated that up to 2 million fish per year escape from salmon farms into the North Atlantic, and for some populations most of the spawning fish are escapees (McGinnity *et al.*, 2003). Farmed fish, bred primarily for fast growth, may have reduced genetic diversity and be less well adapted to local conditions. Hybridization with wild populations may result in offspring of reduced fitness, potentially leading to the extinction of vulnerable populations, and farmed salmon can also pass on pathogens and parasites to wild salmon

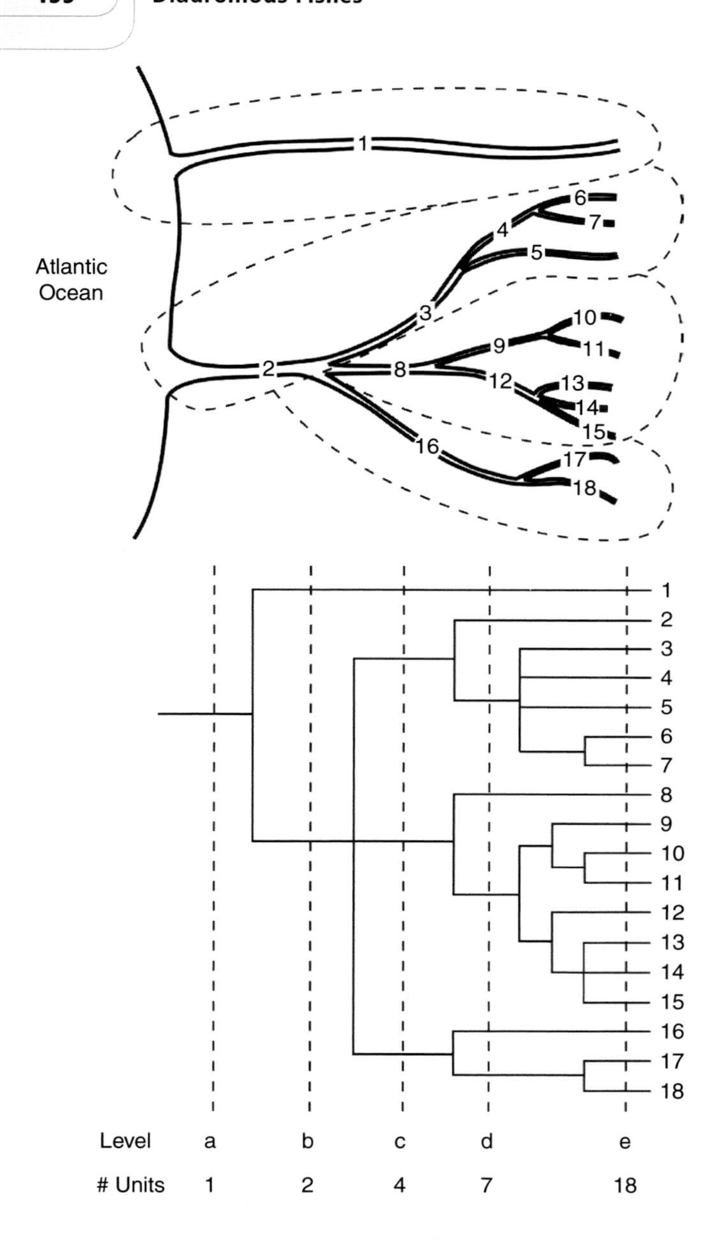


**Fig. 6.8** Identification of evolutionary significant units (ESUs) in hypothetical populations of Atlantic salmon. Genetic and phenotypic data are used to identify distinct populations, as illustrated by the river catchments and phylogenetic tree. Here the decision was made to identify four ESUs (level c), as shown by the dotted lines around the catchments, on the basis of associated socio-economic and legal considerations. From Dodson, J.J., Gibson, R.J., Cunjak, R.A., *et al.* (1998). Elements in the development of conservation plans for Atlantic salmon (*Salmo salar*). *Canadian Journal of Fisheries and Aquatic Sciences*, 55 (Suppl. 1), 312–23. © 2008 Canadian Science Publishing or its licensors. Reproduced with permission.

(McGinnity *et al.*, 2003; Naylor *et al.*, 2005; Naish *et al.*, 2008). From a meta-analysis of existing data, Ford and Myers (2008) found a significant decline in populations of wild salmon in several regions where wild populations were exposed to salmon farms, with in many cases reductions of more than 50%. Reasons for the reduced survival of wild salmon that are migrating past salmon farms may include the impact of escaped salmon on competition for mates and interbreeding and the effect of parasites and diseases. Farmed salmon can be heavily infected with an ectoparasitic copepod, which may pose a threat to wild salmon (Costello, 2009; Krkošek, 2010). Possible measures to reduce the number of escapes and potential harm to ecosystems include more secure nets or closed-wall pens, land-based tanks, raising of fish that are sterile or unable to breed in the wild, restricting the transport of fish, and vaccination programmes to control the spread of diseases. Generally, however, initiatives of salmon-producing countries to prevent or mitigate escapes are relatively weak (Naylor *et al.*, 2005). Salmon farming can also contribute to overfishing of species taken for fishmeal and result in organic waste polluting the seabed under cages (see Chapter 5).

## Seahorses

Many fish species are taken not for food but to meet the increasing global trades in aquarium species and animal products of supposed therapeutic value. Heavily exploited in this regard are coral reef fishes (see Chapter 13) and seahorses.

Seahorses (*Hippocampus* spp.) are among the most distinctive of fish. They have popular appeal and are sometimes used as flagship species in marine conservation. They are, however, the focus of unsustainable international trade, largely for their use in traditional Chinese medicine where dried seahorse is believed to be effective for a range of conditions (Fig. 6.9). Smaller numbers of seahorses are taken for the aquarium and curio trades. Thailand is the main supplier, and the major consumers are Hong Kong SAR, Taiwan, and mainland China. The global trade involves some 10 to 20 million individuals annually, the great majority taken as trawl bycatch (Vincent *et al.*, 2011).

The 47 or so species of seahorses live in sheltered coastal habitats of temperate and tropical regions – particularly in seagrass, algae, and coral reefs. Most of the species are traded but mainly the SE Asian species *Hippocampus trimaculatus, H. spinosissimus, H. kelloggi*, and *H. kuda* (Foster *et al.*, 2016), all of which are threatened (IUCN Red List). Various characteristics render

**Fig. 6.9** Trade in dried seahorses for traditional Chinese medicine has a major impact on several seahorse species, particularly those of SE Asia. Photograph: Getty Images/Stuart Dee.

seahorses vulnerable to intense exploitation and hinder their recolonization, notably their sparse distributions, low mobility and small home ranges, widespread monogamy, lengthy parental care, and low reproductive output, exacerbated by Allee effects at reduced population densities (see Chapter 4). Seahorses are unique in that the female deposits her eggs in the male's abdominal brood pouch to be fertilized, and here they are brooded (about 9–45 days) until the young are released (Foster & Vincent, 2004).

Seahorse populations appear to be declining sharply in many parts of the world. Overfishing – from bycatch as well as targeted fisheries – is mainly to blame, although habitat degradation and pollution are also likely to be significant factors in many areas. Various measures are appropriate to seahorse conservation. Seahorses provide a source of income for many people in poor coastal communities, so developing community-based fisheries management is likely to be a key aspect of seahorse conservation programmes to help foster local responsibility and sustainable management. Projects in the Philippines, for example, have involved engaging with local communities to establish and manage marine protected areas and promote responsible harvesting (Vincent *et al.*, 2011). Options for modifying fishing practices include minimum size limits, leaving pregnant males or holding them until they have given birth, and avoiding destructive fishing methods.

Seahorse aquaculture is also being actively pursued, and now at least 13 species are being raised commercially or researched for their culture potential. But so far captive breeding has contributed animals only to the aquarium trade, and economic large-scale culturing to address the demand from traditional medicine may be some way off (Koldewey & Martin-Smith, 2010). Efforts also need to be directed towards reducing demand by promoting suitable alternatives within traditional medicine.

An important step in regulating trade in seahorses, which came into effect in 2004, was the listing of the genus *Hippocampus* in Appendix II of CITES. This requires seahorse exports by signatory nations to be legal and sustainable. However, given the limited knowledge of the status of most species, the relative impact of exploitation, and the taxonomic challenges, it can be difficult for states to make assessments in order to enforce these regulations. Nevertheless, better compliance with CITES is clearly needed. An analysis of CITES seahorse data by Foster *et al.* (2016) revealed substantial discrepancies between export and import records for species and of the volumes traded. Also, CITES does not cover domestic trade, which may be considerable (Perry *et al.*, 2010).

About a quarter of the seahorse species on the IUCN Red List are categorized as threatened. Most species, however, are listed as data deficient, emphasizing the urgent need for better information on the biology and ecology of seahorses as well as on trading patterns and other threats so as to enable appropriate conservation strategies to be developed.

The fishes examined in this chapter illustrate several key issues in marine conservation, including the underlying importance of life-history traits in the susceptibility of species to overfishing, even for non-target species; the need for a good understanding of the biology and ecology of species, how they will respond to exploitation, and the importance of that information to guide management; the necessity of integrated approaches to management so as to protect all habitat essential to a species' life cycle; and the need for international agreement and cooperation in conserving highly migratory species and in regulating trade in threatened and valuable species.

## REFERENCES

Allsopp, M., Page, R., Johnston, P. & Santillo, D. (2009). *State of the World's Oceans*. London: Springer.

Amemiya, C.T., Alföldi, J., Lee, A.P., *et al.* (2013). The African coelacanth genome provides insights into tetrapod evolution. *Nature*, **496**, 311–16.

Anderson, E.D. (1990). Fishery models as applied to elasmobranch fisheries. In *Elasmobranchs as Living Resources: Advances in the Biology, Ecology, Systematics, and the Status of the Fisheries*, ed. H.L. Pratt Jr, S.H. Gruber & T. Taniuchi, *NOAA Technical Report*, **90**, 473–84.

Andrews, A.H., Tracey, D.M. & Dunn, M.R. (2009). Lead-radium dating of orange roughy (*Hoplostethus atlanticus*): validation of a centenarian life span. *Canadian Journal of Fisheries and Aquatic Sciences*, **66**, 1130–40.

Balon, E.K., Bruton, M.N. & Fricke, H. (1988). A fiftieth anniversary reflection on the living coelacanth, *Latimeria chalumnae*: some new interpretations of its natural history and conservation status. *Environmental Biology of Fishes*, **23**, 241–80.

Barker, M.J. & Schluessel, V. (2005). Managing global shark fisheries: suggestions for prioritizing management strategies. *Aquatic Conservation: Marine and Freshwater Ecosystems*, **15**, 325–47.

Beverton, R.J.H., Cooke, J.G., Csirke, J.B., *et al.* (1984). Dynamics of single species. In *Exploitation of Marine Communities*, ed. R.M. May, pp. 13–58. Berlin: Springer-Verlag.

Billard, R. & Lecointre, G. (2001). Biology and conservation of sturgeon and paddlefish. *Reviews in Fish Biology and Fisheries*, **10**, 355–92.

Bonfil, R. (1994). Overview of world elasmobranch fisheries. *FAO Fisheries Technical Paper*, **341**, 119 pp.

Bonfil, R. (2005). Fishery stock assessment models and their application to sharks. In *Management Techniques for Elasmobranch Fisheries*, ed. J.A. Musick & R. Bonfil. *FAO Fisheries Technical Paper*, **474**, 154–81.

Bronzi, P., Rosenthal, H. & Gessner, J. (2011). Global sturgeon aquaculture production: an overview. *Journal of Applied Ichthyology*, **27**, 169–75.

Bruton, M.N. (1995). Threatened fishes of the world: *Latimeria chalumnae* Smith, 1939 (Latimeriidae). *Environmental Biology of Fishes*, **43**, 104.

Bruton, M.N. & Stobbs, R.E. (1991). The ecology and conservation of the coelacanth *Latimeria chalumnae*. *Environmental Biology of Fishes*, **32**, 313–39.

Burgess, G.H. (2011). *The International Shark Attack File.* Available at www.flmnh.ufl.edu/fish/sharks/ISAF .htm.

Camhi, M., Fowler, S., Musick, J., Bräutigam, A. & Fordham, S. (1998). Sharks and their relatives: ecology and conservation. *Occasional Paper of the IUCN Species Survival Commission No. 20*, 39 pp. Gland, Switzerland: IUCN.

Cheung, W.W.L., Watson, R., Morato, T., Pitcher, T.J. & Pauly, D. (2007). Intrinsic vulnerability in the global fish catch. *Marine Ecology Progress Series*, **333**, 1–12.

Cisneros-Montemayor, A.M., Barnes-Mauthe, M., Al-Abdulrazzak, D., Navarro-Holm, E. & Sumaila, U.R. (2013). Global economic value of shark ecotourism: implications for conservation. *Oryx*, **47**, 381–8.

Clark M. (2001). Are deepwater fisheries sustainable? – the example of orange roughy (*Hoplostethus atlanticus*) in New Zealand. *Fisheries Research*, **51**, 123–35.

Clark, M.R. (2009). Deep-sea seamount fisheries: a review of global status and future prospects. *Latin American Journal of Aquatic Research*, **37**, 501–12.

Clark, M.R., Fincham, D.J. & Tracey, D.M. (1994). Fecundity of orange roughy (*Hoplostethus atlanticus*) in New Zealand waters. *New Zealand Journal of Marine and Freshwater Research*, **28**, 193–200.

Clark, M.R. & Tracey, D.M. (1994). Changes in a population of orange roughy, *Hoplostethus atlanticus*, with commercial exploitation on the Challenger Plateau, New Zealand. *Fishery Bulletin*, **92**, 236–53.

Clarke, J., Milligan, R.J., Bailey, D.M. & Neat, F.C. (2015). A scientific basis for regulating deep-sea fishing by depth. *Current Biology*, **25**, 2425–9.

Clarke, S.C., Magnussen, J.E., Abercrombie, D.L., McAllister, M.K. & Shivji, M.S. (2006a). Identification of shark species composition and proportion in the Hong Kong shark fin market based on molecular genetics and trade records. *Conservation Biology*, **20**, 201–11.

Clarke, S.C., McAllister, M.K., Milner-Gulland, E.J., *et al.* (2006b). Global estimates of shark catches using trade records from commercial markets. *Ecology Letters*, **9**, 1115–26.

Clarke, S., Milner-Gulland, E.J. & Bjørndal, T. (2007). Social, economic, and regulatory drivers of the shark fin trade. *Marine Resource Economics*, **22**, 305–27.

Cliff, G. & Dudley, S.F.J. (2011). Reducing the environmental impact of shark-control programs: a case study from KwaZulu-Natal. *Marine and Freshwater Research* **62**, 700–9.

Compagno, L.J.V. (1991). Government protection for the great white shark (*Carcharadon carcharias*) in South Africa. *South African Journal of Science*, **87**, 284–5.

Costello, M.J. (2009). How sea lice from salmon farms may cause wild salmonid declines in Europe and North America and be a threat to fishes elsewhere. *Proceedings of the Royal Society B*, **276**, 3385–94.

Cullis-Suzuki, S. & Pauly, D. (2010). Failing the high seas: a global evaluation of regional fisheries management organizations. *Marine Policy*, **34**, 1036–42.

Dell'Apa, A., Smith, M.C. & Kaneshiro-Pineiro, M.Y. (2014). The influence of culture on the international management of shark finning. *Environmental Management*, **54**, 151–61.

Devine, J.A., Baker, K.D & Haedrich, R.L. (2006). Deep-sea fishes qualify as endangered. *Nature*, **439**, 29.

Dodson, J.J., Gibson, R.J., Cunjak, R.A., *et al.* (1998). Elements in the development of conservation plans for Atlantic salmon (*Salmo salar*). *Canadian Journal of Fisheries and Aquatic Sciences*, **55** (Suppl. 1), 312–23.

Drazen, J.C. & Haedrich, R.L. (2012). A continuum of life histories in deep-sea demersal fishes. *Deep-Sea Research I*, **61**, 34–42.

Dulvy, N.K., Baum, J.K., Clarke, S., *et al.* (2008). You can swim but you can't hide: the global status and conservation of oceanic pelagic sharks and rays. *Aquatic Conservation: Marine and Freshwater Ecosystems*, **18**, 459–82.

Dulvy, N.K., Fowler, S.L., Musick, J.A., *et al.* (2014). Extinction risk and conservation of the world's sharks and rays. *eLife*, 00590.

Dulvy, N.K. & Reynolds, J.D. (2002). Predicting extinction vulnerability in skates. *Conservation Biology*, **16**, 440–50.

Dulvy, N.K., Sadovy, Y. & Reynolds, J.D. (2003). Extinction vulnerability in marine populations. *Fish and Fisheries*, **4**, 25–64.

Erdmann, M. (2006). Lessons learned from the conservation campaign for the Indonesian coelacanth, *Latimeria menadoensis*. *South African Journal of Science*, **102**, 501–4.

Eschmeyer, W.N., Fricke, R., Fong, J.D. & Polack, D.A. (2010) Marine fish diversity: history of knowledge and discovery (Pisces). *Zootaxa*, **2525**, 19–50.

FAO (2009). Aquaculture production: quantities 1950–2005. FISHSTAT Plus – Food and Agriculture Organization of the United Nations. Universal software for fishery statistical time series (online or CDROM; accessed February 2010).

FAO (2014). *The State of World Fisheries and Aquaculture: Opportunities and Challenges*. Rome: FAO.

Ferretti, F., Worm, B., Britten, G.L., Heithaus, M.R. & Lotze, H.K. (2010). Patterns and ecosystem consequences of shark declines in the ocean. *Ecology Letters*, **13**, 1055–71.

Field, I.C., Meekan, M.G., Buckworth, R.C. & Bradshaw, C.J.A. (2009). Susceptibility of sharks, rays and chimaeras to global extinction. *Advances in Marine Biology*, **56**, 275–363.

Ford, J.S. & Myers, R.A. (2008). A global assessment of salmon aquaculture impacts on wild salmonids. *PLoS Biology*, **6**, e33.

Foster, S.J. & Vincent, A.C.J. (2004). Life history and ecology of seahorses: implications for conservation and management. *Journal of Fish Biology*, **65**, 1–61.

Foster, S., Wiswedel, S. & Vincent, A. (2016). Opportunities and challenges for analysis of wildlife trade using CITES data – seahorses as a case study. *Aquatic Conservation: Marine and Freshwater Ecosystems*, **26**, 154–72.

Fowler, S. (1999). The role of non-governmental organisations in the international conservation of elasmobranchs. In *Case Studies of the Management of Elasmobranch Fisheries*, ed. R. Shotton, *FAO Fisheries Technical Paper*, **378** (2), 880–903.

Fowler, S.L., Cavanagh, R.D., Camhi, M., *et al.* (comp. and ed.). (2005). *Sharks, Rays and Chimaeras: the Status of the Chondrichthyan Fishes. Status Survey.* IUCN/SSC Shark Specialist Group. Gland, Switzerland and Cambridge, UK: IUCN. 461 pp.

Francis, M.P. (2013). Temporal and spatial patterns of habitat use by juveniles of a small coastal shark (*Mustelus lenticulatus*) in an estuarine nursery. *PLoS ONE*, **8**, e57021.

Francis, M.P., Campana, S.E. & Jones, C.M. (2007). Age under-estimation in New Zealand porbeagle sharks (*Lamna nasus*): is there an upper limit to ages that can be determined from shark vertebrae? *Marine and Freshwater Research*, **58**, 10–23.

Fricke, H. (1997). Living coelacanths: values, eco-ethics and human responsibility. *Marine Ecology Progress Series*, **161**, 1–15.

Fricke, H. (2001). Coelacanths: a human responsibility. *Journal of Fish Biology*, **59** (Supplement A), 332–8.

Fricke, H., Hissmann, K., Froese, R., *et al.* (2011). The population biology of the living coelacanth studied over 21 years. *Marine Biology*, **158**, 1511–22.

Gault, A., Meinard, Y. & Courchamp, F. (2008). Consumers' taste for rarity drives sturgeons to extinction. *Conservation Letters*, **1**, 199–207.

Gilman, E.L. (2011). Bycatch governance and best practice mitigation technology in global tuna fisheries. *Marine Policy*, **35**, 590–609.

Goldman, K.J., Cailliet, G.M., Andrews, A.H. & Natanson, L.J. (2012). Assessing the age and growth of chondrichthyan fishes. In *Biology of Sharks and their Relatives*, 2nd edn, ed. J.C. Carrier, J.A. Musick & M.R. Heithaus, pp. 423–51. Boca Raton, FL: CRC Press.

Gruber, S.H. & Fowler, S. (1991). Elasmobranch conservation: the establishment of the IUCN Shark Specialist Group. *Aquatic Conservation: Marine and Freshwater Ecosystems*, **1**, 193–4.

Hamady, L.L., Natanson, L.J., Skomal, G.B. & Thorrold, S.R. (2014). Vertebral bomb radiocarbon suggests extreme longevity in white sharks. *PLoS ONE*, **9**, e84006.

Helfman, G.S (2007). *Fish Conservation: a Guide to Understanding and Restoring Global Aquatic Biodiversity and Fishery Resources.* Washington, DC: Island Press.

Herndon, A., Gallucci, V.F., DeMaster D., *et al.* (2010). The case for an international commission for the conservation and management of sharks (ICCMS). *Marine Policy*, **34**, 1239–48.

Hissmann, K., Fricke, H. & Schauer, J. (1998). Population monitoring of the coelacanth (*Latimeria chalumnae*). *Conservation Biology* **12**, 759–65.

Hoenig, J.M. & Gruber, S.H. (1990). Life-history patterns in the elasmobranchs: implications for fisheries management. In *Elasmobranchs as Living Resources: Advances in the Biology, Ecology, Systematics, and the Status of the Fisheries*, ed. H.L. Pratt Jr, S.H. Gruber & T. Taniuchi, *NOAA Technical Report*, **90**, 1–16.

Hutchings, J.A. (2001). Conservation biology of marine fishes: perceptions and caveats regarding assignment of extinction risk. *Canadian Journal of Fisheries and Aquatic Sciences*, **58**, 108–21.

ICCAT (2011). *Report of the Standing Committee on Research and Statistics (SCRS) (Madrid, Spain, October 3–7, 2011).* Madrid, Spain: International Commission for the Conservation of Atlantic Tunas.

Iglésias, S.P., Toulhoat, L. & Sellos, D.Y. (2010). Taxonomic confusion and market mislabelling of threatened skates: important consequences for their conservation status. *Aquatic Conservation: Marine and Freshwater Ecosystems*, **20**, 319–33.

ISSF (2011). Status of the world fisheries for tuna: stock status ratings – 2011. *ISSF Technical Report 2011-04.* McLean, VA: International Seafood Sustainability Foundation.

Kinney, M.J. & Simpfendorfer, C.A. (2009). Reassessing the value of nursery areas to shark conservation and management. *Conservation Letters*, **2**, 53–60.

Koldewey, H.J. & Martin-Smith, K.M. (2010). A global review of seahorse aquaculture. *Aquaculture*, **302**, 131–52.

Krogh, M. & Reid, D. (1996). Bycatch in the protective shark meshing programme off south-eastern New South Wales, Australia. *Biological Conservation*, **77**, 219–26.

Krkošek, M. (2010). Sea lice and salmon in Pacific Canada: ecology and policy. *Frontiers in Ecology and the Environment*, **8**, 201–9.

Lack, M. (2014). Challenges for international governance. In *Sharks: Conservation, Governance and Management*, ed. E.J. Techera & N. Klein, pp. 46–65. Oxford, UK: Routledge.
Lack, M. & Sant, G. (2009). *Trends in Global Shark Catch and Recent Developments in Management*. Cambridge, UK: TRAFFIC International.
Leidy, R.A. & Moyle, P.B. (1998). Conservation status of the world's fish fauna: an overview. In *Conservation Biology: for the Coming Decade*, 2nd edn, ed. P.L. Fiedler & P.M. Kareiva, pp. 187–227. New York: Chapman & Hall.
Lynch, A-M.J., Sutton, S.G. & Simpfendorfer, C.A. (2010). Implications of recreational fishing for elasmobranch conservation in the Great Barrier Reef Marine Park. *Aquatic Conservation: Marine and Freshwater Ecosystems*, **20**, 312–18.
Mace, G.M. & Hudson, E.J. (1999). Attitudes toward sustainability and extinction. *Conservation Biology*, **13**, 242–6.
Magnuson, J. (1987). The significance of sharks in human psychology. In *Sharks: an Enquiry into Biology, Behavior, Fisheries, and Use*, ed. S. Cook, pp. 85–94. Corvallis, OR: Oregon State University Extension Service.
McDowall, R.M. (1999). Different kinds of diadromy: different kinds of conservation problems. *ICES Journal of Marine Science*, **56**, 410–13.
McGinnity, P., Prodöhl, P., Ferguson, A., *et al.* (2003). Fitness reduction and potential extinction of wild populations of Atlantic salmon, *Salmo salar*, as a result of interactions with escaped farm salmon. *Proceedings of the Royal Society B*, **270**, 2443–50.
Merrett, N.R. & Haedrich, R.L. (1997). *Deep-Sea Demersal Fish and Fisheries*. London: Chapman & Hall.
Moore, J.A. (1999). Deep-sea finfish fisheries: lessons from history. *Fisheries*, **24** (7), 16–21.
Moore, J.A. & Mace, P.M. (1999). Challenges and prospects for deep-sea finfish fisheries. *Fisheries*, **24** (7), 22–3.
Myers, R.A.; Baum, J.K.; Shepherd, T.D., Powers, S.P. & Peterson, C.H. (2007). Cascading effects of the loss of apex predatory sharks from a coastal ocean. *Science*, **315**, 1846–50.
Naish, K.A., Taylor, J.E.; Levin, P.S., *et al.* (2008). An evaluation of the effects of conservation and fishery enhancement hatcheries on wild populations of salmon. *Advances in Marine Biology*, **53**, 61–194.
Naylor, R. & Burke, M. (2005). Aquaculture and ocean resources: raising tigers of the sea. *Annual Review of Environment and Resources*, **30**, 185–218.
Naylor, R., Hindar, K., Fleming, I.A., *et al.* (2005). Fugitive salmon: assessing the risks of escaped fish from net-pen aquaculture. *BioScience*, **55**, 427–37.
Nelson, J.S. (2006). *Fishes of the World*, 4th edn. Hoboken, NJ: John Wiley & Sons Inc.
Nikaido, M., Takeshi, S., Emerson J.J., *et al.* (2011). Genetically distinct coelacanth population off the northern Tanzanian coast. *Proceedings of the National Academy of Sciences*, **108**, 18009–13.
Norse, E.A., Brooke, S., Cheung, W.W.L., *et al.* (2012). Sustainability of deep-sea fisheries. *Marine Policy*, **36**, 307–20.
Oliver, S., Braccini, M., Newman, S.J. & Harvey, E.S. (2015). Global patterns in the bycatch of sharks and rays. *Marine Policy*, **54**, 86–97.
Paterson, R. (1979). Shark meshing takes a heavy toll of harmless marine animals. *Australian Fisheries*, **6** (10), 17–23.
Pauly, D. (1988). Fisheries research and the demersal fisheries of Southeast Asia. In *Fish Population Dynamics: the Implications for Management*, 2nd edn, ed. J.A. Gulland, pp. 329–48. Chichester, UK: John Wiley & Sons Ltd.
Perry, A.L., Lunn, K.E. & Vincent, A.C.J. (2010). Fisheries, large-scale trade, and conservation of seahorses in Malaysia and Thailand. *Aquatic Conservation: Marine and Freshwater Ecosystems*, **20**, 464–75.
Pierce, S.J., Méndez-Jiménez, A., Collins, K., Rosero-Caicedo, M. & Monadjem, A. (2010). Developing a Code of Conduct for whale shark interactions in Mozambique. *Aquatic Conservation: Marine and Freshwater Ecosystems*, **20**, 782–8.
Pikitch, E.K., Doukakis, P., Lauck L., Chakrabarty, P. & Erickson, D.L. (2005). Status, trends and management of sturgeon and paddlefish fisheries. *Fish and Fisheries*, **6**, 233–65.

Pourkazemi, M. (2006). Caspian Sea sturgeon conservation and fisheries: past present and future. *Journal of Applied Ichthyology*, **22** (Suppl. 1), 12–16.

Press, K.M., Mandelman, J., Burgess, E., *et al.* (2016). Catching sharks: recreational saltwater angler behaviours and attitudes regarding shark encounters and conservation. *Aquatic Conservation: Marine and Freshwater Ecosystems*, **26**, 689–702.

Raymakers, C. (2006). CITES, the Convention on International Trade in Endangered Species of Wild Fauna and Flora: its role in the conservation of Acipenseriformes. *Journal of Applied Ichthyology*, **22** (Suppl. 1), 53–65.

Reid, D.D., Robbins, W.D. & Peddemors, V.M. (2011). Decadal trends in shark catches and effort from the New South Wales, Australia, Shark Meshing Program 1950–2010. *Marine and Freshwater Research*, **62**, 676–93.

Reynolds, J.D., Dulvy, N.K., Goodwin, N.B. & Hutchings, J.A. (2005). Biology of extinction risk in marine fishes. *Proceedings of the Royal Society B*, **272**, 2337–44.

Rigby, C. & Simpfendorfer, C.A. (2015). Patterns in life history traits of deep-water chondrichthyans. *Deep-Sea Research II*, **115**, 30–40.

Roberts, C.M. (2002). Deep impact: the rising toll of fishing in the deep sea. *Trends in Ecology & Evolution*, **17**, 242–5.

Rogan, E. & Mackey, M. (2007). Megafauna bycatch in drift nets for albacore tuna (*Thunnus alalunga*) in the NE Atlantic. *Fisheries Research*, **86**, 6–14.

Rosenthal, H., Pourkazemi, M. & participants of the 5th International Symposium on Sturgeons (2006). Ramsar Declaration on Global Sturgeon Conservation. *Journal of Applied Ichthyology*, **22** (Suppl. 1), 5–11.

Safina, C. (2001). Tuna conservation. In *Tuna: Physiology, Ecology, and Evolution*, ed. B.A. Block & E.D. Stevens, pp. 413–59. San Diego, CA: Academic Press.

Safina, C. & Klinger, D.H. (2008). Collapse of bluefin tuna in the western Atlantic. *Conservation Biology*, **22**, 243–6.

Schaefer, K.M. (2001). Reproductive biology of tunas. In *Tuna: Physiology, Ecology, and Evolution*, ed. B.A. Block & E.D. Stevens, pp. 225–70. San Diego, CA: Academic Press

Stevens, J.D., Bonfil, R., Dulvy, N.K. & Walker, P.A. (2000). The effects of fishing on sharks, rays, and chimaeras (chondrichthyans), and the implications for marine ecosystems. *ICES Journal of Marine Science*, **57**, 476–94.

Vianna, G.M.S., Meekan, M.G., Pannell, D.J., Marsh, S.P. & Meeuwig, J.J. (2012). Socio-economic value and community benefits from shark-diving tourism in Palau: a sustainable use of reef shark populations. *Biological Conservation*, **145**, 267–77.

Vincent, A.C.J., Foster, S.J. & Koldewey, H.J. (2011). Conservation and management of seahorses and other Syngnathidae. *Journal of Fish Biology*, **78**, 1681–724.

Vincent, A.C.J., Sadovy de Mitcheson, Y.J., Fowler, S.L. & Lieberman, S. (2014). The role of CITES in the conservation of marine fishes subject to international trade. *Fish and Fisheries*, **15**, 563–92.

Wainwright, T.C. & Kope, R.G. (1999). Methods of extinction risk assessment developed for US West Coast salmon. *ICES Journal of Marine Science*, **56**, 444–8.

Waples, R.S. (1991). Pacific salmon, *Oncorhynchus* spp., and the definition of 'species' under the Endangered Species Act. *Marine Fisheries Review*, **53** (3), 11–22.

Williams, R.N., Bisson, P.A., Bottom, D.L., *et al.* (1999). Scientific issues in the restoration of salmonid fishes in the Columbia River. *Fisheries*, **24** (3), 10–19.

Worm, B., Davis, B., Kettemer, L., *et al.* (2013). Global catches, exploitation rates, and rebuilding options for sharks. *Marine Policy*, **40**, 194–204.

# 7 Marine Turtles

There are some 10 500 known extant species of reptiles, but only about 80 can be considered as primarily marine animals (Rasmussen *et al.*, 2011; Uetz *et al.*, 2016). The majority of these are sea snakes, most of which bear live young and have severed their terrestrial links. Marine turtles and crocodiles, on the other hand, return to land to lay their eggs. So too does the only species of marine lizard, the marine iguana (*Amblyrhynchus cristatus*), endemic to the Galápagos Islands. Reptiles are largely confined to the tropics and subtropics, since as ectotherms they have increasing difficulty surviving in higher latitudes. A number of marine reptile species are seriously threatened by over-exploitation, human disturbance, and destruction of habitat, particularly the turtles.

There are seven extant species of sea turtle: loggerhead (*Caretta caretta*), green (*Chelonia mydas*), hawksbill (*Eretmochelys imbricata*), olive ridley (*Lepidochelys olivacea*), Kemp's ridley (*Lepidochelys kempii*), and flatback turtle (*Natator depressa*), all belonging to the family Cheloniidae; and the leatherback turtle (*Dermochelys coriacea*) of the family Dermochelyidae. The following brief accounts are based largely on reviews by Spotila (2004) and IUCN (www.iucnredlist.org).

The loggerhead turtle is a circumglobal species that occurs from tropical to temperate latitudes, with the adults foraging primarily in nearshore waters on benthic invertebrates, mainly molluscs and crustaceans. The major nesting grounds are in the Atlantic and Indian oceans, notably in Florida and Oman.

Widely distributed in tropical and subtropical waters, the green turtle is the only predominantly herbivorous species, feeding mainly on seagrasses and macroalgae. It has nesting grounds at many locations in the Atlantic, Indian, and western Pacific, but there are only 10–15 populations with 2000 or more nesting females per year. By far the largest nesting colonies are at Tortuguero on the Caribbean coast of Costa Rica (around 27 000 nesting females per year) and Raine Island in the northern Great Barrier Reef (about 18 000).

The hawksbill turtle inhabits coastal waters throughout the tropical Atlantic and Indo-Pacific, often in reef areas. It feeds largely on benthic invertebrates associated with coral reefs but particularly sponges. The main nesting colonies are in the Caribbean and Indian and Pacific Oceans but with few supporting more than 1000 females per year. The largest colony (4000 females per year) is on Milman Island, northern Great Barrier Reef.

The olive ridley occurs mainly in tropical to warm-temperate waters of the Pacific and Indian Oceans. It is the most abundant of the species with major mass nesting sites on the Pacific coasts of Mexico and Costa Rica (300 000–600 000 nests per year) and in Orissa, north-eastern India (150 000–200 000 nests). Adults take a wide variety of food, mainly benthic invertebrates.

Adults of Kemp's ridley are restricted to inshore waters around the Gulf of Mexico, but juveniles and immature individuals range more widely in tropical and temperate coastal areas of the NW Atlantic. Adults frequent sandy and muddy bottoms where they feed mainly on crustaceans. Its nesting sites are along the western Gulf of Mexico, most importantly on beaches near Rancho

Nuevo in north-eastern Mexico. A mass nesting here in 1947 was estimated at more than 40 000 females, but now the total number of adult females may be only some 5000.

Like the Kemp's ridley, the flatback turtle has a restricted geographical distribution, predominantly shelf waters of northern Australia but also eastern Indonesia and southern Papua New Guinea. The adults feed mainly on soft-bodied benthic invertebrates.

The leatherback turtle, the sole living member of the Dermochelyidae, is in many respects unlike the cheloniid species. Its streamlined body lacks scales and is covered by a leathery skin with prominent dorsal keels. It is also by far the largest species, has a thick oily dermis and extensive deposits of blubber, and can regulate its body temperature (Bostrom *et al.*, 2010). Although they nest in the tropics, leatherbacks range into cool temperate and even subpolar latitudes and can remain active in near-freezing waters at the surface but also at depth. They can undertake deep (> 1200 m), prolonged (> 1 hour) dives, most probably to search the water column for prey concentrations (Houghton *et al.*, 2008). Whereas adult cheloniid turtles are primarily benthic feeders of coastal waters, the leatherback is highly pelagic and feeds mainly on jellyfish and other gelatinous zooplankton. It normally occurs in coastal waters only during the breeding season. The largest nesting colonies (2000–4000 females per year) are in the Atlantic, notably along the coasts of French Guiana, Suriname, and Trinidad in the west and in Gabon in the east.

## Life-History Characteristics

Marine turtle species are all large, a feature seemingly necessitated by their high fecundity (at least for reptiles) and long-distance migrations (Hendrickson, 1980). The smallest species, the ridleys, have a maximum carapace length of 75–80 cm, whilst the largest, the leatherback turtle, can attain a carapace length of 1.8 m. The other four species typically reach carapace lengths of around 1 m. Age at sexual maturity and longevity vary greatly. The leatherback, the fastest growing, generally reaches maturity at about 10 years, but green turtles may not attain maturity until they reach 30–40 years and live to 45–60 years (Spotila, 2004).

Several aspects of the life history of marine turtles render them highly vulnerable to human exploitation and disturbance. Females nest on sandy beaches, just above the high-water mark, in tropical to subtropical latitudes. Nesting occurs mostly in spring and summer, the principal exception being the leatherback, which usually nests in autumn and winter. Sea turtles are migratory and can travel extensive, in some cases trans-oceanic, distances between feeding grounds and rookeries. Some green turtle populations migrate from feeding grounds off the coast of Brazil to nest on Ascension Island in the southern mid-Atlantic, a distance of more than 2000 km. Marine turtles in general exhibit a strong fidelity to particular feeding and nesting sites. Some species, most characteristically the ridleys, congregate at nesting beaches en masse in arribadas (arrivals) (Fig. 7.1). During the nesting season, female turtles usually return to their beach with a fortnightly periodicity to lay two to six successive clutches of eggs, averaging about 100 eggs per clutch, but fewer larger eggs in the case of flatback and leatherback turtles. Most species nest mainly at night. Using her flippers the female excavates a hole in the sand to deposit her eggs and backfills the nest. Female ridleys nest every 1–2 years – the remigration interval – but in the other species 2–3 years usually elapses between breeding seasons (Van Buskirk & Crowder, 1994). Incubation time is around 2 months, depending on the species and sand temperature. Hatchlings emerge mostly at night and head immediately for the open sea. There is high natural mortality of eggs and hatchlings as they are taken by a wide variety of predators. Erosion of beaches by cyclones and storms can also cause mass mortality of eggs.

**Fig. 7.1** Mass nesting of olive ridley sea turtles at Ostional Beach, Costa Rica. This arribada in November 2006 was estimated at 62 757 turtles. Republished with permission of Chelonian Research Foundation, from Olive ridley mass nesting ecology and egg harvest at Ostional Beach, Costa Rica, Valverde, R.A., Orrego, C.M., Tordoir, M.T., *et al.*, *Chelonian Conservation and Biology*, 11 (1), pages 1–11, copyright 2012; permission conveyed through Copyright Clearance Center, Inc.

In marine turtles, the sex of the hatchling is not determined genetically but by the incubation temperature during a particular period of embryonic development. There is a pivotal temperature around 28–30°C, depending on the species, that produces a 1:1 sex ratio. A degree or so below produces a predominance of males, whereas a higher incubation temperature results in more females. Since the temperature range that would produce both males and females is narrow, nests usually yield hatchlings of one sex (Davenport, 1997). The sex ratio can thus vary significantly as a result of slight differences in nest temperature not only between sites that are geographically distant but also within rookeries.

For most species there is still a lack of information on the biology of the hatchlings and juveniles, but three main types of life-history pattern can be recognized in terms of the use made of neritic and oceanic waters during development (Bolten, 2003) (Fig. 7.2). The flatback turtle, restricted to the north Australian shelf, is the least migratory, and development appears to be completed in the neritic zone. In the other six species there is an oceanic juvenile stage. This may last for some years, with the juveniles leading a surface pelagic existence and following major currents or gyres. There is evidence that loggerheads, green turtles, hawksbills, and Kemp's ridleys spend their early juvenile period associated with *Sargassum* weed or with other materials concentrated along driftlines and convergences where they seek shelter and food. Eventually they move to coastal waters, become

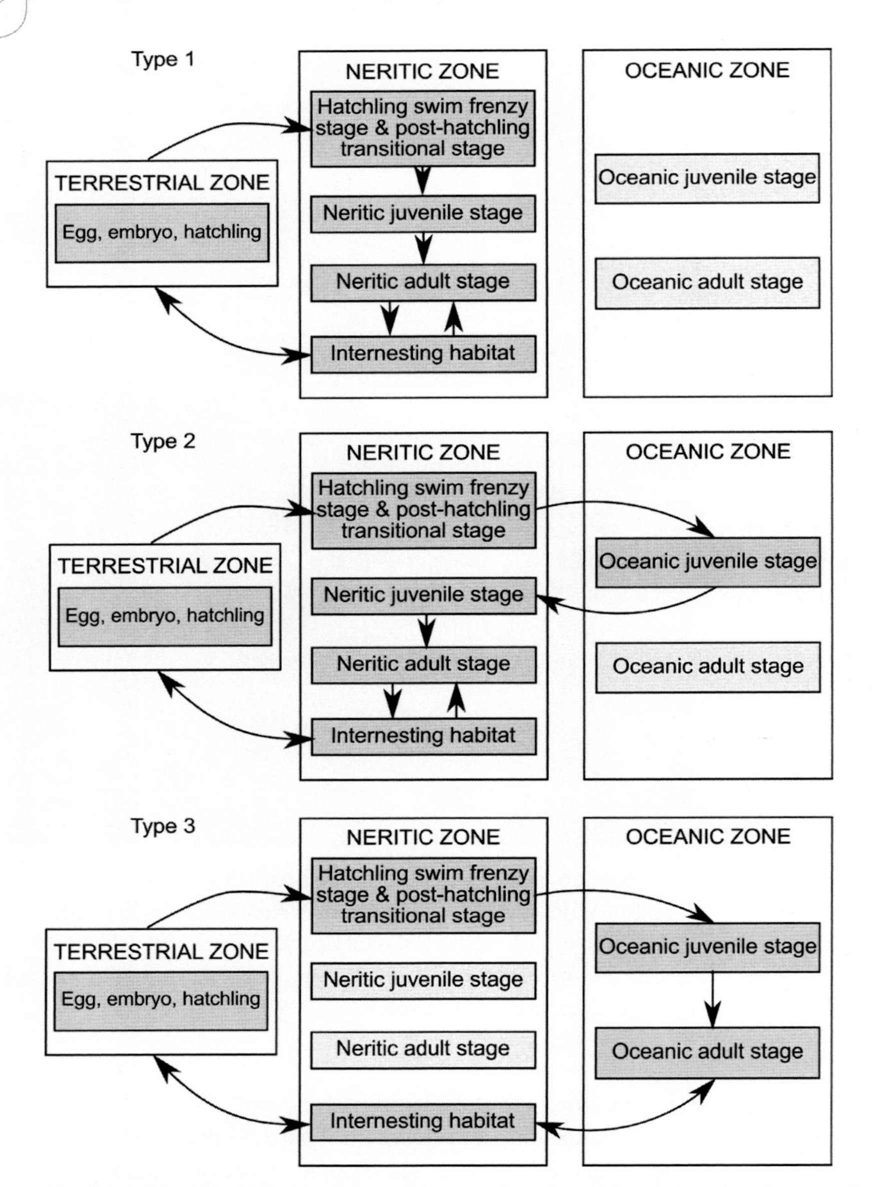


**Fig. 7.2** Major life-history patterns of sea turtles. Type 1 (complete development in the neritic zone) is shown by the flatback. Type 2 (early juvenile development in the oceanic zone and later juvenile development in the neritic zone) is shown by the loggerhead, green, hawksbill, and Kemp's ridley. Type 3 (complete development in the oceanic zone) is shown by the leatherback and olive ridley. Copyright © 2003 from Variation in sea turtle life history patterns: neritic vs. oceanic developmental stages, by A.B. Bolten, in *The Biology of Sea Turtles* Vol. II, ed. P.L. Lutz, J.A. Musick & J. Wyneken. Reproduced by permission of Taylor and Francis Group, LLC, a division of Informa plc.

bottom-feeders, and may remain in the neritic zone for maybe a decade before reaching maturity. With the leatherback and olive ridley, however, development is completed in the oceanic zone, and adults return to the neritic zone only for reproduction.

Sea turtles display strong homing behaviour, returning to their natal region to reproduce. Tagging and mitochondrial DNA (mtDNA) data have shown that natal homing by females predominates in all species, although the geographic precision varies considerably. Green and loggerhead females, for example, are known to return to their natal beach or at least within distances of tens of kilometres to nest, whereas olive ridleys and leatherbacks appear to be less specific. To what extent males undertake migrations and homing is unclear, but nuclear DNA data – as opposed to solely maternally inherited mtDNA – indicate that they can provide significant gene flow between nesting colonies. Nevertheless, homing by females imparts strong population structure (Bowen & Karl, 2007) (Fig. 7.3) and has important conservation implications.

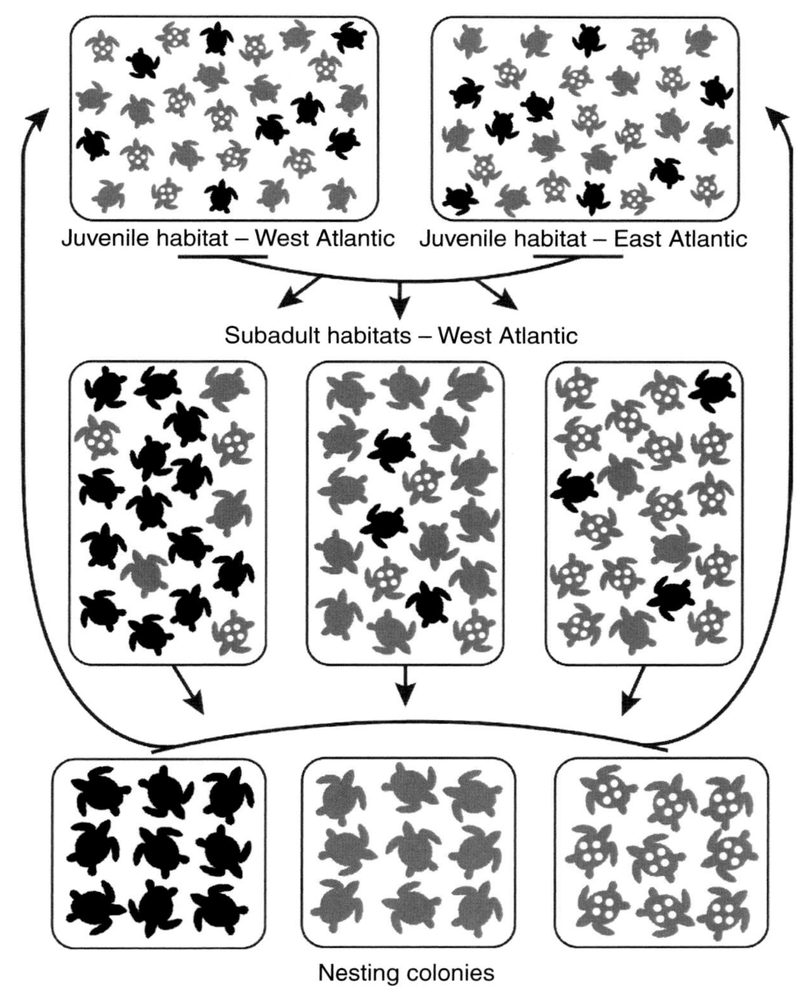


**Fig. 7.3** A model of loggerhead turtle population structure based on three hypothetical nesting colonies and the main life-history stages. Breeding females (and possibly males) show strong natal homing so that the nesting colonies show marked population structure. By contrast, the resulting juveniles intermingle during their oceanic existence and show no such structure. Subadults, however, recruit to coastal habitat near their natal rookery, introducing significant population structure, which is then further increased by the females' fidelity to nesting sites. There is thus increasing population structure from the juvenile to subadult to adult stages. Republished with permission from Bowen, B.W., Bass, A.L., Soares, L. & Toonen, R.J. (2005). Conservation implications of complex population structure: lessons from the loggerhead turtle (*Caretta caretta*). *Molecular Ecology*, 14, 2389–402. © 2005 Blackwell Publishing Ltd.

Life-cycle characteristics make sea turtle populations especially difficult to census and monitor. It is generally feasible to count only nesting females, and even then extensive tag-recapture data may be needed over many years to take sufficient account of the remigration interval and the sometimes huge natural variation in the number of females observed to nest in a given season (Pritchard, 1980). Tagging itself is a problem given the longevity of the animals. Unfortunately, conventional tags of monel alloy or plastic attached to one fore flipper often do not last long enough, and the preferred methods are now to tag both fore flippers with titanium or inconel alloy tags or to use such external tags together with internal microchips inserted into the shoulder muscle (Balazs, 1999). Surveys of nesting females take no account of the adult males and the juveniles of the population. A population may appear relatively stable over a number of years from beach counts, but if there is excessive mortality of eggs or juveniles, this may only become manifest with the crash of nesting females after two or three decades given the late age to maturity for most species (Seminoff & Shankar, 2008; Mortimer, 1991) (Fig. 7.4).

Apart from the flatback, for which population information is inadequate, sea turtle species are categorized by the IUCN as threatened, with the hawksbill and Kemp's ridley assessed as critically endangered, the green turtle as endangered, and the loggerhead, leatherback, and olive ridley as vulnerable (www.iucnredlist.org). The use of such global assessments for sea turtles (and indeed other taxa with comparable life-history and distributional characteristics) has been criticized since individual populations may be considerably worse or better off than the overall categorization indicates and may not as a result attract an appropriate level of conservation effort. Local and regional assessments and monitoring of long-term trends are thus important (Seminoff & Shankar, 2008) (Fig. 7.5).

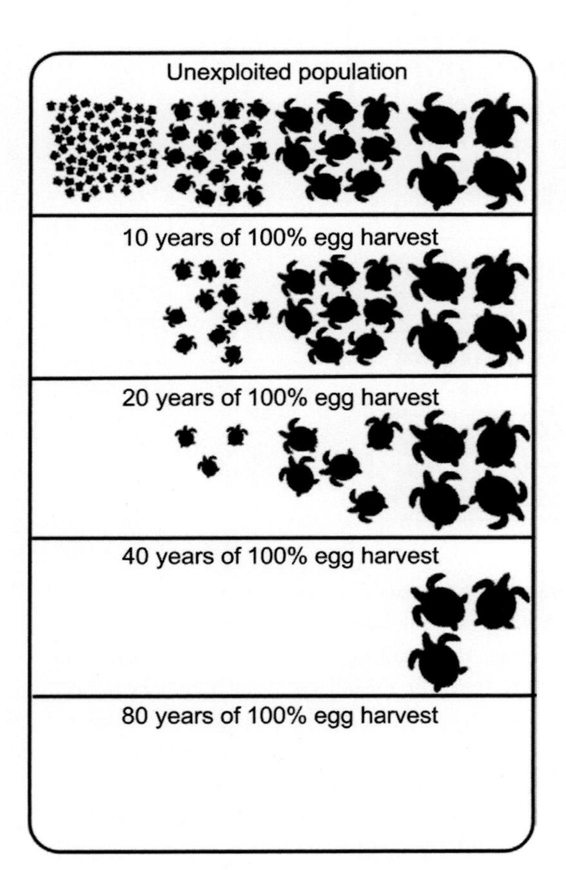


**Fig. 7.4** The effect of excessive egg harvesting on a marine turtle population. Because of the late age at maturity the fate of this population will be hidden for decades. Reprinted from *Journal of Experimental Marine Biology and Ecology*, 356, Seminoff, J.A. & Shankar, K., Marine turtles and IUCN Red Listing: a review of the process, the pitfalls, and novel assessment approaches, pages 52–68, copyright 2008, with permission from Elsevier. Mortimer, J.A. (1991). *Recommendations for the Management of Marine Turtle Populations of Pulau Sipadan, Sabah.* Report to WWF Malaysia, Project No. 3868. 36 pp., with permission from WWF Malaysia.

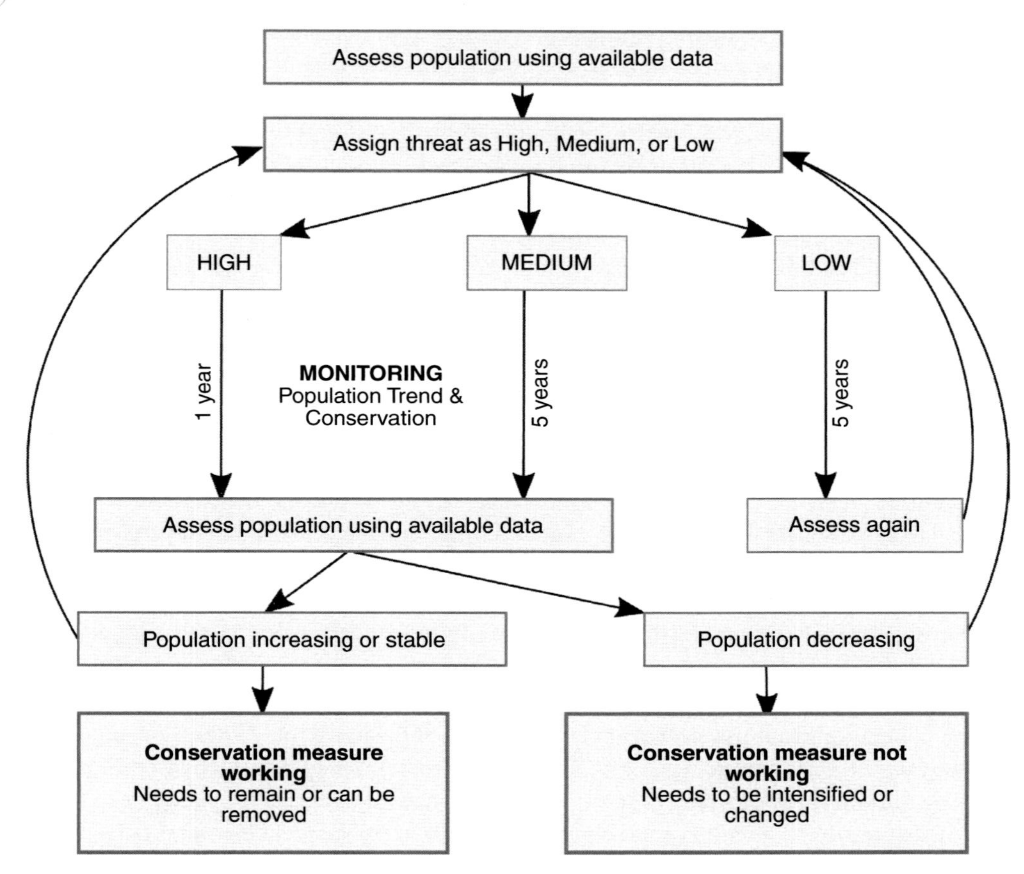


**Fig. 7.5** A proposed conservation assessment framework for focal species such as marine turtles. Reprinted from *Journal of Experimental Marine Biology and Ecology*, 356, Seminoff, J.A. & Shankar, K., Marine turtles and IUCN Red Listing: a review of the process, the pitfalls, and novel assessment approaches, pages 52–68, copyright 2008, with permission from Elsevier.

## Exploitation

Wherever they occur, sea turtles have been used as a source of meat and eggs for human consumption as well as for various other products, and there are innumerable examples of excessive exploitation and the collapse of populations (Thorbjarnarson *et al*., 2000).

Most sought after for its meat is the green turtle. The wider Caribbean provides a notorious example of the systematic destruction of green turtle populations (King, 1995). European exploitation of turtles had already begun there in the sixteenth century, largely for the revictualling of ships, but far more intensive exploitation was to follow, resulting in the serial destruction of large rookeries on Bermuda, the Greater Antilles, Bahamas, Florida, and Dry Tortugas. Even as early as 1620 the Bermuda Assembly had introduced legislation prohibiting the killing of turtles less than 'Eighteen inches in the Breadth or Dyameter' (Carr, 1984). It is believed the largest rookery was on the Cayman Islands, which for many years supported an important turtling industry. Large-scale commercial exploitation of green turtles and eggs commenced in the Cayman Islands in the 1650s, initially in order to supply the new British colony of Jamaica. By the late eighteenth century the turtle populations on the Caymans had been virtually destroyed, and the Cayman turtling fleet moved to southern Cuba. This population too was rapidly depleted, and by 1840 the fleet moved on again, this time to the Nicaraguan coast and cays where extensive turtle feeding grounds occurred. Nicaragua itself established slaughterhouses in the 1970s, processing several thousand turtles annually for several years, although by this stage turtle populations in Central American waters had been effectively decimated. Similarly, massive numbers of olive ridleys were landed in Mexico and Ecuador in the 1960–70s for their hide (turtle leather) and meat, up to tens of thousands per season, in some cases involving the slaughtering of almost entire arribadas (Groombridge & Wright, 1982). The abundance of green turtles in the Caribbean was a key factor that facilitated European exploration and colonization of the New World tropics. The ecological extinction of green turtles in the Caribbean, as the dominant herbivore of seagrass meadows, has undoubtedly resulted in significant changes to these food webs (Bjorndal & Jackson, 2003).

Today, some 40 countries allow for the direct take of marine turtles (Humber *et al.*, 2014). This amounts to an annual legal catch of more than 42 000 animals, most of them (80%) green turtles, with Papua New Guinea and Nicaragua accounting for nearly 60% of the total global take. However, illegal take and bycatch are probably far more significant sources of mortality.

The taking of eggs has also been a major cause of decline in most of the species. There are many beaches where nearly all eggs have been collected, leading to the collapse of populations (Thorbjarnarson *et al.*, 2000). For example, massive exploitation of green turtle eggs has taken place on the Sarawak Turtle Islands where in the late 1920s to mid-1930s up to 2–3 million eggs were collected annually; however, by the mid-1960s the annual take had declined sharply to around 0.3–0.5 million (King, 1995). Limited, strictly managed collecting of eggs by local people may be appropriate as a conservation measure given that a female may only need to nest for a couple of seasons for the value of her eggs to exceed that of her meat (Groombridge & Wright, 1982; Campbell, 1998).

Alternatively, there may be opportunities for non-consumptive use of turtles, in particular through ecotourism. This can provide a sustainable source of income for local communities and raise awareness of conservation issues, importantly among the upcoming generation (Vieitas *et al.*, 1999; Witherington & Frazer, 2003). Turtle watching has focused on beach nesting but may also involve diving and snorkelling in nearshore habitats used by juvenile and adult turtles. Such tourism ventures need to be carefully managed to avoid adverse impacts on the behaviour and physiology of turtles (Landry & Taggart, 2010).

A number of studies have reported recovery of depleted populations once turtles are protected from human exploitation. For instance, six of the world's green turtle nesting populations – four in the Pacific and two in the Atlantic – have shown significant increases over 25 years or more, mainly attributable to protection of eggs and turtles (Chaloupka *et al.*, 2008). The largest green turtle nesting aggregation in the South Atlantic at Ascension Island has, for example, recovered strongly since legal protection and the end of the commercial harvesting that had severely depleted the population (Weber *et al.*, 2014). The average number of nests per year increased from about 3700 in 1977–82 to 23 700 in 2010–13 (Fig. 7.6), although the current population of adult females, about 14 800, is still below conservative estimates of pre-human abundance.

The chief threat to the hawksbill turtle has been the trade in tortoiseshell – the patterned scutes of the carapace and plastron – which are highly valued for making jewellery and other artefacts. Trade in tortoiseshell is known to have been important to a number of ancient cultures in widely separate parts of the world. In modern times, the trade increased immensely, particularly since

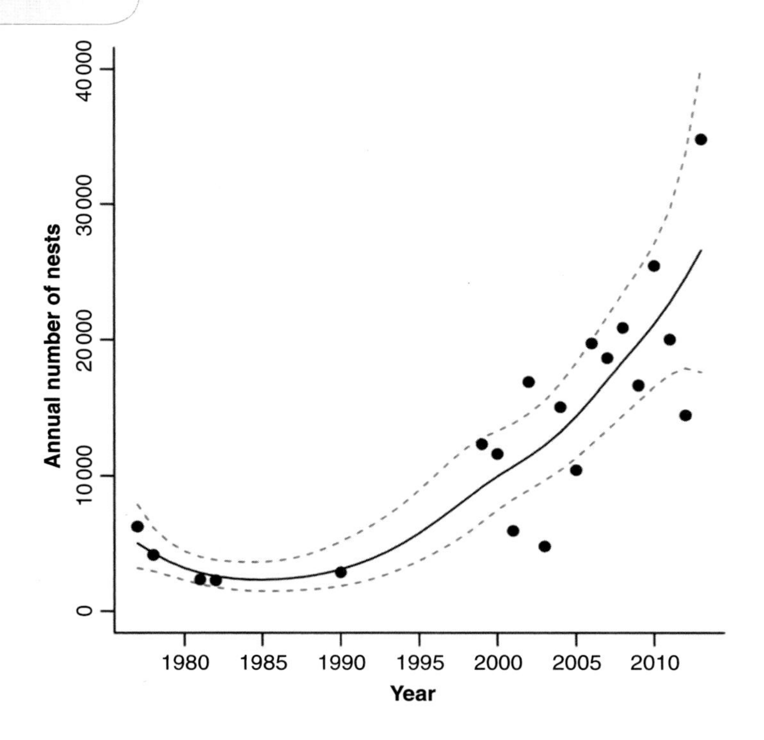


**Fig. 7.6** Long-term trend in the annual number of green turtle nests at Ascension Island. Trend line is a cubic smoothing spline fitted using a generalized additive model. Broken lines show the 95% confidence envelopes for the smoothes. Springer *Biodiversity and Conservation*, Recovery of the South Atlantic's largest green turtle nesting population, volume 23, 2014, pages 3005–18, Weber, S.B., Weber, N., Ellick, J., *et al.*, © Springer Science+Business Media Dordrecht 2014. With permission of Springer.

the early 1970s, but has since come under international and national controls. By far the largest consumer in the twentieth century has been Japan, where hawksbill tortoiseshell, known as bekko, is used for ornamental carvings. Japan, for instance, imported more than 1.3 million large hawksbills between 1950 and 1992 and more than 575 000 juveniles – which are prepared as tourist curios – between 1970 and 1986. Legal Japanese imports ceased at the end of 1992, but in many countries domestic and illegal international trade continues to be a problem (Mortimer & Donnelly, 2008).

All species of marine turtle are listed in Appendix I of the Convention on International Trade in Endangered Species of Wild Fauna and Flora (CITES) (see Chapter 4), so that international commerce in marine turtles and their products is forbidden by signatory nations. But in practice, marine turtles have probably been traded more heavily than any other Appendix I species (Wells & Barzdo, 1991). Some trading countries are not a party to the convention, not all signatories adhere to its conditions, and some have entered reservations on the Appendix I listing for certain species. Japan, which has been by far the largest importer of turtle products, ratified CITES in 1980 but placed reservations on hawksbill, green, and olive ridley turtles, thereby exempting itself from the ban on their trade. But under political pressure, Japan eventually withdrew its CITES reservation in 1994 (Bowman *et al.*, 2010).

## Incidental Capture

Marine turtles commonly occur as bycatch in various fisheries, and many population declines are likely attributable to incidental catch (Lutcavage *et al.*, 1997). Trawl fisheries, gillnets, longlines, and entanglement in lost or discarded fishing gear are mainly responsible. Worldwide bycatch in trawl, gillnet, and longline fisheries from 1990 to 2008 has been reported at about 85 000 turtles (Wallace *et al.*, 2010). However, this is believed to represent less than 1% of the true total bycatch given that the data come mainly from onboard observer reports, which cover only a small proportion of fishing effort, and the dearth of information from small-scale fishing activities. Marine turtle bycatch data vary greatly, and there is a particular need for increased monitoring to better assess impacts and priorities for mitigative measures. There is evidence, for instance, of very high bycatch for some populations (e.g. loggerheads and Kemp's ridleys in NW Atlantic trawls, North Pacific loggerheads in nets and longlines, and olive ridleys in North Indian Ocean trawls) (Wallace *et al.*, 2013).

In US coastal waters, shrimp trawling has been identified as the most important anthropogenic source of mortality for juvenile to adult turtles. From records of sea turtle bycatch for US fisheries between 1990 and 2007, Finkbeiner *et al.* (2011) conservatively estimated an annual mortality of 71 000 turtles before the implementation of bycatch mitigation measures. The great majority (>80%) of these deaths were attributable to the South-east and Gulf of Mexico shrimp trawl fishery and mainly impacted Kemp's ridleys.

Ways of reducing turtle deaths from incidental capture include spatial and temporal restrictions on fishing effort for areas or at times of high turtle concentrations, such as off nesting beaches. There are also ways to modify trawl gear so as to let turtles escape if they are caught, and several turtle excluder devices (TEDs) have been designed. Most incorporate a grille that allows the catch to pass through to the codend but that deflects turtles out through the top or bottom of the net (Fig. 7.7). The use of TEDs has become mandatory in a number of trawl fisheries around the world. All US shrimp trawlers fishing in US waters are required to use approved types of TEDs. The regulation came into effect in 1989 but was amended in 2003 to require TEDs with larger escape openings so as to provide more effective protection for the larger sea turtle species, not just the smaller species such as Kemp's ridley (Mukherjee & Segerson, 2011). Bycatch data indicate that since the 2003 TED enlargement regulation, sea turtle mortality has dropped by more than 90% in the South-east to Gulf of Mexico shrimp trawl fishery (Finkbeiner *et al.*, 2011).

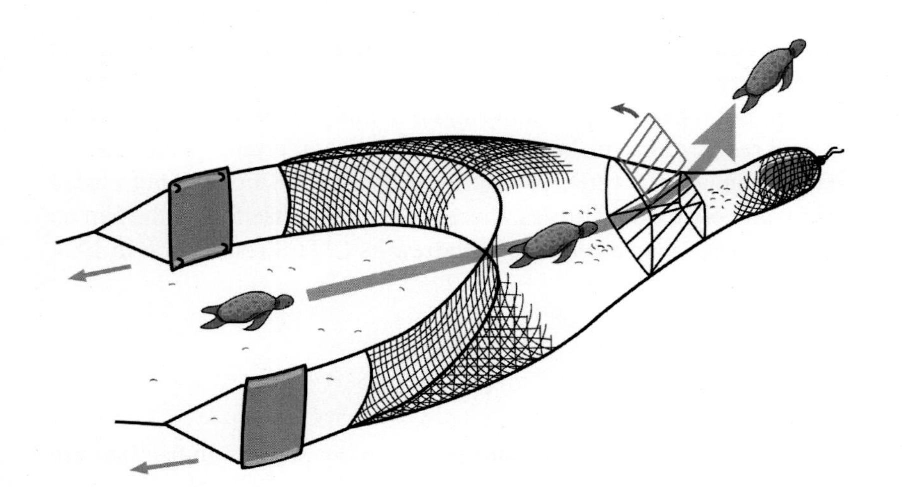

**Fig. 7.7** Turtle excluder device.

## Pollution

Marine turtles are potentially vulnerable to a range of pollutants, including heavy metals, hydrocarbons, and persistent organic pollutants (POPs) (see Chapter 2), and given their long life spans they are liable to bioaccumulate metals and POPs. However, information on tissue burdens of pollutants is relatively sparse, and even less is known about the effects of such substances on turtles and the degree to which fitness might be undermined (Lutcavage *et al.*, 1997).

Ingestion of anthropogenic debris, in particular plastics, and entanglement in ropes and ghost nets appear to be significant sources of mortality for turtles (Schuyler *et al.*, 2014; Gall & Thompson, 2015). Even ingesting small amounts of plastic debris can kill juvenile turtles and may have significant effects on demography through absorption of toxins (Bjorndal *et al.*, 1994; Santos *et al.*, 2015). By their epipelagic habit, turtle hatchlings appear to be especially vulnerable to ingesting debris. Turtles spend their early life in the open ocean, often associated with pelagic driftlines (Witherington *et al.*, 2012). Buoyant debris, including tar balls and styrofoam beads, also tends to concentrate in such zones and may be mistaken as food. Whilst ingestion may not be fatal, there may be significant sublethal effects because of a limited ability of turtles to compensate for inert debris diluting their diet (McCauley & Bjorndal, 1999). Observance of Annex V of MARPOL (see Chapter 2) should reduce the incidence of turtle mortality attributable to plastics, although from gut contents of dead turtles found along the south Texas coast, Shaver and Plotkin (1998) found no significant difference in the proportions of turtles that had ingested debris before and after the implementation of MARPOL.

Sea turtles are potentially very vulnerable to effects of global climate change. A warming of 2°C could result in sex ratios highly skewed towards females at many nesting grounds, in addition to which sea-level rise and increased storm activity associated with global warming could adversely impact the nesting beaches themselves. And natal homing behaviour means that females are unlikely to seek alternative nesting sites (Davenport, 1997; Poloczanska *et al.*, 2009). Projections based on climate change models indicate that survival of eggs and hatchlings of eastern Pacific leatherback turtles will decline by about 50–60% over the next century due to warmer, drier conditions (Santidrián Tomillo *et al.*, 2012).

## Species and Habitat Protection

Many countries have legislation that provides, at least nominally, for total or partial protection of sea turtles, and numerous reserves have been designated to include nesting beaches, although often in practice such provisions cannot be adequately enforced (Groombridge & Wright, 1982). Rigorous enforcement of protection is essential in the case of nesting concentrations. That rookeries can be genetically distinct as a result of natal homing by females indicates that management plans needs to be colony-specific, as protection afforded to one colony may not be of immediate benefit to others. This means that rookeries that have been destroyed may not recolonize by natural recruitment, such as the major green turtle nesting colonies in the Caribbean extirpated centuries ago. It means too that small nesting colonies should not be overlooked as they can contribute significantly to maintaining genetic diversity and thus a species' resilience (Seminoff & Shankar, 2008).

In many parts of the world turtle nesting habitat has been lost or severely degraded as a result of tourist and other coastal development. Bright artificial lights along beachfronts can also discourage females from nesting and disorientate hatchlings as they move towards the sea. The presence of humans on beaches can disturb nesting females. Off-road vehicles can compact the sand over the nests, crush eggs, and leave ruts that trap hatchlings, and power boats and jet skis can disturb and collide with turtles (Lutcavage *et al.*, 1997; Davenport & Davenport, 2006). Also, as available nesting habitat contracts, the carrying capacity of a beach may be exceeded, limiting hatchling production (Mazaris *et al.*, 2009). Predation of eggs and hatchlings can be exacerbated by introduced mammalian predators, such as feral dogs and pigs (Stancyk, 1995).

Whilst much of the focus of sea turtle conservation is on nesting beaches – where management is most practicable – other critical habitats may need protecting, such as feeding grounds, internesting areas, and migration routes. The largest nesting colony of leatherback turtles in the Eastern Pacific is at Playa Grande in Costa Rica. Extensive satellite tracking data show that after nesting the turtles travel south, mainly during February to April, to presumed foraging grounds in the South Pacific gyre. With the identification of such migration corridors and other open sea high-use areas, possible spatio-temporal protection measures can be examined so as to mitigate the impact of fisheries interactions (Shillinger *et al.*, 2008).

Since marine turtles are highly migratory and cross jurisdictional boundaries, conservation initiatives need to be developed in collaboration with both regional and international agencies concerned with open ocean resources, including UN bodies and the various regional fisheries management organizations and tuna commissions (see Chapter 6). A treaty concerned specifically with sea turtle conservation is the Inter-American Convention for the Protection and Conservation of Sea Turtles. It entered into force in 2001, currently has 15 contracting parties, and has potential to strengthen protection of turtles throughout the Americas. Concerns, however, expressed by Campbell *et al.* (2002) include a mismatch in scale between the need for local community-based initiatives through the use of an international instrument and inclusion of an exception clause concerning domestic use. International agreement under the Bonn Convention (see Chapter 4), which deals with the conservation of migratory animals, also has potential to be of considerable benefit to marine turtle conservation, though a number of key countries are not yet parties (Hykle, 2002). All species except the flatback are listed in Appendix I, and the convention includes two memoranda of understanding relating specifically to sea turtles of particular regions (Atlantic coast of Africa and Indian Ocean and SE Asia). Wold (2002) provides a comprehensive review of international instruments relevant to sea turtle conservation. Overall, however, it is questionable how effective international instruments have been in promoting sea turtle conservation. Tiwari (2002) recommends that non-governmental organizations could help significantly in this regard by playing a greater role in awareness and promotion of such agreements.

## Relocation and Captive Rearing

Certain management practices are aimed at increasing the survival of eggs and hatchlings, in particular, egg relocation and captive rearing. Relocation is widely used where eggs are in danger of being washed away or taken by poachers. Such intervention is, however, controversial. A particular concern is that if nest-site preference is an inherited trait, then relocating eggs that are otherwise doomed to a less vulnerable location may distort the gene pool, but to what extent this may be a general issue is unclear. However, for the largest nesting population of loggerheads in eastern Australia, Pfaller *et al.* (2008) found that virtually all females selected at least one unsuccessful nest site and that experienced breeders selected more successful sites, indicating that relocation is unlikely to introduce artificial selection and may be an appropriate conservation tool.

Moving eggs entails some risk. It can fatally disrupt the extra-embryonic membranes if movement occurs at a critical period of development (Blanck & Sawyer, 1981), and because of the effect of incubation temperature on sexual differentiation, relocating eggs may alter the sex ratio. Transplanting eggs to distant beaches in an attempt to establish new breeding sites or to re-establish former nesting populations may not succeed if hatchlings normally return to their natal beach. The very restricted nesting and threatened status of Kemp's ridley renders it especially vulnerable, and since 1978 efforts have been made to re-establish a second nesting site on Padre Island National Seashore, a protected area in SW Texas. The programme has incorporated experimental procedures including artificial imprinting of hatchlings on sand from Padre Island (on the supposition that olfactory imprinting of the natal beach takes place) and head-starting of hatchlings (to 1992). Some imprinted, head-started turtles have returned to nest on Padre Island, but whether the project will significantly contribute to the number of nesting females remains to be seen (Shaver & Caillouet, 2015).

Turtle hatchlings are vulnerable to high mortality from predation and other factors. Head-starting is thus the practice of raising hatchlings in captivity for at least several months so that when released they are at a size where their survival rate may be improved. Critics, however, question whether head-starting is cost-effective and can contribute significantly to breeding populations. Population models indicate that protecting sub-adults and adults is far more beneficial than protecting hatchlings. Also, abnormal conditions of captivity may impair essential behavioural and physiological responses and increase the risk of disease (Burke, 2015).

Other techniques that have generated much debate as to their conservation value are ranching and farming (Ross, 1999). Ranching entails taking eggs or hatchlings from the wild with the aim of raising turtles to a marketable size. Although farming relies initially on wild populations, the goal eventually is for the captive population to be self-sustaining. Starting with eggs taken from the wild, some farming operations have succeeded in raising sexually mature turtles that have produced fertile eggs and hatchlings. However, to rear succeeding generations in captivity and maintain a closed system as a viable commercial concern has yet to be demonstrated. A potential consequence of commercial ranching and farming ventures is that since their success depends on promoting trade and developing new markets, opportunities for illegal trade in wild stocks may emerge, thereby undermining conservation efforts (Dodd, 1995).

Captive rearing techniques have proved popular, but it is debatable whether they can contribute to sea turtle conservation, except maybe as a last resort. Rather than manipulative management, efforts should focus on the underlying causes of turtle declines (Meylan & Ehrenfeld, 2000).

## Key Issues

As we have seen, a number of aspects of the biology of sea turtles have important implications for their conservation – in particular, that they are large, long-lived, and late-maturing, wide-ranging species with complex population structure and with life cycles that involve oceanic and coastal phases and intertidal nesting. There are still, however, major gaps in our knowledge of sea turtle biology and ecology, of the status of populations, and of the effects of human impacts. Research addressing such shortcomings is urgently required to properly inform management and conservation (Hamann *et al.*, 2010). Important too are analyses of where conservation efforts can be most gainfully directed for the species and circumstances in question. To identify global conservation priorities for marine turtles, Wallace *et al.* (2011) evaluated the relative risks and threats facing each biogeographically discrete population segment, or regional management unit (RMU), 58 RMUs having been identified. Risk criteria were population size, recent trend, long-term trend, rookery vulnerability, and genetic diversity; and threat criteria were fisheries bycatch, take, and coastal development. Eleven RMUs scored high risk and high threat, five of them within the Indian Ocean, indicating the most pressing needs for conservation action.

Turtle conservation efforts have largely focused on the eggs. Programmes aimed at reducing egg mortality are essential, especially on beaches where nesting is critically low or where excessive egg collecting is a threat. But also particularly effective are likely to be measures that address mortality of juveniles and adults, such as from direct exploitation, international trade, and fisheries bycatch. Other key conservation issues concern the preservation of nesting habitat and improved public education and involvement of local communities in conservation. Ultimately, sea turtles need to be restored to their roles in marine ecosystems.

## Other Marine Reptiles

### Sea Snakes

Sea snakes inhabit tropical regions of the Pacific and Indian Oceans, mainly in sheltered coastal and estuarine waters. There are some 70 species, all of the family Elapidae, with the highest diversity in the waters of northern Australia, Malaysia, and Indonesia. The great majority (the subfamily Hydrophiinae) are ovoviviparous and entirely aquatic. In addition, there are a few species of sea krait (subfamily Laticaudinae). Sea kraits spend much of their time out of water and are oviparous, laying their eggs on the shore, and demonstrate a strong fidelity to particular stretches of shore (Heatwole, 1999; Brischoux *et al*., 2009). In general appearance, most sea snakes do not differ greatly from their land relatives, except that all species have an oar-like tail, and many are laterally compressed. Most species attain an average length of about 1 m. They tend to be highly venomous and prey on fish. Most species dive to the sea floor to capture prey and are thus limited to relatively shallow inshore waters. The yellow-bellied sea snake (*Pelamis platura*) can, however, feed at the surface and is a truly oceanic species. It has the widest distribution of all sea snakes, through the tropical Pacific and Indian Oceans (Guinea *et al*., 2010).

Sea snakes are absent from Atlantic waters, as they presumably failed to reach the Pacific coast of America before the final emergence of the Central American isthmus in the Pliocene. The cold Humboldt Current prevents sea snakes spreading south along the west coast of South America, and similarly the cold Benguela Current prevents them from rounding southernmost Africa. The yellow-bellied sea snake would probably thrive in the tropical Atlantic, hence a concern of proposals to construct a Central American sea-level canal because it would likely enable the yellow-bellied sea snake to colonize the Caribbean (Rubinoff & Kropach, 1970).

Sea snakes are exploited for their skins. The commercial export of sea snakes from the Philippines began on a large scale in 1934, mainly for *Pseudolaticauda semifasciata, Laticauda laticaudata*, and *Hydrophis ornatus*. Sea snakes are also taken for human consumption, especially in SE Asia. The beaked sea snake, *Enhydrina schistosa*, is the species most prevalent on many SE Asian coasts and is heavily fished, especially in Malaysia. Most of the sea snake species that have been studied tend to be *K*-selected (see Chapter 4) and thus vulnerable to overexploitation (Heatwole, 1999). Sea snake fisheries are in general unregulated, and some populations are likely to be endangered.

Indirect threats to sea snakes arise through fisheries bycatch (notably prawn trawling), loss and degradation of mangrove and coral reef habitat, coastal pollution, and overfishing of their prey species. Particularly concerning are recent dramatic declines in the abundance and diversity of sea snakes that have been reported for reefs protected from obvious anthropogenic impacts, such as in Australia's Timor Sea and Great Barrier Reef, indicating a number of unidentified factors contributing to overall habitat degradation (Lukoschek *et al*., 2013). Sea snakes remain one of the least known major groups of reptiles and have yet to attract strong conservation interest (Rasmussen *et al*., 2011).

## Saltwater Crocodile

Twenty-five species of crocodile occur through the tropics and subtropics, one of which regularly inhabits estuarine and brackish waters, the estuarine or saltwater crocodile (*Crocodylus porosus*) (Rasmussen *et al.*, 2011; Uetz *et al.*, 2016). It is found from the Bay of Bengal through coastal SE Asia to northern Australia and is the largest living reptile, with some individuals exceeding 6 m in length. An opportunistic carnivore, juveniles eat mainly crustaceans, insects, and small fish, whilst adults take an increasing proportion of larger vertebrates. Females become sexually mature at about 12 years, males at about 16 years, and the maximum age appears to be at least 60 years. Nesting usually occurs during the annual wet season. A nest mound is built of vegetation debris, and the clutch size averages 50 eggs. As in marine turtles, the sex of the embryos is governed by the temperature at which the eggs are incubated. Although a widespread species, *C. porosus* is now rare through much of its range, having been severely depleted by hunting and habitat destruction. Loss of coastal wetlands, especially mangroves, is a major threat (Groombridge, 1987).

Management and conservation programmes for saltwater crocodiles are, however, succeeding in some areas, such as in Papua New Guinea and Australia (Ross, 1998). In Australia's Northern Territory, intense commercial hunting for skins began in 1945 and continued to 1971 when the species became legally protected, though by which time its numbers were severely depleted. But since protection it has essentially recovered (Fukuda *et al.*, 2011). Farming of saltwater crocodiles is important in some countries and has contributed to sustainable use.

Much of the range of *C. porosus* in the Northern Territory lies within Aboriginal lands, and in some areas the animals are revered and have a special ceremonial significance (Lanhupuy, 1987). However, public attitude is a significant issue in the recovery of a wild population of a dangerous animal such as *C. porosus*, and many people remain averse to such conservation efforts. A key element of the crocodile conservation programme in the Northern Territory has been a public awareness and education programme, advocating the need to protect whole ecosystems, including crocodiles, but at the same time promoting public safety (Butler, 1987). Translocating 'problem' animals to remote sites is unlikely to be effective as saltwater crocodiles show strong site fidelity and can travel hundreds of kilometres to return directly to their home location (Read *et al.*, 2007).

The saltwater crocodile is included in CITES Appendix I, except for the populations in Australia, Indonesia, and Papua New Guinea, which are in Appendix II so as to permit trade in skins from farming. CITES has recommended a universal tagging system for crocodile skins in international trade so that their origin can be clearly identified and illegal and unregulated exploitation can be curtailed.

Although reptiles are not well represented in marine environments, they present a wide range of challenges for conservation, especially in the case of marine turtles. Several aspects of their life history, migratory nature, and homing behaviour – involving their use of intertidal, neritic, and oceanic habitats – call for strong management programmes at national levels together with international initiatives to protect turtles. Illegal take and bycatch are serious concerns, as is climate change with its implications for reproductive physiology and nesting habitats. Marine turtles can be considered flagship species (see Chapter 4) given the positive public attention they are able to attract and their potential to symbolize marine conservation issues (Frazier, 2005). By contrast, one of the challenges for the conservation of sea snakes and crocodiles is the often poor public awareness and acceptance they engender. Public education has a vital role to play in the conservation of all marine reptiles. But so too does research; we still only poorly understand important aspects of the biology and ecology of these remarkable animals.

## REFERENCES


Balazs, G.H. (1999). Factors to consider in the tagging of sea turtles. In *Research and Management Techniques for the Conservation of Sea Turtles*, ed. K.L. Eckert, K.A. Bjorndal, F.A. Abreu-Grobois & M. Donnelly. IUCN/SSC Marine Turtle Specialist Group Publication, No. 4, 101–9.

Bjorndal, K.A., Bolten, A.B. & Lagueux, C.J. (1994). Ingestion of marine debris by juvenile sea turtles in coastal Florida habitats. *Marine Pollution Bulletin*, **28**, 154–58.

Bjorndal, K.A. & Jackson, J.B.C. (2003). Roles of sea turtles in marine ecosystems: reconstructing the past. In *The Biology of Sea Turtles*, Vol. II, ed. P.L. Lutz, J.A. Musick & J. Wyneken, pp. 259–73. Boca Raton, FL: CRC Press.

Blanck, C.E. & Sawyer, R.H. (1981). Hatchery practices in relation to early embryology of the loggerhead sea turtle, *Caretta caretta* (Linné). *Journal of Experimental Marine Biology and Ecology*, **49**, 163–77.

Bolten, A.B. (2003). Variation in sea turtle life history patterns: neritic vs. oceanic developmental stages. In *The Biology of Sea Turtles*, Vol. II, ed. P.L. Lutz, J.A. Musick & J. Wyneken, pp. 243–57. Boca Raton, FL: CRC Press.

Bostrom, B.L., Jones, T.T., Hastings, M. & Jones, D.R. (2010). Behaviour and physiology: the thermal strategy of leatherback turtles. *PLoS ONE*, **5**, e13925.

Bowen, B.W., Bass, A.L., Soares, L. & Toonen, R.J. (2005). Conservation implications of complex population structure: lessons from the loggerhead turtle (*Caretta caretta*). *Molecular Ecology*, **14**, 2389–402.

Bowen, B.W. & Karl, S.A. (2007). Population genetics and phylogeography of sea turtles. *Molecular Ecology*, **16**, 4886–907.

Bowman, M., Davies, P. & Redgwell, C. (2010). *Lyster's International Wildlife Law*, 2nd edn. Cambridge, UK: Cambridge University Press.

Brischoux, F., Bonnet, X. & Pinaud, D. (2009). Fine scale fidelity in sea kraits: implications for conservation. *Biodiversity and Conservation*, **18**, 2473–81.

Burke, R.L. (2015). Head-starting turtles: learning from experience. *Herpetological Conservation and Biology*, **10**, 299–308.

Butler, W.H. (1987). 'Living with Crocodiles' in the Northern Territory of Australia. In *Wildlife Management: Crocodiles and Alligators*, ed. G.J.W. Webb, S.C. Manolis & P.J. Whitehead, pp. 229–31. Chipping Norton, NSW: Surrey Beatty & Sons Pty Limited.

Campbell, L.M. (1998). Use them or lose them? Conservation and the consumptive use of marine turtle eggs at Ostional, Costa Rica. *Environmental Conservation*, **25**, 305–19.

Campbell, L.M., Godfrey, M.H. & Drif, O. (2002). Community-based conservation via global legislation? Limitations of the Inter-American Convention for the Protection and Conservation of Sea Turtles. *Journal of International Wildlife Law and Policy*, **5**, 121–43.

Carr, A. (1984). *So Excellent a Fishe: A Natural History of Sea Turtles*, revised edn. New York: Charles Scribner's Sons.

Chaloupka, M., Bjorndal, K.A., Balazs, G.H., *et al.* (2008). Encouraging outlook for recovery of a once severely exploited marine megaherbivore. *Global Ecology and Biogeography*, **17**, 297–304.

Davenport, J. (1997). Temperature and the life-history strategies of sea turtles. *Journal of Thermal Biology*, **22**, 479–88.

Davenport, J. & Davenport, J.L. (2006). The impact of tourism and personal leisure transport on coastal environments: a review. *Estuarine, Coastal and Shelf Science*, **67**, 280–92.

Dodd, C.K. Jr (1995). Does sea turtle aquaculture benefit conservation? In *Biology and Conservation of Sea Turtles*, revised edn, ed. K.A. Bjorndal, pp. 473–80. Washington, DC: Smithsonian Institution Press.

Finkbeiner, E.M., Wallace, B.P., Moore, J.E., *et al.* (2011). Cumulative estimates of sea turtle bycatch and mortality in USA fisheries between 1990 and 2007. *Biological Conservation*, **144**, 2719–27.

Frazier, J. (2005). Marine turtles: the role of flagship species in interactions between people and the sea. *Maritime Studies (MAST)*, **3** (2) & **4** (1), 5–38.

Fukuda, Y., Webb, G., Manolis, C., *et al.* (2011). Recovery of saltwater crocodiles following unregulated hunting in tidal rivers of the Northern Territory, Australia. *Journal of Wildlife Management*, **75**, 1253–66.

Gall, S.C. & Thompson, R.C. (2015). The impact of debris on marine life. *Marine Pollution Bulletin*, **92**, 170–9.
Groombridge, B. (1987). The distribution and status of world crocodilians. In *Wildlife Management: Crocodiles and Alligators*, ed. G.J.W. Webb, S.C. Manolis & P.J. Whitehead, pp. 9–21. Chipping Norton, NSW: Surrey Beatty & Sons Pty Limited.
Groombridge, B. & Wright, L. (1982). *The IUCN Amphibia – Reptilia Red Data Book. Part 1: Testudines, Crocodylia, Rhynchocephalia*. Gland, Switzerland: IUCN.
Guinea, M., Lukoschek, V., Cogger, H., *et al*. (2010). *Pelamis platura*. The IUCN Red List of Threatened Species 2010: e.T176738A7293840 (accessed 2 March 2016).
Hamann, M., Godfrey, M.H., Seminoff, J.A., *et al*. (2010). Global research priorities for sea turtles: informing management and conservation in the 21st century. *Endangered Species Research*, **11**, 245–69.
Heatwole, H. (1999). *Sea Snakes*, 2nd edn. Malabar, FL: Krieger Publishing Co.
Hendrickson, J.R. (1980). The ecological strategies of sea turtles. *American Zoologist*, **20**, 597–608.
Houghton, J.D.R., Doyle, T.K., Davenport, J., Wilson, R.P. & Hays, G.C (2008). The role of infrequent and extraordinary deep dives in leatherback turtles (*Dermochelys coriacea*). *Journal of Experimental Biology*, **211**, 2566–75.
Humber, F., Godley, B.J. & Broderick, A.C. (2014). So excellent a fishe: a global overview of legal marine turtle fisheries. *Diversity and Distributions*, **20**, 579–90.
Hykle, D. (2002). The Convention on Migratory Species and other international instruments relevant to marine turtle conservation: pros and cons. *Journal of International Wildlife Law and Policy*, **5**, 105–19.
King, F.W. (1995). Historical review of the decline of the green turtle and the hawksbill. In *Biology and Conservation of Sea Turtles*, revised edn, ed. K.A. Bjorndal, pp. 183–8. Washington, DC: Smithsonian Institution Press.
Landry, M.S. & Taggart, C.T. (2010). 'Turtle watching' conservation and guidelines: green turtle (*Chelonia mydas*) tourism in nearshore coastal environments. *Biodiversity and Conservation*, **19**, 305–12.
Lanhupuy, W. (1987). Australian Aboriginal attitudes to crocodile management. In *Wildlife Management: Crocodiles and Alligators*, ed. G.J.W. Webb, S.C. Manolis & P.J. Whitehead, pp. 145–7. Chipping Norton, NSW: Surrey Beatty & Sons Pty Limited.
Lukoschek, V., Beger, M., Ceccarelli, D., Richards, Z. & Pratchett, M. (2013). Enigmatic declines of Australia's sea snakes from a biodiversity hotspot. *Biological Conservation*, **166**, 191–202.
Lutcavage, M.E., Plotkin, P., Witherington, B. & Lutz, P.L. (1997). Human impacts on sea turtle survival. In *The Biology of Sea Turtles*, ed. P.L. Lutz & J.A. Musick, pp.387–409. Boca Raton, FL: CRC Press.
McCauley, S.J. & Bjorndal, K.A. (1999). Conservation implications of dietary dilution from debris ingestion: sublethal effects in post-hatchling loggerhead sea turtles. *Conservation Biology*, **13**, 925–9.
Mazaris, A.D., Matsinos, G. & Pantis, J.D. (2009). Evaluating the impacts of coastal squeeze on sea turtle nesting. *Ocean & Coastal Management*, **52**, 139–45.
Meylan, A.B. & Ehrenfeld, D. (2000). Conservation of marine turtles. In *Turtle Conservation*, ed. M.W. Klemens, pp. 96–125. Washington, DC: Smithsonian Institution Press.
Mortimer, J.A. (1991). *Recommendations for the Management of Marine Turtle Populations of Pulau Sipadan, Sabah*. Report to WWF Malaysia, Project No. 3868. 36 pp.
Mortimer, J.A. & Donnelly, M. (2008). *Eretmochelys imbricata*. The IUCN Red List of Threatened Species 2008: e.T8005A12881238 (accessed 2 March 2016).
Mukherjee, Z. & Segerson, K. (2011). Turtle excluder device regulation and shrimp harvest: the role of behavioral and market responses. *Marine Resource Economics*, **26**, 173–89.
Pfaller, J.B., Limpus, C.J. & Bjorndal, K.A. (2008). Nest-site selection in individual loggerhead turtles and consequences for doomed-egg relocation. *Conservation Biology*, **23**, 72–80.
Poloczanska, E.S., Limpus, C.J. & Hays, G.C. 2009. Vulnerability of marine turtles to climate change. *Advances in Marine Biology*, **56**, 151–211.
Pritchard, P.C.H. (1980). The conservation of sea turtles: practices and problems. *American Zoologist*, **20**, 609–17.
Rasmussen, A.R., Murphy, J.C., Ompi, M., Gibbons, J.W. & Uetz, P. (2011). Marine reptiles. *PLoS ONE*, **6**, e27373.

Read, M.A., Grigg, G.C., Irwin, S.R., Shanahan, D. & Franklin, C.E. (2007). Satellite tracking reveals long distance coastal travel and homing by translocated estuarine crocodiles, *Crocodylus porosus*. *PLoS ONE*, **2**, e949.

Ross, J.P. (ed.). (1998). *Crocodiles. Status Survey and Conservation Action Plan*, 2nd edn. IUCN/SSC Crocodile Specialist Group. Gland, Switzerland and Cambridge, UK: IUCN.

Ross, J.P. (1999). Ranching and captive breeding sea turtles: evaluation as a conservation strategy. In *Research and Management Techniques for the Conservation of Sea Turtles*, ed. K.L. Eckert, K.A. Bjorndal, F.A. Abreu-Grobois & M. Donnelly. IUCN/SSC Marine Turtle Specialist Group Publication, No. 4, 197–201.

Rubinoff, I. & Kropach, C. (1970). Differential reaction of Atlantic and Pacific predators to sea snakes. *Nature*, **228**, 1288–90.

Santidrián Tomillo, P., Saba, V.S., Blanco, G.S., *et al.* (2012). Climate driven egg and hatchling mortality threatens survival of eastern Pacific leatherback turtles. *PLoS ONE*, **7**, e37602.

Santos, R.S., Andrades, R., Boldrini, M.A. & Agnaldo Silva Martins, A.S. (2015). Debris ingestion by juvenile marine turtles: an underestimated problem. *Marine Pollution Bulletin*, **93**, 37–43.

Schuyler, Q., Hardesty, B.D., Wilcox, C. & Townsend, K. (2014). Global analysis of anthropogenic debris ingestion by sea turtles. *Conservation Biology*, **28**, 129–39.

Seminoff, J.A. & Shankar, K. (2008). Marine turtles and IUCN Red Listing: a review of the process, the pitfalls, and novel assessment approaches. *Journal of Experimental Marine Biology and Ecology*, **356**, 52–68.

Shaver, D.J. & Caillouet, C.W. Jr (2015). Reintroduction of Kemp's ridley (*Lepidochelys kempii*) sea turtle to Padre Island National Seashore, Texas and its connection to head-starting. *Herpetological Conservation and Biology*, **10**, 378–435.

Shaver, D.J. & Plotkin, P.T. (1998). Marine debris ingestion by sea turtles in south Texas: pre- and post-MARPOL Annex V. In *Proceedings of the Sixteenth Annual Symposium on Sea Turtle Biology and Conservation*, comp. R. Byles & Y. Fernandez, NOAA Technical Memorandum NMFS-SEFSC-412, pp. 124.

Shillinger, G.L., Palacios, D.M., Bailey, H., *et al.* (2008). Persistent leatherback turtle migrations present opportunities for conservation. *PLoS Biology*, **6**, e171.

Spotila, J.R. (2004). *Sea Turtles: a Complete Guide to their Biology, Behavior, and Conservation*. Baltimore, MD: The Johns Hopkins University Press.

Stancyk, S.E. (1995). Non-human predators of sea turtles and their control. In *Biology and Conservation of Sea Turtles*, revised edn, ed. K.A. Bjorndal, pp. 139–52. Washington, DC: Smithsonian Institution Press.

Thorbjarnarson, J., Lagueux, C.J., Bolze, D., Klemens, M.W. & Meylan, A.B. (2000). Human use of turtles: a worldwide perspective. In *Turtle Conservation*, ed. M.W. Klemens, pp. 33–84. Washington, DC: Smithsonian Institution Press.

Tiwari, M. (2002). An evaluation of the perceived effectiveness of international instruments for sea turtle conservation. *Journal of International Wildlife Law and Policy*, **5**, 145–56.

Uetz, P., Freed, P. & Hošek, J. (ed.). (2016). *The Reptile Database*. www.reptile-database.org (accessed 6 February 2017).

Valverde, R.A., Orrego, C.M., Tordoir, M.T., *et al.* (2012). Olive ridley mass nesting ecology and egg harvest at Ostional Beach, Costa Rica. *Chelonian Conservation and Biology*, **11**, 1–11

Van Buskirk, J. & Crowder, L.B. (1994). Life-history variation in marine turtles. *Copeia*, 1994 (1), 66–81.

Vieitas, C.F., Lopez, G.G. & Marcovaldi, M.A. (1999). Local community involvement in conservation – the use of mini-guides in a programme for sea turtles in Brazil. *Oryx*, **33**, 127–31.

Wallace, B.P., DiMatteo, A.D., Bolten, A.B., *et al.* (2011). Global conservation priorities for marine turtles. *PLoS ONE*, **6**, e24510.

Wallace, B.P., Kot, C.Y., DiMatteo, A.D., *et al.* (2013). Impacts of fisheries bycatch on marine turtle populations worldwide: toward conservation and research priorities. *Ecosphere*, **4**, article 40.

Wallace, B.P., Lewison, R.L., McDonald, S.L., *et al.* (2010). Global patterns of marine turtle bycatch. *Conservation Letters*, **3**, 131–42.

Weber, S.B., Weber, N., Ellick, J., *et al.* (2014). Recovery of the South Atlantic's largest green turtle nesting population. *Biodiversity and Conservation*, **23**, 3005–18.

Wells, S.M. & Barzdo, J.G. (1991). International trade in marine species: is CITES a useful control mechanism? *Coastal Management*, **19**, 135–54.

Witherington, B.E. & Frazer, N.B. (2003). Social and economic aspects of sea turtle conservation. In *The Biology of Sea Turtles*, Vol. II, ed. P.L. Lutz, J.A. Musick & J. Wyneken, pp. 355–84. Boca Raton, FL: CRC Press.

Witherington, B., Hirama, S. & Hardy, R. (2012). Young sea turtles of the pelagic *Sargassum*-dominated drift community: habitat use, population density, and threats. *Marine Ecology Progress Series*, **463**, 1–22.

Wold, C. (2002). The status of sea turtles under international law and international environmental agreements. *Journal of International Wildlife Law and Policy*, **5**, 11–48.

# 8 Seabirds

The term 'seabird' applies to birds that normally feed out at sea and breed on offshore islands or at coastal sites (Furness & Monaghan, 1987; Schreiber & Burger, 2001) as opposed to waders or shorebirds (see Chapter 11), which typically feed on the shore and migrate to inland breeding sites. Major groups of seabirds include penguins (Sphenisciformes); albatrosses, petrels, shearwaters, and fulmars (Procellariiformes); pelicans, tropicbirds, frigatebirds, gannets, and cormorants (Pelecaniformes); and skuas, gulls, terns, and auks (Charadriiformes) (Table 8.1).

**Table 8.1** Main Groups of Seabirds

| Order | Family | Common name | No. of species |
|---|---|---|---|
| Anseriformes | Anatidae | Ducks | 18 |
| Sphenisciformes | Spheniscidae | Penguins | 18 |
| Procellariiformes | Diomedeidae | Albatrosses | 22 |
| | Procellariidae | Petrels, shearwaters | 82 |
| | Hydrobatidae | Storm-petrels | 23 |
| | Pelecanoididae | Diving petrels | 4 |
| Pelecaniformes | Phaethontidae | Tropicbirds | 3 |
| | Fregatidae | Frigatebirds | 5 |
| | Pelecanidae | Pelicans | 3 |
| | Sulidae | Gannets, boobies | 10 |
| | Phalacrocoracidae | Cormorants, shags | 29 |
| Charadriiformes | Laridae | Gulls, terns | 86 |
| | Stercorariidae | Skuas, jaegers | 8 |
| | Alcidae | Auks | 24 |

From Croxall, J.P., Butchart, S.H.M., Lascelles, B., *et al.* (2012). Seabird conservation status, threats and priority actions: a global assessment. *Bird Conservation International*, **22**, Table S1.

Penguins are a flightless, Southern Hemisphere order and particularly significant in the food web of the Southern Ocean (see Chapter 17). Procellariiforms have a worldwide distribution, but their species richness peaks in southern temperate latitudes, possibly related to the wind energy at these latitudes used for their long-distance foraging (Davies *et al.*, 2010). Pelecaniforms, by contrast, are more characteristic of lower latitudes. Tropicbirds and frigatebirds inhabit tropical to subtropical latitudes, whilst pelicans and sulids (gannets and boobies) range from temperate to tropical regions. Cormorants, the most speciose of the pelecaniforms, are cosmopolitan but occur mainly along temperate and tropical coasts. Among the charadriiforms, skuas breed at high latitudes of both hemispheres, but most migrate towards equatorial regions for the non-breeding period. Gulls and terns comprise a species-rich cosmopolitan family, mainly frequenting coastal and inshore waters, whilst the auks are confined to the Northern Hemisphere from polar to temperate latitudes and are especially well represented in the North Pacific.

There are about 350 species of seabirds, although this is only about 3.5% of known bird species (Croxall *et al.*, 2012). Biodiversity hotspots for seabirds occur mainly in southern cool temperate latitudes, with the highest number of foraging seabird species in the New Zealand-Tasman Sea region and around subantarctic islands. High latitude regions have the lowest number of foraging species but support the highest seabird densities (Karpouzi *et al.*, 2007) (Fig. 8.1).

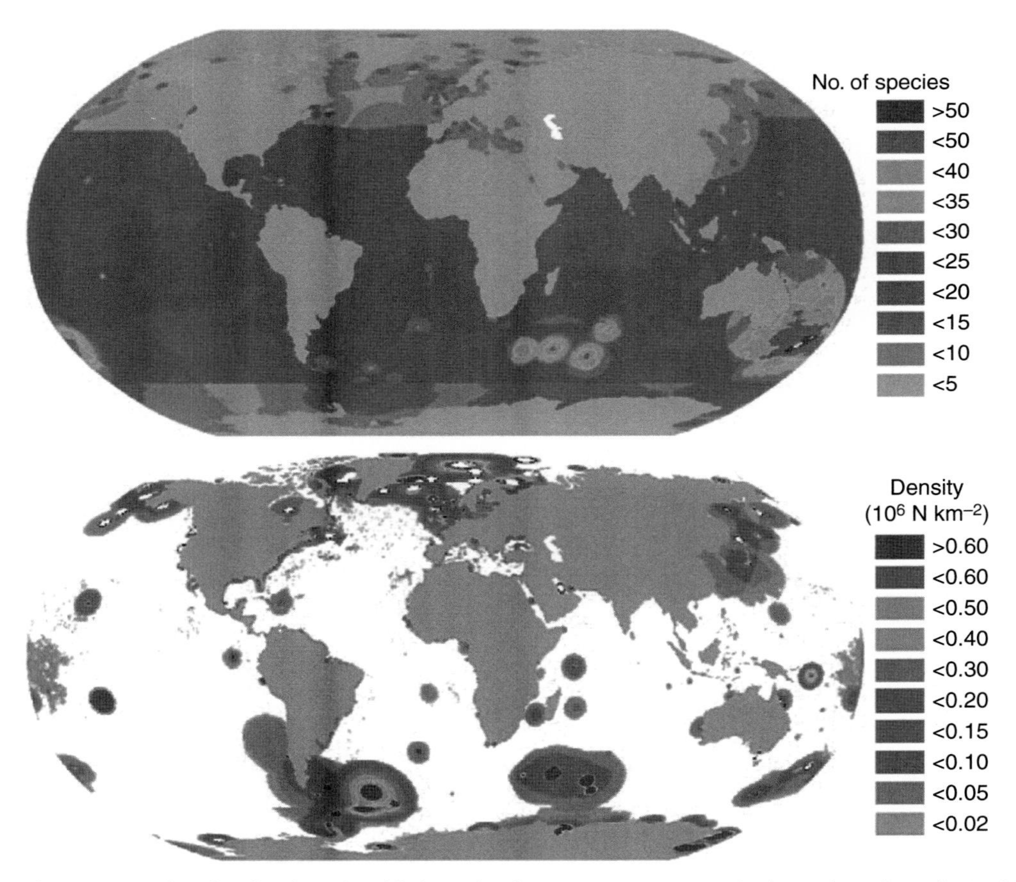


**Fig. 8.1** Foraging distribution of seabird species for an average year, as (top) number of species and (bottom) number (N) of individuals per km$^2$. Reprinted and not modified from Karpouzi, V.S., Watson, R. & Pauly, D. (2007). Modelling and mapping resource overlap between seabirds and fisheries on a global scale: a preliminary assessment. *Marine Ecology Progress Series*, 343, 87–99. With permission from Inter-Research. (A black and white version of this figure will appear in some formats. For the colour version, please refer to the plate section.)

Seabirds are often conspicuous predators in marine ecosystems and have exploited a considerable variety of feeding opportunities in coastal to oceanic habitats (Shealer, 2001) (Fig. 8.2). Most species take small pelagic fishes, crustaceans, and squid, aggregating at localized concentrations of prey such as at areas of enhanced productivity associated with upwellings, frontal regions, and eddies. Their main foraging methods are pursuit diving (notably by penguins, diving petrels, auks, and cormorants), picking prey from the surface (storm-petrels, skuas, gulls, terns, and large petrels), seizing larger prey at the surface (albatrosses and pelicans), and plunge diving (gannets and boobies, tropicbirds, and many terns). Some seabird species feed at night to exploit vertically migrating prey such as squid, lanternfish, and krill. The skuas and gulls include species that prey heavily on other seabirds, mainly during the breeding season. Species may employ different methods opportunistically, and composition of the diet can change seasonally and with foraging area, depending on prey availability.

In their life-history characteristics, seabirds differ markedly from most land birds (Hamer *et al.*, 2001; Schreiber & Burger, 2001). They tend to be much larger bodied and, as such, exhibit characteristics often linked with size. Seabirds are generally long-lived birds – often with a life span of decades – with deferred breeding, small clutch size, and extended periods of incubation and chick

**Fig. 8.2** Seabirds are important predators in marine food webs. The SW Pacific Ocean is a hotspot for seabird abundance and species richness, and included here are several albatross species – wandering (*Diomedea exulans*), northern royal (*D. sanfordi*), southern royal (*D. epomophora*), Chatham (*T. eremita*), shy (*T. cauta*), and black-browed (*T. melanophrys*) – as well as northern giant petrel (*Macronectes halli*) and Cape petrel (*Daption capense*). Photograph: P.K. Probert. (A black and white version of this figure will appear in some formats. For the colour version, please refer to the plate section.)

rearing (Table 8.2). High adult survival rates (typically at least 80%) enable populations to endure extended periods of poor recruitment (Furness & Monaghan, 1987). On the other hand, they are sensitive to changes affecting adult mortality. Seabirds have likely evolved such a lifestyle as pelagic foragers dependent on food sources that can be limited and highly dispersed, and so they need to be adapted to cope with periods of poor food availability and breeding failure. As we have seen with other groups, such *K*-selected traits have crucial conservation implications. So too does the fact that nearly all seabird species are colonial, with breeding colonies often numbering thousands or sometimes millions of birds. Furthermore, many seabird species, especially procellariiforms, return to their natal colony to breed. Scarcity of suitable nesting sites inaccessible to land predators may constrain seabirds to relatively few dense breeding colonies. Such coloniality brings potential disadvantages, such as increased competition for food in the vicinity of the colony and greater vulnerability to human disturbance. But outside the breeding season, seabirds tend to be highly dispersed, spending long periods at sea. A common pattern in procellariiforms is for the young, once they have fledged, to remain at sea until they first breed, and in many seabirds there are well-defined seasonal migrations, in some cases involving movements between opposite hemispheres. This means that outside their breeding aggregations it is difficult to obtain reliable data on the distribution and abundance of seabirds, and even during the nesting period, pre-breeders – which may account for up to half the population – cannot be readily censused.

Large-scale perturbations affecting climate and oceanic productivity can strongly influence seabird populations, which may be compounded by anthropogenic stresses. Dramatic in this respect have been the effects of strong El Niño-Southern Oscillation events. These can have worldwide repercussions (see Chapter 1), including starvation and breeding failure among seabird populations of the central Pacific. High productivity along the Peruvian coast is depressed sharply during an El Niño as upwelling water becomes overlain by an eastward influx of warm nutrient-poor water. Until the late 1960s, the Peruvian upwelling system supported the world's largest fishery, for Peruvian anchovy, a species that is also the main prey of birds responsible for the massive guano deposits of this coast, notably the Guanay cormorant (*Phalacrocorax bougainvillii*), Peruvian booby (*Sula variegata*), and brown pelican (*Pelecanus occidentalis*) (Duffy, 1983; Furness & Monaghan, 1987). Moderate El Niños occur about every 5 years, resulting in average adult mortality of guano birds of

**Table 8.2** Life-History Characteristics of Seabirds and Passerines (Typical Land Birds)

| Life-history characteristic | Seabirds | Passerines |
|---|---|---|
| Body size | Large | Small |
| Age at first breeding | 2–9 years | 1–2 years |
| Clutch size | 1–5 | 4–8 |
| Incubation period | 20–69 days | 12–18 days |
| Chick-rearing period | 30–280 days | 20–35 days |
| Maximum life span | 12–60 years | 5–15 years |

17%, but in severe El Niños, about every 12 years, the adult populations can be nearly halved and with total nesting failure. In the early 1970s, overfishing and recruitment failure of anchovy due to a strong El Niño combined to produce a collapse of both the fishery and the seabird populations. Seemingly, abundance of prey in the past meant the bird populations were able to recover from El Niño crashes, but now the fishery had removed this buffer. The exceptionally severe El Niño of 1982–3 caused enormous upheaval of seabird populations, including more than 80% mortality of the Peruvian guano bird populations, major population decreases of endemic seabirds of the Galápagos, such as the flightless cormorant (*Phalacrocorax harrisi*) and Galápagos penguin (*Spheniscus mendiculus*), and complete reproductive failure of seabirds on Christmas Island in the central Pacific due to major changes in primary productivity and flooding of nest sites by heavy rains (Duffy, 1990). These are extreme examples; nevertheless, seabird populations can fluctuate considerably, and the underlying causes may not be obvious. Some undetected shift in oceanographic conditions may, for example, result in recruitment failure of a key prey species. Mass mortalities of seabirds from apparently natural causes such as from a food shortage or gales occur sporadically, although such events may be aggravated by contaminants or other anthropogenic influences.

Humans can have major impacts on seabird populations, especially at breeding sites through introduction of predators, habitat disturbance, and direct exploitation and at sea from interactions with fisheries and pollution (Boersma *et al.*, 2001; Croxall *et al.*, 2012). The decline and extirpation of populations through human impact can often be due to a combination of threats. This is evident even from the archaeological record, notably with human colonization of islands where effects of predation, introduced species, and removal or alteration of indigenous vegetation have invariably led to large-scale extinctions (Grayson, 2001). On Easter Island in eastern Polynesia, archaeological sites have yielded the bones of 25 species of seabirds, but of these only the red-tailed tropicbird (*Phaethon rubricauda*) nests there today. Easter Island evidently lost more of its indigenous seabirds to prehistoric human activities than other islands of its size in Oceania, but major losses of species have been reported for islands throughout the tropical Pacific as a result of human impact – as indeed for oceanic islands worldwide – with small procellariids hardest hit (Milberg & Tyrberg, 1993; Steadman, 1995).

Numerous seabird populations have been severely depleted by human activities, and some species are now extinct (Table 8.3). In terms of IUCN Red List categories, 28% of seabird species are globally threatened, a higher proportion than for any other comparably speciose group of birds. Pelagic seabirds are particularly imperilled, with penguins and procellariiforms accounting for nearly half of the threatened species. Many of these species are especially vulnerable as they typically have small breeding populations and breed at relatively few sites over a limited range (Croxall *et al.*, 2012).

**Table 8.3** Critically Endangered (CR) and Extinct (EX) Seabirds on the IUCN Red List

| Scientific name | Common name | Status |
|---|---|---|
| Diomedeidae | | |
| *Diomedea amsterdamensis** | Amsterdam albatross | CR |
| *Diomedea dabbenena* | Tristan albatross | CR |
| *Phoebastria irrorata* | Waved albatross | CR |
| Procellariidae | | |

**Table 8.3** (cont.)

| Scientific name | Common name | Status |
|---|---|---|
| *Bulweria bifax* | Small St Helena petrel | EX |
| *Pseudobulweria aterrima* | Mascarene petrel | CR |
| *Pseudobulweria becki* | Beck's petrel | CR |
| *Pseudobulweria macgillivrayi* | Fiji petrel | CR |
| *Pterodroma caribbaea* | Jamaica petrel | CR |
| *Pterodroma magentae* | Magenta petrel | CR |
| *Pterodroma phaeopygia* | Galápagos petrel | CR |
| *Pterodroma rupinarum* | Large St Helena petrel | EX |
| *Puffinus auricularis* | Townsend's shearwater | CR |
| *Puffinus bryani* | Bryan's shearwater | CR |
| *Puffinus mauretanicus* | Balearic shearwater | CR |
| Hydrobatidae | | |
| *Fregetta maoriana* | New Zealand storm-petrel | CR |
| *Oceanodroma macrodactyla* | Guadalupe storm-petrel | CR |
| Fregatidae | | |
| *Fregata andrewsi* | Christmas frigatebird | CR |
| Phalacrocoracidae | | |
| *Phalacrocorax onslowi* | Chatham shag | CR |
| *Phalacrocorax perspicillatus* | Spectacled cormorant | EX |
| Laridae | | |
| *Sterna bernsteini* | Chinese crested tern | CR |
| Alcidae | | |
| *Pinguinus impennis* | Great auk | EX |

*Regarded as a subspecies of *D. exulans* in the World Register of Marine Species.
Extracted from www.iucnredlist.org (accessed 24 June 2016).

## Exploitation

Seabirds tend to be vulnerable to human predation as they are commonly synchronous breeders nesting in dense colonies. Also, once on land they have less mobility, and their often white plumage renders them conspicuous. Seabirds have been exploited mainly for their eggs and meat but have also been killed for feathers and as a source of oil (Feare, 1984). Seabirds have been an important source of food for some, especially more isolated, communities. There is a long tradition of harvesting auks in the North Atlantic, such as in Greenland, Iceland, and the Faeroes, and in some places complex rules were developed that helped sustain the resource (Evans & Nettleship, 1985). The most prevalent form of exploitation has been the taking of eggs (Feare, 1984). Indeed, for reasons of size, palatability, and colonial nesting, seabird eggs have been particularly sought after. Some 80 species of seabirds are known to have been exploited for their eggs, the most heavily targeted being the sooty tern (*Onychoprion fuscatus*), but others include the African penguin (*Spheniscus demersus*), short-tailed shearwater (*Puffinus tenuirostris*), herring gull (*Larus argentatus*), black-headed gull (*L. ridibundus*), Arctic tern (*Sterna paradisaea*), brown noddy (*Anous stolidus*), thick-billed murre (*Uria lomvia*), and common guillemot (*U. aalge*) (Cott, 1953). For the African penguin, which breeds mainly on islands in the productive Benguela Current around southern Africa, massive exploitation of its eggs appears to have contributed significantly to its decline. More than 13 million eggs were removed over the period 1900–30 (Frost *et al.*, 1976). Other egg collecting operations have been similarly staggering (Nettleship & Evans, 1985). For localized populations, removal of even relatively modest numbers of eggs may impact on populations. But removal of eggs is a lesser cost to the population than loss of adults. Eggs will often be replaced within the same breeding season and may provide a sustainable source of food for human communities. The sooty tern is widely exploited for its eggs in parts of the tropics, particularly in the Seychelles where cropping levels of 10–20% of the eggs have been suggested as sustainable (Feare, 1984).

Human exploitation appears to have been directly responsible for many of the known declines of North Atlantic auks, with the common guillemot, thick-billed murre, razorbill (*Alca torda*), and Atlantic puffin (*Fratercula arctica*) suffering heavily. Some colonies had been exterminated by the late 1700s, yet pressure generally increased on these populations from the nineteenth century, as on seabirds generally, in the face of a rapidly increasing human population equipped with more effective boats and firearms (Evans & Nettleship, 1985). Some albatross species were killed in huge numbers during the late nineteenth and early twentieth centuries for their feathers. Feather hunting brought the short-tailed albatross (*Phoebastria albatrus*) of the North Pacific to near extinction by the 1940s with only about 50–60 birds remaining. There are now more than 2500 birds, but more than 80% of the total population breeds on Torishima Island, an active volcano south of Honshu, Japan (Finkelstein *et al.*, 2010).

A notable example of seabird overexploitation is the extinction of the great auk (*Pinguinus impennis*), the largest (< 75 cm) and only flightless member of the recent auks (Bengtson, 1984; Gaston & Jones, 1998). At least in historical times it appears to have occurred mainly in cold-temperate regions of the North Atlantic, particularly on the western side. The largest known colony was on Funk Island, about 60 km north-east of Newfoundland, where an estimated 100 000 pairs bred. From the early sixteenth century, European countries began fishing the rich Newfoundland grounds. Great auk colonies were a convenient source of fresh meat to provision fishing fleets, and a period of mass slaughter ensued. Later, during the eighteenth century, the birds were killed for their feathers, in great demand for stuffing mattresses. The Funk Island colony was extinct by about 1800. Once the species had become rare its demise was hastened by a lucrative trade to secure specimens for collectors. The last reliable sighting was in 1852.

## Disturbance

Humans also intrude upon seabird colonies for reasons other than direct exploitation. Many colonies are the subject of research programmes or are popular tourist attractions. Whilst such activities might ultimately benefit seabird conservation – furnishing data for management, providing revenue for conservation programmes, raising environmental awareness – they nevertheless require careful management.

One area caught up in the rapid growth of wildlife-based tourism is coastal Patagonia, which, since the 1980s, has seen increasing numbers of people visiting its many seabird colonies, with some sites, such as around Península Valdés, attracting more than 100 000 visitors a year (Yorio *et al.*, 2001). Magellanic penguins (*Spheniscus magellanicus*) are one of the main attractions, in this case a species that appears to be relatively tolerant of humans.

The small breeding colony of northern royal albatross (*Diomedea sanfordi*) at Taiaroa Head, south-eastern New Zealand, has also become a major tourist attraction. The colony is within a reserve with strictly controlled public viewing from an observatory. Managed viewing began in 1972, and annual visitor numbers have grown rapidly to some 50 000 by the mid-1990s. At the same time the average number of albatross nests per year has increased to about 20. Although the birds are ostensibly tolerant of tourists, such activities may nevertheless affect breeding behaviour and nesting distribution, and safeguarding both the wildlife resource and the tourist activity requires considerable management intervention (Higham, 1998; Robertson, 1998).

Of particular concern are unregulated recreational activities in coastal areas where people may walk through colonies, on occasions with dogs, or use off-road vehicles. Human intrusions can result in abandonment of nests, making eggs and chicks more vulnerable to predators and eventually reducing breeding success (Boersma *et al.*, 2001). The potential impact of human disturbance on productivity of ground-nesting seabirds is illustrated by a study of brown pelicans (*Pelecanus occidentalis*) in NW Mexico (Anderson & Keith, 1980). In areas disturbed by visitors, on average up to 0.6 young fledged per nest built, in contrast to 1.2–1.5 young produced in undisturbed areas. In this case, effects of disturbance were mostly through losses of eggs and young to predators following short- or long-term nest abandonment.

Many studies have reported negative effects of nature-based recreation on birds. The ecological significance of a disturbance response is likely to depend on the duration, intensity, and extent of the disturbance as well as on the species involved and factors such as age, condition, and reproductive status (Steven *et al.*, 2011). Responses may range from short-term physiological changes to effects at the population level that affect breeding success (Fig. 8.3). Management strategies to minimize disturbance may include the use of sanctuaries, limitation of visitor numbers, and the use of temporal and spatial zoning.

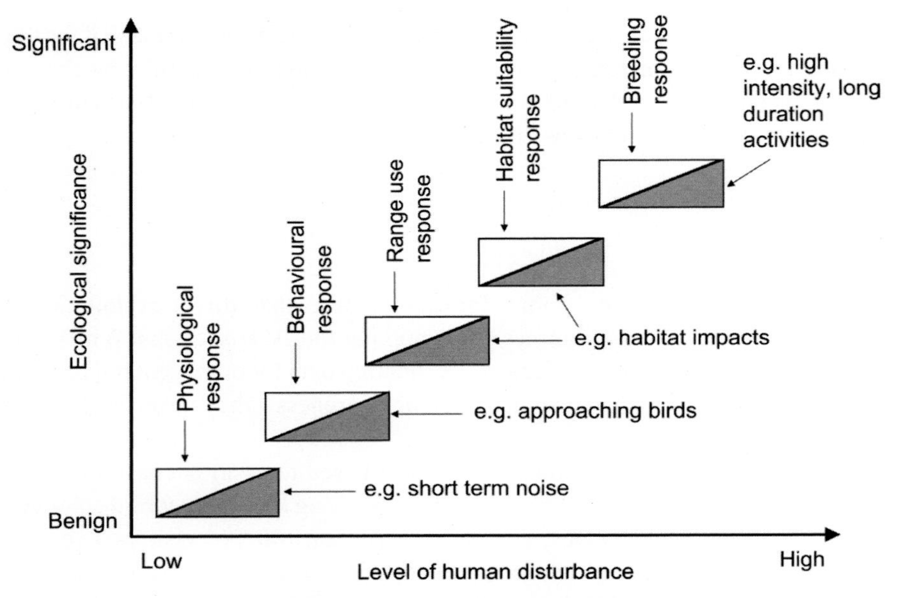


**Fig. 8.3** Conceptual model of the relationship between the level of human disturbance to birds and likely ecological significance. Physiological responses include changes in heart rate or secretion of stress hormone, behavioural responses include changes in foraging and evasion, and population responses include changes in reproductive success (e.g. number of nests built, eggs laid, and chicks hatched or fledged). Reprinted from *Journal of Environmental Management*, Vol. 92, Steven, R., Pickering, C. & Castley, J.G., A review of the impacts of nature based recreation on birds, pages 2287–94, copyright 2011, with permission from Elsevier.

## Introduced Predators

Predators other than humans can also wreak havoc in seabird colonies. Especially damaging are predatory mammals that have been introduced to otherwise inaccessible seabird sites, either intentionally, such as cats, mustelids, mongooses, and foxes, or accidentally, notably rats. Indeed, invasive species at breeding sites represent the main danger to threatened seabirds, impacting some three-quarters of species (Croxall *et al.*, 2012).

Often the most disastrous impacts of invasive species are due to rats and feral cats (Moors & Atkinson, 1984; Nogales *et al.*, 2004; Jones *et al.*, 2008). Such populations can build up rapidly, preying on adults, chicks, and eggs. By 1975, the population of cats on Marion Island in the subantarctic Indian Ocean had reached about 2100 – from the five originally taken there in 1949 to control feral house mice – and was increasing at a rate of 26% per annum. The cats were killing about 450 000 burrowing petrels per year and had already caused the local extinction of the common diving-petrel (*Pelecanoides urinatrix*) (Bester *et al.*, 2002). Cats have also been implicated as mainly responsible for devastating populations of prions and petrels on subantarctic Macquarie Island and of boobies (*Sula dactylatra* and *S. sula*) on Ascension Island as well as for the likely extinction of the Guadalupe storm-petrel (*Oceanodroma macrodactyla*), last sighted in 1912, which bred on Guadalupe Island off Baja California. Species most vulnerable to such predation are small procellariiforms such as prions and petrels and terns, which, apart from being small, generally have inadequate defensive behaviour and often make prolonged feeding flights, leaving eggs and chicks unattended.

Introduced predators have, however, been successfully eradicated from a number of islands. Cats have been removed from about 50 islands, mostly by trapping and hunting (Nogales *et al.*, 2004), and invasive rodents have been eradicated from more than 280 islands, in nearly all cases using poisoned bait (Howald *et al.*, 2007). Such campaigns are costly and can necessitate intensive effort over a number of years, so most such eradications have occurred on small (< 5 $km^2$) islands. In the case of Marion Island (290 $km^2$) – the largest island successfully cleared of cats – a control programme was initiated in 1977 and cats finally eradicated in 1991 using a feline virus followed by hunting, trapping, and poisoning (Bester *et al.*, 2002). Taylor *et al.* (2000) describe the successful eradication of rats from Langara Island, British Columbia. Predation had drastically reduced the island's nesting seabirds, including its population of ancient murrelets (*Synthliboramphus antiquus*) to less than 10% of its historical size. Eradication was effected using bait stations containing an anticoagulant poison (brodifacoum) spaced at a density of about 1 per hectare (equivalent to a rat's home range). Given the island's size (31 $km^2$), this meant nearly 4000 bait stations, indicating the logistical hurdle of tackling larger islands. There are risks associated with such programmes that need to be assessed. In the above case, ravens were secondarily poisoned from scavenging rat carcasses, with probably more than half the local population killed.

Cats were often introduced to islands to control rodent or rabbit populations, and where cats are preying mainly on the rats, not eradicating both simultaneously could in fact increase predation of seabirds; hence the need for an ecosystem-wide approach to such programmes. With the eradication of cats on Little Barrier Island, New Zealand, in 1980, the rat population was released from predation pressure and increased. The breeding success of Cook's petrel (*Pterodroma cookii*) then plummeted as the rats preyed on petrel chicks and eggs, and petrel numbers recovered only after the eradication of rats in 2004 (Rayner *et al.*, 2007).

Other potentially important human impacts on seabird colonies arise from destruction or modification of nesting habitat, such as from coastal residential or tourist developments, changing the coastal vegetation, encroachment of farming, and introducing grazing animals (Vermeer & Rankin, 1984). Deforestation or introduction of exotic plant species may render habitat unsuitable for nesting. A major reason for the decline of sooty terns in the nineteenth century appears to be that many of its nesting islands in the western Indian Ocean were planted with coconut trees (Feare, 1984). Soil erosion following deforestation may destroy sites of burrow-nesting species. Breeding habitat of Atlantic puffin along the Gulf of St Lawrence has been destroyed as burrows were systematically dug up to obtain the adults (Evans & Nettleship, 1985). Habitat alteration especially threatens gadfly petrels (*Pterodroma* spp.), such as the critically endangered magenta petrel (*Pterodroma magentae*) and Chatham petrel (*P. axillaris*) of New Zealand and Bermuda petrel (*P. cahow*) and black-capped petrel (*P. hasitata*) of the Caribbean (Vermeer & Rankin, 1984).

It may be feasible to restore seabird colonies to sites where they have been eliminated. Active restoration methods involve in particular translocation of chicks to new sites and luring adult birds to restoration sites using decoys and sound recordings (Jones & Kress, 2012). Chick translocation can be used for species in which chicks imprint on their natal colony before fledging and are not fed by parents after they leave the colony. This requires translocating downy chicks to release sites and hand-rearing them to fledgling age. Chick translocation was used in the first active restoration project in the 1970s to re-establish Atlantic puffins on two islands in the Gulf of Maine where they had not nested since the late nineteenth century when extirpated by overhunting (Kress, 1997). Chicks were taken from Newfoundland – some 950 to each island – and on relocation were captive-reared in individual artificial burrows before being allowed to depart to sea when ready. The mean age of chicks at relocation was 17 days. It is only after about this time that puffins learn the location of their natal island and so will then return to nest at a transplant site or at a nearby existing puffin colony.

Translocated puffins began to return when 2 years old, and birds began breeding on both islands 8 years after the first translocation. Chick translocation is unlikely to succeed for species that continue to feed their young after leaving the colony. In such cases, methods of luring adult birds to restoration sites may be possible. Colonies of terns have been successfully restored by attracting birds with decoys and sound recordings also to islands off the Maine coast. In both these cases, to facilitate recolonization by puffins and terns, the populations of herring gulls and great black-backed gulls (*Larus marinus*) on the islands were culled (Kress, 1997; Jones & Kress, 2012).

## Interactions with Fisheries

Direct exploitation of seabirds by humans has declined over the past century, although there are still seabird populations significantly impacted by hunting or egg collecting (Boersma *et al.*, 2001). However, other pressures on seabird populations have increased greatly in recent decades, notably from interactions with fisheries. Areas supporting large fish populations can attract both seabirds and fishers, and as seabirds are commonly perceived as competing directly with fisheries, this can lead to calls for culling of seabird populations (Tasker *et al.*, 2000). Globally, the biomass of prey consumed by seabirds each year is of similar magnitude to the total marine fisheries catch. But in terms of composition, only about 40% of the overall food consumption by seabirds comprises fish – and then mostly small species and juveniles – with the other main prey being krill (38%) and squid (20%) (Fig. 8.4). Furthermore, mapping the overlap between seabirds and fisheries on a global scale indicates that most of the food consumed by seabirds is taken from offshore areas where overlap with fisheries is low (Karpouzi *et al.*, 2007). Overall, seabirds are not normally in direct competition with fisheries, although predation on young stages may have the potential to adversely affect recruitment to some fisheries, and progressive 'fishing down the food web' (see Chapter 5) may increase potential competition with fisheries. Large seabird colonies may locally deplete fish populations

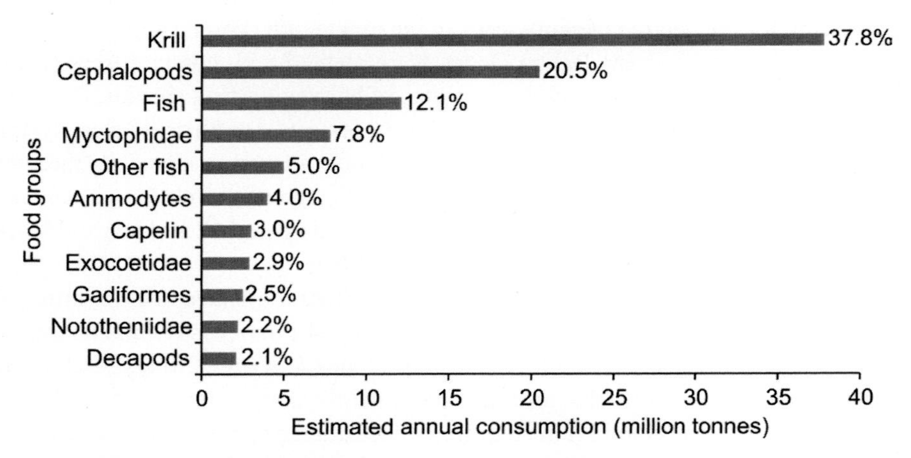


**Fig. 8.4** The contribution of prey groups to the estimated annual global food consumption for all seabird species combined. Krill: Euphausiacea; Cephalopods; squids and octopuses; Myctophidae: lanternfishes; Ammodytes: sandeels; Capelin: *Mallotus villosus*; Exocoetidae: flying fishes; Gadiformes: cods and allies; Nototheniidae: icefishes; Decapods: prawns, crabs, and allies. Redrawn but not modified from Karpouzi, V.S., Watson, R. & Pauly, D. (2007). Modelling and mapping resource overlap between seabirds and fisheries on a global scale: a preliminary assessment. *Marine Ecology Progress Series*, 343, 87–99. With permission from Inter-Research.

(Birt *et al.*, 1987), but depletion of commercial fish species by seabirds is unlikely; predatory fish are typically more important as consumers of food fish than seabirds, likely accounting for an order of magnitude more fish biomass (Bax, 1991). Direct effects of seabirds on fisheries are thus likely to be localized and, overall, minor.

Fisheries, however, can significantly impact on seabird populations in a number of ways (Furness & Monaghan, 1987; Duffy & Schneider, 1994; Tasker *et al.*, 2000; Montevecchi, 2001) (Table 8.4). Fisheries-seabird interactions are seldom straightforward: fisheries may have negative or positive and direct or indirect impacts on seabird populations, seabird species vary considerably in their susceptibility and response to such impacts, and effects are often confounded by oceanographic changes and large variations in fish recruitment. Moreover, fisheries and seabird data are typically examined at different spatial scales: fish populations tend to be assessed for large sea areas whereas seabirds are usually studied at individual colonies (Bailey, 1989). Thus even with adequate demographic data it may be difficult to establish to what extent fisheries are affecting seabird populations. That said, there are areas of particular concern such as the accidental capture of seabirds in fishing gear, notably the mortality of shearwaters and auks in gillnets and albatrosses and other procellariiforms in longline fisheries. Seabird populations can also be indirectly affected where, for example, fisheries alter the structure of a bird's prey community or make food available in the form of discards and offal.

## Bycatch in Fisheries

Of the various methods of fishing (see Chapter 5), two in particular also kill many seabirds – gillnetting and longlining. Synthetic gillnets came into widespread use in the mid-twentieth century and, virtually invisible underwater, present a potentially serious hazard to foraging seabirds. Žydelis *et al.* (2013) estimated that at least 400 000 seabirds die each year as a result of entanglement in gillnets, principally pursuit-diving and benthic-feeding species, with the highest bycatches in the NW Pacific, Iceland, and the Baltic Sea.

Coastal gillnet fisheries are typically small-scale operations, but their cumulative impact on seabird populations can be substantial. For the Baltic and North Sea region, the overall bycatch in gillnets is likely to be of the order of 100 000–200 000 birds per year, mainly various sea ducks, diving ducks, grebes, auks, and cormorants, and at least for some species, where there is adequate information on demographic parameters, this additional mortality is clearly of conservation concern (Žydelis *et al.*, 2009).

**Table 8.4** Main Interactions between Fisheries and Seabirds

| | Negative | Positive |
|---|---|---|
| Direct | Mortality from incidental capture | Provide food via discards and offal |
| | Disturbance | |
| Indirect | Depletion of prey | Removal of competitors |
| | Increased abundance of scavengers/ predators | Increased abundance of small fishes |

Huge numbers of seabirds have drowned in driftnets (set mainly for squid and salmon) in the North Pacific, particularly sooty (*Puffinus griseus*) and short-tailed shearwaters. Uhlmann *et al.* (2005) estimated that the North Pacific driftnet fisheries may in total have killed up to 34 million shearwaters over the period 1952–2001, but deficiencies in the bycatch data and the way they are gathered make it impossible to accurately predict the long-term impacts on shearwater populations. Better standardization of observer data would help significantly in this regard. Although both these shearwater species are very abundant, they are long-lived (21–30 years), so population monitoring over many years is needed to identify trends. Although this mortality declined following the 1992 UN global ban on the use of long driftnets on the high seas (see Chapter 5), much of this fishing effort subsequently transferred to longlining (Montevecchi, 2001).

Possible measures to reduce seabird bycatch in gillnets include increasing the visibility of the net (by incorporating a highly visible mesh panel in the upper portion of the net or using a dye to render the net opaque) and using acoustic alerts (pingers clipped to the corkline) (Løkkeborg, 2011). Melvin *et al.* (1999) examined strategies that would reduce seabird bycatch in a gillnet fishery in north-west USA but without reducing fishing efficiency for the target species (salmon) or increasing the bycatch of other species. The main species impacted is the common guillemot, which is also the seabird most affected worldwide by coastal gillnet fisheries. The greatest reductions in bycatch, up to 70–75%, could be achieved if visual or acoustic alerts were combined with time and area closures to take account of peak abundances of bycatch species. The use of gillnets should be severely controlled or prohibited in the vicinity of important seabird colonies, especially in the case of threatened populations or species. Entanglement in gillnets is, for instance, considered a significant threat to New Zealand mainland populations of the endangered yellow-eyed penguin (*Megadyptes antipodes*) (Darby & Dawson, 2000).

Longlining is a widely used method of fishing for pelagic and demersal species. Pelagic longlining concentrates on tuna and billfish in tropical to temperate waters, whereas demersal longlining mainly targets a variety of cold-water species such as cods, haddock, halibuts, hake, pollock in the northern Atlantic and Pacific, and Patagonian toothfish in the Southern Ocean. The gear comprises a long mainline along which, every several metres, is attached a short branchline with a baited hook. Longlines are paid out from the stern, and each can be many kilometres in length with thousands of hooks. The commonest cause of mortality is when birds take baited hooks as the line enters the water and then drown. Seabirds most often caught are albatrosses and large petrels as they are surface foragers large enough to take longline baits. For most fisheries, catch rates would average less than one bird per 1000 hooks set. However, given that a single vessel may operate up to 35 000 hooks per day, the combined bycatch of a longline fleet over a year can mount up phenomenally. In some cases tens of thousands of birds may be killed in one season of a longline fishery. The scale of such mortality became apparent from the number of birds being hooked by Japanese southern bluefin tuna vessels in southern temperate waters. Based on a mean catch rate of 0.41 birds per 1000 hooks, Brothers (1991) estimated an annual mortality from this fishery of 44 000 birds, with black-browed (*Thalassarche melanophris*), shy (*T. cauta*), and wandering (*Diomedea exulans*) albatrosses being the most abundantly taken species. Also heavily impacted have been Laysan (*Phoebastria immutabilis*) and black-footed (*P. nigripes*) albatrosses taken in the central North Pacific by tuna longlining and albatross and petrel species killed by demersal longlining for Patagonian toothfish in the Southern Ocean, with white-chinned petrel (*Procellaria aequinoctialis*) the most abundantly taken. There is, however, in this area a high level of pirate fishing – probably making little use of mitigative measures to reduce seabird mortality – making it difficult to assess the mortality (Brothers *et al.* 1999). Anderson *et al.* (2011) estimated that globally at least 160 000 and possibly more than 320 000 seabirds are killed annually in longline fisheries. Also contributing to this high level of uncertainty is that seabirds are mainly caught during line setting, whereas bycatch is usually recorded as the line is hauled in. Using data from pelagic longline fisheries over a 15-year period (1988–2003) from four regions, Brothers *et al.* (2010) found that more than half the birds observed to be caught when lines were being set were not retrieved when the gear was hauled in, indicating the potential for significant underestimation of bycatch and emphasizing the need for onboard observers to record birds caught during both setting and hauling.

For a number of seabird populations the level of incidental catch from longlining appears to be contributing significantly to their decline. Of serious concern are the unsustainable losses to fisheries of several species of albatrosses and petrels, with incidental catches amounting to 1–16% of breeding populations. In the case of the wandering albatross, its total population probably decreased some 40% between the early 1960s and early 1990s, and over this period somewhere between 50 000 and 122 000 birds are estimated to have died on Japanese longlines. Populations have shown some recovery since the mid-1980s with shifts in fishing effort, although longlining remains a major threat (Weimerskirch *et al.*, 1997). Especially at risk from fishing mortality are the short-tailed, northern royal, Amsterdam (*Diomedea exulans amsterdamensis*), Tristan (*D. dabbenena*), and Chatham (*Thalassarche eremita)* albatrosses, all threatened species. The Amsterdam albatross is critically endangered and one of the rarest seabirds. It has a single nesting population on Amsterdam Island in the SE Indian Ocean and in the early 1980s had been reduced to fewer than 10 breeding pairs, seemingly because of longlining in the vicinity of Amsterdam Island. Fishing effort subsequently shifted from this sector of the Indian Ocean, and the number of breeding birds increased to 25–30 pairs (Rivalan *et al.*, 2010). But longlining still occurs in the species' range and could account for about 5% of individuals each year depending on how well bycatch mitigation measures are implemented (Thiebot *et al.*, 2016).

Incidental catch of seabirds in longline fisheries can be significantly reduced in a number of ways (Brothers *et al.*, 1999; Bull, 2007; Løkkeborg, 2011). The effectiveness of various mitigation measures depends on the fishery and seabird species concerned, and for best results methods may be used in combination. One of the simplest measures is to set lines only at night, which particularly reduces mortality of albatrosses. Birds can be deterred from taking baited hooks by the use of a bird-scaring line, a line with attached streamers deployed astern above where the line is being set, or to use weighted hooks so that the bait sinks more rapidly. Also, area or seasonal closures are clearly of benefit where fisheries may be close to breeding colonies or undertaken during breeding seasons. Mitigation measures are required by some countries and regulatory authorities. For example, bird-scaring lines, weighted lines, and night-time setting are all prescribed for longline vessels operating in the Southern Ocean under regulations of the Convention on the Conservation of Antarctic Marine Living Resources (CCAMLR) (see Chapter 17).

Trawl fisheries are also responsible for incidental mortality of seabirds, mainly as a result of birds colliding with or becoming entangled in trawl cables. Compared to gillnets and longlining, few of the birds impacted by trawl gear end up on deck. The extent of this impact is therefore less well understood but is likely to have been underestimated (Wagner & Boersma, 2011). Extrapolations by Watkins *et al.* (2008) suggest the impact of the South African hake trawl fishery on seabirds – a mortality of around 18 000 birds per year (mostly albatrosses) – is comparable to that of the bycatch of seabirds from longlining in South African waters. Bird-scaring lines and other measures to deter birds from hitting cables can reduce this source of mortality. Netsonde cables – a cable to the net monitor on the headline of the trawl – are especially hazardous and have been banned in some regions (e.g. CCAMLR, New Zealand). No dumping of offal or, where this is not practicable, careful management of processing waste is also effective in reducing mortality (Løkkeborg, 2011).

Bycatch of seabirds in fisheries is a major threat to seabirds worldwide, with the impact of longlining on albatrosses known to be especially serious (Gales, 1998; Tasker *et al.*, 2000). Implementation and enforcement of mitigative measures at the international and state levels are a priority, such as those required by CCAMLR. There has been further movement on this problem in recent years. In 1999 FAO adopted an International Plan of Action for Reducing Incidental Catch of Seabirds in Longline Fisheries (IPOA – Seabirds). This details ways that seabird bycatch can be reduced, and on this basis a number of countries have produced national plans. All albatross species are listed in Appendix I or II of the Convention on the Conservation of Migratory Species of Wild Animals (CMS) (see Chapter 4). Under CMS, an international Agreement on the Conservation of Albatrosses and Petrels (ACAP) entered into force in 2004, aimed in particular at species vulnerable to longline mortality and establishing a framework to mitigate population declines, although important fishing nations may elect not to be party to the agreement. Also relevant to seabird bycatch are the FAO Code of Conduct for Responsible Fisheries and UN Fish Stocks Agreement, which both set out duties to conserve non-target species (see Chapter 5).

## Altered Community Structure and Prey Availability

Fisheries may impact on seabirds less directly by altering the structure of the community in which they forage. We have seen (Chapter 5) that overfishing typically shifts the age-structure of targeted populations towards younger fish and the community towards more *r*-selected species. As a fishery depletes top predator fish – typically the preferred species – so smaller, lower-trophic-level species become more abundant, to be targeted in turn. Thus, overall, smaller fish become relatively more abundant in the community, which could in theory mean more food for seabirds.

One of the best-documented regions for long-term data on fisheries' catches and seabird populations is the North Sea (Furness, 2002), an area that has shown a typical response to overfishing (e.g. Sherman *et al.*, 1981). Since the 1950s there has been a marked decline in stocks of the traditional commercial species of the North Sea, such as Atlantic cod, whiting, haddock, herring, and mackerel, and over this period there has been a concomitant growth of fisheries for more opportunistic species, notably for sandeels (mainly *Ammodytes marinus*). The sandeel fishery expanded rapidly in the 1970s to reach annual catches averaging around 900 000 t in the late 1990s. Sandeels are, however, a key prey species of seabirds in the North Sea, such as for European shag, terns, skuas, black-legged kittiwake (*Rissa tridactyla*), common guillemot, and northern fulmar (*Fulmarus glacialis*), and Cook *et al.* (2014) found that changes in the breeding failure of seabird species were strongly correlated with pressure from the sandeel fishery.

Fisheries management should take into account the relative dependence of seabird populations on different prey species. Seabird species are, however, likely to differ markedly in their susceptibility to changes in abundance of particular prey species, and this may vary within their geographic range and with life-history stage. The relative vulnerability of seabird breeding success to reduced food abundance is likely to depend on such factors as body size, cost of foraging, potential foraging range, ability to dive, amount of 'spare' time in the daily budget, and ability to switch diet (Furness & Tasker, 2000). Also, mapping the foraging distribution of seabirds and the prey taken by fisheries enables the identification of areas of high conservation value for seabirds (e.g. for high densities or species richness) that are likely to conflict with fisheries (Karpouzi *et al.*, 2007). Management measures can then be examined depending, for instance, on the jurisdictional status of the waters, any regional frameworks for marine resource management, and appropriate mitigative tools.

The often-large fluctuations in abundance exhibited by the more opportunistic fish species, such as sandeels, highlight the need for a precautionary approach to setting catch limits to minimize the risk of a fish stock and its dependent seabirds from collapsing, as we saw earlier with the anchovy and guano birds. A basic management objective should be to ensure sufficient abundance of prey for seabird reproduction. Cury *et al.* (2011) suggest that seabird prey biomass needs to be kept above one-third of the maximum prey biomass in order to sustain long-term productivity of seabird populations (Fig. 8.5). Below this threshold seabird breeding success declines sharply, a response that was found for a range of ecosystems and seabird species.

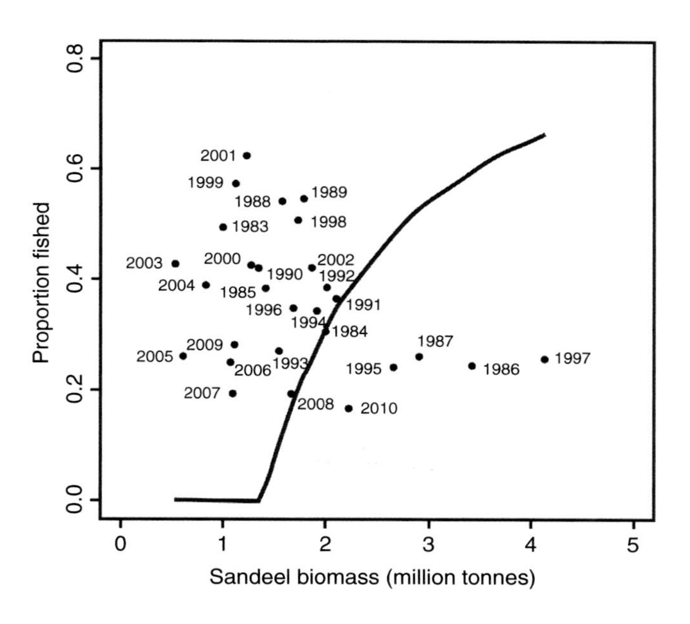


**Fig. 8.5** The proportion of the sandeel stock fished each year between 1983 and 2010 in relation to the total sandeel biomass in the North Sea. The line shows the maximum proportion of the sandeel stock that can be fished if one-third of the long-term maximum biomass is left for birds each year, as suggested as a general rule by Cury *et al.* (2011), a target that was met in only 5 years. Reprinted from *Ecological Indicators*, 38, Cook, A.S.C.P., Dadam, D., Mitchell, I., Ross-Smith, V.H. & Robinson, R.A., Indicators of seabird reproductive performance demonstrate the impact of commercial fisheries on seabird populations in the North Sea, pages 1–11, copyright 2014, with permission from Elsevier.

## Fisheries Discards

Discards and offal from fisheries can be an important source of food for seabirds. In the North Sea, for example, about 1 million t is discarded annually – enough to support 6 million birds – and taken mainly by northern fulmar, northern gannet, great skua, and several gull species (Tasker *et al.*, 2000). Marked increases in populations of scavenging seabirds have been observed in the Atlantic and elsewhere, directly attributable to the provision of waste generated by fisheries. Species reduced to critically low numbers may benefit from fishery discards. A colony of the then-threatened Audouin's gull (*Larus audouinii*) became established at the Ebro Delta, NW Mediterranean, in 1981 and increased dramatically so that by 1997 the colony held several thousand pairs and most of the species' world population. Growth of the colony has been attributed to the gull behaving partly as a scavenger and exploiting a high availability of trawler discards. However, to allow fish populations around the Ebro Delta to recover, there has been a 2-month trawling moratorium annually since 1991, but as this coincides with the gulls' breeding season it has significantly decreased breeding success. On the other hand, the gull would be threatened too by overfishing (Oro *et al.*, 1996).

Development of a hake fishery on the Patagonian shelf appears to have significantly impacted the world's largest colony of Magellanic penguins (> 200 000 active nests) at Punta Tombo, Argentina, where the number of nests has fallen by more than 20% since 1987. In particular, the discarding of undersized hake has benefited scavenging seabirds at the expense of penguins, resulting in starvation of penguin chicks, and waste from hake processing plants supports a large population of kelp gulls (*Larus dominicanus*), which are important predators of penguin eggs and chicks (Wagner & Boersma, 2011). The fishery also impacts other seabirds. The effects of fisheries on seabirds are thus unlikely to be straightforward given that the community as a whole responds to the disturbance as well as to natural variability.

## Pollution

Seabirds tend to be long-lived, wide-ranging top predators and may potentially suffer from chronic as well as acute impacts from a variety of pollutants, including metals, organochlorines, petroleum hydrocarbons, and plastic debris. Here we can look only briefly at major areas likely to be important to conservation of seabird populations. Reviews providing more detailed coverage include those of Nisbet (1994) and Burger and Gochfield (2001).

### Metals and Organochlorines

With metals, the main concern has focused on those that have no known metabolic function, such as mercury, lead, and cadmium, and that can have a range of effects on development and the nervous system. In the case of mercury, high levels of inorganic mercury have been recorded in the liver of some seabirds. More toxic, however, is methylmercury. But whilst seabirds can accumulate this organic mercury in their feathers, they can then excrete it during moulting (Thompson & Furness, 1989). Toxicity is likely to depend on a range of factors, and to what extent sublethal effects of mercury and other metals may significantly harm seabird populations is unclear.

Adverse impacts are evident in the case of organochlorine compounds such as DDT and PCBs (see Chapter 2). Among other effects, DDT disrupts calcium deposition in eggs, and through the 1960s and 1970s, when DDT was more widely used, this resulted in seabirds laying thin-shelled eggs that broke under incubation (Burger & Gochfield, 2001). There is still evidence of such contaminants accumulating in seabirds and contributing to population declines. The ivory gull (*Pagophila eburnea*) is a high Arctic species at risk of extinction across much of its breeding range. Miljeteig *et al.* (2012) found high levels of halogenated contaminants (organochlorine pesticides, PCBs, and brominated flame retardants) in the eggs of ivory gulls, levels associated with a degree of eggshell thinning (of up to 17%) that has been linked to declines in bird populations.

### Plastics

Whereas a number of studies indicate that organochlorine levels in seabirds have declined in recent decades, hazards associated with other synthetic compounds have increased. Plastics and other persistent debris that float pose a danger to many seabirds through entanglement (such as from discarded fishing nets or packaging materials) and from ingestion. Seabirds can mistake pieces of plastic debris for prey, their prey may already contain ingested particles, or parent birds may feed bits of plastic to their chicks in regurgitated food. Seabirds ingest fragments of discarded products as well as industrial pellets. Ingestion of marine debris has been documented for more than 120 seabird species, with the highest incidence among surface-feeding species, especially procellariiforms (Laist, 1997; Gall & Thompson, 2015). The incidence of ingestion can be at least 80–90% of individuals, as for instance in the case of Laysan albatrosses in the central North Pacific (Robards *et al.*, 1997; Gray *et al.*, 2012). A spatial risk analysis of plastic ingestion by seabirds indicates that the northern fringe of the Southern Ocean is the area of highest expected impact and that globally by mid-century at least 95% of seabirds species and of individuals within these species will have ingested plastic (Wilcox *et al.*, 2015).

Ingested material may block or damage the gut or suppress feeding. Of 50 Laysan albatross chicks examined on Hawaiian islands, 45 had plastic items lodged in the upper digestive tract, with three of the chicks having impactions or lesions (Fry *et al.*, 1987). Spear *et al.* (1995) reported a significant negative correlation between the number of plastic particles ingested and body weight for several procellariiform species, indicating that physical condition may be impaired by ingestion of plastic. Ingested plastics may be an additional source of persistent organic pollutants such as organochlorine pesticides and PCBs (Colabuono *et al.*, 2010). MARPOL Annex V prohibiting the disposal of plastics at sea from ships came in to force in 1988 (see Chapter 2), but a number of studies indicate that the incidence of plastic ingestion has not abated since then or has even increased (Robards *et al.*, 1995; Vlietstra & Parga, 2002). The analysis by Wilcox *et al.* (2015) shows that exposure of seabirds to plastic and ingestion rate are directly related, underlining the importance of effective measures to reduce waste.

## Oil

The most obvious effects of pollution on seabirds tend to be those due to oil spills, where seabirds are conspicuous casualties and sometimes killed in large numbers (Fig. 8.6). Seabird populations are potentially very vulnerable to oil spills, particularly where colonies are close to major shipping lanes and oil industry facilities. Such mortalities raise concerns about possible long-term effects on survival of these populations with their naturally low rates of reproduction.

**Fig. 8.6** Seabirds are often among the most significant casualties of oil spills. A northern gannet on a north Devon beach, UK, following the *Christos Bitas* spill, October 1978. Photograph: P.K. Probert.

Birds that encounter floating oil are affected primarily by the plumage losing its natural water repellence and becoming clogged, resulting in reduced buoyancy and insulation. Even if waterlogging does not result in drowning, the increased metabolism needed to maintain body temperature is likely to rapidly deplete energy reserves. Birds may also ingest oil through preening and inhale toxic fumes, particularly from a fresh spill where the oil still contains much of its more volatile fractions. Oiled birds may exhibit a number of abnormal conditions affecting, for instance, the lungs, kidneys, liver, and nasal salt glands, but most birds are likely to die from drowning or hypothermia, compounded by stress and shock, before pathological changes occur. Exposure to oil may have sublethal effects of long-term significance, such as inducing anaemia and immunosuppression, predisposing birds to infections (Briggs *et al.*, 1997). Ingestion of oil during the breeding season, even in relatively small amounts, can temporarily depress egg laying and reduce hatching success, and oil transferred from plumage to incubating eggs can be lethal to the developing embryo (Clark, 1984; Nisbet, 1994).

Seabirds vary in their susceptibility to floating oil depending on their lifestyle. Most at risk are species that spend much of their time on the sea surface and dive for food, such as penguins, auks, cormorants, gannets, and sea ducks. A number of such species also aggregate at sea in dense rafts. Vulnerability is also greater where colonies are under other stresses, for example, at the edge of their distributional range. Most large kills due to oil spills have occurred in boreal waters, where there are large populations of surface-diving birds, high shipping density, and often harsh weather conditions that put shipping at greater risk and place additional stress on seabirds. Sea surface temperature is also important since in warm waters volatile toxic fractions evaporate more rapidly and slicks are less persistent. Thus overall, oil spills in the tropics have been less damaging to seabirds than those at higher latitudes (Clark, 1984; Evans & Nettleship, 1985).

The extent to which seabirds are affected by an oil spill depends on such factors as the volume and type of oil, weather and sea conditions, current velocity and direction, and seabird species in the affected area and their abundance. Under particular circumstances, even relatively small spills can result in heavy mortality if there is a concentration of vulnerable birds, and conversely, some massive spills have resulted in few recorded deaths. In the largest incidents, estimated mortality can range from 10 000 to 500 000 birds, although no consistent relationship has been found between the volume of oil spilled and the resultant bird mortality. From an analysis of 45 spills, Burger (1993) found only a weak correlation between spill volume and number of birds killed, which was not significant if the data point for the *Exxon Valdez* spill was excluded. The grounding of this tanker in March 1989 in Prince William Sound, Alaska, resulted in one of the largest spills in US maritime history (some 37 000 t of crude oil) and a huge bird mortality. More than 30 000 dead birds were recovered, of which three-quarters were guillemots, but the total number of deaths has been estimated at up to ten times this number (Piatt *et al.*, 1990). How readily seabird populations impacted by the *Exxon Valdez* spill recovered is, however, disputed (Lance *et al.*, 2001; Wiens *et al.*, 2001).

In many cases, the occurrence of oiled birds cannot be attributed to a particular incident, and potentially more significant overall than spills from tanker accidents are many small-scale releases of oily wastes from ships' bilges and tanks and illegal discharges. Wiese and Robertson (2004) conservatively estimated that between 1998 and 2000 an average of 315 000 auks were killed annually off south-eastern Newfoundland due to chronic oil pollution from ships – in other words, an annual mortality similar to that of the *Exxon Valdez* kill.

It is difficult to arrive at estimates of seabird mortality attributable to oil pollution. Estimates are based on the numbers of oiled birds found on the shore, but such counts may be unreliable.

What proportion died at sea and were not washed up can be only roughly assessed, but overall mortality may typically be about four to five times the body count (Burger, 1993). Posthumous oiling may also be a source of error. Systematic surveys of beached birds can, however, indicate long-term trends in oiling rates. Standardization of such methods (Camphuysen & Heubeck, 2001) enables comparisons to be made between regions and over time, providing other sources of mortality (such as hunting and loss in gillnets) remain relatively constant (see Wilhelm *et al.*, 2009). Data for the southern North Sea show a gradual decline consistent with measures to reduce oil discharges (Fig. 8.7).

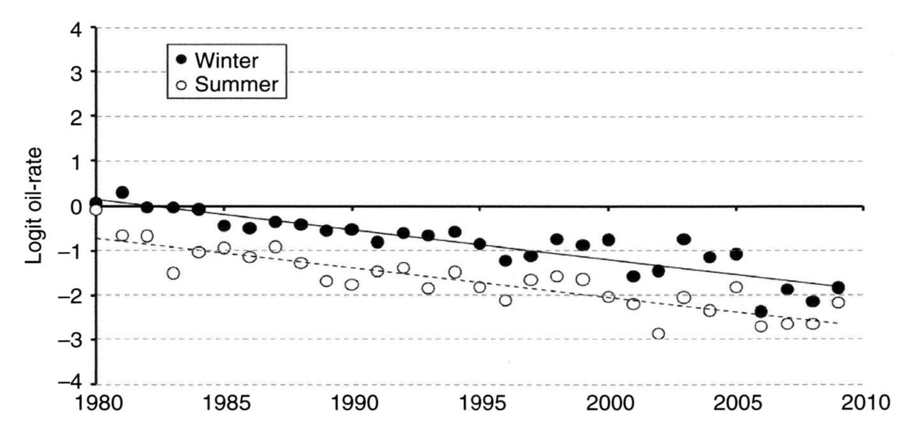


**Fig. 8.7** Oil rates of beached birds (*Larus* gulls) in the Netherlands in winter (November–April) and summer (May–October) from 1980 to 2009. The logit-transformed rates (% of birds found oiled of all intact carcasses found) indicate a gradual decline in oil pollution in the southern North Sea but also the greater vulnerability of birds in winter. Reprinted from *Marine Pollution Bulletin*, Vol. 62, Camphuysen, C.J., Seabirds and chronic oil pollution: self-cleaning properties of gulls, Laridae, as revealed from colour-ring sightings, pages 514–19, copyright 2011, with permission from Elsevier.

But whether mortality from oil pollution significantly affects seabird populations is rarely beyond doubt. Clark (1984), from a North Sea perspective, concluded that in general oil pollution had not damaged seabird populations and that even the largest losses from spills are comparable to natural catastrophes that seabird populations sustain from time to time. Similarly, Gaston and Jones (1998) could identify few cases where oil pollution could be linked to reductions or extirpations of auk populations despite massive mortalities; nevertheless, oiling most likely explains the disappearance of Atlantic puffins and common guillemot from the English Channel.

Potentially valuable as conservation tools in this regard are models to forecast how seabird populations are likely to respond to spills and techniques for assessing the vulnerability of populations. These can be used in oil spill contingency planning and in the management of offshore oil exploration and production. Various indices have been proposed to assess the potential vulnerability of seabirds to oil and other surface pollutants (Box 8.1). Such methods depend on detailed seabird data being available for the region in question, but bird density data may not be available at an appropriate spatial scale (Begg *et al.*, 1997). Also, whilst seabirds may have reasonably predictable broad-scale distributions from year to year, this may not be the case at smaller spatial scales (Fauchald *et al.*, 2002). The problem remains that a small spill in the wrong place at the wrong time can cause a disproportionate mortality.

## Box 8.1 Assessing the Vulnerability of Seabird Populations to Oil Pollution

Indices to assess the vulnerability of seabird populations to oil pollution are used as conservation tools to evaluate, for instance, the risk of oil exploration at sea and of shipping routes (Begg *et al.*, 1997; Renner & Kuletz, 2015). Several such tools have been developed. A method described by Williams *et al.* (1995) combines an:

- Oil vulnerability index (OVI) for each seabird species (to quantify its risk to oil pollution) with an
- Area vulnerability score (AVS) (to quantify the impact of oil pollution for a given area).

The OVI is based on four scored factors:

(a) proportion of time spent on the sea surface by the species
(b) size of the biogeographical population of the species
(c) potential rate of recovery of the species after a reduction in numbers (based on average clutch size and age at first breeding)
(d) reliance on the marine environment by the species

Factors are scored between 1 and 5 (low to high vulnerability), with the first two factors, considered the most important, given double weighting:

$$\mathrm{OVI} = 2a + 2b + c + d$$

The AVS is derived by combining the OVI with the density of a species in an area for each month and mapped for latitude x longitude rectangles (e.g. cells measuring 15' latitude × 30' longitude or at higher resolution for areas of high density and coverage (Begg *et al.*, 1997)):

$$\mathrm{AVS} = \sum_{\mathrm{species}} \ln(p + 1) \times \mathrm{OVI}$$

where $p$ is the density calculated for a species in the area.

Using this method, White *et al.* (2001) produced a vulnerability atlas for the waters of the Falkland Islands (petroleum exploration having begun there). The islands' coastal waters support highly vulnerable concentrations of seabirds all year, but especially between October and April coinciding with the breeding season for most of the species. Thus over summer, the numbers of resident species, such as gentoo penguin (*Pygoscelis papua*) and imperial shag (*Phalacrocorax atriceps*), are augmented by a large influx of visitors, including southern rockhopper (*Eudyptes chrysocome*) and Magellanic (*Spheniscus magellanicus*) penguins, all of which are highly vulnerable to surface pollution. AVSs are lower from April to September. Although significant numbers of non-breeding birds arrive over this period, they tend to be less vulnerable species, such as southern fulmar (*Fulmarus glacialoides*) and Cape petrel (*Daption capense*).

Cleaning and rehabilitating oiled seabirds have met with varying success. For common guillemots that had been oiled and cleaned, Sharp (1996) reported a post-release life expectancy of only about 10 days, with a negligible number surviving long enough to breed. Such results would indicate that whilst cleaning may help address an animal welfare issue it does not contribute to conservation of impacted populations. In some cases, however, successful de-oiling of threatened seabirds is probably making a significant contribution to their conservation. The African penguin, an endangered species (IUCN Red List) endemic to southern Africa, has decreased in abundance by more than 85% in the past century. A combination of factors has contributed to this decline, including oiling. Particularly since the late 1960s, African penguins have been hit by a series of large oil spills, the most serious incidents in recent years being from the sinkings of the *Apollo Sea* in 1994 and the *Treasure* in 2000. Nearly half the 10 000 oiled penguins collected after the *Apollo Sea* spill were successfully cleaned and released back into the wild, and after the *Treasure* spill more than 90% of the 19 000 oiled penguins that were cleaned were able to be released. There appeared to be no large-scale mortality following these releases and, most importantly, evidence that a considerable number of treated birds subsequently bred (Wolfaardt *et al.*, 2009). Treating oiled birds is time-consuming, and few cleaning centres could cope with incidents of the above magnitude. Results from other spills have often been disappointing, but if successful, the procedure could be crucial for rare species.

## Marine Protected Areas

In terms of possible marine protected areas, seabirds pose the challenge of requirements at different spatial scales. Breeding colonies and, potentially, coastal feeding grounds could be catered for under domestic legislation, but for the many highly mobile seabird species, protection beyond the breeding colony needs other measures (Yorio, 2009). One approach to evaluate candidate areas for seabird conservation would be to adopt a criterion of a certain proportion of the population of a species (e.g. 1%, as for waterfowl in the Ramsar Convention – see Chapter 11) for an area that has a density that exceeds the average regional density by a certain amount (e.g. four times) (Skov *et al.*, 2007). Important bird areas (IBAs) are sites that have been identified on the basis of standardized quantitative criteria as critically important for the conservation of bird populations (www.birdlife.org). IBA criteria were originally developed for land and freshwater habitats but are now being adapted for the marine environment. Criteria needed for marine IBAs include such situations as seaward extensions to breeding colonies, coastal congregations of non-breeding seabirds, migration bottlenecks, and high-seas sites.

## Research for Conservation

Birds are at least fortunate in the amount of public interest they attract. Bird-watching is a popular recreation, and countless organizations around the world are devoted to the study and conservation of birds. Major non-governmental bodies that work particularly for bird conservation include the National Audubon Society in the USA and the Royal Society for the Protection of Birds in the UK, whilst at the global level, BirdLife International is a worldwide partnership of NGOs. Such bodies are very active in scientific research, advocacy, and conservation programmes relevant to seabirds. Nevertheless, an international commission for seabirds is perhaps needed to help facilitate information exchange on conservation and management of seabirds, monitor data, co-ordinate programmes, ensure the inclusion of seabirds in international legislation, and secure appropriate resources (Duffy, 1994).

Anthropogenic pressures on seabird populations have shifted in historical times. Overall, human predation on seabirds has declined in recent centuries, although unsustainable practices persist and the impact of feral animals on seabird colonies continues to be a major problem. Some species are facing new threats, notably procellariiforms taken as fisheries bycatch, but for many species human influences are less well defined. We clearly need a better understanding of the threats to seabirds and the vulnerability of species in terms of their biology and ecology (Croxall *et al.*, 2012). Pressing research requirements to address management and conservation particularly concern aspects of population dynamics, at-sea spatial ecology, trophic dynamics and the community roles of seabirds, direct effects of fisheries, global change and population response to environmental variability, and management of anthropogenic impacts (Lewison *et al.*, 2012). On the face of it, pollution does not appear to have substantially harmed many seabird populations. Nor can indirect effects of fisheries be clearly linked to adverse impacts on seabird populations; indeed, some indirect effects of fisheries may have benefited seabirds. Even less clear is the extent to which a combination of chronic influences may be of significance. For instance, does a seabird population exposed to low-level but persistent contaminants – such as PAHs and endocrine-disrupting organochlorines – and where the structure of its prey community is gradually shifting as a result of exploitation still have the same capacity to sustain natural disasters? The greatest impacts on seabird populations may ultimately derive from global climate change, particularly by way of changes to water column characteristics, pelagic productivity, and food-web structure. We urgently need better data on the status of seabird species around the world to keep track of population trends and distinguish between natural long-term variability and anthropogenic effects and identify critical sites in need of protection (Wooller *et al.*, 1992; Sydeman *et al.*, 2012). It is essential that such programmes are sustained for at least decades. Given the typical life-history characteristics of seabirds there can be a long delay before impacts on juveniles show up in the breeding population; some albatross species do not start breeding until they are 10–15 years old. Not only is such information essential for effective conservation of seabirds, it is also likely to prove valuable in monitoring the state of marine ecosystems. As long-lived, wide-ranging top predators, seabirds are good candidates as marine ecological indicators (Durant *et al.*, 2009).

# REFERENCES


Anderson, D.W. & Keith, J.O. (1980). The human influence on seabird nesting success: conservation implications. *Biological Conservation*, **18**, 65–80.

Anderson, O.R.J., Small, C.J., Croxall, J.P., *et al.* (2011). Global seabird bycatch in longline fisheries. *Endangered Species Research*, **14**, 91–106.

Bailey, R.S. (1989). Interactions between fisheries, fish stocks and seabirds. *Marine Pollution Bulletin*, **20**, 427–30.

Bax, N.J. (1991). A comparison of the fish biomass flow to fish, fisheries, and mammals in six marine ecosystems. *ICES Marine Science Symposia*, **193**, 217–24.

Begg, G.S., Reid, J.B., Tasker, M.L. & Webb, A. (1997). Assessing the vulnerability of seabirds to oil pollution: sensitivity to spatial scale. *Colonial Waterbirds*, **20**, 339–52.

Bengtson, S.-A. (1984). Breeding ecology and extinction of the great auk (*Pinguinus impennis*): anecdotal evidence and conjectures. *Auk*, **101**, 1–12.

Bester, M.N., Bloomer, J.P., van Aarde, R.J., *et al.* (2002). A review of the successful eradication of feral cats from sub-Antarctic Marion Island, Southern Indian Ocean. *South African Journal of Wildlife Research*, **32**, 65–73.

Birt, V.L., Birt, T.P., Goulet, D., Cairns, D.K. & Montevecchi, W.A. (1987). Ashmole's halo: direct evidence for prey depletion by a seabird. *Marine Ecology Progress Series*, **40**, 205–8.

Boersma, P.D., Clark, J.A. & Hillgarth, N. (2001). Seabird conservation. In *Biology of Marine Birds*, ed. E.A. Schreiber & J. Burger, pp. 559–79. Boca Raton, FL: CRC Press.

Briggs, K.T., Gershwin, M.E. & Anderson, D.W. (1997). Consequences of petrochemical ingestion and stress on the immune system of seabirds. *ICES Journal of Marine Science*, **54**, 718–25.

Brothers, N. (1991). Albatross mortality and associated bait loss in the Japanese longline fishery in the Southern Ocean. *Biological Conservation*, **55**, 255–68.

Brothers, N.P., Cooper, J. & Løkkeborg, S. (1999). The incidental catch of seabirds by longline fisheries: worldwide review and technical guidelines for mitigation. *FAO Fisheries Circular*, No. 937, 100 p.

Brothers, N., Duckworth, A.R., Safina, C. & Gilman, E.L. (2010). Seabird bycatch in pelagic longline fisheries is grossly underestimated when using only haul data. *PLoS ONE*, **5**, e12491.

Bull, L.S. (2007). Reducing seabird bycatch in longline, trawl and gillnet fisheries. *Fish and Fisheries*, **8**, 31–56.

Burger, A.E. (1993). Estimating the mortality of seabirds following oil spills: effects of spill volume. *Marine Pollution Bulletin*, **26**, 140–3.

Burger, J. & Gochfield, M. (2001). Effects of chemicals and pollution on seabirds. In *Biology of Marine Birds*, ed. E.A. Schreiber & J. Burger, pp. 485–525. Boca Raton, FL: CRC Press

Camphuysen, C.J. (2011). Seabirds and chronic oil pollution: self-cleaning properties of gulls, Laridae, as revealed from colour-ring sightings. *Marine Pollution Bulletin*, **62**, 514–19.

Camphuysen, C.J. & Heubeck, M. (2001). Marine oil pollution and beached bird surveys: the development of a sensitive monitoring instrument. *Environmental Pollution*, **112**, 443–61.

Clark, R.B. (1984). Impact of oil pollution on seabirds. *Environmental Pollution (Series A)*, **33**, 1–22.

Colabuono, F.I., Taniguchi, S. & Montone, R.C. (2010). Polychlorinated biphenyls and organochlorine pesticides in plastics ingested by seabirds. *Marine Pollution Bulletin*, **60**, 630–4.

Cook, A.S.C.P., Dadam, D., Mitchell, I., Ross-Smith, V.H. & Robinson, R.A. (2014). Indicators of seabird reproductive performance demonstrate the impact of commercial fisheries on seabird populations in the North Sea. *Ecological Indicators*, **38**, 1–11.

Cott, H.B. (1953). The exploitation of wild birds for their eggs. *Ibis*, **95**, 409–49.

Croxall, J.P., Butchart, S.H.M., Lascelles, B., *et al.* (2012). Seabird conservation status, threats and priority actions: a global assessment. *Bird Conservation International*, **22**, 1–34.

Cury, P.M., Boyd, I.L., Bonhommeau, S., *et al.* (2011). Global seabird response to forage fish depletion—one-third for the birds. *Science*, **334**, 1703–6.

Darby, J.T. & Dawson, S.M. (2000). Bycatch of yellow-eyed penguins (*Megadyptes antipodes*) in gillnets in New Zealand waters 1979–1997. *Biological Conservation*, **93**, 327–32.

Davies, R.G., Irlich, U.M., Chown, S.L. & Gaston, K.J. (2010). Ambient, productive and wind energy, and ocean extent predict global species richness of procellariiform seabirds. *Global Ecology and Biogeography*, **19**, 98–110.

Duffy, D.C. (1983). Environmental uncertainty and commercial fishing: effects on Peruvian guano birds. *Biological Conservation*, **26**, 227–38.

Duffy, D.C. (1990). Seabirds and the 1982–1984 El Niño-Southern Oscillation. In *Global Ecological Consequences of the 1982–83 El Nino-Southern Oscillation*, ed. P.W. Glynn, pp. 395–415. Amsterdam: Elsevier.

Duffy, D.C. (1994). Afterwards: an agenda for managing seabirds and islands. In *Seabirds on Islands: Threats, Case Studies and Action Plans*, ed. D.N. Nettleship, J. Burger & M. Gochfield, *BirdLife Conservation Series*, **1**, 311–18. Cambridge, UK: BirdLife International.

Duffy, D.C. & Schneider, D.C. (1994). Seabird-fishery interactions: a manager's guide. In *Seabirds on Islands: Threats, Case Studies and Action Plans*, ed. D.N. Nettleship, J. Burger & M. Gochfield, *BirdLife Conservation Series*, **1**, 26–38. Cambridge, UK: BirdLife International.

Durant, J.M., Hjermann, D.Ø., Frederiksen, M., *et al*. (2009). Pros and cons of using seabirds as ecological indicators. *Climate Research*, **39**, 115–29.

Evans, P.G.H. & Nettleship, D.N. (1985). Conservation of the Atlantic Alcidae. In *The Atlantic Alcidae: The Evolution, Distribution and Biology of the Auks Inhabiting the Atlantic Ocean and Adjacent Water Areas*, ed. D.N. Nettleship & T.R. Birkhead, pp. 427–88. London: Academic Press.

Fauchald, P., Erikstad, K.E. & Systad, G.H. (2002). Seabirds and marine oil incidents: is it possible to predict the spatial distribution of pelagic seabirds? *Journal of Applied Ecology*, **39**, 349–60.

Feare, C.J. (1984). Human exploitation. In *Status and Conservation of the World's Seabirds*, ed. J.P. Croxall, P.G.H. Evans & R.W. Schreiber, *ICBP Technical Publication*, **2**, 691–9. Cambridge, UK: International Council for Bird Preservation.

Finkelstein, M.E., Wolf, S., Goldman, M., *et al*. (2010). The anatomy of a (potential) disaster: volcanoes, behavior, and population viability of the short-tailed albatross (*Phoebastria albatrus*). *Biological Conservation*, **143**, 321–31.

Frost, P.G.H., Siegfried, W.R. & Cooper, J. (1976). Conservation of the jackass penguin (*Spheniscus demersus* (L.)). *Biological Conservation*, **9**, 79–99.

Fry, D.M., Fefer, S.I. & Sileo, L. (1987). Ingestion of plastic debris by Laysan albatrosses and wedge-tailed shearwaters in the Hawaiian Islands. *Marine Pollution Bulletin*, 18, 339–43.

Furness, R.W. (2002). Management implications of interactions between fisheries and sandeel-dependent seabirds and seals in the North Sea. *ICES Journal of Marine Science*, **59**, 261–9.

Furness, R.W. & Monaghan, P. (1987). *Seabird Ecology*. Glasgow, UK: Blackie.

Furness, R.W. & Tasker, M.L. (2000). Seabird-fishery interactions: quantifying the sensitivity of seabirds to reductions in sandeel abundance, and identification of key areas for sensitive seabirds in the North Sea. *Marine Ecology Progress Series*, **202**, 253–64.

Gales, R. (1998). Albatross populations: status and threats. In *Albatross Biology and Conservation*, ed. G. Robertson & R. Gales, pp. 20–45. Chipping Norton, NSW: Surrey Beatty & Sons.

Gall, S.C. & Thompson, R.C. (2015). The impact of debris on marine life. *Marine Pollution Bulletin*, **92**, 170–9.

Gaston, A.J. & Jones, I.L. (1998). *The Auks: Alcidae*. Oxford, UK: Oxford University Press.

Gray, H., Lattin, G.L. & Moore, C.J. (2012). Incidence, mass and variety of plastics ingested by Laysan (*Phoebastria immutabilis*) and black-footed albatrosses (*P. nigripes*) recovered as by-catch in the North Pacific Ocean. *Marine Pollution Bulletin*, **64**, 2190–2.

Grayson, D.K. (2001). The archaeological record of human impacts on animal populations. *Journal of World Prehistory*, **15**, 1–68.

Hamer, K.C., Schreiber, E.A. & Burger, J. (2001). Breeding biology, life histories, and life history-environment interactions in seabirds. In *Biology of Marine Birds*, ed. E.A. Schreiber & J. Burger, pp. 217–61. Boca Raton, FL: CRC Press.

Higham, J.E.S. (1998). Tourists and albatrosses: the dynamics of tourism at the Northern Royal Albatross Colony, Taiaroa Head, New Zealand. *Tourism Management*, **19**, 521–31.

Howald, G., Donlan, C.J., Galván, J.B., *et al*. (2007). Invasive rodent eradication on islands. *Conservation*

*Biology*, **21**, 1258–68.

Jones, H.P., Tershy, B.R., Zavaleta, E.S., *et al.* (2008). Severity of the effects of invasive rats on seabirds: a global review. *Conservation Biology*, **22**, 16–26.

Jones, H.P. & Kress, S.W. (2012). A review of the world's active seabird restoration projects. *Journal of Wildlife Management*, **76**, 2–9.

Karpouzi, V.S., Watson, R. & Pauly, D. (2007). Modelling and mapping resource overlap between seabirds and fisheries on a global scale: a preliminary assessment. *Marine Ecology Progress Series*, **343**, 87–99.

Kress, S.W. (1997). Using animal behavior for conservation: case studies in seabird restoration from the Maine coast, USA. *Journal of the Yamashina Institute for Ornithology*, **29**, 1–26.

Laist, D.W. (1997). Impacts of marine debris: entanglement of marine life in marine debris including a comprehensive list of species with entanglement and ingestion records. In *Marine Debris: Sources, Impacts, and Solutions*, ed. J.M. Coe & D.B. Rogers, pp. 99–139. New York: Springer.

Lance, B.K., Irons, D.B., Kendall, S.J. & McDonald, L.L. (2001). An evaluation of marine bird population trends following the *Exxon Valdez* oil spill, Prince William Sound, Alaska. *Marine Pollution Bulletin*, **42**, 298–309.

Lewison, R., Oro, D., Godley, B.J., *et al.* (2012). Research priorities for seabirds: improving conservation and management in the 21st century. *Endangered Species Research*, **17**, 93–121.

Løkkeborg, S. (2011). Best practices to mitigate seabird bycatch in longline, trawl and gillnet fisheries – efficiency and practical applicability. *Marine Ecology Progress Series*, **435**, 285–303.

Melvin, E.F., Parrish, J.K. & Conquest, L.L. (1999). Novel tools to reduce seabird bycatch in coastal gillnet fisheries. *Conservation Biology*, **13**, 1386–97.

Milberg, P. & Tyrberg, T. (1993). Naïve birds and noble savages – a review of man-caused prehistoric extinctions of island birds. *Ecography*, **16**, 229–50.

Miljeteig, C., Gabrielsen, G.W., Strøm, H., *et al.* (2012). Eggshell thinning and decreased concentrations of vitamin E are associated with contaminants in eggs of ivory gulls. *Science of the Total Environment*, **431**, 92–9.

Montevecchi, W.A. (2001). Interactions between fisheries and seabirds. In *Biology of Marine Birds*, ed. E.A. Schreiber & J. Burger, pp. 527–57. Boca Raton, FL: CRC Press.

Moors, P.J. & Atkinson, I.A.E. (1984). Predation on seabirds by introduced animals, and factors affecting its severity. In *Status and Conservation of the World's Seabirds*, ed. J.P. Croxall, P.G.H. Evans & R.W. Schreiber, *ICBP Technical Publication*, **2**, 667–90. Cambridge, UK: International Council for Bird Preservation.

Nettleship, D.N. & Evans, P.G.H. (1985). Distribution and status of the Atlantic Alcidae. In *The Atlantic Alcidae: the Evolution, Distribution and Biology of the Auks Inhabiting the Atlantic Ocean and Adjacent Water Areas*, ed. D.N. Nettleship & T.R. Birkhead, pp. 53–154. London: Academic Press.

Nisbet, I.C.T. (1994). Effects of pollution on marine birds. In *Seabirds on Islands: Threats, Case Studies and Action Plans*, ed. D.N. Nettleship, J. Burger & M. Gochfield, *BirdLife Conservation Series*, **1**, 8–25. Cambridge, UK: BirdLife International.

Nogales, N., Martín, A., Tershy, B.R., *et al.* (2004). A review of feral cat eradication on islands. *Conservation Biology*, **18**, 310–19.

Oro, D., Jover, L. & Ruiz, X. (1996). Influence of trawling activity on the breeding ecology of a threatened seabird, Audouin's gull *Larus audouinii*. *Marine Ecology Progress Series*, **139**, 19–29.

Piatt, J.F., Lensink, C.J., Butler, W., Kendziorek, M. & Nysewander, D.R. (1990). Immediate impact of the 'Exxon Valdez' oil spill on marine birds. *Auk*, **107**, 387–97.

Rayner, M.J., Hauber, M.E., Imber, M.J., Stamp, R.K. & Clout, M.N. (2007). Spatial heterogeneity of mesopredator release within an oceanic island system. *Proceedings of the National Academy of Sciences*, **104**, 20862–5.

Renner, M. & Kuletz, K.J. (2015). A spatial–seasonal analysis of the oiling risk from shipping traffic to seabirds in the Aleutian Archipelago. *Marine Pollution Bulletin*, **101**, 127–36.

Rivalan, P., Barbraud, C., Inchausti, P. & Weimerskirch, H. (2010). Combined impacts of longline fisheries and climate on the persistence of the Amsterdam albatross *Diomedea amsterdamensis*. *Ibis*, **152**, 6–18.

Robards, M.D., Gould, P.J. & Piatt, J.F. (1997). The highest global concentrations and increased abundance of oceanic plastic debris in the North Pacific: evidence from seabirds. In *Marine Debris: Sources, Impacts,*

*and Solutions*, ed. J.M. Coe & D.B. Rogers, pp. 71–80. New York: Springer.
Robards, M.D., Piatt, J.F. & Wohl, K.D. (1995). Increasing frequency of plastic particles ingested by seabirds in the subarctic North Pacific. *Marine Pollution Bulletin*, **30**, 151–7.
Robertson, C.J.R. (1998). Factors influencing the breeding performance of the Northern Royal Albatross. In *Albatross Biology and Conservation*, ed. G. Robertson & R. Gales, pp. 99–104. Chipping Norton, NSW: Surrey Beatty & Sons.
Schreiber, E.A. & Burger, J. (2001). Seabirds in the marine environment. In *Biology of Marine Birds*, ed. E.A. Schreiber & J. Burger, pp. 1–15. Boca Raton, FL: CRC Press.
Sharp, B.E. (1996). Post-release survival of oiled, cleaned seabirds in North America. *Ibis*, **138**, 222–8.
Shealer, D.A. (2001). Foraging behavior and food of seabirds. In *Biology of Marine Birds*, ed. E.A. Schreiber & J. Burger, pp. 137–77. Boca Raton, FL: CRC Press.
Sherman, K., Jones, C., Sullivan, L., Smith, W., Berrien, P. & Ejsymont, L. (1981). Congruent shifts in sand eel abundance in western and eastern North Atlantic ecosystems. *Nature*, **291**, 486–9.
Skov, H., Durinck, J., Leopold, M.F. & Tasker, M.L. (2007). A quantitative method for evaluating the importance of marine areas for conservation of birds. *Biological Conservation*, **136**, 362–71.
Spear, L.B., Ainley, D.G. & Ribic, C.A. (1995). Incidence of plastic in seabirds from the tropical Pacific, 1984–91: relation with distribution of species, sex, age, season, year and body-weight. *Marine Environmental Research*, **40**, 123–46.
Steadman, D.W. (1995). Prehistoric extinctions of Pacific island birds: biodiversity meets zooarchaeology. *Science*, **267**, 1123–31.
Steven, R., Pickering, C. & Castley, J.G. (2011). A review of the impacts of nature based recreation on birds. *Journal of Environmental Management*, **92**, 2287–94.
Sydeman, W.J., Thompson, S.A. & Kitaysky, A. (2012). Seabirds and climate change: roadmap for the future. *Marine Ecology Progress Series*, **454**, 107–17.
Tasker, M.L., Camphuysen, C.J., Cooper, J., *et al.* (2000). The impacts of fishing on marine birds. *ICES Journal of Marine Science*, **57**, 531–47.
Taylor, R.H., Kaiser, G.W. & Drever, M.C. (2000). Eradication of Norway rats for recovery of seabird habitat on Langara Island, British Columbia. *Restoration Ecology*, **8**, 151–60.
Thiebot, J.-B., Delord, K., Barbraud, C., Marteau, C. & Weimerskirch, H. (2016). 167 individuals versus millions of hooks: bycatch mitigation in longline fisheries underlies conservation of Amsterdam albatrosses. *Aquatic Conservation: Marine and Freshwater Ecosystems*, **26**, 674–88.
Thompson, D.R. & Furness, R.W. (1989). Comparison of the levels of total and organic mercury in seabird feathers. *Marine Pollution Bulletin*, **20**, 577–9.
Uhlmann, S., Fletcher, D. & Moller, H. (2005). Estimating incidental takes of shearwaters in driftnet fisheries: lessons for the conservation of seabirds. *Biological Conservation*, **123**, 151–63.
Vermeer, K. & Rankin, L. (1984). Influence of habitat destruction and disturbance on nesting seabirds. In *Status and Conservation of the World's Seabirds*, ed. J.P. Croxall, P.G.H. Evans & R.W. Schreiber, *ICBP Technical Publication*, **2**, 723–36. Cambridge, UK: International Council for Bird Preservation.
Vlietstra, L.S. & Parga, J.A. (2002). Long-term changes in the type, but not amount, of ingested plastic particles in short-tailed shearwaters in the southeastern Bering Sea. *Marine Pollution Bulletin*, **44**, 945–55.
Wagner, E.L. & Boersma, P.D. (2011). Effects of fisheries on seabird community ecology. *Reviews in Fisheries Science*, **19**, 157–67.
Watkins, B.P., Petersen, S.L. & Ryan, P.G. (2008). Interactions between seabirds and deep-water hake trawl gear: an assessment of impacts in South African waters. *Animal Conservation*, **11**, 247–54.
Weimerskirch, H., Brothers, N. & Jouventin, P. (1997). Population dynamics of wandering albatross *Diomedea exulans* and Amsterdam albatross *D. amsterdamensis* in the Indian Ocean and their relationships with long-line fisheries: conservation implications. *Biological Conservation*, **79**, 257–70.
White, R.W., Gillon, K.W., Black, A.D. & Reid, J.B. (2001). *Vulnerable Concentrations of Seabirds in Falkland Islands Waters*. Peterborough, UK: Joint Nature Conservation Committee.

Wiens, J.A., Day, R.H., Murphy, S.M. & Parker, K.R. (2001). On drawing conclusions nine years after the *Exxon Valdez* oil spill. *The Condor*, **103**, 886–92.

Wiese, F.K. & Robertson, G.J. (2004). Assessing seabird mortality from chronic oil discharges at sea. *Journal of Wildlife Management*, **68**, 627–38.

Wilcox, C., Van Sebille, E. & Hardesty, B.D. (2015). Threat of plastic pollution to seabirds is global, pervasive, and increasing. *Proceedings of the National Academy of Sciences*, **112**, 11899–904.

Wilhelm, S.I., Robertson, G.J., Ryan, P.C., Tobin, S.F. & Elliot, R.D. (2009). Re-evaluating the use of beached bird oiling rates to assess long-term trends in chronic oil pollution. *Marine Pollution Bulletin*, **58**, 249–55.

Williams, J.M., Tasker, M.L., Carter, I.C. & Webb, A. (1995). A method of assessing seabird vulnerability to surface pollutants. *Ibis*, **137**, S147–52.

Wolfaardt, A.C., Williams, A.J., Underhill, L.G.,Crawford, R.J.M. & Whittington, P.A. (2009). Review of the rescue, rehabilitation and restoration of oiled seabirds in South Africa, especially African penguins *Spheniscus demersus* and Cape gannets *Morus capensis*, 1983–2005. *African Journal of Marine Science*, **31**, 31–54.

Wooller, R.D., Bradley, J.S. & Croxall, J.P. (1992). Long-term population studies of seabirds. *Trends in Ecology & Evolution*, **7**, 111–14.

Yorio, P. (2009). Marine protected areas, spatial scales, and governance: implications for the conservation of breeding seabirds. *Conservation Letters*, **2**, 171–8.

Yorio, P., Frere, E., Gandini, P. & Schiavini, A. (2001). Tourism and recreation at seabird breeding sites in Patagonia, Argentina: current concerns and future prospects. *Bird Conservation International*, **11**, 231–45.

Žydelis, R., Bellebaum, J., Österblom, H., *et al.* (2009). Bycatch in gillnet fisheries – an overlooked threat to waterbird populations. *Biological Conservation*, **142**, 1269–81.

Žydelis, R., Small, C. & French, G. (2013). The incidental catch of seabirds in gillnet fisheries: a global review. *Biological Conservation*, **162**, 76–88.

# 9 Marine Mammals

Whales, seals, and other marine mammals have for millennia been important to many maritime societies as a source of food and other natural products – in some cases continuing to be so – and in various ways contributing richly to human traditions and cultures. Marine mammals feature prominently in marine conservation endeavours on both professional and public fronts, and many conservation organizations have the protection of marine mammals as a primary focus. Marine mammal management and conservation engender a high level of agency and political activity from the national to the international level.

Mammals that have taken up a marine existence number some 120 species and occur in three major groups (Table 9.1). Two of these are entirely aquatic: the Cetacea (about 84 marine species), comprising the whales, dolphins, and porpoises; and the Sirenia (three marine species), comprising the manatees and dugong. The third group, belonging to the Carnivora, includes the seals, sea lions, and walrus – the Pinnipedia – nearly all of which (33 species) depend on marine waters for food but also make use of land or sea-ice habitat. Also within the Carnivora are two species of otter that feed only in the sea and the polar bear. Overall, species richness in marine mammals is highest in temperate waters of both hemispheres (Fig. 9.1), although for pinnipeds the peaks in species richness are more polar (Kaschner *et al.*, 2011; Pompa *et al.*, 2011).

**Table 9.1** Groups of Marine Mammals

| Major Group | | Family | |
|---|---|---|---|
| Sirenia | | | |
| | | Dugongidae | dugong |
| | | Trichechidae | manatees |
| Cetacea | | | |
| | Mysticeti | | baleen whales |
| | | Balaenidae | bowhead, right whales |
| | | Balaenopteridae | rorquals |
| | | Eschrichtiidae | gray whale |
| | | Neobalaenidae | pygmy right whale |
| | Odontoceti | | toothed whales |
| | | Delphinidae | dolphins, pilot whales, killer whale |
| | | Kogiidae | pygmy sperm whales |
| | | Monodontidae | beluga, narwhal |
| | | Phocoenidae | porpoises |

**Table 9.1** (cont.)

| Major Group | Family | |
|---|---|---|
| | Physeteridae | sperm whale |
| | Ziphiidae | beaked whales, bottlenose whales |
| Carnivora | | |
| | Mustelidae | otters |
| | Ursidae | polar bear |
| Pinnipedia | | |
| | Odobeniidae | walrus |
| | Otariidae | fur seals, sea lions |
| | Phocidae | seals |

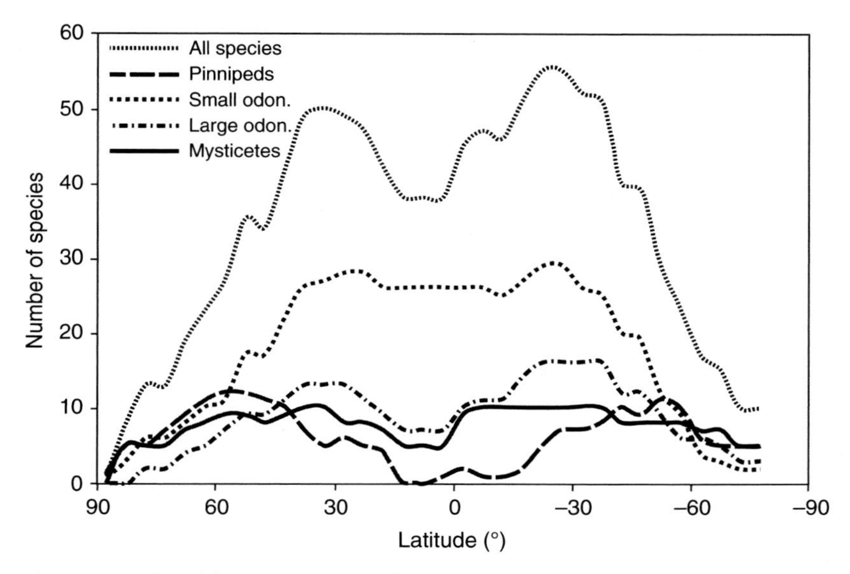


**Fig. 9.1** Species richness by latitude for all extant pinnipeds and cetaceans (small and large odontocetes and mysticetes). Reproduced with minor changes from Kaschner, K., Tittensor, D.P., Ready, J., Gerrodette, T. & Worm B. (2011). Current and future patterns of global marine mammal biodiversity. *PLoS ONE*, 6, e19653. Licensed under a Creative Commons Attribution 4.0 International License: https://creativecommons.org/licenses/by/4.0/.

Globally, marine mammals require about 10% of total primary production to sustain consumption, and they play important roles in structuring food webs (Morissette *et al.*, 2012). Feeding activity, for example, has a key role in the case of sea otters as predators in kelp forests, sirenians as grazers of seagrass beds, and gray whale and walrus as physical perturbators of sediment benthos (Bowen, 1997). Walruses disturb thousands of square kilometres of seabed each year by their feeding activity. This results in large-scale changes in benthic community structure and, by greatly increasing nutrient flux from the sediment over wide areas, is likely to enhance productivity (Ray *et al.*, 2006).

Marine mammals uniquely illustrate a number of issues in marine conservation. Our affinity with these animals and level of concern, at least in recent decades, have resulted in legislation in various countries aimed to protect marine mammals, legislation that may extend beyond conservation focused on populations to include measures concerned with the welfare of individual animals. The Marine Mammal Protection Act of 1972 of the USA prohibits the taking of marine mammals, where the term 'take' means to harass, hunt, capture, or kill or attempt to (Benson, 2012). Among marine mammals, cetaceans at least have evolved advanced cognitive abilities and a sense of self-awareness (Bekoff, 2003), contributing to a heightened debate on the ethics of how we interact with whales and dolphins. Indeed, a Declaration of Rights for Cetaceans drawn up in 2010 affirms 'that all cetaceans as persons have the right to life, liberty and wellbeing' (www.cetaceanrights.org).

Of all marine organisms, marine mammals, particularly whales and other cetaceans, now arouse the most widespread interest and concern, although it is easy to forget how quickly public attitudes can change. A survey in 1978 showed that three-quarters of Americans endorsed the hunting of non-endangered whales if it resulted in useful products, whereas a similar survey just 15 years later showed that only a quarter approved of whaling under any circumstance (Lavigne *et al.*, 1999). Similarly, a survey in Scotland in 2003 demonstrated a high level of public concern about cetacean conservation with most wanting improved legislation to protect cetaceans (Howard & Parsons, 2006). For many, the fate of whales has now almost come to symbolize the state of the world's oceans in general. In that regard, cetaceans represent effective flagship species in helping to raise public interest in marine life and an awareness of marine conservation.

Major conservation issues relating to marine mammals stem in particular from direct overexploitation and heavy incidental mortality associated with fisheries. In many instances this has led to drastic depletion of populations, loss of genetic diversity, and even extinctions. Three marine mammal species have been brought to extinction, mainly as a result of hunting – Steller's sea cow (*Hydrodamalis gigas*), Caribbean monk seal (*Monachus tropicalis*), and Japanese sea lion (*Zalophus japonicus*) – and more than 20% of extant species are currently assessed as threatened with global extinction (Table 9.2). Several species, whilst not globally at risk, have subspecies or subpopulations that are threatened, such as the beluga in Cook Inlet (Alaska) and the harbour porpoise in the Baltic, which are both critically endangered subpopulations. Several of these threatened populations are in enclosed waters, for instance, certain cetacean populations in the Mediterranean (Table 16.2). National and international laws relating to marine mammals include provisions for conservation of subspecies and subpopulations, yet many species are still too poorly known to enable such units to be identified (Taylor, 2005). Even for such high-profile marine animals as whales there is an urgent need for research on taxonomy, distribution, and abundance. Indeed, the conservation status of many species – most of the beaked whales, for instance – is unknown because of lack of data.

Given their life-history traits as *K*-strategists, marine mammals are liable to be impacted dramatically by any extrinsic mortality. The fastest breeders produce no more than a single young per year (except for the semi-aquatic marine otter and polar bear), and most cetaceans produce a single calf every 2 to 5 years. By contrast, the major groups of terrestrial mammals have species that produce litters of varying size. This evolution of a litter size of one appears to be a physiological constraint on mammals that returned to the sea, a result of the energetic demands of parental care (Estes, 1979). As long-lived top predators, marine mammals are also susceptible to bioaccumulation of pollutants. The literature on marine mammals and their management and conservation is huge, and here we can review only major issues. Perrin *et al.* (2009) provide a comprehensive treatment of all marine mammals, on which the following overview of the major groups is based.

**Table 9.2** Marine Mammal Species on the IUCN Red List That Are Categorized as Globally Threatened

CR: critically endangered, EN: endangered, VU: vulnerable

| Species | Common name | Red List category |
|---|---|---|
| Pinnipeds | | |
| *Arctocephalus galapagoensis* | Galápagos fur seal | EN |
| *Callorhinus ursinus* | Northern fur seal | VU |
| *Neophoca cinerea* | Australian sea lion | EN |
| *Phocarctos hookeri* | New Zealand sea lion | EN |
| *Zalophus wollebaeki* | Galápagos sea lion | EN |
| *Cystophora cristata* | Hooded seal | VU |
| *Monachus monachus* | Mediterranean monk seal | EN |
| *Monachus schauinslandi* | Hawaiian monk seal | EN |
| *Pusa caspica* | Caspian seal | EN |
| Otters | | |
| *Enhydra lutris* | Sea otter | EN |
| *Lutra felina* | Marine otter | EN |
| Polar bear | | |
| *Ursus maritimus* | Polar bear | VU |
| Sirenians | | |
| *Dugong dugon* | Dugong | VU |
| *Trichechus manatus* | West Indian manatee | VU |
| *Trichechus senegalensis* | West African manatee | VU |
| Baleen whales | | |
| *Eubalaena glacialis* | North Atlantic right whale | EN |

**Table 9.2** (cont.)

| Species | Common name | Red List category |
|---|---|---|
| *Eubalaena japonica* | North Pacific right whale | EN |
| *Balaenoptera borealis* | Sei whale | EN |
| *Balaenoptera musculus* | Blue whale | EN |
| *Balaenoptera physalus* | Fin whale | EN |
| Toothed whales, dolphins, porpoises | | |
| *Physeter macrocephalus* | Sperm whale | VU |
| *Cephalorhynchus hectori* | Hector's dolphin | EN |
| *Sousa teuszii* | Atlantic humpbacked dolphin | VU |
| *Neophocaena asiaeorientalis* | Finless porpoise | VU |
| *Neophocaena phocaenoides* | Indo-Pacific finless porpoise | VU |
| *Phocoena sinus* | Vaquita | CR |
| *Pontoporia blainvillei* | Franciscana | VU |

Extracted from www.iucnredlist.org (accessed 24 June 2016).

## Cetaceans

Living cetaceans fall into two distinct groups: the whalebone or baleen whales, the Mysticeti (15 species); and the toothed whales, the Odontoceti, which include the dolphins and porpoises (about 70 marine species). The baleen whales are all large; most species have a maximum length in the range of 15–30 m. The toothed whales are small to medium-size cetaceans (< 1.5–12 m), apart from the sperm whale (*Physeter macrocephalus*) (< 20 m). The principal species exploited by traditional whaling methods comprise the majority of baleen whales and the sperm whale, the so-called great whales. But as we shall see, the smaller cetaceans have not escaped harm, either deliberate or accidental, from human activities.

Although much is known about the natural history and habits of the commercially exploited whales, our knowledge is fragmentary for many of the smaller cetaceans. Some are known only from a few stranded specimens, and this, coupled with the difficulties of identification at sea – even for experienced observers – means that we have little idea of their abundance and distribution. Because of their differing habits, frequency of observation at sea may be a poor guide to relative abundance of species. Some, such as the northern bottlenose whale (*Hyperoodon ampullatus*) and common bottlenose dolphin (*Tursiops truncatus*), appear to approach and be fascinated by ships, and riding bow waves by many dolphin species is well known. Others, such as beaked whales and most porpoises, are by contrast shy and difficult to approach, and it is possible that many of the other smaller cetaceans about which we know so little fall into this category.

Cetaceans are all predators, but the two major groups differ markedly in their methods of feeding and type of prey taken. Baleen whales are essentially ‘strainers’. Suspended from the upper jaw are hundreds of fringed keratinous plates, the baleen or whalebone, which they use for sifting food from near-surface water. Right (*Eubalaena* spp.) and bowhead whales (*Balaena mysticetus*) have baleen with finely spaced fringes and are adapted to feed mainly on small planktonic crustaceans, notably copepods, which they obtain by swimming with their mouths open and filtering the water continuously over their baleen plates. Rorquals (Balaenopteridae), on the other hand, have coarser fringes for taking somewhat larger prey. They gulp large volumes of water, which they then expel from the sides of their mouth whilst the baleen retains the food. The blue whale (*Balaenoptera musculus*) feeds almost entirely on krill – large planktonic crustaceans – but the other rorqual species take krill as well as small schooling fish. The gray whale (*Eschrichtius robustus*) has relatively coarse fringes and feeds primarily on benthic amphipods, which it sifts from seabed sediment. The enormous bulk and filter-feeding habit of the baleen whales make them reliant on productive regions to fulfil their nutritional needs. This involves some species making extensive seasonal migrations to high latitudes in both northern and southern hemispheres.

Toothed whales typically feed on fish or squid. The killer whale (*Orcinus orca*) feeds on seals and sea lions and by biting chunks from baleen whales. Many of the toothed whales feed in surface waters, but the sperm whale descends to depths of more than 1000 m for prey. Odontocetes produce a variety of sounds, including high-frequency clicks for prey detection as well as for navigation and communication. Mysticetes produce low-frequency signals, sometimes organized into ‘songs’, for communication.

Cetaceans exhibit strongly *K*-selected life-history characteristics. For many whale species, puberty occurs between 5 and 12 years, and a single calf is born every 2 or 3 years. Gestation is typically 10–12 months and weaning 6–7 months, but lactation is often longer in odontocetes (< 2–5 years). Maximum age is typically 40–100 years in mysticetes and 12–50 years in odontocetes, although sperm whales can attain at least 70 years and killer whales may live past 100 years. So although long-lived, females generally have a low reproductive output. Typical rates of population increase are 2–4% per year for most dolphin species and 2–10% for most baleen whales, but rates of increase for severely depleted baleen stocks can range up to 14% (Best, 1993; Bowen & Siniff, 1999).

Some whales, notably the fin (*Balaenoptera physalus*), blue, common minke (*B. acutorostrata*), sei (*B. borealis*), and humpback (*Megaptera novaeangliae*), occur in all oceans including the tropics, but other species have more restricted distributions. The bowhead, narwhal, and beluga or white whale (*Delphinapterus leucas*) are confined to arctic and subarctic regions, and the gray whale today lives only in the North Pacific; its North Atlantic populations were hunted to extinction 300 to 400 years ago. Some cetaceans, such as Bryde’s whale (*Balaenoptera edeni*), pantropical spotted dolphin (*Stenella attenuata*), and spinner dolphin (*S. longirostris*), are essentially tropical in distribution. There are, however, many small cetaceans with apparently very restricted distributions. Some are essentially estuarine (and there are also freshwater dolphins in tropical river systems of Asia and South America). Whales that frequent temperate or polar regions tend to be isolated by continental land masses into North Atlantic, North Pacific, and Southern Ocean populations or even separate species as with the three right whale species (*Eubalaena glacialis, E. japonica*, and *E. australis*, respectively) (Churchill *et al.*, 2012).

Many whales show more or less distinct seasonal migration patterns, moving into high-latitude waters in summer to feed on the rich zooplankton and returning to low latitudes in the winter to breed. Such a pattern of migration is shown by the gray, fin, blue, and humpback whales. Because the winter migrations of baleen whales towards the tropics occur at different times in the Northern and Southern Hemispheres, the populations north and south of the equator remain distinct. Even for wide-ranging whales species, stocks within ocean basins may remain relatively isolated. Mitochondrial DNA analysis indicates that humpback whales are divided into genetically distinct regional populations, seemingly because of a strong fidelity to traditional migratory routes that develops when a calf first accompanies its mother on her annual migration. The existence of stocks indicates that such relatively self-contained populations should be managed separately (Baker & Palumbi, 1996).

## Pinnipeds

The seals, sea lions, and walrus comprise the Pinnipedia. Thirty-three living marine pinniped species are recognized, belonging to three families (Perrin *et al.*, 2009). The 'true' seals, the Phocidae, comprise a group of Northern Hemisphere seals (nine species), the monk seals (two species), Antarctic seals (four species), and elephant seals (two species). The family Otariidae contains the fur seals (nine species) and the usually somewhat larger sea lions (six species), whilst the Odobeniidae contains the walrus (*Odobenus rosmarus*).

Phocids, unlike otariids, cannot rotate their hind flippers under their body for walking. Otariids can thus move on land more effectively than phocids and swim mainly using their large front flippers. Phocids are more cumbersome on land but more streamlined and effective swimmers, using their hind flippers and lower body to scull through the water. The walrus also mainly uses its hind flippers for propulsion but is a slow swimmer compared to phocids. Most distinctive are its upper canines, which in both sexes are enlarged to form tusks, used to indicate social status and for hauling out on ice (Riedman, 1990).

Apart from the monk seals, phocids inhabit temperate to polar waters, and walruses have a northern circumpolar distribution. On the other hand, none of the otariids is truly polar, and some have a tropical or subtropical distribution. Fur seals are predominantly a Southern Hemisphere group, but sea lions are common in both hemispheres. Pinnipeds are notably absent from some parts of the world, in particular from the tropical Indian Ocean and Asian waters, possibly for reasons of palaeo-oceanography and availability of sufficiently productive feeding areas (Riedman, 1990). However, Cairns *et al.* (2008) suggest that the rarity of pinnipeds (and pursuit-diving seabirds) at low latitudes is due to the effect of temperature on the relative swimming speeds of endotherms and ecotherms and hence their predation success. In warm water, fish as are thus more difficult for pinnipeds to catch, whereas large ectothermic predators of pinnipeds, such as sharks, are advantaged.

Most seals and sea lions feed largely on fish and cephalopods, often taking a range of prey species depending on availability. Some also take other invertebrates, especially crustaceans and bivalve molluscs, and a few prey on other pinnipeds and birds. One of the most specialist feeders, the (misnamed) crabeater seal (*Lobodon carcinophagus*), feeds almost exclusively on krill. Walruses are benthic feeders, taking mainly clams.

Pinnipeds are all relatively large mammals. Reducing surface-to-volume ratio helps with heat conservation, and lessening drag also increases potential swimming efficiency, as does their compact streamlined shape (Bonner, 1982). Large body size also has the advantage of reducing the number of potential predators. However, they still need to haul out for breeding, so pinnipeds have not reached the body size of many cetaceans. Most pinniped species have a maximum length of

1.5–3 m; largest are the walrus (males up to 3.6 m in length) and elephant seals (up to at least 5 m) (Jefferson *et al.*, 1993). Loss of body heat is also reduced by a subcutaneous layer of fat, or blubber, which also serves as a food reserve enabling pinnipeds to endure periods of fasting. In addition, in most pinnipeds the body is insulated by a well-developed fur coat that traps a layer of air against the skin. Some, however, have less fur and more blubber, notably the largest species, the walrus and elephant seals.

Otariids are all highly polygynous – a male mating with several females during a breeding season – with the usually territorial males being considerably larger than the females. Phocids are more variable in their breeding systems. Some, like the elephant seals, are also highly polygynous and sexually dimorphic, whereas ice-breeding seals, such as the Arctic ringed seal (*Pusa hispida*), tend to be serial monogamists – they stay with one partner for the breeding season but typically find a new one next season – and the males and females are of similar size. Pinnipeds have to haul out on land (all otariids) or on ice (most phocids and walrus) to give birth and rear their young and, particularly in some phocids, to moult. Also, some species are migratory and make annual movements between breeding sites (rookeries) and feeding grounds, such as phocids that breed on seasonal pack ice (Riedman, 1990).

## Otters

Several species of otter make use of marine or estuarine environments, but only two forage entirely in the sea. The marine otter (*Lutra felina*), the smallest marine mammal (< 115 cm total length), is an elusive, poorly known species with a disjunct distribution along the coast of Peru and Chile. It frequents open rocky coasts, feeding mainly on fish, crustaceans, and molluscs, but is only semi-aquatic, spending most of its time ashore in caves. Intensively hunted in the twentieth century, it is now threatened more by increasing human impact along the coast, further fragmenting its habitat (Valqui, 2012), and is categorized as endangered on the IUCN Red List.

The sea otter (*Enhydra lutris*) is far better known and fully aquatic, not needing to come ashore. Its ranges around the North Pacific rim from Baja California to northern Japan and is associated with kelp forest habitat where it feeds mainly on benthic invertebrates, in particular sea urchins, molluscs, and crustaceans. Its populations have recovered markedly from intense exploitation for its valuable pelt, but it is still endangered (Estes *et al.*, 2009a).

## Sirenians

The three extant species of marine sirenians are coastal animals, tropical to subtropical in distribution and – unlike all other marine mammals – primarily herbivorous, feeding on aquatic vegetation (Marsh *et al.*, 2011). The dugong (*Dugong dugon*) (Fig. 9.2) inhabits coastal waters of the Indo-West Pacific, with the largest populations in northern Australia (probably around 70 000 individuals), although many populations are now extinct or close to extinction. The West Indian manatee (*Trichechus manatus*) ranges from the south-eastern USA and Gulf of Mexico through the greater Caribbean to northern South America, and the West African manatee (*T. senegalensis*) occurs from Senegal to Angola. Manatees extend into estuarine and riverine habitats, whereas the dugong is more strictly marine. They are all large (up to at least 3 m in length), without hindlimbs, and with paddle-like forelimbs. They swim using their large tail fluke, rounded in manatees and whale-like

**Fig. 9.2** A dugong feeding on seagrass in the Red Sea, Egypt. Dugong inhabit coastal and estuarine waters from east Africa to Vanuatu in the western Pacific. Although hunting is now banned in most of their range, illegal poaching is widespread, and incidental capture in fishing gear is also a major cause of mortality. Dugong feed primarily on seagrasses, so protection of seagrass beds is also important. Photograph: Getty Images/David Peart. (A black and white version of this figure will appear in some formats. For the colour version, please refer to the plate section.)

in the dugong. Sirenians have long been sought for their meat, hides, and blubber, and hunting is still significant for many communities. Fishing-related mortality and loss of habitat are also important in some populations. In the USA, the main identified cause of death of manatees is from collisions with boats, but boat strikes are likely to be a growing problem impacting other sirenian populations (Marsh *et al.*, 2011). All three species are categorized as vulnerable to extinction on the IUCN Red List.

Steller's sea cow was a giant relative of the dugong, at least 7–8 m in length, that lived along the cold North Pacific coast where it foraged on kelp. Relict populations around the Commander Islands in the Bering Sea came to the attention of scientists in 1741, but the species was extinct less than 30 years later as a result of excessive exploitation by whalers and fur hunters provisioning ships, a striking example of the vulnerability of a *K*-selected species to even pre-industrial hunting (Turvey & Risley, 2006; Anderson & Domning, 2009).

# Hunting

Marine mammals have been hunted for a wide range of products, notably for meat, blubber, oil, hides, fur, baleen, bones, and ivory. Large-scale commercial hunting has particularly impacted species of whales (for oil, meat, and baleen), pinnipeds (for pelts and oil), and the sea otter (for fur).

## Origins of Commercial Whaling

Whales have been exploited for their meat and oil from earliest times, as stranded animals were used by coastal communities. Early targeted whaling was from open boats, using hand-held spears or harpoons, methods that survive today among some peoples in the form of subsistence whaling. Directed hunting of large whales appears to date back about a thousand years to the Basques' hunting of right whales in the Bay of Biscay. Over the next several centuries whaling spread northward through Europe to Arctic waters, and other European countries became involved. The Basques, however, also appear to be the first to have undertaken commercial whaling in the western North Atlantic. They were active in the Gulf of St Lawrence area in the sixteenth to seventeenth centuries, taking an estimated 25 000–40 000 whales, mostly bowhead whales – a major impact on the species (McLeod *et al.*, 2008).

The continued exploitation of northern waters, from North America and Europe, was such that by the late eighteenth century stocks of both bowheads and right whales had been severely depleted. By this time commercial whaling had spread to the Pacific and to other parts of the world including the Indian Ocean, South Africa, South America, and the Falkland Islands, as the nations that had been exploiting North Atlantic whales extended their activities worldwide and local industries developed using similar whaling methods. A major development during the eighteenth century was the New England-based fishery for sperm whales. Sperm whale oil was in demand, particularly for use in lamps and candles, and by the late eighteenth century Yankee whalers were operating far afield (Ellis, 2009).

Key developments in the mid-nineteenth century allowed greater exploitation of the faster-swimming rorquals, which had previously been largely beyond the reach of whalers. These innovations were the use of steam-powered catcher vessels, invention of the explosive harpoon (contrived to kill a whale on impact), and an inflation lance to pump air into the carcass as, unlike right whales, dead rorquals do not float. As a result, there followed an extension of commercial whaling to the most distant parts of the world, including the Southern Ocean, where land bases were established in South Georgia and the South Shetland Islands. Exploitation of blue, fin, and sei whales in these waters was now feasible (Fig. 17.6).

The expansion of Antarctic whaling was aided enormously by the introduction in the 1920s of factory ships with stern ramps, as these were independent of shore stations and supplied by catcher boats could work at sea continuously for months (Clapham & Baker, 2009) (Fig. 9.3). Whale stocks declined rapidly. Humpback catches had already dropped drastically by 1920, and attention was successively directed to blue, fin, and sei whales, each showing equally dramatic reductions. In later years the smaller minke whales were in turn exploited (see Chapter 17).

The intensive whaling in Antarctic waters caused an overproduction of oil to the extent that international agreement was reached to restrict further whaling. Following a Convention for the Regulation of Whaling drawn up in 1931, the concept of the blue whale unit (BWU) was introduced. One BWU was equivalent to 2 fin, 2.5 humpback, or 6 sei whales on the basis of average oil yields

**Fig. 9.3** Catcher boats working with factory ships enabled whaling to be carried out at sea for months at a time, severely depleting many whale stocks from about the 1920s. Photograph: Scott Polar Research Institute, University of Cambridge, with permission.

(Clapham & Baker, 2009). This attempt to control catches helped stabilize oil production but was disastrous for whale stocks, as it encouraged the taking of the largest whales and did not discriminate between species in terms of their individual management needs. Further restrictions introduced in the late 1930s included a minimum length for blue and fin whales and opening dates for the whaling season. International agreement in 1937 extended the above restrictions to humpback and sperm whales. It also prohibited the taking of females accompanied by calves, prohibited whaling in certain areas, and protected right and gray whales and subsequently the humpback, which was protected in the Southern Ocean.

## Post-war Whaling and the International Whaling Commission

Since the mid-1940s, the history of commercial whaling has been closely linked with the International Whaling Commission (IWC) (Gambell, 1977, 1999). The need to regulate post-war whaling led in 1946 to interested nations signing the International Convention for the Regulation of Whaling (ICRW) 'to provide for the proper conservation of whale stocks and thus make possible the orderly development of the whaling industry'. One of the main outcomes of the convention was the establishment of the IWC, which meets annually and sets catch quotas and other management measures. At this time, IWC members sought to strike a balance between the conservation of whale stocks and the pressing demand for fats and oils in the aftermath of the war. As with many such organizations, the strength of the IWC depended on the goodwill and cooperation of the nations concerned, the IWC having no power to enforce its regulations. Membership of the IWC is voluntary, and some states have always declined to join. Also, any member state can object to a decision of the commission and not be bound by it. Other loopholes have tended to frustrate the intentions of the commission, as examined later.

Initially, the IWC continued to use the BWU for the regulation of Antarctic catches. Pre-war catches, amounting to about 24 000 BWU per season, were recognized as too high, and over the next decade or so quotas were cut back, albeit – in the face of industry resistance – far less than needed. By the early 1960s, whale stocks had dropped to such an extent that whaling was for some nations becoming uneconomic, and many factory ships were scrapped, laid up, or sold.

The IWC had by now seen the need for an independent assessment of Antarctic whale stocks. This led to the formation of a committee of scientists charged with assessing whale stocks and reporting on sustainable yields. The committee's 1964 report recommended total protection for all blue and humpback whales in the Antarctic and a major reduction in catch of fin whales and that regulation should in future be on the basis of individual species. There followed a major crisis for the IWC in that no quota could be agreed for the 1964–5 season, the few remaining whaling nations – Japan, Norway, and the USSR – again setting their own limits. However, in the late 1960s agreements were reached to reduce catches to below the sustainable yield level for each stock based on the best scientific estimates. Finally, for the 1972–3 season, the BWU was phased out and quotas were set for individual species.

The basis for setting catch quotas depends on knowledge of the number of whales in a stock and the rates of recruitment and natural mortality. These parameters are, however, seldom known with any accuracy. The concept of maximum sustainable yield (MSY) was developed as a method for the management of fish populations (see Chapter 5). It is based on the assumption that, within limits, per capita reproductive rate increases as population size decreases, allowing a relationship between population size, recruitment level, and natural mortality to be formulated. The excess of recruitment over natural mortality gives the potential for exploitation. At a certain population level, typically rather above 50% of the original figure for an unexploited population, yield is at a maximum for that stock. The concept of MSY as a means of regulating whaling has been criticized in that even a small error in the parameters adopted for a stock could lead to an accelerated decline through exploitation at a level that the population cannot sustain. The IUCN at its General Assembly in 1975 concluded that MSY and other single-species management concepts were not adequate as a basis for the management of wild living resources, including whales. Instead, it proposed that management should be based on ecological relationships according to a number of principles incorporating the ideas of the maintenance of present and future options, the inclusion of a safety factor, reduction of waste, and public review of information and assessments before and during utilization of a resource. Undoubtedly, the history of the exploitation of whales, as of so many other living resources of the sea, points to the need for more conservative management.

It became apparent in the early 1990s that the catastrophic decline of some stocks had been hastened because of illegal whaling carried out on a massive scale by the then-USSR from shortly after World War II until the early 1970s when international observers were introduced. Much of this had been carried out in the Southern Hemisphere where, between 1947 and 1972, unreported catches totalled about 100 000 whales, almost half of which were humpbacks. Other countries have engaged in illegal whaling and falsification of records and in pirate whaling conducted by non-IWC countries under flags of convenience (Brownell & Yablokov, 2009).

In 1972, the UN Conference on the Human Environment passed a resolution on whaling. This called for a 10-year moratorium on all commercial whaling as well as increased scientific research into whale stocks and a strengthening of the IWC secretariat and its capabilities. The IWC was unable to accept the proposal for a moratorium since it considered that stocks that had become depleted had already been given total protection, whilst the more abundant stocks could be safely exploited at controlled levels without causing further depletion. A new procedure for setting quotas was, however, developed and later revised (Box 9.1). Nevertheless, the IWC finally agreed in 1982 to an indefinite postponement of all commercial whaling, with full effect from 1986, a moratorium that is still in place. This change in attitude in part reflected the altered composition of the IWC. When it was formed in 1946, all members of the commission were actively involved in commercial whaling, whereas by the 1980s most members no longer had whaling interests. The commission now has some 90 member nations, with most fully committed to conservation of whale stocks. Indeed, the period of modern industrial whaling may represent the largest hunt in human history, at least in terms of biomass. Rocha *et al.* (2014) estimate that over the twentieth century nearly 2.9 million

## Box 9.1 IWC Management Procedures

### New Management Procedure

Under the IWC's New Management Procedure (NMP) adopted in 1975, stocks were to be managed under three categories according to their abundance relative to the level providing the maximum sustainable yield (MSY). The most depleted stocks were afforded protection, which resulted in the end of fin and sei whaling in the Antarctic. However, in the case of those stocks categorized as still able to be exploited, determining sustainable catch levels proved problematic – in most cases the data needed to estimate MSY were simply inadequate. This led to the IWC suspending commercial whaling – from 1985 to 1986 in Antarctic waters and from 1986 elsewhere – and developing a revised method of setting catch limits (Gambell, 1993; Cooke, 1995).

### Revised Management Procedure

The Revised Management Procedure (RMP) was accepted by IWC in 1994. This more precautionary approach calculates catch limits on the basis of two elements: the time series of annual catches and estimates of current abundance by area. The RMP aims to maintain whale stocks at 72% of their original unexploited abundance. Below 54% no whaling is allowed, and quotas progressively increase as populations recover towards their original abundance. Also, quotas are linked to the quality of the abundance data: where uncertainty in abundance is high, the RMP model gives a lower quota, and vice versa. The RMP is part of a larger Revised Management Scheme that includes other compliance components. If the IWC voted to resume commercial whaling, then the RMP would be implemented. Elements of the RMP are, however, used in managing other non-whale stocks (Butterworth & Punt, 1999; Kock, 2007).

large whales were killed and processed. About 70% were taken in the Southern Hemisphere, with catches peaking in the 1950s and 1960s. Acceptance of a moratorium by the IWC has not, however, marked the end of whaling. Various provisions enable whaling to continue under different guises. Parties can avoid having to abide by an amendment in the Schedule to the ICRW by registering an objection. Norway resumed commercial whaling of minke whales in the North Atlantic in 1994 using this procedure.

Also under the ICRW there is provision for whaling 'to satisfy aboriginal subsistence need'. The IWC sets these catch limits, although this can include quotas for stocks so depleted that commercial whaling would be forbidden. Aboriginal subsistence whaling has in recent years mainly involved the taking of bowhead whales from the Bering-Chukchi-Beaufort Seas, gray whales from the eastern North Pacific, minke and fin whales from Greenland, and humpback whales from St Vincent and the Grenadines in the Caribbean. But in most cases the whaling is not for subsistence nor of long tradition and can involve modern methods. Holt (2000) considered the only traditional, truly aboriginal subsistence whaling left to be the hunting of sperm whales by villagers on the Indonesian island of Lembata.

Also contentious is the issue of scientific whaling. Under the ICRW, contracting governments can grant themselves special permits to kill an unregulated number of whales for scientific research, and once the scientific data have been collected, the meat can be sold on the open market (Gales *et al.*, 2005). Since the moratorium, Norway, Iceland, and Japan have used this provision but drawn widespread criticism that they are exploiting a loophole to continue commercial whaling. Japan in particular has used such scientific permits, killing 14 660 whales between 1987 and 2012, a level of exploitation comparable to commercial whaling operations (Clapham, 2015). The vast majority have been minke whales from the Southern Ocean, in fact within the IWC's Southern Ocean Sanctuary (see below). Stated objectives of Japan's scientific whaling programmes are mainly to better understand stock structure for management purposes. But in response to a case filed by Australia, the International Court of Justice ruled in 2014 that Japan's Antarctic whaling programme was not being conducted for scientific purposes. Japan as a result suspended this particular operation (JARPA II) but so as to reformulate its scientific objectives and continue whaling (Clapham, 2015).

There appears to be no justification to use lethal sampling to obtain such data. Abundance estimates can, for instance, be derived from visual surveys, and where tissue is required for stock identification or ageing, non-lethal methods are available. Biopsy darts can be used to obtain skin samples for DNA fingerprinting, and DNA methylation (a method that cells use to control gene expression) can be used to estimate age (Hoelzel *et al.*, 2002; Polanowski et al., 2014). Molecular genetic analyses, with the results in a formal DNA register, would also provide an effective means of monitoring and verifying catch records. For example, analysis of whale products from retail markets in Japan and South Korea has revealed meat from several whale species protected by the moratorium, suggesting illicit hunting (Baker *et al.*, 2010). Such methods also indicate illegal trade in protected whales in contravention to CITES (Baker *et al.*, 2010).

Global estimates for the great whales indicate that current populations of baleen whales are around 10–20% of their pre-exploitation levels and odontocetes are at about one-third (Ruzicka *et al.*, 2013). But there is now evidence of population increases. The eastern North Pacific gray whales, North Atlantic humpbacks, and the subpopulation of bowheads in the Bering, Chukchi, and Beaufort seas are all estimated to have annual rates of population increase of around 3%, and Southern Hemisphere humpback stocks have estimated rates of increase of 5–10% per annum (www.iucnredlist.org).

## Small Cetaceans

Although the ICRW does not define the term 'whale', it lists a dozen species in an annex to the convention, namely the baleen, sperm, and bottlenose whales. Pro-whaling governments have taken the view that these are the species that come under the IWC's jurisdiction, whereas others maintain that the convention is not necessarily limited in this regard. So to what extent the IWC should be actively involved in the management of the unlisted and mostly smaller cetaceans is unclear, though there is agreement that the IWC's Scientific Committee should be concerned with threats to small cetaceans and provide advice, as indeed it does.

Abundance estimates may be poor for many baleen whales, but the situation is far worse for most small cetaceans where the great majority are categorized by IUCN as data deficient, meaning there is inadequate information to assess their risk of extinction (see Chapter 4). There is, however, evidence that various species of small cetaceans face very significant threats, particularly through interactions with fisheries. Indeed, whilst prospects for the great whales have improved in recent years, especially since the moratorium, the situation for smaller species has in general deteriorated.

Small cetaceans are taken for human consumption or as fishing bait in many countries. Directed hunts include those for beluga and narwhal (*Monodon monoceros*) in the Arctic, long-finned pilot whale (*Globicephala melas*) in the Faroe Islands, and short-finned pilot whale (*Globicephala macrorhynchus*) in Japan and for various species in Chile, Peru, and Sri Lanka (Mulvaney & McKay, 2000; Olson, 2009). Several species of small cetaceans are caught for human consumption by artisanal fishermen along the central Peruvian coast. Monitoring these catches in the mid-1980s, Read *et al.* (1988) found that more than 90% of the small cetaceans landed were dusky dolphins (*Lagenorhynchus obscurus*) taken in drift gillnets. Their estimates (for 1985) indicated a total catch of nearly 10 000 dusky dolphins for the entire Peruvian coast, but in the absence of abundance estimates for any of the small cetaceans exploited in coastal Peru, they were unable to assess the impact on populations. Another heavily exploited species has been Dall's porpoise (*Phocoenoides dalli*), hunted in Japanese waters. In 1988 the take peaked at more than 40 000 animals, but catches have since more than halved.

## Sealing

Highly gregarious large mammals that are cumbersome and vulnerable out of water and have thick pelts and abundant blubber and meat were also no doubt hunted by the earliest humans that encountered them. Archaeological evidence from sites in NW Europe, notably in the Baltic region, shows that seals were being hunted from Stone Age times. Seals have been central to the economy and culture of some peoples, such as Inuit, and subsistence and small-scale commercial sealing is documented for various cultures over the centuries. In many instances colonies were no doubt eradicated, particularly with the advent of firearms, although some societies, such as on islands in the Kattegat, had ancient regulations, persisting at least until the end of the eighteenth century, that enabled a sustained exploitation. Large-scale commercial sealing was virtually unknown before the late eighteenth century, but from then until the early twentieth century many pinniped species have been commercially exploited, often on a huge scale (Bonner, 1982).

In the Northern Hemisphere, the species most heavily impacted by commercial sealing have been the walrus, northern fur seal (*Callorhinus ursinus*), and harp seal (*Pagophilus groenlandicus*) (Bonner, 1982). Walruses live in Arctic coastal regions, usually in association with pack ice in relatively shallow areas (< 100 m) where they can forage for clams. They are no longer commercially exploited, but the Atlantic population has yet to recover fully. Significant numbers are still shot each year by indigenous peoples.

The northern fur seal occurs in the North Pacific and Bering Sea and became the target of intense exploitation, particularly following the discovery of the huge Pribilof Islands' population. Given the severe depletion, only limited hunting was agreed on under an international convention of 1911 and its subsequent extensions. Commercial hunting has since ceased, but as the latest convention lapsed in 1984 the species is not now under international agreement regulating its exploitation. Nevertheless, the largest breeding population in the Pribilof Islands (half the total population) is still declining, at about 6% annually. Possible reasons include reduced prey availability through competition with fisheries and entanglement in fishing gear (Lee *et al.*, 2014).

The harp seal, by contrast, is still hunted commercially. Indeed, the Canadian harp seal hunt remains the largest for any marine mammal. The species is widely distributed in the Arctic and subarctic North Atlantic and is separable into three distinct breeding populations. The NW Atlantic population is the largest – it has increased from 1.1 million animals in the early 1970s to now more than 7 million – and is the most heavily exploited. The seals are taken mainly for their pelts. Over the past 60 years the annual commercial catch has been around 200 000 animals, the great majority of them pups (Hammill *et al.*, 2015). The hunt has proved highly controversial in terms of the setting of catch quotas and methods of slaughter (e.g. Hammill & Stenson, 2010; Daoust *et al.*, 2014).

Monk seals, unusual in being seals of warm water regions, have fared particularly badly as a result of hunting, human disturbance, and persecution. The Caribbean monk seal (*Monachus tropicalis*) is now extinct – the last confirmed sighting was in 1952 – and the two remaining species, the Mediterranean monk seal (*M. monachus*) and Hawaiian monk seal (*M. schauinslandi*), are endangered (www.iucnredlist.org). The Mediterranean species, once common throughout the Mediterranean and SE North Atlantic, now has a total population of merely 350 to 450 animals, most of them in the eastern Mediterranean (see Chapter 16). The situation is marginally better for the Hawaiian monk seal. Although the total population is only about 1000 individuals in the northwestern Hawaiian Islands, its main breeding sites are within protected areas (Lowry *et al.*, 2011).

The two species of elephant seal have been heavily exploited, the blubber being processed for its oil. The northern elephant seal (*Mirounga angustirostris*) is a species of the NE Pacific but with a restricted breeding distribution from northern California to the Baja California, mainly on offshore islands (Fig. 9.4). Intensive hunting began around 1810, and by the 1880s the species was all but extinct. A herd of maybe 10 to 30 seals survived on remote Guadalupe Island (Hoelzel *et al.*, 2002). The species has since rebounded dramatically, recolonizing much of its former range and with a total population (estimated in 2010) of between 210 000 and 239 000 individuals (Lowry *et al.*, 2014). As a legacy of this population bottleneck, the species exhibits very reduced genetic variation (Hindell & Perrin, 2009).

The southern elephant seal (*Mirounga leonina*) has a virtually circumpolar distribution with breeding colonies on most of the subantarctic islands. Commercial hunting at South Georgia continued until 1964, and some rookeries were extirpated and have yet to recolonize (Jefferson *et al.*, 1993). But the species' wider and less accessible distribution meant it was spared the near extinction and loss of genetic variation of its northern counterpart. Its population is believed not to have fallen below 1000 animals and was last estimated (in 2000) at about 640 000 (Hindell & Perrin, 2009).

Sealing in the Southern Hemisphere also targeted the *Arctocephalus* fur seals. All but one of the eight species occur in the Southern Hemisphere, and all were heavily exploited in the eighteenth

**Fig. 9.4** A northern elephant seal rookery at Piedras Blancas, California. The species was hunted to near extinction by the late nineteenth century but recovered markedly following legal protection and now has a total population in the NE Pacific of more than 210 000 individuals. Photograph: P.K. Probert.

and nineteenth centuries, in some cases almost to extinction, such as the Antarctic fur seal (Bonner, 1982) (see Chapter 17). It is difficult to judge what the pre-exploitation abundance and distribution of these species would have been, especially as sealers were secretive about the location of good sealing grounds. Only the Cape fur seal (*A. pusillus*) is still legally hunted. Around 50 000 animals, nearly all pups, have been taken annually in Namibia in recent years (Japp *et al.*, 2012).

Released from overexploitation and with a shift to conservation, pinnipeds overall show good evidence of recovery. An analysis by Magera *et al.* (2013) of population abundance time series for 20 pinniped species showed half the populations to be significantly increasing. Nevertheless, as numbers increase and species recolonize their former range, management issues arise, such as those relating to fisheries, tourism, and coastal habitat use.

## Effects of Hunting

Depletion of marine mammal populations may strongly affect food-web dynamics. A striking example is the role of sea otters in structuring nearshore communities of the North Pacific rim (Estes & Palmisano, 1974). Sea otters feed predominantly on benthic invertebrates, including sea urchins (*Strongylocentrotus* spp.), which in turn are major consumers of kelp. By limiting

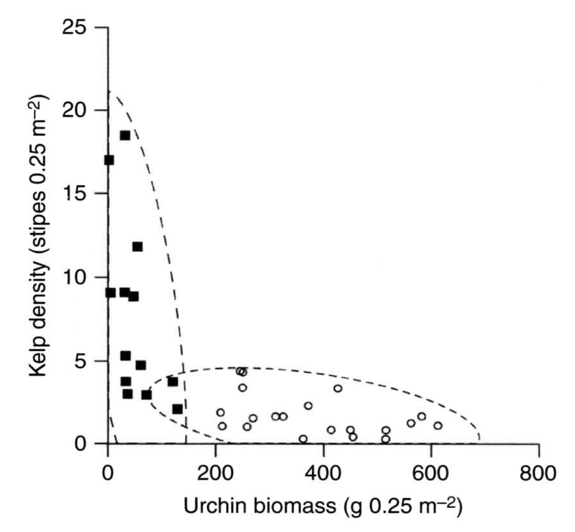


**Fig. 9.5** Kelp density versus urchin biomass in the Aleutian archipelago measured at 19 islands ($n = 463$ sites) between 1987 and 2006, with data averaged by island-year combination. Dashed lines are 90% confidence ellipses around the two aggregates identified by cluster analysis, which reflect the kelp-dominated (solid squares) and urchin-dominated (open circles) phase states. Each data point represents the average of all sites sampled at one island during 1 year ($n = 34$ island-year combinations). The kelp-dominated state is associated with a high population density of sea otters (> 6 otters per km surveyed), and the urchin-dominated state with a low density of otters (< 6 otters per km surveyed). Reprinted with permission from Estes, J.A., Tinker, M.T. & Bodkin, J.L. (2010). Using ecological function to develop recovery criteria for depleted species: sea otters and kelp forests in the Aleutian archipelago. *Conservation Biology*, 24, 852–60, published by Wiley. © 2010,Society for Conservation Biology.

the grazing pressure exerted by urchin populations, sea otters help maintain well-developed kelp beds and their many associated species. A lack of sea otters, on the other hand, typically leads to overgrazing by sea urchins. These nearshore reef systems tend to exist in one of two states – kelp forest or sea urchin barrens – that are strongly indicative of high- and low-density populations of sea otters (Fig. 9.5).

Sea otters were once widespread in nearshore waters from northern Japan to Baja California, but the fur trade from the mid-eighteenth to early twentieth century brought the species to near extinction. An international treaty of 1911 ended commercial hunting of sea otters, allowing remnant populations to recover over the following several decades. This was aided by reintroduction programmes in the 1960s and 1970s (Estes *et al.*, 2009a). Some populations, however, have failed to recover or have since crashed in recent decades. Predation by killer whales might be implicated in the decline of the sea otter population in the Bering Sea-Aleutian Islands region over recent decades, which would have implications for ecosystem-based management (DeMaster *et al.*, 2006) (Fig. 9.6).

For the Southern Ocean there is ongoing debate on the extent to which food-web structure may have changed in the wake of intensive whaling – notably the removal of more than a million baleen whales and a possible ‘krill surplus’ – or whether climate-change related processes are primarily responsible (e.g. Ainley *et al.*, 2007) (see Chapter 17).

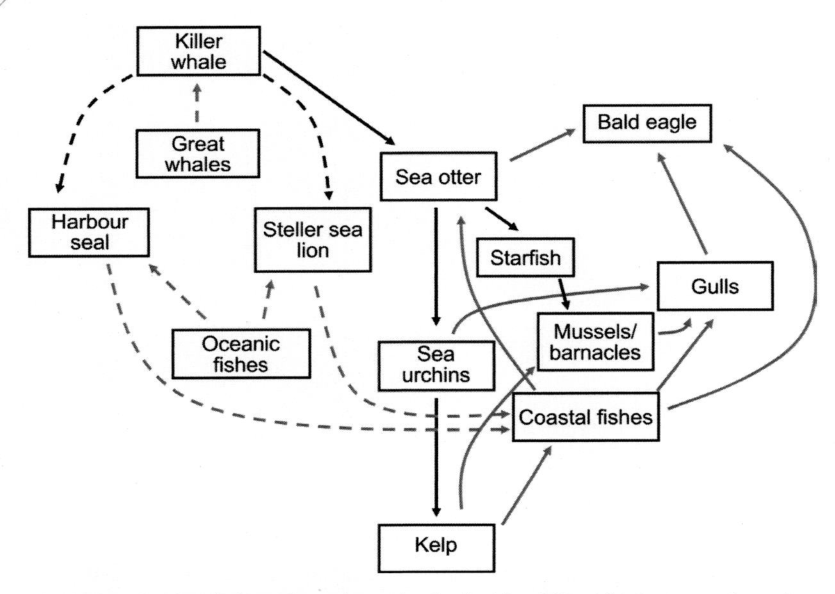


**Fig. 9.6** A food web for selected species in the North Pacific Ocean and southern Bering Sea. Solid and dashed arrows represent known and suspected linkages, and black and grey arrows top-down and bottom-up forcing. Sea otters strongly influence kelp forest community structure and may in turn be predated by killer whales. From Estes, J.A., Doak, D.F., Springer, A.M. & Williams, T.M. (2009b). Causes and consequences of marine mammal population declines in southwest Alaska: a food-web perspective, *Philosophical Transactions of the Royal Society B*, 364, 1647–58, by permission of the Royal Society.

## Marine Mammal Watching

With the demise of industrial whaling, the most significant 'use' of whales is now whale watching, which has grown dramatically worldwide since the late 1980s. Projections indicate that further increases in tourism based on viewing marine mammals from a boat or shore have the potential to generate more than $2.5 billion a year and support some 19 000 jobs worldwide (Cisneros-Montemayor *et al.*, 2010). Marine mammal watching has the potential to strengthen public interest in marine mammals as well as marine conservation in general and provide educational opportunities.

Marine mammal watching may disturb animals and therefore needs to be carefully managed. Whale-watching boats can elicit a variety of responses, but most consistently when boats are present animals spend more time travelling and less time resting and take more circuitous paths (Senigaglia *et al.*, 2016). This indicates a greater expenditure of energy, which may ultimately affect reproductive success. Codes of conduct for whale watching have been developed around the world, covering such issues as minimum approach distances of boats and aircraft, the number of boats at a time, the nature and degree of interaction (swimming, touching, feeding, noise), and particular restrictions (e.g. viewing pods with calves) (Garrod & Fennell, 2004). Such codes vary greatly, and most are voluntary. Ideally, an internationally recognized code of conduct is needed. The IWC has, for instance, drawn up general principles to minimize adverse impacts of whale watching.

Other marine mammals have also become the focus of tourist operations. In winter, Florida manatees (*Trichechus manatus latirostris*) aggregate at warm-water sites fed by natural springs or power-plant discharges. The largest aggregation – about 350 manatees – occurs at Crystal River at the head of a spring-fed estuarine system on the Gulf coast and attracts each year some 100 000 visitors (Sorice *et al.*, 2006).

## Interactions with Fisheries

Marine mammals and fisheries might impinge on one another in various ways, but most relevant here are the potential for both to be competing for a common food resource and the direct impact of fishing operations on marine mammals.

### Competition with Fisheries

There is a view that marine mammals compete with fisheries and that reducing their populations would thereby help fish stocks; such claims have been made for 20% of those marine mammal species reviewed (Plagányi & Butterworth, 2005). Culls have been carried out on this basis, but it appears that none has clearly benefited fisheries' catches. Ecosystem models indicate that even if marine mammals were completely removed from the world's oceans, any increase in fisheries catches would be no more than minor (Morissette *et al.*, 2012; Ruzicka *et al.*, 2013). There are a number of reasons why a straightforward outcome would be unlikely (Northridge & Hofman, 1999; Yodzis, 2001). Food webs tend to be highly complex, with many marine mammals taking a range of prey species, just as the target species of a fishery will have a number of predators.

The importance of understanding major predatory links is illustrated by the interaction between the fishery for hake off the South African west coast and the Cape fur seal, which as mentioned earlier is still hunted, supposedly for fisheries management. This fishery targets *Merluccius capensis* and *M. paradoxus*, which are mainly shelf and upper slope species, respectively. Where they overlap, large *M. capensis* prey heavily on juvenile *M. paradoxus. M. capensis* is important in the diet of fur seals, but modelling indicates that a seal cull would have minimal impact on the hake fishery and could even be detrimental since, under reduced fur seal predation, *M. capensis* would feed more heavily on *M. paradoxus*, and overall hake catch would decline (Punt & Butterworth, 1995).

The main predators of fish tend to be other fish rather than marine mammals. Furthermore, most fish stocks exhibit high recruitment variability and are subject to large-scale physical forcing (see Chapter 5), factors that may be far more significant than predation. Appropriate multi-species models to help us better understand and manage such interactions are still some way off, as is our understanding of physical forcing on ecosystem structure and functioning.

Conversely, extraction of prey by fisheries might be harmful to marine mammal populations – such an impact has been suggested for about a dozen species (Plagányi & Butterworth, 2005) – but again there is little clear-cut evidence, probably also for reasons of food-web complexity and large-scale oceanographic variability. Heavy fishing in the Bering Sea and Gulf of Alaska since the 1970s has, for instance, been implicated in the decline of populations of the northern fur seal, harbour seal (*Phoca vitulina*), and Steller sea lion (*Eumetopias jubatus*) as a result of shifts in ecosystem structure. Over recent decades the Steller sea lion has undergone a marked decline throughout much of its range in the North Pacific rim; its current total population of about 143 000 represents a decline of 28% since 1981 (Gelatt & Lowry, 2012). But an analysis of Steller sea lions in Alaska indicates that populations were largely unaffected by removals of the commercial fish species that dominate their diet and suggests that the overall decline is the result of several factors, such as killer whale predation, large-scale shifts in ocean climate, fisheries, and competition with other species (Hui *et al.*, 2015). Plagányi and Butterworth (2009) concluded that, overall, 'it is currently virtually impossible to wholly substantiate claims that predation by marine mammals is adversely impacting a fishery or *vice versa*'.

## Bycatches of Marine Mammals

Unintended capture in fisheries is an important source of mortality for some marine mammal populations, particularly among pinnipeds and small cetaceans. In the case of large whales, high rates of entanglement in fishing gear are known for some populations, notably North Atlantic right whales and certain humpback populations (Knowlton *et al.*, 2012). For small cetaceans in general, incidental mortality from fisheries probably accounts for greater losses than targeted exploitation, although there is not always a clear distinction between direct and indirect takes. At least in artisanal fisheries, cetaceans caught incidentally in gillnets may be kept for human consumption or used as bait. As a result of such bycatch, small cetaceans may then be targeted, a trend that appears to have increased in recent decades (Robards & Reeves, 2011). The global fisheries bycatch of cetaceans and pinnipeds has been estimated at about 650 000 animals annually, split roughly equally between the two groups, and taken very largely in gillnet fisheries (Read *et al.*, 2006). Gillnets also account for most of the incidental capture of sirenians (Marsh *et al.*, 2011). For most parts of the world, however, the impact of this mortality on marine mammal populations is poorly known, and adequate statistics are urgently needed.

### The Tuna-Dolphin Problem

The issue of marine mammal bycatch came to prominence when the scale of mortality being inflicted on dolphins and porpoises by certain high-seas fisheries became widely known. Concern over dolphin mortality associated with the purse seine fishery for yellowfin tuna in the eastern tropical Pacific generated considerable public concern and was a significant factor leading to the enactment of the US Marine Mammal Protection Act. Gosliner (1999) and Gerrodette (2009) provide overviews of this tuna-dolphin problem.

There is often a close association between schools of large yellowfin tuna (*Thunnus albacares*) and certain dolphin species, although reasons for the relationship are still unclear. From the late 1950s, pole-and-line fishing for tuna was being superseded by the use of large (< 2 km long) purse seines, and the common practice was to use dolphin schools to locate the tuna and encircle both with a purse seine. Initially, as there was no reliable way of releasing the dolphins without risking loss of the tuna, the entire catch was hauled in and the dolphins then discarded. Through the 1960s the incidental kill of dolphins in this US fishery ranged from 200 000 to 600 000 animals per year, mostly pantropical spotted, spinner, and short-beaked common dolphin (*Delphinus delphis*) (Fig. 9.7). Subsequently, with restrictions placed on US tuna boats, considerable reductions in incidental mortality were achieved. Important was the development of a 'backdown' procedure to enable dolphins to escape from the net. This involves the vessel going astern once about half the net has been hauled in so as to create a channel for the dolphins and submerging the cork-line at the apex of the channel where they can be herded out. From the 1970s an increasing proportion of the tuna was being taken by non-US vessels, mainly from Mexico and Central America. Dolphin mortality rose again in the mid-1980s to more than 100 000 animals per year, and the USA subsequently banned the import of tuna from countries that did not use fishing practices that reduced dolphin mortality. However,

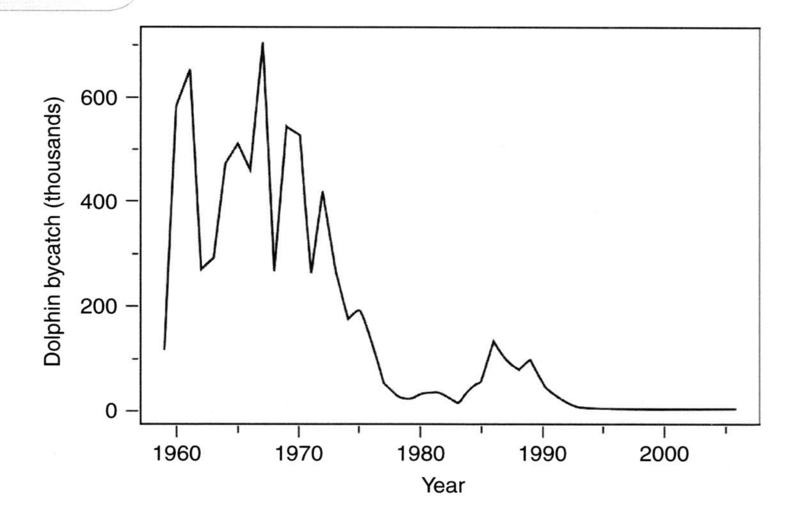


**Fig. 9.7** Estimated mortality of dolphins (mainly spotted and spinner dolphins) in the eastern tropical Pacific purse seine fishery for yellowfin tuna. Reprinted from *Encyclopedia of Marine Mammals*, 2nd edn, Gerrodette, T., The tuna-dolphin issue, pages 1192–5, copyright 2009, with permission from Elsevier.

in 1992 through an international initiative of the Inter-American Tropical Tuna Commission – the La Jolla Agreement – dolphin mortality was greatly reduced by the introduction of bycatch quotas. Under this system, a total allowable dolphin mortality for the fishery is divided among the participating vessels, and if a vessel reaches its individual dolphin mortality limit, it must, for the rest of the year, cease targeting dolphin schools in order to fish for tuna. There is thus an incentive for vessel owners to reduce incidental mortality, and dolphin deaths dropped dramatically even though fishing effort on tuna associated with dolphins did not decrease. This was, however, a voluntary accord. A binding international agreement has since been negotiated – the Agreement on the International Dolphin Conservation Program – which entered into force in 1999. Based on the La Jolla Agreement and subsequent legislation, it aims to reduce dolphin mortalities in the tuna purse seine fishery in the eastern Pacific to levels approaching zero through the setting of annual limits and to seek ecologically sound means of capturing large yellowfin tuna that are not in association with dolphins. It limits incidental mortality in the fishery to no more than 5000 dolphins annually through the assignment of dolphin mortality limits and requires 100% observer coverage. Since 1998 annual dolphin deaths have been cut to fewer than 2000, though this is still one of the largest documented cetacean bycatches. Whilst dolphin mortalities have been greatly reduced, the populations involved have been severely depleted and will still need decades to recover. It is unlikely that pole-and-line fishing could now meet the demand for tuna by the canning industry, so the fishery would resist a move away from purse seining. The advantage of targeting tuna associated with dolphins ('dolphin sets') is that they are almost exclusively yellowfin, and they are larger than tuna caught by the other methods of purse seine fishing, resulting in few discards. The purse seine fishery can also target tuna that occur as free-swimming schools not associated with dolphins, so-called 'school sets', or schools associated with floating objects, usually tree trunks and branches, or, increasingly, artificial fish aggregating devices, 'log sets'. But school sets and log sets yield smaller tuna and result in higher discard rates. They also result in larger bycatches of other fish and turtles. Sets where a drifting community is targeted produce in particular large bycatches of a wide variety of other fish including marlins, sharks, dolphinfish, wahoo, and triggerfishes. In terms of overall impact, dolphin sets may be the least damaging. But what level of incidental mortality in the tuna fishery is acceptable? The tuna-dolphin issue illustrates another contentious matter: that of international free trade versus environmental protection (Box 9.2).

## Box 9.2 World Trade, Marine Conservation, and Eco-labelling

The World Trade Organization regulates international trade between participating countries. A potential concern is that trade agreements may override provisions of environmental agreements and lead to an overall decline in environmental standards. This was highlighted by a dispute between the USA and Mexico in the early 1990s over tuna. As required under its Marine Mammal Protection Act, the USA imposed a ban on yellowfin tuna from Mexico because of the high incidental kill of dolphins in its tuna purse seine fishery. The dispute resolution panel ruled, however, that such a ban by the USA contravened its trade agreement obligations. The decision was not eventually adopted as the parties agreed on dolphin protection measures. So whilst trade restriction provisions might be effective instruments to advance marine conservation policy, such rulings may conflict with trade agreements.

In the tuna-dolphin dispute, US tuna packers brought in labelling to indicate 'dolphin-safe' tuna. It may, however, be difficult to verify that appropriate fishing practices have been adhered to, and such eco-labels are not necessarily a certification of sustainability (Joyner & Tyler, 2000; Kirby *et al.*, 2014). As Clover (2004) discusses, dolphin-safe labelling diverts attention from the wider problem of bycatch in tuna fisheries. It may in fact be possible 'to reduce nearly all bycatch in tuna fisheries to nominal levels' given the range of gear technology solutions and other initiatives that could be used (Gilman, 2011).

## Gillnet Fisheries

Other high-seas fisheries that have generated much controversy are large-scale driftnet fisheries, with their impact on small cetaceans (Northridge & Hofman, 1999). Whilst long driftnets had been used in international waters since the 1940s, it was not until the 1980s that the effect of their much larger-scale use aroused an international outcry to a great extent because of their impact on marine mammals and other non-target species. Particularly serious was the bycatch problem associated with driftnet fisheries in the North Pacific for squid, tuna, and salmon. By the mid-1980s this involved more than 800 vessels deploying in total up to 40 000 km of net each night and each year killing some tens of thousands of small cetaceans, mainly northern right whale dolphin (*Lissodelphis borealis*), Pacific white-sided dolphin (*Lagenorhynchus obliquidens*), and Dall's porpoise. The impact of the fishery led to its closure by UN resolution at the end of 1992 (see Chapter 5).

Nevertheless, the high mortality of small cetaceans caused by entanglement in drift and set gillnets continues to be a major problem. The scale of gillnet fisheries in many parts of the world is enormous. Coastal gillnet fisheries can account for mortalities of small cetaceans comparable to those inflicted by high-seas driftnet fisheries. In the NE Indian Ocean region there are, for instance, some 2.5 million fishers, most of whom are engaged in small-scale fisheries using gillnets, and each year a total of about 21 million km of net is deployed. In Sri Lanka alone this probably results in a kill of more than 40 000 cetaceans per year. For several stocks of marine cetaceans it appears that the rate of incidental mortality from gillnets has been, or continues to be, unsustainable. These include those for the vaquita (*Phocoena sinus*) in the Gulf of California, Hector's dolphin (*Cephalorhynchus hectori*) in New Zealand, Indo-Pacific humpbacked dolphin (*Sousa chinensis*) and bottlenose dolphin taken in anti-shark nets off Natal, striped dolphin (*Stenella coeruleoalba*) in the Mediterranean, harbour porpoise in the western North Atlantic, Burmeister's porpoise (*Phocoena spinipinnis*) in Peru and Chile, and Indo-Pacific finless porpoise (*Neophocaena phocaenoides*) in Chinese waters (IWC, 1994; Jefferson & Curry, 1994; Slooten & Dawson, 2010).

The harbour porpoise is widely distributed in cool temperate and subpolar waters of the Northern Hemisphere, but throughout its range it suffers considerable mortality through incidental capture in fisheries, with several populations at least partly threatened by gillnet entanglement (Jefferson *et al.*, 1993; Jefferson & Curry, 1994). In the USA, harbour porpoises have been subject to high levels of bycatch in the bottom gillnet fishery in the Gulf of Maine. Whilst annual bycatch of harbour porpoises in this fishery fell dramatically during the 1990s, from about 3000 to 300 animals, this reflected the fall in cod landings rather than conservation efforts (Read *et al.*, 2006; Geijer & Read, 2013). Subsequently, steps taken in 1998 under the Marine Mammal Protection Act (MMPA) have had some effect in reducing bycatch.

The MMPA mandates that a population should not be allowed to decline below its 'optimum sustainable population' (OSP) level – a population size between the levels of maximum net productivity and carrying capacity – and that incidental mortality is to be less than the potential biological removal (PBR) level. PBR is defined as 'the maximum number of animals, not including natural mortalities, that may be removed from a marine mammal stock whilst allowing that stock to reach or maintain its optimum sustainable population'. The PBR is the product of the minimum population estimate, one-half the maximum net productivity rate, and a recovery factor (Wade, 1998). If incidental mortality exceeds the PBR, the MMPA requires that a Take Reduction Plan be developed to reduce bycatch to levels less than the PBR – within 6 months as an immediate goal and to insignificant levels within 5 years.

The Take Reduction Plan for the Gulf of Maine harbour porpoises involved time-area closures and the use of acoustic deterrent devices (pingers). Partial success is indicated by the landings and bycatch no longer being so strongly linked (Fig. 9.8). Nevertheless, bycatch levels have exceeded PBR since 2004 (Geijer & Read, 2013). For some reason there is poor compliance with bycatch mitigation measures in this fishery. But that is not always the case. To address the problem of common dolphin (*Delphinus delphis* and *D. capensis*) bycatch in drift gillnet fisheries along the US Pacific coast, a Take Reduction Plan requiring the use of pingers and modifications to fishing gear was implemented in 1997. In this fishery compliance has been high, highlighting the importance of

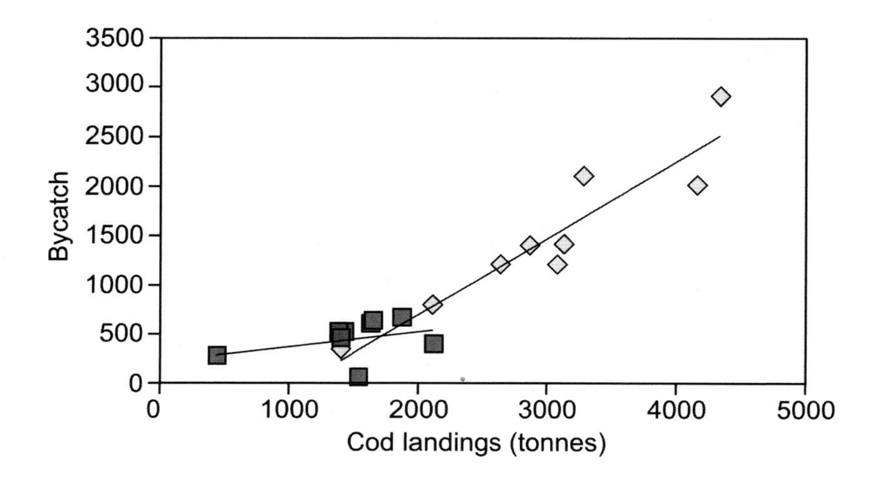


**Fig. 9.8** Relationships between cod landings and bycatch of harbour porpoise in the Gulf of Maine bottom gillnet fishery before (diamonds) and after (squares) implementation of a Take Reduction Plan in 1998. Reprinted from *Biological Conservation*, Vol. 159, Geijer, C.K.A. & Read, A.J., Mitigation of marine mammal bycatch in U.S. fisheries since 1994, pages 54–60, copyright 2013, with permission from Elsevier.

needing to understand factors that may facilitate acceptance of such conservation measures (Geijer & Read, 2013).

Widely distributed species such as the harbour porpoise are probably not in danger of global extinction from fishing mortality. But for cetaceans with very restricted distributions, loss through bycatch may represent a critical threat. A perilous situation is faced by the vaquita, a small porpoise found only in the northern Gulf of California, Mexico, that is threatened by incidental capture from gillnetting and shrimp trawling. Only a few tens of animals remain – a decline of about 90% since the late 1990s – and the species could soon be beyond recovery unless a gillnet ban can be enforced and fishers compensated for switching to vaquita-safe fishing methods (Rojas-Bracho & Reeves, 2013; IUCN SSC – Cetacean Specialist Group, 2016).

Fisheries interactions, including bycatch mortality, seriously impact a number of pinnipeds; in fact they represent the primary threat to most of the taxa categorized as threatened by IUCN (Kovacs *et al.*, 2012). One species of much concern in this regard is the New Zealand sea lion (*Phocarctos hookeri*), now one of the rarest pinnipeds and with a very restricted distribution. Its total population is about 10 000 animals (Geschke & Chilvers, 2009). Archaeological evidence indicates that the species was once widely distributed around New Zealand, but hunting by early Maori and commercial sealing brought the species close to extinction, and by 1830 the Auckland Islands, 500 km south of New Zealand, held the only known remaining breeding colony (Childerhouse & Gales, 1998). Although New Zealand prohibited sealing in 1893, the species is still largely confined to subantarctic islands, with rookeries at the Auckland Islands accounting for about 70% of total pup production. However, since the late 1990s pup production at the Auckland Islands has halved, a decline linked to females failing to return to breeding areas (Robertson & Chilvers, 2011; Chilvers, 2012). Trawl fisheries on the Auckland Islands shelf, particularly for squid, developed in the late 1970s and resulted in sea lions drowning in trawls, most of them females, and modelling indicates that the survival of adult females most strongly affects the population (Meyer *et al.*, 2015). Since 1993 the fishery has had a sea lion bycatch limit imposed for each fishing season, which, if exceeded, requires the fishery to close early, as has occurred several times. Other measures have been the establishment of a 12-nm fishing exclusion zone around the Auckland Islands and the deployment of observers on trawlers. Also, since 2004, sea lion exclusion devices (SLEDs) have regularly been used. These are designed to divert caught sea lions out through an escape hatch in the trawl net, similar to the devices used to reduce the incidental catch of turtles (Fig. 7.7). The mean estimated number of captures peaked in 1996–7 at 142 individuals but fell to less than 20 a decade later (Hamilton & Baker, 2015). The use of SLEDs appears to have contributed to this decline, but whether some animals that escape have been seriously injured and subsequently die is not clear. Trawling-related mortality of sea lions would be reduced to negligible levels if this fishery were able to switch to jigging, as used in other squid fisheries (Chilvers, 2012). Other factors may, however, be contributing to the decline of the Auckland Islands' population, including changes in the availability of prey species (natural and/or fishing-related shifts) and bacterial diseases of pups (Meyer *et al.*, 2015). A combination of factors appears most likely.

## Bycatch Mitigation

Some strategies have proved effective in addressing the problems of marine mammal bycatch in commercial fisheries, as illustrated by Read (2010). Crucially important is having independent observers on vessels to assess the rate of bycatch of marine mammals (and other animals) in relation to fishing practices and to provide feedback on how practices can be improved. Exploring options to reduce bycatch to sustainable levels then calls for formal discussions between the fishing industry, environmental agencies, and other stakeholders to negotiate appropriate measures that can be applied. Having well-defined targets, such as a PBR level, are important. Once mitigation measures are agreed on and implemented, compliance needs to be monitored by observers.

### Gillnet Modifications

Modifications to fishing gear to reduce incidental capture have focused on measures to make gillnets more obvious to marine mammals (Dawson, 1991). The abilities of odontocetes and pinnipeds to detect gillnets and factors that make them vulnerable to entanglement are still poorly understood. It is likely that dolphins and porpoises can usually detect monofilament gillnets by echolocation at sufficient distance to avoid entanglement, but since the animals may spend considerable amounts of time not using their sonar, a net may remain undetected. In this case, modifying nets to make them acoustically more reflective may be of limited value. Even if a net is detected by sonar, its diffuse echo may not be interpreted as a barrier or may be ignored by an animal that is focusing on prey. Another modification, mandatory in some fisheries, is to attach to the net a device that emits a deterrent noise – a pinger that produces a high-pitched noise every few seconds. However, for reasons that are not yet understood, pingers appear to work well in some situations but are ineffective in others. Their use can, for instance, significantly reduce bycatch of harbour porpoise, short-beaked common dolphin, striped dolphin, fransicana (*Pontoporia blainvillei*), and beaked whales, whereas bottlenose dolphin appear to be undeterred (Dawson *et al.*, 2013). There are questions that need to be addressed in this regard, such as how taxon-specific repellent sounds might be, the mechanism of deterrence, whether animals become habituated, and to what extent the costs involved might limit application, especially given the widespread use of gillnets in many less developed countries (Dawson *et al.*, 1998). Otherwise, the most effective ways to reduce the incidence of marine mammal entanglement in gillnets are time and area restrictions and area closures.

## Pollution

The extent to which pollution may adversely affect marine mammals is still poorly understood, and seldom has exposure to an environmental contaminant been shown to result directly in a harmful effect on a population (O'Hara & O'Shea, 2005). There are, however, contaminants of particular concern, including organochlorines, polycyclic aromatic hydrocarbons (PAHs), and metals (see Chapter 2), sublethal effects of which may ultimately affect populations. Many such studies have focused on pinnipeds. Elevated levels of DDT and PCBs have, for instance, been implicated in reproductive impairment, reduced immune response, and lesions in various pinniped species (Reijnders *et al.*, 2009). Immuno-suppression induced by contaminants may be a factor contributing to mass die-offs of seals and dolphins that have occurred at various locations since the late 1980s. Deaths have in most cases been due to viral infection, but animals would be more susceptible if their immune system had been compromised. This serves to remind us that pollutants and other environmental factors, natural as well as anthropogenic, are not acting in isolation and that a marine mammal, like any organism, is continuously coping with a range of interacting influences, some of which may reinforce or offset each other. As often as not the decline of a marine mammal population will be the result of the cumulative impact of stressors, making it difficult to tease out the impact of a particular factor.

Only low levels of organochlorine and metal contaminants have been recorded in tissues of baleen whales, probably because of their generally open ocean distributions and low trophic level, indicating that contaminants are unlikely to significantly affect populations (O'Shea & Brownell, 1994). Much higher contaminant levels have been recorded in toothed whales, which is to be expected given that they feed at a higher trophic level and many are small (higher metabolic rate) and coastal. Also, cetaceans in particular have a low capacity to degrade organochlorines, thereby favouring their accumulation. In some cases, lipophilic contaminants can reach high levels, mostly in blubber. As marine mammals are often long-lived top predators with fat stores that accumulate lipophilic pollutants, they may serve as effective sentinel species for evaluating marine ecosystem health (Bossart, 2011).

Marine mammals are potentially highly vulnerable to oil pollution, particularly those living in cold waters that spend a lot of time at or near the surface, have a well-developed coat (as oiled fur loses its insulating capacity and buoyancy), and groom themselves (thereby ingesting oil). This makes sea otters among the most vulnerable. The *Exxon Valdez* oil spill in Prince William Sound, Alaska, in 1989 (see Chapter 8) severely impacted sea otters, with probably at least a quarter of the population in the western sound killed within months of the spill. Recovery of the population in this heavily impacted area has been slow, probably because otters are still being exposed to residual oil in intertidal sediments when they dig for clams (Monson *et al.*, 2011; Bodkin *et al.*, 2012).

Debris is a potential hazard for marine mammals through entanglement and/or ingestion. Entanglement in marine debris has been documented for 45% of the world's marine mammal species and ingestion for 26% (Gall & Thompson, 2015). Particularly high is the incidence of entanglement in fur seals and sea lions (~80% of species), a problem caused mainly by plastic packaging bands or pieces of fishing net or line around the animal's neck, which may result in life-threatening wounds (Laist, 1997). Levels of entanglement of up to 6–8% have been reported at certain sites for the Hawaiian monk seal, California sea lion (*Zalophus californianus*), and New Zealand fur seal (*Arctocephalus australis forsteri*) (Laist, 1997; Boren *et al.*, 2006). Most reported entanglement rates for pinnipeds are less than 1%, but such rates may still be sufficient to adversely affect populations.

A species particularly vulnerable to entanglement in debris is the Hawaiian monk seal. Threatened with extinction, its current population of about 1000 individuals is dwindling by about 4.5% per year. Several factors appear to be contributing to this decline including entanglement (Lowry *et al.*, 2011). Hawaiian monk seals seldom migrate between islands, so subpopulations may be severely impacted by high localized rates of entanglement. Annex V of MARPOL prohibits the disposal at sea of plastics (see Chapter 2), but the number of monk seal entanglements did not diminish from 1989 after MARPOL came into effect, nor was the amount of driftnet on beaches found to be lower from 1993 following implementation of the ban on high-seas driftnet fishing (Henderson, 2001). Later surveys in 1999–2001 of monk seal nursery zones also indicated that debris accumulation has not diminished (Boland & Donohue, 2003).

It may be possible to successfully release entangled seals – Boren *et al.* (2006) were able to free 43% of the entangled New Zealand fur seals – but for other species or situations this may not be feasible to achieve safely. Outreach campaigns can help address the root of the problem, raising awareness of the issue, the requirements of MARPOL, the importance of cutting used packaging loops, and the need to minimize the use of disposable plastics in the fishing industry (Raum-Suryan *et al.*, 2009).

## Noise Pollution

Over the past century or so humans have greatly increased the amount of noise in the oceans, such as from shipping, sonar, seismic surveys, offshore oil rigs, dredging, and trawling. Marine mammals – or at least cetaceans, pinnipeds, and sirenians – depend on underwater sound for a variety of purposes, notably communication, locating prey, avoiding predators, and, probably, navigation (Würsig & Richardson, 2009). Anthropogenic noise at relevant frequencies can elicit behavioural changes in cetaceans, including changes in diving, swimming, and vocalization, and there is evidence linking naval sonar or seismic surveys to mass strandings and mortalities of cetaceans, especially beaked whales (Weilgart, 2007). Probably more often, noise pollution has chronic, sublethal effects, such as by increasing stress, masking important natural sounds, or leading to abandonment of habitat. The issue of noise looks set to become increasingly significant, and opportunities to reduce noise levels need to be pursued, especially in areas important for marine mammals. Marine protected areas with noise buffer zones may be effective in this regard (Weilgart, 2007).

# Global Climate Change

Global climate change associated with increased greenhouse gas emissions has a multitude of implications for the structure and functioning of marine ecosystems (see Chapter 2). Climate change models indicate high-latitude regions to be among those most affected, regions that are also important for marine mammals in terms of biodiversity and ecological role. The Arctic has seen a rapid loss of multi-year sea ice over recent decades, and models suggest that it will be nearly ice free (less than 1 million $km^2$) in summer in the 2030s (Wang & Overland, 2012). Arctic sea ice provides habitat for 11 marine mammal species. Likely to be among the most severely impacted are those that depend on sea ice as a platform for hunting, breeding, and resting, such as the polar bear (*Ursus maritimus*), walrus, bearded seal (*Erignathus barbatus*), and ringed seal (Moore & Huntington, 2008; Kovacs *et al.*, 2011). Polar bears depend on sea ice areas for hunting ringed seals, their primary prey. But given the projected global warming and loss of sea ice, by mid-century polar bears may be confined to northern portions of their range with a total population of only about a third of today's (20 000–25 000), raising serious doubts about the species' longer-term viability (Stirling & Derocher, 2012).

# Marine Protected Areas

The ICRW provides for the designation of whale sanctuary areas, and two major ocean areas are currently designated as such by the IWC where commercial whaling is prohibited. The Indian Ocean Sanctuary was established in 1979. The original proposal, from the Republic of Seychelles, was for the sanctuary to extend to Antarctica, but IWC members could agree to it extending only as far as 55° S so as to exclude minke whale grounds of the Southern Ocean (Holt, 1983). However, in 1994 the Southern Ocean Sanctuary was adopted, with its northern boundary at 55° S in the Indian Ocean, thus joining the Indian Ocean Sanctuary, and at 40° S in the Atlantic, 60° S in the Pacific, and around South America (Fig. 9.9). Many cetacean species occur within these areas, but the sanctuary

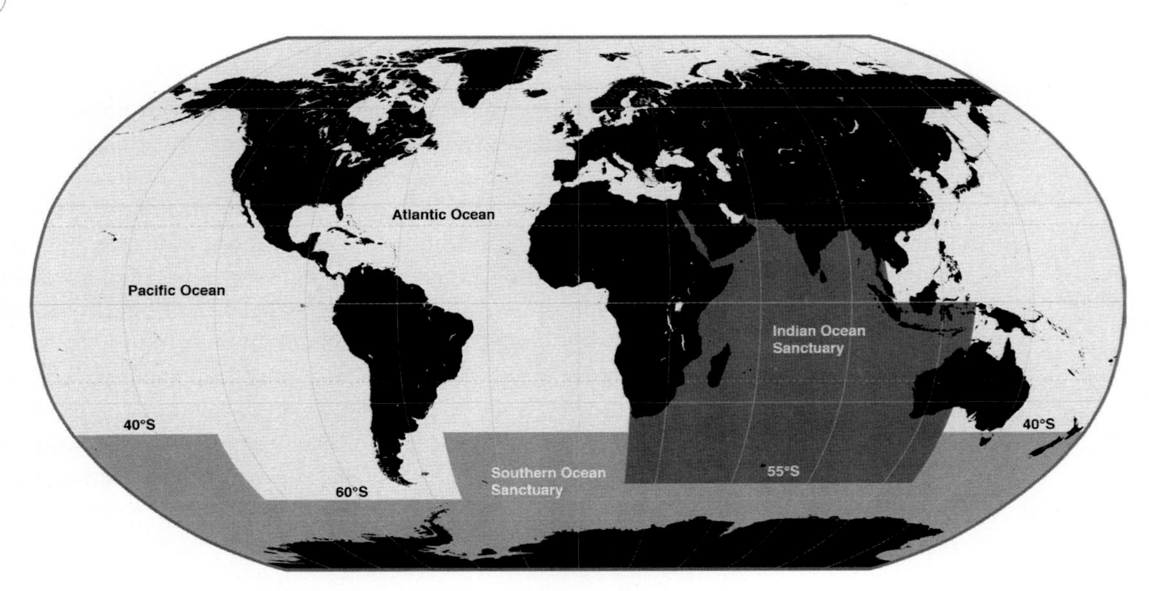


**Fig. 9.9** Whale sanctuaries established by the International Whaling Commission. From coordinates given in the IWC schedule, paragraphs 7 (a) and (b). Map template: www.FreeVectorMaps.com.

designation is generally regarded as applying only to the large whale species. Further IWC sanctuaries have been proposed, including ones immediately north of the Southern Ocean Sanctuary to cover the South Atlantic and much of the South Pacific, but have failed to win adequate support at IWC meetings. Various countries have, however, designated their EEZs as marine mammal sanctuaries. These are mostly in the Southern Hemisphere, but neither these national nor IWC sanctuaries generally meet IUCN criteria for marine protected areas in terms of regulation and management (Hoyt, 2011).

Marine protected areas (MPAs) are potentially a key tool in marine mammal conservation but pose a number of challenges, particularly given the range and mobility of many of the species. The use of MPAs in marine mammal conservation has been reviewed by Reeves (2000) and Hoyt (2009). Crucial to establishing MPAs for marine mammals is defining the threats they face and identifying critical habitat, or those parts of the animal's range 'that are essential for day-to-day well-being and survival, as well as for maintaining a healthy population growth rate' (Hoyt, 2011). Typically these will be areas essential to a population for such activities as feeding, breeding, calving or pupping, hauling out, or possibly for migration. This is, however, problematic; in most cases we do not know exactly what habitat animals are using and why. Identifying critical habitat requires survey data on at least distribution and abundance over several years. Where adequate sightings and associated environmental data are available, spatial modelling may then be used to characterize critical habitat. Zoning may be an appropriate approach, with critical habitat incorporated in core no-take areas affording the highest level of protection and surrounding buffer zones where carefully managed activities such as ecotourism can be undertaken. Flexibility of zone boundaries may need to be considered to account for temporal and spatial variability of populations, such as from seasonal aggregations or to accommodate oceanographic features that are important to marine mammals but that are not stationary. Also, MPAs are likely to be most effective when organized into networks that link areas of critical habitat. Such areas will often be large and in waters beyond national jurisdiction, raising issues with regard to designation and management.

Modelling by Davidson *et al.* (2012) identified the main predictors of extinction risk in marine mammals to be those related to speed of life history and thereby a species' ability to recover from human impacts. Hotspots for species most at risk were found to have little overlap with MPAs. Such approaches are valuable in identifying vulnerable taxa and key areas and in helping prioritize conservation efforts. In terms of potential priority for protection, Pompa *et al.* (2011) identified nine key areas on the basis of their marine mammal species richness (representing more than 80% of all species) – often related to upwelling regions – plus six marine or brackish sites deemed irreplaceable because of their endemic species (Hawaiian Islands, Galápagos Islands, San Felix and Juan Fernández Islands, Kerguelen Islands, Mediterranean Sea, Caspian Sea).

As far as cetaceans are concerned, Hoyt (2011) documents 570 protected areas with cetacean habitat, although relatively few have been designated expressly to protect whales and dolphins – most would in any case be too small to be fully effective – and for cetaceans there are only some 20 MPA networks. So despite widespread interest in whale and dolphin conservation, provision of protected areas is still limited. When MPAs are established, effective management must be ensured; setting up an MPA does not necessarily mean that adequate protection is in place. The Pelagos Sanctuary for Mediterranean marine mammals, which came into force in 2002, encompasses some 87 500 km$^2$ of the Ligurian Sea (Fig. 16.1) and includes a productive frontal system. In summer, about 1000 fin whales and 20 000–30 000 striped dolphin aggregate within the sanctuary, and six other cetacean species are regularly seen (Notarbartolo di Sciara *et al.*, 2008). However, zoning plans have yet to be instituted and the level of resourcing and management raises doubts about the effectiveness of the sanctuary for cetacean conservation (Hoyt, 2011).

Surprisingly little attention has in fact been given to examining the possible effectiveness of area-based protection measures for marine mammals. An exception concerns the value of an MPA to Hector's dolphin, a species endemic to coastal waters of New Zealand with a total population of around 7000 animals (Fig. 4.1). One of the highest-density areas is around Banks Peninsula on the South Island's east coast. This is also, however, an area where there has been intensive gillnetting, and catch data from the mid-1980s indicate that the number of dolphins being killed in gillnets exceeded even the maximum projected rates of population growth. The unsustainable level of gillnet mortality led to the establishment in 1988 of a sanctuary around Banks Peninsula – out to 4 nm (7.4 km) and with an area of 1170 km$^2$ – where commercial gillnetting is prohibited and recreational gillnetting restricted (Dawson & Slooten, 1993). Analysis of more than 20 years of data from before and after the sanctuary's establishment shows an improvement in adult survival rate of just over 5% (Gormley *et al.*, 2012). This change is biologically significant and provides evidence that an MPA approach can improve survival in a marine mammal. Nevertheless, the population's trajectory remains slightly negative. The reason for this appears to be insufficient protection of the population offshore. The animals tend to remain close inshore in summer where they are within the sanctuary, but in winter many range farther offshore outside the sanctuary boundary (Rayment *et al.*, 2010).

Populations of Hector's dolphins around New Zealand are genetically distinct (Fig. 9.10). Gene flow is likely hindered by the species' small home range (typically 30–50 km), exacerbating the conservation challenge. Management plans must focus not only on protecting these populations but also on protecting corridors to facilitate movement of individuals between local populations – and the occasional exchange between regions – in order to avoid further fragmentation and loss of genetic diversity (Hamner *et al.*, 2012b). The isolated population of Hector's dolphin confined to the west coast of New Zealand's North Island, a subspecies known as Maui's dolphin, currently (2010–11) totals only about 55 individuals (Hamner *et al.*, 2012a) and is likely heading for extinction given the inadequate restrictions on gillnet use in this area.

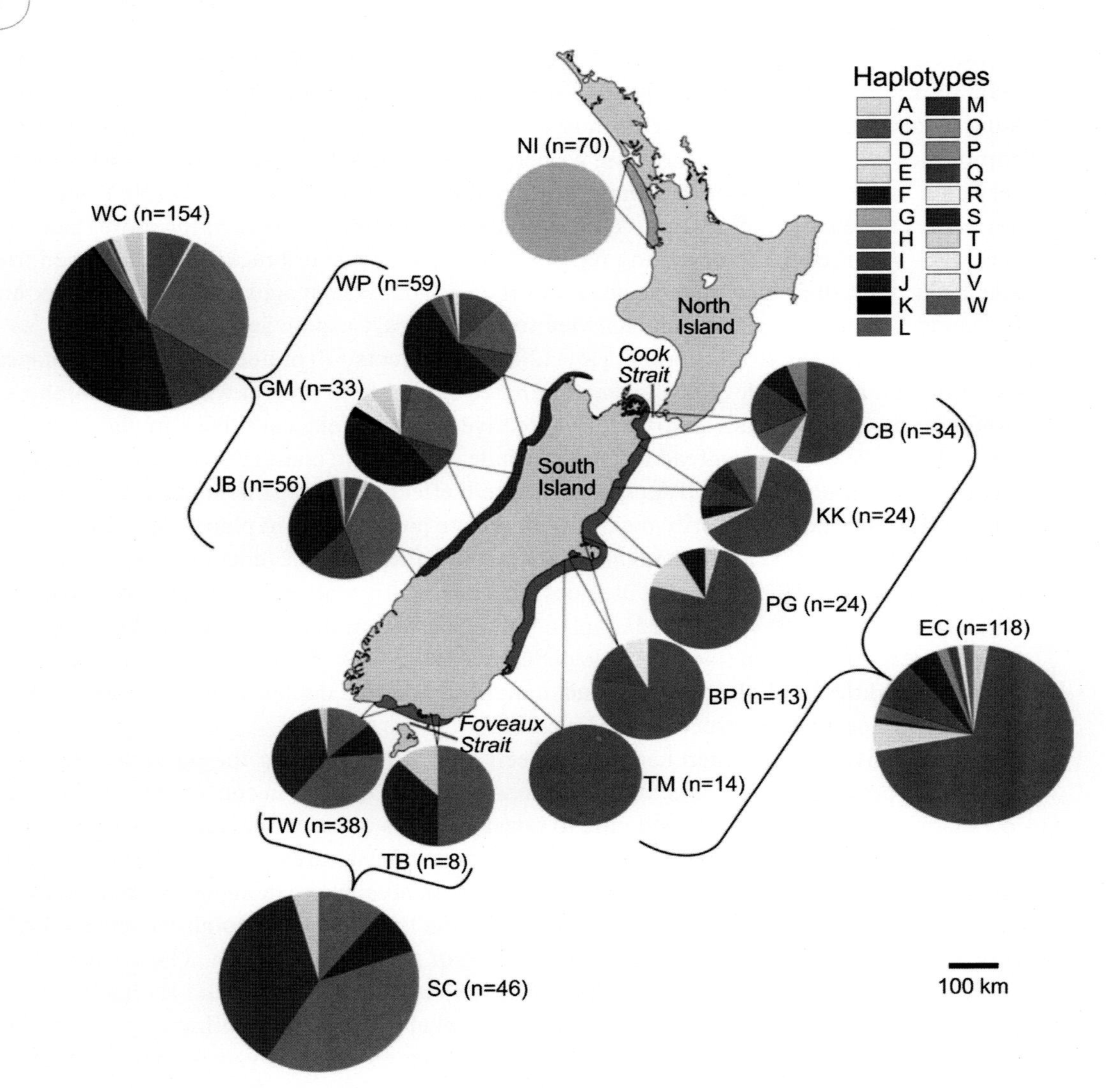


**Fig. 9.10** Genetic differentiation (distribution of haplotypes) of populations of Hector's dolphin (*Cephalorhynchus hectori hectori*) and Maui's dolphin (*C. hectori maui*). Mitochondrial control region haplotypes (360 base pairs) for the Maui's dolphin (NI); regional populations of the Hector's dolphin: East Coast (EC), West Coast (WC), South Coast (SC); and local populations of the Hector's dolphin: Cloudy Bay (CB), Kaikoura (KK), Pegasus Bay (PG), Banks Peninsula (BP), Timaru (TM), Westport (WP), Greymouth (GM), Jackson Bay (JB), Te Waewae Bay (TW), Toetoe Bay (TB). From *Conservation Genetics*, Genetic differentiation and limited gene flow among fragmented populations of New Zealand endemic Hector's and Maui's dolphins, Vol. 13, 2012, pages 987–1002, Hamner, R.M., Pichler, F.B., Heimeier, D., Constantine, R. & Baker, C.S., © Springer Science+Business Media B.V. 2012. With permission of Springer. (A black and white version of this figure will appear in some formats. For the colour version, please refer to the plate section.)

## Translocation

A further tool that may be used to help conserve endangered populations is translocation: 'the human-mediated movement of living organisms from one area, with release in another' (IUCN, 2012). This might be used to reintroduce a species to an area of its former range or to aid the recovery of a severely depleted population. Translocations have been used in marine mammal conservation, for instance, in the recovery of sea otter populations, as mentioned earlier. Success may depend on the degree of genetic differentiation between populations. The Hawaiian and Mediterranean monk seals are both highly endangered and with low genetic diversity. The Hawaiian monk seal exhibits similar genetic variation throughout its range so that translocating individuals within the archipelago may be a viable option. But in the case of the Mediterranean monk seal, the Eastern Mediterranean and Western Saharan populations are reproductively isolated, and translocation attempts between the two are less likely to succeed because of genetic incompatibility (Baker *et al.*, 2011; Schultz, 2011).

In this chapter we have looked at direct and indirect threats facing marine mammals and at various measures to mitigate these pressures. The effective management and conservation of marine mammals present a range of major challenges. Many populations are still severely depleted following gross overexploitation or face new or continuing pressure as a result of targeted hunting or incidental mortality, particularly from fisheries. Other stresses arise from human disturbance, pollutants, and global climate change. Life-history characteristics of marine mammals render them especially susceptible to additional extrinsic sources of mortality, and the range or migratory behaviour of many species means that they often span national and international waters. We still, however, lack even basic information on the biology and ecology of many marine mammal species to assess their conservation status; about 40% of species are categorized as data deficient in the IUCN Red List. An important priority, therefore, is for improved data on species, their life-history parameters, population estimates, distribution, and use of key habitats, such as for feeding, breeding, and migration. Incomplete information should not, however, be a reason for delaying conservation measures where a marine mammal population is seriously at risk (the precautionary principle; see Chapter 3). That we have often failed to avert crises in marine mammal conservation – despite our avowed interest in these animals – perhaps illustrates that ultimately we have yet to properly acknowledge and take account of ourselves as a component in ecosystems and that effective conservation requires a shift in social and economic values (Reynolds *et al.*, 2009). Nevertheless, there are grounds for cautious optimism. There is increasing awareness of the problems facing marine mammals and of the need for effective conservation measures worldwide. Abundance trends indicate that many formerly depleted marine mammal populations are recovering; overall, more than 40% of populations for which there are adequate data are reported to be significantly increasing (Magera *et al.*, 2013). Whilst there are still huge gaps in such data, these trends are encouraging.

## REFERENCES

Ainley, D., Ballard, G., Ackley, S., *et al.* (2007). Paradigm lost, or is top-down forcing no longer significant in the Antarctic marine ecosystem? *Antarctic Science*, **19**, 283–90.

Anderson, P.K. & Domning, D.P. (2009). Steller's sea cow *Hydrodamalis gigas*. In *Encyclopedia of Marine Mammals*, 2nd edn, ed. W.F. Perrin, B. Würsig & J.G.M. Thewissen, pp. 1103–6. London: Academic Press.

Baker, C.S. & Palumbi, S.R. (1996). Population structure, molecular systematics, and forensic identification of whales and dolphins. In *Conservation Genetics: Case Histories from Nature*, ed. J.C. Avise & J.L. Hamrick, pp. 10–49. New York: Chapman & Hall.

Baker, C.S., Steel, D., Choi, Y., *et al.* (2010). Genetic evidence of illegal trade in protected whales links Japan with the US and South Korea. *Biology Letters*, **6**, 647–50.

Baker, J.D., Becker, B.L., Wurth, T.A., *et al.* (2011). Translocation as a tool for conservation of the Hawaiian monk seal. *Biological Conservation*, **144**, 2692–701.

Bekoff, M. (2003). Consciousness and self in animals: some reflections. *Zygon*, **38**, 229–45.

Benson, E. (2012). Endangered science: the regulation of research by the U.S. Marine Mammal Protection and Endangered Species Acts. *Historical Studies in the Natural Sciences*, **42**, 30–61.

Best, P.B. (1993). Increase rates in severely depleted stocks of baleen whales. *ICES Journal of Marine Science*, **50**, 169–86.

Bodkin, J.L., Ballachey, B.E., Coletti, H.A., *et al.* (2012). Long-term effects of the 'Exxon Valdez' oil spill: sea otter foraging in the intertidal as a pathway of exposure to lingering oil. *Marine Ecology Progress Series*, **447**, 273–87.

Boland, R.C. & Donohue, M.J. (2003). Marine debris accumulation in the nearshore marine habitat of the endangered Hawaiian monk seal, *Monachus schauinslandi* 1999–2001. *Marine Pollution Bulletin*, **46**, 1385–94.

Bonner, W.N. (1982). *Seals and Man: A Study of Interactions*. Seattle: Washington Sea Grant.

Boren, L.J., Morrissey, M., Mull, C.G. & Gemmell, N.J. (2006). Entanglement of New Zealand fur seals in man-made debris at Kaikoura, New Zealand. *Marine Pollution Bulletin*, **52**, 442–6.

Bossart, G.D. (2011). Marine mammals as sentinel species for oceans and human health. *Veterinary Pathology*, **48**, 676–90.

Bowen, W.D. (1997). Role of marine mammals in aquatic ecosystems. *Marine Ecology Progress Series*, **158**, 267–74.

Bowen, W.D. & Siniff, D.B. (1999). Distribution, population biology, and feeding ecology of marine mammals. In *Biology of Marine Mammals*, ed. J.E. Reynolds III & S.A. Rommel, pp. 423–84. Washington, DC: Smithsonian Institution Press.

Brownell, R.L. Jr & Yablokov, A.V. (2009). Whaling, illegal and pirate. In *Encyclopedia of Marine Mammals*, 2nd edn, ed. W.F. Perrin, B. Würsig & J.G.M. Thewissen, pp. 1235–9. London: Academic Press.

Butterworth, D.S. & Punt, A.E. (1999). Experiences in the evaluation and implementation of management procedures. *ICES Journal of Marine Science*, **56**, 985–98.

Cairns, D.K., Gaston, A.J. & Huettmann, F. (2008). Endothermy, ecothermy and the global structure of marine vertebrate communities. *Marine Ecology Progress Series* **356**, 239–50.

Childerhouse, S. & Gales, N. (1998). Historical and modern distribution and abundance of the New Zealand sea lion *Phocarctos hookeri*. *New Zealand Journal of Zoology*, **25**, 1–16.

Chilvers, B.L. (2012). Population viability analysis of New Zealand sea lions, Auckland Islands, New Zealand's sub-Antarctics: assessing relative impacts and uncertainty. *Polar Biology*, **35**, 1607–15.

Churchill, M., Berta, A. & Deméré, T. (2012). The systematics of right whales (Mysticeti: Balaenidae). *Marine Mammal Science*, **28**, 497–521.

Cisneros-Montemayor, A.M., Sumaila, U.R., Kaschner, K. & Pauly, D. (2010). The global potential for whale watching. *Marine Policy*, **34**, 1273–8.

Clapham, P.J. (2015). Japan's whaling following the International Court of Justice ruling: Brave New World – or business as usual? *Marine Policy*, **51**, 238–41.

Clapham, P.J. & Baker, C.S. (2009). Whaling, modern, In *Encyclopedia of Marine Mammals*, 2nd edn, ed. W.F. Perrin, B. Würsig & JGM. Thewissen, pp. 1239–43. London: Academic Press.

Clover, C. (2004). *The End of the Line*. London: Ebury Press.

Cooke, J.G. (1995). The International Whaling Commission's Revised Management Procedure as an example of a new approach to fishery management. In *Whales, Seals, Fish and Man*, ed. A.S. Blix, L. Walløe & Ø. Ulltang, pp. 647–57. Burlington, MA: Elsevier.

Daoust, P.-Y., Hammill, M., Stenson, G. & Caraguel, C. (2014). A review of animal welfare implications of the Canadian commercial seal hunt: a critique. *Marine Policy*, **43**, 367–71.

Davidson, A.D., Boyer, A.G., Kim, H., *et al.* (2012). Drivers and hotspots of extinction risk in marine mammals. *Proceedings of the National Academy of Sciences*, **109**, 3395–400.

Dawson, S.M. (1991). Modifying gillnets to reduce entanglement of cetaceans. *Marine Mammal Science*, **7**, 274–82.

Dawson, S.M., Northridge, S., Waples, D. & Read, A.J. (2013). To ping or not to ping: the use of active acoustic devices in mitigating interactions between small cetaceans and gillnet fisheries. *Endangered Species Research*, **19**, 201–21.

Dawson, S.M., Read, A. & Slooten, E. (1998). Pingers, porpoises and power: uncertainties with using pingers to reduce bycatch of small cetaceans. *Biological Conservation*, **84**, 141–16.

Dawson, S.M. & Slooten, E. (1993). Conservation of Hector's dolphins: the case and process which led to establishment of the Banks Peninsula Marine Mammal Sanctuary. *Aquatic Conservation: Marine and Freshwater Ecosystems*, **3**, 207–21.

DeMaster, D.P., Trites, A.W., Clapham, P., *et al.* (2006). The sequential megafaunal collapse hypothesis: testing with existing data. *Progress in Oceanography*, **68**, 329–42.

Ellis, R. (2009). Whaling, traditional. In *Encyclopedia of Marine Mammals*, 2nd edn, ed. W.F. Perrin, B. Würsig & J.G.M. Thewissen, pp. 1243–54. London: Academic Press.

Estes, J.A. (1979). Exploitation of marine mammals: *r*-selection of *K*-strategists? *Journal of the Fisheries Research Board of Canada*, **36**, 1009–17.

Estes, J.A., Bodkin, J.L. & Ben-David, M. (2009a). Otters, marine. In *Encyclopedia of Marine Mammals*, 2nd edn, ed. W.F. Perrin, B. Würsig & J.G.M. Thewissen, pp. 807–16. London: Academic Press.

Estes, J.A., Doak, D.F., Springer, A.M. & Williams, T.M. (2009b). Causes and consequences of marine mammal population declines in southwest Alaska: a food-web perspective. *Philosophical Transactions of the Royal Society B*, **364**, 1647–58.

Estes, J.A. & Palmisano, J.F. (1974). Sea otters: their role in structuring nearshore communities. *Science*, **185**, 1058–60.

Estes, J.A., Tinker, M.T. & Bodkin, J.L. (2010). Using ecological function to develop recovery criteria for depleted species: sea otters and kelp forests in the Aleutian archipelago. *Conservation Biology*, **24**, 852–60.

Gales, N.J., Kasuya, T., Clapham, P.J. & Brownell, R.L. Jr (2005). Japan's whaling plan under scrutiny. *Nature*, **435**, 883–4.

Gall, S.C. & Thompson, R.C. (2015). The impact of debris on marine life. *Marine Pollution Bulletin*, **92**, 170–9.

Gambell, R. (1977). Whale conservation: role of the International Whaling Commission. *Marine Policy*, **1**, 301–10.

Gambell, R. (1993). International management of whales and whaling: an historical review of the regulation of commercial and aboriginal subsistence whaling. *Arctic*, **46**, 97–107.

Gambell, R. (1999). The International Whaling Commission and the contemporary whaling debate. In *Conservation and Management of Marine Mammals*, ed. J.R. Twiss Jr & R.R. Reeves, pp. 179–98. Washington, DC: Smithsonian Institution Press.

Garrod, B. & Fennell, D.A. (2004). An analysis of whalewatching codes of conduct. *Annals of Tourism Research*, **31**, 334–52.

Geijer, C.K.A. & Read, A.J. (2013). Mitigation of marine mammal bycatch in U.S. fisheries since 1994. *Biological Conservation*, **159**, 54–60.

Gelatt, T. & Lowry, L. (2012). *Eumetopias jubatus*. The IUCN Red List of Threatened Species 2012: e.T8239A17463451 (accessed 19 February 2016).

Gerrodette, T. (2009). The tuna-dolphin issue. In *Encyclopedia of Marine Mammals*, 2nd edn, ed. W.F. Perrin, B. Würsig & J.G.M. Thewissen, pp. 1192–5. London: Academic Press.

Geschke, K. & Chilvers, B.L. (2009). Managing big boys: a case study on remote anaesthesia and satellite tracking of adult male New Zealand sea lions (*Phocarctos hookeri*). *Wildlife Research*, **36**, 666–74.

Gilman, E.L. (2011). Bycatch governance and best practice mitigation technology in global tuna fisheries. *Marine Policy*, **35**, 590–609.

Gormley, A.M., Slooten, E., Dawson, S., *et al.* (2012). First evidence that marine protected areas can work for marine mammals. *Journal of Applied Ecology*, **49**, 474–80.

Gosliner, M.L. (1999). The tuna-dolphin controversy. In *Conservation and Management of Marine Mammals*, ed. J.R. Twiss Jr & R.R. Reeves, pp. 120–55. Washington, DC: Smithsonian Institution Press.

Hamilton, S. & Baker, G.B. (2015). Review of research and assessments on the efficacy of sea lion exclusion devices in reducing the incidental mortality of New Zealand sea lions *Phocarctos hookeri* in the Auckland Islands squid trawl fishery. *Fisheries Research*, **161**, 200–6.

Hammill, M.O. & Stenson, G.B. (2010). Comment on 'Towards a precautionary approach to managing Canada's commercial harp seal hunt' by Leaper et al. *ICES Journal of Marine Science*, **67**, 321–2.

Hammill, M.O., Stenson, G.B., Doniol-Valcroze, T. & Mosnier, A. (2015). Conservation of northwest Atlantic harp seals: past success, future uncertainty? *Biological Conservation*, **192**, 181–91.

Hamner, R.M., Oremus, M., Stanley, M., *et al.* (2012a). *Estimating the Abundance and Effective Population Size of Maui's Dolphins Using Microsatellite Genotypes in 2010–11, with Retrospective Matching to 2001–07.* Auckland, New Zealand: Department of Conservation, 44 pp.

Hamner, R.M., Pichler, F.B., Heimeier, D., Constantine, R. & Baker, C.S. (2012b). Genetic differentiation and limited gene flow among fragmented populations of New Zealand endemic Hector's and Maui's dolphins. *Conservation Genetics*, **13**, 987–1002.

Henderson, J.R. (2001). A pre- and post-MARPOL Annex V summary of Hawaiian monk seal entanglements and marine debris accumulation in the northwestern Hawaiian Islands, 1982–1998. *Marine Pollution Bulletin*, **42**, 584–9.

Hindell, M.A. & Perrin, W. F. (2009). Elephant seals: *Mirounga angustirostris* and *M. leonina*. In *Encyclopedia of Marine Mammals*, 2nd edn, ed. W.F. Perrin, B. Würsig & J.G.M. Thewissen, pp. 364–8. London: Academic Press.

Hoelzel, A.R., Goldsworthy, S.D. & Fleischer, R.C. (2002). Population genetic structure. In *Marine Mammal Biology: an Evolutionary Approach*, ed. A.R. Hoelzel, pp. 325–52. Oxford, UK: Blackwell.

Holt, S.J. (1983). The Indian Ocean Whale Sanctuary. *Ambio*, **12**, 345–7.

Holt, S. (2000). Whales and whaling. In *Seas at the Millennium: An Environmental Evaluation*, vol. III, ed. C. Sheppard, pp. 73–88. Oxford, UK: Elsevier Science Ltd.

Howard, C. & Parsons, E.C.M. (2006). Attitudes of Scottish city inhabitants to cetacean conservation. *Biodiversity and Conservation*, **15**, 4335–56.

Hoyt, E. (2009). Marine protected areas. In *Encyclopedia of Marine Mammals*, 2nd edn, W.F. Perrin, B. Würsig & J.G.M. Thewissen, pp. 696–705. London: Academic Press.

Hoyt, E. (2011). *Marine Protected Areas for Whales, Dolphins and Porpoises*, 2nd edn. Abingdon, UK: Earthscan.

Hui, T.C.Y., Gryba, R., Gregr, E.J. & Trites, A.W. (2015). Assessment of competition between fisheries and Steller sea lions in Alaska based on estimated prey biomass, fisheries removals and predator foraging behaviour. *PLoS ONE*, **10**, e0123786.

IUCN (2012). *IUCN Guidelines for Reintroductions and Other Conservation Translocations August 2012.* Adopted by Species Survival Commission Steering Committee at Meeting SC 4 6, 5th September 2012. www.issg.org/pdf/publications/Translocation-Guidelines-2012.pdf

IUCN SSC – Cetacean Specialist Group (2016). www.iucn-csg.org (accessed 15 May 2016).

IWC (1994). Report of the workshop on mortality of cetaceans in passive fishing nets and traps. *Report of the International Whaling Commission Special Issue* **15**, 6–57.

Japp, D.W., Purves, M.G. & Wilkinson, S. (2012). *Benguela Current Large Marine Ecosystem: State of Stocks Review.* Report No. 2 (2012). Cape Town, South Africa: Capricorn Fisheries Monitoring.

Jefferson, T.A. & Curry, B.E. (1994). A global review of porpoise (Cetacea: Phocoenidae) mortality in gillnets. *Biological Conservation*, **67**, 167–83.

Jefferson, T.A., Leatherwood, S. & Webber, M.A. (1993). *FAO Species Identification Guide. Marine Mammals of the World.* Rome: FAO.

Joyner, C.C. & Tyler, Z. (2000). Marine conservation versus international free trade: reconciling dolphins with tuna and sea turtles with shrimp. *Ocean Development & International Law*, **31**, 127–50.

Kaschner, K., Tittensor, D.P., Ready, J., Gerrodette, T. & Worm B. (2011). Current and future patterns of global marine mammal biodiversity. *PLoS ONE* **6**, e19653.

Kirby, D.S., Visser, C. & Hanich, Q. (2014). Assessment of eco-labelling schemes for Pacific tuna fisheries. *Marine Policy*, **43**, 132–42.

Knowlton, A.R., Hamilton, P.K., Marx, M.K., Pettis, H.M. & Kraus, S.D. (2012). Monitoring North Atlantic right whale *Eubalaena glacialis* entanglement rates: a 30 yr retrospective. *Marine Ecology Progress Series*, **466**, 293–302.

Kock, K.-H. (2007). Antarctic marine living resources – exploitation and its management in the Southern Ocean. *Antarctic Science*, **19**, 231–8.

Kovacs, K.M, Aguilar A, Aurioles D, *et al.* (2012). Global threats to pinnipeds. *Marine Mammal Science*, **28**, 414–36.

Kovacs, K.M., Lydersen, C., Overland, J.E. & Moore, S.E. (2011). Impacts of changing sea-ice conditions on Arctic marine mammals. *Marine Biodiversity*, **41**, 181–94.

Laist, D.W. (1997). Impacts of marine debris: entanglement of marine life in marine debris including a comprehensive list of species with entanglement and ingestion records. In *Marine Debris: Sources, Impacts, and Solutions*, ed. J.M. Coe & D.B. Rogers, pp. 99–139. New York: Springer.

Lavigne, D.M., Scheffer, V.B. & Kellert, S.R. (1999). The evolution of North American attitudes toward marine mammals. In *Conservation and Management of Marine Mammals*, ed. J.R. Twiss Jr & R.R. Reeves, pp. 10–47. Washington, DC: Smithsonian Institution Press.

Lee, O.A., Burkanov, V. & Neill, W.H. (2014). Population trends of northern fur seals (*Callorhinus ursinus*) from a metapopulation perspective. *Journal of Experimental Marine Biology and Ecology*, **451**, 25–34.

Lowry, L.F., Laist, D.W., Gilmartin, W.G. & Antonelis, G.A. (2011). Recovery of the Hawaiian monk seal (*Monachus schauinslandi*): a review of conservation efforts, 1972 to 2010, and thoughts for the future. *Aquatic Mammals*, **37**, 397–419.

Lowry, M.S., Condit, R., Hatfield, B., *et al.* (2014). Abundance, distribution, and population growth of the northern elephant seal (*Mirounga angustirostris*) in the United States from 1991 to 2010. *Aquatic Mammals*, **40**, 20–31.

Magera, A.M., Mills Flemming, J.E., Kaschner, K., Christensen, L.B. & Lotze, H.K. (2013). Recovery trends in marine mammal populations. *PLoS ONE*, **8**, e77908.

Marsh, H., O'Shea, T.J. & Reynolds, J.E. III (2011). *Ecology and Conservation of the Sirenia: Dugongs and Manatees*. Cambridge, UK: Cambridge University Press.

McLeod, B.A., Brown, M.W., Moore, M.J., *et al.* (2008). Bowhead whales, not right whales, were the primary target of 16th- to 17th-century whalers in the western North Atlantic. *Arctic*, **61**, 61–75.

Meyer, S., Robertson, B.C., Chilvers, B.L. & Krkošek, M. (2015). Population dynamics reveal conservation priorities of the threatened New Zealand sea lion *Phocarctos hookeri*. *Marine Biology*, **162**, 1587–96.

Monson, D.H., Doak, D.F., Ballachey, B.E. & Bodkin, J.L. (2011). Could residual oil from the *Exxon Valdez* spill create a long-term population 'sink' for sea otters in Alaska? *Ecological Applications*, **21**, 2917–32.

Moore, S.E. & Huntington, H.P. (2008). Arctic marine mammals and climate change: impacts and resilience. *Ecological Applications*, **18** (Suppl.), S157–65.

Morissette, L., Christensen. V. & Pauly, D. (2012). Marine mammal impacts in exploited ecosystems: would large scale culling benefit fisheries? *PLoS ONE*, **7**, e43966.

Mulvaney, K. & Mckay, B. (2000). Small cetaceans: small whales, dolphins and porpoises. In *Seas at the Millennium: an Environmental Evaluation*, Vol.III, ed. C. Sheppard, pp. 89–103. Oxford, UK: Elsevier Science Ltd.

Northridge, S.P. & Hofman, R.J. (1999). Marine mammal interactions with fisheries. In *Conservation and Management of Marine Mammals*, ed. J.R. Twiss Jr & R.R. Reeves, pp. 99–119. Washington, DC: Smithsonian Institution Press.

Notarbartolo di Sciara, G., Agardy, T., Hyrenbach, D., Scovazzi, T. & van Klaveren, P. (2008). The Pelagos Sanctuary for Mediterranean marine mammals. *Aquatic Conservation: Marine and Freshwater Ecosystems*, **18**, 367–91.

O'Hara, T.M. & O'Shea, T.J. (2005). Assessing impacts of environmental contaminants. In *Marine Mammal Research: Conservation beyond Crisis*, ed. J.E. Reynolds III, W.F. Perrin, R.R. Reeves, S. Montgomery & T.J. Ragen, pp. 63–83. Baltimore, MD: The Johns Hopkins University Press.

Olson, P.A. (2009). Pilot whales *Globicephala melas* and *G. macrorhynchus*. In *Encyclopedia of Marine Mammals*, 2nd edn, ed. W.F. Perrin, B. Würsig & J.G.M. Thewissen, pp. 847–52. London: Academic Press.

O'Shea, T.J. & Brownell, R.L. Jr (1994). Organochlorine and metal contaminants in baleen whales: a review and evaluation of conservation implications. *Science of the Total Environment*, **154**, 179–200.

Perrin, W.F., Würsig, B. & Thewissen, J.G.M. (ed.) (2009). *Encyclopedia of Marine Mammals*, 2nd edn. London: Academic Press.

Plagányi, E.E. & Butterworth, D.S. (2005). Indirect fishery interactions. In *Marine Mammal Research: Conservation beyond Crisis*, ed. J.E. Reynolds III, W.F. Perrin, R.R. Reeves, S. Montgomery & T.J. Ragen, pp. 19–45. Baltimore, MD: The Johns Hopkins University Press.

Plagányi, E.E. & Butterworth, D.S. (2009). Competition with fisheries. In *Encyclopedia of Marine Mammals*, 2nd edn, ed. W.F. Perrin, B. Würsig & J.G.M. Thewissen, pp. 269–75. London: Academic Press.

Polanowski, A.M., Robbins, J., Chandler, D. & Jarman, S.N. (2014). Epigenetic estimation of age in humpback whales. *Molecular Ecology Resources*, **14**, 976–87.

Pompa, S., Ehrlich, P.R. & Ceballos, G. (2011). Global distribution and conservation of marine mammals. *Proceedings of the National Academy of Sciences*, **108**, 13600–5.

Punt, A.E. & Butterworth, D.S. (1995). The effects of future consumption by the Cape fur seal on catches and catch rates of the Cape hakes. 4. Modelling the biological interaction between Cape fur seals *Arctocephalus pusillus pusillus* and the Cape hakes *Merluccius capensis* and *M. paradoxus*.*South African Journal of Marine Science*, **16**, 255–85.

Raum-Suryan, K.L., Jemison, L.A. & Pitcher, K.W. (2009). Entanglement of Steller sea lions (*Eumetopias jubatus*) in marine debris: identifying causes and finding solutions. *Marine Pollution Bulletin*, **58**, 1487–95.

Ray, G.C., McCormick-Ray, J., Berg, P. & Epstein, H.E. (2006). Pacific walrus: benthic bioturbator of Beringia. *Journal of Experimental Marine Biology and Ecology*, **330**, 403–19.

Rayment, W., Dawson, S. & Slooten, E. (2010). Seasonal changes in distribution of Hector's dolphin at Banks Peninsula, New Zealand: implications for protected area design. *Aquatic Conservation: Marine and Freshwater Ecosystems*, **20**, 106–16.

Read, A.J. (2010). Conservation biology. In *Marine Mammal Ecology and Conservation: A Handbook of Techniques*, ed. I.L. Boyd, W.D. Bowen & S.J. Iverson, pp. 340–59. Oxford, UK: Oxford University Press.

Read, A.J., Drinker, P. & Northridge, S. (2006). Bycatch of marine mammals in U.S. and global fisheries. *Conservation Biology*, **20**, 163–9.

Read, A.J., van Waerebeek, K., Reyes, J.C., McKinnon, J.S. & Lehman, L.C. (1988). The exploitation of small cetaceans in coastal Peru. *Biological Conservation*, **46**, 53–70.

Reeves, R.R. (2000). *The Value of Sanctuaries, Parks, and Reserves (Protected Areas) as Tools for Conserving Marine Mammals*. Final Report to the Marine Mammal Commission, contract number T74465385. Bethesda, MD: Marine Mammal Commission, 50 pp.

Reijnders, P.J.H., Aguilar, A. & Borrell, A. (2009). Pollution and marine mammals. In *Encyclopedia of Marine Mammals*, 2nd edn, ed. W.F. Perrin, B. Würsig & J.G.M. Thewissen, pp. 890–8. London: Academic Press.

Reynolds III, J.E., Marsh, H. & Ragen, T.J. (2009). Marine mammal conservation. *Endangered Species Research*, **7**, 23–8.

Riedman, M. (1990). *The Pinnipeds: Seals, Sea Lions, and Walruses*. Berkeley, CA: University of California Press.

Robards, M.D. & Reeves, R.R. (2011). The global extent and character of marine mammal consumption by humans: 1970–2009. *Biological Conservation*, **144**, 2770–86.

Robertson, B.C. & Chilvers, B.L. (2011). The population decline of the New Zealand sea lion *Phocarctos hookeri*: a review of possible causes. *Mammal Review*, **41**, 253–75.

Rocha, R.C. Jr, Clapham, P.J. & Ivashchenko, Y.V. (2014). Emptying the oceans: a summary of industrial whaling catches in the 20th century. *Marine Fisheries Review*, **76** (4), 37–48.

Rojas-Bracho, L. & Reeves, R.R. (2013). Vaquitas and gillnets: Mexico's ultimate cetacean conservation challenge. *Endangered Species Research*, **21**, 77–87.

Ruzicka, J.J., Steele, J.H., Ballerini, T., Gaichas, S.K. & Ainley, D.G. (2013). Dividing up the pie: whales, fish, and humans as competitors. *Progress in Oceanography*, **116**, 207–19.

Schultz, J.K. (2011). Population genetics of the monk seals (genus *Monachus*): a review. *Aquatic Mammals*, **37**, 227–35.

Senigaglia, V., Christiansen, F., Bejder, L., *et al.* (2016). Meta-analyses of whale-watching impact studies: comparisons of cetacean responses to disturbance. *Marine Ecology Progress Series*, **542**, 251–63.

Slooten, E. & Dawson, S.M. (2010). Assessing the effectiveness of conservation management decisions: likely effects of new protection measures for Hector's dolphin (*Cephalorhynchus hectori*). *Aquatic Conservation: Marine and Freshwater Ecosystems*, **20**, 334–47.

Sorice, M.G., Shafer, C.S. & Ditton, R.B. (2006). Managing endangered species within the use–preservation paradox: the Florida manatee (*Trichechus manatus latirostris*) as a tourism attraction. *Environmental Management*, **37**, 69–83.

Stirling, I. & Derocher, A.E. (2012). Effects of climate warming on polar bears: a review of the evidence. *Global Change Biology*, **18**, 2694–706.

Taylor, B.L. (2005). Identifying units to conserve. In *Marine Mammal Research: Conservation beyond Crisis*, ed. J.E. Reynolds III, W.F. Perrin, R.R. Reeves,S. Montgomery & T.J. Ragen, pp. 149–62. Baltimore, MD: The Johns Hopkins University Press.

Turvey, S.T. & Risley, C.L. (2006). Modelling the extinction of Steller's sea cow. *Biology Letters*, **2**, 94–7.

Valqui, J. (2012). The marine otter *Lontra felina* (Molnia, 1782): a review of its present status and implications for future conservation. *Mammalian Biology*, **77**, 75–83.

Wade, P.R. (1998). Calculating limits to allowable human-caused mortality of cetaceans and pinnipeds. *Marine Mammal Science*, **14**, 1–37.

Wang, M. & Overland, J.E. (2012). A sea ice free summer Arctic within 30 years: an update from CMIP5 models, *Geophysical Research Letters*, **39**, L18501.

Weilgart, L.S. (2007). The impacts of anthropogenic ocean noise on cetaceans and implications for management. *Canadian Journal of Zoology*, **85**, 1091–116.

Würsig, B. & Richardson, W.J. (2009). Noise, effects of. In *Encyclopedia of Marine Mammals*, 2nd edn, ed. W.F. Perrin, B. Würsig & J.G.M. Thewissen, pp. 765–73. London: Academic Press.

Yodzis, P. (2001). Must top predators be culled for the sake of fisheries? *Trends in Ecology & Evolution*, **16**, 78–84.

# 10 Coastal Waters

In the next few chapters we look at major marine ecosystems, the human activities affecting them, and approaches to management and conservation. This chapter provides an introduction to coastal waters. We then examine particular coastal, open ocean, and deep-sea habitats and the use of protected areas in marine conservation.

Coastal waters encompass a huge diversity of habitats, including wave-exposed rocky shores, sheltered mudflats, wetlands, tropical and temperate reefs, and benthic and pelagic habitats of the open continental shelf (Box 10.1). Although coastal waters comprise only 9% of the world ocean by area (Harris *et al.*, 2014), they are of enormous importance ecologically and economically as productive regions that provide 80% of the world's wild-caught seafood (Watson *et al.*, 2015). They are also of immense value for the ecosystem services they provide, such as coastal protection, nutrient cycling, and carbon sequestration (Barbier *et al.*, 2011). Coastal ecosystems are, however, especially impacted by human activities. In a global assessment, Halpern *et al.* (2008) used indices of anthropogenic impact for a number of marine ecosystems. More than 40% of the world's ocean had medium to very high impact scores, and mostly these were areas of shelf and slope. Ecosystems with the highest impact scores included coral reefs, seagrass beds, mangroves, rocky reefs, and continental shelves.

### Box 10.1 Major Coastal Habitats

Cliffs
Rocky/boulder shores
Shingle beaches
Sandy beaches
Sand dunes
Sand- and mudflats
Mangroves
Saltmarsh
Seagrass beds
Coral reefs
Temperate reefs
Kelp forests
Subtidal sediment habitats
    pebble, gravel, sand, mud, biogenic
Estuaries, lagoons
Brackish seas
Pelagic habitats
    fronts, eddies, upwellings

The term 'coastal waters' is used here to mean those sea areas significantly influenced by their proximity to land. One can recognize a zone where oceanic circulation and tides are modified by their interaction with continental topography and coastal winds and where coastal bathymetry and land runoff modify physico-chemical parameters such as temperature, salinity and nutrient status. These factors strongly influence ecosystem structure and function, such that coastal waters tend to be distinguishable by their biota and productivity (Longhurst, 2007). Typically, this coastal area corresponds to waters of the continental shelf, or neritic waters, plus any landward enclosed waters, so essentially encompassing the area from high water mark to the shelf break. But given the degree of land-sea interaction in this zone, effective coastal management and conservation call for strategies that take account of natural processes and human activities in adjacent land margins and catchments. This integrative approach is the basis for coastal zone management (examined later), 'which treats shorelands and coastal waters as a single interacting unit' and aims for their unified management (Clark, 1998).

The continental shelf, the submerged border of a continent, extends seawards from the shore to the shelf break, at which point the seabed descends more steeply down the continental slope (Fig. 1.6). The edge of the shelf is usually at water depths of between 100 and 200 m, the global average being about 130 m. Shelf width varies greatly. Some coasts have virtually no shelf, the seabed giving way almost immediately to deep-sea depths, whereas off other coasts the shelf extends for more than 1000 km. Worldwide, shelf width averages 70 km (Eisma, 1988). Much of the world's shelf area is in temperate to polar latitudes of the Northern Hemisphere. There are also large shelf areas in SE Asia. The Southern Hemisphere has far less landmass, but among the major southern shelf areas are those off Australia and Argentina. Coastal habitats and communities are thus far more extensive in some regions than others.

Shelf ecosystems also vary considerably in terms of climate, circulation, and other oceanographic processes. In turn, these factors give rise to differences in primary productivity and its seasonality, benthic biomass, and the relative importance of demersal and pelagic fish and marine mammals. The fate of this production differs considerably between regions. Shelf ecosystems tend to show a latitudinal difference in the proportion of primary production transferred to pelagic and benthic components. For tropical shelf ecosystems, pelagic components typically account for a greater share of energy flow, whereas benthic production is proportionately more significant at mid- to high latitudes. Pelagic fisheries thus tend to be more important at low latitudes and demersal fisheries at mid-latitudes (Petersen & Curtis, 1980; Alongi, 2005). Management models developed for temperate waters may require substantial modification to be useful in high-latitude or tropical seas.

The degree to which coastal waters are enclosed by landmasses and separated from the open ocean has implications for their susceptibility to terrestrial influences and vulnerability to anthropogenic impacts. Many of the world's most populous coasts border semi-enclosed bodies of water that have restricted circulation and receive major rivers. Important enclosed seas include the North, Baltic, Mediterranean, Red, and South China seas; Persian Gulf; Bay of Bengal; and Gulf of Mexico. Nearly all such basins are in the Northern Hemisphere. Enclosed seas can differ considerably in salinity depending on the relative importance of evaporation and freshwater inflow. For instance, the average salinity of the Mediterranean Sea is about 38.5, but it is less than 20 for the Baltic Sea (Ketchum, 1983), a difference that strongly influences biodiversity.

East of the Mediterranean Sea lie three brackish seas: the Black Sea, with only restricted exchange of water with the Mediterranean; and the Caspian and Aral seas, both completely isolated from the open ocean. This region, together with the Mediterranean basin, was once occupied by part of the ancient Tethys Ocean that separated Laurasia from Gondwana. The region's modern biota appears to be a mixture of elements from various sources, notably the Mediterranean Sea, brackish and freshwater assemblages of the late Tertiary, glacial relicts, and modern immigrants. Each component is biogeographically distinct, but the impoverished Aral Sea fauna is mainly of freshwater rather than of marine origin (Briggs, 1974). Human activities have drastically altered these seas. Pollution, alterations to freshwater inflow, overfishing, and invasive species have had major and, in some cases, catastrophic impacts.

Estuarine and biogenic habitats are a feature of many coasts. An estuary is a semi-enclosed coastal area, typically a river mouth, where salinity is markedly lowered by freshwater inflow, but usually with large fluctuations. Sediment shores of estuaries and of other sheltered coasts often support important wetland habitat, notably saltmarshes in temperate latitudes and mangroves in the tropics and subtropics. Also highly significant as biogenic habitat of tropical waters are coral reefs. Management and conservation of estuaries, wetlands, and coral reefs warrant separate consideration and are examined in the next three chapters.

Coastal ecosystems are among the world's most productive. Nutrients derived from landmasses are in good supply and kept in the euphotic zone by mixing processes. Phytoplankton productivity for shelf waters is typically of the order of 200–400 g C $m^{-2}$ $y^{-1}$, which is at least twice that of the open ocean. In wind-driven coastal upwelling systems well supplied with nutrients, productivity can reach 700–800 g C $m^{-2}$ $y^{-1}$. And even higher rates, exceeding 1000 g C $m^{-2}$ $y^{-1}$, are seen in coastal vegetated habitats where rooted macrophytes tap into nutrient-rich sediments (Mann, 2000). By comparison, the primary production of croplands is about 300–500 g C $m^{-2}$ $y^{-1}$ (Huston & Wolverton, 2009).

## Benthic Habitats

Major benthic habitat types are briefly considered here as they are often the most appropriate by which to assess the impact of human activities in coastal waters and to frame management and conservation strategies.

### Intertidal Zone

The most obvious difference between shores is the nature of the substratum, whether, for instance, it consists of rock, boulders, shingle, sand, or mud. Such categories form a gradation, and shores of mixed particle size are common; nevertheless, it is convenient to broadly distinguish between rocky and soft sediment shores. Other major environmental gradients that strongly affect the biota of shores are the local climate, the vertical gradient from sea to land (amplified in particular by waves and tides), the degree of wave exposure, and the salinity gradient from marine to freshwater (Raffaelli & Hawkins, 1996).

Shore organisms display zonation patterns in response to these gradients as well as to biotic interactions, such as competition for space and food. Zonation is particularly evident on rocky shores, where similar major zones can be found worldwide. Typically there is a low-shore zone (around the level of low-water spring tides) dominated by large brown macroalgae, or kelps; then a mid-shore zone characterized by barnacles and mussels; and finally a high-shore zone (around high-water spring tides) characterized by black lichen and periwinkles (small snails of the family Littorinidae). Zonation is less obvious on sediment shores. Many of these animals burrow into the sediment where they are less affected by stresses associated with emersion and where competition for space is less intense than on two-dimensional rocky substrata. Nevertheless, zones characterized mainly by amphipod and isopod crustaceans are recognizable on wave-exposed sandy shores. Sediment shores of moderate wave exposure support more diverse macrofaunas, usually dominated by bivalve molluscs, polychaete worms, and crustaceans. Sheltered shores, such as in estuaries and inlets, can be especially productive and important as feeding grounds for waders and support wetland vegetation such as mangrove and saltmarsh.

## Nearshore Subtidal Zone

Organisms that play a major role in structuring habitats are also often important in nearshore subtidal zones; seagrass, for example, on sediment bottoms, and reef-building corals on rocky areas in the tropics. In temperate and higher-latitude areas, a zone of large macroalgae often extends from the low shore into the rocky subtidal. Here we find extensive kelp beds where the rocky substratum is at most only gently shelving, whilst epifauna tend to dominate steeply shelving or vertical surfaces (Witman & Dayton, 2001).

Kelp beds are of considerable ecological importance and conservation interest. Kelps are large brown seaweeds, particularly those of the orders Laminariales (*Laminaria, Macrocystis, Nereocystis, Alaria, Ecklonia, Undaria*) and Durvillaeales (*Durvillaea*). In clear water, kelps can extend to water depths of 20–30 m. Kelps, seagrasses, and other macrophytes are significant sources of organic detritus to coastal systems. Average detrital production by kelps alone is estimated at some 700 g C $m^{-2}$ $y^{-1}$ (Krumhansl & Scheibling, 2012).

Extensive kelp-forest communities occur along the NE Pacific coast, where major surface-canopy species are *Eualaria fistulosa, Macrocystis pyrifera,* and *Nereocystis luetkeana* between Alaska and California. Kelp forests form complex three-dimensional habitats with different canopy layers characterized by different algal groups. In kelp forests of southern California, for example, Dayton *et al.* (1984) describe five main layers: a floating canopy (*Macrocystis pyrifera*), an erect understorey (*Pterygophora californica* and *Eisenia arborea*), a prostrate canopy (*Laminaria farlowii*), a turf of articulated coralline algae, and encrusting coralline algae. Hundreds of associated species inhabit kelp forests, in particular other algae, invertebrates, and fishes as well as seabirds such as cormorants, pelicans and gulls, and sea otters and harbour seals (Foster *et al.*, 1991; Graham, 2004).

The biomass of kelp beds can be dramatically affected by sea urchins, which are major herbivores of kelps worldwide (Fig. 10.1). There has been considerable debate on kelp-urchin interactions and possible controlling factors (e.g. Mann, 2000). Studies on North Pacific and NW Atlantic coasts indicate that sea urchins in kelp beds can switch between two very different feeding behaviours and that, consequently, these systems can alternate between two stable states. When kelp is abundant and urchin numbers are relatively low, the urchins adopt a cryptic mode of life, living under rocks and in crevices and intercepting drift algae and detritus for food. But if urchin abundance increases and they can no longer be supported by detrital material, they emerge to graze directly on live kelp. Dense aggregations of urchins in this feeding mode can quickly destroy whole kelp beds to produce 'urchin barrens'. In Southern California, Graham (2004) found the urchin barrens to have low diversity: about a third of the 275 commonly recorded species occurred significantly more often in the *Macrocystis pyrifera* kelp forests. Various factors may contribute to the wide variability in urchin and kelp abundances, such as favourable environmental conditions for heavy recruitment, mass mortality from storms or outbreaks of disease, large-scale oceanographic changes (e.g. El Niño events), or a decline in urchin predators. In many cases, changes in abundance are probably due to a combination of factors. Particular attention has, however, focused on the role of urchin predators, notably fish and lobsters in the NW Atlantic and sea otters in the North Pacific (Fig. 9.5), with their overexploitation linked to population explosions of urchins (Tegner & Dayton, 2000).

**Fig. 10.1** A kelp forest dominated by *Macrocystis pyrifera* and an aggregation of the sea urchin *Evechinus chloroticus* in Paterson Inlet, Stewart Island, New Zealand. The interaction between grazing sea urchins and kelp can result in the community structure shifting between two states: kelp forest and urchin barrens. Photograph: S. Wing. (A black and white version of this figure will appear in some formats. For the colour version, please refer to the plate section.)

Where a rocky substratum extends deeper than the macroalgal zone, it is characterized mainly by epifaunal invertebrates including sponges, cnidarians, bryozoans, and ascidians. Some of these epifauna may themselves contribute importantly to the structural complexity of the habitat, particularly gorgonian, antipatharian, and hydrozoan corals (Hiscock & Mitchell, 1980; Witman & Dayton, 2001).

## Subtidal Sediment Habitats

Rocky subtidal habitats of coastal waters are mostly confined to the nearshore zone. Seaward, the seafloor of the continental shelf is predominantly sedimentary. The type of sediment depends on such factors as the source and supply of modern terrigenous material, whether the shelf was formerly glaciated (contributing relict gravel and stones), and the importance of calcareous skeletal material (e.g. from shells). Very broadly, muddy shelf sediments tend to be commonest in the wet tropics, sandy sediments on temperate shelves, and gravelly sediments in high-latitude areas, reflecting the relative influence of climate, riverine inputs, and glaciation.

Benthic assemblages of similar structure and function occur in shelf sediments worldwide, their specific composition depending mainly on biogeography, factors related to bathymetry, and the nature of the sediment. Macrofaunal communities of fine-grained sediments are typically dominated numerically by polychaete worms (Fig. 10.2). Other important infaunal groups include

**Fig. 10.2** Polychaete worms are usually the dominant macrofauna of soft-sediment communities in terms of number of species and individuals. This polychaete (a species of *Perinereis*) is a member of the family Nereididae, commonly known as ragworms. Some ragworm species are dug for bait. Photograph: © Rod Morris/www.rodmorris.co.nz. (A black and white version of this figure will appear in some formats. For the colour version, please refer to the plate section.)

amphipod and isopod crustaceans, bivalve molluscs, brittlestars, and heart urchins. Often important among the mobile epifauna are gastropods, anomuran and brachyuran crabs, sea stars, brittlestars, and regular urchins. Sessile epifauna become especially conspicuous on gravelly and stony bottoms where hard surfaces for attachment are plentiful. On such shelves, sponges, hydroids, anemones, octocorals, bryozoans, and ascidians can comprise a major component of the benthos. Sessile epifauna can markedly increase seabed habitat complexity and, thereby, benthic biodiversity. Loss of benthic habitat-forming species can thus severely impact associated species. In the Wadden Sea, for example, the destruction of oyster banks, tube-building polychaete reefs, seagrass beds, and saltmarshes has been a major factor in the extinction or severe depletion of species (Lotze *et al.*, 2005).

Although macrofaunal groups dominate the biomass of coastal soft-bottom habitats – and are used almost exclusively to assess the conservation value of such areas – much or most of the benthic secondary productivity is likely to be contributed by the meiofauna, usually defined as animals that pass through a 0.5 mm mesh sieve but are retained on a finer mesh (usually 30–60 μm). Nematodes are the most abundant members of the meiofauna (Fig. 4.4), usually followed by harpacticoid copepods. Smaller still is the microbenthos, which includes the benthic bacteria, vitally important, among other things, as food for deposit-feeding invertebrates and in the decomposition of organic matter and regeneration of nutrients into the water column. The benthos thus occupies a central role in shelf ecosystems, converting the flux of organic matter from the water column into animal and microbial biomass of the seabed. In turn these components provide food for demersal fish and other bottom-feeding animals and release nutrients back into the water column for reuse by phytoplankton.

## Habitat Mapping

An essential prerequisite for effective management and conservation of coastal as well as offshore waters is adequate information on the distribution of habitats and communities. This information is needed, for example, in designating marine protected areas, in identifying areas potentially vulnerable to anthropogenic disturbance, and in guiding fisheries and other extractive uses. Distributional patterns in the marine environment can be examined across the spatial spectrum, from global and regional levels down to specific sites and communities, and different approaches can be taken depending upon the purpose of the analysis and type and availability of abiotic and biotic data (Roff & Zacharias, 2011). As we have seen (Chapter 1), broad-scale biogeographic patterns can be delineated, including realms, provinces, and ecoregions for coastal areas (Spalding *et al.*, 2007). Such classifications can then be used as the basis for finer-scale characterization to inform local management and conservation. Benthic environments tend to be the focus of these assessments as they are generally more suitable than pelagic environments for categorizing protected areas and assessing environmental impacts.

A wide range of methods is available for mapping seabed habitat, including satellite and aerial imagery, acoustic survey techniques, underwater photography, and direct benthic sampling. Methods vary in terms of the area that can be mapped per unit effort and the degree of resolution, from very large scale (10–100s $km^2$ per hour) but of low horizontal resolution (10–100 m) for remote sensing to very detailed but labour intensive for direct sampling methods (Kenny *et al.*, 2003). Aerial and satellite imaging methods for seabed mapping are limited to very shallow waters, typically less than 10 m water depth. They have, for example, been used for mapping intertidal habitats, seagrass beds, kelp beds, and reef habitats. An important advance in habitat mapping has come from the development of multibeam echo-sounding, enabling broad-scale surveys with complete seafloor coverage to be undertaken. High-resolution seabed imagery and associated environmental data, complemented by benthic sampling (e.g. grab, epibenthic sled, video), can be used to produce benthic habitat maps (Brown *et al.*, 2011) (Fig. 10.3).

Adequate biological data are often, however, not available for benthic mapping. Alternatively, abiotic variables may be used as surrogates to describe and classify patterns of benthic biodiversity (McArthur *et al.*, 2010). This approach has the advantage that whereas sampling the benthos, directly or as seabed images, and processing those samples are labour intensive and costly, geophysical and other abiotic data can be gathered and analysed more quickly, enabling large areas to be covered. In many situations, using abiotic surrogates may be the only practical option for areas where the biota is poorly known and management decisions need to be made without delay. Abiotic surrogates tend to perform best at large spatial scales, but – particularly for surveys at smaller spatial scales – habitat classifications need to be validated where possible using biological data. In a study of continental shelf rocky reefs of south-east Queensland, Australia, Richmond and Stevens (2014) identified distinct benthic assemblages. But their classification, based on commonly used physical surrogates available at the local scale for MPA planning, had a relatively poor relationship with the benthic assemblage structure, explaining only 22% of variation. Also, physical surrogates may not reflect the degree of anthropogenic impact on the benthos, such as from fishing and introduced species (McArthur *et al.*, 2010).

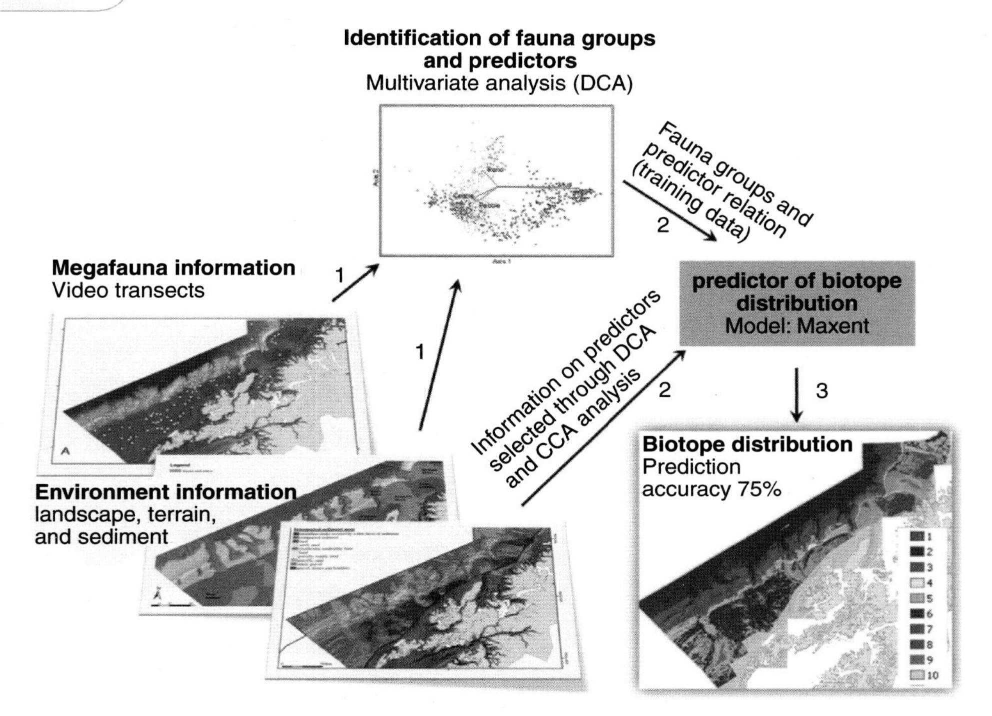


**Fig. 10.3** General steps used to identify and predict the distribution of benthic habitats/biotopes: (1) collecting information on the fauna and environmental factors to identify faunal groups and related environmental descriptors, (2) using the fauna-environmental factor relationship for construction of a model, and (3) producing predictive biotope maps based on environmental information and the model. In this example, for the shelf and upper slope off northern Norway, biotopes are based on the composition of benthic megafauna sampled by video transects. This component of the benthos includes large habitat-forming organisms that are especially vulnerable to physical disturbance. Detrended correspondence analysis (DCA) and canonical correspondence analysis (CCA) are the multivariate techniques used here to identify groups of samples and the relationships between biota and environmental gradients. Maxent is a modelling tool used here to predict biotope distribution. Reprinted from *Journal of Sea Research,* Vol. 100, Buhl-Mortensen, L., Buhl-Mortensen, P., Dolan, M.J.F. & Gonzalez-Mirelis, G., Habitat mapping as a tool for conservation and sustainable use of marine resources: some perspectives from the MAREANO Programme, Norway, pages 46–61, copyright 2014, with permission from Elsevier. (A black and white version of this figure will appear in some formats. For the colour version, please refer to the plate section.)

Benthic habitat mapping is increasingly being used to help inform management and conservation of coastal waters, with differing use of abiotic and biotic variables. Roff *et al.* (2003) propose, for example, a hierarchical approach to habitat classification based on features of oceanography (e.g. temperature, stratification, exposure) and physiography (e.g. bottom relief, substratum type) (Box 10.2). They applied this classification to the Canadian exclusive economic zone (EEZ) to examine the broad-scale distribution of marine natural regions and to the shelf off Nova Scotia to examine the more detailed distribution of pelagic and benthic seascapes. Huang *et al.* (2011) developed this approach to habitat classification for the Australian EEZ by using strata (three levels with nine environmental variables) informed by knowledge of benthic ecology to optimize the choice of environmental variables and their threshold values. Also their classifications, based on fuzzy logic, recognize the typically gradual transitions between habitats.

## Box 10.2 Geophysical Factors That Can Be Used to Classify Habitat Types

### Oceanographic Factors

Ice cover
Temperature
Salinity
Water masses (temperature and salinity signatures)
Temperature anomalies
Temperature gradients (e.g. fronts)
Light penetration
Water column stratification
Nutrient concentrations
Tidal amplitude
Tidal currents
Exposure (to wave action or desiccation)
Oxygen concentration

### Physiographic Factors

Tectonic motion
Latitude
Depth (bathymetry)
Relief/bottom slope
Rate of change of slope/heterogeneity
Substratum particle size
Rock type

With permission from Roff, J.C., Taylor, M.E. & Laughren, J. (2003). Geophysical approaches to the classification, delineation and monitoring of marine habitats and their communities. *Aquatic Conservation: Marine and Freshwater Ecosystems*, 13, 77–90. Copyright © 2003 John Wiley & Sons, Ltd.

It is important for management and conservation purposes that the results from such mapping exercises can be incorporated into agreed-upon frameworks for classifying habitats so that outputs from surveys are comparable, at least at a regional level. One such scheme increasingly being used is the European Union Nature Information System (EUNIS) habitat classification (see Galparsoro *et al.* (2012) for a recent overview and appraisal). EUNIS includes as one of its primary categories a hierarchical classification of European marine habitats, essentially intertidal areas and those seaward, including fully marine to brackish habitats. The classification recognizes eight major categories of marine habitat based on substratum type and water depth, each of which is further subdivided in terms of physical and then biological factors to individual biotopes (a physical habitat and its associated species) (Table 10.1). For marine and coastal waters under US jurisdiction, the Coastal and Marine Ecological Classification Standard has a similar aim of providing a standard format for classifying the marine environment to its constituent biotopes (Federal Geographic Data Committee, 2012).

**Table 10.1** European Union Nature Information System (EUNIS) Hierarchical Classification of Marine Habitats with Examples of Successive Levels under Littoral Rock and Sublittoral Sediment

In most cases habitats are defined up to Level 5. There is also a description of each biotope. For example, A5.3741: 'In deep offshore sandy mud adjacent to oil or gas platforms, organic enrichment from drill cuttings leads to the development of communities dominated by the pollution tolerant opportunist *Capitella capitata* and *Ophryotrocha dubia* (or other species of *Ophryotrocha*) [polychaetes]. These species are generally found in extremely high abundances and accompanied by *Thyasira* spp. [bivalves], *Raricirrus beryli, Paramphinome jeffreysii,* and *Chaetozone setosa* [polychaetes]. Other taxa including *Exogone verugera, Pholoe inornata* [polychaetes], and *Idas simpsoni* [bivalve] may also be present.' For A1.1122, *Chthamalus* is a genus of acorn barnacles, and *Lichina pygmaea* is a lichen.

| Level | 1 | 2 | 3 | 4 | 5 | 6 |
|---|---|---|---|---|---|---|
| | A: Marine habitats | | | | | |
| | | A1: Littoral rock and other hard substrata | | | | |
| | | | A1.1: High energy littoral rock | | | |
| | | | | A1.11: Mussel and/or barnacle communities | | |
| | | | | | A1.112: *Chthamalus* spp. on exposed upper eulittoral rock | |
| | | | | | | A1.1122: *Chthamalus* spp. and *Lichina pygmaea* on steep exposed upper eulittoral rock |
| | | A2: Littoral sediment | | | | |
| | | A3: Infralittoral rock and other hard substrata | | | | |
| | | A4: Circalittoral rock and other hard substrata | | | | |
| | | A5: Sublittoral sediment | | | | |
| | | | A5.3: Sublittoral mud | | | |
| | | | | A5.37: Deep circalittoral mud | | |
| | | | | | A5.374: *Capitella capitata* and *Thyasira* spp. in organically enriched offshore circalittoral mud and sandy mud | |
| | | | | | | A5.3741: *Capitella capitata, Thyasira* spp. and *Ophryotrocha dubia* inorganically enriched offshore circalittoral mud or sandy mud |
| | | A6: Deep-sea bed | | | | |
| | | A7: Pelagic water column | | | | |
| | | A8: Ice-associated marine habitats | | | | |

From www.eea.europa.eu/themes/biodiversity/eunis/eunis-habitat-classification. Copyright holder: European Environment Agency (EEA).

To evaluate the range and distribution of habitats and species within a large area requires a comprehensive and systematic approach. An example of survey and assessment of a coastal zone for marine conservation purposes is the Marine Nature Conservation Review (MNCR) of Great Britain, carried out between 1987 and 1998. The review provides a comprehensive baseline of information on marine habitats and species to aid coastal zone management and to help identify sites and species of nature conservation importance (Hiscock, 1996). For this purpose, the coastal zone of Great Britain (to about 5 km offshore) was divided into 15 sectors based on major features of geomorphology, biogeography, and oceanography. For each sector, existing information on physical characteristics and on the distribution of species, communities, and habitats was collated. Large stretches of coast were found to be poorly known, or existing information proved to be of limited use given the variety of methods used and level of detail involved in data collections. Thus, a substantial element of the MNCR was carrying out new field surveys to fill gaps in knowledge, and a phased approach to surveys was used to provide increasing levels of detail. An essential aspect of such widespread surveys is the establishment from the outset of standardized methods of field survey and of data collection, storage, and analysis. A classification of marine benthic biotopes was developed for the MNCR that is compatible with and contributes to the EUNIS classification.

## Human Activities

Human pressure on coastal areas has increased steadily with growth of the world population and the increasing proportion of people living near the coast. About 40% of the world's population now lives within 100 km of a coast (Small & Cohen, 2004). In many parts of the world, this pressure is exacerbated by tourism, particularly on coasts with beaches and coral reefs. We looked earlier (Chapter 2) at major types of human impact on the marine environment, such as from resource exploitation, disturbance and physical alteration of habitat, biological invasions, and various types of pollutants, including organic wastes, oil, persistent organic pollutants, plastics, and heavy metals. These impacts are often most pronounced in coastal waters given the intensity of human activity and sources of disturbance and pollutants and the vulnerability of many coastal habitats. Among the most serious threats are habitat loss, the overexploitation of living resources, various pollutants (including sewage, nutrients, and marine litter), invasive species, and altered sedimentation patterns. The relative significance of such factors for coastal areas varies widely at local to regional scales, but there are also threats with ecosystem-wide implications, notably those associated with climate change (UNEP/GPA, 2006; Crain *et al.*, 2009). These anthropogenic stresses are examined here and in later chapters in relation to specific habitats.

### Nutrient Enrichment

The situation with sewage discharge is generally worsening worldwide as coastal areas become more populous, and the building of treatment facilities and sewerage infrastructure lags behind. In developing countries, most sewage is discharged untreated. And tourism developments can lead to reduced water quality, particularly from sewage inputs. Apart from the implications for human health, sewage introduces organic matter to coastal ecosystems, the degradation of which can lead to oxygen depletion and increased nutrient loading. Importantly, nutrient enrichment of coastal waters occurs too from agricultural runoff containing fertilizers and animal wastes. Such nutrient

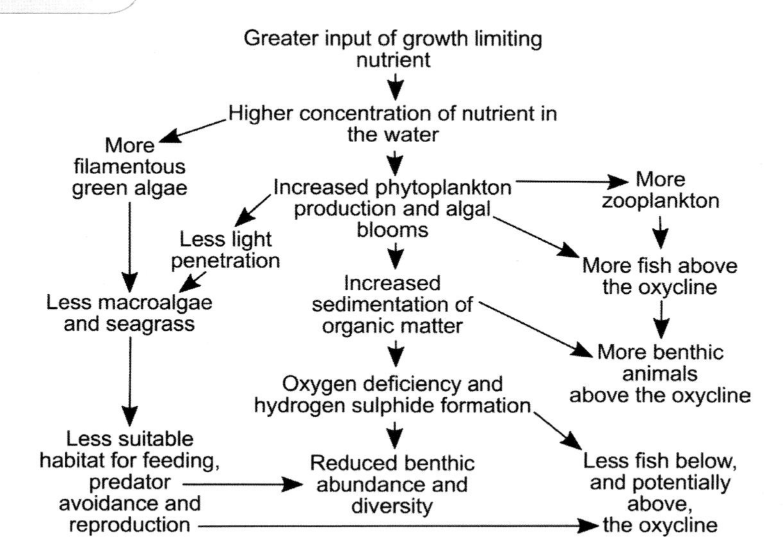


**Fig. 10.4** Effects of increased nutrient input in a coastal ecosystem. Certain effects of enhanced primary production, notably an increased biomass of fish above the sharp drop in oxygen (oxycline), might be deemed beneficial. But once increased production exceeds the assimilative capacity of the system, nutrient over-enrichment will result in oxygen deficiency and loss of macrobiota. From Rabalais, N.N., Turner, R.E., Díaz, R.J. & Justić, D. (2009). Global change and eutrophication of coastal waters. *ICES Journal of Marine Science,* 66, 1528–37, by permission of the International Council for the Exploration of the Sea. Published by Oxford Journals.

inputs can promote excessive algal growth, resulting in eutrophication (see Chapter 2) and a succession of effects in coastal ecosystems (Fig. 10.4).

The sedimentation and aerobic decomposition of this increased production lead to oxygen depletion of the bottom water – if oxygen is being consumed faster than it is being replenished – and mass mortality of benthos. Waters with low and minimal dissolved oxygen are termed 'hypoxic' and 'anoxic' (see Chapter 2). Stratification of the water column impedes the flux of oxygen from the surface mixed layer to beneath the thermocline, so the occurrence of hypoxic or anoxic bottom water is often more a summer phenomenon. Oxygen depletion can arise naturally, such as in upwelling areas and fjord basins, but the frequency and intensity of anthropogenic eutrophication and hypoxia in coastal waters are increasing worldwide such that many coastal areas now have dead zones. Díaz and Rosenberg (2008) describe an exponential rise since the 1960s in the occurrence of hypoxic zones, which now affect more than 400 coastal areas and some 245 000 km$^2$. In most cases, hypoxia is intermittent, with events occurring from several to many times a year or seasonally (usually from summer to autumn) to less than once a year. Other areas, however, are continuously hypoxic. Included in the latter are most of the deep basins of the Baltic Sea, where on occasion the low oxygen zone has extended upwards to make one huge area of up to 70 000 km$^2$. Other major coastal areas subject to anthropogenic hypoxia include the northern Gulf of Mexico off the Mississippi and Atchafalaya rivers, the East China Sea off the Yangtze River, the northern Adriatic Sea (see Chapter 16), Chesapeake Bay, and the northwestern Black Sea (Rabalais *et al.*, 2010).

Increased nutrient loading of coastal waters from human activities has been identified as the most likely reason for the increased reported incidence of algal blooms over recent decades, including the occurrence of harmful blooms (Heisler *et al.*, 2008). (Under bloom conditions phytoplankton populations can increase from background levels of $10^3$ cells L$^{-1}$ to $10^7$ cells L$^{-1}$ or more (Steidinger & Vargo, 1988).) A number of these harmful algae produce toxins, including potent neurotoxins, that can kill animals either directly or through affected prey. Harmful algal blooms have been linked with numerous large-scale die-offs, including of benthic communities, fish, seabirds, and marine mammals (Verity *et al.*, 2002).

Semi-enclosed waters subject to well-developed stratification and receiving large nutrient inputs from adjacent densely populated and/or agricultural areas are highly susceptible to hypoxia and the development of dead zones. A crucial step in reducing hypoxic zones is management of nutrient inputs, in particular through improved and more appropriate agricultural practices and suitable treatment of urban wastewater. For a number of locations, eutrophic conditions have eased following reductions in nutrient loading (Rabalais *et al.*, 2010). In the case of Tampa Bay, Florida, various actions have been taken since the mid-1970s to tackle the problem of nutrient enrichment and eutrophic conditions. These resulted in a 60% reduction in estimated nitrogen loading by 1985–2003, a significant decline in chlorophyll *a* concentrations, and an increase in light penetration. The areal coverage of seagrass subsequently increased, and hypoxic events have decreased in extent and duration (Greening & Janicki, 2006). The biological recovery of hypoxic systems may take decades once oxygen conditions have returned to normal (Steckbauer *et al.*, 2011). However, such systems often have associated problems, such as habitat destruction, overfishing, and invasive species, so reducing nutrient inputs may not return a system to its original state, and restoration efforts may need to be carried out on a number of fronts (Rabalais *et al.*, 2010).

## Riverine Inputs

Human activities also strongly affect the supply of sediment from land to sea, and changes in sediment input have major implications for coastal ecosystems. Increased sediment load can severely impact benthic communities dependent on low levels of turbidity, such as coral reefs, whereas coastal wetlands need depositional environments.

Globally, soil erosion is increasing, particularly as a result of deforestation and agricultural practices. At the same time, however, diversion and damming of rivers are reducing sediment input to the coastal ocean. The net result is that the global modern sediment flux (12.8 billion tonnes per year) is now about 15% less than the pre-human load (Syvitski & Kettner, 2011). There are more than 48 000 large dams around the world (dams more than 15 m high with an average reservoir area of 23 $km^2$), and many more under construction (Syvitski & Kettner, 2011). Among the best known is the Aswan High Dam, which dramatically affected productivity of the south-eastern Mediterranean (see Chapter 16). Most of these large dams are, however, in China. Thousands of dams have been built in the Yangtze River basin, including the world's largest, the Three Gorges Dam. Shortly after filling of the Three Gorges Dam began in 2003, primary production in the large diluted zone off the Yangtze River mouth dropped by more than 80%, accompanied by a shift in the composition of the phytoplankton community from diatoms to flagellates, seemingly because of a change in the flux of nutrients (Gong *et al.*, 2006). Such changes have important repercussions for ecosystem functioning and fisheries of the East China Sea. Since 1950, the rate of sediment accumulation in reservoirs in the Yangtze basin has increased to more than 850 million tonnes per year. As a result, the growth of intertidal areas at the delta has declined dramatically, from about 12 to 3 $km^{-2}$ $y^{-1}$ between the early 1970s and 1998 (Yang *et al.*, 2005). Within this delta region are highly significant wetland sites, including Chongming Dongtan Nature Reserve (326 $km^2$), a site internationally important as a staging and wintering site for millions of waders and wildfowl. Human-induced sea-level rise will exacerbate coastal erosion at such sites deprived of an adequate sediment supply by dams.

In other areas where sediment runoff is unimpeded, changes in land use can result in sedimentation rates in coastal waters increasing by one to two orders of magnitude. Heavy siltation tends to depress biodiversity of inshore benthic communities, with suspension feeders such as bivalves likely to be particularly adversely affected due to clogging of their gills (Thrush *et al.*, 2004).

## Physical Alteration of Habitat

Humans have physically modified the coastline and coastal seabed, often drastically. This results most obviously from urban, tourist, and industrial developments of the coast and associated infrastructure. Coastal developments often involve reclamation and building of protection measures such as seawalls, riprap, groynes, and breakwaters. These constructions can result in radical transformation of the shore and shallow subtidal zones such that little remains of the original habitats. Along many stretches of heavily urbanized coast, soft shorelines have been armoured using such measures so that the sediment habitats have been replaced by artificial hard substrata. In San Diego Bay, California, three-quarters of the shore is now artificial hard substrata (Davis *et al.*, 2002), whilst in Sydney Harbour, Australia, about half the intertidal habitat consists of seawalls or similar artificial habitat (Chapman & Bulleri, 2003). Such situations are typical of many populous coasts. Coastal armouring is likely to increase dramatically, given the ongoing littoralization of the world population, projected sea-level rise, and worsening coastal erosion. It should be remembered that coastal ecosystems themselves, particularly mangroves, saltmarshes, and coral reefs, play a key role in coastal protection (as well as providing many other benefits), underlining the importance of maintaining the health and integrity of these ecosystems (Spalding *et al.*, 2014).

Coastal protection structures, as well as wharves, jetties, and marinas, do provide habitat and may considerably increase the amount of hard substratum available for colonization by benthic organisms. However, structures such as seawalls, even when built using local stone, tend to provide a more uniform habitat than natural shores and support less diverse biotas. Habitat complexity of seawalls can be increased, for example, by incorporating cavities, sediment traps, and tide pools in their construction. Browne and Chapman (2011) found that adding rock pool mimics to a seawall doubled the number of species within months, particularly of mobile animals.

Sources of physical disturbance to the seabed include dredging and spoil disposal for port development and maintenance and dredging of sand and gravel for building aggregate. Highly significant in this regard is bottom fishing using trawls and dredges, which is estimated to occur on about three-quarters of the global continental shelf (Kaiser *et al.*, 2005). Fishing effort is very patchy, with some grounds being fished repeatedly. For the southern North Sea, Rijnsdorp *et al.* (1998) estimated that 30% of the seabed is trawled up to twice a year, and 9% is trawled more than five times by the Dutch beam trawl fleet, which makes up 50–70% of the total beam trawl effort in the North Sea. Chronic trawling pressure is likely to reduce the complexity of seabed habitat and alter benthic community structure. For certain areas of the North Sea, there is evidence of a shift in the composition of the benthos between the 1920s and 1980s, consistent with increased mechanization of the fishing fleet (Frid *et al.*, 2000). Exploitation of coastal fish and shellfish can also seriously impact target species but also in many cases non-target species and the composition of fish communities as a whole (see Chapter 5). And aquaculture operations can result in loss and degradation of habitat and appropriation of coastal waters by seaweed, fish, and shellfish farms.

## Human Pressure on Shores

Among the best-studied effects of human exploitation on coastal marine communities are those on the rocky shores of central and southern Chile, where the impact of food gathering has been assessed by comparing shore community structure inside and outside marine protected areas (Castilla, 1999; Moreno, 2001). The effect of exploitation depends on the trophic role of the targeted species. Heavy collecting of herbivorous keyhole limpets (*Fissurella* spp.) results in a red alga (*Mazzaella laminarioides*) dominating the midlittoral zone. Also gathered is a carnivorous muricid gastropod (*Concholepas concholepas*), its depletion allowing mussels (*Perumytilus purpuratus*) to cover the rocks and in turn ousting barnacles (*Notochthamalus scabrosus* and *Jehlius cirratus*) and several algal species. In many locations, exploitation is much less selective, resulting in general impoverishment of shore communities. Recreational gleaning in which a wide range of shore invertebrates are taken for human consumption is a type of overfishing reported from much of southern Europe and parts of Asia (Jamieson, 1993).

Shore species are collected not only for food. Even in Victorian Britain, the craze for home aquariums probably had significant impacts at coastal resorts. Gosse (1907) recalls in the mid-1850s rock pools in SW England 'thronged with beautiful sensitive forms of life,—they exist no longer … An army of "collectors" has passed over them, and ravaged every corner of them.' A number of conservation organizations produce seashore codes with advice on how best to enjoy a visit to the shore and minimize harm to organisms, such as not collecting live animals to take away, carefully replacing any rocks that have been moved to look for animals sheltering underneath, and not leaving any rubbish behind, especially persistent materials such as plastics. Shallow coastal zones have particular potential for promoting marine conservation, being those areas that people can most readily engage with. The health of the oceans as a whole would for most people be seen to have little or no bearing on their activities, but accessible coastal fringes with their often-rich biodiversity and range of resources and where human impacts can be all too evident offer an opportunity to raise public awareness and concern about the marine environment and the need for conservation action (Vincent, 2011).

Visitor pressure on shores is intensifying as the world population grows, and more people have personal transport and increased leisure time (Thompson *et al.*, 2002). Coastal areas with sandy beaches come under particular pressure in many countries as a result of tourist development, often resulting in essentially irreversible damage from resort development (Davenport & Davenport, 2006). Human trampling in itself can effect compositional changes to rocky shore communities (e.g. Fletcher & Frid, 1996). On sandy shores, trampling and the use of off-road vehicles can be particularly destabilizing to dune systems by reducing vegetation cover and promoting erosion (French, 1997; Brown & McLachlan, 2002). Recreational use of beaches is also potentially disruptive to nesting and foraging birds. Migratory waders, for example, are vulnerable when they have limited time to feed at a stopover location. Weight gain during this period can affect their onward migration and subsequent breeding. Delaware Bay supports the largest concentration of waders in the continental USA. More than a million birds, including knot, sanderling, and turnstone, stop off here during late May to early June to feed, in particular on the eggs of horseshoe crabs, as the crabs congregate to spawn at this time of year. The spectacle also attracts large numbers of bird-watchers, such that measures have been taken to reduce disruption (e.g. restricting access, providing viewing platforms) (Burger *et al.*, 2004). Human disturbance can also severely affect nesting turtles (see Chapter 7).

## Marine Debris

Marine debris is a particularly conspicuous pollutant of many coastal areas, and the situation worldwide is generally worsening. Coastal clean-up surveys indicate the enormous quantities of marine debris, although amounts vary greatly. Results from some 80 surveys carried out in 12 coastal regions around the world showed that three-quarters of sites yielded up to 200 kg of debris per km of shoreline, but on some shores 1–11 tonnes per km were collected (UNEP, 2009). The problem lies chiefly with plastics, by far the commonest type of marine debris (Derraik, 2002). Plastics break down far more slowly than their rate of input to the environment despite international initiatives to reduce the amount of plastics entering the sea. Most plastic waste in the marine environment is from land-based sources, but harm to marine animals, especially through entanglement and ingestion, is not restricted to those of coastal waters (see Chapter 2). Coastal clean-up programmes are organized regularly in many parts of the world, and public awareness of the problem is gaining ground through such activities. However, the solution lies primarily in waste reduction and recycling (Allsopp *et al.*, 2006).

## Oil Pollution

Oil inputs to the marine environment resulting from human activities have decreased since the 1980s; nevertheless, major oil spills remain a considerable risk with the potential for severe disruption to coastal habitats (UNEP/GPA, 2006). Among the largest and best-studied spills of crude oil from tanker accidents (from groundings) have been those from the *Torrey Canyon* off SW England in 1967 (119 000 tonnes of oil spilled), the *Amoco Cadiz* on the Brittany coast in 1978 (223 000 t), and the *Exxon Valdez* in Alaska's Prince William Sound in 1989 (37 000 t). Even larger inputs occurred during the 1991 Gulf War, a release of some 1.6 million t, and as a result of a well blowout at the *Deepwater Horizon* rig in the Gulf of Mexico that released some 700 000 t in 2010 (Farrington, 2013).

The effect of oil spills on coastal habitats has been the subject of a multitude of studies (see, for example, Raffaellii and Hawkins (1996) and Clark (2001) for overviews). Coastal environments can be broadly categorized in terms of their potential vulnerability to oil spills, based largely on coastal geomorphology, degree of wave exposure, and oil residence time. Gundlach and Hayes (1978) developed a vulnerability index that has been widely applied (Table 10.2). Exposed rocky headlands and wave-cut platforms generally recover most rapidly from oil spills, typically within a few years. Coarse-grained sandy and gravel beaches, which are subject to oil penetration and burial, are assigned intermediate index values. Sheltered environments, particularly mudflats, saltmarshes, and mangroves, are the environments that take the longest to recover as once oil becomes incorporated into fine-grained anaerobic sediment, it can take decades to degrade. Recovery can be delayed considerably by inappropriate clean-up techniques. Much of the disturbance to rocky shore communities of SW England following the *Torrey Canyon* spill was due to the liberal use of toxic dispersants. As a result, these shores took about ten years to regain their former community structure.

A key component of any oil spill contingency plan is the identification of ecologically sensitive areas, such as coastal wetlands, seabird colonies, or major spawning grounds, and often this identification exercise will include important seasonal considerations. Coastal sensitivity mapping can be used to show such ecological features (as well as amenity, fisheries, or industrial areas that may be affected), the priorities for protection, and appropriate preventive or clean-up action (e.g. Petersen

**Table 10.2** Potential Vulnerability of Coastal Environments to Oil Spill Damage

| Vulnerability index | Shoreline type | Comments |
|---|---|---|
| 1 | Exposed rocky headlands | Wave reflection keeps most of the oil offshore. No clean-up is necessary. |
| 2 | Eroding wave-cut platforms | Wave swept. Most oil removed by natural processes within weeks. |
| 3 | Fine-grained sand beaches | Oil does not penetrate into the sediment, facilitating mechanical removal if necessary. Otherwise, oil may persist for several months. |
| 4 | Coarse-grained sand beaches | Oil may sink and/or be buried rapidly making clean-up difficult. Under moderate to high energy conditions, oil will be removed naturally within months from most of the beachface |
| 5 | Exposed, compacted tidal flats | Most oil will not adhere to, nor penetrate into, the compacted tidal flat. Clean-up is usually unnecessary. |
| 6 | Mixed sand and gravel beaches | Oil may undergo rapid penetration and burial. Under moderate to low energy conditions, oil may persist for years. |
| 7 | Gravel beaches | Same as above. Clean-up should concentrate on the high-tide swash area. A solid asphalt pavement may form under heavy oil accumulations. |
| 8 | Sheltered rocky coasts | Areas of reduced wave action. Oil may persist for many years. Clean-up is not recommended unless oil concentration is very heavy. |
| 9 | Sheltered tidal flats | Areas of low wave energy. Oil may persist for years. Clean-up is not recommended unless oil accumulation is very heavy. These areas should receive priority protection. |
| 10 | Salt marshes and mangroves | Oil may persist for years. Cleaning of salt marshes by burning or cutting only if heavily oiled. Mangroves should not be altered. Protection of these environments should receive first priority. |

Reproduced with permission from: Gundlach, E.R. & Hayes, M.O. (1978). Vulnerability of coastal environments to oil spill impacts. *Marine Technology Society Journal*, 12 (4), 18–27.

*et al.*, 2002). Computer models are valuable for oil spill management and response to forecast the drift and dispersion of spills using information on the type of oil and environmental factors (wind, currents, tide, temperature, physical and chemical characteristics of oil, and amount and rate of spill) (van Bernem *et al.*, 2008).

## Global Climate Change

Human-induced climate change from increased greenhouse gas emissions is projected to have a range of profound impacts on marine ecosystems (see Chapter 2). Whilst our understanding of these changes and their consequences is still limited, coastal systems already exposed to

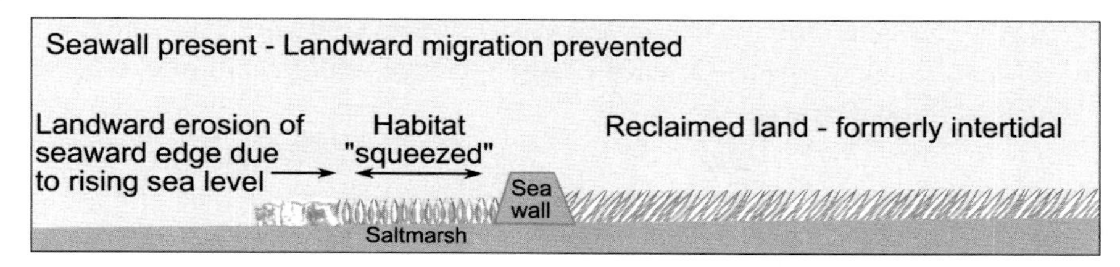


**Fig. 10.5** An example of coastal squeeze where a sea defence prevents the landward migration of habitat, in this case saltmarsh, in response to rising sea level. Reprinted from *Ocean & Coastal Management*, Vol. 84, Pontee, N., Defining coastal squeeze: a discussion, pages 204–7, copyright 2013, with permission from Elsevier.

such stresses as fishing pressure, pollution, and habitat destruction are especially vulnerable to warmer, more acidic seas and rising sea levels (Scavia *et al.*, 2002; Harley *et al.*, 2006). A warmer ocean, for example, is predicted to generate more intense storms – their impact on the coast exacerbated by a rising sea level – including more frequent extreme rainfall events, significantly increasing the delivery of nutrients and sediment to coastal waters. Global warming will also in general enhance stratification, rendering coastal waters more susceptible to hypoxia (Rabalais *et al.*, 2009).

Coastal armouring may result in the loss of intertidal habitat in the face of global climate change. Whilst the low-water mark will migrate landwards in response to sea-level rise, the presence of sea defences or similar fixed structures will prevent a corresponding migration of high-water mark, a so-called coastal squeeze resulting in a loss of intertidal habitat (Pontee, 2013) (Fig. 10.5). Dugan *et al.* (2008) examined habitat loss on open coast beaches of southern California resulting from the narrowing of beaches in front of coastal armouring. They reported loss of upper intertidal zones with loss of habitat for invertebrates, foraging shorebirds, and roosting gulls.

The impact of sea-level rise from global warming on coastal regions will, however, vary greatly from place to place depending upon coastal physiography and the already-prevailing patterns of eustatic and tectonic movement. Some coastal margins are subsiding naturally, such as those of the Gulf of Mexico and much of the US Atlantic coast, Amazon River delta, southern and eastern England, northern Adriatic Sea, and the Ganges-Brahamputra delta (Bird, 1993). Clearly, low-lying areas that are already subsiding will be most severely affected. Some of these areas are densely populated, and inundation will bring massive social and economic upheaval.

A number of studies have already reported poleward shifts in species' distributions coincident with a move in isotherms due to global warming. Coastal species may be able to migrate on north-south aligned coasts, but the problem will be more severe for species on east-west coastlines (Burrows *et al.*, 2011). Using more than 20 000 herbarium records of seaweeds collected in Australia since the 1940s, Wernberg *et al.* (2011) found a pronounced poleward shift in the distribution of temperate species on both the Indian and Pacific Ocean coasts, with median rates of retreat of about 0.7° and 1.7° latitude $°C^{-1}$, respectively. Given projected increases in sea surface temperature over the next several decades, such rates indicate that some hundreds of species will be driven to the edge of southern Australia – a hotspot of seaweed biodiversity and endemism – from where no further retreat is possible. Large seaweeds also provide resources for numerous associated species, so their loss – like that of other macrophytes and habitat-forming invertebrates – may have community-wide implications. One of the seaweeds shifting polewards on the Australian west coast, *Scytothalia dorycarpa*, is a large brown alga that is important as a habitat-forming species. As a result of a period of unusually high sea temperature in 2011, its warm distribution limit was abruptly shifted southwards by about 100 km with associated loss of understorey community structure (Smale

& Wernberg, 2013). Extreme climatic events are forecast to become more common as a result of anthropogenic climate change. Species already close to their physiological limits will be especially vulnerable to such additional stress.

## Coastal Zone Management

An environmental overview of the world's coastal waters by Hinrichsen (1998) shows that despite their diversity, they have a number of similar threats and management needs. Regional management programmes, such as those under UNEP (see Chapter 16) or other forums, need in general to be strengthened, properly supported, and more fully implemented. But management objectives also need to be incorporated into domestic legislation. Many coastal areas suffer from inadequate fisheries regulations, resulting in overexploitation of stocks, excessive bycatch, and damage to benthic habitats as well as from poor control of pollution because of insufficient investment in sewage and industrial wastewater treatment plants, and deficient management of catchments. Also crucial is sound coastal planning that addresses future population growth, urban and tourism development, and protection of vulnerable ecosystems, where often a top priority is conservation of wetland and coral reef habitat. Importantly, coastal management needs to involve local communities and tap in to their expertise. There may be opportunities to adapt traditional methods of resource management to modern programmes (see Chapter 13).

An important step in recognizing the need for effective management of the coastal zone in the face of increasing degradation was the Coastal Zone Management Act of 1972 in the USA 'to preserve, protect, develop, and where possible, to restore or enhance, the resources of the Nation's coastal zone for this and succeeding generations', to which end states should develop and implement 'management programs to achieve wise use of the land and water resources of the coastal zone, giving full consideration to ecological, cultural, historic, and esthetic values as well as the needs for compatible economic development …' Key elements of coastal zone management were further developed under Agenda 21 of the United Nations Conference on Environment and Development (see Chapter 3), which called on coastal states to commit themselves to integrated management and sustainable development of coastal areas under their national jurisdiction. By this stage, leading up to the UN Convention on the Law of the Sea, many countries had declared an exclusive economic zone and assumed rights and responsibilities for management of resources in their coastal waters. Aims of integrated coastal zone management (ICZM) have subsequently been incorporated into a number of international agreements relevant to the management and conservation of coastal resources. The Intergovernmental Panel on Climate Change calls on countries to develop coastal zone management plans, given the potentially immense impact of sea-level rise.

What is meant by ICZM may differ somewhat depending on specific aims of the programme in question, but as generally applied, the term incorporates certain key elements. The European Commission defines ICZM as 'a dynamic, multi-disciplinary and iterative process to promote sustainable management of coastal zones. … ICZM uses the informed participation and co-operation of all stakeholders to assess the societal goals in a given coastal area, and to take actions towards meeting these objectives. ICZM seeks, over the long-term, to balance environmental, economic, social, cultural and recreational objectives, all within the limits set by natural dynamics.' 'Integrated' means integration of objectives and of the instruments needed to meet these objectives, of the policy and administration, and of the terrestrial and marine components in time and space (Commission of the European Communities, 2000). The European approach identifies several key principles for successful ICZM (Box 10.3).

## Box 10.3 Principles of Integrated Coastal Zone Management

1. **A Broad Holistic Perspective** – Coastal zones are influenced by many inter-related forces related to hydrological, geomorphological, socio-economic, institutional, and cultural systems. Successful planning and management must eschew piecemeal decision making in favour of more strategic approaches that look at the bigger picture. The close links (through human and physical processes) between components imply that ICZM should always consider both the marine and terrestrial portions of the coastal zone as well as the river basins draining into it.
2. **A Long-Term Perspective** – The needs of both present and future generations must be considered, ensuring that decisions respect the 'precautionary principle'. Successful planning and management must acknowledge the inherent uncertainty of the future.
3. **Adaptive Management** – Integrated planning and management develop and evolve over the course of years or decades. ICZM does not guarantee the immediate resolution of coastal zone problems. The process requires good information and monitoring so it can be adjusted as problems and knowledge evolve.
4. **Reflect Local Specificity** – There is a wide diversity among the coastal zones, including variations in physical, ecological, social, cultural, institutional, and economic characteristics. ICZM must be based on a thorough understanding of the specific characteristics of the area in question.
5. **Work with Natural Processes** – Natural processes and dynamics of coastal systems are in continual, and sometimes sudden, flux. By working with these processes and respecting the limits imposed by them, activities are more environmentally sustainable and more economically profitable in the long run.
6. **Participatory Planning** – To incorporate the perspectives of all the relevant stakeholders in the planning process.
7. **Support and Involvement of All Relevant Administrative Bodies** – Administrative policies, programmes, and plans set the context for the management of coastal areas and their natural resources.
8. **Use of a Combination of Instruments** – Coastal zone management requires multiple instruments, including a mix of law, economic instruments, voluntary agreements, information provision, technological solutions, research, and education. The correct mix in a specific area will depend on the problems at hand and the institutional and cultural context.

Adapted from the Commission of the European Communities,2000.

ICZM recognizes that the coastal zone straddles an interface encompassing areas of hinterland, shore, and adjacent sea and that effective management must take account of their interactions. What actually constitutes 'the coastal zone' will vary depending on environmental factors but also on the management plan and its goals. The landward and seaward limits may be considered as the respective reach of significant marine and terrestrial influences, with such limits specific to the area under consideration. Where possible, boundaries should be governed primarily by environmental criteria, of which geomorphology is likely to be a prime factor (McGlashan, 2000). Administrative practicalities also need to be considered. In terms of seaward extent, a jurisdictional boundary such as the territorial sea may, for instance, make sense.

To be effective, however, ICZM has significant challenges to overcome. Agencies involved can have conflicting interests on the conservation to exploitation spectrum, and whereas ocean governance generally comes under the wing of central government agencies, land-based activities often come under regional control. The number of agencies involved, the complexities of their responsibilities, and the volume of legislation and policies can make it difficult to achieve integration across the sectors (Forrest, 2006; Shipman & Stojanovic, 2007). Ballinger *et al.* (2010), with reference to the European ICZM principles, found the main difficulties were in promoting a holistic approach that considered the land-sea area as an interacting unit, taking account of long-term processes and trends, and the need for adaptive management, or acknowledging uncertainties and adjusting management as information comes to hand.

ICZM is a valuable approach in focusing on the need for unified management to address degradation of coastal environments. It is hard to say, however, to what extent ICZM initiatives are succeeding as few programmes have been evaluated for their effectiveness, in part due to the difficulty of selecting performance criteria. Such measures are also needed to set improved programme goals (Nobre, 2011). In the next chapter we look at the conservation of estuaries, where human impacts and land-sea interactions in the coastal zone are often the most pronounced.

## REFERENCES


Allsopp, M., Walters, A., Santillo, D. & Johnston, P. (2006). *Plastic Debris in the World's Oceans*. Amsterdam: Greenpeace International.

Alongi, D.M. (2005). Ecosystem types and processes. In *The Sea, Volume 13, The Global Coastal Ocean: Multiscale Interdisciplinary Processes*, ed. A.R. Robinson & K.H. Brink, pp. 317–51. Cambridge, MA: Harvard University Press.

Ballinger, R., Pickaver, A., Lymbery, G. & Ferreria, M. (2010). An evaluation of the implementation of the European ICZM principles. *Ocean & Coastal Management*, **53**, 738–49.

Barbier, E.B., Hacker, S.D., Kennedy, C., *et al.* (2011). The value of estuarine and coastal ecosystem services. *Ecological Monographs*, **81**, 169–93.

Bird, E.C.F. (1993). *Submerging Coasts: The Effects of a Rising Sea Level on Coastal Environments*. Chichester, UK: John Wiley & Sons.

Briggs, J.C. (1974). *Marine Zoogeography*. New York, NY: McGraw-Hill.

Brown, A.C. & McLachlan, A. (2002). Sandy shore ecosystems and the threats facing them: some predictions for the year 2025. *Environmental Conservation*, **29**, 62–77.

Brown, C.J., Smith, S.J., Lawton, P. & Anderson, J.T. (2011). Benthic habitat mapping: a review of progress towards improved understanding of the spatial ecology of the seafloor using acoustic techniques. *Estuarine, Coastal and Shelf Science*, **92**, 502–20.

Browne, M.A. & Chapman, M.G. (2011). Ecologically informed engineering reduces loss of intertidal biodiversity on artificial shorelines. *Environmental Science & Technology*, **45**, 8204–7.

Buhl-Mortensen, L., Buhl-Mortensen, P., Dolan, M.J.F. & Gonzalez-Mirelis, G. (2015). Habitat mapping as a tool for conservation and sustainable use of marine resources: some perspectives from the MAREANO Programme, Norway. *Journal of Sea Research*, **100**, 46–61.

Burger, J., Jeitner, C., Clark, K. & Niles, L.J. (2004). The effect of human activities on migrant shorebirds: successful adaptive management. *Environmental Conservation*, **31**, 283–8.

Burrows, M.T., Schoeman, D.S., Buckley, L.B., *et al.* (2011). The pace of shifting climate in marine and terrestrial ecosystems. *Science*, **334**, 652–5.

Castilla, J.C. (1999). Coastal marine communities: trends and perspectives from human-exclusion experiments. *Trends in Ecology and Evolution*, **14**, 280–3.

Chapman, M.G. & Bulleri, F. (2003). Intertidal seawalls – new features of landscape in intertidal environments. *Landscape and Urban Planning*, **62**, 159–72.

Clark, J.R. (1998). *Coastal Seas: the Conservation Challenge*. Oxford, UK: Blackwell.

Clark, R.B. (2001). *Marine Pollution*, 5th edn. Oxford, UK: Clarendon Press.

Commission of the European Communities. (2000). *Communication from the Commission to the Council and the European Parliament on Integrated Coastal Zone Management: A Strategy for Europe*. 27.09.2000, Brussels, COM(2000) 547 final.

Crain, C.M., Halpern, B.S., Beck, M.W. & Kappel, C.V. (2009). Understanding and managing human threats to the coastal marine environment. *Annals of the New York Academy of Sciences*, **1162**, 39–62.

Davenport, J. & Davenport, J.L. (2006). The impact of tourism and personal leisure transport on coastal environments: a review. *Estuarine, Coastal and Shelf Science*, **67**, 280–92.

Davis, J.L.D., Levin, L.A. & Walther, S.M. (2002). Artificial armored shorelines: sites for open-coast species in a southern California bay. *Marine Biology*, **140**, 1249–62.

Dayton, P.K., Currie, V., Gerrodette, T., *et al.* (1984). Patch dynamics and stability of some California kelp communities. *Ecological Monographs*, **54**, 253–89.

Derraik, J.G.B. (2002). The pollution of the marine environment by plastic debris: a review. *Marine Pollution Bulletin*, **44**, 842–52.

Díaz, R.J. & Rosenberg, R. (2008). Spreading dead zones and consequences for marine ecosystems. *Science*, **321**, 926–9.

Dugan, J.E., Hubbard, D.M., Rodil, I.F., Revell, D.L. & Schroeter, S. (2008). Ecological effects of coastal armoring on sandy beaches. *Marine Ecology*, **29** (Suppl. 1), 160–70.

Eisma, D. (1988). An introduction to the geology of continental shelves. In *Continental Shelves*, ed. H. Postma & J.J. Zijlstra, pp. 39–91. Amsterdam: Elsevier.

Farrington, J.W. (2013). Oil pollution in the marine environment I: inputs, big spills, small spills, and dribbles. *Environment: Science and Policy for Sustainable Development*, **55** (6), 3–13.

Federal Geographic Data Committee (2012). *Coastal and Marine Ecological Classification Standard*. Federal Geographic Data Committee. FGDC-STD–018–2012.

Fletcher, H. & Frid, C.L.J. (1996). Impact and management of visitor pressure on rocky intertidal algal communities. *Aquatic Conservation: Marine and Freshwater Ecosystems*, **6**, 287–97.

Forrest, C. (2006). Integrated coastal zone management: a critical overview. *WMU Journal of Maritime Affairs*, **5**, 207–22.

Foster, M.S., De Vogelaere, A.P., Oliver, J.S., Pearse, J.S. & Harrold, C. (1991). Open coast intertidal and shallow subtidal ecosystems of the northeast Pacific. In *Intertidal and Littoral Ecosystems*, ed. A.C. Mathieson & P.H. Nienhuis, pp. 235–72. Amsterdam: Elsevier.

French, P.W. (1997). *Coastal and Estuarine Management*. London: Routledge.

Frid, C.L.J., Harwood, K.G., Hall, S.J. & Hall, J.A. (2000). Long-term changes in the benthic communities on North Sea fishing grounds. *ICES Journal of Marine Science*, **57**, 1303–9.

Galparsoro, I., Connor, D.W., Borja, A., *et al.* (2012). Using EUNIS habitat classification for benthic mapping in European seas: present concerns and future needs. *Marine Pollution Bulletin*, **64**, 2630–8.

Gong, G.-C., Chang, J., Chiang, K.-P., *et al.* (2006). Reduction of primary production and changing of nutrient ratio in the East China Sea: effect of the Three Gorges Dam? *Geophysical Research Letters*, **33**, L07610.

Gosse, E. (1907). *Father and Son*. London: William Heinemann.

Graham, M.H. (2004). Effects of local deforestation on the diversity and structure of Southern California giant kelp forest food webs. *Ecosystems*, **7**, 341–57.

Greening, H. & Janicki, A. (2006). Toward reversal of eutrophic conditions in a subtropical estuary: water quality and seagrass response to nitrogen loading reductions in Tampa Bay, Florida, USA. *Environmental Management*, **38**, 163–78.

Gundlach, E.R. & Hayes, M.O. (1978). Vulnerability of coastal environments to oil spill impacts. *Marine Technology Society Journal*, **12** (4), 18–27.

Halpern, B.S., Walbridge, S., Selkoe, K.A., *et al.* (2008). A global map of human impact on marine ecosystems. *Science*, **319**, 948–52.

Harley, C.D.G., Hughes, A.R., Hultgren, K.M., *et al.* (2006). The impacts of climate change in coastal marine systems. *Ecology Letters*, **9**, 228–41.

Harris, P.T., Macmillan-Lawler, M., Rupp, J. & Baker, E.K. (2014). Geomorphology of the oceans. *Marine Geology*, **352**, 4–24.

Heisler, J., Glibert, P.M., Burkholder, J.M., *et al.* (2008). Eutrophication and harmful algal blooms: a scientific consensus. *Harmful Algae*, **8**, 3–13.

Hinrichsen, D. (1998). *Coastal Waters of the World: Trends, Threats, and Strategies*. Washington, DC: Island Press.

Hiscock, K. (ed.) (1996). *Marine Nature Conservation Review: Rationale and Methods*. Peterborough, UK: Joint Nature Conservation Committee.

Hiscock, K. & Mitchell, R. (1980). The description and classification of sublittoral epibenthic ecosystems. In *The Shore Environment: 2. Ecosystems*, ed. J.H. Price, D.E.G. Irvine, & W.F. Farnham, *Systematics Association Special Volume*, 17(b), 323–70. London: Academic Press.

Huang, Z., Brooke, B.P. & Harris, P.T. (2011). A new approach to mapping marine benthic habitats using physical environmental data. *Continental Shelf Research*, **31**, S4–16.

Huston, M.A. & Wolverton, S. (2009). The global distribution of net primary production: resolving the paradox. *Ecological Monographs*, **79**, 343–77.

Jamieson, G.S. (1993). Marine invertebrate conservation: evaluation of fisheries over-exploitation concerns. *American Zoologist*, **33**, 551–67.

Kaiser, M.J., Hall, S.J. & Thomas, D.N. (2005). Habitat modification. In *The Sea, Volume 13, The Global Coastal Ocean: Multiscale Interdisciplinary Processes*, ed. A.R. Robinson & K.H. Brink, pp. 927–70. Cambridge, MA: Harvard University Press.

Kenny, A.J., Cato, I., Desprez, M., Fader, G., Schüttenhelm, R.T.E. & Side, J. (2003). An overview of seabed-mapping technologies in the context of marine habitat classification. *ICES Journal of Marine Science*, **60**, 411–18.

Ketchum, B.H. (1983). Enclosed seas – introduction. In *Estuaries and Enclosed Seas*, ed. B.H. Ketchum. pp. 209–18. Amsterdam: Elsevier.

Krumhansl, K.A. & Scheibling, R.E. (2012). Production and fate of kelp detritus. *Marine Ecology Progress Series*, **467**, 281–302.

Longhurst, A.R. (2007). *Ecological Geography of the Sea*, 2nd edn. Burlington, MA: Academic Press.

Lotze, H.K., Reise, K., Worm, B., *et al.* (2005). Human transformations of the Wadden Sea ecosystem through time: a synthesis. *Helgoland Marine Research*, **59**, 84–95.

Mann, K.H. (2000). *Ecology of Coastal Waters: With Implications for Management*, 2nd edn. Malden, MA: Blackwell.

McArthur, M.A., Brooke, B.P., Przeslawski, R., *et al.* (2010). On the use of abiotic surrogates to describe marine benthic biodiversity. *Estuarine, Coastal and Shelf Science*, **88**, 21–32.

McGlashan, D.J. (2000). Coastal management in the future. In *Seas at the Millennium: an Environmental Evaluation*, ed. C.R.C. Sheppard, pp. 349–58. Oxford, UK: Elsevier Science Ltd.

Moreno, C.A. (2001). Community patterns generated by human harvesting on Chilean shores: a review. *Aquatic Conservation: Marine and Freshwater Ecosystems*, **11**, 19–30.

Nobre, A.M. (2011). Scientific approaches to address challenges in coastal management. *Marine Ecology Progress Series*, **434**, 279–89.

Petersen, G.H. & Curtis, M.A. (1980). Differences in energy flow through major components of subarctic, temperate and tropical marine shelf ecosystems. *Dana*, **1**, 53–64.

Petersen, J., Michel, J., Zengel, S., White, M., Lord, C. & Plank, C. (2002). *Environmental Sensitivity Index Guidelines*, version 3.0. NOAA Technical Memorandum NOS OR&R 11.

Pontee, N. (2013). Defining coastal squeeze: a discussion. *Ocean & Coastal Management*, **84**, 204–7.

Rabalais, N.N., Díaz, R.J., Levin, L.A., *et al.* (2010). Dynamics and distribution of natural and human-caused hypoxia. *Biogeosciences*, **7**, 585–619.

Rabalais, N.N., Turner, R.E., Díaz, R.J. & Justić, D. (2009). Global change and eutrophication of coastal waters. *ICES Journal of Marine Science*, **66**, 1528–37.

Raffaelli, D. & Hawkins, S. (1996). *Intertidal Ecology*. London: Chapman & Hall.

Richmond, S. & Stevens, T. (2014). Classifying benthic biotopes on sub-tropical continental shelf reefs: how useful are abiotic surrogates? *Estuarine, Coastal and Shelf Science*, **138**, 79–89.

Rijnsdorp, A.D., Buys, A.M., Storbeck, F. & Visser, E.G. (1998). Micro-scale distribution of beam trawl effort in the southern North Sea between 1993 and 1996 in relation to the trawling frequency of the sea bed and the impact on benthic organisms. *ICES Journal of Marine Science*, **55**, 403–19.

Roff, J.C., Taylor, M.E. & Laughren, J. (2003). Geophysical approaches to the classification, delineation and monitoring of marine habitats and their communities. *Aquatic Conservation: Marine and Freshwater Ecosystems*, **13**, 77–90.

Roff, J. & Zacharias, M. (2011). *Marine Conservation Ecology*. London: Earthscan.

Scavia, D., Field, J.C., Boesch, D.F., *et al.* (2002). Climate change impacts on U.S. coastal and marine ecosystems. *Estuaries*, **25**, 149–64.

Shipman, B. & Stojanovic, T. (2007). Facts, fictions, and failures of Integrated Coastal Zone Management in Europe. *Coastal Management*, **35**, 375–98.

Smale, D.A. & Wernberg, T. (2013). Extreme climatic event drives range contraction of a habitat-forming species. *Proceedings of the Royal Society B*, **280**, article 20122829.

Small, C. & Cohen, J.E. (2004). Continental physiography, climate, and the global distribution of human population. *Current Anthropology*, **45**, 269–77.

Spalding, M.D., Fox, H.E., Allen, G.R., *et al.* (2007). Marine ecoregions of the world: a bioregionalization of coastal and shelf areas. *BioScience*, **57**, 573–83.

Spalding, M.D., Ruffo, S., Lacambra, C., *et al.* (2014). The role of ecosystems in coastal protection: adapting to climate change and coastal hazards. *Ocean & Coastal Management*, **90**, 50–7.

Steckbauer, A., Duarte, C.M., Carstensen J., Vaquer-Sunyer, R. & Conley, D.J. (2011). Ecosystem impacts of hypoxia: thresholds of hypoxia and pathways to recovery. *Environmental Research Letters*, **6**, 025003.

Steidinger, K.A. & Vargo, G.A. (1988). Marine dinoflagellate blooms: dynamics and impacts. In *Algae and Human Affairs*, ed. C.A. Lembi & J.R. Waaland, pp. 373–401. Cambridge, UK: Cambridge University Press.

Syvitski, J.P.M. & Kettner, A. (2011). Sediment flux and the Anthropocene. *Philosophical Transactions of the Royal Society A*, **369**, 957–75.

Tegner, M.J. & Dayton, P.K. (2000). Ecosystem effects of fishing in kelp forest communities. *ICES Journal of Marine Science*, **57**, 579–89.

Thompson, R.C., Crowe, T.P. & Hawkins, S.J. (2002). Rocky intertidal communities: past environmental changes, present status and predictions for the next 25 years. *Environmental Conservation*, **29**, 168–91.

Thrush, S.F., Hewitt, J.E., Cummings, V.J., *et al.* (2004). Muddy waters: elevating sediment input to coastal and estuarine habitats. *Frontiers in Ecology and the Environment*, **2**, 299–306.

UNEP (2009). *Marine Litter: A Global Challenge*. Nairobi: United Nations Environment Programme. 232 pp.

UNEP/GPA (2006). *The State of the Marine Environment: Trends and Processes*. The Hague: Coordination Office of the Global Programme of Action for the Protection of the Marine Environment from Land-based Activities of the United Nations Environment Programme.

van Bernem, C., Wesnigk, J.B., Wunderlich, M., Adam, S. & Callies, U. (2008). Oil pollution in marine ecosystems – policy, fate, effects and response. In *Environmental Crises*, ed. H. von Storch, R.S.J. Tol & G. Flösser, pp. 101–39. Berlin: Springer.

Verity, P.G., Smetacek, V. & Smayda, T.J. (2002). Status, trends and the future of the marine pelagic ecosystem. *Environmental Conservation*, **29**, 207–37.

Vincent, A.C.J. (2011). Saving the shallows: focusing marine conservation where people might care. *Aquatic Conservation: Marine and Freshwater Ecosystems*, **21**, 495–9.

Watson, R.A., Nowara, G.B., Hartmann, K., *et al.* (2015). Marine foods sourced from farther as their use of global ocean primary production increases. *Nature Communications*, **6**, article 7365.

Wernberg, T., Russell, B.D., Thomsen, M.S., *et al.* (2011). Seaweed communities in retreat from ocean warming. *Current Biology*, **21**, 1828–32.

Witman, J.D. & Dayton, P.K. (2001). Rocky subtidal communities. In *Marine Community Ecology*, ed. M.D. Bertness, S.D. Gaines & M.E. Hay, pp. 339–66. Sunderland, MA: Sinauer Associates, Inc.

Yang, S.L., Zhang, J., Zhu, J., *et al.* (2005). Impact of dams on Yangtze River sediment supply to the sea and delta intertidal wetland response. *Journal of Geophysical Research*, **110**, F03006.

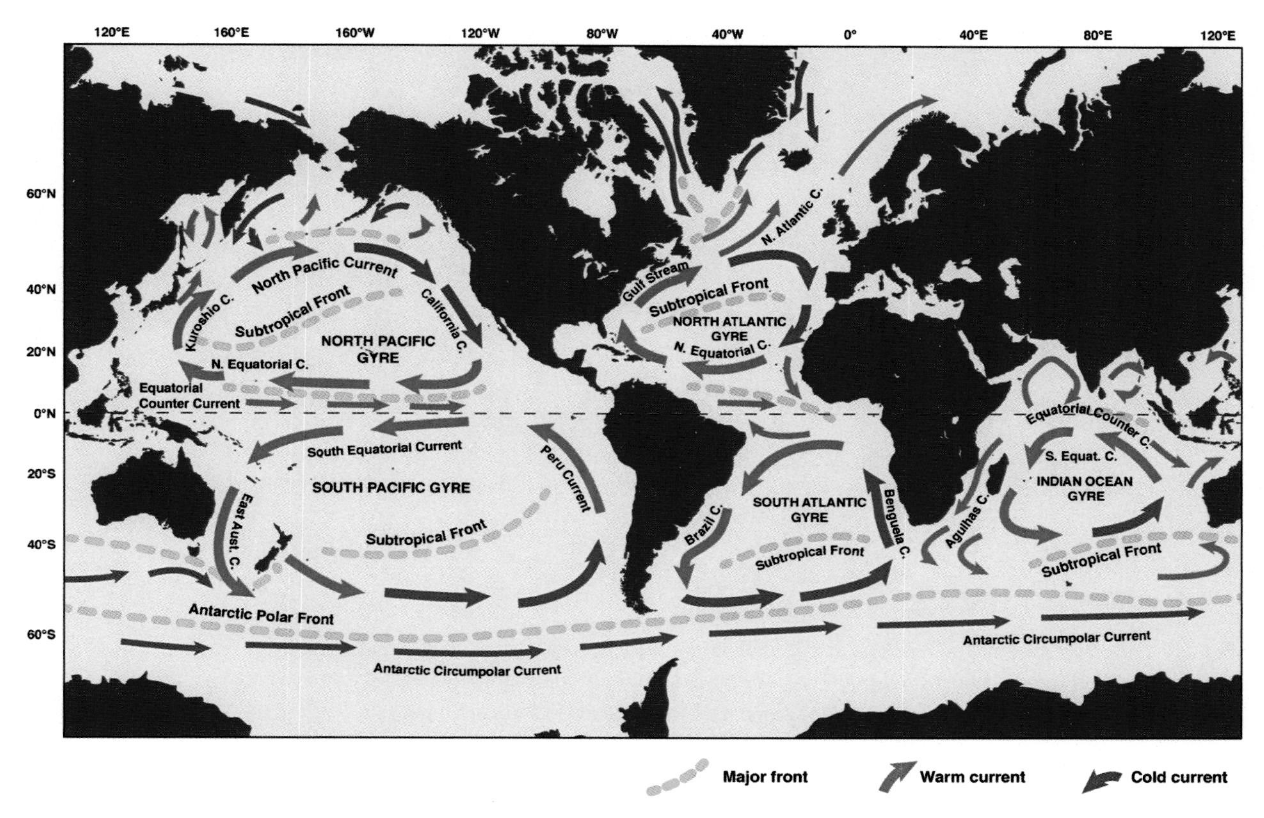


**Fig. 1.2** The major surface currents and fronts of the oceans. The pattern of circulation is dominated by the subtropical gyres, apart from the circumglobal eastward flow in the Southern Ocean.

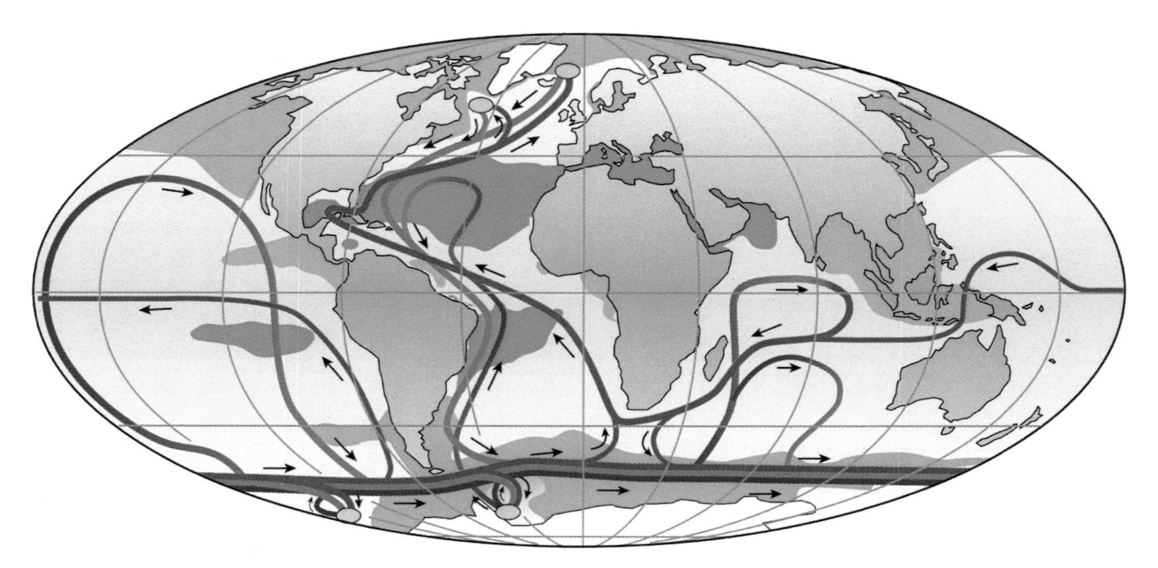

**Fig. 1.3** A simplified diagram of the global thermohaline circulation. Near-surface waters (red lines) flow towards the main regions of deep-water formation (yellow ovals) – in the northern North Atlantic, the Ross Sea, and the Weddell Sea – and recirculate at depth as deep currents (blue lines) and bottom currents (purple lines). Green shading, salinity above 36; blue shading, salinity below 34. Reprinted by permission from Macmillan Publishers Ltd: *Nature*. Rahmstorf, S. (2002). Ocean circulation and climate during the past 120,000 years. *Nature*, 419, 207–14, copyright 2002.

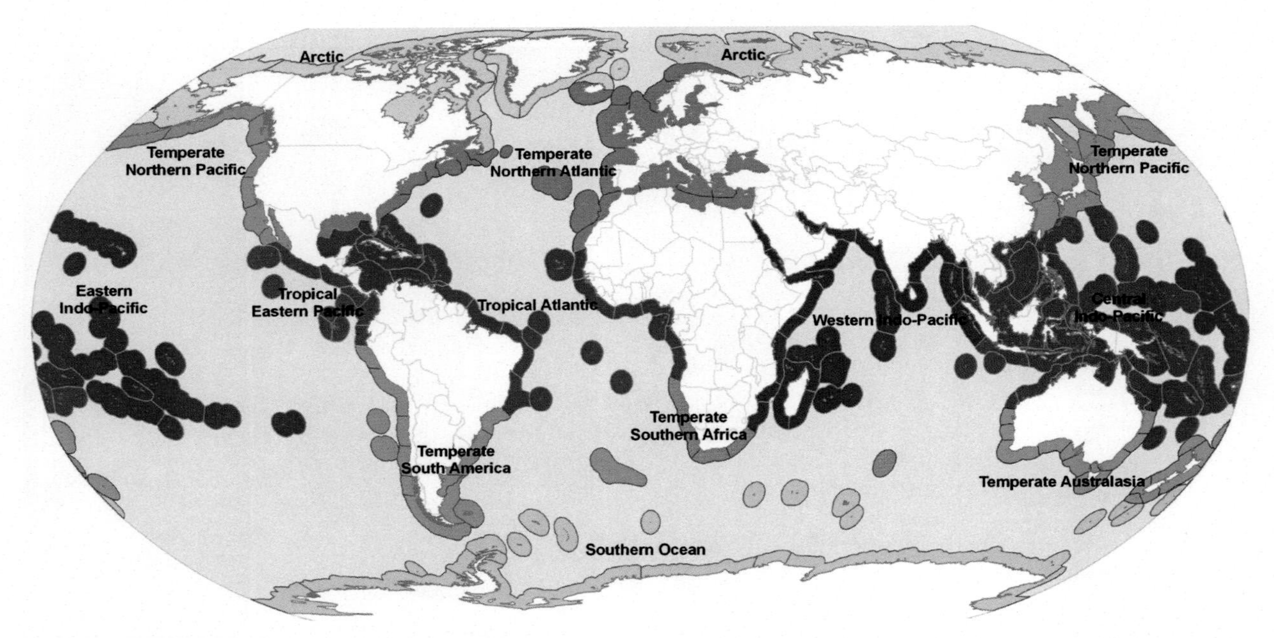


**Fig. 1.4** Biogeographic realms of coastal and shelf areas, with ecoregion boundaries outlined. From Spalding, M.D., Fox, H.E., Allen, G.R., *et al.*, Marine ecoregions of the world: a bioregionalization of coastal and shelf areas, *BioScience*, 2007, Vol. 57 (7), pages 573–83. By permission of American Institute of Biological Sciences.

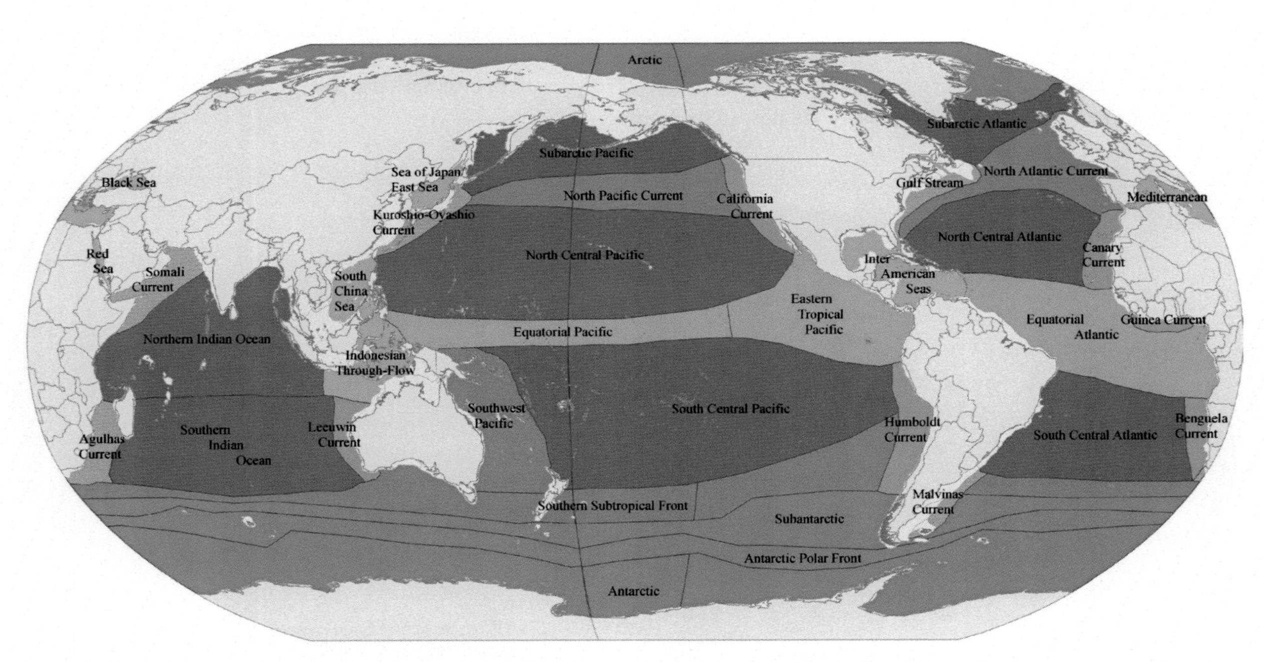


**Fig. 1.5** Biogeographic provinces of surface pelagic waters. The colours represent the different biomes, for example: polar (Arctic), gyre (Subarctic Pacific), transitional (North Pacific Current), eastern boundary current (California Current), western boundary current (Kuroshio-Oyashio Current), equatorial (Equatorial Pacific), and semi-enclosed seas (South China Sea). Reprinted from *Ocean & Coastal Management*, Vol. 60, Spalding, M.D., Agostini, V.N., Rice, J. & Grant, S.M, Pelagic provinces of the world: a biogeographic classification of the world's surface pelagic waters, pages 19–30, copyright 2012, with permission from Elsevier.

**Fig. 2.6** The green shore-crab (*Carcinus maenas*) is native to NW Europe, where it occurs on a variety of open coast to estuarine shores, but has been introduced to many other temperate locations around the world. It is a highly successful invader, tolerant of a wide range of habitats, and an important predator. Photograph: Vital Signs Program at the Gulf of Maine Research Institute.

**Fig. 2.7** Waste outfall from an abattoir. Pollutants enter the marine environment primarily from land-based sources, with organic wastes and nutrients being the main contributors. Anthropogenic nutrient enrichment affects coastal waters in many parts of the world, resulting in degradation of ecosystem structure and function. Photograph: P.K. Probert.

**Fig. 5.2** The concept of fishing down marine food webs. As fisheries (blue arrow) deplete large, high-trophic-level species, they then target smaller, lower-trophic-level species. The accompanying decline in large epibenthic invertebrates represents the impact of bottom trawling. From Pauly, D. (2007). The *Sea Around Us* Project: documenting and communicating global fisheries impacts on marine ecosystems. *Ambio*, 36, 290–5. © Royal Swedish Academy of Sciences 2007.

**Fig. 6.2** The blue shark (*Prionace glauca*), heavily impacted by shark finning and as bycatch. Photograph: M. Francis.

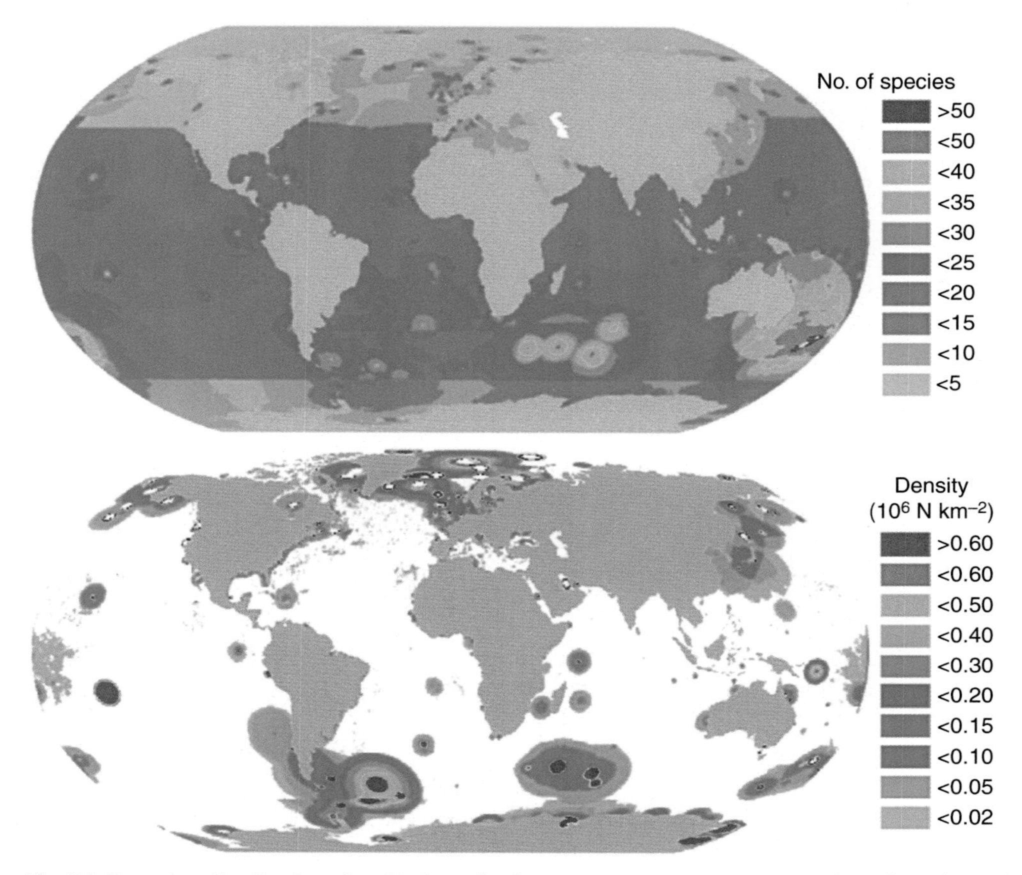


**Fig. 8.1** Foraging distribution of seabird species for an average year, as (top) number of species and (bottom) number (N) of individuals per $km^2$. Reprinted and not modified from Karpouzi, V.S., Watson, R. & Pauly, D. (2007). Modelling and mapping resource overlap between seabirds and fisheries on a global scale: a preliminary assessment. *Marine Ecology Progress Series*, 343, 87–99. With permission from Inter-Research.

**Fig. 8.2** Seabirds are important predators in marine food webs. The SW Pacific Ocean is a hotspot for seabird abundance and species richness, and included here are several albatross species – wandering (*Diomedea exulans*), northern royal (*D. sanfordi*), southern royal (*D. epomophora*), Chatham (*T. eremita*), shy (*T. cauta*), and black-browed (*T. melanophrys*) – as well as northern giant petrel (*Macronectes halli*) and Cape petrel (*Daption capense*). Photograph: P.K. Probert.

**Fig. 9.2** A dugong feeding on seagrass in the Red Sea, Egypt. Dugong inhabit coastal and estuarine waters from east Africa to Vanuatu in the western Pacific. Although hunting is now banned in most of their range, illegal poaching is widespread, and incidental capture in fishing gear is also a major cause of mortality. Dugong feed primarily on seagrasses, so protection of seagrass beds is also important. Photograph: Getty Images/David Peart.

**Fig. 4.1** Hector's dolphin, a small (up to 1.5 m in length) dolphin endemic to New Zealand's coastal waters that is threatened with extinction. Photograph: S. Dawson.

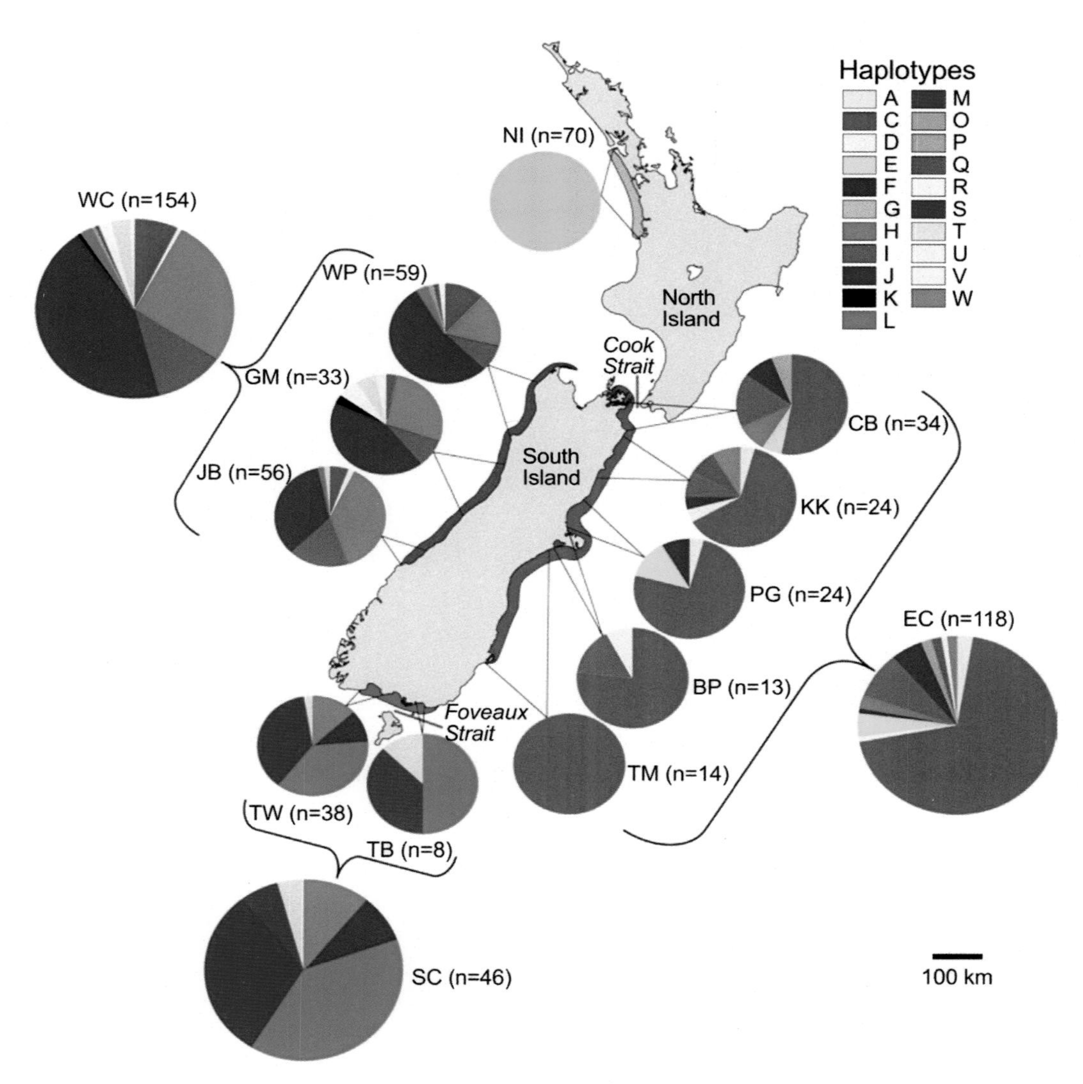


**Fig. 9.10** Genetic differentiation (distribution of haplotypes) of populations of Hector's dolphin (*Cephalorhynchus hectori hectori*) and Maui's dolphin (*C. hectori maui*). Mitochondrial control region haplotypes (360 base pairs) for the Maui's dolphin (NI); regional populations of the Hector's dolphin: East Coast (EC), West Coast (WC), South Coast (SC); and local populations of the Hector's dolphin: Cloudy Bay (CB), Kaikoura (KK), Pegasus Bay (PG), Banks Peninsula (BP), Timaru (TM), Westport (WP), Greymouth (GM), Jackson Bay (JB), Te Waewae Bay (TW), Toetoe Bay (TB). From *Conservation Genetics*, Genetic differentiation and limited gene flow among fragmented populations of New Zealand endemic Hector's and Maui's dolphins, Vol. 13, 2012, pages 987–1002, Hamner, R.M., Pichler, F.B., Heimeier, D., Constantine, R. & Baker, C.S., © Springer Science+Business Media B.V. 2012. With permission of Springer.

**Fig. 10.1** A kelp forest dominated by *Macrocystis pyrifera* and an aggregation of the sea urchin *Evechinus chloroticus* in Paterson Inlet, Stewart Island, New Zealand. The interaction between grazing sea urchins and kelp can result in the community structure shifting between two states: kelp forest and urchin barrens. Photograph: S. Wing.

**Fig. 4.5** A European maerl bed. The unattached coralline algae provide a complex habitat supporting a diverse community. Photograph: Getty Images/Mark Webster.

**Fig. 10.2** Polychaete worms are usually the dominant macrofauna of soft-sediment communities in terms of number of species and individuals. This polychaete (a species of *Perinereis*) is a member of the family Nereididae, commonly known as ragworms. Some ragworm species are dug for bait. Photograph: © Rod Morris/www.rodmorris.co.nz.

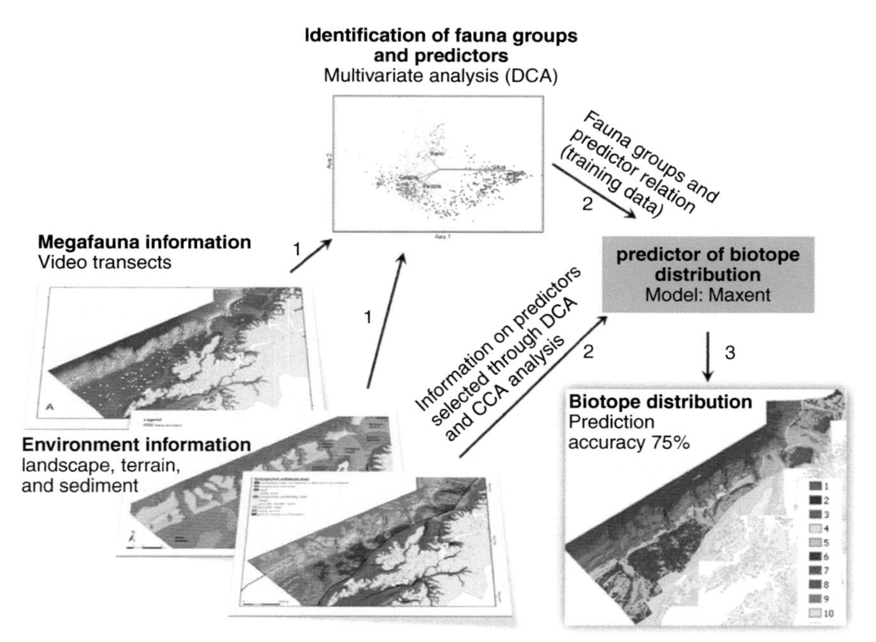


**Fig. 10.3** General steps used to identify and predict the distribution of benthic habitats/biotopes: (1) collecting information on the fauna and environmental factors to identify faunal groups and related environmental descriptors, (2) using the fauna-environmental factor relationship for construction of a model, and (3) producing predictive biotope maps based on environmental information and the model. In this example, for the shelf and upper slope off northern Norway, biotopes are based on the composition of benthic megafauna sampled by video transects. This component of the benthos includes large habitat-forming organisms that are especially vulnerable to physical disturbance. Detrended correspondence analysis (DCA) and canonical correspondence analysis (CCA) are the multivariate techniques used here to identify groups of samples and the relationships between biota and environmental gradients. Maxent is a modelling tool used here to predict biotope distribution. Reprinted from *Journal of Sea Research,* Vol. 100, Buhl-Mortensen, L., Buhl-Mortensen, P., Dolan, M.J.F. & Gonzalez-Mirelis, G., Habitat mapping as a tool for conservation and sustainable use of marine resources: some perspectives from the MAREANO Programme, Norway, pages 46–61, copyright 2014, with permission from Elsevier.)

**Fig. 11.3** Estuarine and other sheltered sediment shores provide essential feeding grounds for migratory waders. Bar-tailed godwits (*Limosa lapponica*) have breeding grounds in northern Scandinavia to Alaska and migrate to southern wintering grounds. Some migrate to Australasia, undertaking the longest non-stop flight of any bird. Photograph: Raewyn Adams.

**Fig. 11.1** A drowned river valley, or ria, is a common form of estuary. The close links between marine and terrestrial environments and competing uses in estuaries underline the need for integrated management and legislation. Note the extensive mudflats and seagrass (*Zostera* spp.) beds exposed here at low water – Kingsbridge estuary, South Devon, UK. Photograph: Getty Images/Dan Burton/robertharding.

**Fig 12.1** A grazed saltmarsh dominated by the alkali grass *Puccinellia maritima* (Northton, Outer Hebrides, Scotland). Saltmarshes are productive systems, important as carbon sinks and well developed in temperate latitudes. Large areas of saltmarsh have been lost to agricultural, urban, and industrial use. Photograph: P.K. Probert.

**Fig. 12.3** Mangroves provide important ecosystem services in tropical to subtropical latitudes, including coastal protection and migratory and nursery habitat. Note the aerial prop roots and pneumatophores. *Rhizophora* in the Solomon Islands. Photograph: Getty Images/Sune Wendelboe.

**Fig. 13.3** Stony corals (Scleractinia) provide the main framework of coral reefs – structurally complex habitats that support highly diverse communities. A reef in Fiji. Photograph: Kim Westerskov.

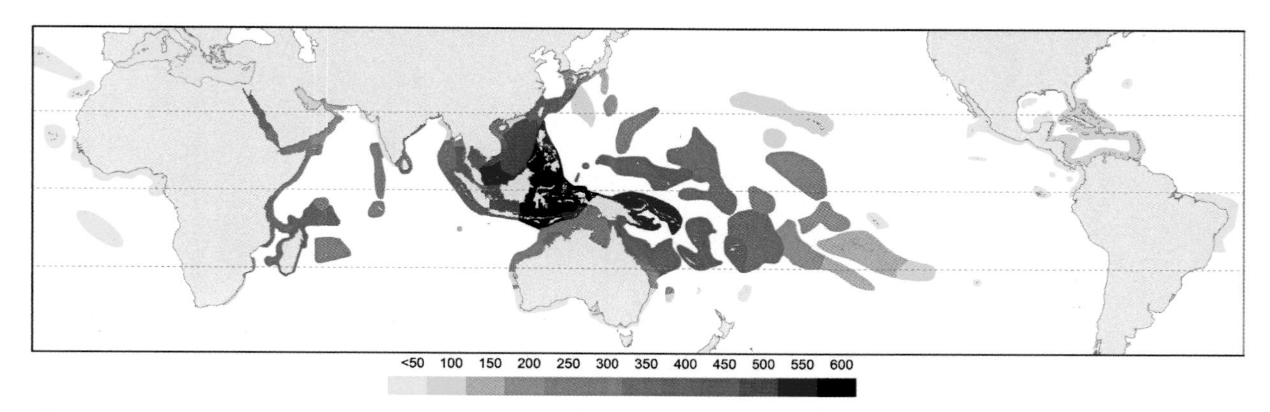


**Fig. 13.4** Contours of species richness for warm-water stony corals. The dashed lines are the equator and the tropics of Cancer and Capricorn (23° 26' N and S). From Veron, J., Stafford-Smith, M., DeVantier, L. & Turak, E. (2015). Overview of distribution patterns of zooxanthellate Scleractinia. Frontiers in Marine Science, 1, article 81. Copyright © 2015 Veron, Stafford-Smith, DeVantier and Turak.

**Fig. 13.5** A bleached *Acropora* sp. in the Wakatobi Marine National Park in south-east Sulawesi, Indonesia. Photograph: J.J. Bell.

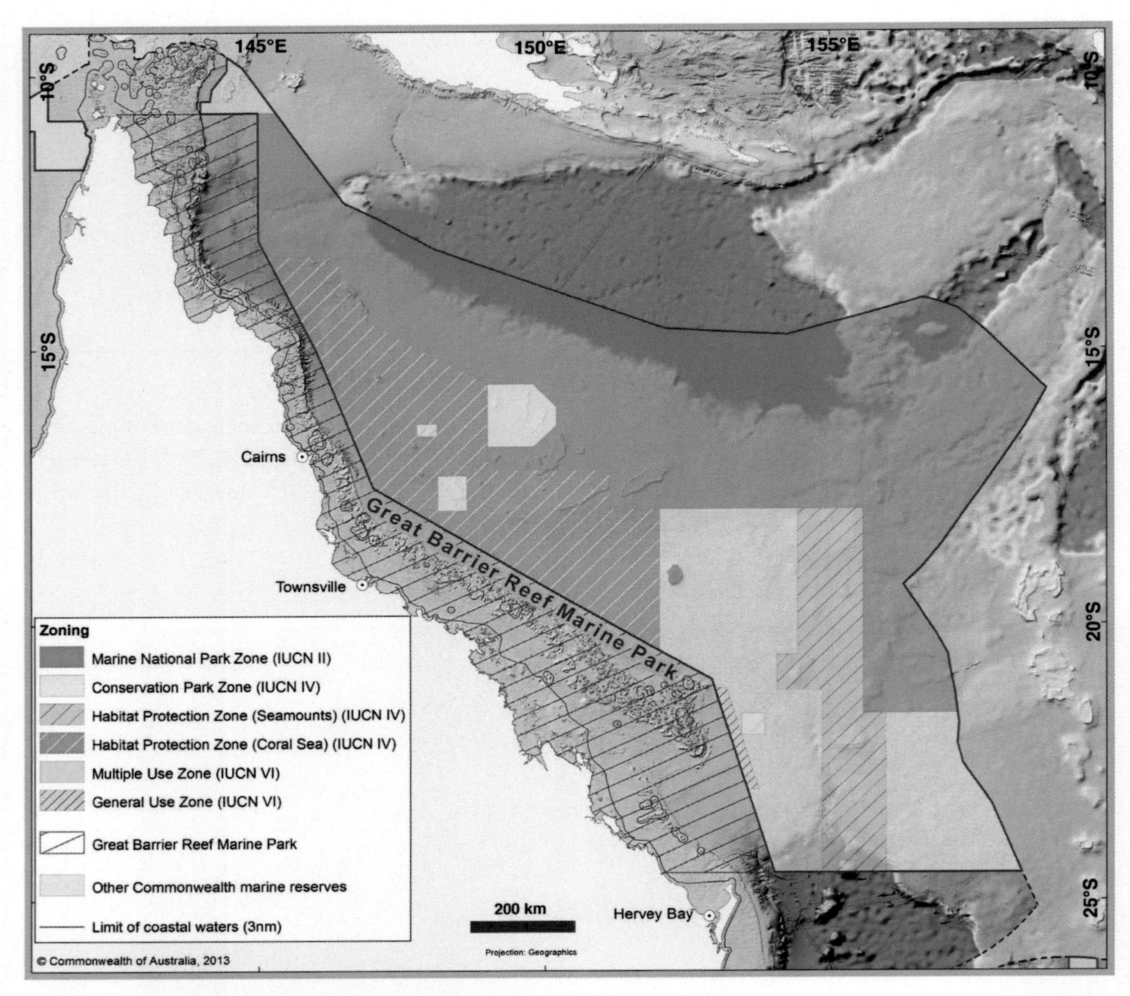


**Fig. 13.9** Boundaries of the Great Barrier Reef Marine Park and Coral Sea Commonwealth Marine Reserve. (The internal zones of the reserve, shown here as of March 2014, are being reviewed and may change.) © Commonwealth of Australia 2016.

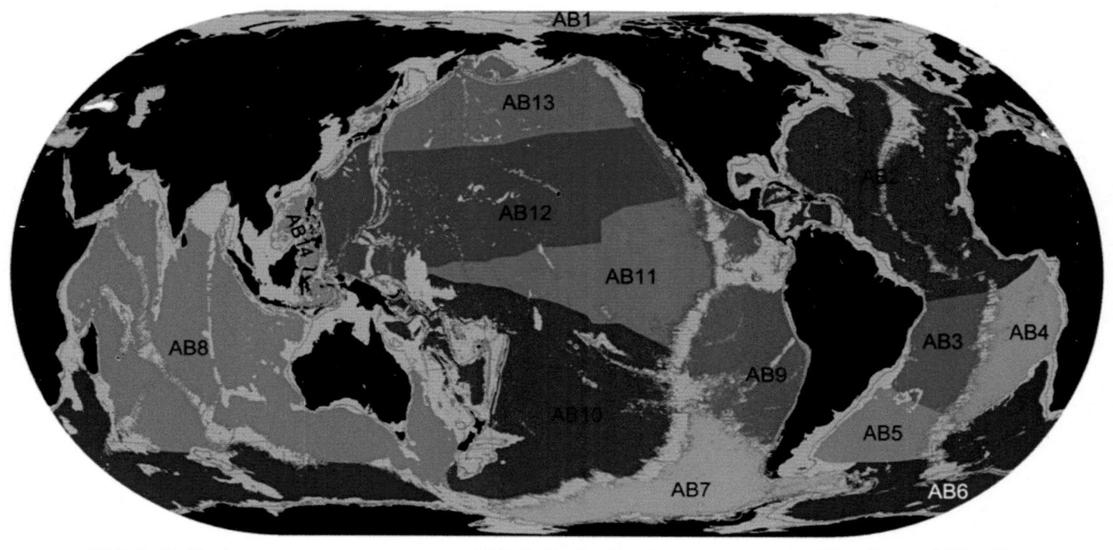


AB1: Arctic Basin
AB2:North Atlantic
AB3: Brazil Basin
AB4: Angola, Guinea, Sierra Leone Basins
AB5: Argentine Basin
AB6: Antarctica East
AB7: Antarctica West
AB8: Indian
AB9: Chile, Peru, Guatemala Basins
AB10: South Pacific
AB11: Equatorial Pacific
AB12: North Central Pacific
AB13: North Pacific
AB14: West Pacific Basins

**Fig. 14.1** Biogeographic provinces of the abyssal zone (3500–6500 m). Reprinted from *Progress in Oceanography*, Vol. 111, Watling, L., Guinotte, J., Clark, M.R. & Smith, C.R., A proposed biogeography of the deep ocean floor, pages 91–112, copyright 2012, with permission from Elsevier.

**Fig. 14.3** Reef-like matrix of the branching stony coral *Solenosmilia variabilis* together with abundant crinoids, a venus flower basket sponge (*Symplectella* sp.), white stylasterid hydrocorals, and two orange roughy (*Hoplostethus atlanticus*), on a seamount at about 1000 m water depth, Chatham Rise, New Zealand. Image provided by the National Institute of Water and Atmospheric Research (NIWA), New Zealand, and captured by NIWA's Deep Towed Imaging System (DTIS). Photograph: NIWA.

**Fig. 17.3** Antarctic krill (*Euphausia superba*), a key species in the Southern Ocean food web. Photograph: Getty Images/Gerald and Buff Corsi/Visuals Unlimited, Inc.

**Fig. 17.4** Adélie penguins are considered vulnerable to effects of climate change, including loss of sea ice and a decline in krill abundance. Parent birds at Cape Bird, Ross Island, returning to sea to forage for their chicks. Photograph: Kim Westerskov.

# 11 Estuaries and Lagoons

Among marine environments most vulnerable to human disturbance are enclosed bodies of water that have limited or no exchange with the open sea. The salinity of such waters is often significantly lower or higher than that of the open sea because of enclosure and the strong influence that the prevailing climate can exert. Various schemes have been advanced to categorize these environments, particularly in terms of physical geography and the salinity and its variability. A common convention is to assign a salinity of $< 0.5$ for freshwater and 30 for the upper level of diluted seawater. 'Brackish' and 'estuarine' are terms applied to waters in this salinity range. Brackish is most usefully applied to water with a relatively stable salinity that is lower than that of adjacent ocean waters, whereas estuarine implies a marked variability of salinity, usually as a result of tidal action. Fully saline (euhaline) water is usually accorded a salinity range of 30–40, and waters with salinities of $> 40$ are termed 'hypersaline'. Estuarine and brackish water environments are better represented in the Northern Hemisphere, which has the greater share of the world's land mass (see Chapter 1), and are a feature particularly of broad, gently sloping continental margins. The world's large low-salinity seas lie in the Northern Hemisphere: Hudson Bay (1.23 million $km^2$), Black Sea (436 000 $km^2$), Baltic Sea (377 000 $km^2$), and Caspian Sea (371 000 $km^2$) (www.worldatlas.com). Although this chapter largely concerns estuaries, many of the environmental issues raised apply also to non-estuarine coastal inlets. We also look briefly at lagoons, anchialine pools, and high-salinity habitats and their particular conservation issues.

## Estuaries

Estuarine systems are a distinctive feature of many coastlines, totalling of the order of 1.2–1.8 million $km^2$ (depending on which large low-salinity sea areas are included), or 0.3–0.5% of the global marine area (Woodwell *et al.*, 1973). In contrast to their small global extent, estuaries have been assessed as among the most valuable ecosystem in terms of services they provide, particularly for nutrient cycling (Costanza *et al.*, 1997).

Various definitions of an estuary have been proposed. But typically, an estuary is a partly enclosed tidal inlet of the sea that has a significant freshwater input and marked cyclical fluctuations in salinity, which usually means that an estuary is a river mouth and its immediate upstream reaches (Lincoln & Boxshall, 1987; Little, 2000). Different types of estuary can be recognized on the basis of their geomorphology, salinity stratification, and circulation, although these characteristics are often strongly related. Most common are drowned river valleys, or coastal plain estuaries, formed by the flooding of river valleys at the end of the last glaciation (e.g. Thames Estuary, Chesapeake Bay) (Fig. 11.1). Fjords too are flooded valleys, but as they have been gouged out by glaciers they are restricted to cool temperate and higher latitudes; fjords can have deep basins, in some cases exceeding 1000 m water depth, in contrast to more typical shallow-water estuaries. Bar-built estuaries,

**Fig. 11.1** A drowned river valley, or ria, is a common form of estuary. The close links between marine and terrestrial environments and competing uses in estuaries underline the need for integrated management and legislation. Note the extensive mudflats and seagrass (*Zostera* spp.) beds exposed here at low water – Kingsbridge estuary, South Devon, UK. Photograph: Getty Images/Dan Burton/ robertharding. (A black and white version of this figure will appear in some formats. For the colour version, please refer to the plate section.)

more common in the tropics, occur where sand or shingle, deposited parallel to the coast, forms spits and barrier islands to enclose a usually shallow basin that may receive a number of rivers. Examples are Albermarle Sound and Pamlico Sound (North Carolina), Wadden Sea (southern North Sea), and Moreton Bay (Queensland). Deltaic estuaries occur where large rivers fan out in a network of channels across extensive deltas, such as those off the Ganges-Brahmaputra, Nile, and Mississippi. Occasionally, estuaries are produced by tectonic processes, notably as a result of land subsidence along fault lines, San Francisco Bay being a classic example.

The volume of freshwater entering an estuary usually exceeds that lost to surface evaporation, so that less dense low-salinity water leaves the estuary in a seaward surface flow, whilst denser water from the sea enters the estuary at depth. This pattern is typical of temperate regions. Depending upon such factors as tidal amplitude, river flow, and estuarine morphology, the degree of mixing varies from estuaries that show strong stratification, with a pronounced low-salinity surface flow, to those that are only weakly stratified or well mixed. Bar-built estuaries are usually shallow, where wind-induced mixing tends to prevail. In some parts of the world, notably the tropics, evaporation can be greater than the freshwater inflow. In this case, denser higher-salinity water produced at the surface sinks to form an outgoing bottom current, whereas water from the sea enters the estuary as a surface flow.

Estuaries typically exhibit a suite of characteristic environmental features. Given the freshwater inflow and tidal mixing, there are generally large fluctuations of salinity. Indeed, many of the world's estuaries are in regions subject to heavy rainfall from cyclones and monsoons and can experience severe freshwater flooding. Temperature variability also tends to be more marked in estuaries than in the adjacent sea because estuaries are often shallow and the inflowing river is more readily influenced by the air temperature than is the sea. Importantly too, since they are partially enclosed, estuaries are sheltered environments subject to only limited wave action. This shelter, together with the supply of fine-grained sediment from the adjacent sea and the catchment, means that estuaries are normally regions where fine sediments are deposited. Estuarine sediments are characteristically muddy, and at low tide extensive mudflats may be exposed in the middle reaches. Such sediments are often rich in organic detritus, particularly if the estuary supports saltmarsh, mangrove, or seagrass beds. High turbidity is a common feature of estuarine waters as a result of this input of fine material and the resuspension of deposited sediment. Dissolved organic compounds introduced by river runoff can also limit the light penetration. Estuarine waters tend to be rich in nutrients derived from the sea as well as from riverine and catchment sources. Dissolved nutrients and particulate organic matter contribute to the high productivity of estuaries.

## Biota of Estuaries

From an ecological viewpoint it is convenient to distinguish a sequence of estuarine zones based on salinity. In a typical estuary, these zones range from the estuary head, where riverine influences prevail (salinities $< 5$), to upper (5–18) and middle reaches (18–25) characterized by wide ranges of salinity and the deposition of muddy sediments to the lower reaches and estuary mouth subject to a narrower range of salinity (25–30) and where stronger tidal currents favour substrata of sandier sediments or rock.

Estuaries typically exhibit strong gradients and variability in physico-chemical factors, which can make for a challenging environment for estuarine organisms and strongly influence their distribution. Salinity is the most obvious of these factors. Species of the adjacent open sea that are tolerant of a narrow salinity range (i.e. stenohaline) are prevented from advancing beyond the estuary's lower reaches by salinities of less than about 25. On the other hand, marine species tolerant of wide changes of salinity (i.e. euryhaline) are typically able to tolerate salinities down to 18 or, in the case of some more opportunistic species, 5. Such species may have to contend with dramatic salinity changes over a tidal cycle. In turn, these species give way to estuarine endemics at salinities of around 2–15. Making their appearance at the head of the estuary are the few freshwater species able to tolerate salinities greater than 5. In a summary of the distribution of benthic macrofauna recorded along four estuaries in southern Africa, Day (1981) reported ranges of 84 to 310 for the total numbers of species, of which the majority were categorized as stenohaline or euryhaline marine and less than 10% as true estuarine species (Table 11.1).

Other environmental factors may be as important as or more important than salinity as potential stressors for estuarine organisms (Kaiser *et al.*, 2011). The deposition of fine-grained sediments in the middle and upper reaches of estuaries and conditions of high turbidity pose problems,

**Table 11.1** The Relative Abundance of Benthic Macrofaunal Species of Differing Salinity Tolerance Recorded in Four Estuaries in Southern Africa

| Stenohaline marine | Euryhaline marine | True estuarine | Freshwater |
|---|---|---|---|
| 11–45% | 44–80% | 6–9.5% | 0.5–5% |

Derived from Day, J.H. (1981). The estuarine fauna. In *Estuarine Ecology with Particular Reference to Southern Africa*, ed. J.H. Day, pp. 147–78. Rotterdam: A.A. Balkema.

especially for animals with feeding and respiratory structures that could become clogged. Estuarine waters and sediments are rich in organic matter, the bacterial degradation of which can lead to oxygen depletion. This is most evident in estuarine mudflats where anaerobic conditions prevail just beneath the sediment surface. Seasonal changes in temperature can be very marked in estuaries at higher latitudes, with potentially important implications for the activity of estuarine organisms, including levels of bacterial degradation.

Estuarine biotas tend to be of relatively low species diversity compared to those of fully marine environments. This low diversity is due to a combination of factors. There are the considerable physiological challenges for estuarine organisms to overcome, such as the problem of osmotic stress faced by freshwater and marine species invading variable estuarine salinities. In addition, estuaries are young environments in geological terms, more so in cold temperate regions where colonization has occurred only since the last glaciation. Also, the predominant benthic habitat of estuaries is mudflat, often a rather uniform substratum offering limited microhabitat diversity. Mudflats can be very extensive in estuaries with a large tidal range. Estuarine biotas are noted for their often-high biomass and productivity fuelled by plentiful organic and nutrient inputs. However, they vary greatly in this respect, not only in their overall productivity but also in the relative contributions of the different components in energy flow.

## Primary Producers

Despite nutrient-rich conditions, phytoplankton production in estuaries is often curtailed by turbidity and flushing time, so that net production may be less than for shelf waters, with maximum values of up to 100–200 g C $m^{-2}$ $y^{-1}$ (McLusky & Elliott, 2004). Often making an equal or greater contribution to estuarine primary production are benthic diatoms and other microalgae on the surface of mudflats. Macroalgal production is typically far less important, although fucoid algae can for instance be abundant in northern temperate areas. However, these sources of production can be considerably less than those contributed by flowering plants, notably by seagrass, saltmarsh, and mangrove communities (see Chapter 12). With their contribution, the total net primary production of an estuary can reach 1600 g C $m^{-2}$ $y^{-1}$. An important feature of the energy flow in estuaries is that the great majority of the production by macrophytes is not consumed directly but enters the food web as organic detritus. Detrital particles and associated microbes are used by zooplankton, some fishes, and, in particular, benthic invertebrates.

## Benthos

In general, the sediment macrofauna of estuaries mainly comprises polychaetes, molluscs (especially bivalves), and crustaceans (chiefly amphipods and isopods). Dense populations of suspension-feeding bivalves, such as mussels and oysters, can significantly influence water quality of estuaries by reducing phytoplankton and suspended solids as well providing other ecosystem services (Coen *et al.*, 2007; Beck *et al.*, 2011). Deposit feeders, however, typically dominate the estuarine benthos. They exploit the detritus and microalgae in the sediment, and the burrowers play an important role in reworking the sediment. Well-known representatives include arenicolid polychaetes (lugworms), tellinid bivalves, corophiid amphipods, and hydrobiid snails. Benthic productivity of estuaries is in general high compared with that of other ecosystems, typically of the order of 10–20 g C $m^{-2}$ $y^{-1}$ (Wilson & Fleeger, 2013). This level of productivity is reflected by the large populations of vertebrate consumers that estuaries conspicuously support, particularly their fish and bird populations.

## Fish

Important fish families in boreal estuaries include salmon and trout (Salmonidae) and smelt and capelin (Osmeridae), whilst in cold-water estuaries of the Southern Hemisphere, galaxiids are a significant component of the endemic fauna. Members of the herring family (Clupeidae) and pleuronectid flounders tend to dominate more temperate regions. Many families are important in the tropics and subtropics, such as anchovies (Engraulidae), lizardfish (Synodontidae), mullets (Mugilidae), croakers (Sciaenidae), gobies (Gobidae), and soleid flounders (Haedrich, 1983).

Fish use estuaries in various ways, notably as feeding, nursery, and refuge areas and for migrating between marine and freshwater habitats. Relatively few species live in estuaries throughout the year and can be regarded as strictly estuarine. But notable among these are the mummichog (*Fundulus heteroclitus heteroclitus*), various gobies, pipefishes, centropomids, antherinids, and engraulids. Many essentially marine or freshwater fishes make only occasional use of estuaries, but others need estuarine habitat for at least part of their life history. The extent to which fish depend on estuaries varies considerably, not only among species but also at geographical, annual, and cohort-specific scales (Fig. 11.2).

Many fish that use estuaries spawn at sea but reside in estuaries for part of their life, a pattern most evident in temperate regions. Examples are mullets such as *Mugil cephalus*, menhaden (*Brevoortia* spp.), anchovies, croakers, milkfish (Chanidae), and flounders. Many species enter estuaries as post-larvae or early juveniles and move first into low salinity areas where there is greater protection from predators. As they grow they gradually move downstream into more saline reaches, often making use of seagrass or mangrove areas, until finally as adults they leave the estuary to spawn. Estuaries can thus provide a nursery habitat where fish can benefit from high estuarine productivity and grow rapidly. A species may use a variety of estuarine habitats, such as seagrass beds, mangroves, marshes, creeks, and open water areas, during its life history.

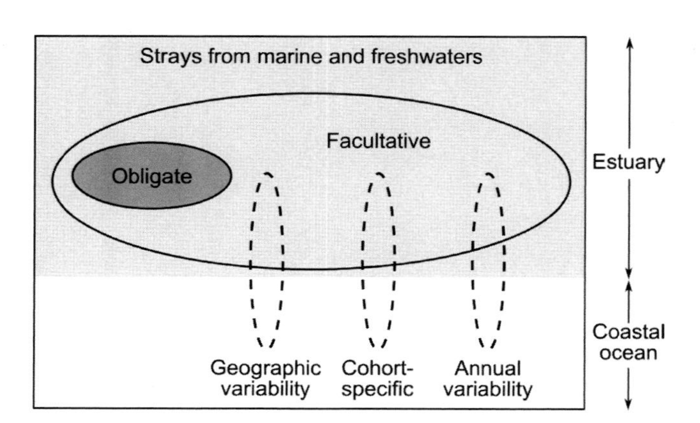


**Fig. 11.2** Estuarine dependence of fishes, which can range from species that have obligate life-history stages, to those that are facultative users or occasional strays. Facultative users may differ in terms of geographic, cohort-specific, or annual variability. Reprinted from *Estuarine, Coastal and Shelf Science*, Vol. 64, Able, K.W., A re-examination of fish estuarine dependence: evidence for connectivity between estuarine and ocean habitats, pages 5–17, copyright 2005, with permission from Elsevier.

Other fish species use estuaries during spawning migrations. Prominent among anadromous fish (those that move from saltwater to freshwater to spawn) are sturgeons and salmonids (see Chapter 6), clupeids, smelts, basses, and lampreys. For adults of anadromous species, the estuary provides them with the opportunity to acclimate before they ascend to freshwater. And for the descending juveniles, the estuary is commonly an important feeding ground. Anadromy is rare at tropical latitudes but becomes increasingly common towards boreal regions. By contrast, at temperate latitudes of the Southern Hemisphere, endemic anadromous fish are almost non-existent, and catadromous species (those that move from fresh- to saltwater to spawn) are better represented. On a worldwide basis, however, catadromy is uncommon. The best-known examples are the freshwater eels (*Anguilla* spp.), which spawn in certain deep-sea areas, the juveniles, or elvers, eventually entering estuaries on their way to freshwater habitats.

## Birds

Birds are often the most conspicuous consumers in estuaries. The composition, seasonal abundance, and trophic role of estuarine bird faunas show latitudinal differences. Temperate estuaries tend to be characterized by migrant populations of waders or shorebirds (Charadrii) and wildfowl (Anatidae). Wildfowl includes ducks and geese, some species of which feed on estuarine macrophytes, whereas others consume both plants and benthic invertebrates. Waders mainly take benthic invertebrates. Tropical estuaries are more obviously characterized by resident populations of herons, egrets, storks, ibises, spoonbills, and flamingos (Ciconiiformes) and pelicans and cormorants (Pelecaniformes). Many of these birds are at least tertiary consumers preying on crabs and other large crustaceans, large gastropods, large insects, and fish (Siegfried, 1981).

### Waders

There are some 200 species of waders, although not all of them make use of coastal habitat. The great majority are sandpipers (Scolopacidae), a family of highly migratory birds that breed mainly in northerly latitudes, and plovers (Charadriidae), which are more temperate and tropical in distribution, tend not to make such long migrations between breeding and feeding grounds, and are less strongly associated with the shore than sandpipers (Hale, 1980; Colwell, 2010). Other well-known wader families are the oystercatchers (Haematopodidae) and the avocets and stilts (Recurvirostridae) (Table 11.2). Most species of waders disperse widely for the summer breeding season, occupying inland tundras and grasslands to feed on insects and earthworms. This dispersion contrasts with their far more restricted distribution outside the breeding season – during migration and on their wintering grounds – when large numbers of birds aggregate in smaller areas

**Table 11.2** The Main Families of Waders That Use Coastal Habitat

| Family | Common name | No. of species |
|---|---|---|
| Recurvirostridae | Avocets, stilts | 11 |
| Haematopodidae | Oystercatchers | 11 |
| Charadriidae | Plovers, dotterels | ~65 |
| Scolopacidae | Sandpipers, curlews, godwits, knots, turnstones | ~88 |

From various sources.

**Fig. 11.3** Estuarine and other sheltered sediment shores provide essential feeding grounds for migratory waders. Bar-tailed godwits (*Limosa lapponica*) have breeding grounds in northern Scandinavia to Alaska and migrate to southern wintering grounds. Some migrate to Australasia, undertaking the longest non-stop flight of any bird. Photograph: Raewyn Adams. (A black and white version of this figure will appear in some formats. For the colour version, please refer to the plate section.)

(Myers *et al.*, 1987). Most species migrate to estuaries and other coastal habitats and exploit intertidal areas as feeding grounds (Fig. 11.3). Here their food consists primarily of benthic invertebrates, especially molluscs, small crustaceans, and polychaetes. Sandpipers in particular have evolved a wide diversity of bill forms and feeding strategies, including deep probing of the sediment for prey. Plovers have short bills and feed mainly by sight on prey at or near the sediment surface. Avocets and stilts have long slender bills and feed chiefly on small invertebrates, whereas oystercatchers have powerful chisel-like bills that enable them to open cockles and other bivalves.

Estuaries provide feeding grounds for overwintering waders, but they are also used by many wader species as staging areas where they can accumulate fat reserves in autumn and spring on their migrations between breeding and wintering grounds. At any one estuary there may be large year-to-year variation in the prey available and in the use of intertidal areas by migratory birds. A considerable proportion of benthic invertebrate production can be taken by estuarine birds, typically around 20–25% (Baird *et al.*, 1985; McLusky & Elliott, 2004). But whether an estuarine shore can provide enough food for wintering waders depends on a range of factors, including the extent of available mudflat, the production and availability of acceptable prey, the size and period of residency of the wader populations, and behavioural interactions between birds that may limit their foraging (Piersma, 1987; Zharikov & Skilleter, 2003). Availability and predictability of food supply may be critical in cold temperate estuaries due to severe winter conditions. Some invertebrates burrow more deeply in winter, beyond the reach of waders. Knot (*Calidris canutus*), for instance, feed mainly on the bivalve *Macoma balthica* during winter, but studies on intertidal flats in eastern England show that for a mean bill length of 33 mm, only 4% of the *Macoma* biomass is accessible to knot in winter compared with 90% in mid-summer (Reading & McGrorty, 1978). Even if prey invertebrates are within bill length, their activity may be so reduced at low temperatures that they provide insufficient visual clues and remain undetected by waders. During such mid-winter conditions, wader density may approach the carrying capacity of feeding grounds.

Waders are often only seasonally abundant in estuaries. Gulls, on the other hand, can be important year-round predators on estuarine flats, feeding on ragworms (nereidid polychaetes) and a variety of other invertebrates (Ambrose, 1986; Moreira, 1995).

Various aquatic reptiles and mammals of high conservation interest also frequent estuaries. Several species of small cetaceans occur in estuarine waters, phocid seals and otters can be important in estuaries and brackish seas of northern temperate and arctic regions, whilst sirenians, estuarine terrapins, and crocodilians occur in many tropical estuaries.

## Human Impacts

Early humans would have been drawn to estuaries as sheltered coastal sites that provided plentiful supplies of fish, shellfish, and wildfowl. In recent centuries, with the development of coastal shipping and fisheries, many early settlements have grown to become busy ports, an expansion that gained further momentum with the Industrial Revolution and resulted in heavy industries located on estuaries. Further industrialization has seen facilities such as oil refineries, chemical and petrochemical complexes, power stations, and container ports often sited on estuaries. As a result, many of the world's major cities have grown up around estuaries. Examples abound of estuaries now heavily polluted by urban and industrial wastes and where there has been major physical alteration of the shore and estuary bed. We need first to examine the main anthropogenic pressures on estuarine ecosystems in order to consider appropriate management and conservation strategies.

### Pollution

As centres of population and by virtue of their environmental characteristics, estuaries are often among the most polluted of coastal environments. The main culprits are usually organic wastes, oil, heavy metals, and persistent organic pollutants. Physico-chemical features of estuaries strongly influence the behaviour and fate of pollutants. Often there is a long residence time of water in an estuary and an extensive mixing zone that can extend downstream and upstream from the point of discharge as the tide ebbs and flows. Also, high rates of sedimentation in estuaries and the adsorption of contaminants onto particles can lead to raised pollutant concentrations in muddy estuarine sediments. Pollution, therefore, is often a key issue in the management and conservation of estuaries. We looked earlier (Chapters 2 and 10) at major types of pollutants and their impacts on coastal waters. Here we briefly examine pollution problems that particularly affect estuaries.

#### Organic Wastes and Nutrients

By volume, the main pollutants entering estuaries are usually sewage (often untreated) and other organic wastes, such as effluents from pulp mills, abattoirs, and other food industries. Organic particles and fine sediments settle out under similar hydrodynamic conditions, so estuarine muds tend naturally to have a relatively high organic content. The consequent level of bacterial degradation results in only a thin surface layer of oxic sediment where aerobic bacteria can break down organic matter. Oxygen can diffuse only a short distance into these muddy sediments, so that beneath this surface oxic layer anaerobic bacteria take over the breakdown of material. Infaunal species of such sediments may have higher tolerances to anoxia or means of accessing overlying oxygenated water, such as by siphons or irrigation of burrows. But estuarine faunas are vulnerable to increased organic inputs if these are beyond the assimilative capacity of the system, in particular its flushing rate, leading to oxygen depletion.

Gradients of organic enrichment in estuarine systems typically display a well-defined pattern of response by benthic fauna (Pearson & Rosenberg, 1978) (Fig. 11.4). Where there is a heavy input of organic material, the sediment and overlying water will be anoxic (unless the system is well flushed) and devoid of macrofauna, and anaerobic degradation can release toxic hydrogen sulphide and methane. Beyond this grossly polluted zone occurs an impoverished community consisting of a few small opportunistic species occurring at high densities and living at or close to the sediment surface. Characteristic pollution-tolerant species include small annelid worms such as capitellids, spionids, and oligochaetes. With further reductions of organic input, species diversity gradually increases as the surface oxic layer of sediment deepens and larger deeper-burrowing species characteristic of the undisturbed community of the area make their appearance. Overall, as the level of disturbance decreases, the community changes from one dominated by *r*-selected species to one where *K*-selected species are important (see Chapter 4). A similar pattern is seen along temporal gradients as pollution lessens. Ivanov *et al.* (2013) followed the recovery of a benthic community in an inlet of the White Sea following the cessation of mussel farming (see Chapter 5). At this Arctic site, where biological processes are slow, it took about 15 years for the community to regain its undisturbed state.

Organic wastes and, particularly, agricultural and urban runoff are major sources of nutrients to estuaries and can result in eutrophic conditions. The overproduction of phytoplankton and macroalgae, decomposition of the resulting organic detritus, and associated oxygen depletion can severely impact estuarine biodiversity. Eutrophication is a major factor in the degradation of estuaries worldwide. An assessment of US estuaries in 2007 showed two-thirds to be moderately to highly eutrophic based on a number of criteria. And given the growing coastal population, conditions are predicted to worsen by 2020 for a similar proportion of estuaries – population density being a reliable indicator of nitrogen load (Bricker *et al.*, 2008).

Macroalgal production from nutrient enrichment of estuaries may result in conspicuous blooms of opportunistic green seaweeds, notably species of *Ulva, Chaetomorpha*, and *Cladophora* (Raffaelli *et al.*, 1998; Smetacek & Zingone, 2013). In spring and summer, these macroalgae can form dense mats covering large areas of intertidal sediment, with generally adverse ecological effects. For example, algal blooms can affect the availability of food for birds. Grazers such as mute swans (*Cygnus olor*) and brent geese (*Branta bernicla*) could conceivably benefit from increased macroalgal production, but species that feed on infaunal invertebrates are disadvantaged by extensive algal mats. The effect of algal mats on the underlying benthos and consequences for feeding birds is evident from studies in the Ythan estuary, part of a nature reserve on the east coast of Scotland noted for its bird populations (Raffaelli *et al.*, 1989; Raffaelli, 2000). Data over a 25-year period for the Ythan catchment indicated an increasing proportion of land used for cereal crops and a corresponding increase in the use of nitrogenous fertilizers. During this time, there was a two- to three-fold increase of nitrogen in the river water, and the biomass of green seaweed (chiefly *Ulva*) in the estuary increased from only a few hundred g $m^{-2}$ to exceed 2–3 kg $m^{-2}$, forming dense mats over much of the estuary. The blanketing effect of the weed alters the underlying benthic fauna, notably by dramatically reducing the density of the burrowing amphipod *Corophium volutator*,

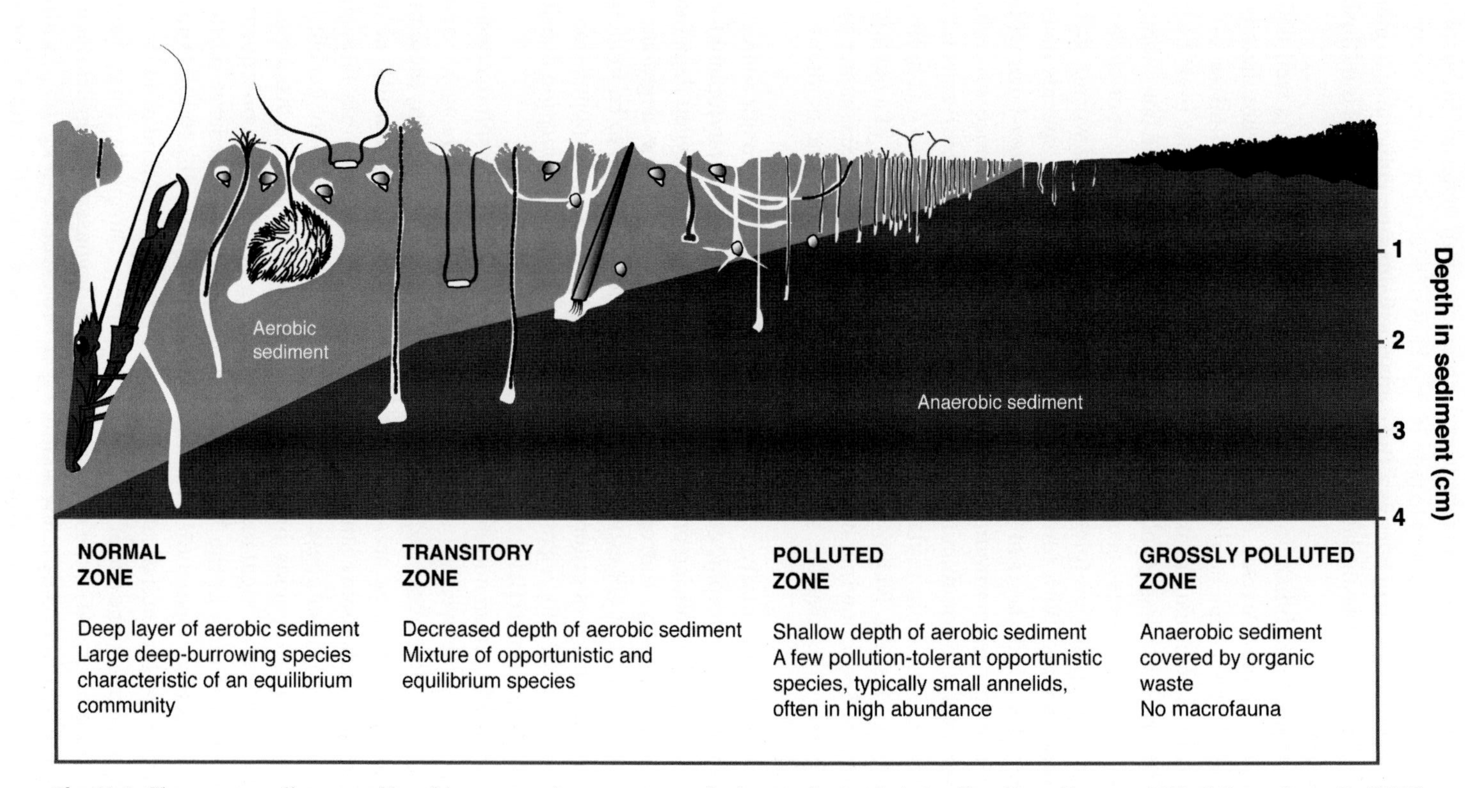


**Fig. 11.4** Changes to sediment and benthic community structure under increasing organic loading. From Pearson, T.H. & Rosenberg, R. (1976). A comparative study of the effects on the marine environment of wastes from cellulose industries in Scotland and Sweden. *Ambio*, 5, 77–9. © Royal Swedish Academy of Sciences.

a major prey item for most of the estuary's bird and fish populations, species such as redshank (*Tringa totanus*), common shelduck (*Tadorna tadorna*), and European flounder (*Platichthys flesus*). Associated with the spread of algal mats in the Ythan estuary has been a decline in several of its shorebird species.

Nutrient enrichment can also result in intense blooms of phytoplankton (including toxic dinoflagellate species) and lead to mass mortality of fish and benthic populations. Estuaries can support large populations of suspension-feeding bivalves, and these can strongly influence the particulate load in the water column. But if these suspension feeders have been overfished, the capacity of the estuary to consume phytoplankton will have been reduced. So eutrophication may also be a symptom of overfishing (Jackson *et al.*, 2001).

## Oil

Although difficult to quantify, estuaries in general receive a disproportionate amount of petroleum hydrocarbons from human activities. Land-based inputs, notably from rivers and urban and industrial runoff, much of which enters estuaries, are estimated to account for 20% of the total anthropogenic input of oil to the oceans (National Research Council, 2003). Many estuaries also receive significant inputs as a result of port activities, including from tanker operations, other shipping, and accidental spills.

Inputs of oil can readily overload estuarine systems, resulting in significant amounts accumulating in sediments and serious contamination and impoverishment of biota. The effects of oil on estuarine biota, however, vary greatly depending, for instance, on the composition and duration of the input, the fate and toxicity of the component hydrocarbons, and the relative vulnerability of the impacted species.

Oil enters estuaries chiefly as a result of chronic, diffuse inputs. But accidental spills result in the most obvious impacts. The low-energy environments of estuaries and associated wetlands are especially sensitive to oil spills (Gundlach & Hayes, 1978; O'Sullivan & Jacques, 2001). Physicochemical breakdown and dispersal are slow, and once oil becomes incorporated into subsurface anoxic sediment, the opportunity for microbial degradation is limited. Hydrocarbon residues may then persist in estuarine mud for decades and continue to affect key functional species (e.g. Culbertson *et al.*, 2008).

## Metals

Rivers normally provide the main route for the natural transport of heavy metals to coastal waters, so certain levels of metals are to be expected in estuaries, depending on the geochemistry of the catchment. Often, however, natural background levels are substantially elevated as a result of urban and industrial inputs, leading in some cases to gross contamination by mercury, cadmium, lead, copper, zinc, chromium, and other heavy metals (Clark, 2001).

Estuaries act as efficient traps for heavy metals. Metals entering estuaries tend to bind to particles, and the mixing of fresh and saline waters promotes flocculation and settlement of particles together with their adsorbed metals. Biological processes, such as biodeposition of faeces and pseudofaeces by filter feeders, assist in this transfer. Metal concentrations in estuarine sediments can thus greatly exceed those in the overlying water, often by three to five orders of magnitude. In terms of potential toxicity, however, what matters most is not the amount of metal in the sediment or water but the fraction that can be taken up by an organism – the amount that is bioavailable. Bioavailability depends on what form the metal is in, environmental factors, and the biology of the organism. Species and life-history stages vary greatly in sensitivity. Broadly speaking, the heavy metals that most often contaminate estuarine sediments rank in toxicity from mercury (most toxic) to cadmium, copper, zinc, chromium, nickel, lead, and arsenic (least toxic), with toxicity increasing as salinity decreases and as temperature increases (McLusky *et al.*, 1986). Estuarine muds can, nevertheless, contain high levels of heavy metals, and correspondingly elevated levels in benthic species, yet still support productive benthic communities. It has also been difficult to establish obvious adverse effects of heavy metals on estuarine birds, but waders feeding in contaminated estuaries may be at risk (Smith *et al.*, 2009).

Among the most obvious effects of metals on estuarine biotas are those attributable to organometals – compounds with metal–carbon bonds. These are usually more toxic than the inorganic forms and include such compounds as methylmercury, alkyl lead, and tributyltin. Marine anti-fouling paints containing organic tin compounds, in particular tributyltin (TBT), have been a significant source of pollution in many estuarine and enclosed waters that support high levels of shipping and boating activity. Anti-fouling paints containing TBT were first introduced in the mid-1960s and came into widespread use in the 1970s to succeed copper-based paints. TBT is an effective anti-fouling compound but is highly toxic with lethal and sub-lethal effects on a wide variety of non-target organisms. Deleterious effects on marine invertebrates can be initiated at remarkably low TBT concentrations ($<1$ ng $L^{-1}$) (Alzieu, 1998).

Molluscs are especially susceptible to TBT. Harmful effects became evident in the 1970s from shell deformation and reproductive failure of oysters in France and in the UK and the USA from abnormalities of the reproductive system of gastropods. TBT was subsequently implicated in similar abnormalities in more than 100 species of gastropods (mainly neogastropods) in many parts of the world. Most obviously, the female snail develops male characteristics, a condition known as imposex, which can result in sterilization. In some regions, a high proportion of snails have been affected, especially in estuaries with a high level of boating and shipping activity, and their populations found to be declining (Evans, 2000).

Widespread concern about the impact of TBT on non-target organisms resulted, from the 1980s, in many industrialized countries restricting the use of TBT-based paints, and in 2001 the International Maritime Organization (IMO) adopted a convention to prohibit the application of toxic anti-fouling paints after 2003, with a complete ban by 2008. Overall, TBT contamination has been declining as a result, with evidence of severely impacted species recovering, such as the dogwhelk (*Nucella lapillus*) around the UK (Langston *et al.*, 2015). But the problem of TBT is likely to persist for many years yet. TBT-based paints are still being used in countries with limited or no TBT legislation and are used illegally. Also, contaminated sediments represent a chronic source of the pollutant. TBT adsorbs strongly to particulate matter in the water and to bottom sediments, and in anaerobic sediments its degradation to inorganic tin may take decades (Antizar-Ladislao, 2008).

It has been argued that restrictions on TBT should be delayed until effective and environmentally safer products are fully developed as TBT-based paints have some environmental benefits. Being so effective in controlling hull fouling, TBT reduces the fuel consumption of ships and thereby the emissions of greenhouse gases and sulphur dioxide (acid rain) as well as the opportunities for alien species to be transported on hulls (Evans, 1999; Sonak *et al.*, 2009).

## Persistent Organic Pollutants

Organochlorine pesticides (notably DDT) and polychlorinated biphenyls (PCBs) (see Chapter 2) have become ubiquitous in estuarine environments and can commonly occur in appreciable levels in sediments and in invertebrates, fishes, birds, and mammals. Like metals, they readily adsorb onto particles and thereby accumulate in estuarine sediments. Organochlorines are very resistant to degradation so that their build up in sediments can be a chronic source of contamination to estuarine biota.

Sediments and fish of San Francisco Bay contain residues of organochlorine pesticides, notably of DDT, chlordanes, and dieldrin, as a legacy of their heavy use in the catchment. They still occur at levels likely to be harmful to marine life even though the pesticides were banned by the late 1980s. Model predictions indicate that if no more pesticides enter the bay, the active sediment layer would be free of residues in 10–30 years; however, given continuing input, recovery is projected to take considerably longer or may not occur at all. Management actions to hasten recovery, such as the remediation of pesticide hotspots, need to be examined (Connor *et al.*, 2007).

Of particular concern is the potential for organochlorines to bioaccumulate in top predators, including estuarine birds and mammals. A striking example concerns the threatened population of beluga (*Delphinapterus leucas*) resident in the St Lawrence Estuary in eastern Canada. This southernmost and isolated population of some 1000 animals has been chronically exposed to a mixture of agricultural and industrial contaminants including heavy metals, PAHs, and organochlorines. Elevated levels of contaminants have been found in beluga tissues, which may help explain the lack of recovery of this population since its protection from hunting in 1979. The cocktail of contaminants includes known carcinogens, which may be linked to an unusually high incidence of cancers in this population (Newman & Smith, 2006). Encouragingly, blubber levels of various organochlorines declined significantly between 1987 and 2002, likely related to a reduction of inputs, among other factors (Lebeuf *et al.*, 2007).

### Thermal and Radioactive Discharges

Coastal power stations, many of them sited on estuaries, discharge large volumes of heated effluent. In temperate regions, although the effluent can be about 10°C warmer than the intake water, the heat is likely to dissipate relatively quickly and have only localized biological effects. The situation is potentially more critical where heated effluents enter tropical waters as many organisms are already near their upper thermal limit. Cooling water discharged from nuclear power stations also contains low-level radioactive wastes, although such inputs would normally be minute compared to natural levels of radioactivity (see Chapter 2). Some genetic aberration may occur in individual organisms at maximum dose rates, but there appears to be no evidence that adverse effects would become apparent at population, community, or ecosystem levels (McLusky & Elliott, 2004).

## Physical Alterations

### Dredging

Reclamation, dredging, and port development are among the most widespread physical impacts, particularly for large estuaries.

Maintenance dredging is undertaken to keep open navigable harbours and channels. In addition, capital works dredging may occasionally be undertaken to enlarge existing channels or for new port developments. Dredging for building aggregates (sands and gravels) is also often carried out in estuarine areas. Mechanical dredges use grabs, buckets or scoops to remove bottom material, whilst hydraulic dredges extract seabed sediment as a slurry.

The removal of substratum entails a high mortality of benthic biota. In the immediate vicinity of the dredging there will also be smothering of benthos and clogging of filter-feeding and respiratory mechanisms due to high suspended solid loads. Smothering and burial will be a major cause of mortality for benthos at the disposal site. How readily the benthos recovers – regains a community structure comparable to its pre-impact state (e.g. in terms of species composition, diversity, and biomass) – depends on various factors, but in particular on the spatial extent and intensity of

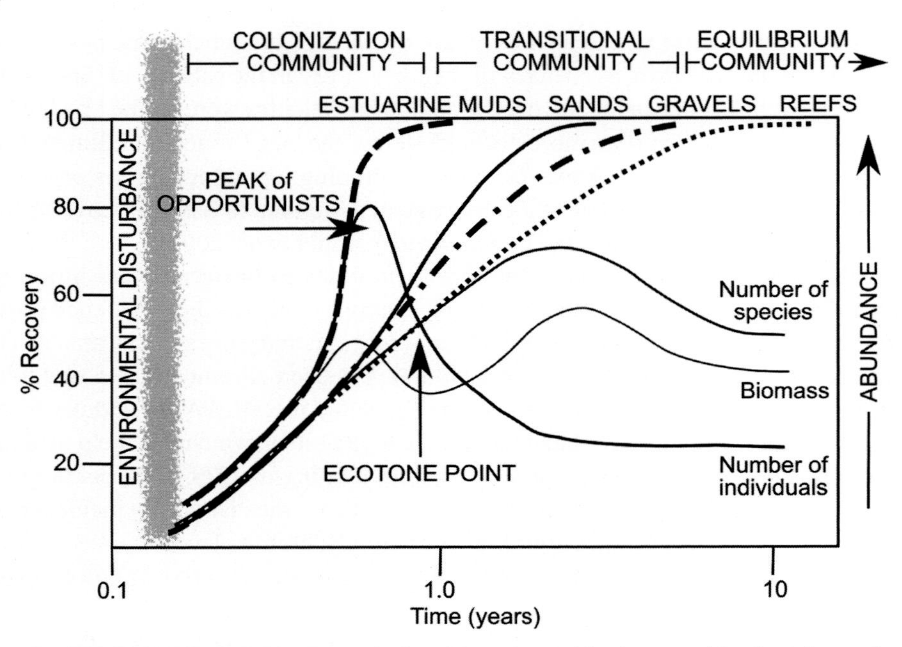


**Fig. 11.5** Typical rates of recovery for benthic communities impacted by dredging and community structure in terms of number of individuals, number of species, and biomass. Mobile estuarine muds, characterized by communities of relatively low diversity dominated by opportunistic (*r*-selected) species, tend to recover quickly. On the other hand, coarse sediments less subject to natural disturbance are often characterized by communities containing slow-growing (*K*-selected) species and take years to re-establish. The 'ecotone point' represents the region between the colonization community, dominated by a restricted number of small opportunistic species, and the transitional community. The transitional community has an elevated number of species as it contains both opportunistic and equilibrium species. From The impact of dredging works in coastal waters: a review of the sensitivity to disturbance and subsequent recovery of biological resources on the sea bed, *Oceanography and Marine Biology: An Annual Review*, Vol. 36, pages 127–78, Newell, R.C., Seiderer, L.J. & Hitchcock, D.R., copyright © 1998, UCL Press. Reproduced by permission of Taylor & Francis Books UK.

the dredging, the nature of the substratum, and its benthic community. Many estuarine sediments are muds, subject to frequent natural disturbance, and characterized by opportunistic species. Such communities may recover quickly – in less than a year – from dredging activity. By comparison, communities of coarser sediments with longer-lived and habitat-forming species may take several years to at least a decade to recover (Newell *et al.*, 1998; Foden *et al.*, 2009) (Fig. 11.5).

The other main environmental aspect of dredging concerns water quality. Increased turbidity during dredging can be marked, although usually short lived, so that any light attenuation and limitation of primary production are localized and temporary. Since many estuaries are naturally turbid, this effect is likely to assume significance only where natural turbidity is low. Disturbance of sediment from dredging operations can also facilitate the mobilization of nutrients, heavy metals, and synthetic organic compounds, an important consideration in heavily industrialized estuaries. In the case of urban estuaries and harbours, it is important to determine the levels of contaminants in the sediment to be dredged as well as their chemical state and the quantities exchanged between sediment and aqueous phases. Various agencies have developed bioassay criteria to evaluate the toxicity of dredge spoil to marine and estuarine organisms.

For the disposal of unpolluted sediments at open-water sites, the main options are to select either a high-energy dispersive site where dumped sediments are carried away or a containment site in a low-energy depositional environment where the spoil is likely to remain within the dump site. Options for the disposal of contaminated sediment include capping the dumped spoil with clean sediment or constructing artificial islands. When dredge spoil is released from the vessel at open-water sites, the great majority descends rapidly through the water column as a high-density flow, and typically less than 5% of the sediment volume is transported away from the site. So there is normally only limited potential for contamination of a wider area (Kennish, 1992). There can, however, be considerable difficulties in predicting the fate of spoil dumped at open-water sites. Dredge spoil is often used for reclamation (see below), and opportunities may arise to use uncontaminated dredge spoil for wetland creation and beach nourishment.

By altering the geometry of an estuary, dredging and reclamation can affect the hydraulic and sedimentological regimes and, in turn, the benthic communities. Dredging increases the volume of the estuary, thereby reducing tidal velocities and favouring sediment deposition – and hence further dredging.

There has been considerable research into environmental effects of dredging and spoil disposal, and various protocols and guidelines have been developed for environmental protection, from international (e.g. London Convention, Chapter 2) to regional and national levels. At the regional level, for example, the OSPAR Convention for the Protection of the Marine Environment of the NE Atlantic has guidelines for the management of dredged material. These include evaluating the need for dredging and disposal, physical, and chemical characterization of the material to be dredged including its toxicity, survey and sampling requirements of the area to be dredged, evaluating the sea disposal site, assessment of potential effects, and associated environmental monitoring (OSPAR Commission, 2014).

## Reclamation

Many estuaries have undergone major physical alteration as a result of extensive areas lost to infilling or reclamation. Typically this change occurs gradually over many years in a piecemeal fashion, with the loss of fringing wetland areas followed by successive downshore encroachments. At least a quarter of Britain's total estuarine area has been lost to reclamation since Roman times, initially to provide farmland but in recent decades largely to satisfy port development, urban and industrial expansion, and recreational developments such as marinas. Intensive development can mean the original foreshore is entirely eradicated or reduced to small fragments (Davidson, 1991). In the Netherlands, the estuarine and coastal geography has been radically altered over the centuries with the construction of protective seawalls and the conversion of enclosed areas to polders and freshwater reservoirs.

As a specific example, Sydney Harbour estuary, Australia, has undergone extensive modification with nearly a quarter of the estuary having been infilled since European colonization and with the loss of extensive areas of mudflat, mangrove, and saltmarsh (Fig. 11.6). This modification mostly occurred during the period 1922–55 for port and residential developments, but an additional incentive for reclamation prior to that period was the removal of unsanitary mudflats that had been used for sewage and garbage disposal. Only minor reclamation has occurred since 1980 given changes in environmental opinion and legislation and the likelihood that the estuary has reached its capacity for such development (Birch *et al.*, 2009).

Invertebrate production can vary considerably within an estuary, and simply assessing the area lost to reclamation is not a reliable indicator of the loss of invertebrate production and the impact on dependent predators. Within the past 200 years nearly half the intertidal area of the Forth estuary in eastern Scotland has been destroyed. Demersal fish in the estuary appear to feed predominantly in

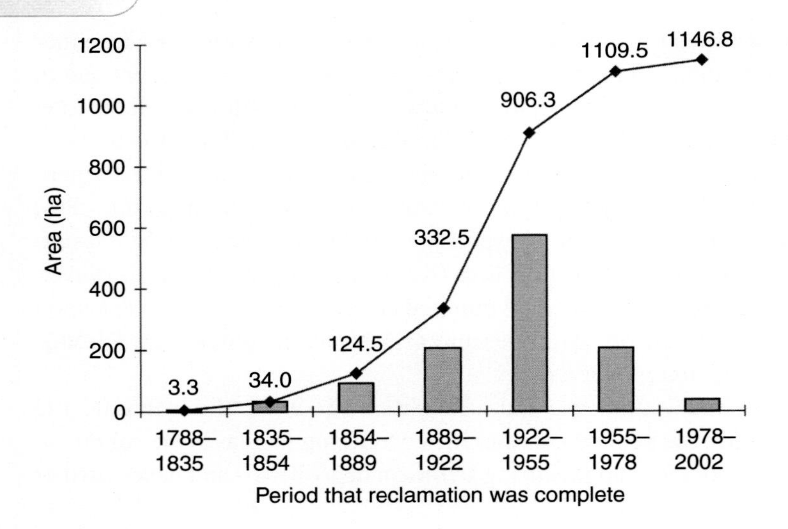


**Fig. 11.6** Reclamation in Sydney Harbour estuary since European colonization. From Birch, G.F., Murray, O., Johnson, I. & Wilson, A. (2009). Reclamation in Sydney estuary, 1788–2002. *Australian Geographer*, 40, 347–68. Reproduced by permission of Taylor & Francis Ltd, www.tandfonline.com.

intertidal areas, which are more productive than the more extensive subtidal areas. So whilst reclamation has removed 24% of the natural fish habitat of the Forth estuary, it has removed 40% of the fishes' food supply (McLusky *et al.*, 1992).

Reclamations may also significantly reduce the food available for birds. Critically, although the carrying capacity of an impacted site may not be exceeded (at least in terms of the number of bird-days it can support or the number that can survive overwintering in good condition), increased competition may mean that birds do not build up adequate body reserves to migrate to their breeding grounds. For *K*-selected wader and wildfowl species, even a slight decrease in survivorship may significantly impact population size (Goss-Custard *et al.*, 2002). Habitat loss can affect the fitness of displaced birds as evidenced by the impoundment of Cardiff Bay, UK, with the loss of 200 ha of intertidal habitat that supported up to 300 wintering redshank. Displaced birds, probably coming under strong competition at adjacent feeding areas, had poor body condition and a 44% increase in mortality rate (Burton *et al.*, 2006).

## Hydraulic Engineering Schemes

In addition to reclamation, other engineering schemes can alter estuarine hydraulics (Mitchell, 1978). Rivers have long been regulated for irrigation, flood control, navigational purposes, and, more recently, hydroelectric generation. Regulation and diversion of freshwater flows can affect estuarine and coastal species in a number of ways by altering migration and spawning patterns (particularly of anadromous species), survival of young, general productivity, and water quality (Drinkwater & Frank, 1994).

Barrages have been built across numerous estuaries for freshwater storage, coastal defence, or harnessing tidal energy. As part of the Netherland's flood protection scheme, the southern part of the former Zuiderzee was closed off by a dam in 1932 to become a huge freshwater lake, the IJsselmeer (3440 $km^2$ before subsequent reclamations). More recently, the Delta Plan was completed, a massive scheme involving the mouths of the estuaries of the Rhine, Meuse, and Scheldt; the construction of dams and storm surge barriers; and the creation of lakes. Such schemes raise major issues of nature conservation, particularly in relation to saltmarshes, intertidal invertebrates, waders, wildfowl, and commercial fisheries.

Tidal movement in estuaries and inlets has for centuries been used as a source of energy to drive mills. Over the past century, consideration has been given to harnessing tidal energy to generate electricity, and potential sites around the world have been identified. In the case of tidal barrages, suitable inlets need to have a large tidal range (> 4 m) and be narrow enough for barrage construction to be practicable. The first commercial tidal power station, at the mouth of the Rance estuary, France, was completed in 1966, and other tidal power stations have since been constructed. Environmental concerns of such schemes include the effects on estuarine circulation and sediment dynamics, water quality, salinity regime, pelagic and benthic communities, and the intertidal area available to feeding and roosting birds.

## Living Resources

### Fisheries and Aquaculture

Few species of importance to commercial or recreational fisheries are restricted to estuaries, but many species use estuaries at some stage in their life cycle. Estuaries can provide suitable nursery areas, protection from predation, and high food availability. In addition, many diadromous fishes depend on estuaries in migrating between marine and freshwater habitats, including a number that are highly sought after and of conservation concern, including species of salmon and sturgeon (see Chapter 6). An analysis of US commercial fishery landings for 2000–4 indicates that species that use estuaries during their life cycle account for about 46% by weight and 68% by value of the fish and shellfish landed nationwide. Among the most important species are Atlantic menhaden (*Brevoortia tyrannus*), shrimps, Pacific salmon (*Oncorhynchus* spp.), crabs, and lobsters (Lellis-Dibble *et al.*, 2008). Fishery yields are commonly enhanced in coastal areas that have large estuaries, high river discharges, and extensive wetlands, seemingly a reflection of the typically high productivity and other benefits that estuarine habitats confer.

Fishing can impact estuarine and associated coastal ecosystems in various ways, including effects on target and non-target organisms, nursery functions, community and habitat structure, and the potential for local extinctions (Blaber *et al.*, 2000), examples of which we examined in earlier chapters. One of the major impacts of fishing on estuarine habitats has been the loss of shellfish reefs and beds, in particular the loss of oyster reefs (Fig. 11.7). Such reefs once dominated many temperate and subtropical estuaries. Beck *et al.* (2011) estimate an 85% loss of these ecosystems globally, due mainly to overharvesting, introductions of non-native oysters, and associated oyster diseases as well as other stresses from human activities. Shellfish reefs provide important ecosystem services. As dense beds of filter feeders, they can control the suspended particulate load in the water column, thereby increasing water clarity and growth of benthic macrophytes. They also protect shores from erosion and provide habitat for many associated species. Shellfish reefs are in many ways the temperate equivalent of coral reefs but have received far less attention in terms of their management and conservation. However, restoration programmes, as at a number of US sites, are encouraging (Beck *et al.*, 2011).

In general, conventional capture fisheries of estuarine and coastal waters have declined, and there has been an increasing use of estuaries and brackish waters for aquaculture, such as for various species of fish, shrimps, prawns, and bivalves. Aquaculture can have significant environmental impacts (see Chapter 5). One of the main aquaculture industries in estuaries is oyster farming, in particular cultivation of the Pacific oyster (*Crassostrea gigas*). This commonly involves holding the oysters in trays or mesh bags or on sticks set out on the estuarine tidal flats. In this case, environmental issues include the appropriation of intertidal space, biodeposition and changes to the sediments and benthos beneath oyster farms, the depletion of phytoplankton, and the introduction of pests such as fouling species and nuisance microalgae (Forrest *et al.*, 2009). Total world production from brackish water aquaculture stands at about 5 million t, of which penaeid shrimps account for more than half (FAO, 2012). Construction of coastal ponds for aquaculture is a serious threat to tropical wetlands (see Chapter 12).

**Fig. 11.7** The eastern oyster (*Crassostrea virginica*), native to estuaries along the US Atlantic coast and Gulf of Mexico, can form extensive reef habitat. In many locations, however, oyster reefs have almost disappeared as a result of overharvesting, sedimentation, pollution, and eutrophication. Photograph: Michael Beck.

## Bait Harvesting

Collecting bait is carried out on many estuarine and other sheltered sediment shores. Cockles (*Cerastoderma edule*), for example, are gathered on many eastern Atlantic estuarine shores, and soft-shell clams (*Mya arenaria*) are taken on western Atlantic shores (where they are native) as well as on shores in Europe and the NE Pacific (where they have been introduced). Among the animals most often sought by bait diggers are large polychaetes, notably species of lugworm (Arenicolidae), ragworm (Nereididae), and bloodworm (Glyceridae), and ghost shrimps (Callianassidae). These activities are undertaken both recreationally and commercially and can be on a surprisingly large scale, resulting in considerable physical disturbance of the shore. For the bloodworm (*Glycera dibranchiata*) fishery in Maine, Sypitkowski *et al.* (2009) estimate that on average nearly a quarter of the state's mudflats (about 1800 ha) are turned over annually. Diggers use tined hoes to expose the worms, which live at a sediment depth of about 10 cm. Harvesting can also be carried out mechanically, such as in the Dutch Wadden Sea where digging machines cut 40 cm–deep gullies for harvesting lugworms (*Arenicola marina*) and cockles are collected by suction dredging, both of which seriously impact the populations of the target and associated species (Beukema, 1995; Piersma *et al*., 2001). On the UK's Norfolk coast (North Sea), bait digging for lugworms and the ragworm *Hediste diversicolor* – which involves disturbing the sediment to a depth of 30–40 cm – has been implicated in a marked decline of cockle populations (Jackson & James, 1979). Cockles are a major prey species of oystercatchers (*Haematopus ostralegus*), and birds could be deprived of food in areas of high bait-digging activity.

### Disturbance to Estuarine Birds

Bait digging may disturb feeding waders. The Bay of Fundy in eastern Canada is a major stopover site for migratory waders. During their southward migration, more than 500 000 semipalmated sandpipers (*Calidris pusilla*) (most of the world population) use the bay to feed, particularly on the amphipod *Corophium volutator*. Bait digging for bloodworms is, however, carried out in the bay, and Shepherd and Boates (1999) found that in dug areas the overall density of *C. volutator* decreased by 39% and the foraging efficiency of sandpipers by 69%. Birds foraging in harvested areas may take longer to deposit the fat needed for migration, thereby delaying their arrival at wintering grounds, or depart with insufficient fat.

The presence of people as a source of disturbance to feeding or roosting shorebirds needs to be taken into account in site management – such as in the provision and location of access and pathways – particularly with increasing use of coastal areas for recreational and ecotourism activities. Birds may react to people walking over intertidal flats by moving to a new site. Limiting access to feeding grounds, especially during cold weather, could significantly affect a bird's energy balance and, ultimately, its survival or reproductive success. In addition to adequate feeding habitat, estuarine birds need suitable areas for roosting when they are excluded from their feeding grounds at periods of high tide. Adjacent saltmarshes are often important in this respect. Roosts also need to be free from undue disturbance. Mitchell *et al*. (1988) recorded a 57% decline in the total number of waders roosting on Britain's Dee estuary (a Ramsar site) between 1975–6 and 1984–5 and attributed this decrease mainly to increased levels of disturbance from dogs, horse riders, and walkers. Bar-tailed godwits (*Limosa lapponica*) and knots that used feeding grounds on the Dee and formerly roosted there switched to a neighbouring estuary to roost. In the case of knots, the return trip of 40 km accounted for 14% of the bird's daily energy expenditure. Even a seemingly low-magnitude human disturbance – individual pedestrians on a track alongside an estuary at high tide – may flush significant numbers of birds or result in a roost being abandoned (Navedo & Herrera, 2012).

It may, however, be difficult to distinguish natural population changes and the impact of human activities. It is often unclear to what extent such declines are natural population phenomena or the consequences of human-induced environmental perturbations, such as degradation of habitat in wintering and staging areas due to development, human disturbance, pesticide use, and water pollution.

## Introduced Species

A common, often-conspicuous feature of estuarine biotas is the presence of non-indigenous species that have been introduced by human activities (see Chapter 2; Box 11.1). Indeed, there may be few, if any, estuaries that are now without introduced species. Many estuaries have ports that have been in use for centuries, providing opportunities for exotic species to be introduced by vessels as fouling

## Box 11.1 Invasive Species of Estuarine Systems

Invasive species are a conspicuous component of the biota of many estuaries, with a range of impacts on ecosystem structure and function. Some of the best-known invasive species are summarized below. Detailed information is available from the Global Invasive Species Database (www.issg.org/gisd).

### *Mnemiopsis leidyi*

*M. leidyi* is a ctenophore native to temperate to subtropical estuaries of the western Atlantic that was introduced, probably via ballast water, to the Black Sea in the early 1980s. It subsequently invaded adjacent seas and, from about 2005, the North Sea and Baltic Sea. *M. leidyi* is an important predator of zooplankton, and its population explosion in the Black Sea resulted in dramatic declines in zooplankton, fish eggs and larvae, and zooplankton-feeding fish in the 1990s. The pelagic ecosystem and fisheries of the Black Sea were profoundly impacted. In the late 1990s, however, another introduced ctenophore was found in the Black Sea, *Beroe ovata*. This species preys on *M. leidyi*, and the food web structure of the Black Sea began to recover (Purcell *et al.*, 2001; Oliveira, 2007; Katsanevakis *et al.*, 2014).

### *Ficopomatus enigmaticus*

Possibly native to Australia, *F. enigmaticus* is a serpulid polychaete that has invaded brackish waters in many parts of the world since at least the 1920s. Its dense aggregations of calcareous tubes form reefs that can markedly alter the physical environment. The reefs provide new habitat, including for other non-native species (Heiman & Micheli, 2010; McQuaid & Griffiths, 2014).

### *Musculista senhousia* Asian Date Mussel

This west Pacific species now occurs in many parts of the world, including Australasia, Mediterranean Sea, NE Pacific, and east Africa, with the main introduction pathways being aquaculture, ballast water, and hull fouling. It mainly colonizes soft substrata where it can form dense byssal mats on the sediment surface (densities of up to 10 000 individuals $m^{-2}$ or more) that can exclude native species and displace seagrass (Sousa *et al.*, 2009).

### *Potamocorbula amurensis* Asian Clam

Native to China, Japan and Korea, this clam arrived in San Francisco Bay in the mid-1980s, most likely in ballast water. It underwent a population explosion, with densities reaching 10 000 individuals $m^{-2}$, to become the dominant soft-bottom macrobenthic species in the estuary. Its suspension feeding appears to have dramatically reduced plankton biomass and substantially altered the pelagic food web (Carlton *et al.*, 1990; Greene *et al.*, 2011).

### *Carcinus maenas* European Green Shore-crab (Fig. 2.6)

This crab has become widely distributed in temperate regions outside its native range. Since the early nineteenth century it has established populations in Australia, South Africa, Japan, and North and South America involving a variety of dispersal mechanisms. The crab is a significant predator that takes a wide range of benthic prey and potentially competes with native crabs (Leignel *et al.*, 2014).

***Eriocheir sinensis* Chinese Mitten Crab**

The Chinese mitten crab was discovered in Germany in the early twentieth century, and from there it spread rapidly throughout northern Europe. More recently it has become established on the west and east coasts of North America. Several dispersal pathways have been responsible. Its burrowing activity results in the collapse of estuarine banks, erosion of marsh sediments, habitat alteration, and decreased biodiversity (Dittel & Epifanio, 2009).

***Spartina* spp. cordgrasses (see Chapter 12)**

and boring organisms or in ballast. In addition, efforts to develop fisheries have led to the intentional introduction of many fish and shellfish species to estuaries as well as the accidental introduction of numerous attendant species.

Some estuaries are heavily invaded, in many cases with introduced species becoming dominant elements of the biota and altering the structure and function of the impacted communities. San Francisco Bay has the highest documented number of non-native species of any estuary in the world, about 240 species (Lee *et al.*, 2003), and two-thirds of the introduced species in San Francisco Bay assessed by Molnar *et al.* (2008) were evaluated as harmful in terms of their ecological impact.

Estuaries harbour more invasive species than open coast habitats, a susceptibility that is probably due to various factors. Shipping and aquaculture, the most important vectors (Fig. 2.5), are often concentrated in estuaries, thereby increasing the chance of transfers, and being relatively enclosed, estuaries tend to retain dispersive stages. Estuaries are also geologically young systems with relatively low species richness and may offer more ‘empty’ niches (Preisler *et al.*, 2009).

Reducing the impact of invasive species on estuaries requires action on several fronts to help restore ecosystem structure and function. Strict control of ballast water release is needed to minimize the inflow of exotic species, whilst pollution control and habitat improvement help maintain biodiversity and lessen the success of invaders. To fully regain balanced estuarine ecosystems that are resistant to invasion, as well as being resilient overall, may require the reintroduction of top-level predators (Briggs, 2012).

## Integrated Management

Apart from their fish and bird life, estuaries have not traditionally attracted high nature conservation interest. More often they have been viewed as convenient watercourses for receiving sewage and industrial wastes or of economic benefit if infilled. Given strong commercial pressures, loss and degradation of estuarine habitat continue, although the potential economic implications of safeguarding estuarine productivity and estuary-dependent fisheries and aquaculture are now better appreciated. Recent decades have witnessed increasing concern about the plight of estuaries worldwide, and in many countries the importance of conserving estuarine habitats and wildlife is now recognized. Estuaries are scarce environments in areal extent, but being highly productive, their influence can be disproportionately far-reaching, especially in the case of the highly migratory species they can support.

There are, however, major difficulties to be addressed to achieve effective management and conservation of estuarine habitats. A key problem is the multiplicity of uses often concentrated in estuaries, many of which may conflict. Also, developments tend to be pursued in a piecemeal fashion without regard to the ecosystem as a whole. Small developments, assessed individually, may be unlikely to result in measurable impacts on the system as a whole and thus deemed to be acceptable, but over decades the cumulative effect of this approach can be profound. This problem is not of course peculiar to estuaries; it besets many areas of coastal planning. But such difficulties tend to be accentuated in estuaries where there can be a concentration of groups with overlapping and competing interests, such as port and local authorities, industries, recreational and commercial fishing interests, boating clubs, amenity groups, and conservation organizations. Furthermore, and crucially, estuarine health depends on activities throughout the catchment as well as processes occurring within the estuary itself, which management, to be effective, must take into account. Commonly, for instance, marine environmental impacts have been ignored in projects involving regulation of freshwater flows, even though such modifications may be various and substantial (Drinkwater & Frank, 1994). A further obstacle to more holistic management of estuarine systems is that the terrestrial, freshwater, and marine components generally fall under different national policies and legislation. Estuaries illustrate in a nutshell the importance of integrated coastal zone management (see Chapter 10).

As we have seen, estuaries are often assailed by multiple anthropogenic factors. Included here are numerous pollutants, various physical changes and disturbances affecting an estuary's topography and hydrodynamics, biological invasions, and impacts related to fisheries and aquaculture. The difficulty of identifying the effect of a particular anthropogenic factor on estuarine degradation is further compounded by estuaries being highly variable systems. As a result, they tend naturally to exhibit indices of community structure – based in particular on abundance, species richness, and biomass – that are similar to those of anthropogenically stressed systems (Elliott & Quintino, 2007). These considerations indicate the need for more integrated approaches to assessing estuarine 'health', approaches that reflect functionality – the capacity of the estuary to support processes and provide ecosystem services. For example, as a means of integrated assessment, Van Niekerk *et al.* (2013) used a number of abiotic and biotic measures of South African estuaries to derive an estuarine health index (EHI). EHI scores were then translated into health categories reflecting overall ecological state, functionality, and management (Fig. 11.8).

Estuarine systems vary in their inherent vulnerability to anthropogenic impacts, which is important for setting management and conservation priorities. Engle *et al.* (2007), for instance, classified US estuaries based on physical and hydrological parameters likely to be important in modulating pollutant stress on estuarine biota, in particular factors influencing the residence time of water in an estuary. From a cluster analysis of 138 estuarine systems, they identified nine classes. Estuaries within a group would thus be expected to respond similarly to a similar pollutant stress from the catchment. Such a classification can be used to assess the susceptibility of estuaries to pollutants.

Some degree of contamination is inevitable for developed estuaries. The challenge is in controlling contaminant levels so as to optimize the resource for conflicting human uses without unduly jeopardizing ecological integrity. Maximum allowable concentrations need to be established for the various contaminants on the basis of their toxicity and amounts being discharged and the assimilative capacity of the estuary (Wilson, 1988). An associated monitoring programme should be in place to ensure that contaminant concentrations in the estuary and in selected indicator species do not exceed levels permitted by the consenting authority.

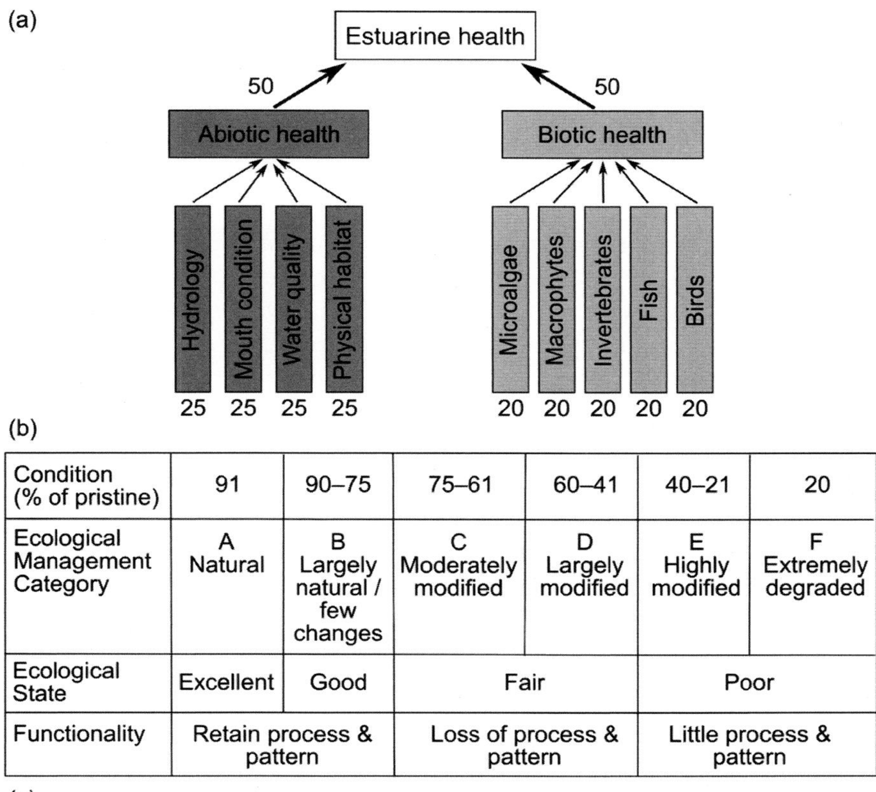


| Condition (% of pristine) | 91 | 90–75 | 75–61 | 60–41 | 40–21 | 20 |
|---|---|---|---|---|---|---|
| Ecological Management Category | A Natural | B Largely natural / few changes | C Moderately modified | D Largely modified | E Highly modified | F Extremely degraded |
| Ecological State | Excellent | Good | Fair | | Poor | |
| Functionality | Retain process & pattern | | Loss of process & pattern | | Little process & pattern | |

(c)
Description of the Ecological Management Categories

A: Unmodified, natural.

B: Largely natural with few modifications. A small change in natural habitats and biota may have taken place but the ecosystem functions and processes are essentially unchanged.

C: Moderately modified. A loss and change of natural habitat and biota have occurred but the basic ecosystem functions and processes are still predominantly unchanged.

D: Largely modified. A large loss of natural habitat, biota and basic ecosystem functions and processes have occurred.

E: Seriously modified. The loss of natural habitat, biota and basic ecosystem functions and processes are extensive.

F: Critically/Extremely modified. Modifications have reached a critical level and the system has been modified completely with an almost complete loss of natural habitat and biota. In the worst instances the basic ecosystem functions and processes have been destroyed and the changes are irreversible.

**Fig. 11.8** The components and weightings of an estuarine health index (a), the derived health categories (b), and details of the categories (c). Reprinted from *Estuarine, Coastal and Shelf Science*, Vol. 130, Van Niekerk, L., Adams, J.B., Bate, G.C., et al., Country-wide assessment of estuary health: an approach for integrating pressures and ecosystem response in a data limited environment, pages 239–51, copyright 2013, with permission from Elsevier.

## Amelioration

Reclamations, port and industrial expansion, flood protection, and hydraulic engineering works result in losses of estuarine habitat that are effectively irreversible. But for other forms of habitat degradation, notably poor water quality, ameliorative measures are likely to be possible. Water quality criteria can be applied to protect certain features of estuaries, such as the passage of migratory fish, resident biotic communities, and public health standards. To realize such criteria requires compliance monitoring to ensure that physico-chemical parameters of effluents are being met and monitoring of the receiving system – the water, sediment, and biota – to keep a check on any environmental impacts. As well as local and national water quality standards, there may be regional and international agreements on marine pollution that are applicable.

The UK's Thames Estuary is a notable example of rehabilitation following improved water quality. The estuary became grossly polluted during the nineteenth century but was restored to aerobic conditions in the 1960s following major improvements in sewage treatment. The number of fish species recorded in upper reaches of the estuary increased from fewer than 10 in the early 1960s to some 120 species by 40 years later, including a number of species of conservation interest, such as the smelt (*Osmerus eperlanus*), twaite shad (*Alosa fallax*), and sea lamprey (*Petromyzon marinus*) (Thomas, 1998; Colclough *et al.*, 2002). Other changes have included the development of a more diverse benthic community (from one dominated by tubificid worms characteristic of organically polluted sediments), the recovery of the benthic algal flora, and the return in winter of thousands of migratory wildfowl and waders (Andrews & Rickard, 1980).

Opportunities may arise for the restoration and creation of new estuarine habitat, usually so as to provide compensatory feeding and roosting areas for waders and wildfowl. Habitat creation should not, however, be regarded as an alternative to the safeguarding of existing nature conservation sites, given the uncertainties of achieving a particular habitat type and the long lead time needed to ascertain if the venture has been successful (Davidson, 1990). Davidson and Evans (1986) suggest general criteria for the creation and improvement of habitat to help ensure such sites are appropriate for the birds for which they are intended. They recommend, most importantly, that: (1) such activities do not compromise any existing conservation importance of the site; (2) any created habitat provides substrata and invertebrate fauna similar to those of the area lost so that it will likely support the displaced shorebirds; (3) satisfactory compensation is achieved only if all displaced birds can settle and survive on the new site; and (4) compensatory habitat is established well in advance of the area lost, usually within a minimum of five years.

A national initiative for estuarine management and conservation is the USA's National Estuary Program (NEP), authorized under an amendment to the Clean Water Quality Act of 1987 and managed by the US Environmental Protection Agency (EPA). The mission of the NEP is to identify nationally significant estuaries and to establish and oversee a process for improving and protecting their water quality and enhancing their living resources. The focus of the NEP strategy is on ecosystem management and taking account of the human communities, both as part of the ecosystem and in the decision making (Poole, 1996). States initially nominate estuaries, and if the estuary meets the programme guidelines, the EPA convenes a management conference of interested parties to characterize the estuary, define its problems, and develop a conservation and management plan that establishes goals and objectives. The management plan addresses three general areas: water and sediment quality, focusing on pollution abatement and control; living resources, including restoration as well as protection of special habitat areas; and land use and water resources, using regulatory and non-regulatory means to conserve land and water. A comparative analysis by Lubell (2004) indicates that the NEP's inclusive and collaborative style of governance involving multiple stakeholders is effective at resolving conflict and increasing the level of cooperation on estuary restoration projects. The resulting network of contacts between stakeholder organizations in NEP areas is far denser than that in non-NEP areas (Schneider *et al.*, 2003). The NEP currently includes 28 estuary projects. These include Puget Sound and San Francisco estuary on the Pacific coast; Galveston Bay and the huge Barataria-Terrebonne estuarine complex on the Gulf of Mexico; and, on the east coast, Buzzards Bay, Narragansett Bay, Long Island Sound, Delaware Estuary, and Albemarle-Pamlico Sound.

## Protected Areas

Estuaries may appear to feature prominently in conservation efforts, at least for their intertidal and fringing habitats. British estuaries, for instance, are particularly diverse in terms of their geomorphology, habitats, and biota, and about 90% of them have protected areas (as sites of special scientific interest) (Davidson, 1991). Often, however, such areas have been designated primarily for their bird life. Although the potential conservation importance of other biota is increasingly being recognized, subtidal estuarine habitats still tend to be poorly represented in marine conservation efforts worldwide.

Estuaries pose considerable challenges for protected area designation. They typically exhibit wide environmental gradients, many support a high level of human activity and are often significantly degraded as a result, and apart from their fringing habitats and vertebrates, they tend to lack the popular appeal of other coastal environments. A key objective of any marine protected area (MPA) programme is to establish a comprehensive network that captures the representative and unique habitats and communities for that biogeographical area. And, particularly in the case of estuaries, such a programme should be part of an integrated coastal zone management plan (see Chapter 10), given the importance of catchment influences. Indeed, it may be important to seek protection of a whole catchment. For estuaries in tropical rain forest regions, where annual rainfall may be several metres, appropriate protection or management of the forest catchment may be essential to ameliorate water runoff and prevent disastrous soil erosion (Carr, 1982).

To develop a network of estuarine protected areas in Tasmania, Edgar *et al.* (2000) assessed the conservation significance of all 111 estuaries of moderate or large size. These ranged from near-pristine estuaries to ones heavily degraded due to such impacts as heavy metal pollution, eutrophication, siltation, urban development, dam construction, foreshore reclamation, channel dredging, and introduced species. Estuaries were categorized into a limited number of groups based on a series of variables relating to geomorphology, hydrology, and catchment area, and the biological validity of these groups was assessed by comparison with assemblages of macrobenthos and fish. The level of anthropogenic disturbance was assessed by data on human population, land use, and land tenure, and the estuary with least disturbance in each group was assigned the highest conservation significance (Fig. 11.9). The estuarine groups identified by Edgar *et al.* (2000) differed in a number of ways from the usual geomorphological categories because of the effects of tidal range, salinity, and runoff, illustrating that classifications based solely on physico-chemical data may not closely reflect biotic patterns.

For large estuaries, especially those with commercial development, an MPA for the whole system may not be achievable or appropriate, but the high degree of spatial variability typical of large estuaries does need to be taken into account in developing an effective MPA network. Using conservation planning software to explore MPA options for Long Island Sound, north-eastern USA, Neely

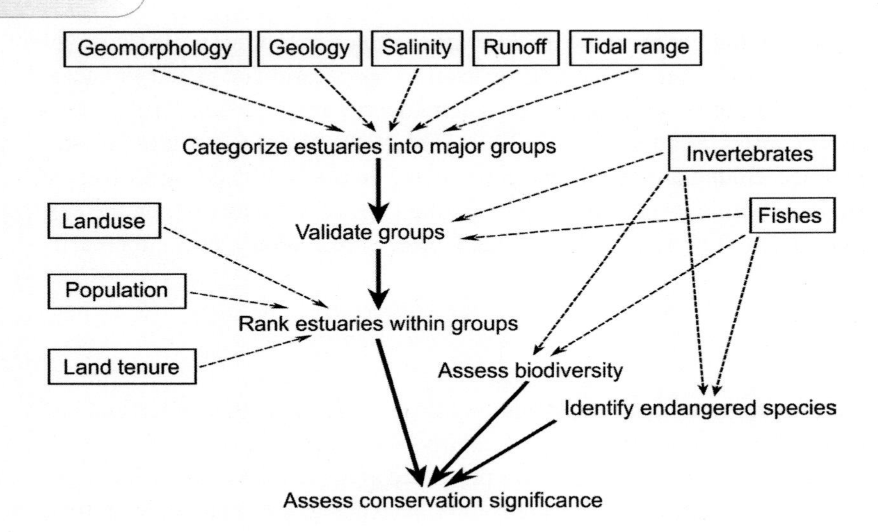


**Fig. 11.9** A process to identify the most appropriate estuaries to be included in a system of estuarine protected areas. The data sets used are in rectangles. Reprinted from *Biological Conservation*, Vol. 92, Edgar, G.J., Barrett, N.S., Graddon, D.J. & Last, P.R., The conservation significance of estuaries: a classification of Tasmanian estuaries using ecological, physical and demographic attributes as a case study, pages 383–97, copyright 2000, with permission from Elsevier.

and Zajac (2008) proposed a network consisting of areas in the main body of the estuary, together with a set of more peripheral areas to protect the full range of the estuary's biodiversity and function. Additionally, marine zoning might be used to confine certain activities to particular parts of a large estuary and to reduce conflicts between user groups.

## The Ramsar Convention

An important international agreement relevant to the conservation of estuarine and coastal wetland habitats is the 1971 Convention on Wetlands of International Importance especially as Waterfowl Habitat, known as the Ramsar Convention. The convention came into force in 1975 and was the first international initiative concerned with habitat conservation. 'Wetland' is defined very broadly under the convention to include (in addition to freshwater wetlands) saltmarsh and mangrove but also unvegetated tidal flats, beaches, and coastal areas of shallow water to depths of 6 m at low tide; and 'waterfowl' is taken to mean birds ecologically dependent on such areas, including waders. A major aim of the convention is 'to stem the progressive encroachment on and loss of wetlands, now and in the future'. Also central to the convention is the 'wise use of wetlands' defined as 'the maintenance of their ecological character, achieved through the implementation of ecosystem approaches, within the context of sustainable development' (Ramsar Convention Secretariat, 2010). The convention requires each contracting party to designate suitable wetlands within its territory for inclusion in a list of wetlands of international importance (Article 2.1). Contracting parties are obliged 'to promote the conservation of the wetlands included in the List, and as far as possible the wise use of wetlands in their territory' (Article 3.1) but also to 'promote the conservation of wetlands and waterfowl by establishing nature reserves on wetlands, whether they are included in the List or not' (Article 4.1). Whilst the convention highlights waterfowl – and this group was initially a major focus – the criteria for selection of important wetlands enable the convention to be used for the conservation of other wetland biota. There are now nine criteria to guide implementation (Box 11.2). Group A sites are internationally important as rare or unique examples of wetland, and Group B sites are internationally important for conserving biodiversity. Eight criteria are available to judge the latter based on species and ecological communities, waterbirds, fish, and other taxa (www.ramsar.org).

## Box 11.2 Ramsar Convention Criteria for Identifying Wetlands of International Importance

### Group A: Sites Containing Representative, Rare, or Unique Wetland Types

Criterion 1: A wetland should be considered internationally important if it contains a representative, rare, or unique example of a natural or near-natural wetland type found within the appropriate biogeographic region.

### Group B: Sites of International Importance for Conserving Biological Diversity

#### Criteria Based on Species and Ecological Communities

Criterion 2: Supports vulnerable, endangered, or critically endangered species or threatened ecological communities.
Criterion 3: Supports populations of plant and/or animal species important for maintaining the biological diversity of a particular biogeographic region.
Criterion 4: Supports plant and/or animal species at a critical stage in their life cycles, or provides refuge during adverse conditions.

#### Specific Criteria Based on Waterbirds

Criterion 5: Regularly supports 20 000 or more waterbirds.
Criterion 6: Regularly supports 1% of the individuals in a population of one species or subspecies of waterbird.

#### Specific Criteria Based on Fish

Criterion 7: Supports a significant proportion of indigenous fish subspecies, species or families, life-history stages, species interactions and/or populations that are representative of wetland benefits and/or values and thereby contributes to global biological diversity.
Criterion 8: Is an important source of food for fishes, spawning ground, nursery and/or migration path on which fish stocks, either within the wetland or elsewhere, depend.

### Specific Criteria Based on Other Taxa

Criterion 9: Regularly supports 1% of the individuals in a population of one species or subspecies of wetland-dependent non-avian animal species.

From www.ramsar.org.

There are now some 170 contracting parties to the convention and more than 2200 sites totalling 2 million km$^2$ on the Ramsar list. More than 900 sites totalling 650 000 km$^2$ are listed as marine or coastal wetlands (www.ramsar.org). Among the most significant sites are those located at the head of the Bay of Fundy (between New Brunswick and Nova Scotia), where the mudflats and tidal marshes form one of the most important resting and feeding areas for waders in eastern North America; the

Banc d'Arguin in Mauritania, the most important site for waders in the east Atlantic, attracting more than two million waders each winter; and the Wadden Sea, western Europe's most important tidal wetland for birds. Whilst the Ramsar network is extensive, it is dominated by sites in the temperate North Atlantic with few sites in some other regions, such as the eastern Indo-Pacific (Spalding *et al.*, 2007).

### Shorebird Reserve Networks

Greater international coordination in conservation is needed in the case of waders and wildfowl that depend on stopover sites in different countries to undertake their migrations. Often these sites are estuaries, but rarely are interconnected sites adequately protected. In the case of the great knot (*Calidris tenuirostris*), a threatened shorebird, only 7% of its distribution is covered by protected areas during migration (Runge *et al.*, 2015).

One example of cooperation in strengthening such protected areas is the Western Hemisphere Shorebird Reserve Network (WHSRN), a collaboration of organizations committed to protecting habitats across the Americas that are of critical importance to migratory waders. Launched in 1985, the network now has more than 90 sites encompassing about 130 000 $km^2$ (www.whsrn.org). Member sites are ranked as being of 'hemispheric' significance if they host more than 500 000 waders annually or 30% of a species' biogeographic population. Coastal sites in this category include the Copper River Delta and Yukon Delta (Alaska), San Francisco Bay, Bay of Fundy, Delaware Bay, Reentrâncias Maranhenses (NE Brazil), and the Atlantic Coast Reserve of Tierra del Fuego. There are also sites of 'international' significance (>100 000 waders annually or > 10% of a species' biogeographic population) and 'regional' significance (>20 000 waders annually or >1% of a species' biogeographic population). Colwell (2010) concludes that both Ramsar and WHSRN have heightened awareness of the importance of staging and wintering habitats for the conservation of waders. However, it is still unclear to what extent this awareness has translated into more effective management.

## Coastal Lagoons

The difference between an estuary and a lagoon is not necessarily clear-cut. Many estuaries have an intermittent (often seasonal) connection with the open sea depending on the build-up or erosion of a sandbar at the mouth and the input of freshwater. At times of low rainfall a closed estuary may then become lagoonal in character. However, a coastal lagoon is typically a shallow, more permanent body of saline water that is separated from the adjacent sea by a low-lying depositional barrier, usually of sand or shingle (Barnes, 1980). Most commonly, the enclosing barrier results from the development of a spit or bar or from a chain of barrier islands so that the lagoon is an elongated body of water lying parallel to the coastline. Lagoons are generally shallow but retain water at times of low tide on the adjacent sea. Exchange of water with the sea is by connecting channels, by percolation through the barrier, or by overtopping during storms. Tidal fluctuations are small or absent so that exposure of intertidal sediments is limited. Typically, lagoonal sediments are organically rich sands and muds. Whilst lagoon salinity often shows a relatively stable longitudinal profile, there can be marked seasonal (or shorter) variations in salinity as a result of weather patterns. Salinity can range from < 1 to extremes of > 200. Since lagoons are generally small and shallow (often only 1–2 m deep), their physical environment can be highly variable. Barrier/lagoonal systems, a feature of an estimated 13% of the world's coastline, range from tropical to arctic environments, especially where there is a small (< 2 m) tidal range. Particularly well-developed lagoon systems occur along the east coast of the USA and Gulf of Mexico, northern Alaska, Siberia, southern Baltic, Mediterranean Sea, Black Sea, Caspian Sea, eastern Asia (Kamchatka, Sakhalin, and northern Hokkaido), Gulf of Guinea, Sri Lanka and SE India, Brazil, South Africa, and southern Australia. There are some large lagoon systems (> 1000 $km^2$), like those on the eastern seaboard of the USA, but many are only a few hectares (Colombo, 1977; Barnes, 1980; Bamber *et al.*, 1992; Bird, 1994).

Lagoonal biotas are typically of low diversity and mixed origin, in many instances resembling those of estuaries in being characterized by relatively few species able to tolerate the physical variability. The biota may include an important component of species from freshwater habitats that are able to withstand brackish conditions. Whilst the number of species unique to lagoons is not large, they are often of high nature conservation interest on account of their restricted distributions and vulnerability. Lagoons are often important too for migratory species, including various fishes and birds. Again, as in estuaries, detrital food webs usually predominate, with the main sources of detritus being fringing and submerged vegetation, notably mangrove, saltmarsh, seagrasses, and, in lagoons of low salinity, reeds, pondweeds, (Potamogetonaceae), and tasselweeds (Ruppiaceae).

Human use of coastal lagoons takes a wide variety of forms. Most important is the exploitation of species for food. Lagoons are among the most productive habitats, with fisheries yields per unit area (e.g. 100–150 kg ha $y^{-1}$, excluding aquaculture) that can appreciably exceed those of adjacent coastal areas. Fisheries mostly target migratory fish that move between the lagoon and adjacent sea, but crustaceans and bivalves can also feature in catches. Since prehistoric times lagoons have also been used for raising and culturing fish, and lagoon aquaculture now accounts for at least 6% of world aquaculture production, primarily of fishes, prawns, and bivalves (Pérez-Ruzafa & Marcos, 2012).

Many of the human impacts on lagoons are similar to those that beset estuaries close to centres of population. Lagoons of sufficient size have long attracted residential development and been used as harbours, resulting in major modifications of the shoreline and the dredging of navigable channels. At some coastal resorts, small lagoons have been developed for recreation, such as boating and bathing, largely destroying any natural values. Pressure from agriculture and from industrial and urban growth has also meant that many lagoons have been lost through infilling. Lagoons are highly vulnerable to contamination, in part because of long residence times and their tendency to retain substances. Nutrient enrichment and eutrophication resulting from agricultural runoff, other forms of pollution, and climate change impacts are of particular concern (Beer & Joyce, 2013).

Lagoons are transient features geologically, becoming either completely isolated and evolving into freshwater marshes and ultimately land or reverting to coastal embayments on erosion of the barrier. Small lagoons may persist no more than a few hundred years, and even within a human lifetime a lagoon may, as a result of natural causes, alter substantially or perhaps disappear. The question then arises to what extent a lagoon should be actively managed, in an attempt to preserve a particular community structure, or allowed to evolve naturally (Colombo, 1977). There is no simple answer. In terms of conservation and management, lagoons are among the habitats that most need to be assessed individually. Lagoons within a particular region can differ considerably in their environmental characteristics and biotic structure and in the natural and human forces bearing upon them. Of paramount importance are alterations to the morphology and hydraulics of the lagoon and the use and management of the surrounding land, particularly as it affects the quality and quantity of freshwater inflow and sources of agricultural, urban, and industrial wastes. A further consideration is the exploitation of the living resources. The use of lagoons for some fisheries and non-intensive forms of aquaculture might be appropriate and consistent with conservation goals, at least in ensuring that water quality is safeguarded.

Lagoons have been systematically researched in Britain, which has some 40 predominantly small (< 5 ha) coastal lagoons concentrated along the lower-lying southern and eastern coasts. In world terms they are, however, atypical in occurring on macrotidal coasts and in being confined to shingle foreshores. Barnes (1989) and Bamber *et al.* (1992) have carried out surveys and conservation assessments of the British lagoons to identify those deemed nationally significant. Important considerations include geomorphological type and size, geographical representativeness, species richness and number of specialist lagoonal species, naturalness, ecological setting, and vulnerability. Among the lagoons judged to be of high ecological and conservation importance are those with communities dominated by specialist lagoonal species. In Britain, about 40 species are known only from lagoon-like habitats. Given their very restricted distributions, some are among the country's rarest marine species. The small mud-burrowing anemone *Edwardsia ivelli*, known only from a single brackish lagoon in West Sussex, may now be extinct. A number of British lagoons are statutory reserves.

Perhaps more than any other marine habitat, lagoons offer opportunities for creative conservation and rehabilitation. Along the English North Sea coast are numerous small brackish lagoons that owe their origin directly or indirectly to human activities and that have a biota similar to that occurring in natural systems (Barnes, 1991). However, although such lagoons are relatively easy to recreate behind longshore shingle barriers, the opportunities for natural colonization may now be curtailed by coastal defences built to prevent flooding. Disused dock basins can, with suitable management of the physical environment leading to improved water quality, support diverse and abundant communities in inner-city areas and potentially contribute to the conservation of lagoonal and estuarine species (Hawkins *et al.*, 1999). Nevertheless, the extent to which such communities can be artificially introduced and restructured is hindered by a lack of knowledge of the factors regulating lagoonal communities (Beer & Joyce, 2013).

## Anchialine and Hypersaline Environments

An anchialine ecosystem is a special type of estuary occupying subterranean crevices and caverns on limestone and lava-flow shores. As anchialine pools and caves have underground connections with the open sea, they are tidally influenced. They are typically small, up to a few hundred metres in diameter, and occur mostly in the tropics and subtropics (Bishop *et al.*, 2015). Remarkable crustacean-dominated faunas have been described from anchialine environments and many new taxa discovered, including Remipedia, a class of predatory, seemingly ancient, crustaceans found almost exclusively in anchialine systems (Neiber *et al.*, 2011) (Fig. 11.10). As anchialine species have restricted geographical ranges, high habitat specificity, and small population sizes, they are highly vulnerable to human activities (Cabezas *et al.*, 2012). Anchialine environments can be common where there is suitable terrain. There are, for instance, several hundred anchialine pools on Hawaii. Some anchialine sites around the world are protected, but many have been degraded or lost as a result of pollution, introduction of exotic species, and coastal development (Sket, 1996).

The salinity of lagoons and pools in arid parts of the world can reach at least 100, or more in extreme cases, and result in the precipitation of sulphate and eventually other more soluble salts. The fauna of such hypersaline systems is generally very impoverished, which appears to allow for the development of complex associations of benthic microbes, especially mats produced by cyanobacteria. Spectacular in this regard are stromatolites, laminated structures accreted as a result of a benthic microbial community trapping and binding sediment. They are common in the geological record, dating back some 3.5 billion years and among the earliest evidence of life on Earth

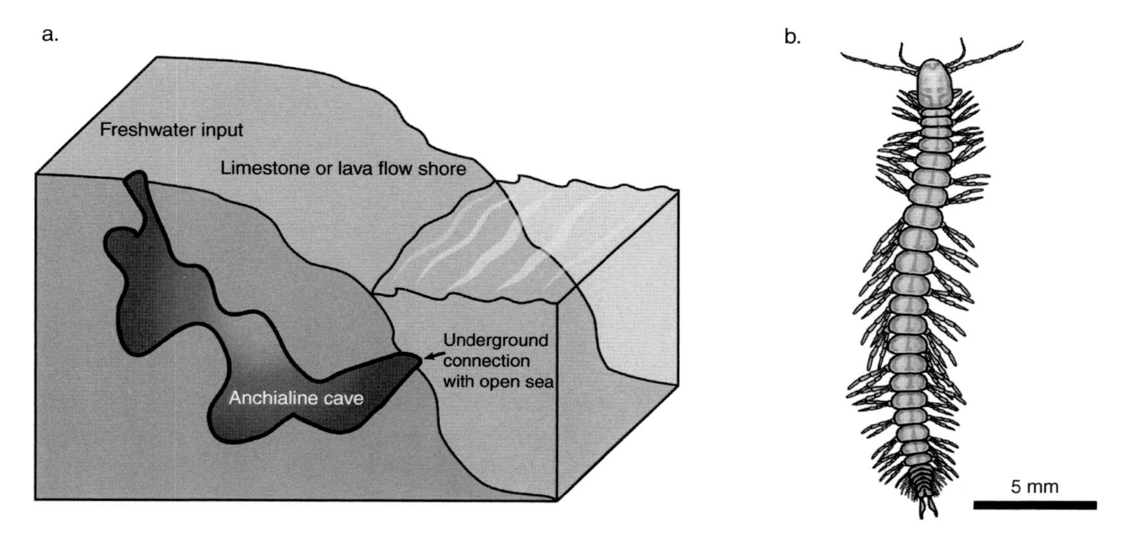


**Fig. 11.10** Diagram of (a) an anchialine ecosystem and (b) a remipede.

(Allwood *et al.*, 2006). Modern stromatolites, of great importance in palaeobiological interpretation, are however very limited in geographical extent.

Western Australia is noted for its contemporary stromatolite formations. Those at Shark Bay cover about 100 km of shoreline around the hypersaline Hamelin Pool to water depths of 3–4 m (Chivas *et al.*, 1990). They are arguably the most important living examples known and one of the reasons for Shark Bay being a UNESCO World Heritage site. The opening up of the Shark Bay region as a tourist destination has highlighted issues relating to the management of visitors and public education. A key issue is ensuring adequate protection for these fragile structures. Individual subtidal stromatolites in Shark Bay take up to 1000 years to reach their present heights of about 350 mm and are extremely vulnerable to physical damage.

Estuaries, lagoons, and related habitats highlight many key conservation issues, in particular the close links between marine and terrestrial environments in these coastal habitats and, correspondingly, the need for integrated management and legislation. Vegetated wetlands are often a conspicuous component at these interfaces, contributing importantly to their biodiversity and productivity. They warrant separate consideration, and we look at them in the next chapter.

## REFERENCES

Able, K.W. (2005). A re-examination of fish estuarine dependence: evidence for connectivity between estuarine and ocean habitats. *Estuarine, Coastal and Shelf Science*, **64**, 5–17.

Allwood, A.C., Walter, M.R., Kamber, B.S., Marshall, C.P. & Burch, I.W. (2006). Stromatolite reef from the early Archaean era of Australia. *Nature* **441**, 714–18.

Alzieu, C. (1998). Tributyltin: case study of a chronic contaminant in the coastal environment. *Ocean & Coastal Management*, **40**, 23–36.

Ambrose, W.G. Jr (1986). Estimate of removal rate of *Nereis virens* (Polychaeta: Nereidae) from an intertidal mudflat by gulls (*Larus* spp.). *Marine Biology*, **90**, 243–7.

Andrews, M.J. & Rickard, D.G. (1980). Rehabilitation of the inner Thames estuary. *Marine Pollution Bulletin*, **11**, 327–32.

Antizar-Ladislao, B. (2008). Environmental levels, toxicity and human exposure to tributyltin (TBT)-contaminated marine environment: a review. *Environment International*, **34**, 292–308.

Baird, D., Evans, P.R., Milne, H. & Pienkowski, M.W. (1985). Utilization by shorebirds of benthic invertebrate production in intertidal areas. *Oceanography and Marine Biology: An Annual Review*, **23**, 573–97.

Bamber, R.N., Batten, S.D., Sheader, M. & Bridgwater, N.D. (1992). On the ecology of brackish water lagoons in Great Britain. *Aquatic Conservation: Marine and Freshwater Ecosystems*, **2**, 65–94.

Barnes, R.S.K. (1980). *Coastal Lagoons. The Natural History of a Neglected Habitat*. Cambridge, UK: Cambridge University Press.

Barnes, R.S.K. (1989). The coastal lagoons of Britain: an overview and conservation appraisal. *Biological Conservation*, **49**, 295–313.

Barnes, R.S.K. (1991). European estuaries and lagoons: a personal overview of problems and possibilities for conservation and management. *Aquatic Conservation: Marine and Freshwater Ecosystems*, **1**, 79–87.

Beck, M.W., Brumbaugh, R.D., Airoldi, L., *et al.* (2011). Oyster reefs at risk and recommendations for conservation, restoration, and management. *BioScience*, **61**, 107–16.

Beer, N.A. &. Joyce, C.B. (2013). North Atlantic coastal lagoons: conservation, management and research challenges in the twenty-first century. *Hydrobiologia*, **701**, 1–11.

Beukema, J.J. (1995). Long-term effects of mechanical harvesting of lugworms *Arenicola marina* on the zoobenthic community of a tidal flat in the Wadden Sea. *Netherlands Journal of Sea Research*, **33**, 219–27.

Birch, G.F., Murray, O., Johnson, I. & Wilson, A. (2009). Reclamation in Sydney estuary, 1788–2002. *Australian Geographer*, **40**, 347–68.

Bird, E.C.F. (1994). Physical setting and geomorphology of coastal lagoons. In *Coastal Lagoon Processes*, ed. B. Kjerfve, pp. 9–39. Amsterdam: Elsevier.

Bishop, R.E., Humphreys, W.F., Cukrov, N., *et al.* (2015). 'Anchialine' redefined as a subterranean estuary in a crevicular or cavernous geological setting. *Journal of Crustacean Biology*, **35**, 511–14.

Blaber, S.J.M., Cyrus, D.P., Albaret, J.-J., *et al.* (2000). Effects of fishing on the structure and functioning of estuarine and nearshore ecosystems. *ICES Journal of Marine Science*, **57**, 590–602.

Bricker, S.B., Longstaff, B., Dennison, W., *et al.* (2008). Effects of nutrient enrichment in the nation's estuaries: a decade of change. *Harmful Algae*, **8**, 21–32.

Briggs, J.C. (2012). Marine species invasions in estuaries and harbors. *Marine Ecology Progress Series*, **449**, 297–302.

Burton, N.H.K., Rehfisch, M.N., Clark, N.A. & Dodd, S.G. (2006). Impacts of sudden winter habitat loss on the body condition and survival of redshank *Tringa totanus*. *Journal of Applied Ecology*, **43**, 464–73.

Cabezas, P., Alda, F., Macpherson, E. & Machordom, A. (2012). Genetic characterization of the endangered and endemic anchialine squat lobster *Munidopsis polymorpha* from Lanzarote (Canary Islands): management implications. *ICES Journal of Marine Science*, **69**, 1030–7.

Carlton, J.T., Thompson, J.K., Schemel, L.E. & Nichols, F.H. (1990). Remarkable invasion of San Francisco Bay (California, USA) by the Asian clam *Potamocorbula amurensis*. I. Introduction and dispersal. *Marine Ecology Progress Series*, **66**, 81–94.

Carr, A. III (1982). Tropical forest conservation and estuarine ecology. *Biological Conservation*, **23**, 247–59.

Chivas, A.R., Torgersen, T. & Polach, H.A. (1990). Growth rates and Holocene development of stromatolites from Shark Bay, Western Australia. *Australian Journal of Earth Sciences*, **37**, 113–21.

Clark, R.B. (2001). *Marine Pollution*, 5th edn. Oxford, UK: Clarendon Press.
Coen, L.D., Brumbaugh, R.D., Bushek, D., *et al.* (2007). Ecosystem services related to oyster restoration. *Marine Ecology Progress Series*, **341**, 303–7.
Colclough, S.R., Gray, G., Bark, A. & Knights, B. (2002). Fish and fisheries of the tidal Thames: management of the modern resource, research aims and future pressures. *Journal of Fish Biology*, **61** (Suppl. A), 64–73.
Colombo, G (1977). Lagoons. In *The Coastline*, ed. R.S.K. Barnes, pp. 63–81. Chichester, UK: John Wiley & Sons.
Colwell, M.A. (2010). *Shorebird Ecology, Conservation, and Management*. Berkeley, CA: University of California Press.
Connor, M.S., Davis, J.A., Leatherbarrow, J., *et al.* (2007). The slow recovery of San Francisco Bay from the legacy of organochlorine pesticides. *Environmental Research*, **105**, 87–100.
Costanza, R., d'Arge, R., de Groot, R., *et al.* (1997). The value of the world's ecosystem services and natural capital. *Nature*, **387**, 253–60.
Culbertson, J.B., Valiela, I., Olsen, Y.S. & Reddy, C.M. (2008). Effect of field exposure to 38-year-old residual petroleum hydrocarbons on growth, condition index, and filtration rate of the ribbed mussel, *Geukensia demissa*. *Environmental Pollution*, **154**, 312–19.
Davidson, N.C. (1990). The conservation of British North Sea estuaries. *Hydrobiologia*, **195**, 145–62.
Davidson, N.C. (1991). Human activities and wildlife conservation on estuaries of different sizes: a comment. *Aquatic Conservation: Marine and Freshwater Ecosystems*, **1**, 89–92.
Davidson, N.C. & Evans, P.R. (1986). The role of potential man-made and man-modified wetlands in the enhancement of the survival of overwintering shorebirds. *Colonial Waterbirds*, **9**, 176–88.
Day, J.H. (1981). The estuarine fauna. In *Estuarine Ecology with Particular Reference to Southern Africa*, ed. J.H. Day, pp. 147–78. Rotterdam: A.A. Balkema.
Dittel, A.I. & Epifanio, C.E. (2009). Invasion biology of the Chinese mitten crab *Eriochier sinensis*: a brief review. *Journal of Experimental Marine Biology and Ecology*, **374**, 79–92.
Drinkwater, K.F. & Frank, K.T. (1994). Effects of river regulation and diversion on marine fish and invertebrates. *Aquatic Conservation: Freshwater and Marine Ecosystems*, **4**, 135–51.
Edgar, G.J., Barrett, N.S., Graddon, D.J. & Last, P.R. (2000). The conservation significance of estuaries: a classification of Tasmanian estuaries using ecological, physical and demographic attributes as a case study. *Biological Conservation*, **92**, 383–97.
Elliott, M. & Quintino, V. (2007). The Estuarine Quality Paradox, environmental homeostasis and the difficulty of detecting anthropogenic stress in naturally stressed areas. *Marine Pollution Bulletin* **54**, 640–5.
Engle, V.D., Kurtz, J.C., Smith, L.M., Chancy, C. & Bourgeois, P. (2007). A classification of U.S. estuaries based on physical and hydrologic attributes. *Environmental Monitoring and Assessment*, **129**, 397–412.
Evans, S.M. (1999). Tributyltin pollution: the catastrophe that never happened. *Marine Pollution Bulletin*, **38**, 629–36.
Evans, S.M. (2000). Marine antifoulants. In *Seas at the Millennium: an Environmental Evaluation*, ed. C. Sheppard, pp. 247–56. Oxford, UK: Elsevier Science Ltd.
FAO (2012). *The State of World Fisheries and Aquaculture 2012*. Rome: FAO.
Foden, J., Rogers, S.I. & Jones, A.P. (2009). Recovery rates of UK seabed habitats after cessation of aggregate extraction. *Marine Ecology Progress Series*, **390**, 15–26.
Forrest, B.M., Keeley, N.B., Hopkins, G.A., Webb, S.C. & Clement, D.M. (2009). Bivalve aquaculture in estuaries: review and synthesis of oyster cultivation effects. *Aquaculture*, **298**, 1–15.
Goss-Custard, J.D., Stillman, R.A., West, A.D., Caldow, R.W.G. & McGrorty, S. (2002). Carrying capacity in overwintering migratory birds. *Biological Conservation*, **105**, 27–41.
Greene, V.E., Sullivan, L.J., Thompson, J.K. & Kimmerer, W.J. (2011). Grazing impact of the invasive clam *Corbula amurensis* on the microplankton assemblage of the northern San Francisco Estuary. *Marine Ecology Progress Series*, **431**, 183–93.
Gundlach, E.R. & Hayes, M.O. (1978). Vulnerability of coastal environments to oil spill impacts. *Marine Technology Society Journal*, **12** (4), 18–27.
Haedrich, R.L. (1983). Estuarine fishes. In *Estuaries and Enclosed Seas*, ed. B.H. Ketchum, pp. 183–207. Amsterdam: Elsevier.

Hale, W.G. (1980). *Waders*. London: Collins.

Hawkins, S.J., Allen, J.R. & Bray, S. (1999). Restoration of temperate marine and coastal ecosystems: nudging nature. *Aquatic Conservation: Marine and Freshwater Ecosystems*, **9**, 23–46.

Heiman, K.W. & Micheli, F. (2010). Non-native ecosystem engineer alters estuarine communities. *Integrative and Comparative Biology*, **50**, 226–36.

Ivanov, M.V., Smagina, D.S., Chivilev, S.M. & Kruglikov, O.E. (2013). Degradation and recovery of an Arctic benthic community under organic enrichment. *Hydrobiologia*, **706**, 191–204.

Jackson, J.B.C., Kirby, M.X., Berger, W.H., *et al.* (2001). Historical overfishing and the recent collapse of coastal ecosystems. *Science*, **293**, 629–38.

Jackson, M.J. & James, R. (1979). The influence of bait digging on cockle, *Cerastoderma edule* populations in north Norfolk. *Journal of Applied Ecology*, **16**, 671–9.

Kaiser, M.J., Attrill, M.J., Jennings, S., *et al.* (2011). *Marine Ecology: Processes, Systems, and Impacts*, 2nd edn. Oxford, UK: Oxford University Press.

Katsanevakis, S., Wallentinus, I.W., Zenetos, A., *et al.* (2014). Impacts of invasive alien marine species on ecosystem services and biodiversity: a pan-European review. *Aquatic Invasions*, **9**, 391–423.

Kennish, M.J. (1992). *Ecology of Estuaries: Anthropogenic Effects*. Boca Raton, FL: CRC Press Inc.

Langston, W.J., Pope, N.D., Davey, M., *et al.* (2015). Recovery from TBT pollution in English Channel environments: a problem solved? *Marine Pollution Bulletin*, **95**, 551–64.

Lebeuf, M., Noël, M., Trottier, S. & Measures, L. (2007). Temporal trends (1987–2002) of persistent, bioaccumulative and toxic (PBT) chemicals in beluga whales (*Delphinapterus leucas*) from the St. Lawrence Estuary, Canada. *Science of the Total Environment*, **383**, 216–31.

Lee, H. II, Thompson, B. Lowe, S. (2003). Estuarine and scalar patterns of invasion in the soft-bottom benthic communities of the San Francisco Estuary. *Biological Invasions*, **5**, 85–102.

Leignel, V., Stillman, J.H., Baringou, S., Thabet, R. & Metais, I. (2014). Overview on the European green crab *Carcinus* spp. (Portunidae, Decapoda), one of the most famous marine invaders and ecotoxicological models. *Environmental Science and Pollution Research*, **21**, 9129–44.

Lellis-Dibble, K.A., McGlynn, K.E. & Bigford, T.E. (2008). *Estuarine Fish and Shellfish Species in U.S. Commercial and Recreational Fisheries: Economic Value as an Incentive to Protect and Restore Estuarine Habitat*. U.S. Department of Commerce, NOAA Technical Memorandum NMFS-F/SPO-90, 94 pp.

Lincoln, R.J. & Boxshall, G.A. (1987). *The Cambridge Illustrated Dictionary of Natural History*. Cambridge, UK: Cambridge University Press.

Little, C. (2000). *The Biology of Soft Shores and Estuaries*. Oxford, UK: Oxford University Press.

Lubell, M. (2004). Resolving conflict and building cooperation in the National Estuary Program. *Environmental Management*, **33**, 677–91.

McLusky, D.S., Bryant, D.M. & Elliott, M. (1992). The impact of land-claim on macrobenthos, fish and shorebirds on the Forth Estuary, eastern Scotland. *Aquatic Conservation: Marine and Freshwater Ecosystems*, **2**, 211–22.

McLusky, D.S., Bryant, V. & Campbell, R. (1986). The effects of temperature and salinity on the toxicity of heavy metals to marine and estuarine invertebrates. *Oceanography and Marine Biology: An Annual Review*, **24**, 481–520.

McLusky, D.S. & Elliott, M. (2004). *The Estuarine Ecosystem: Ecology, Threats and Management*, 3rd edn. Oxford, UK: Oxford University Press.

McQuaid, K.A. & Griffiths, C.L. (2014). Alien reef-building polychaete drives long-term changes in invertebrate biomass and diversity in a small, urban estuary. *Estuarine, Coastal and Shelf Science*, **138**, 101–6.

Mitchell, J.R., Moser, M.E. & Kirby, J.S. (1988). Declines in midwinter counts of waders roosting on the Dee estuary. *Bird Study*, **35**, 191–8.

Mitchell, R. (1978). Nature conservation implications of hydraulic engineering schemes affecting British estuaries. *Hydrobiological Bulletin*, **12**, 333–50.

Molnar, J.L. Gamboa, R.L.; Revenga, Carmen, R. & Spalding, M.D. (2008). Assessing the global threat of invasive species to marine biodiversity. *Frontiers in Ecology and the Environment*, **6**, 485–92.

Moreira, F. (1995). Diet of black-headed gulls *Larus ridibundus* on emerged intertidal areas in the Tagus estuary (Portugal): predation or grazing? *Journal of Avian Biology*, **26**, 277–82.

Myers, J.P., Morrison, R.I.G., Antas, P.Z., *et al.* (1987). Conservation strategy for migratory species. *American Scientist*, **75**, 18–26.
National Research Council (2003). *Oil in the Sea III: Inputs, Fates, and Effects*. Washington, DC: National Academies Press.
Navedo, J.G. & Herrera, A.G. (2012). Effects of recreational disturbance on tidal wetlands: supporting the importance of undisturbed roosting sites for waterbird conservation. *Journal of Coastal Conservation*, **16**, 373–81.
Neely, A.E. & Zajac, R.N. (2008). Applying marine protected area design models in large estuarine systems. *Marine Ecology Progress Series*, **373**, 11–23.
Neiber, M.T., Hartke, T.R., Stemme, T., *et al.* (2011) Global biodiversity and phylogenetic evaluation of Remipedia (Crustacea). *PLoS ONE*, **6**, e19627.
Newell, R.C., Seiderer, L.J. & Hitchcock, D.R. (1998). The impact of dredging works in coastal waters: a review of the sensitivity to disturbance and subsequent recovery of biological resources on the sea bed. *Oceanography and Marine Biology: An Annual Review*, **36**, 127–78.
Newman, S. J. & Smith, S. A. (2006). Marine mammal neoplasia: a review. *Veterinary Pathology*, **43**, 865–80.
Oliveira, O.M.P. (2007). The presence of the ctenophore *Mnemiopsis leidyi* in the Oslofjorden and considerations on the initial invasion pathways to the North and Baltic Seas. *Aquatic Invasions*, **2**, 185–9.
OSPAR Commission (2014). *OSPAR Guidelines for the Management of Dredged Material at Sea*. Agreement 2014–06. 39 pp. London: OSPAR.
O'Sullivan, A.J. & Jacques, T.G. (2001). *Impact Reference System. Effects of Oil in the Marine Environment: Impact of Hydrocarbons on Fauna and Flora. European Commission, Directorate General Environment*, original edn 1991, revised edn 1998, Internet edn 2001. www.europa.eu.int/comm/environment/civil/index.htm, p. 79.
Pearson, T.H. & Rosenberg, R. (1976). A comparative study of the effects on the marine environment of wastes from cellulose industries in Scotland and Sweden. *Ambio*, **5**, 77–9.
Pearson, T.H. & Rosenberg, R. (1978). Macrobenthic succession in relation to organic enrichment and pollution of the marine environment. *Oceanography and Marine Biology: An Annual Review*, **16**, 229–311.
Pérez-Ruzafa, A. & Marcos, C. (2012). Fisheries in coastal lagoons: an assumed but poorly researched aspect of the ecology and functioning of coastal lagoons. *Estuarine, Coastal and Shelf Science*, **110**, 15–31.
Piersma, T (1987). Production by intertidal benthic animals and limits to their predation by shorebirds: a heuristic model. *Marine Ecology Progress Series*, **38**, 187–96.
Piersma, T., Koolhaas, A., Dekinga, A., *et al.* (2001). Long-term indirect effects of mechanical cockle-dredging on intertidal bivalve stocks in the Wadden Sea. *Journal of Applied Ecology*, **38**, 976–90.
Poole, S. (1996). The United States National Estuary Program. *Ocean & Coastal Management*, **30**, 63–7.
Preisler, R.K., Wasson, K., Wolff, W.J. & Tyrrell, M.C. (2009). Invasions of estuaries vs. the adjacent open coast: a global perspective. In *Biological Invasions in Marine Ecosystems: Ecological, Management and Geographic Perspectives*, ed. G. Rilov & J.A. Crooks, pp. 587–617. Berlin: Springer-Verlag.
Purcell, J.E., Shiganova, T.A., Decker, M.B. & Houde, E.D. (2001). The ctenophore *Mnemiopsis* in native and exotic habitats: U.S. estuaries versus the Black Sea basin. *Hydrobiologia*, **451**, 145–76.
Raffaelli, D. (2000). Interactions between macro-algal mats and invertebrates in the Ythan estuary, Aberdeenshire, Scotland. *Helgoland Marine Research*, **54**, 71–9.
Raffaelli, D., Hull, S. & Milne, H. (1989). Long-term changes in nutrients, weed mats and shorebirds in an estuarine system. *Cahiers de Biologie Marine*, **30**, 259–70.
Raffaelli, D.G., Raven, J.A. & Poole, L. J. (1998). Ecological impact of green macroalgal blooms. *Oceanography and Marine Biology: An Annual Review*, **36**, 97–125.
Ramsar Convention Secretariat (2010). *Wise Use of Wetlands: Concepts and Approaches for the Wise Use of Wetlands*. Ramsar Handbooks for the Wise Use of Wetlands, 4th edn, vol. 1. Gland, Switzerland: Ramsar Convention Secretariat.
Reading, C.J. & McGrorty, S. (1978). Seasonal variations in the burying depth of *Macoma balthica* (L.) and its accessibility to wading birds. *Estuarine and Coastal Marine Science*, **6**, 135–44.
Runge, C.A., Watson, J.E.M., Butchart, S.H.M., *et al.* (2015). Protected areas and global conservation of migratory birds. *Science*, **350**, 1255–8.

Schneider, M., Scholz, J., Lubell, M., Mindruta, D. & Edwardsen, M. (2003). Building consensual institutions: networks and the National Estuary Program. *American Journal of Political Science*, **47**, 143–58.
Shepherd, P.C.F. & Boates, J.S. (1999). Effects of a commercial baitworm harvest on semipalmated sandpipers and their prey in the Bay of Fundy Hemispheric Shorebird Reserve. *Conservation Biology*, **13**, 347–56.
Siegfried, W.R. (1981). The estuarine avifauna of southern Africa. In *Estuarine Ecology with Particular Reference to Southern Africa*, ed. J.H. Day, pp. 223–50. Rotterdam: A.A. Balkema.
Sket, B. (1996). The ecology of anchihaline caves. *Trends in Ecology and Evolution*, **11**, 221–5.
Smetacek, V. & Zingone, A. (2013). Green and golden seaweed tides on the rise. *Nature*, **504**, 84–8.
Smith, J.T., Walker, L.A., Shore, R.F., *et al.* (2009). Do estuaries pose a toxic contamination risk for wading birds? *Ecotoxicology*, **18**, 906–17.
Sonak, S., Pangam, P., Giriyan, A. & Hawaldar, K. (2009). Implications of the ban on organotins for protection of global coastal and marine ecology. *Journal of Environmental Management*, **90**, S96–S108.
Sousa, R., Gutiérrez, J.L. & Aldridge, D.C. (2009). Non-indigenous invasive bivalves as ecosystem engineers. *Biological Invasions*, **11**, 2367–85.
Spalding, M.D., Fox, H.E., Allen, G.R., *et al.* (2007). Marine ecoregions of the world: a bioregionalization of coastal and shelf areas. *BioScience*, **57**, 573–83.
Sypitkowski, E., Ambrose, W.G. Jr, Bohlen, C. & Warren, J. (2009). Catch statistics in the bloodworm fishery in Maine. *Fisheries Research*, **96**, 303–7.
Thomas, M. (1998). Temporal changes in the movements and abundance of Thames estuary fish populations. In *A Rehabilitated Estuarine Ecosystem: The Environment and Ecology of the Thames Estuary*, ed. M.J. Attrill, pp. 115–39. Dordrecht, Netherlands: Kluwer Academic Publishers.
Van Niekerk, L., Adams, J.B., Bate, G.C., *et al.* (2013). Country-wide assessment of estuary health: an approach for integrating pressures and ecosystem response in a data limited environment. *Estuarine, Coastal and Shelf Science* **130**, 239–51.
Wilson, J.G. (1988). *The Biology of Estuarine Management*. London: Croom Helm.
Wilson, J.G. & Fleeger, J.W. (2013). Estuarine benthos. In *Estuarine Ecology*, 2nd edn, ed. J.W. Day Jr, B.C. Crump, W.M. Kemp & A. Yáñez-Arancibia, pp. 303–25. Hoboken, NJ: John Wiley & Sons, Inc.
Woodwell, G.M., Rich, P.H. & Hall, C.A.S. (1973). Carbon in estuaries. In *Carbon and the Biosphere*, ed. G.M. Woodwell & E.V. Pecan, Proceedings of the 24th Brookhaven Symposium in Biology, pp. 221–40. Springfield, VA: Technical Information Center, Office of Information Services, U.S. Atomic Energy Commission.
Zharikov, Y. & Skilleter, G.A. (2003). Depletion of benthic invertebrates by bar-tailed godwits *Limosa lapponica* in a subtropical estuary. *Marine Ecology Progress Series*, **254**, 151–62.

# 12 Coastal Wetlands

The term 'wetlands' is used here for low-lying areas that are submerged or periodically inundated and where rooted plants form a conspicuous component of the biota (Mitsch & Gosselink, 2000). Most are freshwater ecosystems, but wetlands also include important coastal habitats, notably saltmarshes, mangroves, and seagrass beds, that often characterize marine, estuarine, and brackish environments. Like their freshwater counterparts, the plants themselves provide much of the habitat, increasing its complexity and biodiversity. Coastal wetlands can also stabilize sediments, provide shoreline protection, and are often highly productive systems, important in nutrient cycling and with key functional links to other coastal and marine ecosystems. They are among the most valuable habitats in terms of the resources and ecosystem services they provide (Costanza *et al.*, 2014) but also very vulnerable to anthropogenic disturbance and among the most threatened of habitats (Barbier *et al.*, 2011). The vital importance of conserving coastal wetlands has been reiterated by a number of international forums, including the Ramsar Convention (1971), Agenda 21 of the UN Conference on Environment and Development (1992), and the World Summit on Sustainable Development (2002).

## Saltmarshes

Saltmarshes are areas vegetated by herbs, grasses, or low shrubs that border saline water bodies and are periodically subject to tidal flooding (Adam, 1990). They develop particularly in sheltered sites where fine sediment is accumulating, allowing saltmarsh plants to take root. Saltmarshes develop on the upper shore, normally from about mean high-water neap up to the level of the highest tides. Downshore, saltmarsh typically gives way to unvegetated tidal flats, seagrass beds, or mangroves, whilst landward and up-estuary, saltmarsh merges into terrestrial and freshwater marsh assemblages. Saltmarshes are best developed in temperate latitudes and can extend to Arctic regions as long as ice abrasion is not limiting. Saltmarshes are more restricted in distribution in the subtropics to tropics, where they are largely replaced by mangroves. The global extent of saltmarshes is estimated to be of the order of 400 000 $km^2$ (Nellemann *et al.*, 2009).

A range of flowering plants contribute to saltmarsh assemblages around the world. Among the more widespread or cosmopolitan genera are *Salicornia* and *Sarcocornia* (glassworts), *Suaeda* (seablites), *Limonium* (sea lavenders), *Puccinellia* (alkali grasses), *Distichlis* (spikegrasses), *Spartina* (cordgrasses), *Juncus* (rushes), and *Carex* (sedges) (Fig. 12.1). The composition of the vegetation varies within biogeographical regions depending on a range of physical and chemical variables, so that a number of major types of saltmarsh are recognized, and usually a clear zonation is evident related to elevation. Species richness of the saltmarsh flora increases from the tropics towards temperate latitudes, contrary to the usual pattern for flowering plants (Adam, 2002). Major types of saltmarsh are outlined in Table 12.1.

**Table 12.1** Major Types of Saltmarsh

| Distribution | Characteristics |
|---|---|
| Arctic | Species poor, lowest zone dominated by *Puccinellia phryganodes*. (No Antarctic analogue) |
| Boreal | More species rich than Arctic marshes. *Triglochin maritimum* (sea arrow-grass) and *Salicornia europaea* widespread. Often brackish, with extensive *Carex*-dominated communities. (Limited subantarctic equivalent) |
| Temperate | Lower zones (before spread of *Spartina anglica*) dominated by *Puccinellia maritima* |
| Europe | Upper-marsh dominated by *Juncus maritimus. Atriplex portulacoides* (sea purslane) occurs widely |
| Western North America | *Distichlis spicata, Carex lyngbei* in brackish sites |
| Japan | *Zoysia sinica* (zoysia grass) dominant in mid-marsh |
| Australasia | *Sarcocornia quinqueflora* dominant in low-marsh, *Juncus kraussii* in upper-marsh |
| South Africa | *Sarcocornia* spp. in lower-marsh, *Juncus kraussii* in upper-marsh, *Spartina maritima* present |
| West Atlantic | Extensive dominance of *Spartina alterniflora* |
| Dry coast | Dwarf shrubs dominant, including *Sarcocornia, Suaeda, Limoniastrum, Frankenia* (seaheaths) |
| Tropical | Very species poor. Extensive *Sporobolus virginicus* and *Paspalum vaginatum* grasslands. *Batis maritima* (saltwort), *Sesuvium portulacastrum* (shoreline purslane), and *Cressa cretica* (littoral bindweed) |

From Adam, P., Saltmarshes in a time of change, *Environmental Conservation*, 29 (1), 39–61, reproduced with permission.

**Fig 12.1** A grazed saltmarsh dominated by the alkali grass *Puccinellia maritima* (Northton, Outer Hebrides, Scotland). Saltmarshes are productive systems, important as carbon sinks and well developed in temperate latitudes. Large areas of saltmarsh have been lost to agricultural, urban, and industrial use. Photograph: P.K. Probert. (A black and white version of this figure will appear in some formats. For the colour version, please refer to the plate section.)

## Mangroves

Mangroves are tropical to subtropical in distribution (roughly delimited by the winter 20°C seawater isotherm) where they can form extensive stands. They flourish in soft sediment along marine and estuarine coasts, typically between mean sea level and high spring tide (Alongi, 2002). Estimates put the global extent of mangroves at around 138 000–152 000 km$^2$, with the greatest extent (about 40%) in Asia (Spalding *et al.*, 2010; Giri *et al.*, 2011). Some 70 species of trees and shrubs from a range of families are generally recognized as characteristic of mangrove communities, although two families generally dominate: Rhizophoraceae (*Bruguiera, Ceriops, Kandelia, Rhizophora*) and Avicenniaceae (*Avicennia*). There are two main biogeographic regions, the Indo-West Pacific and the Atlantic-Eastern Pacific, the former having five to six times as many species (Spalding *et al.*, 2010) (Fig. 12.2). Mangroves range from open patchy vegetation only a metre or so high to closed forests 30 m or more high, depending on abiotic factors and species composition, and show marked zonation parallel to the shoreline (Mitsch & Gosselink, 2000). The principal mangrove species have various adaptations to cope with the demanding conditions, including mechanisms to exclude or excrete salt and, for living in anoxic sediment, root structures to obtain oxygen, notably aerial prop roots and pneumatophores that poke out of the sediment (Fig. 12.3). In many species the seed germinates whilst still on the parent tree so that a seedling is eventually released, which may then float for a prolonged period.

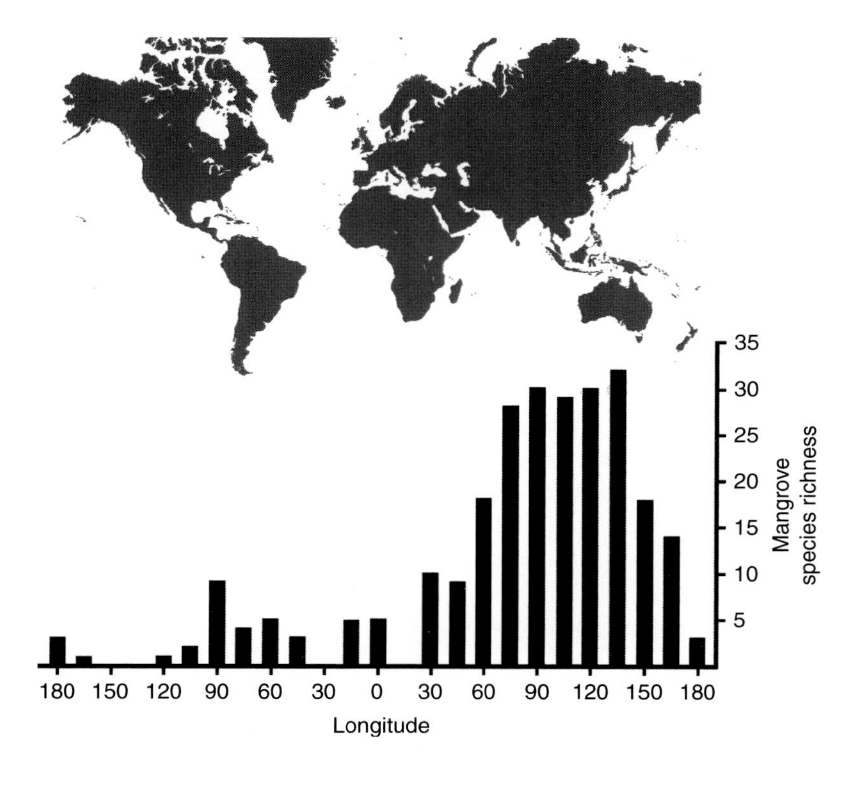


**Fig. 12.2** Species richness of mangroves in relation to longitude. From Ellison, A.M., Farnsworth, E.J. & Merkt, R.E. (1999). Origins of mangrove ecosystems and the mangrove biodiversity anomaly. *Global Ecology and Biogeography*, 8, 95–115. With permission from Wiley. © 1999 Blackwell Science Ltd.

**Fig. 12.3** Mangroves provide important ecosystem services in tropical to subtropical latitudes, including coastal protection and migratory and nursery habitat. Note the aerial prop roots and pneumatophores. *Rhizophora* in the Solomon Islands. Photograph: Getty Images/Sune Wendelboe. (A black and white version of this figure will appear in some formats. For the colour version, please refer to the plate section.)

## Seagrasses

Seagrasses comprise a number of flowering plants adapted to living in shallow subtidal to intertidal environments. They anchor themselves in the substratum with an extensive root and rhizome system, and most have blade-like leaves – up to 2 m long in some species – and underwater pollination. There are some 70 species in four main families: Zosteraceae (*Zostera, Phyllospadix*), Cymodoceaceae (*Amphibolis, Halodule, Cymodocea, Syringodium, Thalassodendron*), Posidoniaceae (*Posidonia*), and Hydrocharitaceae (*Enhalus, Halophila, Thalassia*) (den Hartog & Kuo, 2006; Short *et al.*, 2011). Seven of these genera are mainly tropical: *Halodule, Cymodocea, Syringodium, Thalassodendron, Enhalus, Halophila,* and *Thalassia*; and the other four mainly temperate: *Zostera, Phyllospadix, Amphibolis,* and *Posidonia* (Fig. 16.2). Species richness is highest in tropical regions, particularly in the Indo-Pacific.

Seagrass beds occur worldwide, from the tropics to subpolar latitudes. The extent of seagrass habitat is still poorly known for many regions, but the global area is around 177 000–330 000 km$^2$ (Spalding *et al.*, 2003; Nellemann *et al.*, 2009). Seagrasses occur mainly on sandy to muddy sediments from the intertidal zone to depths at which photosynthesis is still feasible – usually to a few metres but to more than 50 m in the clearest waters – and they flourish over a wide range of salinity, from brackish to hypersaline. Most seagrass beds are monospecific, even in regions of high species richness (Duarte, 2001). Reproduction is predominantly vegetative, and a meadow may be composed of distinct clones, each hundreds to thousands of years old (Migliaccio *et al.*, 2005). Seagrass coverage can be very extensive where there is suitable habitat. One of the largest areas, in the Great Barrier Reef, has a total extent of about 35 000 km$^2$ (Coles *et al.*, 2015).

# Functional Role

## Productivity

Productivity of saltmarshes, mangroves, and seagrass beds varies greatly, although all can be remarkably productive systems on a par with or exceeding intensive agriculture. Above-ground net production for seagrasses is of the order of 300–1500 g C m$^{-2}$ y$^{-1}$, but below-ground production can be even greater. In addition, there is the productivity of associated epiphytes and benthic macro- and microalgae (Mateo *et al.*, 2006). Similarly, rates of 800–1000 g C m$^{-2}$ y$^{-1}$ for the above-ground biomass have been estimated for saltmarshes and mangroves (Little, 2000; Bouillon *et al.*, 2008; Laffoley & Grimsditch, 2009). Vascular plant tissues are not readily broken down by marine animals – much less so than algae – and much of the primary production from coastal wetlands enters food webs as organic detritus or as dissolved organic matter available for microbial use. Large amounts of detritus are exported to adjacent waters and incorporated into coastal sediments. The export of particulate organic carbon from mangroves accounts for as much as 10–11% of the total input of terrestrial carbon to the world ocean (Alongi, 2014).

Vegetated coastal ecosystems are highly significant as carbon sinks, storing carbon within their underlying sediments, in above- and below-ground biomass, and within leaf litter and dead wood. Whilst comprising less than 0.5% of the world's seabed area, coastal wetlands account for some 50–70% of all carbon storage in ocean sediments (Nellemann *et al.*, 2009). The mean long-term rates of carbon sequestration in sediments of vegetated coastal ecosystems are in fact far higher than those of terrestrial forests (Fig. 12.4).

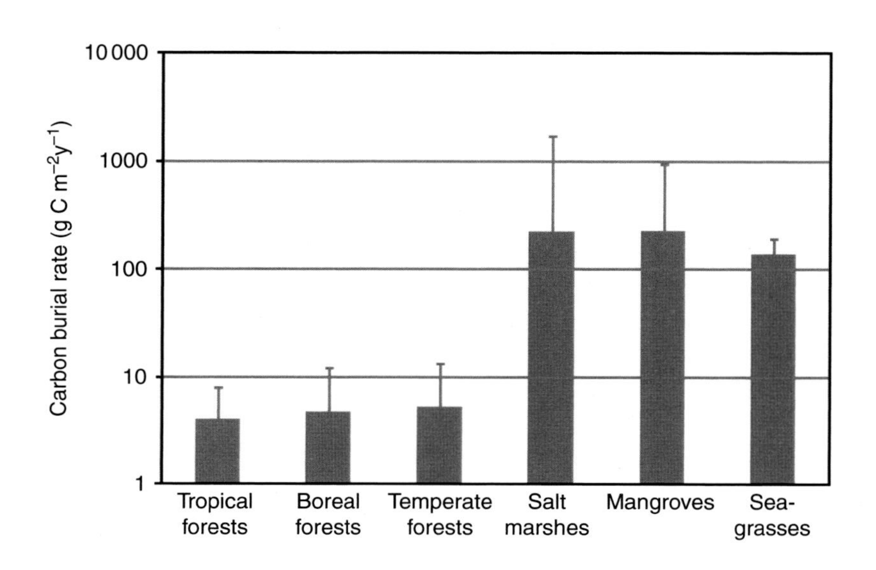


**Fig. 12.4** Mean long-term rates of carbon sequestration in soils in terrestrial forests and sediments in vegetated coastal ecosystems. Error bars indicate maximum rates of accumulation. Republished with permission of the Ecological Society of America from A blueprint for blue carbon: towards an improved understanding of the role of vegetated coastal habitats in sequestering $CO_2$, *Frontiers in Ecology and the Environment*, Mcleod, E., Chmura, G.L., Bouillon, S., et al., Vol. 9 (10), pages 552–60, copyright 2011; permission conveyed through Copyright Clearance Center, Inc.

## Biodiversity

Coastal wetlands are structurally diverse habitats that provide for a wide range and abundance of associated organisms from microbes to vertebrates. A seagrass bed, for instance, provides surfaces that can be used as substrata by a variety of organisms, a source of food, and a refuge within the leaf canopy. Numerous studies have found the fauna of seagrass beds to be more abundant and diverse than that of adjacent unvegetated areas (Hemminga & Duarte, 2000).

The marine invertebrate fauna of saltmarshes is generally of low diversity, although high population densities can occur. The macrobenthos is dominated by detritus feeders, where many of the species are also those of unvegetated tidal flats. Saltmarshes also provide essential habitat for birds, which is often a reason for their designation as sites of nature conservation importance. Wildfowl can be conspicuous on saltmarshes, using them as sites for roosting and feeding (e.g. on *Puccinellia* grass). Waders too use saltmarsh for roosting but generally feed on adjacent unvegetated tidal flats. The upper marsh, when not inundated by the highest tides over summer, can provide nesting habitat, such as for various species of waders, gulls, and terns. Throughout Europe, for example, saltmarshes are important as breeding habitat for redshank. Nearly half the population of redshank breeding in Britain nest on saltmarshes, although surveys since 1985 indicate a serious decline in breeding density of this wader and a need to assess grazing practices on saltmarshes (Malpas *et al.*, 2013).

Mangroves too provide habitat for a wide range of organisms. The emergent vegetation can be important habitat for terrestrial animals, including insects, amphibians, reptiles, birds, and mammals (Kathiresan & Bingham, 2001; Nagelkerken *et al.*, 2008). Among the more obvious marine animals in mangroves are crustaceans, molluscs, and teleost fish. Crabs of the families Grapsidae and Ocypodidae are often conspicuous in Indo-West Pacific mangroves and have an important role in processing mangrove detritus. Mangrove roots and pneumatophores provide a substratum for attached algae and invertebrates, which can be an important source of food for fishes (Blaber, 2013).

Mangroves tend to support distinct fish assemblages, although species richness and abundance can vary considerably between regions (Bloomfield & Gillanders, 2005). Many of the fish and decapod crustaceans of mangroves are fishery species, and there is much debate on the extent to which mangroves support fish populations. It appears that for some areas, mangrove-related species can contribute very significantly to fisheries landings – at least one-third to two-thirds of the entire commercial catch (Walters *et al.*, 2008). Aburto-Oropeza *et al.* (2008) found the landings for fish and blue crab in the Gulf of California to be positively related to the local area of mangroves, or more especially to the length of the mangrove fringe, supporting evidence that many species use the edge of mangroves as nursery or feeding grounds (Fig. 12.5).

Likewise, shelter and abundance of prey make seagrass beds important as fish habitat, and they may serve too as nursery areas for fish and shellfish species (Jackson *et al.*, 2001; Heck *et al.*, 2003). A nursery habitat is one that enhances recruitment to adult populations when compared to adjacent habitats. This function may arise from a combination of factors, notably the density, survival, and growth of juveniles and their movement to adult habitat (Beck *et al.*, 2001; Sheridan & Hays, 2003). The nursery role of seagrass meadows appears to be due mainly to the shelter they provide from predators, although in this regard they are probably similar to other structured habitats (Heck *et al.*, 2003). Studies indicate potential links between loss of seagrass and fisheries, but the relationship between fisheries catches and seagrass cover is still far from clear (Gillanders, 2006).

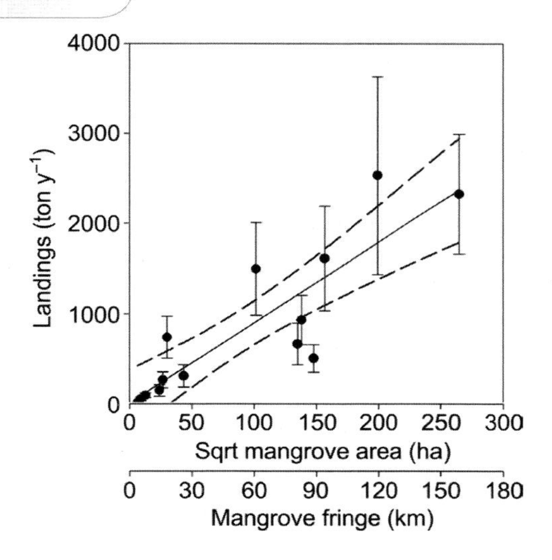


**Fig. 12.5** Relationship between fisheries landings (fish and blue crab) and mangrove area in the Gulf of California. The square root of the mangrove area is directly related to the length of the mangrove fringe (fringe = square root of area × 6.13). Data are average ± standard error (2001–5); solid line, model; dashed line, 95% confidence intervals. From Aburto-Oropeza, O., Ezcurra, E., Danemann, G., *et al.* (2008). Mangroves in the Gulf of California increase fishery yields. *Proceedings of the National Academy of Sciences*, 105, 10456–9. Copyright (2008) National Academy of Sciences, U.S.A.

## Habitat Connectivity

There is, nevertheless, mounting evidence of important faunal linkages between wetlands and adjacent habitats, with species migrating between them at different stages of their life cycle and possibly on a daily or tidal basis. In a study in the Caribbean, Mumby *et al.* (2004) demonstrated that mangroves strongly influence the fish community structure on neighbouring reefs. They compared reefs with associated mangrove cover against reefs with extremely limited or no mangrove cover and found that mangroves provided an important nursery function. Juvenile bluestriped grunt (*Haemulon sciurus*) migrate from seagrass to mangroves before eventually migrating to reefs. The mangroves provide a refuge from predators, thus increasing the survivorship of juveniles. But if mangroves are absent, the juveniles migrate directly to patch reefs but at a smaller size and lower density. Juveniles of some fish, however, appear to depend on mangroves, such as the rainbow parrotfish (*Scarus guacamaia*), which is now locally extinct following mangrove removal (Fig. 12.6). Similarly, Unsworth *et al.* (2008) found the fish fauna of seagrass beds in the Wakatobi Marine National Park, Indonesia, to be significantly influenced by the proximity of adjacent mangrove and coral reef habitats (in terms of abundance, species richness, trophic structure, and assemblage composition).

## Coastal Protection

Coastal wetlands can help protect coastlines, in particular by attenuating wave energy, promoting sedimentation, and/or reducing erosion and movement of sediments (Spalding *et al.*, 2014). In the case of catastrophic events, there is evidence that mangrove forests provided protection along coasts impacted by the 2004 Indian Ocean tsunami (Alongi, 2008) and that the death toll from the cyclone that hit the coast of Orissa, eastern India, in 1999 was higher in villages that had only narrow or no mangroves between them and the coast than in villages with a wider mangrove buffer (Das & Vincent, 2009). The amount of energy absorbed in such instances depends on a host of factors, but mangrove forests probably need to be at least 100 m wide to significantly reduce wave energy (Alongi, 2008).

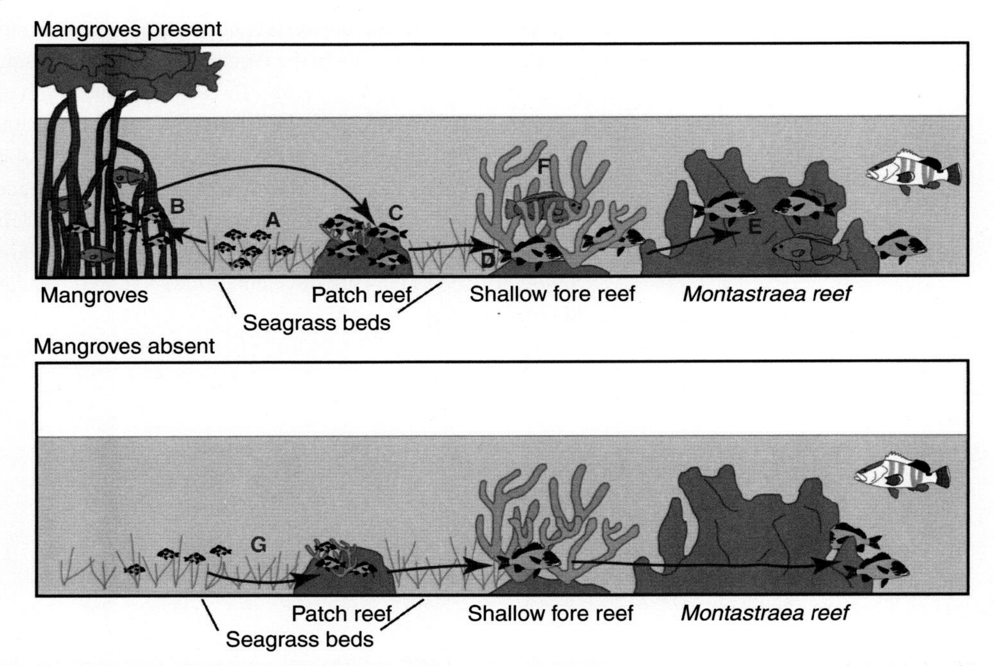


**Fig. 12.6** Diagram showing habitat shifts in the life history of the Caribbean reef fish, bluestriped grunt, and rainbow parrotfish when mangroves are present and absent. Juveniles inhabit seagrass beds (A) before moving to mangroves (B) as intermediate nursery habitat and then to patch reefs, shallow fore reefs, and *Montastraea* reefs (C, D, E). If mangroves are absent, bluestriped grunt migrate directly from seagrass to reefs (G). Rainbow parrotfish (F), however, depend on mangroves and so are rarely seen on reefs when mangroves are absent. From Mumby, P.J. & Harborne, A.R. (2006). A seascape-level perspective of coral reef ecosystems. In *Coral Reef Conservation*, ed. I.M. Coté & J.D. Reynolds, pp. 78–114. Cambridge, UK: Cambridge University Press. With permission from Cambridge University Press.

## Loss and Degradation

Saltmarsh, mangrove, and seagrass habitats worldwide have suffered degradation and major losses in areal extent as a result of increasing human use of the coastal zone. Global estimates indicate annual rates of loss typically in the range of 1–5% over recent decades, with about a third of each habitat having disappeared since the 1940s (Nellemann *et al.*, 2009; Mcleod *et al.*, 2011). Many developed estuaries have lost most or all of their wetland. An assessment by Lotze *et al.* (2006) of 12 temperate estuarine and coastal ecosystems in Europe, North America, and Australia indicated that they had lost more than two-thirds of their wetlands since human settlement. And since 1949, China has lost about 22 000 km$^2$ of coastal wetland, or about half the country's total area of coastal wetland, with the rate of loss accelerating in recent decades in line with a sharp increase in gross domestic product (Tian *et al.*, 2016). Coastal wetlands are among the world's most threatened ecosystems, with serious implications for the biodiversity and functioning of coastal ecosystems. Several factors are responsible, often in combination (Table 12.2).

Saltmarshes have been used and modified over many centuries. A major impact has been the enclosing and development of saltmarsh for agricultural use, in particular the building of embankments to exclude the tide and provide grazing or arable land (Doody, 2001). Piecemeal enclosing of saltmarsh has taken place in Europe since at least Roman times, and where there is sediment

**Table 12.2** Anthropogenic Factors Mainly Responsible for Loss and Degradation of Coastal Wetlands

Climate change and attendant sea-level rise represent a major threat to wetland ecosystems generally, but the implications are still poorly understood.

| Saltmarsh | Mangrove | Seagrass |
|---|---|---|
| Invasive species | Aquaculture (mainly shrimp culture) | Eutrophication |
| Human disturbance and changes in consumer control | Forest products | Sedimentation |
| Agricultural use, reclamation, and coastal engineering | Freshwater diversions | Pollution (biocides, petrochemicals, metals) |
| Pollution and eutrophication | Reclamation | Bottom fishing |
| | | Dredging |

Based on Bromberg Gedan, K., Silliman, B.R. & Mertness, M.D. (2009). Centuries of human-driven change in salt marsh ecosystems. *Annual Review of Marine Science*, **1**, 117–41; Ralph, P.J., Tomasko, D., Moore, K., Seddon, S. & Macinnis-Ng, C.M.O. (2006). Human impacts on seagrasses: eutrophication, sedimentation, and contamination. In *Seagrasses: Biology, Ecology and Conservation*, ed. A.W.D. Larkum, R.J. Orth & C.M. Duarte, pp. 567–93, Dordrecht, Netherlands: Springer; Valiela, I., Bowen, J.L. & York, J.K. (2001). Mangrove forests: one of the world's threatened major tropical environments. *BioScience*, **51**, 807–15.

accretion and new marsh can develop seaward of the enclosure, land claim can be an ongoing process. In the Netherlands, about 4000 $km^2$ of saltmarsh were lost by embanking between the thirteenth and mid-twentieth centuries (Wolff, 1992). Marsh enclosed for grazing may still have nature conservation interest, but large areas of saltmarsh have also been destroyed for urban and industrial development and engineered to preclude seaward development of new marsh (Adam, 1990). Saltmarshes have also been systematically drained to eradicate breeding habitat of mosquitoes and other insects that pose a risk to human health, as in the eastern USA. Overall, in Europe and the USA, possibly up to a half of all saltmarsh has been destroyed (Doody, 2001).

Mangroves too are being lost at a startling rate. Major causes of loss are from deforestation, particularly for aquaculture but also for timber and wood-chip production, urban development, overexploitation of fisheries resources, pollution, and damming and diversion of rivers preventing freshwater from reaching mangroves (Field, 2000; Valiela *et al.*, 2001; Alongi, 2002). Alongi (2008) puts the average rate of loss from deforestation at 1–2% of the total area per year. The loss from aquaculture results from the large-scale conversion of mangroves to ponds for farming of penaeid shrimp and milkfish. At least half the mangrove loss in the Philippines and SE Asia has been attributed to pond culture (Valiela *et al.*, 2001). However, shrimp ponds have a short life of only 5–10 years before the water becomes too acidic and polluted and the ponds are unusable. And abandoned ponds recover slowly, far more so than the rate at which new areas of mangrove are converted. The loss of ecosystem services that the mangroves provide includes the support to wild fisheries stocks and those for sustaining the farming operations. Shrimp farms depend on the mangrove ecosystem for the supply of the wild-caught juveniles to seed the ponds as well as for clean water and mangrove detritus. In more intensive operations, the addition of food pellets containing fishmeal (from offshore fisheries) and the effluent from shrimp ponds – rich in nutrients and organic matter – may adversely affect receiving waters.

Seagrass beds are particularly threatened by nutrient enrichment, such as from sewage effluent, urban and industrial outfalls, and agricultural runoff, resulting in eutrophication. Increased nutrients impact seagrass mainly by promoting the growth of phytoplankton and epiphytes, both of which reduce the amount of light reaching the seagrass itself (Burkholder *et al.*, 2007). Light penetration may be further attenuated by increased suspended sediment levels from runoff, outfalls, and dredging operations (Ralph *et al.*, 2006). As seagrasses require particularly high light levels for photosynthesis, they are especially susceptible to reduced water clarity (Orth *et al.*, 2006). Seagrass species vary considerably, however, in their life-history traits, which influence how a seagrass bed responds to particular environmental factors. Walker *et al.* (2006) describe neighbouring seagrass beds in Western Australia, both subject to eutrophication, one of which has consistently declined since the 1960s whereas the other has expanded, a difference probably due to the species present, their growth rates, and the hydrodynamic environment.

Coastal wetlands, especially those close to urban and industrial areas, can be exposed to a range of pollutants. An area of particular focus is the impact of oil pollution, as sheltered habitats with emergent vegetation are especially vulnerable in this regard (Table 10.2) and are often close to oil production facilities, refineries, and tanker operations. A striking example was the blowout of the *Deepwater Horizon* well in 2010, resulting in moderate to heavy oiling along an estimated 75 km of saltmarsh in the northern Gulf of Mexico (Silliman *et al.*, 2012). The impact of an oil spill on a wetland depends, however, on a range of factors, so few generalizations can be made. In the case of saltmarshes, the main factors affecting impacts are the amount and type of oil (and hence its behaviour, persistence, and toxicity), the location of the oiling and its extent on the sediments and vegetation, the vegetation type and species' sensitivities to oil, climate and time of year, the presence of wildlife of concern, and the exposure to waves and currents (as natural removal processes) (Michel & Rutherford, 2014). Prevailing circumstances thus strongly affect the length of recovery (Table 12.3). For most spills involving light to moderate oiling, recovery may occur within one or two growing seasons, but if a thick layer of surface oil persists, recovery may take up to a few decades. Burns *et al.* (1993) conclude that mangroves may require at least 20 years to recover from catastrophic oil spills once toxic residues are trapped in the deep mud. In terms of spill response, relying on natural recovery is generally the best option provided other flora and fauna are not unduly harmed during that period. But some intervention is often called for in cases of heavy oiling of marshes so as to remove oil and hasten recovery, as long as this can be done without exacerbating the damage. The appropriate response option depends on the oiling conditions (Michel & Rutherford, 2014). A supplementary clean-up approach may be the use of bioremediation: applying microbial populations that can break down the remaining petroleum compounds. These may be indigenous microorganisms, those from other sites, or genetically modified strains (Santos *et al.*, 2011).

The largest-known loss of seagrass occurred in the 1930s as a result of the 'wasting disease' of the eelgrass *Zostera marina* (Muehlstein, 1989). The disease, caused by a marine slime mould, reduced

**Table 12.3** Conditions Affecting the Length of Recovery of Saltmarshes to Oil Spills

| Shortest recovery | Longest recovery |
|---|---|
| Warm climate | Cold climate |
| Some exposure to waves and currents | Sheltered settings |
| Light to heavy oiling of vegetation only | Thick oil on the marsh surface |
| Medium crude oils | Light refined products with heavy loading |
| Less intensive treatment | Heavy oils forming persistent thick residues |
| | Intensive treatment |

Derived from Michel, J. & Rutherford, N. (2014). Impacts, recovery rates, and treatment options for spilled oil in marshes. *Marine Pollution Bulletin*, 82, 19–25.

seagrass populations along the Atlantic coasts of North America and Europe by more than 90%. The disease recurred briefly in 1944, but by the early 1950s seagrass had largely re-established along the Atlantic coast of North America. Further localized outbreaks have occurred since the 1980s (den Hartog, 1996). The 1930s episode severely affected North Atlantic populations of brent geese, for which seagrass is a major food, as well as fish and invertebrate species associated with the seagrass beds (Muehlstein, 1989). It also brought about the extinction of the eelgrass limpet *Lottia alveus* (see Chapter 4).

Habitat loss and degradation have serious implications for individual wetland species. Ten seagrass species are categorized as threatened on the IUCN Red List. They have restricted or highly fragmented distributions and are threatened in particular by shoreline development, pollution, and aquaculture (Short *et al.*, 2011). Similarly, 11 species of mangrove are assessed as threatened, primarily as a result of habitat destruction, with the highest proportion of threatened species (up to 40%) occurring along the Atlantic and Pacific coasts of Central America (Polidoro *et al.*, 2010).

## Introduced Species

Introduced species can substantially modify wetland habitats and communities, most obviously in the case of *Spartina* species, the saltmarsh cordgrasses (Strong & Ayres, 2009, 2013). *Spartina* is a genus of tall (< 2 m) robust grasses, most of them native to temperate Atlantic coasts. They typically colonize muddy estuarine shores and, spreading by rhizomes, can form extensive beds. Spartinas are effective ecosystem engineers as their dense colonies baffle water flow and promote sediment deposition. This capacity has resulted in their being deliberately planted in many estuarine mudflats around the world as a means of accelerating sediment accretion and, thereby, land claim. Human introductions of *Spartina* have resulted in various hybrids between native and introduced species. In some cases, the introduced species, or self-pollinating hybrids in particular, have proved highly successful colonizers, trapping sediment at an impressive rate and transforming large expanses of open mudflat to saltmarsh (Fig. 12.7).

China's coast, for example, is extensively invaded by *Spartina*. Since 1963, four species have been introduced for ecological engineering purposes. Of these, *S. alterniflora* (a western Atlantic species) has spread rapidly since 1981 to cover some 345 km$^2$ by 2007. It occurs in all the coastal provinces, although more than half the total cover is in Jiangsu Province north of Shanghai (Zuo *et al.*, 2012).

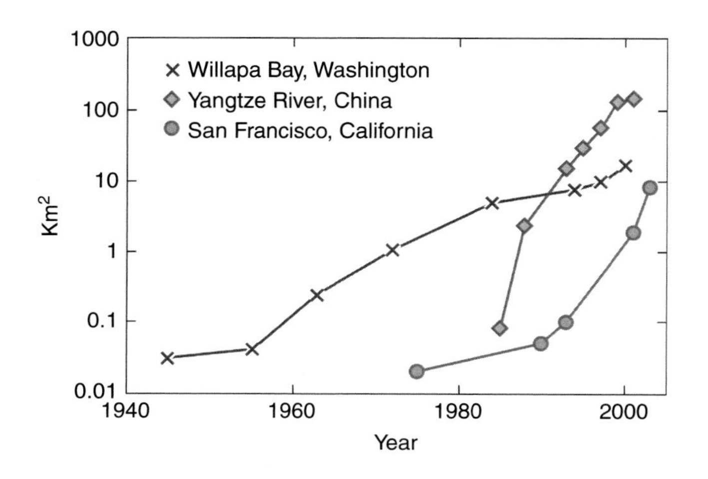


**Fig. 12.7** The spread of three invasive *Spartina* populations: *S. alterniflora* (Willapa Bay and Yangtze River) and *S. alterniflora* x *foliosa* hybrid (San Francisco Bay) that has highly invasive populations. All the invasions were aided by human dispersal. From Strong, D.R. & Ayres, D.R. (2013). Ecological and evolutionary misadventures of *Spartina*. *Annual Review of Ecology, Evolution, and Systematics*, 44, 389–410. Modified with permission from the *Annual Review of Ecology, Evolution, and Systematics*, Volume 44 © by Annual Reviews, www.annualreviews.org.

**Fig. 12.8** *Spartina anglica* is a highly invasive cordgrass that rapidly colonizes tidal mudflats. Orwell estuary, Suffolk, England. Photograph: P.K. Probert.

This level of incursion is likely to strongly impact China's coastal biodiversity, fisheries, aquaculture, and mangroves (Strong & Ayres, 2013).

The most widely introduced cordgrass is *Spartina anglica*, a hybrid species that arose in Southampton Water, southern England, in the late 1880s (Strong & Ayres, 2009) (Fig. 12.8). (It derives from *S. alterniflora*, introduced to the Southampton area in the early nineteenth century, hybridizing with the native *S. maritima*. This resulted in a sterile hybrid that, by chromosome doubling, gave rise to the fertile *S. anglica*.) *S. anglica* is also highly invasive, able to colonize tidal mudflats rapidly. From the site of its original discovery it has spread widely around the British coast. It has been deliberately planted at many sites throughout Europe as well as farther afield, including in Australasia and China.

*Spartina anglica* is able to grow seaward of the normal saltmarsh zone in Europe and at places around Britain has encroached upon or replaced seagrass (*Zostera* spp.) beds (Adam, 1990). Such loss of *Zostera* may depress populations of wildfowl that feed on seagrass, such as brent geese and wigeon (Percival *et al.*, 1998). Similarly, encroachment of *S. anglica* over formerly unvegetated intertidal flats where waders feed has been implicated in the decline of internationally important wintering populations of dunlin in British estuaries (Goss-Custard & Moser, 1988). The presence of *S. anglica* may have some benefits, such as increased coastal protection, the development of new saltmarsh areas, and the provision of habitat for other coastal plants and animals. But, on balance, its spread is generally regarded as harmful to conservation interests, and efforts have been made to control it. Herbicide application is the most commonly used and effective method of control (Doody, 2001).

Other species of *Spartina* are invading tidal flats at numerous sites around the world. Along the Pacific coast of North America are three other introduced species in addition to *S. anglica*.

By using physical characteristics, Daehler and Strong (1996) identified estuaries likely to be vulnerable to future *Spartina* invasion along this coast, information that is useful in prioritizing sites for protection. Plants such as *Spartina* can significantly influence coastal habitat structure, but what factors determine whether a state change in ecosystem functioning occurs? Neira *et al.* (2006) predict that system restructuring occurs when an invader: (1) appears in a setting where comparable structural forms are absent, thereby modifying the hydrodynamic and sedimentary conditions; (2) facilitates key consumers; and/or (3) introduces large quantities of a new food source, such as detritus.

Many other exotic species have become established in coastal wetlands. For seagrass ecosystems, Williams (2007) found records for at least 56 non-native species that have been introduced, mostly invertebrates (63%) and seaweeds (25%), with the main vectors being shipping and boating and aquaculture. Where the ecological effects of introduced species on seagrass habitats have been assessed, they are mainly adverse and affecting seagrass growth or density. Introduced species are often more common in habitats that are under environmental stress, and experiments in seagrass ecosystems also confirm this relationship: disturbance to seagrass beds increases their susceptibility to invasion.

Four seagrass species have been introduced to various sites around the world (Williams, 2007). These include the Asian seagrass *Zostera japonica*, probably introduced from the NW Pacific to the NE Pacific with shipments of oysters from Japan early in the twentieth century. *Z. japonica* co-occurs with the native *Z. marina* on the NE Pacific coast, though not in direct competition; the introduced species mainly colonizes the mid- to upper intertidal zones, whereas *Z. marina* occupies the lower intertidal area. However, the ecological implications of the introduction of *Z. japonica* are poorly understood, and within its established range there is no consistent management approach among the various federal and state agencies (Shafer *et al.*, 2014).

Mangroves were deliberately introduced to Hawaii for the purpose of stabilizing coastal mudflats (Allen, 1998). Although it has suitable environment, the Hawaiian archipelago has no native mangrove species, presumably because of its isolation. However, at least six mangrove species were introduced to Hawaii during the early twentieth century, of which the red mangrove (*Rhizophora mangle*) has become well established on most of the main islands. Its spread is causing a number of problems, including a reduction in the habitat of some endemic and endangered waterbird species.

## Climate Change

Climate change represents one of the biggest threats to wetlands, but there remains much uncertainty as to how its various aspects might play out (Bromberg Gedan *et al.*, 2009). As we saw earlier, coastal wetlands play a major role as carbon sinks. Despite comprising a much smaller area than terrestrial forests, these coastal systems are of comparable importance in carbon sequestration, hence the vital importance of wetland conservation in mitigating climate change (Mcleod *et al.*, 2011). Protecting wetlands can in fact be a cost-effective way of avoiding carbon dioxide emissions (Siikamäki *et al.*, 2012). The Blue Carbon Initiative is an international programme that seeks to mitigate climate change through the management of coastal wetlands (http://thebluecarboninitiative.org).

The geographical range of wetland species could shift as a result of global warming, and sea-level rise may produce a landward retreat of vegetation if the rate is compatible with accretionary processes and if upward migration is not impeded by shoreline structures – the problem of 'coastal squeeze' we met earlier (Chapter 10). Sediment supply may, however, be inadequate for many tidal wetlands to keep pace with sea-level rise (Stevenson & Kearney, 2009). Lovelock *et al.* (2015) found this to be the case for many mangrove forests across the Indo-Pacific region, where sediment delivery is declining due largely to the damming of rivers. Their modelling indicates that at current rates of sea-level rise, mangrove forests at some locations could be submerged by 2070.

## Public Awareness

Humans have generally not valued coastal wetlands, often viewing them as better suited for agricultural, urban, and industrial development. Huge areas have consequently been lost or severely degraded. Coastal wetlands tend to be especially vulnerable habitats as they are frequently in estuaries or bays close to centres of population and seen as areas ripe for waterside expansion. This proximity to human populations means too that wetlands are often close to sources of contamination. Fundamental to any wetland management programme is the need for an overarching plan for coastal zone management (see Chapter 10) that addresses the question of coastal water quality.

We now have a far better understanding of the contribution of wetlands to the functioning of coastal systems as a whole. Nevertheless, there is still a great need for public education in this area so that society at large better appreciates the importance of these habitats – the benefits they provide in terms of resources and ecosystem services. Raised public awareness is particularly needed for seagrasses, as they attract less research effort and being mainly subtidal are less conspicuous (Orth *et al.*, 2006). Without that awareness, wetlands will continue to be threatened and lost by inappropriate developments that fail to take account of the wider costs.

## Habitat Mapping

Conservation and management of coastal wetlands are also hampered by the often-inadequate information on the type and extent of habitat within a region. There is a crucial need for detailed mapping and monitoring of coastal wetlands. To inform management, information is needed on the range of habitat types and their distribution and representation, human impacts, and conservation significance. Glenn *et al.* (2006), for instance, conducted an inventory of coastal wetlands of the northern Gulf of California, which could then be used to assess conservation status and management options. This unique area in a desert setting is characterized by numerous 'esteros', or negative estuaries, which, because of limited freshwater inflow, tend to be saltier at their head than at their mouth. The major types of wetland are mangroves and saltgrass (*Distichlis palmeri*) marshes. The intertidal vascular plants are all native, and four of them are endemic, including *D. palmeri*. The esteros are significant too as nursery areas for penaeid shrimp and as habitat for nearly 500 bird species, including many migratory species. Human impacts here derive mainly from resorts and vacation homes, shrimp farming, and diversion of freshwater for agriculture; nevertheless, much of the wetland habitat is still intact. Essential information for such assessments, as in this case, can be obtained from satellite and aircraft imagery. Remote sensing data are particularly valuable for mapping and monitoring coastal wetlands and in many cases are the only practicable option. Klemas (2013), for example, shows how such data, combined with limited ground-truth observations, can be used to map biomass and determine long-term changes in tidal marshes. Using remote sensing data (Landsat imagery), Torres-Pulliza *et al.* (2013) mapped seagrass distribution for the Lesser Sunda ecoregion (from Bali to Timor-Leste), with the resulting maps being used to help design a marine protected area network.

## Protection and Management

It would appear that a considerable number of coastal wetlands are protected. Green and Short (2003) list nearly 250 protected areas worldwide that include seagrass ecosystems, whilst Spalding *et al.* (2010) estimate that some 1200 protected areas include mangroves. But in many cases, protected areas have not been designated primarily to conserve wetland habitat – the inclusion of wetland is often incidental – and the size of such areas and level of protection can vary greatly, strongly influencing the likely conservation value. There is, consequently, a pressing need to determine the types, coverage, and degree of protection of wetlands within biogeographic regions to enable effective networks of protected areas to be developed.

Often a prime focus of wetland conservation is protection of habitat for particular animal species. A number of species that regularly occur in seagrass beds are listed as threatened by IUCN, such as dugong, West Indian manatee, green turtle, and several species of seahorse (Green & Short, 2003). Dugongs feed almost exclusively on seagrass (Fig. 9.2), so the optimal conservation strategy is to provide a high level of protection in areas of suitable grazing that still support large numbers of animals (Marsh *et al.*, 1999). Seagrass is important too in the diet of adult green turtles (Lanyon *et al.*, 1989). Dugongs and green turtles are large herbivores, and their grazing can influence the structure of seagrass communities.

Many organisms make use of different wetland habitats during their lifetime. Some, as we have seen, make ontogenetic migrations, including migrating between wetlands and other habitats. An important consideration, therefore, in designing effective networks of marine protected areas (MPAs) is the degree of connectivity between habitats for particular organisms (Sobel & Dahlgren, 2004). Based on the study described earlier, on the use of seagrass and mangrove habitat by populations of coral reef fishes in the Caribbean, Mumby (2006) developed algorithms applicable to conservation planning so as to identify the relative importance of mangrove nursery sites, the connectivity between reefs and mangrove nurseries, nursery habitat areas that are critical to particular reefs, and mangrove sites that are strategically important to restore. The algorithms provide indices of connectivity between habitats, which could be used to identify habitat corridors for incorporating into the design of MPAs. In a study of fish assemblages at sites in the western Pacific, Olds *et al.* (2013) found that connectivity between mangroves and coral reefs can improve the effectiveness of no-take MPAs, increasing the abundance and species richness of fishes.

The principal international initiative to promote the conservation of wetlands is the Ramsar Convention (see Chapter 11). Vegetated coastal wetlands tend to be under-represented, although some major sites are included, such as the Sundarbans Reserved Forest (6000 km$^2$) in Bangladesh. This is the world's largest area of contiguous mangrove forest and supports many threatened plants and animals. The convention as originally drafted was considered to have various deficiencies, although there has been considerable effort to address these, and Ramsar has no doubt become a significant instrument in advancing wetland conservation (Bowman *et al.*, 2010). The number of contracting parties and sites has grown considerably, but management plans still need to be developed for many sites. This comes down to national responsibility and action – the success of the convention depends ultimately on the extent to which parties incorporate Ramsar objectives into national policies and laws.

A number of regions and countries have strategies or legislation intended to protect wetlands. The principal US law regulating water pollution, the Clean Water Act (CWA), aims 'to restore and maintain the chemical, physical, and biological integrity of the Nation's waters', including its wetlands. Section 404 requires applicants seeking to discharge fill material to wetlands to demonstrate that they have used all means to mitigate damage. This requirement means first seeking to avoid damage, secondly minimizing damage, or compensating to provide replacement wetland, so there should be no net loss of wetland area and function. Other countries have a similar mitigation sequence. Compensation could be restoration of a former or degraded wetland, or creation of a new wetland. In response to Section 404 – and comparable legislation in other countries – there is increasing interest in so-called 'mitigation banking' when off-site compensation is to be undertaken, with typically a third party responsible for producing wetland and liable for its ecological success. Coastal wetlands are not yet common in such schemes, and how effective this mechanism might prove in the long term remains to be seen. There are concerns, for instance, that the replacement wetland may be of a different type and of lesser wildlife value (Zedler, 2004; Burgin, 2010). Robertson and Hough (2011) conclude that there has unfortunately been undue emphasis on the compensatory component of mitigation under Section 404 and that for effective wetland conservation far more attention needs to be paid to avoiding and minimizing damage.

In 2015 Sri Lanka embarked on a programme to protect all its mangrove forests (8815 ha), the first country to take such comprehensive action. The joint project, involving the government and NGOs, provides microloans and training to some 15 000 women living in villages in mangrove areas, enabling them to set up small businesses that do not necessitate destroying mangrove forest. The project also includes raising mangrove seedlings in nurseries for replanting nearly 40 $km^2$ of forest that have been cut down and educating the communities on the importance of mangroves. The communities in turn have a responsibility to protect the forest. This type of model has potential worldwide to assist impoverished coastal communities to safeguard vulnerable habitat (Seacology, 2015).

## Restoration

Wetland restoration has become an active area of coastal management over recent decades. Such activity has been driven by increasing concern over the degradation and rate of loss of wetlands worldwide, particularly as we better understand the importance of wetlands in coastal ecological processes and in supporting biodiversity, including species of economic interest. As a result, many programmes are aiming to restore damaged wetlands or reinstate them at sites where they have been lost. Also, there is growing interest in projects to establish wetlands in new areas that are deemed suitable as compensation for loss of wetland elsewhere.

Whilst such techniques have a part to play in coastal management, the prospect of restoration should not undermine the pre-eminence of preventative measures in minimizing degradation or loss of wetland. There may be a number of reasons for embarking on a restoration project, such as maintaining ecosystem services, conservation of biodiversity and landscape, providing habitat for particular wetland-dependent species, sustainable production of natural resources, and coastal protection. The aims need to be clear from the outset, and there must be involvement and support of the local community (Field, 1998). Another primary consideration in any restoration proposal is the suitability of the site: can it be expected to support the desired wetland community in terms of its topography, hydrology, and other environmental conditions? Are factors that contributed initially to the site's degradation still in force? If so, can this impact be remedied?

There are several main approaches to restoring or re-creating saltmarsh (Doody, 2008). The extensive use of introduced Spartinas shows that saltmarsh can readily be established given suitable conditions. The aim of these projects, however, has usually been to promote sediment accretion and facilitate reclamation, whereas more recent saltmarsh creation and restoration schemes generally have conservation objectives (Adam, 1990). Many such restorations in Europe and North America have involved the breaching of old embankments so as to re-establish tidal access and allow degraded marshes to recover. For instance, Warren *et al.* (2002) describe a programme along the Connecticut coast of Long Island Sound where a series of degraded marshes, impounded and dominated by the common reed (*Phragmites australis*) and bulrush or cattail (*Typha angustifolia* and *T. latifolia*), were reconnected to the sound to provide for the restoration of *Spartina*-dominated saltmarsh vegetation.

A major wetland project undertaken in Delaware Bay on the US east coast involves the restoration of about 40 km$^2$ of saltmarsh (Teal & Peterson, 2005). It was undertaken to offset possible effects of the cooling-water intake of a power station on fish populations in the bay. The area of marsh needed to enhance fish production to compensate for potential losses was calculated from a food-chain model (and multiplied by four as a safety factor). Sites chosen for restoration were former saltmarshes at appropriate tidal elevations, with two types of marsh being restored. First, in the lower euryhaline parts of the bay, are old diked meadows, originally harvested for 'salt hay' (*Spartina patens* and *Distichlis spicata*). In these areas, through breaching of the dikes and engineering of channels, the natural tidal regime has been restored. This intervention has enabled *Spartina alterniflora*-dominated marsh to develop, increased the production of detritus, and provided fish habitat. Secondly, there are also marshes, mainly in the bay's low-salinity regions, that had become overrun with *Phragmites australis*. Herbicides and burning have been used to reduce *Phragmites* domination, but this component of the restoration appears to be harder to achieve.

The largest estuarine system on the USA's west coast, San Francisco Bay, originally had some 2200 km$^2$ of tidal marsh, but as a result of diking and filling only about 5% remained by the 1970s (Nichols *et al.*, 1986). Numerous projects have since been under way aimed at restoring thousands of hectares of wetland, but long-term restoration progress is likely to depend on climate change and decisions related to California's water policy (Callaway & Parker, 2012).

Seagrass beds are slow to recover from disturbance and may take decades to recover naturally from a large impact (Thorhaug, 1986) or centuries in the case of *Posidonia* meadows (see Chapter 16). Restorations have been attempted to accelerate recovery, usually by transplanting plugs of sediment (plus shoots, rhizomes, and roots) from a donor site, a method used for a number of species (e.g. *Zostera marina, Thalassia testudinum,* and *Halodule wrightii*). Raising seagrasses from seed is rarely feasible given the difficulty of harvesting seed in sufficient quantity and the often-low viability of seed and survival of seedlings (Thorhaug, 1986; Hemminga & Duarte, 2000). The first large-scale seagrass restoration was carried out in Biscayne Bay, Florida, and involved more than one million plants planted over 60 ha (Thorhaug, 1987). An examination of European seagrass restoration projects reveals, however, a low success rate, and the reported success of programmes is often based on limited evidence: on short monitoring periods (< 1 year) and projects of small spatial scale (< 10 m$^2$) (Cunha *et al.*, 2012). Also, studies have generally focused on the above-ground communities, but the effect of restoration methods on sediment structure and nutrient pools may limit development of the seagrass community and its associated infauna (Bourque & Fourqurean, 2014). Overall, the evidence again emphasizes the importance of avoiding damage to existing seagrass beds, facilitating natural recovery where possible, and that compensatory creation of seagrass beds should be seen only as a last resort (Cunha *et al.*, 2012).

Mangrove restoration has been more intensively researched and pursued. A major objective of most such projects has, however, been to provide a harvestable resource – plantations for wood products – rather than attempting to regain a mature community (Ellison, 2000). A huge afforestation programme has been carried out in Bangladesh. Over the period 1960 to 2001, 1485 km$^2$ of mangroves were planted on newly accreted areas, of which 905 km$^2$ (61%) have succeeded in not being lost to erosion and encroachment. The plantations, dominated by *Sonneratia apetala* and *Avicennia officinalis*, were initially developed mainly to improve coastal protection – this area being prone to severe cyclone damage – but later objectives included accelerating accretion for eventual transfer to agricultural land and to provide timber, employment, and wildlife habitat (Iftekhar & Islam, 2004). Other countries that have been active in large-scale mangrove replanting include Vietnam (670 $km^2$), primarily in the Mekong delta, and Indonesia, Philippines, and Cuba (each 400–500 $km^2$) (Spalding *et al.*, 2010). Mangroves managed for forestry tend to be of low diversity (often, for economic reasons, monocultures of the faster-growing species) and subject to rapid rotational felling, which prevents a mature mangrove community from developing. They may, however, provide a basis for fully restoring a mangrove area, providing that stands are left unharvested so that succession can occur (Ellison, 2000).

Techniques commonly used for re-establishing mangroves include hand sowing or scattering of propagules, the transplanting of seedlings or young trees from donor sites, or the culturing of seedlings in nurseries (Kaly & Jones, 1998). If, however, there is a good local supply of propagules or seeds, then natural regeneration is to be preferred as it will result in vegetation that is most similar to the original (Field, 1998). In fast-growing systems with good rainfall, vegetation and much of the associated biota may have largely re-established in 5–10 years, and by 20 years planted areas can be of similar biomass and productivity to adjacent natural forest (Kaly & Jones, 1998; Ellison, 2000). But this does not necessarily mean that the system has been successfully restored; as with other wetland restoration projects, few studies have also monitored the return of functional characteristics. A meta-analysis by Moreno-Mateos *et al.* (2012) indicates that, following restoration efforts, tidal wetlands may take at least 30 years to recover their biological structure (driven mostly by plant assemblages) and biogeochemical functioning (driven primarily by the storage of carbon in wetland soils).

Silviculture has been the main driver of mangrove restoration efforts, and only more recently has the focus shifted to the important but more difficult task of restoring functionality – the ecosystem goods and services provided by natural mangroves. In this context, Bosire *et al.* (2008) present possible pathways to achieve ecological restoration of mangroves (Fig. 12.9). They emphasize that the primary aim should be natural recovery rather than the planting of mangroves – assuming an adequate supply of seeds from neighbouring areas – and determining if there are factors that are hindering regeneration. Often critical in this regard are the hydrological regime (such as the flooding depth, duration, and frequency) and the soil conditions. They stress too the importance of community involvement, given that the livelihood of coastal populations in such countries often depends on mangroves.

## Sustainable Use

Humans have for millennia used coastal wetlands as sources of food and natural materials. Wetland plants are used in a wealth of traditional crafts. In addition to a wide variety of woods, wetland plants yield tannins and dyes and materials for thatching, matting, basketry, and cordage (Morton, 1976). Wood from mangroves has long been used as an important source of charcoal for cooking and heating as well as for building (Field, 2000). There is also a long history of using saltmarshes for grazing (mainly sheep and cattle), particularly in Europe, where they have been modified by centuries of human use (Adam, 2002). Saltmarshes are often also important for wildfowling, a significant factor in their protection. However, with increasing human pressure on coastal areas, accommodating sustainable use of wetland resources is a major challenge for management and conservation.

The issue is illustrated in the case of mangroves given, in particular, the demands of aquaculture and forestry. There is considerable debate about sustainable mangrove forestry, but one of the best-known successful examples is the mangrove forest of Matang in Peninsular Malaysia. Some 350 $km^2$ of this forest have been managed for wood production for more than a century (using rotational

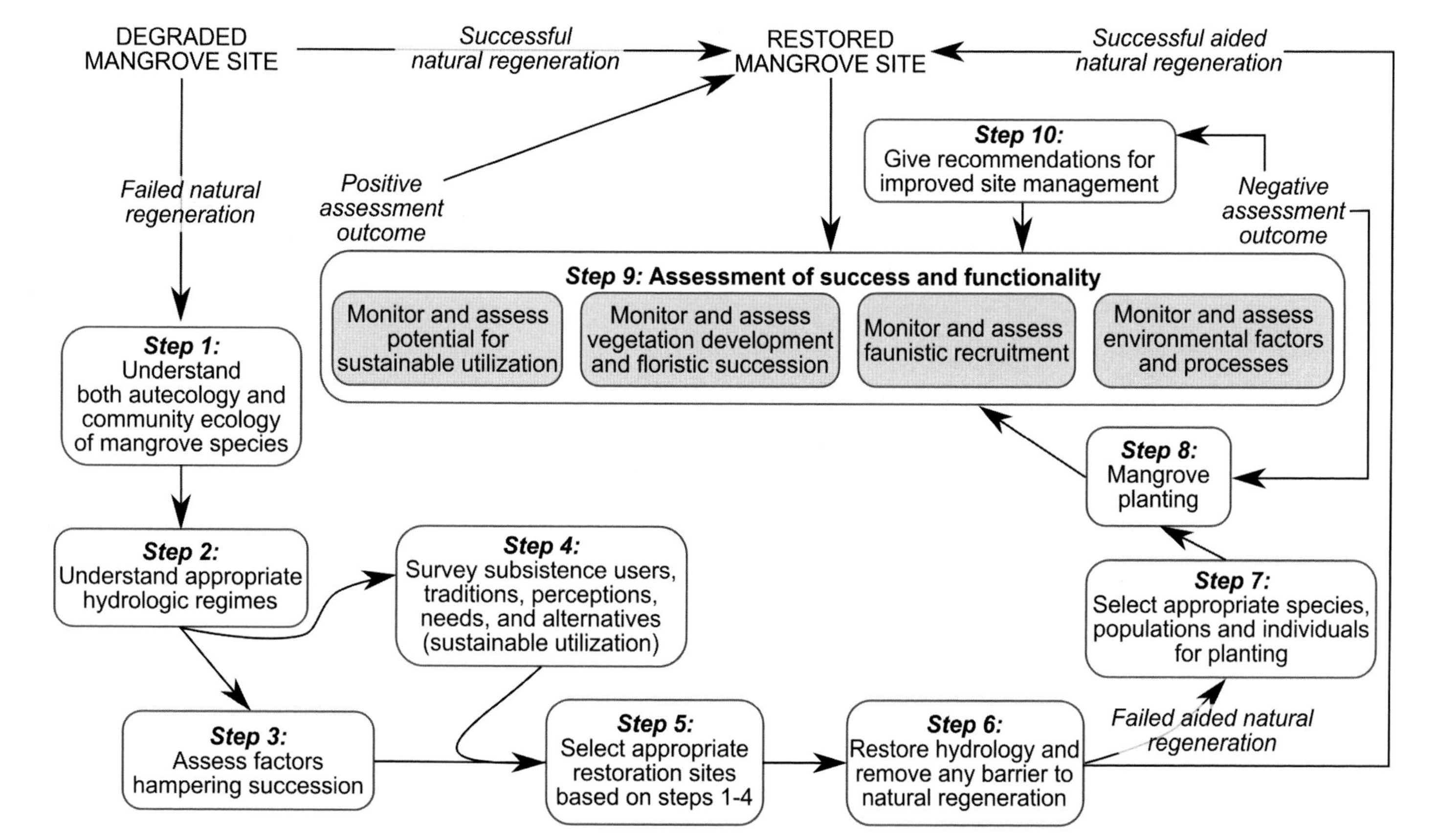


**Fig. 12.9** Possible pathways to achieve the ecological restoration of mangroves. Restoration should be based on the premise of the potential for natural regeneration and removing stresses that are hindering this process. If natural regeneration fails, then the ecology of the species concerned and site conditions (notably the hydrologic regime) need to be examined to assess factors that are hampering succession (Steps 1–3). This will likely involve input of indigenous knowledge (Step 4). The resulting socio-ecological information can then be used to select appropriate restoration sites (Step 5) and lead to the removal of the obstacles to natural regeneration (Step 6). If natural recovery still fails, then appropriate plants are selected for planting (Steps 7–8). The success of replanting should be monitored using indicators of functionality (Step 9). But if the assessment has a negative outcome, recommendations should be given for improved site management (Step 10). When the assessment is positive, i.e. the site has restored, further monitoring may still be appropriate. Reprinted from *Aquatic Botany*, Vol. 89, Bosire, J.O., Dahdouh-Guebas, F., Walton, M., *et al.*, Functionality of restored mangroves: a review, pages 251–9, copyright 2008, with permission from Elsevier.

harvesting) and provided employment for the local population, although there is concern over declining yields in recent decades and the intensive use of herbicides (Ellison & Farnsworth, 2001). There may also be opportunities for aquaculture in mangroves that do not require forest clearance (Walters *et al.*, 2008). Integrated mangrove-aquaculture systems can, for example, combine forest management and low-density fish production (various species of finfish, shrimps, and crabs cultured either singly or in mixed systems), although there is still much research and development work needed in this area. But it may be feasible to accommodate various uses in a mangrove area on a sustainable basis, such as aquaculture, rotational forestry, rice paddies, ecotourism, open-water fishing, protected core areas, and buffer zones (Ellison & Farnsworth, 2001). The International Society for Mangrove Ecosystems has adopted a Charter for Mangroves and developed a global Mangrove Action Plan (ITTO/ISME, 2004). These initiatives recognize the biological, ecological, and socio-economic roles of mangroves as well as the major problems they face and set out a number of guiding principles, responses, and activities to achieve protection and sustainable management of mangroves.

There remains an urgent need for effective management and conservation of coastal wetlands in general. They are vulnerable habitats at the interface of sea and land, caught in the crossfire between marine and terrestrial jurisdictions and all too often regarded as expendable. Saltmarshes, mangroves, and seagrass beds continue to be degraded and lost worldwide at an alarming rate. Yet the crucial role these ecosystems play is becoming increasingly apparent, such as the biodiversity they support – directly and by way of their links with other habitats – as well as their importance to coastal protection, as carbon sinks, and in underpinning fisheries. Their protection as healthy, fully functional ecosystems is one of the most pressing issues in conservation.

## REFERENCES


Aburto-Oropeza, O., Ezcurra, E., Danemann, G., *et al.* (2008). Mangroves in the Gulf of California increase fishery yields. *Proceedings of the National Academy of Sciences*, **105**, 10456–9.

Adam, P. (1990). *Saltmarsh Ecology*. Cambridge, UK: Cambridge University Press.

Adam, P. (2002). Saltmarshes in a time of change. *Environmental Conservation*, **29**, 39–61.

Allen, J.A. (1998). Mangroves as alien species: the case of Hawaii. *Global Ecology and Biogeography Letters*, **7**, 61–71.

Alongi, D.M. (2002). Present state and future of the world's mangrove forests. *Environmental Conservation*, **29**, 331–49.

Alongi, D.M. (2008). Mangrove forests: resilience, protection from tsunamis, and responses to global climate change. *Estuarine, Coastal and Shelf Science*, **76**, 1–13.

Alongi, D.M. (2014). Carbon cycling and storage in mangrove forests. *Annual Review of Marine Science*, **6**, 195–219.

Barbier, E.B., Hacker, S.D., Kennedy, C., *et al.* (2011). The value of estuarine and coastal ecosystem services. *Ecological Monographs*, **81**, 169–93.

Beck, M.W., Heck, K.L. Jr, Able, K.W., *et al.* (2001). The identification, conservation, and management of estuarine and marine nurseries for fish and invertebrates. *BioScience*, **51**, 633–41.

Blaber, S.J.M. (2013). Fishes and fisheries in tropical estuaries: the last 10 years. *Estuarine, Coastal and Shelf Science*, **135**, 57–65.

Bloomfield, A.L. & Gillanders, B.M. (2005). Fish and invertebrate assemblages in seagrass, mangrove, saltmarsh, and nonvegetated habitats. *Estuaries*, **28**, 63–77.

Bosire, J.O., Dahdouh-Guebas, F., Walton, M., *et al.* (2008). Functionality of restored mangroves: a review. *Aquatic Botany*, **89**, 251–9.

Bouillon, S., Borges, A.V., Castañeda-Moya, E., *et al.* (2008). Mangrove production and carbon sinks: a revision of global budget estimates. *Global Biogeochemical Cycles*, **22**, GB2013.

Bourque, A.S. & Fourqurean, J.W. (2014). Effects of common seagrass restoration methods on ecosystem structure in subtropical seagrass meadows. *Marine Environmental Research*, **97**, 67–78.

Bowman, M., Davies, P. & Redgwell, C. (2010). *Lyster's International Wildlife Law*, 2nd edn. Cambridge, UK: Cambridge University Press.

Bromberg Gedan, K., Silliman, B.R. & Mertness, M.D. (2009). Centuries of human-driven change in salt marsh ecosystems. *Annual Review of Marine Science*, **1**, 117–41.

Burgin, S. (2010). 'Mitigation banks' for wetland conservation: a major success or an unmitigated disaster? *Wetlands Ecology and Management*, **18**, 49–55.

Burkholder, J.M., Tomasko, D.A. & Touchette, B.W. (2007). Seagrasses and eutrophication. *Journal of Experimental Marine Biology and Ecology*, **350**, 46–72.

Burns, K.A., Garrity, S.D. & Levings, S.C. (1993). How many years until mangrove ecosystems recover from catastrophic oil spills? *Marine Pollution Bulletin*, **26**, 239–48.

Callaway, J.C. & Parker, V.T. (2012). Current issues in tidal marsh restoration. In *Ecology, Conservation, and Restoration of Tidal Marshes: The San Francisco Estuary*, ed. A. Palaima, pp. 253–62. Berkeley, CA: University of California Press.

Coles, R.G., Rasheed, M.A., McKenzie, L.J., *et al.* (2015). The Great Barrier Reef World Heritage Area seagrasses: managing this iconic Australian ecosystem resource for the future. *Estuarine, Coastal and Shelf Science*, **153**, A1–12.

Costanza, R., de Groot, R., Sutton, P., *et al.* (2014). Changes in the global value of ecosystem services. *Global Environmental Change*, **26**, 152–8.

Cunha, A.H., Marbá, N.N., van Katwijk, M.M., *et al.* (2012). Changing paradigms in seagrass restoration. *Restoration Ecology*, **20**, 427–30.

Daehler, C.C. & Strong, D.R. (1996). Status, prediction and prevention of introduced cordgrass *Spartina* spp. invasions in Pacific estuaries, USA. *Biological Conservation*, **78**, 51–8.

Das, S. & Vincent, J.R. (2009). Mangroves protected villages and reduced death toll during Indian super cyclone. *Proceedings of the National Academy of Sciences*, **106**, 7357–60.

den Hartog, C. (1996). Sudden declines of seagrass beds: 'wasting disease' and other disasters. In *Seagrass Biology: Proceedings of an International Workshop, Rottnest Island, Western Australia, 25–29 January 1996*, ed. J. Kuo, R. C. Phillips, D. I. Walker & H. Kirkman, pp. 307–14. Nedlands, Western Australia: University of Western Australia.

den Hartog, C. & Kuo, J. (2006). Taxonomy and biogeography of seagrasses. In *Seagrasses: Biology, Ecology and Conservation*, ed. A.W.D. Larkum, R.J. Orth & C.M. Duarte, pp. 1–23, Dordrecht, Netherlands: Springer.

Doody, J.P. (2001). *Coastal Conservation and Management: An Ecological Perspective*. Dordrecht, Netherlands: Kluwer Academic Publishers.

Doody, J.P. (2008). *Saltmarsh Conservation, Management and Restoration*. Dordrecht, Netherlands: Springer.

Duarte, C.M. (2001). Seagrasses. In *Encyclopedia of Biodiversity*, ed. S.A. Levin, vol. **5**, pp. 255–68. San Diego, California & London: Academic Press.

Ellison, A.M. (2000). Mangrove restoration: do we know enough? *Restoration Ecology*, **8**, 219–29.

Ellison, A.M. & Farnsworth, E.J. (2001). Mangrove communities. In *Marine Community Ecology*, ed. M.D. Bertness, S.D. Gaines & M.E. Hay, pp. 423–42. Sunderland, MA: Sinauer Associates, Inc.

Ellison, A.M., Farnsworth, E.J. & Merkt, R.E. (1999). Origins of mangrove ecosystems and the mangrove biodiversity anomaly. *Global Ecology and Biogeography*, **8**, 95–115.

Field, C.D. (1998). Rehabilitation of mangrove ecosystems: an overview. *Marine Pollution Bulletin*, **37**, 383–92.

Field, C.D. (2000). Mangroves. In *Seas at the Millennium: An Environmental Evaluation, Vol. III Global Issues and Processes*, ed. C.R.C. Sheppard, pp. 17–32. Oxford, UK: Elsevier Science Ltd.

Gillanders, B.M. (2006). Seagrasses, fish, and fisheries. In *Seagrasses: Biology, Ecology and Conservation*, ed. A.W.D. Larkum, R.J. Orth & C.M. Duarte, pp. 503–36, Dordrecht, Netherlands: Springer.

Giri, C., Ochieng, E., Tieszen, L.L., *et al.* (2011). Status and distribution of mangrove forests of the world using earth observation satellite data. *Global Ecology and Biogeography*, **20**, 154–9.

Glenn, E.P., Nagler, P.L., Brusca, R.C. & Hinojosa-Huerta, O. (2006). Coastal wetlands of the northern Gulf of California: inventory and conservation status. *Aquatic Conservation: Marine and Freshwater Ecosystems*, **16**, 5–28.

Goss-Custard, J.D. & Moser, M.E. (1988). Rates of change in numbers of dunlin, *Calidris alpina*, wintering in British estuaries in relation to the spread of *Spartina anglica*. *Journal of Applied Ecology*, **25**, 95–109.

Green, E.P. & Short, F.T. (2003). *World Atlas of Seagrasses*. Berkeley, CA: University of California Press.

Heck, K.L. Jr, Hays, G. & Orth, R.J. (2003). Critical evaluation of the nursery role hypothesis for seagrass meadows. *Marine Ecology Progress Series*, **253**, 123–36.

Hemminga, M.A. & Duarte, C.M. (2000). *Seagrass Ecology*. Cambridge, UK: Cambridge University Press.

Iftekhar, M.S. & Islam, M.R. (2004). Managing mangroves in Bangladesh: a strategy analysis. *Journal of Coastal Conservation*, **10**, 139–46.

ITTO/ISME (2004). *ISME Mangrove Action Plan for the Sustainable Management of Mangroves 2004–2009*. Okinawa, Japan: International Society for Mangrove Ecosystems and International Tropical Timber Organization. 21 pp.

Jackson. E.L., Rowden, A.A., Attrill, M.J., Bossey, S.J. & Jones, M.B. (2001). The importance of seagrass beds as a habitat for fishery species. *Oceanography and Marine Biology: An Annual Review*, **39**, 269–303.

Kaly, U.L. & Jones, G.P. (1998). Mangrove restoration: a potential tool for coastal management in tropical developing countries. *Ambio*, **27**, 656–61.

Kathiresan, K. & Bingham, B.L. (2001). Biology of mangroves and mangrove ecosystems. *Advances in Marine Biology*, **40**, 81–251.

Klemas, V. (2013). Remote sensing of coastal wetland biomass: an overview. *Journal of Coastal Research*, **29**, 1016–28.

Laffoley, D.d'A. & Grimsditch, G. (eds.). (2009). *The Management of Natural Coastal Carbon Sinks*. Gland, Switzerland: IUCN. 53 pp.

Lanyon, J., Limpus, C.J. & Marsh, H. (1989). Dugongs and turtles: grazers in the seagrass system. In *Biology of Seagrasses: A Treatise on the Biology of Seagrasses with Special Reference to the Australian Region*, ed. A.W.D. Larkum, A.J. McComb & S.A. Shepherd, pp. 610–34, Amsterdam: Elsevier.

Little, C. (2000). *The Biology of Soft Shores and Estuaries*. Oxford, UK: Oxford University Press.

Lotze, H.K., Lenihan, H.S., Bourque, B.J., *et al.* (2006). Depletion, degradation, and recovery potential of estuaries and coastal seas. *Science*, **312**, 1806–9.

Lovelock, C.E., Cahoon, D.R., Friess, D.A., *et al.* (2015). The vulnerability of Indo-Pacific mangrove forests to sea-level rise. *Nature*, **526**, 559–63.

Malpas, L.R., Smart, J., Drewitt, A., Sharps, E. & Garbutt, A. (2013). Continued declines of Redshank *Tringa totanus* breeding on saltmarsh in Great Britain: is there a solution to this conservation problem? *Bird Study*, **60**, 370–83.

Marsh, H., Eros, C., Corkeron, P. & Breen, B. (1999). A conservation strategy for dugongs: implications of Australian research. *Marine and Freshwater Research*, **50**, 979–90.

Mateo, M.A., Cebrián, J., Dunton, K. & Mutchler, T. (2006). Carbon flux in seagrass ecosystems. In *Seagrasses: Biology, Ecology and Conservation*, ed. A.W.D. Larkum, R.J. Orth & C.M. Duarte, pp. 159–92, Dordrecht, Netherlands: Springer.

Mcleod, E., Chmura, G.L., Bouillon, S., *et al.* (2011). A blueprint for blue carbon: toward an improved understanding of the role of vegetated coastal habitats in sequestering $CO_2$. *Frontiers in Ecology and the Environment*, **9**, 552–60.

Michel, J. & Rutherford, N. (2014). Impacts, recovery rates, and treatment options for spilled oil in marshes. *Marine Pollution Bulletin*, **82**, 19–25.

Migliaccio, M., De Martino, F., Silvestre, F. & Procaccini, G. (2005). Meadow-scale genetic structure in *Posidonia oceanica*. *Marine Ecology Progress Series*, **304**, 55–65.

Mitsch, W.J. & Gosselink, J.G. (2000). *Wetlands*, 3rd edn. New York: John Wiley & Sons, Inc.

Moreno-Mateos, D., Power, M.E., Comín, F.A. & Yockteng, R. (2012). Structural and functional loss in restored wetland ecosystems. *PLoS Biology*, **10**, e1001247.

Morton, J.F. (1976). Craft industries from coastal wetland vegetation. In *Estuarine Processes*, vol. **1**, ed. M. Wiley, pp. 254–66. New York: Academic Press.

Muehlstein, L.K. (1989). Perspectives on the wasting disease of eelgrass *Zostera marina*. *Diseases of Aquatic Organisms*, **7**, 211–21.

Mumby, P.J. (2006). Connectivity of reef fish between mangroves and coral reefs: algorithms for the design of marine reserves at seascape scales. *Biological Conservation*, **128**, 215–22.

Mumby, P.J., Edwards, A.J., Arias-González, J.E., *et al.* (2004). Mangroves enhance the biomass of coral reef fish communities in the Caribbean. *Nature*, **427**, 533–6.

Mumby, P.J. & Harborne, A.R. (2006). A seascape-level perspective of coral reef ecosystems. In *Coral Reef Conservation*, ed. I.M. Coté & J.D. Reynolds, pp. 78–114. Cambridge, UK: Cambridge University Press.

Nagelkerken, I., Blaber, S.J.M., Bouillon, S., *et al.* (2008). The habitat function of mangroves for terrestrial and marine fauna: a review. *Aquatic Botany*, **89**, 155–85.

Neira, C., Grosholz, E.D., Levin, L.A. & Blake, R. (2006). Mechanisms generating modification of benthos following tidal flat invasion by a *Spartina* hybrid. *Ecological Applications*, **16**, 1391–404.

Nellemann, C., Corcoran, E., Duarte, C.M., *et al.* (eds.). (2009). *Blue Carbon: the Role of Healthy Oceans in Binding Carbon*. A Rapid Response Assessment. United Nations Environment Programme. Norway: GRID-Arendal.

Nichols, F.H., Cloern, J.E., Luoma, S.N. & Peterson, D.H. (1986). The modification of an estuary. *Science*, **231**, 567–73.

Olds, A.D., Albert, S., Maxwell, P.S., Pitt, K.A. & Connolly, R.M. (2013). Mangrove-reef connectivity promotes the effectiveness of marine reserves across the western Pacific. *Global Ecology and Biogeography*, **22**, 1040–9.

Orth, R.J., Carruthers, T.J.B., Dennison, W.C., *et al.* (2006). A global crisis for seagrass ecosystems. *BioScience*, **56**, 987–96.

Percival, S.M., Sutherland, W.J. & Evans, P.R. (1998). Intertidal habitat loss and wildfowl numbers: applications of a spatial depletion model. *Journal of Applied Ecology*, **35**, 57–63.

Polidoro, B.A., Carpenter, K.E., Collins, L., *et al.* (2010) The loss of species: mangrove extinction risk and geographic areas of global concern. *PLoS ONE*, **5**, e10095.

Ralph, P.J., Tomasko, D., Moore, K., Seddon, S. & Macinnis-Ng, C.M.O. (2006). Human impacts on seagrasses: eutrophication, sedimentation, and contamination. In *Seagrasses: Biology, Ecology and*

*Conservation*, ed. A.W.D. Larkum, R.J. Orth & C.M. Duarte, pp. 567–93, Dordrecht, Netherlands: Springer.

Robertson, M. & Hough, P. (2011). Wetlands regulation: the case of mitigation under Section 404 of the Clean Water Act. In *Wetlands: Integrating Multidisciplinary Concepts*, ed. B.A. LePage, pp. 171–87. Dordrecht, Netherlands: Springer.

Santos, H.F., Carmo, F.L., Paes, J.E.S., Rosado, A.S. & Peixoto, R.S. (2011). Bioremediation of mangroves impacted by petroleum. *Water, Air, & Soil Pollution*, **216**, 329–50.

Seacology (2015). www.seacology.org/project/sri-lanka-mangrove-conservation-project/ (accessed 31 July 2016).

Shafer, D.J., Kaldy, J.E. & Gaeckle, J.L. (2014). Science and management of the introduced seagrass *Zostera japonica* in North America. *Environmental Management*, **53**, 147–62.

Sheridan, P. & Hays, C. (2003). Are mangroves nursery habitat for transient fishes and decapods? *Wetlands*, **23**, 449–58.

Short, F.T., Polidoro, B., Livingstone, S.R., *et al.* (2011). Extinction risk assessment of the world's seagrass species. *Biological Conservation*, **144**, 1961–71.

Siikamäki, J., Sanchirico, J.N. & Jardine, S.L. (2012). Global economic potential for reducing carbon dioxide emissions from mangrove loss. *Proceedings of the National Academy of Sciences*, **109**, 14369–74.

Silliman, B.R., van de Koppel, J., McCoy, M.W., *et al.* (2012). Degradation and resilience in Louisiana salt marshes after the BP–*Deepwater Horizon* oil spill. *Proceedings of the National Academy of Sciences*, **109**, 11234–9.

Sobel, J.A. & Dahlgren, C.P. (2004). *Marine Reserves: a Guide to Science, Design, and Use.* Washington, DC: Island Press.

Spalding, M., Kainuma, M. & Collins, L. (2010). *World Atlas of Mangroves*. London: Earthscan.

Spalding, M., Taylor, M., Ravilious, C., Short, F. & Green, E. (2003). The distribution and status of seagrasses. In *World Atlas of Seagrasses*, ed. E.P. Green & F.T. Short, pp. 5–26. Berkeley, CA: University of California Press.

Spalding, M.D., Ruffo, S., Lacambra, C., *et al.* (2014). The role of ecosystems in coastal protection: adapting to climate change and coastal hazards. *Ocean & Coastal Management*, **90**, 50–7.

Stevenson, J.C. & Kearney, M.S. (2009). Impact of global climate change and sea-level rise on tidal wetlands. In *Human Impacts on Salt Marshes: a Global Perspective*, ed. B.R. Silliman, E.D. Grosholz & M.D. Bertness, pp. 171–206. Berkeley, CA: University of California Press.

Strong, D.R. & Ayres, D.R. (2009). *Spartina* introductions and consequences in salt marshes: arrive, survive, thrive, and sometimes hybridize. In *Human Impacts on Salt Marshes: A Global Perspective*, ed. B.R. Silliman, E.D. Grosholz & M.D. Bertness, pp. 3–22. Berkeley, CA: University of California Press.

Strong, D.R. & Ayres, D.R. (2013). Ecological and evolutionary misadventures of *Spartina*. *Annual Review of Ecology, Evolution, and Systematics*, **44**, 389–410.

Teal, J.M. & Peterson, S.B. (2005). Introduction to the Delaware Bay salt marsh restoration. *Ecological Engineering*, **25**, 199–203.

Thorhaug, A. (1986). Review of seagrass restoration efforts. *Ambio*, **15**, 110–17.

Thorhaug, A. (1987). Large-scale seagrass restoration in a damaged estuary. *Marine Pollution Bulletin*, **18**, 442–6.

Tian, B., Wu, W., Yang, Z. & Zhou, Y. (2016). Drivers, trends, and potential impacts of long-term coastal reclamation in China from 1985 to 2010. *Estuarine, Coastal and Shelf Science*, **170**, 83–90.

Torres-Pulliza, D., Wilson, J.R., Darmawan, A., Campbell, S.J. & Andréfouët, S. (2013). Ecoregional scale seagrass mapping: a tool to support resilient MPA network design in the Coral Triangle. *Ocean & Coastal Management*, **80**, 55–64.

Unsworth, R.K.F., De León, P.S., Garrard, S.L., *et al.* (2008). High connectivity of Indo-Pacific seagrass fish assemblages with mangrove and coral reef habitats. *Marine Ecology Progress Series*, **353**, 213–24.

Valiela, I., Bowen, J.L. & York, J.K. (2001). Mangrove forests: one of the world's threatened major tropical environments. *BioScience*, **51**, 807–15.

Walker, D.I., Kendrick, G.A. & McComb, A.J. (2006). Decline and recovery of seagrass ecosystems – the dynamics of change. In *Seagrasses: Biology, Ecology and Conservation*, ed. A.W.D. Larkum, R.J. Orth & C.M. Duarte, pp. 551–65, Dordrecht, Netherlands: Springer.

Walters, B.B., Rönnback, P., Kovacs, J.M., *et al.* (2008). Ethnobiology, socio-economics and management of mangrove forests: a review. *Aquatic Botany*, **89**, 220–36.
Warren, R.S., Fell, P.E., Rozsa, R., *et al.* (2002). Salt marsh restoration in Connecticut: 20 years of science and management. *Restoration Ecology*, **10**, 497–513.
Williams, S.L. (2007). Introduced species in seagrass ecosystems: status and concerns. *Journal of Experimental Marine Biology and Ecology*, **350**, 89–110.
Wolff, W.J. (1992). The end of a tradition: 1000 years of embankment and reclamation of wetlands in the Netherlands. *Ambio*, **21**, 287–91.
Zedler, J.B. (2004). Compensating for wetland losses in the United States. *Ibis*, **146** (Suppl. 1), 92–100.
Zuo, P., Zhao, S., Liu, C., Wang, C. & Liang. Y. (2012). Distribution of *Spartina* spp. along China's coast. *Ecological Engineering*, **40**, 160–6.

# 13 Coral Reefs

Coral reefs of shallow tropical waters engender more public interest than probably any other marine environment. They have enormous aesthetic appeal with their arresting richness of form and profusion of organisms in warm and usually clear waters. Not surprisingly, coral reef tourism is a huge industry. But coral reefs are also a vital source of seafood for many communities and provide a wide range of ecosystem services. They function as spawning, nursery, and feeding grounds for many organisms; act as natural breakwaters; and play an important role in the global calcium budget (Spurgeon, 1992; Moberg & Folke, 1999). The value of their ecosystem services is enormous (Table 3.2). Cruz-Trinidad *et al.* (2011) estimated the total economic value of a coral reef area of 200 $km^2$ on the west coast of Luzon, Philippines, at $38 million per year, from fisheries, aquaculture, and tourism but mainly from shoreline protection. Coral reefs reduce wave energy by an estimated 97%, a natural defence that may benefit at least 100 million coastal inhabitants (Ferrario *et al.*, 2014). Destruction wrought by the tsunami of December 2004 along the coast of south-western Sri Lanka correlated with the extent of reef cover, inundation being far worse where illegal coral mining allowed water to surge inland (Fernando *et al.*, 2005). Many of the world's coral reefs are now severely degraded, and the conservation of coral reefs is of critical importance. For a recent overview of tropical coral reef biology see Sheppard *et al.* (2009). Cold-water corals are discussed in Chapter 14.

## Distribution and Reef Types

The global area of coral reefs is estimated to be about 250 000 $km^2$ (Burke *et al.*, 2011). However, the total extent may be at least twice this, depending on how coverage is estimated and what is categorized as 'coral reef' (Sheppard *et al.*, 2009). Coral reefs probably occupy only around 0.1% of the world's ocean area, with more than half the total reef area occurring in the Pacific and SE Asia (Table 13.1). Coral reefs are not, therefore, an abundant habitat in the world ocean. They are largely restricted to waters between 30° N and 30° S and the minimum monthly isotherm of 18°C (Kleypas *et al.*, 1999; Sheppard *et al.*, 2009). Major warm and cold currents modify the typical latitudinal range however. The cool California and Humboldt currents of the eastern Pacific constrict coral reef development along the west coast of America, whereas the warm Gulf Stream and Kuroshio Current extend the occurrence of reefs to Bermuda and southern Japanese islands respectively (Fig. 1.2). Reef-forming corals are also restricted by lowered salinity. A salinity range of 23–42 has been reported for reef sites, but typically salinities are 34–35. Water turbidity is an important factor too, high sedimentation having a smothering effect and reducing light penetration. Hermatypic (i.e. reef-forming) corals need light and so tend to thrive in clear waters. This is because most of them contain zooxanthellae, symbiotic single-celled algae (dinoflagellates) that provide nutrition to the polyps through photosynthesis, facilitate nutrient recycling and conservation, and help in calcification of the skeleton (Davy *et al.*, 2012). Coral reefs occur in oligotrophic (i.e. nutrient-poor) waters but make

**Table 13.1** Estimates of Coral Reef Area

| Region | Reef area (km²) | Percent of reef area |
|---|---|---|
| Atlantic | 25 850 | 10.4 |
| Australia | 42 310 | 16.9 |
| Indian Ocean | 31 540 | 12.6 |
| Middle East | 14 400 | 5.8 |
| Pacific | 65 970 | 26.4 |
| South-east Asia | 69 640 | 27.9 |
| Global total | 249 710 | |

Extracted from Burke, L., Reytar, K., Spalding, M. & Perry, A. (2011). *Reefs at Risk Revisited.* Washington, DC: World Resources Institute (Table 4.1). http://www.wri.org/publication/reefs-risk-revisited.

very efficient use of nutrients, particularly through this coral-zooxanthella symbiosis. The lower limit of coral reef growth is usually about 50–70 m water depth, with the most luxuriant growth from the level of the lowest tides to about 25–30 m, providing there is a hard substratum for attachment and strong wave action.

In terms of geomorphology, three main reef types are recognized, although they are not always clearly separable (Sheppard *et al.*, 2009) (Fig. 13.1). Fringing reefs develop on shallow rocky substrata adjacent to islands or landmasses, the coral growth eventually forming a shallow platform up

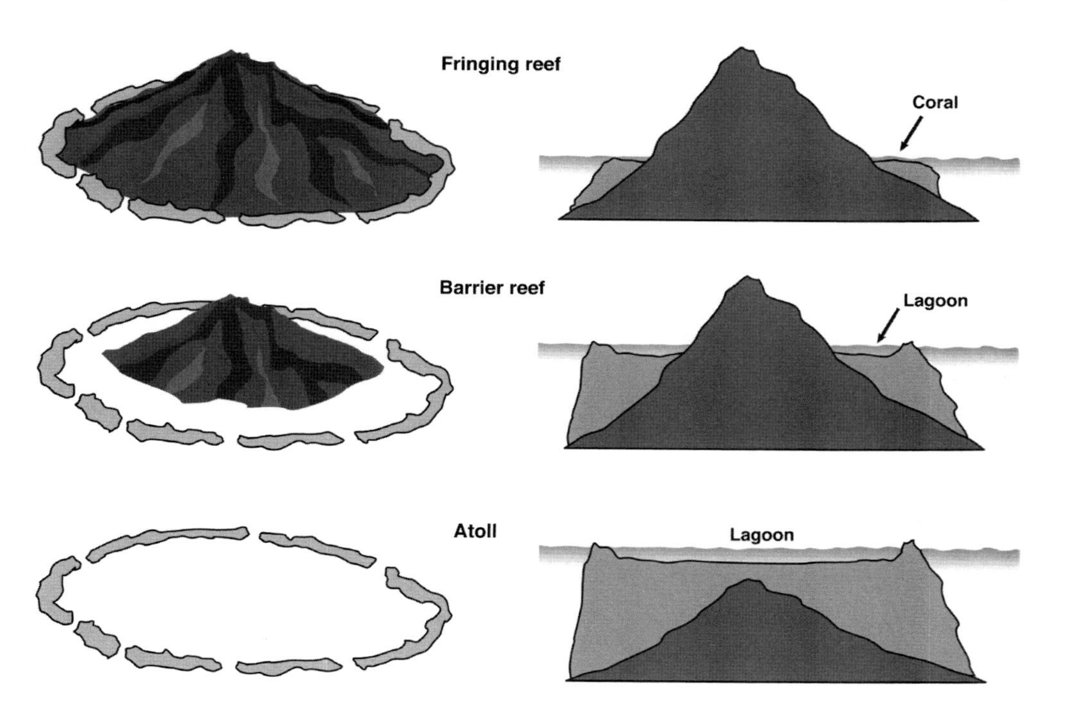


**Fig. 13.1** The main types of coral reef.

to low tide level and with the seaward edge marked by a reef crest and a steep slope to the reef front. There is often a channel between the reef and adjacent land because of sediment input from the land preventing coral growth. Barrier reefs, however, are separated from the land by a lagoon often up to several kilometres wide. They can be continuous for many kilometres or divided into segments and may develop around volcanic islands that are subsiding or subject to sea-level rise. In this case, what was a fringing reef continues to grow upwards and most rapidly around its outer edge, so that in time it becomes separated from the coastline by a lagoon. An atoll is a ring-shaped reef or string of low-lying coral islands enclosing a lagoon, formed where an island has subsided beneath the sea surface, so that what was once a barrier reef now no longer encircles an emergent island. Atolls can be very old structures. The coral limestone above the basement rock can be more than 1 km thick and represent growth over more than 50 million years. Most atolls are oceanic in location and in the Indo-Pacific.

## Biodiversity and Productivity

Coral reefs encompass a huge complexity of features and diversity of habitats that help explain their high species richness (e.g. Arias-González *et al.*, 2008). Important physical factors influencing large-scale vertical and horizontal zonation include water depth (e.g. affecting illumination, turbulence, and turbidity), substratum type (e.g. from intact coral beds to patches of coral rubble and sand), and wave action (e.g. between the windward and leeward sides of an atoll and between the seaward and lagoon sides). We see a range of features in a typical cross-section of a reef (Guilcher, 1988; Sheppard *et al.*, 2009) (Fig. 13.2). On the seaward side is the steep outer slope, often terraced towards its upper part and characterized at the surface by coral-algal spurs or buttresses alternating with deep surge channels. It is here along these spurs on the outer edge of the reef that coral growth is most rapid. But where the tops of the ridges receive the full brunt of wave exposure, they are coral free and covered instead by encrusting coralline algae. Inside this algal ridge is the shallow reef flat, normally the widest part of the reef. The reef flat is usually exposed on low spring tides and may be dotted with pools. It is typically a mix of rubble and sand patches, calcareous algal pavement, and scattered corals including micro-atolls (basin-like structures comprising the outer surfaces of massive corals, typically *Porites*). The heterogeneity of the reef flat provides for a high diversity of associated organisms. Behind the reef flat of barrier reefs and atolls is the lagoon, usually up to some tens

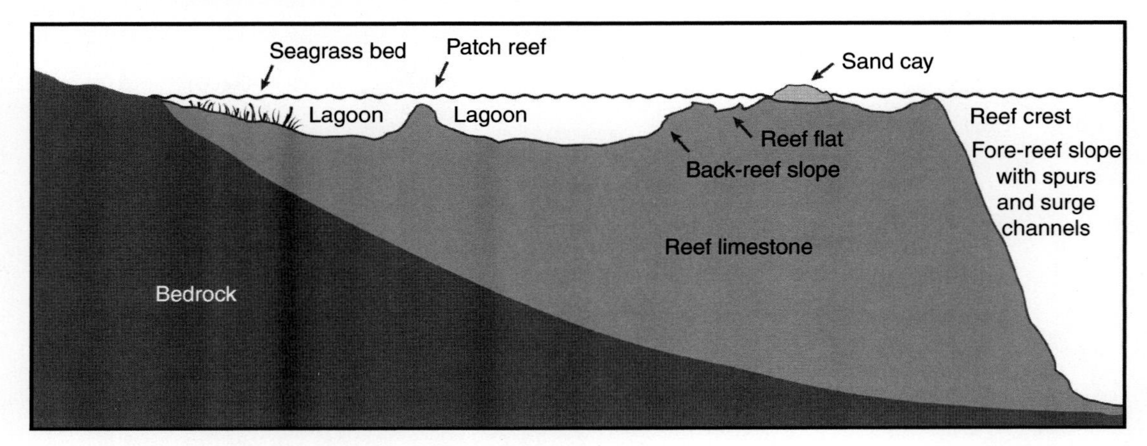


**Fig. 13.2** Cross-section through a typical reef.

of metres deep. Often rising from the lagoon are small patch reefs, but otherwise the lagoon floor is covered by calcareous sediment. Beds of seagrass or of the calcareous green alga *Halimeda* can be conspicuous here. Mangroves may make an appearance in more sheltered intertidal areas, as on the inner shores of fringing reefs. And where sediment accumulates on top of a reef it can give rise to small sandy islands (cays), which in time may support terrestrial vegetation. Such islands can be important as nesting sites for pelagic seabirds such as boobies, tropicbirds, terns, frigatebirds, and shearwaters as well as for marine turtles.

The calcareous framework of coral reefs is due primarily to the growth of hermatypic corals and coralline red algae (Fig. 13.3). Best known of the reef-forming corals are the stony corals, or Scleractinia, of which there are some 800 hermatypic species. The dominant and most

**Fig. 13.3** Stony corals (Scleractinia) provide the main framework of coral reefs – structurally complex habitats that support highly diverse communities. A reef in Fiji. Photograph: Kim Westerskov. (A black and white version of this figure will appear in some formats. For the colour version, please refer to the plate section.)

speciose family is the Acroporidae, which includes the staghorn and elkhorn corals. Many species of non-reef-building (ahermatypic) sclertactinian corals also occur on reefs, such as cup corals (Caryophyllidae) and mushroom corals (Fungiidae). Another group of corals that can be locally important as reef builders are millepores, colonial hydrozoans commonly known as stinging corals or fire corals on account or their toxic defensive polyps. Other coral-like cnidarians important on reefs include gorgonians (e.g. sea whips, sea fans) and soft corals.

Coral reefs harbour a wealth of other invertebrates, including polychaete worms, decapod crustaceans, gastropods, and echinoderms. Many small invertebrates of reefs are cryptic, exploiting the multitude of available microhabitats, including species that burrow into the coral itself, thereby acting as important bioeroders that contribute sediment to the reef system. Coral reefs are possibly the most species-rich of all marine habitats; about one-third of all known multicellular marine species occur on coral reefs, but an estimated three-quarters of coral reef species remain to be discovered (Fisher *et al.*, 2015). Many groups are poorly known taxonomically, not just on coral reefs but throughout the tropics. But among the better-known groups, corals and coral-like cnidarians, gastropods, mantis shrimps, sea cucumbers, and fish attain their greatest diversity on coral reefs (Paulay, 1997). In the case of fish, some 4000 species occur on coral reefs, or about a quarter of all marine fish species. Conspicuous families include groupers (Serrandiae), snappers (Lutjanidae), grunts and sweetlips (Haemulidae), butterflyfish (Chaetodontidae), angelfish (Pomacanthidae), damselfish (Pomacentridae), wrasse (Labridae), parrotfish (Scaridae), and surgeonfish (Acanthuridae) (Spalding *et al.*, 2001). And at least 320 species of fish depend on live coral habitat, a third of them damselfishes and gobies (Gobiidae) (Coker *et al.*, 2014). Important pelagic predators of reefs include Carangidae (including jacks and trevallies), barracudas (*Sphyraena*), and sharks (Longhurst & Pauly, 1987). The exceptional diversity of coral reefs is probably due to a combination of factors, including evolutionary history, the development of tight recycling and intricate food webs in oligotrophic waters, and the structural complexity of reefs and opportunity for niche diversification, including many symbiotic interactions. In terms of latitudinal gradients of diversity (see Chapter 1), the higher diversity of a number of taxa at lower latitudes can be attributed largely to their association with coral reefs.

Nevertheless, there are marked regional differences in coral reef diversity, most obviously between the high and low diversities of the Indo-Pacific and the Atlantic. This is very evident from contours of species richness for the Scleractinia (Fig. 13.4). More than 700 scleractinian corals are known from the Indo-West Pacific as opposed to some 60 from the Western Atlantic. Within the Indo-Pacific, diversity peaks in the Indonesian-Philippines archipelago, a region referred to as the Coral Triangle. Many other groups are similarly more diverse on Indo-West Pacific reefs, such as sponges, cowries and cone snails, mantis and caridean shrimps, echinoderms, butterflyfish, and angelfish (Spalding *et al.*, 2001). The lower diversity of Atlantic reefs dates from the emergence of the isthmus of Panama some 3–3.5 million years ago and subsequent mass extinctions during the Pliocene-Pleistocene glaciations, events that affected the Indo-Pacific far less severely.

Coral reefs are remarkable in being highly productive systems in otherwise-oligotrophic tropical waters. The productivity depends on what one includes in a reef ecosystem and the nature of the reef itself. Primary productivity is typically about 1.5–14.0 g C $m^{-2}$ $d^{-1}$, contributed mainly by benthic microalgae and cyanobacteria, zooxanthellae, fleshy and calcareous macroalgae, and seagrasses. This is typically one to two orders of magnitude greater than the phytoplankton productivity of the surrounding ocean and comparable to estimates for the most productive terrestrial systems (Longhurst & Pauly, 1987). Such productivity is due largely to the consistently high light availability and the tight recycling and retention of nutrients within reef ecosystems.

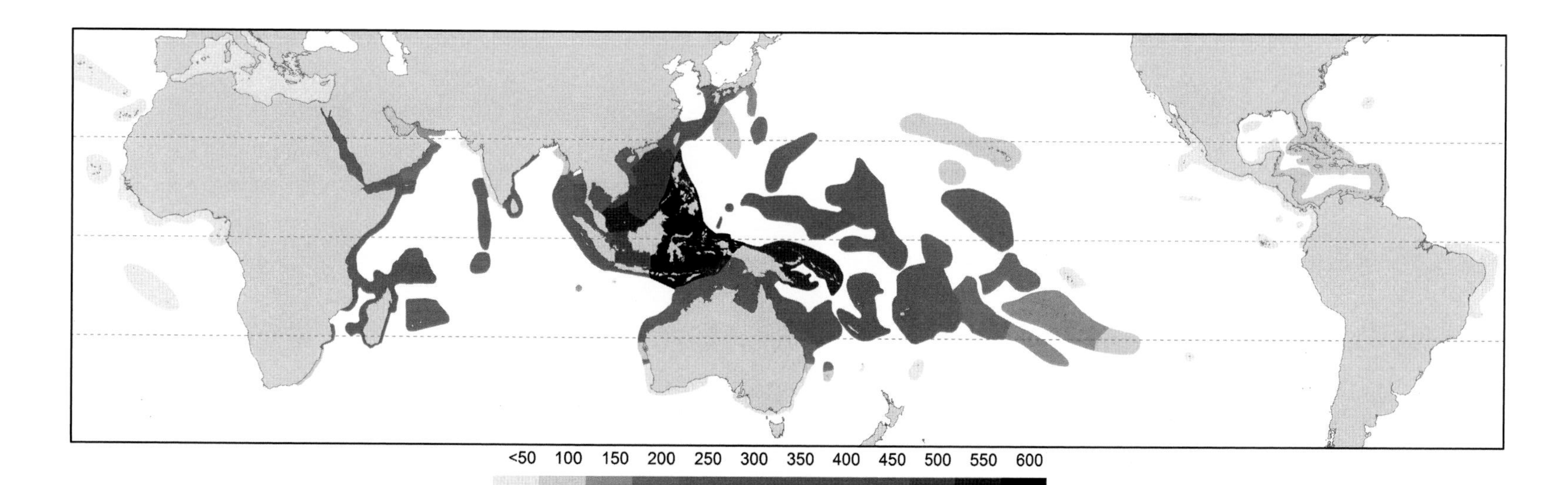


**Fig. 13.4** Contours of species richness for warm-water stony corals. The dashed lines are the equator and the tropics of Cancer and Capricorn (23° 26' N and S). From Veron, J., Stafford-Smith, M., DeVantier, L. & Turak, E. (2015). Overview of distribution patterns of zooxanthellate Scleractinia. Frontiers in Marine Science, 1, article 81. Copyright © 2015 Veron, Stafford-Smith, DeVantier and Turak. (A black and white version of this figure will appear in some formats. For the colour version, please refer to the plate section.)

## Disturbances

Coral reefs are highly dynamic systems, structured by a multiplicity of biotic interactions (such as competition for resources, grazing, predation, and bioerosion) and variability of the physical environment. The latter can include large-scale natural disturbances. Hurricanes, for example, can severely impact coral reefs, depending on such factors as hurricane intensity and frequency and the location and structure of the reef. For Caribbean reefs, Gardner *et al* (2005) found relative coral cover to be on average about 17% less in the year after a hurricane. Widespread damage to reefs can also result from prolonged emersion during unusually low tides that occur during the heat of the day. But as with all disturbances, corals vary considerably in their susceptibility and response.

Higher-than-normal sea temperatures, as well as various other stresses, can cause zooxanthellate corals to lose their symbiotic algae and/or the algal pigments, a response known as bleaching because it results in a marked paling of the coral (Baker *et al.*, 2008) (Fig 13.5). If the stress abates, corals may regain their zooxanthellae and recover, but if it is more severe or persistent or compounded by other stresses, coral mortality is likely. Bleaching and loss of live coral can occur over huge areas of reef. Several mass bleachings documented in the past 30 years have been linked to strong El Niño events (see Chapter 1) where higher-than-normal sea temperatures have occurred (at least 1°C higher than the summer maximum), often for prolonged periods. Global bleaching events triggered by El Niños have occurred in 1998, 2010, and 2015–16. The 1998 event, where sea surface temperatures exceeded the thermal tolerance of corals for months, resulted in coral mortalities of more than 80% on many reefs (Westmacott *et al.*, 2000; Wilkinson, 2000). Particularly

**Fig. 13.5** A bleached *Acropora* sp. in the Wakatobi Marine National Park in south-east Sulawesi, Indonesia. Photograph: J.J. Bell. (A black and white version of this figure will appear in some formats. For the colour version, please refer to the plate section.)

affected by this event were large areas of the Indian Ocean where El Niño coincided with the positive phase of an oscillation (the Indian Ocean dipole) characterized by warmer-than-average sea surface temperatures in the western Indian Ocean. Graham *et al.* (2015) examined the ecosystem dynamics of reefs in the Seychelles, in the western Indian Ocean, severely affected by the 1998 coral bleaching event. The reefs they examined had lost > 90% of their live coral cover. Most of the reefs had recovered towards their pre-disturbance state, with live coral cover reaching an average of 23% by 2011 and with macroalgal cover remaining low (< 1%). By contrast, for the remaining reefs over this period, coral cover remained low (< 3%), but macroalgal cover increased (to 42%). Reefs respond similarly to nutrient enrichment or loss of herbivores (see below).

Coral reefs are also subject to major biological disturbances. Particularly dramatic are the effects of population explosions or mass die-offs of key echinoderms. The crown-of-thorns starfish (*Acanthaster planci*) is a large species (adults are normally 25–40 cm in diameter) of the Indo-Pacific that feeds mainly on hard corals, particularly faster-growing acroporids (Moran, 1990). Crown-of-thorns starfish are not usually common (typically up to 1 per hectare), but on occasions outbreaks occur (more than about 15 per hectare), resulting in extensive coral mortality. Such outbreaks have been recorded since the late 1950s (Ryukyu Islands, Japan) and early 1960s (Great Barrier Reef, Australia). (*A. planci* is in fact a species complex (Vogler *et al.*, 2008), and outbreak patterns may differ between the sibling species.) By contrast, the long-spined sea urchin *Diadema antillarum*, a keystone herbivore of Caribbean coral reefs, suffered a mass die-off throughout the Caribbean in 1983–4. More than 90% of urchins were killed, and after 30 years the average population densities were still only about 12% of those before the die-off (Lessios, 2016). As we shall see, such disturbances have major implications for the structure and functioning of coral reef ecosystems.

It is not clear, however, if these physical and biological disturbances are entirely natural or if human activities are implicated. For example, global warming may increase hurricane intensity (IPCC, 2014) and bleaching events (Baker *et al.*, 2008; Hoegh-Guldberg, 2011). Also, outbreaks of the crown-of-thorns starfish on the Great Barrier Reef may be linked to increased nutrient discharge from land runoff, thereby enhancing larval survival (Brodie *et al.*, 2005), and the *Diadema* mass mortality was probably due to a bacterial pathogen, the spread of which was possibly facilitated by ballast water exchange (Phinney *et al.*, 2001).

## Human Impacts

Human activities directly impact coral reefs in a variety of ways, notably through fishing, degradation of coastal habitat, pollution, and mining (Hatcher *et al.*, 1989; Dubinsky & Stambler, 1996; Brown, 1997) (Box 13.1).

### Fishing

Although coral-reef fisheries account for only about 2–5% of world fisheries catches, they are a vital source of food, particularly in many developing countries. For some island nations, reef seafood may provide more than half the protein for the local people (Pauly *et al.*, 2002; Sheppard *et al.*, 2009). Worldwide there are some 5–7 million coral reef fishers, the great majority of them small-scale fishers and about half of them in SE Asia (Teh *et al.*, 2013). Important fisheries species include pelagic fish that migrate through reef areas, such as mackerels (*Scomberomorus* spp.), tunas, and trevallies as well as reef-associated species including groupers, snappers, and emperors (Lethrinidae). Invertebrates that are exploited in many reef areas include trochus (*Tectus*

## Box 13.1 Threats and Stresses to Coral Reefs

The order does not necessarily reflect the degree of damage caused by threats. The ranking of threats will change for different areas of the world.

### Threats from Global Change

- **Coral Bleaching** – caused by elevated sea surface temperatures due to global climate change.
- **Rising Levels of $CO_2$** – increased concentrations of $CO_2$ in seawater decrease calcification rates in coral reef organisms.
- **Diseases, Plagues, and Invasives** – increases in diseases and plagues of coral reef organisms that are increasingly linked to human disturbances.

### Direct Human Pressures

- **Overfishing (and Global Market Pressures)** – exploitation of fishes and invertebrates beyond sustainable levels, including the use of damaging practices (blast and cyanide fishing).
- **Sediments** – from poor land use, deforestation, dredging.
- **Nutrients and Chemical Pollution** – organic and inorganic chemicals carried with sediments in untreated sewage, waste from agriculture, animal husbandry, and industry; includes complex organics and heavy metals.
- **Development of Coastal Areas** – modification of coral reefs for urban, industrial, transport, and tourism developments, including reclamation and the mining of coral reef rock and sand beyond sustainable limits.

### The Human Dimension – Governance, Awareness, and Political Will

- **Rising Poverty, Increasing Populations, Alienation from the Land** – increasing human populations put increasing pressures on coral reef resources beyond sustainable limits.
- **Poor Capacity for Management and Lack of Resources** – most coral reef countries lack trained personnel for coral reef management, raising awareness, enforcement, and monitoring; also a lack of adequate funding and logistic resources to implement effective conservation.
- **Lack of Political Will and Ocean Governance** – most problems facing coral reefs are manageable if there is political will and effective and non-corrupt governance of resources. Interventions by and inertia in global and regional organizations can impede national action to conserve coral reefs.

From Wilkinson, C. (2006). Status of coral reefs of the world: summary of threats and remedial action. In *Coral Reef Conservation*, ed. I.M. Coté & J.D. Reynolds, pp. 3–39. Cambridge: Cambridge University Press. With permission from Cambridge University Press.

*niloticus*) and giant clams (Tridacninae) of the Indo-Pacific, queen conch (*Lobatus gigas*) in the Caribbean, sea cucumbers (bêche-de-mer or trepang), shrimps, spiny lobsters, and octopuses (Craik *et al.*, 1990).

A rapidly growing area of exploitation is the extraction of live reef fish to supply the lucrative Asian restaurant trade, with a total take of probably at least 60 000 tonnes per year. A number of high-value species are targeted, mainly groupers, snappers, and wrasses (Labridae). These are sourced from many countries in the Indo-Pacific but destined mainly for Hong Kong and mainland China (Vincent, 2006).

One obvious effect of overfishing on coral reef fish communities is the depletion of top predators, such as groupers, jacks, and sharks (e.g. Sandin *et al.*, 2008), *K*-selected species (see Chapter 4) that are vulnerable to overexploitation. Grouper species can form large spawning aggregations or have a high degree of site specificity, making them easily targeted, and many of them are threatened species (Morris *et al.*, 2000).

## Destructive Fishing Practices

In the face of declining yields, highly destructive fishing methods that physically destroy reefs have been employed, such as blast fishing and muro-ami. These techniques have been used particularly in SE Asia. In blast fishing, schooling reef fish are targeted so as to maximize the catch. Homemade bombs are dropped on the reef, killing or stunning fish, as well as many other animals in the vicinity of the explosion, and shattering the coral. The typical bombs that are used can reduce coral to rubble in an area up to 10 m in diameter, and with heavy use whole reefs can be destroyed (Alcala & Gomez, 1987). This has been evident in the Philippines where, although illegal, blast fishing has continued to cause widespread damage to reefs (Spalding *et al.*, 2001). Even if left alone, bombed reefs may take up to 40 years to regain just half their cover of hard coral (Alcala & Gomez, 1987). In muro-ami fishing, swimmers move in a line across a reef, pounding it with stone weights on ropes to drive the fish into a net (Wells & Hanna, 1992). The method has also been common in the Philippines and continues despite being banned.

Trade in marine species for aquariums began on a small scale in the mid-twentieth century but has boomed over the past few decades and is now a multi-million-dollar international industry (Wabnitz *et al.*, 2003). The traffic is mainly in small colourful coral reef fishes, especially damselfishes and clownfishes (Pomacentridae), and to a lesser extent in reef invertebrates, mainly scleractinian corals, molluscs, shrimps, and anemones. These ornamental species are collected mainly in the Philippines and Indonesia for export in particular to the USA and western Europe. Import records for the USA, the main importer, show that in one year a total of 11 million marine aquarium fish comprising 1800 species entered the country (Rhyne *et al.*, 2012). The trade is highly wasteful with cumulative mortality of up to 80% of the fish between collector and retailer, and subsequent survival is often low as marine species tend to be hard to keep in home aquariums and some commonly traded species are simply unsuitable (Sadovy & Vincent, 2002). Collecting coral reef species for the aquarium trade appears to have depleted or even locally extirpated some populations, and destructive fishing methods such as cyanide fishing (see below) are sometimes used (Wabnitz *et al.*, 2003; Dee *et al.*, 2014). Likely to be among the most vulnerable species are those with a restricted range and limited dispersal and fecundity (Lunn & Moreau, 2004). Restricted-range species are common on coral reefs. Roberts *et al.* (2002) found that around 10–20% of coral-reef fish, gastropods, and lobsters have geographic ranges of up to only 50 000 km$^2$, and for some species the global range is equivalent to a single atoll.

Of considerable concern is the use of cyanide fishing as a means of collecting coral reef fish for the aquarium trade. This technique began in the Philippines in the early 1960s but has more recently been used to obtain larger reef fish, such as groupers and wrasse, to supply live fish for Asian restaurants (Burke *et al*., 2011). Collectors use plastic bottles to squirt sodium cyanide solution at fish to stun them temporarily. The fish often retreat into holes in the reef, which may then be broken open to recover the fish. Many of the target specimens that initially survive subsequently die. The cyanide exposure also has lethal or sub-lethal effects on many non-target organisms including the corals themselves (Dee *et al*., 2014). Although widely outlawed, the practice continues.

The marine aquarium trade represents, however, an important source of income for many impoverished coastal communities. Ornamentals can be sold for far more than food fish, which may be an incentive to conserve reefs and promote responsible collecting. The trade is potentially sustainable if appropriately monitored and managed to ensure, for example, that only suitable species are targeted and in appropriate numbers and that only non-destructive collecting methods are used (Spalding *et al*., 2001; Rhyne *et al*., 2014). Specific management practices that show promise include moratoriums on particular species, no-take marine protected areas, traditional marine tenure systems, and import and export restrictions (Dee *et al*. 2014). Aquarium hobbyists represent a huge constituency, and the industry is well placed to raise public awareness and support conservation efforts (Rhyne *et al*., 2014).

The aquarium trade can provide an invasion pathway for alien species, indicating the need for good education and enforcement programmes. Semmens *et al*. (2004) report the occurrence of 16 non-native marine fishes on reefs of south-east Florida. The species are all commonly imported in the marine aquarium trade, suggesting that they have escaped or been released from aquariums. Spectacular has been the invasion of lionfish (*Pterois volitans* and *P. miles*) into the NW Atlantic. Lionfish are popular aquarium fishes native to the Indo-Pacific. They were discovered off Florida in the mid-1980s and from there spread rapidly through the Greater Caribbean region, along the eastern coast of the USA, and more recently to south-eastern Brazil. Lionfish are voracious predators and expected to cause major impacts on reef communities (Betancur-R *et al*., 2011; Ferreira *et al*., 2015).

Some coral reef fishes can be reared successfully in captivity, mainly demersal spawners that have clutches of large eggs that they care for. But unlike the trade in freshwater ornamentals, which is based mostly on captive-bred fish, the great majority of marine fish for the aquarium trade continue to be collected from the wild. If aquaculture of marine ornamental fish is to be advanced and contribute to reef conservation, a number of areas need to be researched, such as broodstock diet, spawning induction, and larval rearing and nutrition (Moorhead & Zeng, 2010).

Coral reefs are also a major source of ornamental species for tourist shops and shell collectors, with the Philippines as the major supplier and the USA the major consumer. Molluscs in particular are taken but also hard corals, gorgonacean sea fans, crustaceans, echinoderms, and certain fish (Vincent, 2006). There are probably some 5000 mollusc species involved, but most popular are large colourful gastropods of the Indo-Pacific and Caribbean, including cowries (Cypraeidae), helmet shells (Cassidae), cone snails (Conidae), volutes (Volutidae), and spider shells and conchs (Strombidae) (Wells, 1989). Many are easily overcollected, and those with narrow ranges are especially vulnerable, such as the volute *Cymbiola rossiniana* restricted to south-west New Caledonia. Some rare species are especially prized by shell collectors and command high prices, notably the cone snail *Conus gloriamaris*. Also popular with collectors is the giant triton *Charonia tritonis,* which, although widespread in the Indo-Pacific, occurs naturally at low densities and is vulnerable to overexploitation. It feeds mainly on sea stars, including the crown-of-thorns starfish.

## Coral Mining

Coral has long been used as an important source of material for building houses and roads, land reclamation, and lime production. Mining of reefs has been a major cause of reef degradation, for example in the Maldives (Brown & Dunne, 1988; Dawson Shepherd *et al.*, 1992), Sri Lanka (Öhman *et al.*, 1993), on the Caribbean coast of Panama (Guzmán *et al.*, 2003) and in Tanzania (Dulvy *et al.*, 1995). In the Maldives, coral rock has been extracted from the shallow flats around Malé Atoll. At sites not subject to mining, the cover of living coral was found to range from 11 to 60%, but at sites where mining had occurred, even up to 16 years earlier, only up to 1% of cover was living coral. Also, coral diversity was markedly reduced at sites subject to intensive mining. Mining results in a substratum of loose mobile rubble seemingly inhospitable to settling corals. If recovery does eventually occur once mining has ceased, it is likely to be very slow, at least of the order of decades. The fish community too was markedly affected by mining, with overall abundance and biomass reduced. Loss of topographic diversity from mining removes, in particular, habitat for small planktivorous species that depend on the reef for shelter. Intermediate-sized benthic herbivores are also reduced. From the early 1970s at the start of the building boom in the Maldives through to the mid-1980s, at least 54 000 $m^3$ of coral is estimated to have been mined, and to meet demand living coral was starting to be taken from outer atoll 'faros', ring-shaped reefs that naturally protect the islands from storms (Brown & Dunne, 1988). Given the projected demand for mined coral, the Maldives government introduced measures from the early 1990s to regulate mining, raise public awareness of the environmental implications of mining, and require the use of alternative building materials, and from 1995 coral was no longer to be used for construction (Jaleel, 2013).

## Land-Based Inputs

Coral reefs are highly vulnerable to the influence of land-derived inputs on coastal water quality. Human population pressure in many coastal regions of the tropics has resulted in coral reefs being exposed to increased sedimentation and turbidity as well as to increased anthropogenic inputs of nutrients, organic matter, and other pollutants. Logging, land clearance, and agriculture can markedly increase the erosion of soil into adjacent reef systems, and the use of agricultural fertilizers and discharge of sewage can result in nutrient enrichment and eutrophication of reef waters (Brown, 1997; Fabricius, 2005).

Changes in water quality from terrestrial runoff can have a wide variety of effects on corals and coral reef organisms, but certain processes are identified as particularly significant (Table 13.2). For example, dissolved inorganic nutrients can have important negative effects on coral reproduction and early life stages and promote the growth of macroalgae. The relative importance of macroalgal blooms affecting coral reefs illustrates the potential balance between bottom-up eutrophication and top-down herbivory (Littler *et al.*, 2006) (see Chapter 2). Increases in particulate organic matter also tend to benefit competitors, notably suspension feeders such as sponges, bivalves, and ascidians, more than corals (Fabricius, 2005). Sedimentation and light reduction from increased turbidity have a number of strongly adverse effects on survival and growth of corals, with early life stages of corals being especially sensitive to sedimentation. Deforestation is a major factor increasing sediment supply to coral reefs. A study by Maina *et al.* (2013) of forest catchments in Madagascar indicates that although river flow and reef sedimentation will decrease under climate-change projections, these declines are outweighed by deforestation – sediment supply being forecast to increase further by

**Table 13.2** Effects of Changes in Water Quality from Terrestrial Runoff on Major Processes Affecting Coral Reefs

Only strong increasing (↑) or decreasing (↓) documented responses are shown.

| | | Dissolved inorganic nutrients | Particulate organic matter | Light reduction | Sedimentation |
|---|---|---|---|---|---|
| Growth and survival of adult corals | Calcification | | | ↓ | ↓ |
| | Tissue thickness | | | | ↓ |
| | Zooxanthellae density | ↑ | | | |
| | Photosynthesis | | | ↓ | ↓ |
| | Colony survival | | | | ↓ |
| Coral reproduction and recruitment | Fertilization | ↓ | | | |
| | Embryo development and larval survival | ↓ | | | |
| | Settlement and metamorphosis | | | ↓ | ↓ |
| | Recruit survival | | | | ↓ |
| Organisms interacting with corals | Crustose coralline algae | | | | ↓ |
| | Bioeroders | | ↑ | | |
| | Macroalgae | ↑ | | ↓ | |
| | Filter feeders (non-zooxanthellate taxa) | | ↑ | ↑ | |
| | Coral predators | | ↑ | | |

Adapted from Fabricius, K.E. (2005). Effects of terrestrial runoff on the ecology of corals and coral reefs: review and synthesis. *Marine Pollution Bulletin*, **50**, 125–46, copyright 2004, with permission from Elsevier.

54–64% if 10–50% of the forest is removed – underlying the importance of land-use management in conserving coral reefs. Reduced water quality is implicated too in the increased prevalence of coral disease (e.g. Williams *et al.*, 2010).

## Tourism

An important factor that has increased human pressure along many coral reef coasts is tourism, especially from resort development and associated infrastructure. The huge growth in coastal tourism along Egypt's mainland Red Sea coast and Sinai Peninsula has, for instance, led to extensive degradation and loss of nearshore fringing reefs (Spalding *et al.*, 2001).

Underwater tourism targets coral reefs, and high levels of diving can directly damage reefs. Harriott *et al.* (1997) concluded that at intensively dived coral-dominated sites (designated as marine protected areas) in eastern Australia, there is potential for considerable environmental impact, mainly from divers' fins. Sessile reef invertebrates vary in their vulnerability to structural damage, so that the amount of diving a particular reef can withstand will depend on its community structure, with branching soft and stony corals and erect sponges being more susceptible than more massive stony corals. At the island of Bonaire in the southern Caribbean, one of the world's most popular diving destinations, Lyons *et al.* (2015) found locations with heavy diver traffic to have roughly 10% less structural complexity than lightly used areas. Massive corals were 31% less abundant at the heavily used sites, whereas gorgonians and sponges were less impacted. Chadwick-Furman (1997) estimated that for fringing reefs in the US Virgin Islands, the critical level of diving frequency was only 500 dives per year per site – an order of magnitude lower than that estimated from other studies – probably because of a high abundance of more fragile forms.

Various management strategies can help reduce the effects of scuba divers on reefs. Public education, including pre-dive briefings to raise awareness of reef conservation, can be effective, as can the presence of dive guides (Lyons *et al.*, 2015). It may also be possible to divert divers away from areas of high stony coral abundance to less sensitive areas or to provide attractive substitutes (e.g. van Treeck & Schuhmacher, 1998). At Eilat in the Gulf of Aqaba, another of the world's top dive spots, artificial reefs have been developed, primarily to reduce the diving pressure on natural coral reef areas (Wilhelmsson *et al.*, 1998).

Other direct physical impacts of tourists on coral reefs include trampling by reef walkers (Kay & Liddle, 1989; Leujak & Ormond, 2008) and anchor damage from vessels. The anchor and chain of one cruise ship off Grand Cayman Island, West Indies, was estimated to have destroyed in one day more than 3000 $m^2$ of previously intact reef (Smith, 1988). No-anchoring areas can be designated to reduce damage and mooring areas established where the bottom is less vulnerable (Dinsdale & Harriott, 2004; Beeden *et al.*, 2014).

## Global Climate Change

In the longer term, global climate change is perhaps the greatest threat to coral reefs. The gradual increase in tropical sea surface temperatures over the past century has brought corals that much closer to the upper limits of their thermal tolerance. Thus any significant additional warming, such as by El Niño events, becomes more critical and likely to cause coral bleaching. Projections indicate that by mid-century the world's coral reefs could be experiencing bleaching conditions annually

(Hoegh-Guldberg, 2011). Recovery from bleaching depends on corals being able to re-establish their zooxanthallae, which may depend on the occurrence of heat-tolerant strains of algae repopulating the coral (Coffroth *et al.*, 2010). Sea-level rise from global warming may be a lesser problem as estimates of upward growth indicate that reefs should be able to keep pace, providing they remain healthy (Hoegh-Guldberg, 2011). This may not be the case for reefs already subject to other stresses and unable to maintain normal rates of calcification, an issue of great concern to inhabitants of low-lying tropical islands such as Kiribati and Marshall Islands (Pacific Ocean) and Maldives (Indian Ocean) (Westmacott *et al.*, 2000).

The increasing concentration of carbon dioxide ($CO_2$) in the atmosphere due to human activities has another major implication for coral reefs. Seawater is naturally slightly alkaline (pH of 8.0–8.2), and tropical surface waters are super-saturated with aragonite, the form of calcium carbonate used by corals. Increasing uptake of $CO_2$ by the ocean lowers its pH (by increasing the weak acid $HCO_3$), making it harder for calcifying organisms to secrete skeletal calcium carbonate (Orr *et al.*, 2005; Chapter 2). Albright *et al.* (2016) found that by increasing the alkalinity of seawater flowing over a reef flat so as to restore pre-industrial pH conditions, net calcification of the coral-reef community increased by an average of 7%. Projections indicate that over the coming decades ocean warming, combined with a decline in the aragonite saturation level, will render large areas of the tropics unsuitable as coral reef habitat, particularly in the central Indo-Pacific (Couce *et al.*, 2013). There are other factors that will determine how corals respond to global climate change; for instance, an increase in the rate of calcification with temperature may partly offset ocean acidification. This is a complex issue, but there is a widespread view that the added stress of global climate change will hasten the degradation of coral reef systems worldwide. There may, however, be wide spatial and temporal variability in the impacts of ocean warming and acidification on reefs given variations in calcification and susceptibility to bleaching among species, potential differences in rates of adaptation, and geographical differences in response and recovery of reefs (Pandolfi *et al.*, 2011).

An assessment by Carpenter *et al.* (2008) indicates that corals now face a far higher risk of extinction given, in particular, the various effects of climate change and other anthropogenic disturbances. Of the 704 species of reef-building corals for which there were sufficient data, Carpenter *et al.* (2008) assigned one-third to the threatened categories of the IUCN Red List (see Chapter 4). Most at risk are species in the families Euphyllidae, Dendrophyllidae, and Acroporidae, where about half the species are threatened. Foden *et al.* (2013) identified 6–9% of coral species as being both highly vulnerable to climate change and already listed as threatened on the IUCN list.

Despite the diffuse nature of $CO_2$-emission impacts on the marine environment, Mcleod *et al.* (2013) identify conservation objectives to help address bleaching and ocean acidification:

- Identify areas likely to be protected from bleaching and that can act as sources of coral larvae to reseed affected areas, and where possible include such refugia in marine protected areas (MPAs); similarly identify areas less exposed or less sensitive to changes in seawater carbonate chemistry.
- Maintain connectivity among MPAs to ensure that they are mutually replenishing to facilitate recovery following a disturbance.
- Protect replicates of the full range of major habitat types, including those likely to experience a variety of ocean chemistry patterns, so as to spread the risk within an MPA network.
- Prioritize areas where local threats can be effectively managed. Reducing local stresses (e.g. land-based sources of pollution, sedimentation, fishing pressure) improves the health of reefs and increases their resilience to bleaching and acidification.

## Effects of Disturbance

An overriding problem facing coral reefs is that they are subject to multiple stresses from natural as well as anthropogenic disturbances (Wilkinson, 2006; Sheppard *et al.*, 2009; Veron *et al.*, 2009). These vary in their nature, intensity, and recurrence as well as the extent to which stresses may reinforce one another. How a coral reef responds to a disturbance depends too on the nature of past disturbances and to what extent its health is already compromised. A reef chronically stressed from human impacts is likely to exhibit poor recovery from a natural disturbance (Brown, 1997). However, despite the range of potential factors and their interactions, some major patterns are evident in the response of coral reef assemblages to perturbation.

Human impacts can elicit major shifts in the structure and functioning of coral reef ecosystems. Removal of predatory species through overfishing can have cascading effects (comparable to predator removal in temperate kelp forests; see Chapter 9). Particularly noticeable in Kenyan reef lagoons subject to different levels of fishing pressure are marked differences in the densities of sea urchins (McClanahan & Muthiga, 1988, 1989; McClanahan & Shafir, 1990). One of the commonest sea urchins of these lagoons is *Echinometra mathaei*, a burrowing species that occurs throughout the Indo-Pacific. Finfish, particularly triggerfishes (Balistidae), are its main predators. Removal of these predators appears to enable *E. mathaei* to exclude its weaker competitors (principally *Diadema* spp.) and become the dominant omnivorous echinoid of hard substrata. In the most heavily exploited lagoon off Kenya's most popular tourist beach, the biomass of *E. mathaei* increased five-fold during a 15-year period. Through feeding and burrowing activities, the increased population of *E. mathaei* reduced the live coral cover and increased erosion of the coral substratum. Eventual loss of topographic complexity in turn resulted in a decline in species diversity and the fisheries potential of the reef. On these Kenyan reefs, finfish accounted for 90% of sea urchin predation, asteroids 5%, and gastropods 5%, indicating that the reduction of finfish had a greater effect on sea urchin populations than shell collecting. The helmet shell *Cypraecassis rufa*, a sea urchin predator, is, for instance, traded as an ornamental species. It therefore becomes important in the management of such reefs to balance the direct economic value of predatory species against their efficiency as sea urchin predators and establish harvesting restrictions accordingly. The alternative may be a reef dominated by sea urchins that are largely unusable for local human consumption.

### Caribbean Reefs

Caribbean coral reefs have experienced a long history of human impact and illustrate how reefs react to various disturbances. Analysis of Pleistocene coral communities from Barbados indicates a remarkable stability in community composition of Caribbean reefs for tens of thousands of years despite major episodes of sea level and climate change. But humans have since wrought an unprecedented change to Caribbean reefs (Pandolfi & Jackson, 2006). Overexploitation of Caribbean coral reefs dates back centuries with the virtual ecological extinction of large herbivores and carnivores, notably turtles and large reef fish, so the trophic structure of these reefs may already have been substantially altered long before the first scientific studies (Jackson, 1997). Well documented and particularly startling has been the change in reef communities over recent decades. A massive decline in coral cover has occurred throughout the Caribbean from about 50% cover in 1977 to 10% in 2001. Gardner *et al.* (2003, 2005) attribute this to a combination of natural and anthropogenic factors including overfishing, sedimentation, eutrophication, habitat destruction, storms, temperature stress, coral disease, and mass mortality of the sea urchin *Diadema antillarum*. The sudden loss of this key grazer in 1983–4 enabled macroalgae to increase dramatically in abundance, although this increase was probably facilitated by nutrient enrichment. Food-web modelling suggests also that overfishing of sharks may have contributed to a reduction in the abundance of herbivorous fishes such as parrotfish (Bascompte *et al.*, 2005). Furthermore, during the 1980s there was a massive loss of live coral cover through the Caribbean caused by a bacterial disease affecting acroporid corals (*Acropora palmata* and *A. cervicornis*), which are among the most important of the Caribbean reef-builders (Aronson & Precht, 2001; Patterson *et al.*, 2002). The result of these changes is that Caribbean reefs have undergone a regime shift, from being coral dominated in the 1970s to algal dominated by the 1990s (Fig. 13.6). Caribbean reefs have subsequently been struck by severe bleaching events, notably in 2005 and 2010, associated with warm-water anomalies, with mortalities of up to 50% in many locations (Eakin *et al.*, 2010). Coinciding with this series of events over recent decades has been a marked decline in the architectural complexity of Caribbean reefs, with serious implications for associated biodiversity (Alvarez-Filip *et al.*, 2009).

Mumby *et al.* (2012) describe a shift in fish community structure over just 7 years on Belize reefs in the western Caribbean attributable to fishing impact. In particular, the abundance of large carnivorous grouper and snapper declined sharply, whereas mesopredators (smaller groupers) increased dramatically in abundance, seemingly in response to reduced predation. In turn, abundance of damselfishes, which are prey of these mesopredators, declined markedly. As the preferred grouper and snapper became scarce, fishers started to exploit parrotfishes, which also underwent a major decline. There was evidence of expansion of a calcified macroalga (*Halimeda*), consistent with reduced grazing by herbivorous parrotfishes. This trophic cascade is illustrated in Fig. 13.7.

The Caribbean region illustrates how rapidly, within a few decades, a combination of stresses can lead to widespread degradation of coral reefs. Whilst some of these factors may be largely natural, their impact appears to have been exacerbated by stresses from various human activities (and vice versa) leading to widespread decline. Bellwood *et al.* (2004) argue that Caribbean reefs are more susceptible to being perturbed than those of high biodiversity regions because on Caribbean reefs key functional groups may be represented only by single species, so that a small change in biodiversity may have a large effect on ecosystem processes. Nevertheless, the Caribbean is not unique in the decline of its coral reefs.

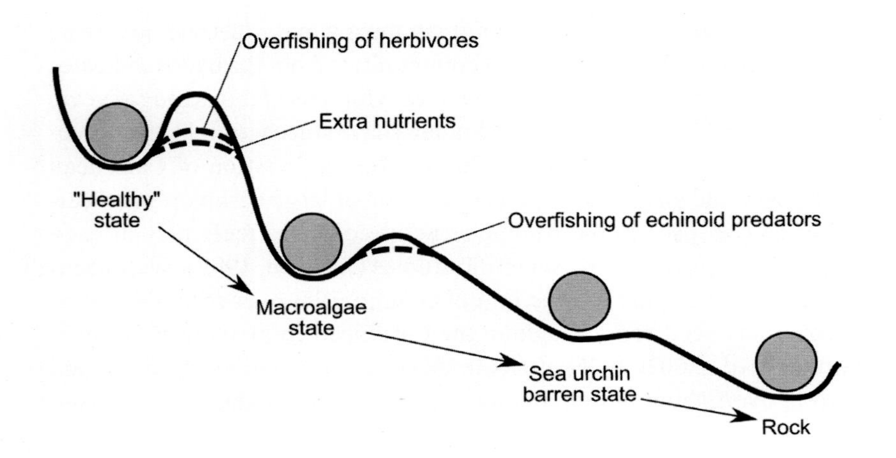


**Fig. 13.6** A conceptual model of the effects of fishing pressure and nutrient enrichment on coral reef ecosystems. Reprinted by permission from Macmillan Publishers Ltd: *Nature* (Bellwood, D.R., Hughes, T.P., Folke, C. & Nyström, M. Confronting the coral reef crisis. Nature, 429, 827–33), copyright (2004).

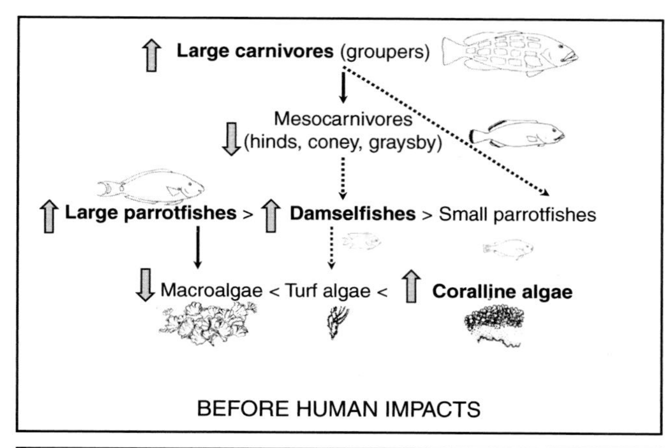


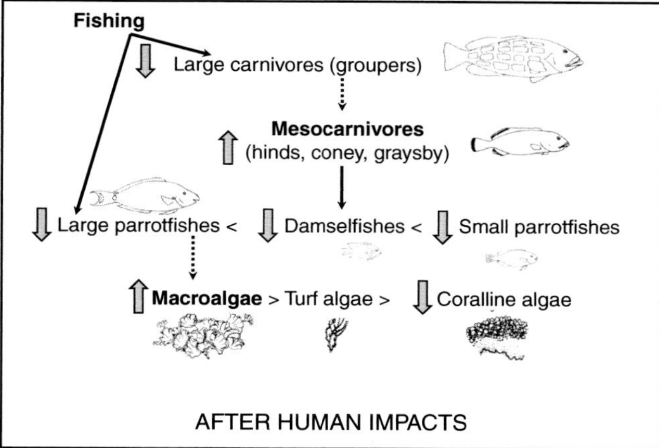


**Fig. 13.7** Impact of fishing on food-web structure of reefs in Belize. In bold: dominant species. Grey arrows: abundance trends. Black arrows: strong interactions. Dotted arrows: weak/non-existent interactions. Reproduced from Mumby, P.J., Steneck, R.S., Edwards, A.J., *et al.* (2012). Fishing down a Caribbean food web relaxes trophic cascades. *Marine Ecology Progress Series*, 445, 13–24. With permission from Inter-Research.

## Worldwide Degradation of Coral Reefs

There is alarming evidence of a decline in coral reef ecosystems in many regions of the world. Pandolfi *et al.* (2003) examined the ecological histories of 14 coral reef ecosystems worldwide. They used the status of seven major categories of biota (essentially carnivores, herbivores, and structural species) for each of seven cultural periods (from prehuman to the present). Their trajectories indicate that most of the reef ecosystems were already significantly degraded before 1900, most obviously from overfishing, and that the pattern of decline was similar worldwide. Large carnivores and herbivores had been substantially depleted almost everywhere by the beginning of the twentieth century, followed by declines in structural biota such as corals and seagrass.

Burke *et al.* (2011) undertook a worldwide analysis of threats to coral reefs based on four major categories of localized human impact: coastal development, land-based pollution and erosion, marine-based pollution and damage, and overfishing and destructive fishing practices; and two global climate-related threats: thermal stress and ocean acidification. For local threats they used risk indicators based largely on human population density and infrastructure features (e.g. size of cities, ports, and hotels), estimates of sediment inputs from rivers, and distance of the threat from the reef. These were modelled separately but also combined in a single integrated index (rated from low to very high) to reflect their cumulative impact on reefs. Local threats were also combined with modelled future estimates of global climate change impacts so as to forecast the threat to reefs in 2030 and 2050. Their assessment indicates that more than 60% of the world's reefs are threatened by one or more local pressures, with overfishing and destructive fishing as the most widespread threat, affecting more than 55% of reefs. Most at risk are reefs of SE Asia, noted for their high biodiversity, where nearly 95% of reefs are threatened. However, when local threats are combined with the global impact of thermal stress, about 75% of all reefs are rated as threatened. The combined effects of ocean warming and acidification are projected to raise the percentage of threatened reefs to more than 90% by 2030 and virtually 100% by 2050, if there is no change in local threats to reefs.

In an assessment of the status of coral reefs, Wilkinson (2008) estimated that, by area, 19% of the world's coral reefs have already been lost (90% of the corals lost and unlikely to recover soon), 15% of reefs are at a critical stage (50–90% loss of corals) and under threat of loss within the next 10–20 years, and 20% show moderate signs of damage (20–50% loss of corals) and threatened within 20–40 years (using business-as-usual projections for climate change). Carlton *et al.* (1999) project that if 5% of the area of coral reefs has been degraded to non-reef state, the number of species brought to extinction as a result could be as high as 50 000–60 000. These studies demonstrate that degradation of the world's coral reefs is one of the most pressing issues of marine conservation, not only for conservation of biodiversity and ecosystem function of reefs but also in turn for their huge socio-economic importance; an estimated 500 million people depend on coral reefs for food, coastal protection, and tourism (Wilkinson, 2008). Sheppard *et al.* (2009) warn of the likelihood that coral reefs 'will be the first major ecosystem in the modern era to become ecologically extinct'. The underlying reasons are familiar: human population growth, exploitation of natural resources for short-term economic gain, and failure to account for ecosystem goods and services.

## Protection of Reefs

### International and Regional Action

Fundamental to safeguarding the long-term health of coral reefs worldwide is the need to address global climate change, particularly in view of the enormous inertia of ocean-atmosphere systems in responding to changes in greenhouse gas emissions. The UN Framework Convention on Climate Change is thus of the utmost importance (see Chapter 2). Coral reefs would similarly benefit from other wide-ranging international agreements, such as the Global Programme of Action for the Protection of the Marine Environment from Land-Based Activities (GPA), the International Convention for the Prevention of Marine Pollution from Ships (MARPOL), and the FAO Code of Conduct for Responsible Fisheries (see Chapters 2 and 5). The importance and value of coral reefs are acknowledged by a number of international initiatives (Box 13.2). There is, however, no international treaty concerned specifically with coral reef conservation.

In terms of regional conservation, the UNEP Regional Seas Programme (see Chapter 16) includes several regions with important coral reefs, notably Eastern Africa, East Asian Seas, Pacific, Red Sea and the Gulf of Aden, the Persian Gulf, South Asia Seas, and Wider Caribbean (Table 16.1). In the case of the Wider Caribbean programme, the island and continental countries of the region have identified priority issues, including land-based sources of pollution and runoff, overexploitation of fish and shellfish, and urbanization and coastal development. An Action Plan for the Caribbean Environment Programme was adopted in 1981 and the legal framework for the plan, the Cartagena

## Box 13.2 International Initiatives Relevant to the Protection of Coral Reefs

### United Nations Conference on Environment and Development (see Chapter 3)

The UNCED action plan for protection of the oceans (Chapter 17 of Agenda 21) identifies coral reefs as a priority area.

### Convention on Biological Diversity (see Chapter 3)

Calls on Parties to manage and protect coastal and marine ecosystems, including coral reefs, and has adopted specific programmes on coral bleaching and the physical degradation and destruction of coral reefs.

### International Coral Reef Initiative

Founded in 1994 as a partnership among governments, international organizations, and NGOs that aims to preserve coral reefs and their related ecosystems in a framework of sustainable use (www.icriforum.org).

### World Summit on Sustainable Development (see Chapter 3)

Endorses Chapter 17 of Agenda 21 and the need to develop national, regional, and international programmes for halting the loss of marine biodiversity, including in coral reefs.

### Convention on International Trade in Endangered Species of Wild Fauna and Flora (see Chapter 4)

Included in Appendix II are all the stony corals (Scleractinia) plus other coral groups – blue coral (*Heliopora coerulea*), organ-pipe corals (Tubiporidae), black corals (Antipatharia), and hydrozoan corals (Milleporidae and Stylasteridae) – and the giant clams (Tridacninae). Only strictly regulated trade in Appendix II species is permitted. Few stony corals are, however, likely to be threatened by international trade, and inclusion of the whole group is more indicative of the desire to safeguard coral reefs and the difficulty of identifying coral species. But none of the coral reef fish species collected for the aquarium trade is included under CITES despite the vulnerability of local populations in the face of the huge, wasteful international trade in ornamental species.

### The Ramsar Convention (see Chapter 11)

The Ramsar Convention includes more than 50 sites with significant coral reefs. Established primarily for the protection of wetlands, it may also be valuable in protecting habitats that have important links with coral reef systems, such as mangroves and seagrass beds.

### World Heritage Convention (see Chapter 15)

Coral reefs form about 40% of the World Heritage marine sites, including the Great Barrier Reef (Abdulla *et al.*, 2013).

Convention (for the protection and development of the marine environment), was adopted in 1983. Supplementing the convention are protocols on oil spills, specially protected areas for wildlife, and pollution from land-based sources. Such programmes have considerable potential to tackle coral reef degradation at a regional scale through collaborative efforts to improve coastal zone management as a whole. Besides the Caribbean, there is a particularly pressing need in SE Asia for coordinated regional programmes of coral reef conservation (Roberts *et al.*, 2002).

## Regional to Local Initiatives

Especially important and potentially effective are measures that can be taken at the regional to local level to limit degradation and overexploitation of coral reef ecosystems. Such measures need to be part of integrated management programmes for the areas in question (see Chapter 10), approaches that take account of the nature and vulnerability of reef systems in relation to human activities and the sustainable use of reef resources. A number of aspects of coral reefs and human use are important in this regard. Coral reefs have high biodiversity, and correspondingly, a wide range of species are exploited. Sought-after species are often strongly *K*-selected, thus highly vulnerable to overexploitation, and may be key species in structuring the reef community. Maintaining the physical integrity of the reef is essential to conserving its biodiversity and ecosystem functioning; so too is the need to manage land-use practices of adjacent coastal areas. Coral reefs are mainly exploited by communities dependent on small-scale subsistence fishing. However, many of these communities are poor and have greatly increased in population size over recent decades, often putting reef systems under severe pressure. In such situations, coastal zone management is unlikely to succeed if not in the context of broader economic programmes to address poverty, employment, and sustainable use of local resources (McClanahan, 2002). Tourism and its associated infrastructure are also a key factor to be considered in economic planning and coastal management for many coral reef coasts.

Although local action cannot alleviate disturbance driven by climate change, factors that tend to reinforce this stress – notably sedimentation and eutrophication – can still be managed locally to help enhance conditions for recruitment and coral recovery. Management strategies need to be tailored to take account of regional differences in the source of stresses and their relative importance as well as the best use of resources (Maina *et al.*, 2011; Mumby & Steneck, 2011).

## Marine Protected Areas

Considerable attention has focused on the use of marine protected areas (MPAs) as a means to help conserve coral reefs. MPAs may be set up for a variety of aims, but in particular for fisheries management and biodiversity conservation (see Chapters 5 and 15). Properly designed and managed MPAs can be effective in conserving target populations, at least for species that are not highly mobile. There is evidence of this from studies of fish populations showing exploited predatory species to be more abundant and of larger average size within no-take MPAs than in neighbouring unregulated areas. A number of such examples are for coral reef MPAs (Roberts & Polunin, 1991; Friedlander *et al.*, 2003; Westera *et al.*, 2003; Pelletier *et al.*, 2005).

MacNeil *et al.* (2015) estimated the fish biomass for more than 800 coral reefs worldwide and found that on protected or near-pristine reefs the resident fish biomass averaged 1 tonne per hectare, whereas the great majority (83%) of fished reefs had less than half this biomass. If protected from fishing, reef fish biomass could potentially recover within 35 years – and within 60 years if heavily depleted. Low-trophic-level species, including herbivores and planktivores, can recover rapidly. Fish predators, on the other hand, are the last to recover. MPAs may not be achievable where local communities depend heavily on reef resources. Encouragingly, however, MacNeil *et al.* (2015) found that even simple restrictions on fishing equipment, species, or access – measures more likely to be observed than MPAs – resulted on average in 27% more fish biomass than on reefs open to fishing.

Given the complex trophic interactions on coral reefs, an MPA may, however, have unexpected results. As we have seen, Caribbean reefs have effectively lost their principal invertebrate grazer, the sea urchin *Diadema*, and parrotfishes have instead become the dominant grazer on most reefs. In this situation, establishing an MPA and restoring the biomass of large predatory fish might be expected to increase the predation pressure on parrotfishes and, thereby, further encourage seaweed growth. In a study of an MPA in the Bahamas archipelago, Mumby *et al.* (2006) found that whilst such predation was enhanced, this was more than offset by the fact that large-bodied parrotfish were more abundant in the MPA. These appear to escape predation by the dominant predator, the Nassau grouper (*Epinephelus striatus*), and as the large parrotfish are the main grazers, macroalgal cover was reduced.

Importantly, MPAs can be effective in preventing the loss of live coral cover – an indicator of reef health (Selig & Bruno, 2010; Hargreaves-Allen *et al.*, 2011) (Fig. 13.8), although there can be a considerable time lag (10–15 years) before coral cover responds to protection given the slow growth of many reef-building corals.

There are nearly 2700 coral reef protected areas worldwide. Whilst this represents about a quarter of the world's reefs located within MPAs, MPAs are poorly represented in areas of greatest threat, and only a small proportion of the world's coral reefs are assessed as being within MPAs where conditions are adequate in terms of extraction and poaching, risks from external threats (sedimentation, pollution, coastal development), and MPA size and connectivity (Mora *et al.*, 2006; Burke *et al.*, 2011). For two of the most diverse groups inhabiting coral reefs, scleractinian corals and wrasses (labrid fishes), only about 6% and 22%, respectively, of the species have 10% of their geographic ranges (as a minimum protection target) within MPAs (Mouillot *et al.*, 2016). Such findings indicate the need for an urgent assessment of coral reef conservation strategies and a huge increase in the establishment of effective coral reef MPAs. Probably at least 30% of the world's coral reefs need to be protected from all exploitation (Sheppard *et al.*, 2009).

Salm *et al.* (2000) provide information on the planning and management of coral reef MPAs. They describe design guidelines based on the use of management zones so as to take account of

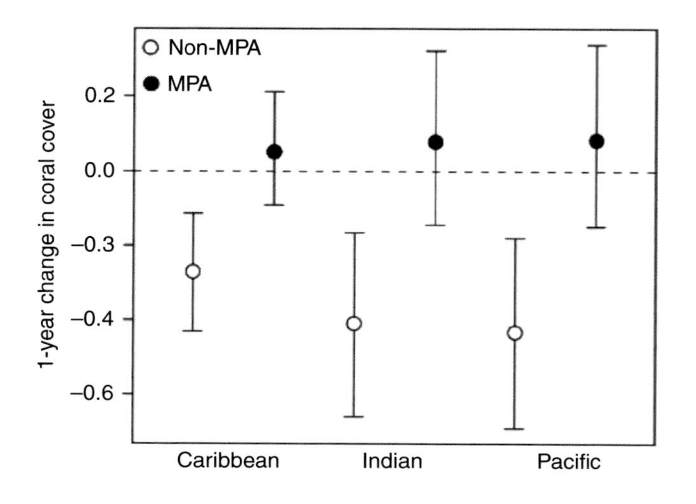


**Fig. 13.8** The change in percent coral cover over 1 year (2004–5) inside and outside MPAs. Coral cover within MPAs increased by 0.05–0.08% and declined by 0.27–0.43% on unprotected reefs. Although these year-to-year changes are small, long-term cumulative effects could be substantial. The error bars are 95% credibility intervals. Reproduced with minor changes from Selig, E.R. & Bruno, J.F. (2010). A global analysis of the effectiveness of marine protected areas in preventing coral loss. *PLoS ONE*, 5, e9278. Licensed under a Creative Commons Attribution 4.0 International License: https://creativecommons.org/licenses/by/4.0/.

the fact that a reef ecosystem extends beyond its physical boundaries and interacts with neighbouring habitats. The basic approach is one commonly advocated for protected areas in general (see Chapter 15). Central to such an MPA is a core area large enough to capture a high diversity of reef morphology and biota. As an example, Salm *et al.* (2000) suggest as a minimum core area one that would include 95% of coral genera and subgenera. Based on reefs of the Chagos archipelago, this could be expected to be an area of at least 4.5 $km^2$, adequate at least to contain propagules of short-distance dispersers (Shanks *et al.*, 2003). Ideally, beyond this core area is a further protected zone delimiting neighbouring coastal habitats that maintain ecological processes of the reef itself (e.g. reef flats, seagrass beds, sand flats, mangroves). This zone may also serve to regulate visitor use by providing recreational opportunities. A further outer buffer zone may encompass more distant linked habitats (notably coastal watershed areas). Potentially damaging activities here might best be controlled by coastal zone management approaches. Essential are monitoring programmes to assess the effectiveness of MPAs linked to adaptive management plans (Wells, 2006). Worrying in this regard is that Burke *et al.* (2012) rated only 6% of the world's coral reefs as being located in MPAs that are effectively managed, whilst for the Coral Triangle, where coral diversity peaks, the proportion is less than 1%.

MPA planning needs to take account of climate warming given the potential significance of its impacts on coral reefs. Magris *et al.* (2015), for example, used historical and projected sea surface temperature data for the Brazilian coast to delineate thermal refugia and thereby configure a coral reef MPA network. Reefs differ in their ability to recover from disturbances, and managing for resilience is seen as an increasingly appropriate strategy. Resilience might be defined, for instance, as the probability of a reef switching from a coral to an algal dominated state (Mumby *et al.* 2014).

Coral reef MPAs are often remote, and enforcing regulations may be difficult, even if the country in question has the necessary resources. Compliance with MPA objectives can be greatly improved if the local community is engaged in the management. This is demonstrated on many Pacific islands where traditional marine tenure systems have operated for centuries. There is a range of such systems where local people have rights to reef areas and can restrict access at particular times or to parts of the reef under their tenure (see also Chapter 5). Such communities, dependent on reef seafood, have typically built up a detailed acquaintance with the natural history of local species, including knowledge of the timing and location of fish movements such as spawning migrations and aggregations, information that is invaluable in management of reef ecosystems (Johannes, 1997). In a study of fish assemblages of the Hawaiian islands, Friedlander *et al.* (2003) found fish biomass at sites under customary stewardship to be equal to or greater than that of no-take MPAs. And Campbell *et al.* (2012) reported reef fish biomass and hard-coral cover to be two to eight times higher at sites under the local customary management system in Aceh, Indonesia. Traditional marine tenure systems have declined with the introduction of Western-style fishing practices. But there is now interest in their re-establishment, at least in modified forms, and integrating them with contemporary approaches to coastal zone management (Cinner & Aswani, 2007).

## Coral Reef Restoration

There is growing interest in active methods for coral reef restoration, in particular direct transplantation of coral – from nubbins (small fragments) to whole colonies – mariculture of coral recruits to adequate size in nurseries for transplant to reef sites, and provision of artificial substrata for natural recruitment. One promising approach that has been trialled in various reefs worldwide and with numerous coral species involves rearing coral nubbins in mid-water floating nurseries and, once they have reached a suitable size, transplanting them into denuded reef areas (Rinkevich, 2014).

## Major Recommendations for Conservation

In brief, conservation of the world's coral reefs requires prompt, effective action at a number of levels. Major recommendations for action have been advanced in various international initiatives. Among these are recommendations for action to conserve coral reefs set out by the Global Coral Reef Monitoring Network (Wilkinson, 2008), which focus on the need to:

- Urgently combat global climate change.
- Maximize coral reef resilience by minimizing direct human pressure on reefs.
- Include more reefs in MPAs.
- Protect remote reefs.
- Improve enforcement of MPA regulations.
- Help improve decision making with better ecological and socio-economic monitoring.

An example of a major coral reef system where many of aspects of conservation have been adopted in an ecosystem-based approach to management is Australia's Great Barrier Reef.

# Great Barrier Reef

The Great Barrier Reef (GBR) off Australia's north-east coast is the world's largest coral reef system. It extends for 2300 km from the Gulf of Papua (~ 9° S) to southern Queensland (~ 24° S), is about 50 km wide in the north to 200 km in the south, and comprises nearly 3000 separate reefs, although only its northern part is more or less continuous barrier reef (Fig. 13.9).

The GBR is a region of high biodiversity. Some 400 species of hermatypic corals have been recorded from the GBR (Veron, 2000). Also important or conspicuous are macroalgae (~ 400–500 species), sponges (~1500 spp.), molluscs (> 4000 spp.), crustaceans (> 1330 spp.), bryozoans (300–500 spp.), echinoderms (~ 800 spp.), fishes (1200–2000 spp.), sea snakes (17 spp.), sea turtles (6 spp.), seabirds (23 spp. with breeding colonies), cetaceans (> 20 spp.), and dugong (*Dugong dugon*). Other important linked ecosystems within the GBR region are extensive areas of mangroves (37 spp.) and seagrass beds (15 spp.), both important as breeding and nursery grounds (Spalding *et al.*, 2001).

In 1975 the Australian Federal Government established the GBR Marine Park to include almost the entire reef, from off Cape York to its southern extremity. This was prompted in particular by concerns about environmental effects of possible oil drilling and coral mining. The GBR Marine Park covers an area of 344 000 km$^2$ of the Australian continental shelf (comparable in size to Germany), of which just over 20 000 km$^2$ is reef. In 1981 the GBR was declared a UNESCO World Heritage site (348 700 km$^2$). The GBR Marine Park and World Heritage area is managed jointly by federal and state agencies (Great Barrier Reef Marine Park Authority (GBRMPA) and Queensland Parks and Wildlife Service, respectively). (See Lawrence *et al.* (2002) for a detailed account of the establishment of the Marine Park and its management.) Waters immediately east of the GBR Marine Park that are within Australia's EEZ were afforded protection in 2012 when an area of nearly 990 000 km$^2$ was designated as the Coral Sea Commonwealth Marine Reserve, although only the outer half is a no-take zone (Fig. 13.9). The GBR is of huge economic value to Australia – largely by way of tourism – estimated to be worth about AUD$5 billion annually and supporting some 66 000 jobs (Day & Dobbs, 2013).

The GBR Marine Park is managed as a multiple-use MPA to cater to a range of conservation objectives and human uses, an example of ecosystem-based management (see Chapter 17). This

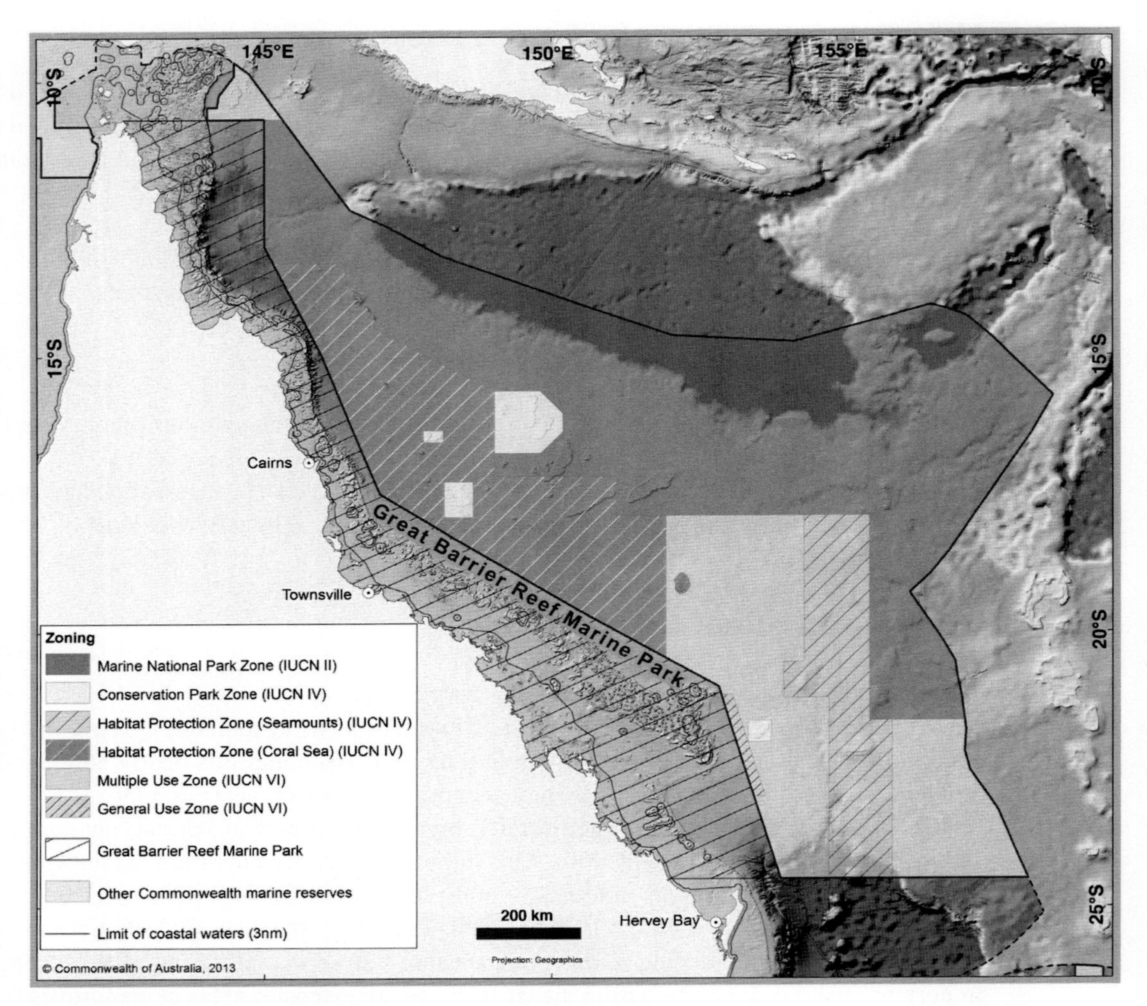


**Fig. 13.9** Boundaries of the Great Barrier Reef Marine Park and Coral Sea Commonwealth Marine Reserve. (The internal zones of the reserve, shown here as of March 2014, are being reviewed and may change.) © Commonwealth of Australia 2016. (A black and white version of this figure will appear in some formats. For the colour version, please refer to the plate section.)

is achieved largely through a zoning plan so as to separate uses (Table 13.3). There are several types of zone, from highly protected areas to ones that provide for a wide range of 'reasonable use' (GBRMPA, 2004). In terms of IUCN protected area categories (see Chapter 15), most of the marine park (62%) corresponds to IUCN Category VI, but more than 33% is now in no-take areas (Categories II and IA). Prior to a rezoning in 2004, less than 5% of the marine park was in no-take areas. As a basis for the rezoning, 70 distinct bioregions were identified and a number of operational principles applied to ensure adequate protection of the full range of habitats and species, such as in terms of size, replication, and spacing of no-take areas representing different bioregions. For instance, no-take areas are at least 20 km across at the smallest dimension (or 10 km for coastal bioregions), and 20% of each bioregion is within no-take areas, with at least three to four no-take areas per bioregion being recommended (Fernandes *et al.*, 2005).

## Table 13.3 Zoning in the Great Barrier Reef Marine Park and Permitted Activities

General use zone: provides opportunities for reasonable use whilst still allowing for the conservation of these areas (34% of GBR Marine Park).

Habitat protection zone: provides for the conservation of areas by protecting and managing sensitive habitats and ensuring they are generally free from potentially damaging activities. Trawling is not permitted (28%).

Conservation park zone: allows for increased protection and conservation whilst providing opportunities for reasonable use and enjoyment, including limited extractive use (2%).

Buffer zone: provides for the protection and conservation of areas in their natural state whilst allowing public access. Trolling for pelagic fish species is allowed, but all other forms of extractive activities are prohibited (3%).

Scientific research zone: allows for research in areas relatively undisturbed by extractive activities. For people not undertaking research, these areas are essentially the same as marine national park zones, where only non-extractive activities are allowed without written permission (< 1%).

Marine national park zone: a 'no-take' area where extractive activities like fishing or collecting are not allowed without a permit. Activities such as boating, swimming, snorkelling, and sailing are allowed (33%).

Preservation zone: no entry without written permission and extractive activities are strictly prohibited (< 1%).

Activities Guide courtesy of the Spatial Data Centre, Great Barrier Reef Marine Park Authority 2016.

**GBRMP Zoning**
(see relevant *Zoning Plans* and *Regulations* for details)

| Activity | General Use Zone | Habitat Protection Zone | Conservation Park Zone | Buffer Zone | Scientific Research Zone * | Marine National Park Zone | Preservation Zone |
|---|---|---|---|---|---|---|---|
| Aquaculture | Permit | Permit | Permit * | ✗ | ✗ | ✗ | ✗ |
| Bait netting | ✓ | ✓ | ✓ * | ✗ | ✗ | ✗ | ✗ |
| Boating, diving, photography | ✓ | ✓ | ✓ | ✓ | ✓ * | ✓ | ✗ |
| Crabbing (trapping) | ✓ | ✓ | ✓ * | ✗ | ✗ | ✗ | ✗ |
| Harvest fishing for aquarium fish, coral and beachworm | Permit | Permit | Permit * | ✗ | ✗ | ✗ | ✗ |
| Harvest fishing for sea cucumber, trochus, tropical rock lobster | Permit | Permit | ✗ | ✗ | ✗ | ✗ | ✗ |
| Limited collecting | ✓ * | ✓ * | ✓ * | ✗ | ✗ | ✗ | ✗ |
| Limited spearfishing (snorkel only) | ✓ | ✓ | ✓ * | ✗ | ✗ | ✗ | ✗ |
| Line fishing | ✓ * | ✓ * | ✓ * | ✗ | ✗ | ✗ | ✗ |
| Netting (other than bait netting) | ✓ | ✓ | ✗ | ✗ | ✗ | ✗ | ✗ |
| Research (other than limited impact research) | Permit | Permit | Permit | Permit | Permit | Permit | Permit |
| Shipping (other than in a designated shipping area) | ✓ | Permit | Permit | Permit | Permit | Permit | ✗ |
| Tourism programme | Permit | Permit | Permit | Permit | Permit | Permit | ✗ |
| Traditional use of marine resources | ✓ * | ✓ * | ✓ * | ✓ * | ✓ * | ✓ * | ✗ |
| Trawling | ✓ | ✗ | ✗ | ✗ | ✗ | ✗ | ✗ |
| Trolling | ✓ * | ✓ * | ✓ * | ✓ * | ✗ | ✗ | ✗ |

**PLEASE NOTE:** This guide provides an introduction to Zoning in the Great Barrier Reef Marine Park. Relevant Great Barrier Reef Marine Park Zoning Plans should be consulted for confirmation of use or entry requirements.

* Additional restrictions / conditions apply.

**ACCESS TO ALL ZONES IS PERMITTED IN AN EMERGENCY.**

The MPA network in the GBR Marine Park provides strong evidence of the benefit of no-take zones, particularly since the re-zoning (McCook *et al.*, 2010). Fish densities on no-take reefs are about twice those of fished reefs, and the fish are also of larger mean size, which is important as larger individuals have a disproportionately greater reproductive output (see Chapter 5). The no-take areas are thus potentially important sources of larvae, including for fished reefs. A similar pattern of response is seen across the whole MPA network of the marine park, encompassing significant latitudinal and inshore-offshore environmental gradients. The effect of protection on abundance and size is seen most strongly in species such as coral grouper (*Plectropomus* spp.), which tend to reside on individual reefs and are the main target of line fishers. But other taxa, such as sharks, dugong, and marine turtles, that range over larger areas spanning protected and open zones appear to benefit less from the MPA network.

Management of the GBR is often regarded as second to none. Particularly concerning, therefore, is evidence of steady environmental deterioration. Average coral cover across the GBR is estimated to have reduced substantially since the 1960s, and populations of species such as dugong and loggerhead turtle have declined sharply. Brodie and Waterhouse (2012) argue that GBRMPA's initial management focus was on zoning and that only since 2000 has there been significant action on important stressors such as water quality and fishing. Among the greatest threats to the GBR are terrestrial runoff, outbreaks of the crown-of-thorns starfish, impacts of fishing, and climate change (Brodie & Waterhouse, 2012). There is also a threat of port development, with associated dredging and spoil disposal (Grech *et al.*, 2013). Management of land runoff and water quality of inshore areas of the GBR presents a particular problem. The catchment area adjacent to the GBR is important for agriculture (principally beef cattle and sugarcane), has undergone considerable urban development, and is a source of sediments, nutrients, pesticides, and other pollutants. There is evidence from inshore reef sites of the central GBR of a historical collapse of *Acropora* assemblages between 1920 and 1955, coinciding with a marked increase in sediment and nutrient loading. Modern monitoring of inshore GBR reefs may thus be based on a significantly shifted baseline (Roff *et al.*, 2013). Indications of increased sedimentation and eutrophication of GBR lagoon waters are evidenced by elevated phytoplankton concentrations, increased growth of macroalgae, and lack of recovery of coral cover following storm and flood damage (Bell & Elmetri, 1995). Since 2003, state and national government programmes have been implemented to improve the quality of water entering the reef, notably in terms of nutrient, sediment, and pesticide loads, and some progress has been made towards load reduction targets (Brodie & Pearson, 2016).

Outbreaks of the crown-of-thorns starfish, a major source of disturbance to the GBR, can cause extensive loss of live coral cover; reefs can lose about 40% of their cover during an outbreak (Osborne *et al.*, 2011). Four major outbreaks have occurred since 1962, mainly impacting reefs in the mid-shelf region (McCook *et al.*, 2010; Brodie & Waterhouse, 2012). Outbreaks appear mainly to be controlled by the availability of phytoplankton food for the starfish larvae and hence by nutrient input from runoff (Fabricius *et al.*, 2010). *A. planci* can be killed by divers using an injection gun to deliver a shot of bile salt solution (which disrupts cell membranes and induces osmotic shock), which has no apparent adverse effects on other reef fauna. To control outbreaks, this needs to be done in conjunction with reducing nutrient inputs and protecting predator species (Rivera-Posada *et al.*, 2014).

Global climate change and effects of ocean warming and acidification represent the most serious long-term threats to the GBR. Coral bleaching has been increasing in frequency and intensity since the 1980s, consistent with an increase in sea surface temperature across the GBR (Brodie & Waterhouse, 2012). In 2016, with ocean warming exacerbated by an unusually long El Niño, only 7% of the reef system was entirely unaffected by bleaching (Normile, 2016). D'Olivo *et al.* (2013) took cores from long-lived (>100 year) *Porites* colonies in a transect across the central GBR to examine coral calcification since the mid-twentieth century. The mid- and outer-shelf reefs now appear to be exhibiting decreasing rates of calcification consistent with effects of climate change. The inner shelf reefs have been undergoing a longer-term decline in calcification, although here largely due to a deterioration in inshore water quality from land-based activities.

De'ath *et al.* (2012) reported a major decline in hard coral cover on the GBR from 28.0 to 13.8% over the period 1985–2012. The loss has particularly affected the central and southern GBR, and the rate of decline has increased since about 2006 (now 1.5% per year). Tropical cyclones, coral predation by crown-of-thorns starfish, and coral bleaching largely account for the losses. Cyclone intensity and the frequency of bleaching events are both increasing with warming ocean temperatures, so strategies to control greenhouse gas emissions are vital to help manage these factors. In the shorter term, the analyses show that in the absence of crown-of-thorns alone, coral cover could increase despite the continuing losses due to cyclones and bleaching. The evidence that improved water quality would reduce the frequency of crown-of-thorns outbreaks in the central and southern GBR also indicates the importance of managing land runoff.

The GBR Marine Park, arguably the most significant MPA in terms of size and biodiversity, faces an uncertain future, especially in view of direct and indirect impacts related to water quality and climate change. Environmental deterioration and industrial development along the coast threaten the GBR's World Heritage status (Grech *et al.*, 2013). Brodie and Pearson (2016) recommend several management actions to safeguard the GBR, including the need for a management province that encompasses not just the GBR but also its entire catchment given the links between land use and water quality in the GBR. Also vital is a commitment to the control of greenhouse gas emissions. At all events, urgent and decisive action is needed.

## REFERENCES

Abdulla, A., Obura, D., Bertzky, B. & Shi, Y. (2013). *Marine Natural Heritage and the World Heritage List*. Gland, Switzerland: IUCN.

Albright, R., Caldeira, L., Hosfelt, J., *et al.* (2016). Reversal of ocean acidification enhances net coral reef calcification. *Nature*, **531**, 362–5.

Alcala, A.C. & Gomez, E.D. (1987). Dynamiting coral reefs for fish: a resource-destructive fishing method. In *Human Impacts on Coral Reefs: Facts and Recommendations*, ed. B. Salvat, pp. 51–60. Moorea, French Polynesia: Antenne Museum EPHE.

Alvarez-Filip, L., Dulvy, N.K., Gill, J.A., Côté, I.M. & Watkinson, A.R. (2009). Flattening of Caribbean coral reefs: region-wide declines in architectural complexity. *Proceedings of the Royal Society B*, **276**, 3019–25.

Arias-González, J.E., Legendre, P. & Rodríguez-Zaragoza, F.A. (2008). Scaling up beta diversity on Caribbean coral reefs. *Journal of Experimental Marine Biology and Ecology*, **366**, 28–36.

Aronson, R.B. & Precht, W.F. (2001). White-band disease and the changing face of Caribbean coral reefs. *Hydrobiologia*, **460**, 25–38.

Baker, A.C., Glynn, P.W. & Riegl, B. (2008). Climate change and coral reef bleaching: an ecological assessment of long-term impacts, recovery trends and future outlook. *Estuarine, Coastal and Shelf Science*, **80**, 435–71.

Bascompte, J., Melián, C.J. & Sala, E. (2005). Interaction strength combinations and the overfishing of a marine food web. *Proceedings of the National Academy of Sciences*, **102**, 5443–7.

Beeden, R., Maynard, J., Johnson, J., *et al.* (2014). No-anchoring areas reduce coral damage in an effort to build resilience in Keppel Bay, southern Great Barrier Reef. *Australasian Journal of Environmental Management*, **21**, 311–19.

Bell, P.R.F. & Elmetri, I. (1995). Ecological indicators of large-scale eutrophication in the Great Barrier Reef lagoon. *Ambio*, **24**, 208–15.

Bellwood, D.R., Hughes, T.P., Folke, C. & Nyström, M. (2004). Confronting the coral reef crisis. *Nature*, **429**, 827–33.

Betancur-R, R., Hines, A., Acero, P.A., *et al.* (2011). Reconstructing the lionfish invasion: insights into Greater Caribbean biogeography. *Journal of Biogeography*, **38**, 1281–93.

Brodie, J., Fabricius, K., De'ath, G. & Okaji. K. (2005). Are increased nutrient inputs responsible for more outbreaks of crown-of-thorns starfish? An appraisal of the evidence. *Marine Pollution Bulletin*, **51**, 266–78.

Brodie, J. & Pearson, R.G. (2016). Ecosystem health of the Great Barrier Reef: time for effective management action based on evidence. *Estuarine, Coastal and Shelf Science*, **183**, 438–51.

Brodie, J. & Waterhouse, J. (2012). A critical review of environmental management of the 'not so Great' Barrier Reef. *Estuarine, Coastal and Shelf Science*, **104–5**, 1–22.

Brown, B.E. (1997). Disturbances to reefs in recent times. In *Life and Death of Coral Reefs*, ed. C. Birkeland, pp. 354–79. New York: Chapman & Hall.

Brown, B.E. & Dunne, R.P. (1988). The environmental impact of coral mining on coral reefs in the Maldives. *Environmental Conservation*, **15**, 159–65.

Burke, L., Reytar, K., Spalding, M. & Perry, A. (2011). *Reefs at Risk Revisited*. Washington, DC: World Resources Institute.

Burke, L., Reytar, K., Spalding, M. & Perry, A. (2012). *Reefs at Risk Revisited in the Coral Triangle*. Washington, DC: World Resources Institute.

Campbell, S.J., Cinner, J.E., Ardiwijaya, R.L., *et al.* (2012). Avoiding conflicts and protecting coral reefs: customary management benefits marine habitats and fish biomass. *Oryx*, **46**, 486–94.

Carlton, J.T., Geller, J.B., Reaka-Kudla, M.L. & Norse, E.A. (1999). Historical extinctions in the sea. *Annual Review of Ecology and Systematics*, **30**, 515–38.

Carpenter, K.E., Abrar, M., Aeby, G., *et al.* (2008). One-third of reef-building corals face elevated extinction risk from climate change and local impacts. *Science*, **321**, 560–3.

Chadwick-Furman, N.E. (1997). Effects of SCUBA diving on coral reef invertebrates in the U.S. Virgin Islands: implications for the management of diving tourism. In *Proceedings of the 6th International Conference on Coelenterate Biology*, ed. J.C. den Hartog, pp. 91–100. Leiden, Netherlands: Nationaal Natuurhistorisch Museum.

Cinner, J.E. & Aswani, S. (2007). Integrating customary management into marine conservation. *Biological Conservation*, **140**, 201–16.

Coffroth, M.A., Poland, D.M., Petrou, E.L., Brazeau, D.A. & Holmberg, J.C. (2010). Environmental symbiont acquisition may not be the solution to warming seas for reef-building corals. *PLoS ONE*, **5**, e13258.

Coker, D.J., Wilson, S.K. & Pratchett, M.S. (2014). Importance of live coral habitat for reef fishes. *Reviews in Fish Biology and Fisheries*, **24**, 89–126.

Couce, E., Ridgwell, A. & Hendy, E.J. (2013). Future habitat suitability for coral reef ecosystems under global warming and ocean acidification. *Global Change Biology*, **19**, 3592–606.

Craik, W., Kenchington, R. & Kelleher, G. 1990. Coral-reef management. In *Coral Reefs*, ed. Z. Dubinsky, pp. 453–67. Amsterdam: Elsevier.

Cruz-Trinidad, A., Geronimo, R.C., Cabral, R.B. & Aliño, P.M. (2011). How much are the Bolinao-Anda coral reefs worth? *Ocean & Coastal Management*, **54**, 696–705.

Davy, S.K., Allemand, D. & Weis, V.M. (2012). Cell biology of cnidarian-dinoflagellate symbiosis. *Microbiology and Molecular Biology Reviews*, **76**, 229–61.

Dawson Shepherd, A.R., Warwick, R.M., Clarke, K.R. & Brown, B.E. (1992). An analysis of fish community responses to coral mining in the Maldives. *Environmental Biology of Fishes*, **33**, 367–80.

Day, J.C. & Dobbs, K. (2013). Effective governance of a large and complex cross-jurisdictional marine protected area: Australia's Great Barrier Reef. *Marine Policy*, **41**, 14–24.

De'ath, G., Fabricius, K.E., Sweatman, H. & Puotinen, M. (2012). The 27–year decline of coral cover on the Great Barrier Reef and its causes. *Proceedings of the National Academy of Sciences*, **109**, 17995–9.

Dee, L.E., Horii, S.S. & Thornhill, D.J. (2014). Conservation and management of ornamental coral reef wildlife: successes, shortcomings, and future directions. *Biological Conservation*, **169**, 225–37.

Dinsdale, E.A. & Harriott, V.J. (2004). Assessing anchor damage on coral reefs: a case study in selection of environmental indicators. *Environmental Management*, **33**, 126–39.

D'Olivo, J.P., McCulloch, M.T. & Judd, K. (2013). Long-term records of coral calcification across the central Great Barrier Reef: assessing the impacts of river runoff and climate change. *Coral Reefs*, **32**, 999–1012.

Dubinsky, Z. & Stambler, N. (1996). Marine pollution and coral reefs. *Global Change Biology* **2**, 511–26.

Dulvy, N.K., Stanwell-smith, D., Darwall, W.R.T. & Horrill, C.J. (1995). Coral mining at Mafia Island, Tanzania: a management dilemma. *Ambio*, **24**, 358–65.

Eakin, C.M., Morgan, J.A., Heron, S.F., *et al.* (2010). Caribbean corals in crisis: record thermal stress, bleaching, and mortality in 2005. *PLoS*, **5**, e13969.

Fabricius, K.E. (2005). Effects of terrestrial runoff on the ecology of corals and coral reefs: review and synthesis. *Marine Pollution Bulletin*, **50**, 125–46.

Fabricius. K.E., Okaji, K. & De'ath, G. (2010). Three lines of evidence to link outbreaks of the crown-of-thorns seastar *Acanthaster planci* to the release of larval food limitation. *Coral Reefs*, **29**, 593–605.

Fernandes, L., Day, J., Lewis, A., *et al.* (2005). Establishing representative no-take areas in the Great Barrier Reef: large-scale implementation of theory on marine protected areas. *Conservation Biology*, **19**, 1733–44.

Fernando, H.J.S., McCulley, J.L., Mendis, S.G. & Perera, K. (2005). Coral poaching worsens tsunami destruction in Sri Lanka. *Eos, Transactions of the American Geophysical Union*, **86** (33), 301, 304.

Ferrario, F., Beck, M.W., Storlazzi, C.D., *et al.* (2014). The effectiveness of coral reefs for coastal hazard risk reduction and adaptation. *Nature Communications*, **5**, 3794.

Ferreira, C.E.L., Luiz, O.J., Floeter, S.R., *et al.* (2015) First record of invasive lionfish (*Pterois volitans*) for the Brazilian coast. *PLoS ONE*, **10**, e0123002.

Fisher, R., O'Leary, R.A., Low-Choy, S., *et al.* (2015). Species richness on coral reefs and the pursuit of convergent global estimates. *Current Biology*, **25**, 500–5.

Foden, W.B., Butchart, S.H.M., Stuart, S.N., *et al.* (2013). Identifying the world's most climate change vulnerable species: a systematic trait-based assessment of all birds, amphibians and corals. *PLoS ONE*, **8**, e65427.

Friedlander, A.M., Brown, E.K., Jokiel, P.L., Smith, W.R. & Rodgers, K.S. (2003). Effects of habitat, wave exposure, and marine protected area status on coral reef fish assemblages in the Hawaiian archipelago. *Coral Reefs*, **22**, 291–305.

Gardner, T.A., Côté, I.M., Gill, J.A., Grant, A. & Watkinson, A.R. (2003). Long-term region-wide declines in Caribbean corals. *Science*, **301**, 958–60.

Gardner, T.A., Côté, I.M., Gill, J.A., Grant, A. & Watkinson, A.R. (2005). Hurricanes, and Caribbean coral reefs: impacts, recovery patterns, and role in long-term decline. *Ecology*, **86**, 174–84.

GBRMPA (2004). *Great Barrier Reef Marine Park Zoning Plan 2003*. Townsville, QLD: Great Barrier Reef Marine Park Authority.

Graham, N.A.J., Jennings, S., MacNeil, A., Mouillot, D. & Wilson, S.K. (2015). Predicting climate-driven regime shifts versus rebound potential in coral reefs. *Nature*, **518**, 94–7.

Grech, A., Bos, M., Brodie, J., *et al.* (2013). Guiding principles for the improved governance of port and shipping impacts in the Great Barrier Reef. *Marine Pollution Bulletin*, **75**, 8–20.

Guilcher, A. (1988). *Coral Reef Geomorphology*. Chichester, UK: John Wiley & Sons.

Guzmán, H.M., Guevara, C. & Castillo, A. (2003). Natural disturbances and mining of Panamanian reefs by indigenous people. *Conservation Biology*, **17**, 1396–1401.

Hargreaves-Allen, V., Mourato, S. & Milner-Gulland, E.J. (2011). A global evaluation of coral reef management performance: are MPAs producing conservation and socio-economic improvements? *Environmental Management*, **47**, 684–700.

Harriott, V.J., Davis, D. & Banks, S.A. (1997). Recreational diving and its impact in marine protected areas in eastern Australia. *Ambio*, **26**, 173–9.

Hatcher, B.G., Johannes, R.E. & Robertson, A.I. (1989). Review of research relevant to the conservation of shallow tropical marine ecosystems. *Oceanography and Marine Biology: An Annual Review*, **27**, 337–414.

Hoegh-Guldberg, O. (2011). The impact of climate change on coral reef ecosystems. In *Coral Reefs: an Ecosystem in Transition*, ed. Z. Dubinsky & N. Stambler, pp. 391–403. New York: Springer-Verlag.

IPCC (2014). *Climate Change 2014: Synthesis Report*. IPCC Fifth Assessment Synthesis Report. Geneva, Switzerland: Intergovernmental Panel on Climate Change.

Jackson, J.B.C. (1997). Reefs since Columbus. *Coral Reefs*, **16** (Suppl.), S23–32.

Jaleel, A. (2013). The status of the coral reefs and the management approaches: the case of the Maldives. *Ocean & Coastal Management*, **82**, 104–18.

Johannes, R.E. (1997). Traditional coral-reef fisheries management. In *Life and Death of Coral Reefs*, ed. C. Birkeland, pp. 380–5. New York: Chapman & Hall.

Kay, A.M. & Liddle, M.J. (1989). Impact of human trampling in different zones of a coral reef flat. *Environmental Management*, **13**, 509–20.

Kleypas, J.A., McManus, J.W. & Meñez, L.A.B. (1999). Environmental limits to coral reef development: where do we draw the line? *American Zoologist*, **39**, 146–59.

Lawrence, D., Kenchington, R. & Woodley, S. (2002). *The Great Barrier Reef: Finding the Right Balance*. Carlton South, Victoria: Melbourne University Press.

Lessios, H.A. (2016). The great *Diadema antillarum* die-off: 30 years later. *Annual Review of Marine Science*, **8**, 267–83.

Leujak, W. & Ormond, R.F.G. (2008). Quantifying acceptable levels of visitor use on Red Sea reef flats. *Aquatic Conservation: Marine and Freshwater Ecosystems*, **18**, 930–44.

Littler, M.M., Littler, D.S. & Brooks, B.L. (2006). Harmful algae on tropical coral reefs: bottom-up eutrophication and top-down herbivory. *Harmful Algae*, **5**, 565–85.

Longhurst, A.R. & Pauly, D. (1987). *Ecology of Tropical Oceans*. San Diego, CA: Academic Press.

Lunn, K.E. & Moreau, M.-A. (2004). Unmonitored trade in marine ornamental fishes: the case of Indonesia's Banggai cardinalfish (*Pterapogon kauderni*). *Coral Reefs*, **23**, 344–51.

Lyons, P.J., Arboleda, E., Benkwitt, C.E., *et al.* (2015). The effect of recreational SCUBA divers on the structural complexity and benthic assemblage of a Caribbean coral reef. *Biodiversity and Conservation*, **24**, 3491–504.

MacNeil, M.A., Graham, N.A.J., Cinner, J.E., *et al.* (2015). Recovery potential of the world's coral reef fishes. *Nature*, **520**, 341–4.

Magris, R.A., Heron, S.F & Pressey, R.L. (2015). Conservation planning for coral reefs accounting for climate warming disturbances. *PLoS ONE*, **10**, e0140828.

Maina, J., de Moel, H., Zinke, J., *et al.* (2013). Human deforestation outweighs future climate change impacts of sedimentation on coral reefs. *Nature Communications*, **4**, article 1986.

Maina, J., McClanahan, T.R., Venus, V., Ateweberhan, M. & Madin, J. (2011). Global gradients of coral exposure to environmental stresses and implications for local management. *PLoS ONE*, **6**, e23064.

McClanahan, T.R. (2002). The near future of coral reefs. *Environmental Conservation*, **29**, 460–83.

McClanahan, T.R. & Muthiga, N.A. (1988). Changes in Kenyan coral reef community structure and function due to exploitation. *Hydrobiologia*, **166**, 269–76.

McClanahan, T.R. & Muthiga, N.A. (1989). Patterns of predation on a sea urchin, *Echinometra mathaei* (de Blainville), on Kenyan coral reefs. *Journal of Experimental Marine Biology and Ecology*, **126**, 77–94.

McClanahan, T.R. & Shafir, S.H. (1990). Causes and consequences of sea urchin abundance and diversity in Kenyan coral reef lagoons. *Oecologia*, **83**, 362–70.

McCook, L.J., Ayling, T., Cappo, M., *et al.* (2010). Adaptive management of the Great Barrier Reef: a globally significant demonstration of the benefits of networks of marine reserves. *Proceedings of the National Academy of Sciences*, **107**, 18278–85.

Mcleod, E., Anthony, K.R.N., Andersson, A., *et al.* (2013). Preparing to manage coral reefs for ocean acidification: lessons from coral bleaching. *Frontiers in Ecology and the Environment*, **11**, 20–7.

Moberg, F. & Folke, C. (1999). Ecological goods and services of coral reef ecosystems. *Ecological Economics*, **29**, 215–33.

Moorhead, J.A. & Zeng, C. (2010). Development of captive breeding techniques for marine ornamental fish: a review. *Reviews in Fisheries Science*, **18**, 315–43.

Mora, C., Andréfouët, S., Costello, M.J., *et al.* (2006). Coral reefs and the global network of marine protected areas. *Science*, **312**, 1750–1.

Moran, P.J. (1990). *Acanthaster planci* (L.): biographical data. *Coral Reefs*, **9**, 95–6.

Morris, A.V., Roberts, C.M. & Hawkins, J.P. (2000). The threatened status of groupers (Epinephelinae). *Biodiversity and Conservation*, **9**, 919–42.

Mouillot, D., Parravicini, V., Bellwood, D.R., *et al.* (2016). Global marine protected areas do not secure the evolutionary history of tropical corals and fishes. *Nature Communications*, **7**, article 10359.

Mumby, P.J., Dahlgren, C.P., Harborne, A.R., *et al.* (2006). Fishing, trophic cascades, and the process of grazing on coral reefs. *Science*, **311**, 98–101.

Mumby, P.J. & Steneck, R.S. (2011). The resilience of coral reefs and its implications for reef management. In *Coral Reefs: an Ecosystem in Transition*, ed. Z. Dubinsky & N. Stambler, pp. 509–19. New York: Springer-Verlag.

Mumby, P.J., Steneck, R.S., Edwards, A.J., *et al.* (2012). Fishing down a Caribbean food web relaxes trophic cascades. *Marine Ecology Progress Series*, **445**, 13–24.

Mumby, P.J., Wolff, N.H., Bozec, Y.-M., Chollett, I. & Halloran, P. (2014). Operationalizing the resilience of coral reefs in an era of climate change. *Conservation Letters*, **7**, 176–87.

Normile, D. (2016). Survey confirms worst-ever coral bleaching at Great Barrier Reef. *Science*, doi:10.1126/science.aaf9933 (accessed 31 May 2016).

Öhman, M.C., Rajasuriya, A. & Lindén, O. (1993). Human disturbances on coral reefs in Sri Lanka: a case study. *Ambio*, **22**, 474–80.

Orr, J.C., Fabry, V.J., Aumont, O., *et al.* (2005). Anthropogenic ocean acidification over the twenty-first century and its impact on calcifying organisms. *Nature*, **437**, 681–6.

Osborne, K., Dolman, A.M., Burgess, S.C. & Johns, K.A. (2011). Disturbance and the dynamics of coral cover on the Great Barrier Reef (1995–2009). *PLoS ONE*, **6**, e17516.

Pandolfi, J.M., Bradbury, R.H., Sala, E., *et al.* (2003). Global trajectories of the long-term decline of coral reef ecosystems. *Science*, **301**, 955–8.

Pandolfi, J.M., Connolly, S.R., Marshall, D.J. & Cohen, A.L. (2011). Projecting coral reef futures under global warming and ocean acidification. *Science*, **333**, 418–22.

Pandolfi, J.M. & Jackson, J.B.C. (2006). Ecological persistence interrupted in Caribbean coral reefs. *Ecology Letters*, **9**, 818–26.

Patterson, K.L., Porter, J.W., Ritchie, K.B., *et al.* (2002). The etiology of white pox, a lethal disease of the Caribbean elkhorn coral, *Acropora palmata*. *Proceedings of the National Academy of Sciences*, **99**, 8725–30.

Paulay, G. (1997). Diversity and distribution of reef organisms. In *Life and Death of Coral Reefs*, ed. C. Birkeland, pp. 298–353. New York: Chapman & Hall.

Pauly, D., Christensen,V., Guénette, S., *et al.* (2002). Towards sustainability in world fisheries. *Nature*, **418**, 689–95.

Pelletier, D, García-Charton, J.A., Ferraris, J., *et al.* (2005). Designing indicators for assessing the effects of marine protected areas on coral reef ecosystems: a multidisciplinary standpoint. *Aquatic Living Resources*, **18**, 15–33.

Phinney, J.T., Muller-Karger, F., Dustan, P. & Sobel, J. (2001). Using remote sensing to reassess the mass mortality of *Diadema antillarum* 1983–1984. *Conservation Biology*, **15**, 885–91.

Rhyne, A.L., Tlusty, M.F. & Kaufman, L. (2014). Is sustainable exploitation of coral reefs possible? A view from the standpoint of the marine aquarium trade. *Current Opinion in Environmental Sustainability*, **7**, 101–7.

Rhyne, A.L., Tlusty, M.F., Schofield, P.J., *et al.* (2012). Revealing the appetite of the marine aquarium fish trade: the volume and biodiversity of fish imported into the United States. *PLoS ONE*, **7**, e35808.

Rinkevich, B. (2014). Rebuilding coral reefs: does active reef restoration lead to sustainable reefs? *Current Opinion in Environmental Sustainability*, **7**, 28–36.

Rivera-Posada, J., Pratchett, M.P., Aguilar, C., Grand, A. & Caballes, C.F. (2014). Bile salts and the single-shot lethal injection method for killing crown-of-thorns sea stars (*Acanthaster planci*). *Ocean & Coastal Management*, **102**, 383–90.

Roberts, C.M., McClean, C.J., Veron, J.E.N., *et al.* (2002). Marine biodiversity hotspots and conservation priorities for tropical reefs. *Science*, **295**, 1280–4.

Roberts, C.M. & Polunin, N.V.C. (1991). Are marine reserves effective in management of reef fisheries? *Reviews in Fish Biology and Fisheries*, **1**, 65–91.

Roff, G., Clark, T.R., Reymond, C.E., *et al.* (2013). Palaeoecological evidence of a historical collapse of corals at Pelorus Island, inshore Great Barrier Reef, following European settlement. *Proceedings of the Royal Society B*, **280**, article 20122100.

Sadovy, Y.J. & Vincent, A.C.J. (2002). Ecological issues and the trades in live reef fishes. In *Coral Reef Fishes: Dynamics and Diversity in a Complex Ecosystem*, ed. P.F. Sale, pp. 391–420. San Diego, CA: Academic Press.

Salm, R.V., Clark, J.R. & Siirila, E. (2000). *Marine and Coastal Protected Areas: A Guide for Planners and Managers*, 3rd edn. Washington, DC: IUCN.

Sandin, S.A., Smith, J.E., DeMartini, E.E., *et al.* (2008) Baselines and degradation of coral reefs in the northern Line Islands. *PLoS ONE* **3**, e1548.

Selig, E.R. & Bruno, J.F. (2010). A global analysis of the effectiveness of marine protected areas in preventing coral loss. *PLoS ONE*, **5**, e9278.

Semmens, B.X., Buhle, E.R., Salomon, A.K. & Pattengill-Semmens, C.V. (2004). A hotspot of non-native marine fishes: evidence for the aquarium trade as an invasion pathway. *Marine Ecology Progress Series*, **266**, 239–44.

Shanks, A.L., Grantham, B.A. & Carr, M.H. (2003). Propagule dispersal distance and the size and spacing of marine reserves. *Ecological Applications*, **13** (Suppl.), S159–69.

Sheppard, C.R.C, Davy, S.K. & Pilling, G.M. (2009). *The Biology of Coral Reefs*. Oxford, UK: Oxford University Press.

Smith, S.H. (1988). Cruise ships: a serious threat to coral reefs and associated organisms. *Ocean & Shoreline Management*, **11**, 231–48.

Spalding, M.D., Ravilious, C. & Green, E.P. (2001). *World Atlas of Coral Reefs*. Berkeley, CA: University of California Press.

Spurgeon, J.P.G. (1992). The economic valuation of coral reefs. *Marine Pollution Bulletin*, **24**, 529–36.

Teh, L.S.L., Teh, L.C.L. & Sumaila, U.R. (2013). A global estimate of the number of coral reef fishers. *PLoS ONE*, **8** (6), e65397.

van Treeck, P. & Schuhmacher, H. (1998). Mass diving tourism – a new dimension calls for new management approaches. *Marine Pollution Bulletin*, **37**, 499–504.

Veron, J.E.N. (2000). *Corals of the World*, Vol. 3. Townsville, QLD: Australian Institute of Marine Science and CRR Qld Pty Ltd.

Veron, J.E.N., Hoegh-Guldberg, O., Lenton, T.M., *et al.* (2009). The coral reef crisis: the critical importance of $< 350$ ppm $CO_2$. *Marine Pollution Bulletin*, **58**, 1428–36.

Veron, J., Stafford-Smith, M., DeVantier, L. & Turak, E. (2015). Overview of distribution patterns of zooxanthellate Scleractinia. *Frontiers in Marine Science*, **1**, article 81.

Vincent, A.C.J. (2006). Live food and non-food fisheries on coral reefs, and their potential management. In *Coral Reef Conservation*, ed. I.M. Coté & J.D. Reynolds, pp. 183–236. Cambridge, UK: Cambridge University Press.

Vogler, C., Benzie, J., Lessios, H., Barber, P. & Wörheide, G. (2008). A threat to coral reefs multiplied? Four species of crown-of-thorns starfish. *Biology Letters*, **4**, 696–9.

Wabnitz, C., Taylor, M., Green, E. & Razak, T. (2003). *From Ocean to Aquarium: The Global Trade in Marine Ornamental Species*. Cambridge, UK: UNEP World Conservation Monitoring Centre.

Wells, S.M. (1989). Impacts of the precious shell harvest and trade: conservation of rare or fragile resources. In *Marine Invertebrate Fisheries: Their Assessment and Management*, ed. J.F. Caddy, pp. 443–54. New York: Wiley.

Wells, S. (2006). Assessing the effectiveness of marine protected areas as a tool for improving coral reef management. In *Coral Reef Conservation*, ed. I.M. Coté & J.D. Reynolds, pp. 314–31. Cambridge, UK: Cambridge University Press.

Wells, S. & Hanna, N. (1992). *The Greenpeace Book of Coral Reefs*. New York: Sterling Publishing Co., Inc.

Westera, M., Lavery, P. & Hyndes, G. (2003). Differences in recreationally targeted fishes between protected and fished areas of a coral reef marine park. *Journal of Experimental Marine Biology and Ecology*, **294**, 145–68.

Westmacott, S., Teleki, K., Wells, S. & West, J. (2000). *Management of Bleached and Severely Damaged Coral Reefs*. Gland, Switzerland: IUCN.

Wilhelmsson, D., Ohman, M.C., Stahl, H. & Shlesinger, Y. (1998). Artificial reefs and dive tourism in Eilat, Israel. *Ambio*, **27**, 764–6.

Wilkinson, C.R. (2000). World-wide coral reef bleaching and mortality during 1998: a global climate change warning for the new Millennium? In *Seas at The Millennium: An Environmental Evaluation* Vol. III, ed. C. Sheppard, pp. 43–57. Oxford, UK: Elsevier Science Ltd.

Wilkinson, C. (2006). Status of coral reefs of the world: summary of threats and remedial action. In *Coral Reef Conservation*, ed. I.M. Coté & J.D. Reynolds, pp. 3–39. Cambridge, UK: Cambridge University Press.

Wilkinson, C. (2008). *Status of Coral Reefs of the World: 2008*. Townsville, QLD: Global Coral Reef Monitoring Network and Reef and Rainforest Research Centre.

Williams, G.J., Aeby, G.S., Cowie, R.O.M. & Davy, S.K. (2010). Predictive modeling of coral disease distribution within a reef system. *PLoS ONE*, **5**, e9264.

# 14 The Deep Sea

A distinction can be made between coastal waters, extending to the edge of the continental shelf, and the oceanic waters beyond. Important linkages between coastal and open-ocean waters are evident in ecological processes and biota. Nevertheless, modification of oceanic circulation by the submerged margin of the continent results in obvious differences in the structure and functioning of shelf and open-ocean ecosystems. The division is also relevant from the point of view of human use and management. For centuries, the open oceans have in effect been invulnerable to human influence, protected by their vastness and a relatively small human population with limited capacity or need to exploit them. There was little incentive to seek new fishing grounds beyond the accessible, productive, and still largely unpolluted coastal waters. Grotius's principle of 'freedom of the seas' of the early seventeenth century recognized that establishing rights to oceanic resources was unnecessary (see Chapter 3). But the pressures of human population and advances in fishing and offshore technology changed this picture dramatically. With the establishment of an international legal framework for the oceans under the 1982 UN Convention on the Law of the Sea, there is a distinction too between a state's exclusive economic zone or legal continental shelf and those seaward areas beyond national jurisdiction. This has major consequences for management and conservation of marine resources.

The deep sea is usually taken to mean the region below 200 m water depth, where the continental shelf gives way to the steeper gradient of the continental slope (Fig. 1.6). Continental slope depths encompass the bathyal zone, which typically extends to water depths of about 3000 m (the lower limit of the bathyal zone being variously recognized as between 2000 and 4000 m). The shelf edge and slope are in places incised by submarine canyons. Together, the shelf and slope comprise the continental margin, beyond which continental crust gives way to oceanic crust, which is denser and mainly comprises basalt. Beyond the base of the slope lies the huge abyssal zone at water depths of about 3000–6000 m, which accounts for more than half of the Earth's surface. Interrupting the flat abyssal topography is the mid-ocean ridge system where adjacent tectonic plates are spreading apart and new crust is being formed. Also rising from the abyssal plains and resulting from former volcanic activity are steep-sided oceanic islands and seamounts. The deepest parts of the seafloor are the ocean trenches at water depths of 6000–11 000 m, the hadal zone, where subduction is occurring – where one tectonic plate is moving under its neighbour and into the mantle as the two plates converge.

By its vast extent, the open ocean is responsible for the major share of global marine productivity. Limited by nutrient input, open-ocean areas are, however, far less productive per unit area than coastal waters (see Chapter 1), and only a fraction of surface production reaches deep-sea organisms. On average about 2–3% of primary production is transported below 1000 m, and only about half of that reaches the abyssal seafloor (Jahnke, 1996). There are important exceptions: deep-sea areas, for instance, that receive organic inputs because of down-slope transport or large pulses of phytodetritus. But, in general, deep-sea communities dependent on particulate flux from surface production are constrained by food input and are consequently of low biomass and productivity. As a result, most deep-sea benthic communities are dominated by invertebrates that are smaller bodied than those of more productive shallow-water environments.

The deep-sea sediment benthos is characterized by the usual major groups, such as deposit-feeding polychaetes, peracarid crustaceans (mainly tanaidaceans, isopods, amphipods, and cumaceans), and bivalve and gastropod molluscs. Studies since the late 1960s indicate that this deep-sea invertebrate benthos, particularly on the continental slope, can be far more diverse than the benthos of coastal sediments. Grassle and Maciolek (1992), for example, examined diversity of the soft-sediment macrobenthos of the continental slope off the north-eastern USA. They took quantitative bottom samples that they sieved on a fine mesh (0.3 mm) to extract the fauna. Their stations at 1500 m and 2100 m water depth each yielded some 280–360 species for seabed surface areas of 1.62 $m^2$. The often-high diversity of the deep-sea sediment benthos appears to be related to a range of factors that promote small-scale patchiness of the habitat, such as fluxes of phytodetritus, sporadic falls of carrion, foraging and sediment reworking activities, and pits, mounds, and minute biogenic structures. The fine-scale environmental heterogeneity that results is relevant to the small body size of deep-sea species (Rex & Etter, 2010). Diversity may, however, be reduced as a result of low oxygen conditions, as under upwelling zones and other areas of high surface productivity; where the seabed is tectonically unstable and susceptible to slumping, such as in trenches; and in deep basins that have become isolated, such as the Sea of Japan and Red Sea (Gage & Tyler, 1991; Rex & Etter, 2010).

We have looked at the broad-scale biogeography of open-ocean surface waters (Chapter 1, Fig. 1.5). Sampling of deep-water pelagic and benthic organisms has by comparison been far more limited, and understanding of their major biogeographical patterns is still poorly developed. Watling *et al.* (2013) proposed a biogeography of the deep-ocean floor based on water-mass characteristics (temperature and salinity) and organic matter flux, factors that are believed to be good predictors of the distribution of organisms. In particular, they proposed provinces for two major depth zones: lower bathyal at 800–3500 m and abyssal at 3500–6500 m. The lower bathyal zone mainly comprises the lower continental margins, seamounts and slopes of oceanic islands, and mid-ocean ridges, whilst the abyssal zone includes most of the deep-sea floor with major basins separated by mid-ocean ridges. Fourteen provinces are proposed for each major zone; those for the most extensive zone, the abyssal, are shown in Fig. 14.1.

## Deep-Sea Habitats

Although deep ocean covers most of the planet, our knowledge of its biology and ecology lags far behind that of other environments. That said, our understanding of oceanic systems has advanced enormously in recent decades. Significantly, for example, deep-ocean ecosystems are known to be far more diverse and dynamic than early studies suggested and to include many unique habitats (Ramirez-Llodra *et al.*, 2010; Levin & Sibuet, 2012). As our understanding of the diversity, structure, and function of deep-sea habitats and ecosystems matures, so must our approaches to their management and conservation be developed. Here we briefly consider the diversity of deep-sea benthic habitats – focusing on those that tend to feature in discussions on deep-sea conservation and management – and look at differences in their geomorphology, oceanography, biodiversity, and vulnerability to human disturbance.

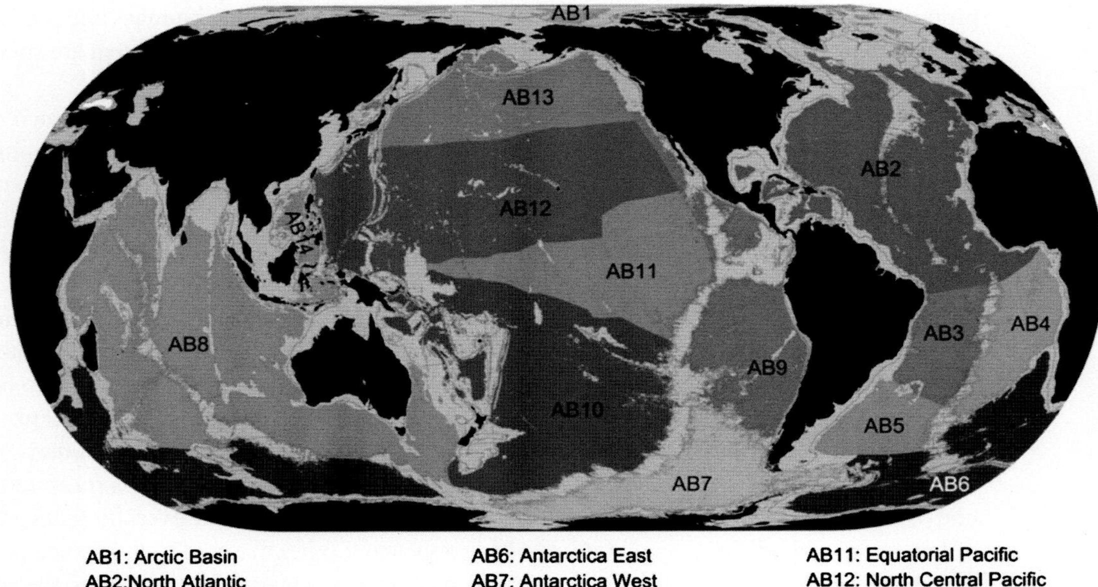


**Fig. 14.1** Biogeographic provinces of the abyssal zone (3500–6500 m). Reprinted from *Progress in Oceanography*, Vol. 111, Watling, L., Guinotte, J., Clark, M.R. & Smith, C.R., A proposed biogeography of the deep ocean floor, pages 91–112, copyright 2012, with permission from Elsevier. (A black and white version of this figure will appear in some formats. For the colour version, please refer to the plate section.)

## Submarine Canyons

A feature of continental margins worldwide, submarine canyons are typically steep-walled, V-shaped valleys that incise the continental slope and sometimes part of the shelf. In a global inventory, Harris *et al.* (2014) identified nearly 9500 large submarine canyons with a mean length of 41 km (large canyons being defined as those spanning a depth range of at least 1000 m and incising at least 100 m into the slope). Most submarine canyons result from erosion by submarine landslides and sediment-laden currents and are important conduits for the transport of coastal sediment to deep water. Canyon heads may also intercept organic-rich material that is then carried to down-slope benthos. In addition, the topography of canyons can modify current flow to produce localized downwelling or upwelling, boosting primary productivity and the flux of food to the benthos. As a result, submarine canyons can be hotspots of productivity. Toothed whales, particularly beaked whales and sperm whales, appear to be strongly associated with a number of submarine canyons, using them as important foraging areas throughout the year (Moors-Murphy, 2014). Kaikoura Canyon, New Zealand, is noted for its sperm whales but also supports a benthos of exceptional biomass and unusual community structure (De Leo *et al.*, 2010; Leduc *et al.*, 2014). The benthic biodiversity of canyons can be enhanced where exposures of bedrock provide hard surfaces for attached epifauna, such as cold-water corals, which in turn provide biogenic habitat for many associated species. Such habitat occurs more commonly in the heads of shelf-incising canyons at water depths shallower than about 1500 m. This is within the depth limit for bottom trawling, making these diverse communities vulnerable to severe physical disturbance (Harris & Whiteway, 2011).

## Seamounts

Seamounts are discrete undersea mountains that rise steeply from the deep-sea floor. Most are volcanic in origin and occur in chains or clusters, many having erupted sequentially where a tectonic plate has passed over a hotspot in the mantle. They are usually roughly conical in shape and range in size from large seamounts rising thousands of metres from the seafloor to subsea knolls and hills a few hundreds of metres high (Yesson *et al.*, 2011).

Seamounts are found in all ocean basins. Estimates of their number worldwide vary widely depending on the size of the feature being considered. Seamounts have traditionally been defined as having a height of more than 1000 m, but as there is no ecological reason to exclude smaller features, many studies now take an elevation of at least 100 m as a threshold. On this basis, the global number of seamounts is estimated to range from some tens of thousands (Kim & Wessel, 2011) to at least one million (Costello *et al.*, 2010). But although numerous, seamounts represent a relatively uncommon habitat compared with the vastness of the deep sea. Large seamounts (> 1000 m high) constitute in areal extent only 2% of the ocean floor (Harris *et al.*, 2014).

Seamounts can include a diversity of topographic features and substrata and often have extensive areas of exposed rock – unusual in the soft sediment-dominated deep sea. These hard surfaces provide habitat for attached epifauna, most conspicuously stony corals (Scleractinia), black corals (Antipatharia), and gorgonian corals (Alcyonacea), in places forming reefs that support a diverse associated fauna (Consalvey *et al.*, 2010). Koslow *et al.* (2001) surveyed the benthic macrofauna of seamounts off southern Tasmania. On unfished to lightly fished seamounts peaking at depths of < 1400 m, they recorded a dense and diverse invertebrate fauna dominated by suspension feeders, including gorgonian and antipatharian corals, hydroids, sponges, and suspension-feeding ophiuroids and sea stars but with up to 90% of the biomass comprising live *Solenosmilia variabilis*, a matrix-forming scleractinian coral.

Seamounts have also attracted attention for their fish and fisheries. About 800 fish species are recognized as seamount species, representing a unique component of fish biodiversity (Morato & Clark, 2007). From catch statistics, Watson *et al.* (2007) identified 13 fish species that are taken primarily from seamounts, dominated by oreos and orange roughy, and a further 29 commercial species associated with seamounts. Annual landings of primary seamount species appear to have peaked by the early 1990s at about 120 000 tonnes. A number of mainly benthopelagic and demersal fishes form dense aggregations on seamounts. The very low productivity and extreme longevity of these species make them especially vulnerable to overfishing (see Chapter 6).

There is evidence too of increased pelagic biodiversity in the vicinity of seamounts. Using tuna longline observer data for areas of the western and central Pacific, Morato *et al.* (2010) found higher species richness within 30–40 km of seamount summits compared with other oceanic and coastal areas, with a number of species of shark, billfish, and other pelagic teleost fishes tending to aggregate near seamounts (Fig. 14.2). Some cetaceans and seabirds are also reported to be significantly more abundant in the vicinity of seamounts (Morato *et al.*, 2008).

Seamounts can be sites of enhanced productivity, manifest for instance by an abundance of epibenthic suspension feeders to increased numbers of pelagic predators. Interactions between seamount topography and the local current regime appear to underpin such productivity, including an intensified flow around seamounts that provides an increased water-borne food supply, and the entrapment of diurnally migrating plankton over seamount summits (Clark *et al.*, 2010).

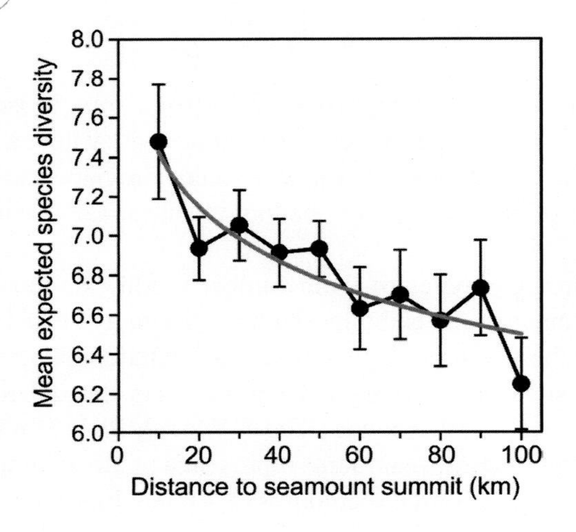


**Fig. 14.2** Pelagic species diversity in relation to distance to seamount summit (with 95% confidence limits and logarithmic regression line). The species diversity is the expected mean species diversity for 40 individuals. Reproduced with permission from Morato, T., Hoyle, S.D., Allain, V. & Nicol, S.J. (2010). Seamounts are hotspots of pelagic biodiversity in the open ocean. *Proceedings of the National Academy of Sciences*, 107, 9707–11.

## Deep-Water Coral Reefs

Deep-water corals are remarkably diverse, and for some groups, such as black corals, octocorals, and stylasterids, most of the species occur in deep water (Table 14.1). Among the stony corals, the Scleractinia, about one-third of species live at upper slope depths. Most of these are solitary species, such as cup corals (Cairns, 2007). However, several species of deep-water scleractinian corals form significant biogenic habitat as reefs or reef-like structures where there is suitable hard substratum, mostly at shelf break to continental slope depths, including on seamounts. The global area of such deep reefs is unknown but may be comparable to that of their shallow warm-water counterparts (Corcoran & Hain, 2006).

Among the most important of the deep-sea habitat-forming corals are the cosmopolitan species *Lophelia pertusa, Madrepora oculata*, and *Solenosmilia variabilis*; the Southern

**Table 14.1** Number of Species of Major Groups of Corals and Number That Occur in Water Deeper than 50 m

| Main groups of corals | Total number of species | Number of species deeper than 50 m (without zooxanthellae) |
|---|---|---|
| Scleractinia stony corals | 1482 | 615 (41%) |
| Antipatharia black corals | 237 | 178 (75%) |
| Octocorallia gorgonians | 3093 | 2320 (75%) |
| Stylasteridae hydrocorals | 246 | 220 (89%) |

Reproduced with permission from Cairns, S.D. (2007). Deep-water corals: an overview with special reference to diversity and distribution of deep-water scleractinian corals. *Bulletin of Marine Science*, 81, 311–22.

**Fig. 14.3** Reef-like matrix of the branching stony coral *Solenosmilia variabilis* together with abundant crinoids, a venus flower basket sponge (*Symplectella* sp.), white stylasterid hydrocorals, and two orange roughy (*Hoplostethus atlanticus*), on a seamount at about 1000 m water depth, Chatham Rise, New Zealand. Image provided by the National Institute of Water and Atmospheric Research (NIWA), New Zealand, and captured by NIWA's Deep Towed Imaging System (DTIS). Photograph: NIWA. (A black and white version of this figure will appear in some formats. For the colour version, please refer to the plate section.)

Hemisphere species *Goniocorella dumosa*; and *Enallopsammia profunda* of the western Atlantic (Roberts *et al.*, 2009) (Fig. 14.3). The best known of these is *Lophelia pertusa*, mainly on account of studies of North Atlantic continental margins where it has often been recorded, from small patch reefs to extensive banks. Particularly massive reefs have been mapped along the shelf break off Norway – complexes up to 35–40 km long and 3 km wide at 300–400 m water depth (Fosså *et al.*, 2005). Habitat is provided not only by the framework of living coral but also by dead coral and accumulated rubble. More than 1800 species have been recorded as living on or in *Lophelia* reefs in the NE Atlantic and 2700 species worldwide, although most are probably facultative associates rather than being dependent on coral habitat (Roberts *et al.*, 2009; Roberts & Cairns, 2014). Nevertheless, this indicates a faunal diversity comparable to that of tropical shallow-water coral reefs, albeit with some obvious differences. The coral diversity itself is much lower in the case of *Lophelia* and other deep-water reefs, with typically only one or two species as the primary habitat formers, and they also harbour far fewer mollusc and fish species than tropical reefs (Rogers, 1999).

Deep-sea coral reefs tend to attract fish, including many commercially important species (Roberts & Hirshfield, 2004). As structurally complex habitats they may provide increased availability of food and shelter. From underwater video observations at water depths of 260–320 m off northern Norway, Kutti *et al.* (2015) found Norway redfish (*Sebastes viviparus*) and tusk (*Brosme brosme*) to be twice as abundant on coral mounds and sponge fields than on unstructured seabed. Deep-water coral reefs may be afforded protection where they are deemed to provide essential fish habitat that is recognized in legislation (see Chapter 5).

## Hydrothermal Vents

Most deep-sea organisms ultimately depend for their food on the euphotic zone and the resulting downward flux of particulate organic matter. Benthic abundance and biomass thus tend to decline markedly with water depth and distance from shore towards the mid-ocean regions of low primary productivity and where sinking particles are more likely to be fully exploited during their long descent. It therefore came as a dramatic discovery when in 1977 submersible investigations of the Galápagos Rift in the eastern Pacific revealed a luxuriant community of large benthic invertebrates at a depth of 2500 m. The community was associated with hot-water vents. These occur where seawater that has permeated newly formed crust issues from the seafloor, but now superheated (up to 400°C or more) and laden with metals and sulphide leached from the hot rock. Subsequent explorations revealed similar vent faunas dotted along mid-oceanic ridges and at other sites of tectonic activity, such as some subduction zones. Vent communities are highly productive. They are fuelled by bacteria able to use the sulphide-rich vent fluid as a substrate rather than by particulate flux from the euphotic zone (although their oxygen still comes ultimately from photosynthesis). Such primary production driven by chemical energy is known as chemosynthesis. Associated invertebrates filter bacteria from the water or graze bacterial films, but a number of them are sustained by symbiotic sulphur-oxidizing bacteria harboured in their tissues.

Hundreds of species have now been described from vent sites, the great majority endemic to these specialized habitats, and studies are revealing a high level of cryptic species diversity (Van Dover, 2011). Most vent species appear to be restricted to vent habitats, with many known so far from only one site (Tunnicliffe *et al.*, 1998; Wolff, 2005). In terms of biogeography, 11 hydrothermal vent provinces have been recognized, related mainly to ocean basins and the relative isolation along mid-ocean ridge systems (Van Dover *et al.*, 2002; Rogers *et al.*, 2012).

Among the most conspicuous invertebrates of the east Pacific vents are giant siboglinid polychaetes (tubeworms up to 1.5 m long), alvinellid polychaetes, mussels and vesicomyid clams, limpets, crabs, galatheids, and anemones. Vent faunas differ considerably, however, between ocean basins as well as between sections of ridge crest. Vent communities in the Atlantic, for instance, differ noticeably from those in the Pacific. They lack the giant tubeworms and alvinellids, and instead shrimps and mussels dominate the biomass (Tunnicliffe *et al.*, 1998). Active vents fields are very localized, only tens to hundreds of metres across and spaced perhaps 10–100 km apart along a ridge crest. They are also ephemeral given that vent activity at a site may persist for only a matter of decades.

Other chemosynthetic systems have been discovered that are likewise supported by microbial primary producers using reduced compounds as energy sources (typically hydrogen sulphide or methane), such as cold seeps, wood falls, and whale skeletons. And the faunas that colonize these habitats include many taxa closely related to those at hydrothermal vents (Smith & Baco, 2003).

## Ocean Trenches

The deepest parts of the ocean, the hadal zone, are at water depths of 6000 to 11 000 m and occur mainly in ocean trenches, formed along active plate margins where subduction is occurring. They thus tend to be narrow, linear basins (rarely more than 2000 km long) that run parallel to continents and island-arc systems. Harris *et al.* (2014) identify a total of 56 trenches, accounting for 0.5% of the ocean floor, with the great majority of trench habitat in the Pacific (80% by area). High seismic activity and proximity to landmasses can result in higher sedimentation rates in trenches. They are also zones of extreme hydrostatic pressure (which increases by 1 atmosphere for every 10 m water depth). This poses physiological and biochemical constraints on organisms, but hadal depths are otherwise not exceptional in terms of deep-sea environmental characteristics (Jamieson *et al.*, 2010). However, as trenches tend to be isolated basins separated by abyssal plains, this strongly affects the composition and endemicity of their faunas. For trench faunas as a whole, about 56% of species are reported to be endemic, with 95% of these found in only one trench or group of neighbouring trenches (Jamieson, 2015), and ten biogeographical provinces are recognized (Watling *et al.*, 2013) (Fig. 14.4). No commercially valuable resources have yet been discovered in trenches. Bioprospecting may yield novel compounds of interest to the biotechnology industry, ones that occur in this extreme high-pressure, low-temperature environment (Jamieson, 2015). But it seems unlikely that trenches would be significantly threatened by extractive activities. They are in any case very difficult to access. However, they could be vulnerable to certain types of human disturbance, such as a trench-wide pollution event (Angel, 1982). Trenches are thus scarce habitats, harbouring unique faunas with sparse and essentially enclosed populations.

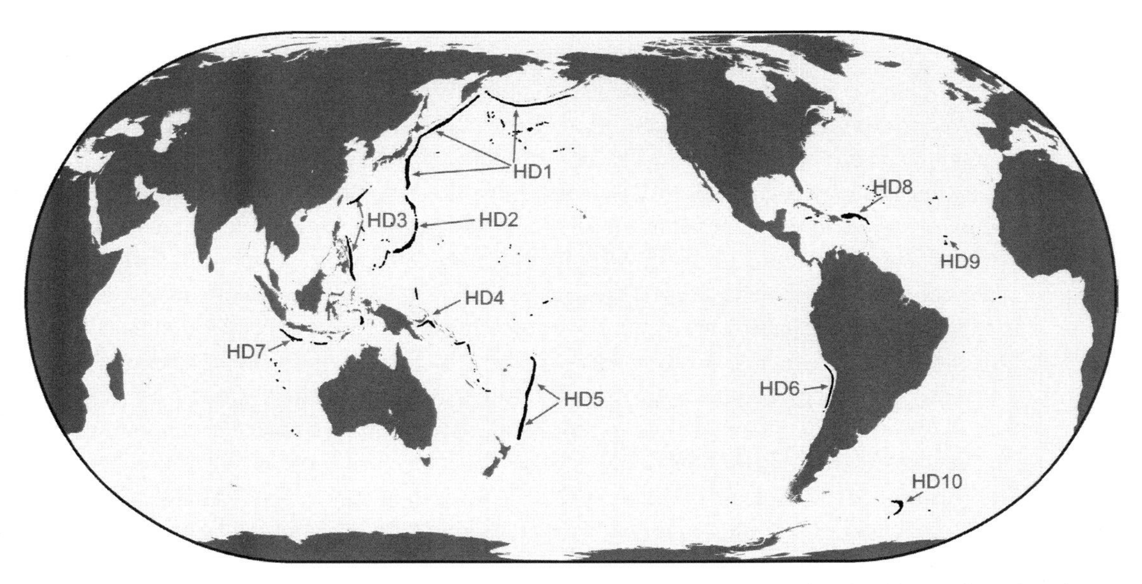


**Fig. 14.4** Location of ocean trench systems and the biogeographic provinces of the hadal zone based on those proposed by Belyaev (1989): HD1 Aleutian-Japan, HD2 Philippine, HD3 Mariana, HD4 Bougainville-New Hebrides, HD5 Tonga–Kermadec, HD6 Peru–Chile, HD7 Java, HD8 Puerto Rico, HD9 Romanche, HD10 Southern Antilles. Reprinted from *Progress in Oceanography*, Vol. 111, Watling, L., Guinotte, J., Clark, M.R. & Smith, C.R., A proposed biogeography of the deep ocean floor, pages 91–112, copyright 2012, with permission from Elsevier.

## Bottom Fisheries

Human use of deep-ocean resources has so far mainly concerned their biota. Whales were among the first animals of the open ocean to be seriously depleted (see Chapter 9). Since then, with the over-exploitation of many inshore stocks and advances in fishing technology, many species of squid and pelagic and demersal fishes have become the focus of important fisheries. Major pelagic resources of the open ocean include tunas, salmon, sharks, billfish, and squid. Deep-water demersal fisheries have also developed, targeting bathyal fishes such as grenadiers and orange roughy. Many of these stocks have been heavily overexploited, although as we have seen the contribution such species can make to sustainable fisheries is very limited given their life-history traits (see Chapter 6).

Also of particular concern is the impact of deep-water bottom fisheries on benthic habitats and communities. Increases in the size and power of vessels and advances in echo-sounding, navigation, and fishing gear technology have greatly increased the capacity, water depth, and efficiency of deep-sea bottom trawlers over recent decades and in turn their potential environmental impact. Bottom trawlers can now fish down to 2000 m water depth. Disturbance of the seabed and benthos by bottom trawling and dredging has attracted much attention in recent decades, particularly in coastal waters where major fishing grounds have traditionally been located. The effects of towed fishing gear on the bottom are likely, however, to be far more severe in deeper water with the potential for major disturbance to benthic communities and where biological characteristics of species indicate that recovery may be measured in decades if not centuries (Jones, 1992; Watling & Norse, 1998).

An overriding effect of bottom trawling is to simplify benthic habitat complexity (Thrush & Dayton, 2002), evident from studies of both soft and hard bottom areas. There is, for instance, a long-established bottom trawl fishery for shrimp on the NW Mediterranean upper continental slope, which intensified markedly from the 1970s. The trawling induces downslope sediment flows that significantly smooth the bottom topography and greatly reduce benthic habitat heterogeneity (Puig *et al.*, 2012). Resuspension of sediment from repeated trawling results in significant loss of surface organic material, notably recently settled phytodetritus, and thereby less food for the benthos. The fauna of chronically trawled sediments is markedly reduced in abundance and diversity, with an assemblage structure dominated by opportunistic species (Pusceddu *et al.*, 2014). Such results indicate that in many parts of the world, intensive bottom trawling may be important in altering the morphology of the upper continental slope and its benthic biodiversity over extensive areas.

Particularly vulnerable to trawling damage are habitats in which much of the structural complexity is provided by large sessile epifauna (see Chapter 5). We see this in the deep sea where hard substrata provide habitat for corals and other attached suspension feeders, such as on seamounts and in the heads of canyons. Fishing these topographically complex areas has been made possible by various technological developments, including trawl gear designed for use on rough seafloor (Clark & Koslow, 2007). Seamounts have attracted attention in this regard given the dense aggregations of commercially valuable fish and well-developed epibenthos that can characterize these features (Probert *et al.*, 1997; Clark & Koslow, 2007). Seamount trawl fisheries date mainly from the 1970s, in many cases developing rapidly but unsustainably with targeted stocks crashing within a matter of years. Such operations can mean intense fishing pressure on individual seamounts. For seamounts in the New Zealand region, O'Driscoll and Clark (2005) estimated the median intensity of fishing to be about 130 km of tows per 1 $km^2$ of seamount, but in some cases fishing intensity was at least an order of magnitude greater. A comparison of fished and unfished seamounts off southern Tasmania indicates the severe impact that trawling can have on the benthos (Althaus *et al.*, 2009). Unfished seamounts supported coral thickets of *Solenosmilia variabilis* covering on average 52% of the bottom, compared to less than 1% coral cover on fished seamounts (Fig. 14.5). And associated

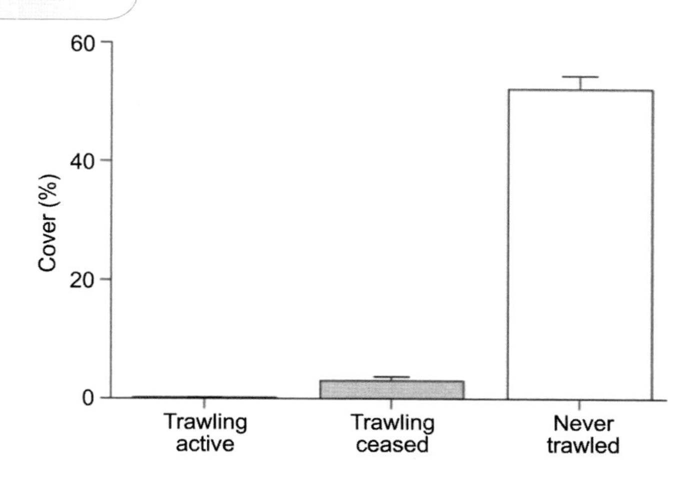


**Fig. 14.5** The percent cover of habitat-forming coral (*Solenosmilia variabilis*) on seamounts off southern Tasmania, for seamounts that were actively trawled, where trawling had ceased 5–10 years ago, and for those that had never been trawled. Modified from Althaus, F., Williams, A., Schlacher, T.A., *et al.* (2009). Impacts of bottom trawling on deep-coral ecosystems of seamounts are long-lasting. *Marine Ecology Progress Series*, 397, 279–94. With permission from Inter-Research.

with the removal of this coral habitat were three-fold declines in the diversity and density of other epibenthos.

Deep coral-dominated assemblages require a long period to recover from trawling damage. Studies of seamounts in this SW Pacific region show no evidence of assemblage change in up to 10 years following the cessation of trawling (Althaus *et al.*, 2009; Williams *et al.*, 2010). Indeed, given estimates of the longevity of deep-water corals, recovery might be expected to take centuries or even millennia. Colonies of deep-sea habitat-forming scleractinian corals can attain ages of at least 600 years (Houlbrèque *et al.*, 2010), and large reefs may take several thousand years to develop (Rogers, 1999). Extreme longevity, of the order of 2000–4000 years, has been determined by radio-carbon dating for some deep-sea non-scleractinian corals (e.g. a zoanthid gold coral (*Gerardia* sp.) and an antipatharian black coral (*Leiopathes* sp.)) (Roark *et al.*, 2009).

So-called precious corals are highly sought after for making jewellery. Included here are deep-sea species of *Corallium, Gerardia*, and *Primnoa*, and many seamounts have been exploited for their precious corals. These are the slowest-growing organisms of any known fishery, and populations have typically collapsed not long after being discovered (Tsounis *et al.*, 2010). Populations of a red coral *Corallium* sp. on the Emperor Seamounts in the North Pacific have, for example, been devastated.

## Minerals and Hydrocarbons

### Minerals

The deep seabed harbours immense mineral and hydrocarbon resources. The principal mineral resources are metal-rich deposits, notably manganese nodules on abyssal plains (valuable mainly for their nickel, cobalt, and copper content); cobalt-rich ferromanganese crusts on seamounts, ocean ridges, and other topographic highs; and seafloor massive sulphide (SMS) deposits enriched with various base and precious metals at hydrothermal sites (Rona, 2008). Whilst many of these deposits are attractive to mining companies, commercial extraction is not yet economically and/or technologically viable. However, there are currently exploration contracts for more than 1.8 million km$^2$ of seabed, with about half of this area in EEZs and the other half in areas beyond national jurisdiction. About 45% of the contracted area is for SMS deposits, mostly within EEZs in the SW Pacific (Hein *et al.*, 2013), and deposits in this region are likely to be among the first to be mined (Boschen *et al.*, 2013, 2015). Deposits within EEZs, being outside the jurisdiction of the International Seabed Authority (see below), may be subject to weaker regulatory regimes and hence more attractive to mining companies (Halfar & Fujita, 2002).

The full environmental implications of seafloor mining are still unclear. Also, impacts are likely to vary considerably depending on the site, its assemblages, and the method of mining. There is, however, the potential for major long-lasting impacts on deep-sea ecosystems. Most studies have focused on the impact of manganese nodule mining. Such nodules are typically 1–10 cm in diameter and range in density from a sparse cover to almost continuous pavements on the sediment surface. They are widely distributed on the abyssal seafloor, but only certain nodule fields would be worth mining in terms of metal content, nodule density, and minimum extent. The highest quality nodule fields are in the north-eastern equatorial Pacific, notably south-east of Hawaii between the Clarion and Clipperton fracture zones. Several mining claims have been registered with the International Seabed Authority in this area, each initially up to 150 000 $km^2$ and for a 15-year contract.

Nodule mining would involve some type of mechanical or hydraulic collector to harvest the nodules from the sediment, a riser to convey them to a surface mining platform for crushing or direct transfer to an ore carrier, and the discharge of tailings into the subsurface water (Oebius *et al.*, 2001). Removal of the nodules means the loss of that hard substratum for epibenthos, whilst other benthos in the path or vicinity of the collector system would largely be destroyed or smothered. Waste material discharged from the mining platform would probably be piped to beneath the euphotic zone. Organisms in these mid-ocean regions experience very low natural levels of suspended sediment, so the resulting turbidity plumes could have widespread effects. A single nodule mining operation could, over its contract period, severely impact benthic communities over tens of thousands of square kilometres (Smith *et al.*, 2008). Benthic recolonization is likely to be slow in this naturally quiescent environment. Miljutin *et al.* (2011) found that the nematode assemblage in a track resulting from experimental nodule dredging had not recovered after 26 years, with nematode density, biomass, and diversity significantly lower within the track than outside.

## International Seabed Authority

Under the UN Convention on the Law of the Sea (UNCLOS), the seafloor beyond national jurisdiction, 'the Area', and its mineral resources are declared the 'common heritage of mankind', with no state able to claim sovereignty over any part of the international seabed or its resources (Articles 136 and 137). Resources under this part of UNCLOS are defined as 'all solid, liquid or gaseous mineral resources *in situ* in the Area at or beneath the sea-bed, including polymetallic nodules'. The convention established the International Seabed Authority (ISA) to administer the resources of the Area and to control their exploitation and mining. The need for the ISA was prompted by the expectation that commercial exploitation of manganese nodule mining was imminent – UNCLOS was adopted in 1982 – whereas industrial-scale mining may not be viable for some years yet. In the interim, the ISA is overseeing exploratory surveys and research on mining impacts as well as associated measures to protect the marine environment (as required under Article 145 of UNCLOS). The ISA is mandated to take a precautionary approach in its activities and to require a contractor applying for nodule mining rights to designate 'reference zones' for protecting representative areas of seabed and their biodiversity and for assessing the impacts of mining. Wedding *et al.* (2013) illustrate the development of a protected area network for the Clarion-Clipperton fracture zone using spatial analysis of biophysical and social datasets to balance protection of biodiversity and resource use. Their nine replicate marine protected areas cover 24% of the region targeted for nodule mining.

### Hydrocarbons

About one-third of the world's crude oil production is now from continental margins. Most offshore wells have been on continental shelves, but increasingly production is moving on to the continental slope, with some fields now at 3000 m water depth. (The absence of sedimentary rocks precludes production beyond continental margins.) Important areas for deep offshore production include the Gulf of Mexico and off Brazil and West Africa (Sandrea & Sandrea, 2007). Environmental concerns include the discharge of cuttings and drilling muds and their impact on benthic communities. Far more damaging would be an accidental blowout, as occurred at the *Deepwater Horizon* well in the Gulf of Mexico in 2010. Some 700 000 tonnes of crude oil were released at the seafloor at a water depth of about 1500 m south-east of the Mississippi Delta. Impacts included a significant reduction in subsurface dissolved oxygen, from the bacterial degradation of hydrocarbons – in an area that already experiences natural large-scale hypoxia – as well as toxic effects of chemical dispersants and the smothering of deep-water corals (Ramirez-Llodra *et al.*, 2011; Fisher *et al.*, 2014).

A potentially much larger hydrocarbon resource – possibly an order of magnitude greater than that of conventional fossil fuels – are methane hydrates that occur widely in upper continental slope sediments (Hester & Brewer, 2009). At low temperature and high pressure, methane from microbial decomposition of organic matter remains at depth as an ice-like solid. To what extent it would be practical and economic to exploit such hydrates remains to be seen. Environmental considerations of exploiting gas hydrates include climate change implications and the disturbance and destabilization of slope sediments (Ramirez-Llodra *et al.*, 2011).

## Pollution and Waste Disposal

The world's coastal waters receive a wide range of pollutants, including organic wastes and nutrients, petroleum hydrocarbons, synthetic organic compounds, and metals, mainly as a result of land-based discharges and in many instances with marked biological impacts (see Chapter 2). By contrast, the open ocean has generally been considered relatively clean, with concentrations of contaminating inputs usually insufficient to cause detectable problems, although there is generally far less information available on substances in the open ocean.

There is, nevertheless, evidence of elevated concentrations of contaminants in the deep ocean. Organochlorines and other persistent organic compounds have penetrated into the deep sea and now occur more or less globally. Likewise, fallout of radionuclides from nuclear tests and the Chernobyl accident has been traced to abyssal depths. The downwelling of dense water masses in high-latitude regions and transport in the thermohaline circulation (Fig. 1.3) are probably important means of entry and dispersal for these contaminants, although a more direct route is by way of contaminants associated with particulate organic matter sinking through the water column. But as yet there appears little evidence of such contaminants adversely impacting deep-sea life (Thiel, 2003; Ramirez-Llodra *et al.*, 2011).

The deep ocean has been, and often still is, considered a suitable repository for various types of waste, the view being that it is a remote, depauperate, monotonous environment with enormous assimilative capacity and with only weak links to surface and coastal waters. But we now know the deep ocean to be a far more heterogeneous, dynamic, and less isolated environment than previously thought. Deep-sea sites have been used for the dumping of wastes, such as sewage sludge, dredge spoil, low-level radioactive wastes, and pharmaceutical products, mainly in the latter decades of the twentieth century (Ramirez-Llodra *et al.*, 2011). Such intentional disposal is now banned under the London Convention and its protocol (see Chapter 2), although leakage from containers, as in the case of radioactive wastes, can be expected in tens to hundreds of years' time (Benn *et al.*, 2010).

There may, however, be increasing pressure to look to the deep ocean for waste disposal, and for certain wastes this may be preferable environmentally than disposal in other environments. Angel and Rice (1996) examined under what circumstances such disposal might be considered, arguing that it is unrealistic to exclude the use of the deep ocean as a potential option for waste disposal. They suggest, given our present knowledge of deep-ocean ecosystems, that disposal of inert, metal-rich, or even organic-rich wastes into accumulative regimes on the floor of the abyssal ocean would not create major deleterious impacts. But they caution that any such proposals must be fully evaluated by appropriately scaled experiments and that there are still biological questions to be resolved. Whilst our understanding of the biology and ecology of the deep sea has advanced enormously in recent decades, we are still some way from being able to forecast the full effects of disposal operations on its ecological processes. A precautionary approach is essential.

In places, oceanic waters show obvious signs of pollution, such as at fronts and convergences where tar balls, plastics, and other persistent wastes become concentrated (Cózar *et al.*, 2014; Eriksen *et al.*, 2014). Persistent debris can accumulate on the seafloor, sometimes in high densities, with potentially harmful effects on marine life, particularly through accidental ingestion and entanglement (see Chapter 2). Pham *et al.* (2014b) reported on the distribution and density of marine litter recorded from video and trawl surveys across 32 sites in European waters. Litter was found at all sites and all depths, from 35 m to 4500 m, with plastic and derelict fishing gear as the most abundant litter items. Locations with the highest litter densities (> 20 items per hectare), mainly plastics, were submarine canyons, underlining their role as conduits for the transport of materials from continental shelves into deeper waters (Fig. 14.6). Derelict fishing gear, on the other hand, was most common at topographic features targeted by commercial fisheries, such as seamounts and banks. Whilst the disposal of plastics and other wastes at sea has been banned under international

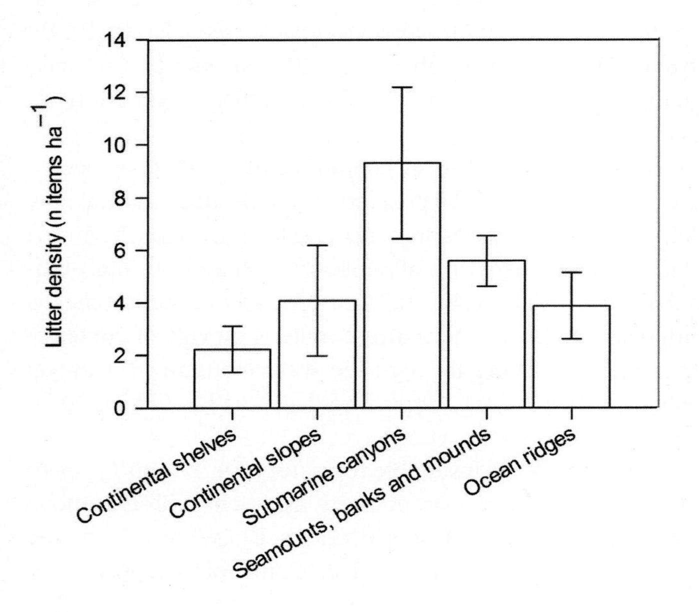


**Fig. 14.6** Mean density of litter items (± standard error) across different physiographic settings in European waters. The groups differ significantly. Reproduced with no changes from Pham, C.K., Ramirez-Llodra, E., Alt, C.H.S., *et al.* (2014b). Marine litter distribution and density in European seas, from the shelves to deep basins. *PLoS ONE*, 9, e95839. Licensed under a Creative Commons Attribution 4.0 International License: https://creativecommons.org/licenses/by/4.0/.

legislation (notably the MARPOL Convention; see Chapter 2), illegal disposal continues, and the persistent nature of the material means that the legacy of earlier disposal is still very evident (Ramirez-Llodra *et al.*, 2011).

Determining the spatial extent of human activities that impact the deep seafloor is important for management purposes but difficult in practice, even for areas where data are likely to be comprehensive. For the NE Atlantic – an area heavily impacted by human activities – Benn *et al.* (2010) estimated the extent of impact on the deep seafloor of submarine communication cables, marine scientific research, oil and gas industry, bottom trawling and the dumping of radioactive waste, munitions, and chemical weapons. The spatial extent of bottom trawling was by far the most significant – at least an order of magnitude greater than the other activities combined. For all activities, the total area impacted was about 28 000 $km^2$, which represents less than 1% of deep seafloor (deeper than 200 m) of the area examined. However, activities are concentrated in certain, mainly shallower, areas and likely to be impacting some habitats more than others.

## Ocean Warming and Acidification

Anthropogenic emissions of carbon dioxide perhaps represent the greatest threat to the health of the deep oceans worldwide, given the implications they have for climate change, ocean acidification, and ecosystem function (GESAMP, 2009; Ramirez-Llodra *et al.*, 2011). Global atmospheric changes are likely to have far-reaching repercussions for open-ocean ecosystems (see Chapter 2). The full extent of such changes is still unclear, but climate warming has the potential, for instance, to modify global ocean circulation, decrease oxygen solubility, increase water column stratification – and thereby reduce vertical mixing – and promote the release of methane from gas hydrates (methane being far more efficient as a greenhouse gas than carbon dioxide).

One of the major consequences of climate change for deep-sea communities is projected to be a reduction in the supply of organic matter to the seabed. Surface warming of the ocean, and thereby more pronounced stratification, will reduce the supply of nutrients to phytoplankton and in turn the flux of particulate organic carbon (POC) to the benthos. Projections by Jones *et al.* (2014) suggest that a reduced export of POC to the seafloor will result in an overall decrease in global benthic biomass of about 5% in this century. Most of the world's seamounts and deep-water coral reefs are forecast to experience reductions in biomass from this reduced flux.

The reduced availability of dissolved oxygen in deep-water masses as a result of ocean warming is forecast to significantly increase the extent of regions naturally susceptible to hypoxia. This is mainly expected to affect mesopelagic depths in the world's subtropical and tropical oceans and result in major shifts in the structure and function of these ecosystems (Ramirez-Llodra *et al.*, 2011).

Ocean acidification, from increased absorption of carbon dioxide by seawater, makes it increasingly difficult for marine organisms to build skeletons of calcium carbonate (see Chapter 2). Deep-water corals are likely to be particularly affected in this respect as they use aragonite for their skeletons. Aragonite is considerably more soluble than calcite, the other form of calcium carbonate used by calcifying organisms. With a lowering of seawater pH, the depth of the aragonite saturation horizon – the depth at which aragonite starts to dissolve – rises closer to the surface. The shallowing of the aragonite horizon over this century is predicted to expose most deep-sea stony corals to undersaturated water, thereby inhibiting their ability to build skeletons (Guinotte *et al.*, 2006; Ramirez-Llodra *et al.*, 2011). Summits and upper flanks of seamounts, lying in waters with a higher aragonite saturation state and less susceptible to projected changes in ocean chemistry than the surrounding seafloor, may act as temporary refugia in need of particular protection (Tittensor *et al.*, 2010).

It is important to bear in mind that habitats may be subject to more than one significant anthropogenic stressor. Different impacts may interact, possibly with a combined effect that is magnified. There are various ways such synergistic interactions may operate in deep-sea ecosystems. In particular, global climate change and its associated effects would be expected to exacerbate impacts of fishing, physical disturbance, and persistent pollutants (Ramirez-Llodra *et al.*, 2011).

## Protecting Deep-Sea Habitats

Human activities have significantly impacted deep-sea communities and habitats in many parts of the world. Overall, human impacts on deep-sea ecosystems have shifted historically from those associated with disposal and dumping to those resulting from exploitation; however, effects of anthropogenic climate change are expected to become increasingly important. Ramirez-Llodra *et al.* (2011) identify deep-sea habitats and communities that currently and in the near future appear most at risk from human impacts, notably benthic communities of upper continental slope sediments, deep-water corals, submarine canyon benthic communities, oxygen minimum zones, and pelagic and benthic communities of seamounts. Sediment slopes, canyons, deep-water corals, and seamounts are particularly impacted by fishing activities; and climate change and ocean acidification have major implications for sediment slope and deep coral communities. Climate change is the main threat for oxygen minimum zones as a result of increasing hypoxia. Canyons, as conduits, are impacted too by the accumulation of debris and by other pollutants. And vulnerable to major impacts from mineral extraction are hydrothermal vents, manganese nodule fields and seamounts, and methane seeps for their hydrocarbon resources.

Fishing is now the main human activity impacting deep-sea ecosystems and is the focus of much of the discussion about deep-sea conservation. We looked earlier (Chapter 5) at the major problems related to fisheries and international instruments (e.g. the UN Fish Stocks Agreement and FAO Code of Conduct for Responsible Fisheries) that aim to promote sustainable management and lessen environmental impact. Deep-sea bottom trawl fisheries are of particular concern in this regard. They impact large areas of seabed and have the potential to substantially alter benthic habitats. The damaged communities could take centuries to recover given the extremely slow growth of many habitat-forming species. This issue has been highlighted in resolutions on sustainable fisheries adopted by the UN General Assembly that draw attention to the adverse impacts of bottom trawling on vulnerable marine ecosystems and call on states to take appropriate conservation and management action (Resolutions 59/25 (2004), 61/105 (2006), and 64/72 (2009)).

### Vulnerable Marine Ecosystems

To facilitate action in the wake of these resolutions, the FAO developed guidelines for the management of deep-sea fisheries. These include measures to help protect deep-sea vulnerable marine ecosystems (VMEs) (FAO, 2009). VMEs are defined in terms of general characteristics related to uniqueness or rarity, functional significance of the habitat, fragility, life-history traits, and structural complexity (Box 14.1). The guidelines also provide examples of taxa, communities, and habitat-forming species that may contribute to forming VMEs as well as the physical features that may support them (Box 14.2). The FAO guidelines call on states and regional fisheries management organizations (RFMOs) to undertake assessments to establish whether deep-sea fishing activities are likely to produce 'significant adverse impacts' (defined in terms of impacts that compromise ecosystem structure or function) and to adopt measures for the prevention of such impacts on VMEs and for protecting their biodiversity (FAO, 2009).

## Box 14.1 Criteria for the Identification of Vulnerable Marine Ecosystems

1. **Uniqueness or Rarity** – an area or ecosystem that is unique or that contains rare species whose loss could not be compensated for by similar areas or ecosystems. These include:
   - habitats that contain endemic species;
   - habitats of rare, threatened, or endangered species that occur only in discrete areas; or
   - nurseries or discrete feeding, breeding, or spawning areas.
2. **Functional Significance of the Habitat** – discrete areas or habitats that are necessary for the survival, function, spawning/reproduction, or recovery of fish stocks, particular life-history stages (e.g. nursery grounds or rearing areas), or rare, threatened, or endangered marine species.
3. **Fragility** – an ecosystem that is highly susceptible to degradation by anthropogenic activities.
4. **Life-History Traits of Component Species That Make Recovery Difficult** – ecosystems that are characterized by populations or assemblages of species with one or more of the following characteristics:
   - slow growth rates;
   - late age of maturity;
   - low or unpredictable recruitment; or
   - long-lived.
5. **Structural Complexity** – an ecosystem that is characterized by complex physical structures created by significant concentrations of biotic and abiotic features. In these ecosystems, ecological processes are usually highly dependent on these structured systems. Further, such ecosystems often have high diversity, which is dependent on the structuring organisms.

From FAO (2009). *International Guidelines for Management of Deep-Sea Fisheries in the High Seas.* Rome: FAO. 73 pp.

## Box 14.2 Examples of Potentially Vulnerable Species Groups, Communities, and Habitats and Features That Potentially Support Them

These often display characteristics consistent with possible VMEs.

**Species Groups, Communities, and Habitat-Forming Species** that are potentially vulnerable to destructive fishing practices and that may contribute to forming VMEs:

1. certain coldwater corals and hydroids, e.g. reef builders and coral forest including: stony corals (Scleractinia), alcyonaceans and gorgonians (Octocorallia), black corals (Antipatharia), and hydrocorals (Stylasteridae);
2. some types of sponge dominated communities;
3. communities composed of dense emergent fauna where large sessile foraminiferans (xenophyophores) and invertebrates (e.g. hydroids and bryozoans) form an important structural component of habitat; and

4. seep and vent communities composed of endemic invertebrate and microbial species.

**Topographical, Hydrophysical, or Geological Features** that potentially support the species groups or communities referred to above:

1. submerged edges and slopes (e.g. corals and sponges);
2. summits and flanks of seamounts, guyots, banks, knolls, and hills (e.g. corals, sponges, xenophyphores);
3. canyons and trenches (e.g. burrowed clay outcrops, corals);
4. hydrothermal vents (e.g. microbial communities and endemic invertebrates); and
5. cold seeps (e.g. mud volcanoes for microbes, hard substrates for sessile invertebrates).

From FAO (2009). *International Guidelines for Management of Deep-Sea Fisheries in the High Seas.* Rome: FAO. 73 pp.

The response of RFMOs to protecting vulnerable marine ecosystems has been varied (Rogers & Gianni, 2010; Weaver *et al*., 2011). Some, such as the General Fisheries Commission for the Mediterranean and the Commission for the Conservation of Antarctic Marine Living Resources (see Chapters 16 and 17), have closed large areas to bottom trawling. Overall, however, Weaver *et al*. (2011) conclude that the resolutions of the UN General Assembly have been only partially applied and that VMEs are not being adequately protected. Admittedly, there are difficulties in putting such provisions into practice. In particular, there is inadequate information on the distribution of VMEs, questions as to what exactly constitutes a VME, the spatial patchiness and size of ecosystem components, and the metrics that are appropriate for requiring action (FAO, 2008; Auster *et al*., 2011). Also, the identification of VMEs has been up to individual RFMOs, resulting in different approaches among regions. An important step would be to have a consistent approach to identifying VMEs that is relevant to all regions, such as that developed by Ardron *et al*. (2014).

A key requirement for protecting VMEs is determining where they occur or are likely to occur. Benthic community data are very limited for most deep-sea areas and rarely adequate for mapping the distribution of VMEs. To overcome this limitation, species distribution models (SDMs) are increasingly being used to predict where VME taxa are likely to occur in areas that have not been sampled. On the basis of records of a species and associated environmental information, SDMs such as Maxent (Phillips *et al*., 2006) can be used to predict the potential distribution of a species over unsampled areas where appropriate environmental data are available. The availability of databases covering a range of parameters relevant to characterizing benthic environments has enabled SDMs to be applied at regional, if not global, scales and at relatively fine resolution (about 1 $km^2$). There are, nevertheless, potential uncertainties and errors inherent in SDMs, and seabed surveys for ground-truthing are needed to improve model accuracy (Vierod *et al*., 2014). An important consideration is that the distribution of a known habitat-forming species may not necessarily be a good proxy for the habitat it forms and thereby of a VME. Predictive modelling of the distribution of *Lophelia pertusa* on banks in the NE Atlantic showed that whilst 7% of the study area was highly suitable for the species, less than 1% was suitable for *Lophelia* reefs, indicating that mapping of VMEs should focus on potential habitat, not just on the species (Howell *et al*., 2011).

Several studies have used SDMs to predict the distribution of deep-sea corals, information that can then inform fisheries management and conservation objectives. Davies and Guinotte (2011) mapped the predicted distribution of five species of reef-forming deep-water corals. They used some 30 environmental layers, covering a range of bathymetric, hydrographic, chemical, and biological variables, to characterize seafloor conditions and predict habitat suitability at a global scale. The most important factors in determining habitat suitability were found to be water depth, temperature, aragonite saturation state, and salinity. Using the habitat suitability values from this study, Penney and Guinotte (2013) generated indices of the likelihood of occurrence of VMEs within the New Zealand high seas region and plotted these against data on bottom trawling effort to quantify the risk of significant impacts on VMEs. Estimates of seabed areas swept by bottom trawlers were used as a measure of fishing intensity, combined with a discounting factor to take account of impacts in previously fished areas. Such an approach could be used to examine different trawl closure scenarios and optimize the protection of VMEs. Ultimately it would be appropriate for closed areas to be incorporated into protected area networks (see below).

The approach, described by Penney *et al.* (2009), to protecting high-seas VMEs off New Zealand – which is within the South Pacific RFMO – is similar to that used by other RFMOs. This recognizes the importance of safeguarding less impacted areas, applying a 'move-on' rule where the likelihood of encountering a VME is unknown, and restricting fishing (if it is to continue) to those areas already most damaged (Box 14.3). Interim measures within the New Zealand high seas region may, however, be suboptimal for protecting VMEs (Penney & Guinotte, 2013).

## Box 14.3 Protection Measures for Vulnerable Marine Ecosystems (VMEs)

Measures described by Penney *et al.* (2009) to protect high-seas VMEs incorporate a three-tier management approach and a move-on rule.

Fisheries catch and effort data are used to classify blocks (20-minute latitude/longitude blocks) into three tiers:

- Tier 1 – lightly trawled blocks, which are closed to further fishing;
- Tier 2 – moderately trawled blocks, where a 'move-on' rule applies (using the protocol described by Parker *et al.*, 2009);
- Tier 3 – heavily trawled blocks, which are open to fishing.

The move-on rule requires vessels to cease fishing if VMEs are encountered and move away before they resume fishing (resolutions of the UN General Assembly).

- A move-on rule is triggered when the benthic bycatch contains a certain biomass or volume of species indicative of a VME.
- RFMOs have stipulated minimum move-on distances, for example 2 nm (~3 km).

There are, however, uncertainties relating to move-on rules, such as how bycatch thresholds relate to the presence of VMEs and how such levels relate to the biomass of VME taxa that are actually impacted, given that bottom-fishing gear is inefficient at sampling sessile benthos. Requiring vessels to move away may even spread the impact on VMEs. Also, RFMOs may apply the rule only to limited types of VMEs and set bycatch thresholds that are too high (Auster *et al.*, 2011; Weaver *et al.*, 2011; Penney & Guinotte, 2013).

Switching to methods of fishing that are less damaging to benthic communities than bottom trawling may be an important way to help protect deep-sea coral reefs and other vulnerable habitats. Pham *et al.* (2014a) demonstrated that deep-sea bottom longline fishing has little impact on cold-water corals compared to trawling, resulting in low bycatch and in situ damage. For their study site in the Azores, they estimated that between 4000 and 23 000 longline deployments would be necessary to remove 90% of the density of corals, compared to only 13 trawls. On the other hand, longlines can be responsible for high bycatches of sharks (Chapter 6) and seabirds (Chapter 8). Longlines are inefficient for catching some species targeted by trawl fisheries, so their use as an alternative might be resisted by the industry.

Proposals have been made for a moratorium on high-seas bottom trawling. The target fishes are highly vulnerable to overexploitation, and the trawling severely damages seabed habitat. Moreover, such fisheries account for only a small fraction of the global marine fish catch and would be uneconomic without government subsidies (Sumaila *et al.*, 2010).

## Marine Protected Areas

The great majority of marine protected areas (MPAs) worldwide are located in coastal areas. These areas continue to come under intense pressure from effects of human activity, and there is a clear need to establish representative networks of MPAs in coastal waters. However, such networks should extend into offshore waters, particularly as we become increasingly aware of the threats to deep-sea ecosystems and their biodiversity. This has been recognized, for example, by Parties to the Convention on Biological Diversity. In 2008, the parties adopted criteria for identifying ecologically or biologically significant marine areas (EBSAs) in open-ocean waters and deep-sea habitats (Table 14.2) and guidance for establishing representative networks of marine protected areas (Dunn *et al.*, 2014). Among the deep-sea habitats and communities cited as meeting the criteria are hydrothermal vents, seamounts, and deep-water coral and sponge communities. EBSAs have to date been identified largely as a result of expert opinion. Clark *et al.* (2014) proposed a more objective method to complement this process involving four main steps: (1) identify the area to be examined, (2) determine appropriate datasets and thresholds to use in the evaluation, (3) evaluate data for each area/habitat against a set of criteria, and (4) identify and assess candidate EBSAs. To evaluate the

**Table 14.2** Scientific Criteria for Identifying Ecologically or Biologically Significant Marine Areas in Need of Protection in Open-Ocean Waters and Deep-Sea Habitats

| | |
|---|---|
| 1 | Uniqueness or rarity |
| 2 | Special importance for life history stages of species |
| 3 | Importance for threatened, endangered or declining species and/or habitats |
| 4 | Vulnerability, fragility, sensitivity, or slow recovery |
| 5 | Biological productivity |
| 6 | Biological diversity |
| 7 | Naturalness |

From Annex I of CBD Decision IX/20.

method, they used data on offshore seamounts in the South Pacific Ocean, examining options for different combinations of criteria to arrive at a selection process that satisfies the EBSA criteria and delivers a suitable number of seamounts. These may then be considered for some form of management, including possible designation as MPAs.

Seamounts illustrate some of the main issues in using MPAs and other instruments to protect deep-sea ecosystems (Probert *et al.*, 2007). Like other deep-sea habitats, seamounts are poorly represented in MPA statistics. Of seamounts with an elevation greater than 500 m, fewer than 1% are located within MPAs; however, only about one-third of such seamounts are in areas under national jurisdiction, and relatively few countries have used legislation to protect them (Alder & Wood, 2004; Yesson *et al.*, 2011). One of the first seamount MPAs was designated by the Azores government in 1988. The Formigas Bank MPA in the south-east of the archipelago, which extends to 1800 m water depth, was initially 377 $km^2$ but enlarged in 2003 to 595 $km^2$ with further restrictions on fisheries. Within Australia's EEZ, the Huon Commonwealth Marine Reserve south of Tasmania includes a 'habitat protection zone' of 389 $km^2$ established to protect the benthic communities of a cluster of seamounts.

Several international instruments are relevant in managing and protecting seamounts and other deep-sea ecosystems (Alder & Wood, 2004). RFMOs mainly cover areas beyond national jurisdiction and can provide a potential avenue for protecting seamounts on the high seas. Certain areas vulnerable to bottom trawling have been closed by RFMOs, although in some cases only temporarily. Enforcement of closures needs, however, to be considered. Many seamounts are remote features, and illegal fishing is widespread. The problems of managing exploitation in international waters and the vulnerability of seamount species have led to calls for a global ban on high seas bottom trawling until a suitable management regime is in place (Alder & Wood, 2004). Possible conservation measures, including MPAs in areas beyond national jurisdiction, are examined in the following chapter.

To achieve representative coverage of habitats and communities in MPA networks requires good information on patterns of biodiversity from global to regional scales. Whilst an understanding of global biogeographic patterns for the deep seabed is emerging (Watling *et al.*, 2013), finer-scale information is generally inadequate as a basis for developing MPA networks that are representative of deep-sea habitats. It is rarely practicable to rely on faunal data from direct sampling for such information: few areas have been surveyed in sufficient detail, and generating new data that are adequate is unlikely to be feasible given the huge areas in question and attendant effort and cost. For these reasons, the main approach to classifying deep-sea marine environments to inform MPA selection has been to select physical variables considered the most biologically relevant, and thereby likely to predict distinct habitats or communities, and that are already available in databases (but see also Chapter 10 and Williams *et al.*, 2009).

Clark *et al.* (2011), for example, outline a hierarchical classification of large seamounts (> 1500 m elevation) at bathyal depths in the world's high seas (a total of nearly 11 000 seamounts). They used four variables: organic matter flux (i.e. potential food supply), summit depth (important for instance if the summit reaches the photic zone or migrating plankton layer), dissolved oxygen, and distance between a seamount and its nearest neighbour (as a measure of connectivity and likely 'uniqueness'). Their analysis produced up to 20 seamount classes for each of the biogeographic provinces of the lower bathyal zone (using the global classification of the deep seafloor described earlier (Watling *et al.*, 2013)). How many and which particular seamounts in each class should be selected for a network will depend on a number of other factors, such as the percentage area of habitat to be protected (recommended targets are usually in the 10–50% range), the existence of particular habitats or VMEs, and whether seamounts have already been heavily trawled and should be excluded.

There may, on the other hand, be a case to protect certain sites from further disturbance to allow them to recover, although this is likely to be more easily achieved in the case of habitats that are expected to recover relatively quickly, such as active vents. Alternatively, restoring or actively assisting in the recovery of a deep-sea site may become an option to consider. Ecological restoration is an important tool in management of coastal habitats such as wetlands (see Chapter 12). Restoration techniques for deep-sea habitats have yet to be developed – there are considerable socio-economic, ecological, and technological issues to address – but the scope and options need to be examined as pressure on deep-sea resources increases (Van Dover *et al.*, 2014).

Whilst we have focused on protection of benthic habitats, we must not overlook the need to manage whole ecosystems. Seamounts are again a good example in this regard (Probert, 1999). Flow-topography interactions and enhancement of productivity can make seamounts biodiversity hotspots that support aggregations of seabirds, marine mammals, and fish as well as a diversity of benthic communities including those dominated by habitat-forming epifauna. Consequently, a management plan for a seamount will often need to encompass not only the seafloor and its biota but also that of the overlying water column and, as for other deep-sea habitats, be suitably precautionary in its approach.

Strategies for protecting open-ocean and deep-sea habitats are still being developed, but establishing MPAs is seen as one of the key ways for conserving their marine biodiversity. In the next chapter we examine in more detail the use of MPAs.

# REFERENCES


Alder, J. & Wood, L. (2004). Managing and protecting seamount ecosystems. In *Seamounts: Biodiversity and Fisheries*, ed. T. Morato & D. Pauly. Fisheries Centre, University of British Columbia, Canada. Fisheries Centre Research Report, **12** (5), 67–75.

Althaus, F., Williams, A., Schlacher, T.A., *et al.* (2009). Impacts of bottom trawling on deep-coral ecosystems of seamounts are long-lasting. *Marine Ecology Progress Series*, **397**, 279–94.

Angel, M.V. (1982). Ocean trench conservation. *The Environmentalist*, **2** (Suppl. 1), 1–17.

Angel, M.V. & Rice, A.L. (1996). The ecology of the deep ocean and its relevance to global waste management. *Journal of Applied Ecology*, **33**, 915–26.

Ardron, J.A., Clark, M.R., Penney, A.J., *et al.* (2014). A systematic approach towards the identification and protection of vulnerable marine ecosystems. *Marine Policy*, **49**, 146–54.

Auster, P.J., Gjerde, K., Heupel, E., *et al.* (2011). Definition and detection of vulnerable marine ecosystems on the high seas: problems with the 'move-on' rule. *ICES Journal of Marine Science*, **68**, 254–64.

Belyaev, G.M. (1989). *Deep-Sea Ocean Trenches and Their Fauna*. Moscow: Nauka Publishing House. [In Russian, translation by Scripps Institution of Oceanography Library.]

Benn, A.R., Weaver, P.P., Billet, D.S.M., *et al.* (2010). Human activities on the deep seafloor in the North East Atlantic: an assessment of spatial extent. *PLoS ONE*, **5**, e12730.

Boschen, R.E., Rowden, A.A., Clark, M.R. & Gardner, J.P.A. (2013). Mining of deep-sea seafloor massive sulfides: a review of the deposits, their benthic communities, impacts from mining, regulatory frameworks and management strategies. *Ocean & Coastal Management*, **84**, 54–67.

Boschen, R.E., Rowden, A.A., Clark, M.R., *et al.* (2015). Megabenthic assemblage structure on three New Zealand seamounts: implications for seafloor massive sulfide mining. *Marine Ecology Progress Series*, **523**, 1–14.

Cairns, S.D. (2007). Deep-water corals: an overview with special reference to diversity and distribution of deep-water scleractinian corals. *Bulletin of Marine Science*, **81**, 311–22.

Clark, M.R. & Koslow, J.A. (2007). Impacts of fisheries on seamounts. In *Seamounts: Ecology, Fisheries & Conservation*, ed. T.J. Pitcher, T. Morato, P.J.B. Hart, *et al.*, pp. 413–41. Oxford, UK: Blackwell

Clark, M.R., Rowden, A.A., Schlacher, T., *et al.* (2010). The ecology of seamounts: structure, function, and human impacts. *Annual Review of Marine Science*, **2**, 253–78.

Clark, M.R., Rowden, A.A., Schlacher, T.A., *et al.* (2014). Identifying ecologically or biologically significant areas (EBSA): a systematic method and its application to seamounts in the South Pacific Ocean. *Ocean & Coastal Management*, **91**, 65–79.

Clark, M.R., Watling, L., Rowden, A.A., Guinotte, J.M. & Smith, C.R. (2011). A global seamount classification to aid the scientific design of marine protected area networks. *Ocean & Coastal Management*, **54**, 19–36.

Consalvey, M., Clark, M.R., Rowden, A.A. & Stocks, K.I. (2010). Life on seamounts. In *Life in the World's Oceans: Diversity, Distribution, and Abundance*, ed. A.D. McIntyre, pp. 123–38. Chichester, UK: Wiley-Blackwell.

Corcoran, E. & Hain, S. (2006). Cold-water coral reefs: status and conservation. In *Coral Reef Conservation*, ed. I.M. Côté & J.D. Reynolds, pp. 115–44. Cambridge, UK: Cambridge University Press.

Costello, M.J., Cheung, A. & De Hauwere, N. (2010). Surface area and the seabed area, volume, depth, slope, and topographic variation for the world's seas, oceans, and countries. *Environmental Science & Technology*, **44**, 8821–8.

Cózar, A., Echevarría, F., González-Gordillo, J.I., *et al.* (2014). Plastic debris in the open ocean. *Proceedings of the National Academy of Sciences*, **111**, 10239–44.

Davies, A.J. & Guinotte, J.M. (2011). Global habitat suitability for framework-forming cold-water corals. *PLoS ONE*, **6**, e18483.

De Leo, F.C., Smith, C.R., Rowden, A.A., Bowden, D.A. & Clark, M.R. (2010). Submarine canyons: hotspots of benthic biomass and productivity in the deep sea. *Proceedings of the Royal Society B*, **277**, 2783–92.

Dunn, D.C., Ardron, J., Bax, N., *et al.* (2014). The Convention on Biological Diversity's Ecologically or Biologically Significant Areas: origins, development, and current status. *Marine Policy*, **49**, 137–45.

Eriksen, M., Lebreton, L.C.M., Carson, H.S., *et al.* (2014). Plastic pollution in the world's oceans: more than 5 trillion plastic pieces weighing over 250,000 tons afloat at sea. *PLoS ONE*, **9**, e111913.

FAO (2008). Report of the FAO Workshop on Vulnerable Ecosystems and Destructive Fishing in Deep-sea Fisheries. *FAO Fisheries Report*, 829, 18 pp. Rome: FAO.

FAO (2009). *International Guidelines for Management of Deep-Sea Fisheries in the High Seas*. Rome: FAO. 73 pp.

Fisher, C.R., Hsinga, P.-Y., Kaiser, C.L., *et al.* (2014). Footprint of *Deepwater Horizon* blowout impact to deep-water coral communities. *Proceedings of the National Academy of Sciences*, **111**, 11744–9.

Fosså, J.H., Lindberg, B., Christensen, O., *et al.* (2005). Mapping of *Lophelia* reefs in Norway: experiences and survey methods. In *Cold-Water Corals and Ecosystems*, ed. A. Freiwald & J.M. Roberts, pp. 359–91. Berlin: Springer.

Gage, J.D. & Tyler, P.A. (1991). *Deep-Sea Biology: A Natural History of Organisms at the Deep-Sea Floor*. Cambridge, UK: Cambridge University Press.

GESAMP (IMO/FAO/UNESCO-IOC/UNIDO/WMO/IAEA/UN/UNEP Joint Group of Experts on the Scientific Aspects of Marine Environmental Protection) (2009). Pollution in the Open Ocean: a Review of Assessments and Related Studies. *GESAMP Reports and Studies*, **79**, 64 pp.

Grassle, J.F. & Maciolek, N.J. (1992). Deep-sea species richness: regional and local diversity estimates from quantitative bottom samples. *American Naturalist*, **139**, 313–41.

Guinotte, J.M., Orr, J., Cairns, S., *et al.* (2006). Will human-induced changes in seawater chemistry alter the distribution of deep-sea scleractinian corals? *Frontiers in Ecology and the Environment*, **4**, 141–6.

Halfar, J. & Fujita, R.M (2002). Precautionary management of deep-sea mining. *Marine Policy*, **26**, 103–6.

Harris, P.T., Macmillan-Lawler, M., Rupp, J. & Baker, E.K. (2014). Geomorphology of the oceans. *Marine Geology*, **352**, 4–24.

Harris, P.T. & Whiteway, T. (2011). Global distribution of large submarine canyons: geomorphic differences between active and passive continental margins. *Marine Geology*, **285**, 69–86.

Hein, J.R, Mizell, K., Koschinsky, A. & Conrad, T.A. (2013). Deep-ocean mineral deposits as a source of critical metals for high- and green-technology applications: comparison with land-based resources. *Ore Geology Reviews*, **51**, 1–14.

Hester, K.C. & Brewer, P.G. (2009). Clathrate hydrates in nature. *Annual Review of Marine Science*, **1**, 303–27.

Houlbrèque, F., McCulloch, M., Roark, B., *et al.* (2010). Uranium-series dating and growth characteristics of the deep-sea scleractinian coral: *Enallopsammia rostrata* from the Equatorial Pacific. *Geochimica et Cosmochimica Acta*, **74**, 2380–95.

Howell, K.L., Holt, R.D., Pulido Endrino, I. & Stewart, H. (2011). When the species is also a habitat: comparing the predictively modelled distributions of *Lophelia pertusa* and the reef habitat it forms. *Biological Conservation*, **144**, 2656–65.

Jahnke, R.A. (1996). The global ocean flux of particulate organic carbon: areal distribution and magnitude. *Global Biogeochemical Cycles*, **10**, 71–88.

Jamieson, A. (2015). *The Hadal Zone: Life in the Deepest Oceans*. Cambridge, UK: Cambridge University Press.

Jamieson, A.J., Fujii, T., Mayor, D.J., Solan, M. & Priede, I.G. (2010). Hadal trenches: the ecology of the deepest places on Earth. *Trends in Ecology and Evolution*, **25**, 190–7.

Jones, D.O.B., Yool, A., Wei, C.-L., *et al.* (2014). Global reductions in seafloor biomass in response to climate change. *Global Change Biology*, **20**, 1861–72.

Jones, J.B. (1992). Environmental impact of trawling on the seabed: a review. *New Zealand Journal of Marine and Freshwater Research*, **26**, 59–67.

Kim, S.-S. & Wessel, P. (2011). New global seamount census from altimetry-derived gravity data. *Geophysical Journal International*, **186**, 615–31.

Koslow, J.A., Gowlett-Holmes, K., Lowry, J.K., *et al.* (2001). Seamount benthic macrofauna off southern Tasmania: community structure and impacts of trawling. *Marine Ecology Progress Series*, **213**, 111–25.

Kutti, T., Fosså, J.H. & Bergstad, O.A. (2015). Influence of structurally complex benthic habitats on fish distribution. *Marine Ecology Progress Series*, **520**, 175–90.

Leduc, D., Rowden, A.A., Nodder, S.D., *et al.* (2014). Unusually high food availability in Kaikoura Canyon linked to distinct deep-sea nematode community. *Deep-Sea Research II*, **104**, 310–18.

Levin, L.A. & Sibuet, M. (2012). Understanding continental margin biodiversity: a new imperative. *Annual Review of Marine Science*, **4**, 79–112.

Miljutin, D.M., Miljutina, M.A., Arbizu, P.M. & Galèron, J. (2011). Deep-sea nematode assemblage has not recovered 26 years after experimental mining of polymetallic nodules (Clarion-Clipperton Fracture Zone, Tropical Eastern Pacific). *Deep-Sea Research I*, **58**, 885–97.

Moors-Murphy, H.B. (2014). Submarine canyons as important habitat for cetaceans, with special reference to the Gully: a review. *Deep-Sea Research II*, **104**, 6–19.

Morato, T. & Clark, M.R. (2007). Seamount fishes: ecology and life histories. In *Seamounts: Ecology, Fisheries and Conservation*, ed. T.J. Pitcher, T. Morato, P.J.B. Hart, *et al.*, pp. 170–88. Oxford, UK: Blackwell.

Morato, T., Hoyle, S.D., Allain, V. & Nicol, S.J. (2010). Seamounts are hotspots of pelagic biodiversity in the open ocean. *Proceedings of the National Academy of Sciences*, **107**, 9707–11.

Morato, T., Varkey, D.A., Damaso, C., *et al.* (2008). Evidence of a seamount effect on aggregating visitors. *Marine Ecology Progress Series*, **357**, 23–32.

O'Driscoll, R.L. & Clark, M.R. (2005). Quantifying the relative intensity of fishing on New Zealand seamounts. *New Zealand Journal of Marine and Freshwater Research*, **39**, 839–50.

Oebius, H.U., Becker, H.J., Rolinski, S. & Jankowski, J.A. (2001). Parametrization and evaluation of marine environmental impacts produced by deep-sea manganese nodule mining. *Deep-Sea Research II*, **48**, 3453–67.

Parker, S.J., Penney, A.J. & Clark, M.R. (2009). Detection criteria for managing trawl impacts on vulnerable marine ecosystems in high seas fisheries of the South Pacific Ocean. *Marine Ecology Progress Series*, **397**, 309–17.

Penney, A.J. & Guinotte, J.M. (2013). Evaluation of New Zealand's high-seas bottom trawl closures using predictive habitat models and quantitative risk assessment. *PLoS ONE*, **8**, e82273.

Penney, A.J., Parker, S.J. & Brown, J.H. (2009). Protection measures implemented by New Zealand for vulnerable marine ecosystems in the South Pacific Ocean. *Marine Ecology Progress Series*, **397**, 341–54.

Pham, C.K., Diogo, H., Menezes, G., *et al.* (2014a). Deep-water longline fishing has reduced impact on Vulnerable Marine Ecosystems. *Scientific Reports*, **4**, article 04837.

Pham, C.K., Ramirez-Llodra, E., Alt, C.H.S., *et al.* (2014b). Marine litter distribution and density in European seas, from the shelves to deep basins. *PLoS ONE*, **9**, e95839.

Phillips, S.J., Anderson, R.P. & Schapire, R.E. (2006). Maximum entropy modeling of species geographic distributions. *Ecological Modelling*, **190**, 231–59.

Probert, P.K. (1999). Seamounts, sanctuaries and sustainability: moving towards deep-sea conservation. *Aquatic Conservation: Marine and Freshwater Ecosystems*, **9**, 601–5.

Probert, P.K., Christiansen, S., Gjerde, K.M., Gubbay, S. & Santos, R.S. (2007). Management and conservation of seamounts. In *Seamounts: Ecology, Fisheries and Conservation*, ed. T.J. Pitcher, T. Morato, P.J.B. Hart, *et al.*, pp. 442–75. Oxford, UK: Blackwell.

Probert, P.K., McKnight, D.G. & Grove, S.L. (1997). Benthic invertebrate bycatch from a deep-water trawl fishery, Chatham Rise, New Zealand. *Aquatic Conservation: Marine and Freshwater Ecosystems*, **7**, 27–40.

Puig, P., Canals, M., Company, J.B., *et al.* (2012). Ploughing the deep sea floor. *Nature*, **489**, 286–90.

Pusceddu, A., Bianchelli, S., Martín, J., *et al.* (2014). Chronic and intensive bottom trawling impairs deep-sea biodiversity and ecosystem functioning. *Proceedings of the National Academy of Sciences*, **111**, 8861–6.

Ramirez-Llodra, E., Brandt, A., Danovaro. R., *et al.* (2010). Deep, diverse and definitely different: unique attributes of the world's largest ecosystem. *Biogeosciences*, **7**, 2851–99.

Ramirez-Llodra, E., Tyler, P.A., Baker, M.C., *et al.* (2011) Man and the last great wilderness: human impact on the deep sea. *PLoS ONE*, **6**, e22588.

Rex, M.A. & Etter, R.J. (2010). *Deep-Sea Biodiversity: Pattern and Scale*. Cambridge, MA: Harvard University Press.

Roark, E.B., Guilderson, T.P., Dunbar, R.B., Fallon, S.J. & Mucciarone, D.A. (2009). Extreme longevity in proteinaceous deep-sea corals. *Proceedings of the National Academy of Sciences*, **106**, 5204–8.

Roberts, J.M. & Cairns, S.D. (2014). Cold-water corals in a changing ocean. *Current Opinion in Environmental Sustainability*, **7**, 118–26.

Roberts, J.M., Wheeler, A.J., Freiwald, A. & Cairns, S.D. (2009). *Cold-Water Corals: the Biology and Geology of Deep-Sea Coral Habitats*. Cambridge, UK: Cambridge University Press.

Roberts, S. & Hirshfield, M. (2004). Deep-sea corals: out of sight, but no longer out of mind. *Frontiers in Ecology and the Environment*, **2**, 123–30.

Rogers, A.D. (1999). The biology of *Lophelia pertusa* (Linnaeus 1758) and other deep-water reef-forming corals and impacts from human activities. *International Review of Hydrobiology*, **84**, 315–406.

Rogers, A.D. & Gianni, M. (2010). *The Implementation of UNGA Resolutions 61/105 and 64/72 in the Management of Deep-Sea Fisheries on the High Seas*. London: International Programme on the State of the Ocean, 97 pp.

Rogers, A.D., Tyler, P.A., Connelly, D.P., *et al.* (2012). The discovery of new deep-sea hydrothermal vent communities in the Southern Ocean and implications for biogeography. *PLoS Biology*, **10**, e1001234.

Rona, P.A. (2008). The changing vision of marine minerals. *Ore Geology Reviews*, **33**, 618–66.

Sandrea, I. & Sandrea, R. (2007). Exploration trends show continued promise in world's offshore basins. *Oil & Gas Journal*, **105** (9), 34–40.

Smith, C.R. & Baco, A.R. (2003). Ecology of whale falls at the deep-sea floor. *Oceanography and Marine Biology: An Annual Review*, **41**, 311–54.

Smith, C.R., Paterson, G., Lambshead, J., *et al.* (2008). Biodiversity, species ranges, and gene flow in the abyssal Pacific nodule province: predicting and managing the impacts of deep seabed mining. *ISA Technical Study*, 3, 38 pp. Kingston, Jamaica: International Seabed Authority.

Sumaila, U.R., Khan, A., Teh, L., *et al.* (2010). Subsidies to high seas bottom trawl fleets and the sustainability of deep-sea demersal fish stocks. *Marine Policy*, **34**, 495–7.

Thiel, H. (2003). Anthropogenic impacts on the deep sea. In *Ecosystems of the Deep Oceans*, ed. P.A. Tyler, pp. 427–70. Amsterdam: Elsevier.

Thrush, S.F. & Dayton, P.K. (2002). Disturbance to marine benthic habitats by trawling and dredging: implications for marine biodiversity. *Annual Review of Ecology and Systematics*, **33**, 449–73.

Tittensor, D.P., Baco, A.R., Hall-Spencer, J.M., Orr, J.C. & Rogers, A.D. (2010). Seamounts as refugia from ocean acidification for cold-water stony corals. *Marine Ecology*, **31** (Suppl. 1), 212–25.

Tsounis, G., Rossi, S., Grigg, R., *et al.* (2010). The exploitation and conservation of precious corals. *Oceanography and Marine Biology: An Annual Review*, **48**, 161–212.

Tunnicliffe, V., McArthur, A.G. & McHugh, D. (1998). A biogeographical perspective of the deep-sea hydrothermal vent fauna. *Advances in Marine Biology*, **34**, 353–442.

Van Dover, C.L. (2011). Mining seafloor massive sulphides and biodiversity: what is at risk? *ICES Journal of Marine Science*, **68**, 341–8.

Van Dover, C.L., Aronson, J., Pendleton, L., *et al.* (2014). Ecological restoration in the deep sea: desiderata. *Marine Policy*, **44**, 98–106.

Van Dover, C.L., German, C.R., Speer, K.G., Parson, L.M. & Vrijenhoek, R.C. (2002). Evolution and biogeography of deep-sea vent and seep invertebrates. *Science*, **295**, 1253–7.

Vierod, A.D.T., Guinotte, J.M. & Davies, A.J. (2014). Predicting the distribution of vulnerable marine ecosystems in the deep sea using presence-background models. *Deep-Sea Research II*, **99**, 6–18.

Watling, L., Guinotte, J., Clark, M.R. & Smith, C.R. (2013). A proposed biogeography of the deep ocean floor. *Progress in Oceanography*, **111**, 91–112.

Watling, L. & Norse, E.A. (1998). Disturbance of the seabed by mobile fishing gear: a comparison to forest clearcutting. *Conservation Biology*, **12**, 1180–97.

Watson, R., Kitchingman, A. & Cheung, W.W. (2007). Catches from world seamount fisheries. In *Seamounts: Ecology, Fisheries and Conservation*, ed. T.J. Pitcher, T. Morato, P.J.B. Hart, *et al.*, pp. 400–12. Oxford, UK: Blackwell.

Weaver, P.P.E., Benn, A., Arana, P.M., *et al.* (2011). *The impact of deep-sea fisheries and implementation of the UNGA Resolutions 61/105 and 64/72*. Report of an international scientific workshop, National Oceanography Centre, Southampton, 45 pp.

Wedding, L.M., Friedlander, A.M., Kittinger, J.N., *et al.* (2013). From principles to practice: a spatial approach to systematic conservation planning in the deep sea. *Proceedings of the Royal Society B*, **280**, article 20131684.

Williams, A., Bax, N.J., Kloser, R.J., *et al.* (2009) Australia's deep-water reserve network: implications of false homogeneity for classifying abiotic surrogates of biodiversity. *ICES Journal of Marine Science*, **66**, 214–24.

Williams, A., Schlacher, T.A., Rowden, A.A., *et al.* (2010). Seamount megabenthic assemblages fail to recover from trawling impacts. *Marine Ecology*, **31** (Suppl. 1), 183–99.

Wolff, T. (2005). Composition and endemism of the deep-sea hydrothermal vent fauna. *Cahiers de Biologie Marine*, **46**, 97–104.

Yesson, C., Clark, M.R., Taylor, M.L. & Rogers, A.D. (2011). The global distribution of seamounts based on 30 arc seconds bathymetry data. *Deep-Sea Research I*, **58**, 442–53.

# 15 Marine Protected Areas

Setting aside areas to protect marine resources is a practice dating back centuries, as in the systems of customary marine tenure on Pacific islands and the tradition of fishery closures in western cultures (see Chapter 5). Typically, such closures are temporary, often seasonal. More recent is the idea of setting aside areas on a long-term basis to protect marine biodiversity more broadly. Protected areas are now seen as a key component of marine conservation – if not the main tool – as a way to minimize human disturbance and safeguard ecosystem structure and function. There is also increasing interest in the use of protected areas in fisheries management (see Chapter 5). The goals for achieving sustainable fisheries and the wider objectives of marine conservation are, however, converging, more so in the case of artisanal fisheries. In this chapter we take a closer look at protected areas in the marine environment. What, for instance, are the objectives in setting up marine protected areas? What criteria are to be used in establishing them? How should such areas be structured and managed? What type and degree of protection are needed? And does the technique deliver the intended objectives? The topic of marine protected areas is an active field of marine conservation research that has developed a large literature. For overviews see, for instance, Agardy (1997), Sobel and Dahlgren (2004), Claudet (2011), and Roff and Zacharias (2011).

The International Union for the Conservation of Nature (IUCN) defines a protected area as 'a clearly defined geographical space, recognized, dedicated and managed, through legal or other effective means, to achieve the long-term conservation of nature with associated ecosystem services and cultural values' (Dudley, 2008). There exists an array of terms for protected areas in the sea (Al-Abdulrazzak & Trombulak, 2012), but 'marine protected area' (MPA) is an umbrella term covering a range from no-take marine reserves to areas where only mining or oil exploration are restricted.

We are well used to nature reserves on land. Around 15% of the terrestrial realm is now in protected areas, but this compares to only 3.5% for the world's oceans (Lubchenco & Grorud-Colvert, 2015). Designating protected areas in the sea has been much slower to take hold, hindered by a traditional view of unrestricted access to the sea and its resources. The rise in interest in MPAs dates from the early 1960s and in particular from the First World Conference on National Parks in 1962 at which the idea of a comprehensive network of MPAs was advanced. A recommendation of the Conference 'invites the governments of all those countries having marine frontiers, and other appropriate agencies, to examine as a matter of urgency the possibility of creating marine parks or reserves to defend underwater areas of special significance from all forms of human interference, and further recommends the extension of existing national parks and equivalent reserves with shorelines, into the water to the ten fathom [18 m] depth or the territorial limit or some other appropriate off-shore boundary' (Adams, 1962). The use of MPAs has subsequently been endorsed and promoted by other international initiatives including UNCED's Agenda 21, the Convention on Biological Diversity (Article 8), and the World Summit on Sustainable Development. UNCLOS does not refer to MPAs directly, only to general objectives of marine conservation (under Part XII).

These include measures 'to protect and preserve rare or fragile ecosystems as well as the habitat of depleted, threatened or endangered species and other forms of marine life' (Article 195 (5)).

By 1970 there were little more than 100 MPAs worldwide (Kelleher & Kenchington, 1992), whereas now there are at least 9000 (Thomas *et al.*, 2014; Costello & Ballantine, 2015). The great majority are in coastal waters, so that worldwide about 11% of territorial seas and 8% of EEZs are within MPAs, whereas the coverage for areas beyond national jurisdiction is only 0.25%. Also, most MPAs are small; the median size is less than 5 $km^2$ (McCauley *et al.*, 2015).

## Purpose of MPAs

MPAs are promoted for a wide variety of reasons (Jones, 1994; Watson *et al.*, 2014). Conservation of marine biodiversity is generally regarded as their primary role, in particular as areas free from the major impacts of fishing, notably the removal of biomass and the damage to benthic habitats. But they can have various other functions, in particular in meeting fisheries, scientific, social, and cultural objectives. Different goals may be compatible within an MPA. There may be primary and secondary objectives, whilst larger multiple-use areas may aim to meet several objectives, with different parts of an area designated for different goals under an overall management strategy. The IUCN recommends that a national system of MPAs should have the following aims (Resolution 17.38 of the IUCN General Assembly) (Kelleher, 1999):

- To protect and manage substantial examples of marine and estuarine systems to ensure their long-term viability and to maintain genetic diversity.
    - (i) To protect depleted, threatened, rare, or endangered species and populations and, in particular, to preserve habitats considered critical for the survival of such species.
    - (ii) To protect and manage areas of significance to the life cycles of economically important species.
    - (iii) To prevent outside activities from detrimentally affecting the MPAs.
    - (iv) To provide for the continued welfare of people affected by the creation of MPAs.
    - (v) To preserve, protect, and manage historical and cultural sites and natural aesthetic values of marine and estuarine areas, for present and future generations.
    - (vi) To facilitate the interpretation of marine and estuarine systems for the purposes of conservation, education, and tourism.
    - (vii) To accommodate within appropriate management regimes a broad spectrum of human activities compatible with the primary goal in marine and estuarine settings.
    - (viii) To provide for research and training, and for monitoring the environmental effects of human activities, including the direct and indirect effects of development and adjacent land-use practices.

The IUCN has categorized protected areas based upon their management objectives, from areas established primarily for biodiversity conservation and scientific research to areas where sustainable use and maintenance of cultural heritage are the prime objectives. These categories are applicable to MPAs (Table 15.1), and activities that may be appropriate to them are summarized in Table 15.2. (For a specific example, see Table 13.3.) In most cases, the protection of marine ecosystems and their biotic components is the principal goal and the one focused on here. As we have seen in earlier chapters, protected areas are often advocated to help conserve species, communities, and habitats.

**Table 15.1** IUCN Protected Area Categories

| Category | Explanation and application to marine protected areas |
|---|---|
| Ia – strict nature reserve | Strictly protected areas set aside to protect biodiversity and also possibly geological/geomorphological features, where human visitation, use, and impacts are strictly controlled and limited to ensure protection of the conservation values. Such protected areas can serve as reference areas for scientific research and monitoring. No-take areas/marine reserves fit this description. They may be a whole MPA or a separate zone within a multiple-use MPA. |
| Ib – wilderness area | Usually large unmodified or slightly modified areas, retaining their natural character and influence without permanent or significant human habitation, which are protected and managed so as to preserve their natural condition. The issue of 'wilderness' in the marine environment is less clear than for terrestrial protected areas, but these should be strictly protected areas for maintenance of environmental services. |
| II – national park | Large natural or near-natural areas set aside to protect large-scale ecological processes, along with the complement of species and ecosystems characteristic of the area, which also provide a foundation for environmentally and culturally compatible, spiritual, scientific, educational, recreational, and visitor opportunities. Marine national parks fit this description. |
| III – natural monument or feature | Areas are set aside to protect a specific natural monument, which can be a landform, seamount, submarine cavern, geological feature such as a cave, or even a living feature such as an ancient grove. They are generally quite small protected areas and often have high visitor value. Marine national monuments fall into this category, but a relatively uncommon designation for marine ecosystems. |
| IV – habitat/species management area | Areas to protect particular species or habitats and management reflect this priority. Many Category IV protected areas will need regular, active interventions to address the requirements of particular species or to maintain habitats, but this is not a requirement of the category. Includes whale sanctuaries, seabird rookeries, and some managed areas for fisheries. |
| V – protected landscape or seascape | Area where the interaction of people and nature over time has produced an area of distinct character with significant, ecological, biological, cultural, and scenic value and where safeguarding the integrity of this interaction is vital to protecting and sustaining the area and its associated nature conservation and other values. Some coastal biosphere reserves fit this category. |
| VI – protected area with sustainable use of natural resources | Areas that conserve ecosystems and habitats together with associated cultural values and traditional natural resource management systems. They are generally large, with most of the area in a natural condition, where a proportion is under sustainable natural resource management and where low-level non-industrial use of natural resources compatible with nature conservation is seen as one of the main aims of the area. Multiple-use MPAs, including biosphere reserves, fit this category. |

Based on Dudley (2008) and Agardy (1997).

**Table 15.2** Marine Activities That May Be Appropriate for Each IUCN-Protected Area Management category

N = no

N* = generally no unless special circumstances apply

Y = yes

Y* = yes because no alternative exists, but special approval is essential

* = variable, depends on whether this activity can be managed in such a way that it is compatible with the MPA's objectives

| Activities | Ia | Ib | II | III | IV | V | VI |
|---|---|---|---|---|---|---|---|
| Research: non-extractive | Y* | Y | Y | Y | Y | Y | Y |
| Non-extractive traditional use | Y* | Y | Y | Y | Y | Y | Y |
| Restoration/enhancement for conservation (e.g. invasive species control, coral reintroduction) | Y* | * | Y | Y | Y | Y | Y |
| Traditional fishing/collection in accordance with cultural tradition and use | N | Y* | Y | Y | Y | Y | Y |
| Non-extractive recreation (e.g. diving) | N | * | Y | Y | Y | Y | Y |
| Large-scale, low-intensity tourism | N | N | Y | Y | Y | Y | Y |
| Shipping (except as may be unavoidable under international maritime law) | N | N | Y* | Y* | Y | Y | Y |
| Problem wildlife management (e.g. shark control programmes) | N | N | Y* | Y* | Y* | Y | Y |
| Research: extractive | N* | N* | N* | N* | Y | Y | Y |
| Renewable energy generation | N | N | N | N | Y | Y | Y |
| Restoration/enhancement for other reasons (e.g. beach replenishment, fish aggregation, artificial reefs) | N | N | N* | N* | Y | Y | Y |
| Fishing/collection: recreational | N | N | N | N | * | Y | Y |
| Fishing/collection: long-term and sustainable local fishing practices | N | N | N | N | * | Y | Y |
| Aquaculture | N | N | N | N | * | Y | Y |
| Works (e.g. harbours, ports, dredging) | N | N | N | N | * | Y | Y |
| Untreated waste discharge | N | N | N | N | N | Y | Y |
| Mining (seafloor as well as sub-seafloor) | N | N | N | N | N | Y* | Y* |
| Habitation | N | N* | N* | N* | N* | Y | N* |

Reproduced with permission from: Day J., Dudley N., Hockings M., *et al.* (2012). *Guidelines for Applying the IUCN Protected Area Management Categories to Marine Protected Areas.* Gland, Switzerland: IUCN. 36 pp.

## No-Take MPAs

The degree of protection provided by MPAs varies greatly. It may amount to only minor restrictions on fishing or other activities and provide minimal conservation value. On the other hand, in no-take MPAs – often called marine reserves (Lubchenco *et al.*, 2003) – fishing and other extractive activities are prohibited, and where possible, other forms of human disturbance are minimized. The level of protection that can be attained has obvious implications for the extent to which an MPA can be expected to conserve biodiversity and natural ecological function. Fishing in fact is permitted in more than 90% of MPAs (Costello & Ballantine, 2015). But permanent no-take areas have a number of compelling advantages. They are best equipped to conserve biodiversity, can act as refugia for heavily exploited populations, and are straightforward to understand in terms of enforcement. No-take MPAs are also invaluable to marine scientific research by providing sites that are free from direct human disturbance (Ballantine, 2014). They can be used as controls for a wide variety of studies, including fundamental research on species interactions and the functioning of marine ecosystems to studies of the impacts of fishing and the effectiveness of different regimes of resource management.

### Effect of No-Take MPAs

Evidence from numerous studies shows that effective no-take MPAs can significantly benefit the growth and development of enclosed populations and communities. Fish biomass can increase dramatically with protection to at least four to five times that of fished areas (Lester *et al.*, 2009; Edgar *et al.*, 2014). Protection within no-take reserves can also result in significant increases in abundance and size of individuals and in species richness of assemblages (Fig. 15.1). Increases in these biological variables are evident for all latitudes and regions and not due to MPAs being in more favourable locations nor to fishing pressure increasing outside MPAs (Lester *et al.*, 2009). However, an analysis

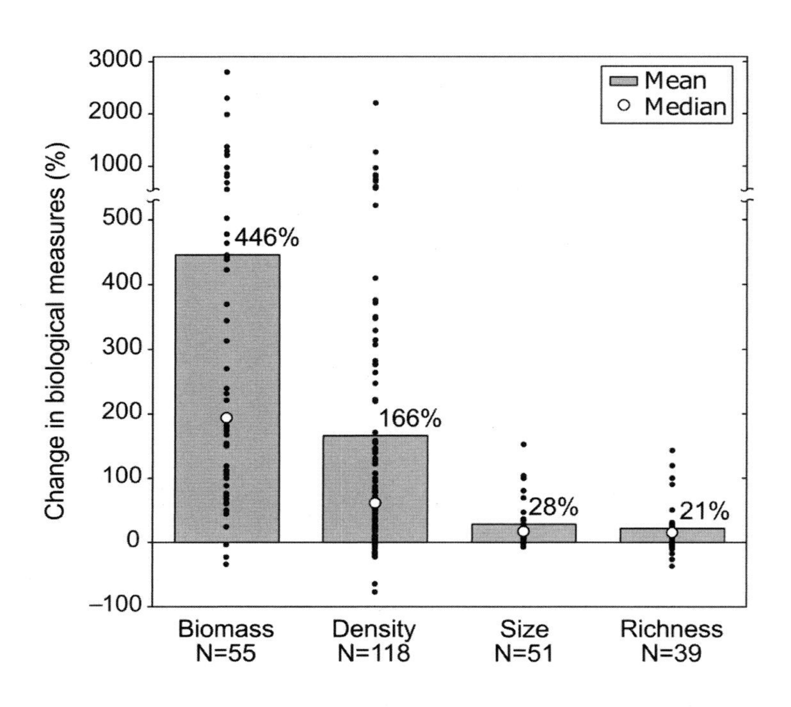


**Fig. 15.1** Mean (grey bars) and median (o) percent change in biomass, numerical density, size of individuals, and species richness within no-take MPAs. All four biological variables show statistically significant increases. (•) Individual reserve responses. N = number of reserves for which each biological variable was measured. Redrawn but not modified from Lester, S.E., Halpern, B.S., Grorud-Colvert, K., *et al.* (2009). Biological effects within no-take marine reserves: a global synthesis. *Marine Ecology Progress Series*, 384, 33–46. With permission from Inter-Research.

by Edgar *et al.* (2014) indicates that the most effective MPAs are not just no-take, they are also well enforced, old (> 10 years), large (> 100 $km^2$), and isolated, thereby reducing emigration of animals into areas where they are vulnerable to fishing because of continuous habitat.

The rebuilding of predatory fish biomass and the attendant levels of top-down control that typically occur within no-take MPAs following their establishment can result in major shifts in community structure and dynamics as the system recovers. Strong evidence for this comes from long-term studies at the Leigh Marine Reserve in northern New Zealand, a 5 $km^2$ no-take MPA with shallow subtidal reef habitat (Shears & Babcock, 2003; Leleu *et al.*, 2012). When the reserve was established in 1977, about one-third of the reef habitat was 'urchin barrens', rock that had been grazed bare by sea urchins (*Evechinus chloroticus*). Twenty years after establishment, however, urchin barrens were virtually absent, whilst the area of kelp forest (*Ecklonia radiata*) had roughly doubled in size to occupy 60% of reef area. The main predators of urchins are snapper *Chrysophrys auratus* and the spiny lobster *Jasus edwardsii*, and both these predators, heavily targeted by fishers, increased in size and abundance within the reserve. Establishing the reserve appears to have set in motion a trophic cascade, with predators indirectly controlling the biomass of primary producers (as sea otters similarly maintain kelp beds; Chapter 10). Long-term studies of multiple marine reserves in both tropical and temperate reef habitats show that direct effects of protection on target species can occur relatively rapidly, becoming evident on average within 5 years. But indirect effects – in particular a decline in prey species – take significantly longer to appear, on average 13 years and in some cases much longer (Babcock *et al.*, 2010). Importantly, communities in fully protected MPAs, where the exploited component of the community has a refuge and the food web is intact, can exhibit increased stability and persistence (Wing & Jack, 2013). MPAs that support such communities and help to maintain ecosystem structure and function are predicted to be more resilient to stress and may, for instance, help mitigate impacts of climate change (CBD, 2003; Green *et al.*, 2014).

An MPA may also enhance production outside its boundaries, notably as a result of the net emigration of juveniles and adults from the protected populations (so-called spillover) and by the net export of eggs and larvae (or recruitment subsidy) (Sale *et al.*, 2005). Whether such movement into adjacent unprotected areas is significant can be difficult to detect but is one of the main reasons why MPAs are advocated for fisheries management (see Chapter 5). If no-take MPAs support higher densities and larger individuals of heavily fished species – and therefore higher reproductive capacity – than areas outside, the spillover of adults and larvae may help maintain adjacent fisheries. Bigger, older individuals in an MPA can make a disproportionate contribution to reproductive output, sometimes up to an order of magnitude more egg production than in adjacent fished areas (e.g. Jack & Wing, 2010). From a study in Guam in the western tropical Pacific, Taylor and McIlwain (2010) found that no-take MPAs accumulate larger, older individuals of *Lethrinus harak*, a heavily targeted emperor fish. The oldest females, up to 11 years, have an ovary about 10 times heavier than those first reaching maturity at around 4 years.

Larval export from no-take MPAs may contribute significantly to juvenile recruitment in fished areas as well to other reserves, as shown by genetic parentage analyses to detect parent-offspring pairs (Harrison *et al.*, 2012). Yellow tang (*Zebrasoma flavescens*), a surgeonfish, is an important herbivore on coral reefs in the western and central Pacific. But the juveniles are heavily exploited for the aquarium trade, especially in Hawaii. To help sustain the fishery, a network of MPAs where collecting for the aquarium trade is banned was established along the west coast of the island of Hawaii in 1999. Using parentage analysis, Christie *et al.* (2010) found direct evidence of larval connectivity in yellow tang among the MPAs as well as successful dispersal of larvae from an MPA to unprotected sites. They detected dispersal distances of up to 184 km, but actual distances may be considerably greater (Fig. 15.2).

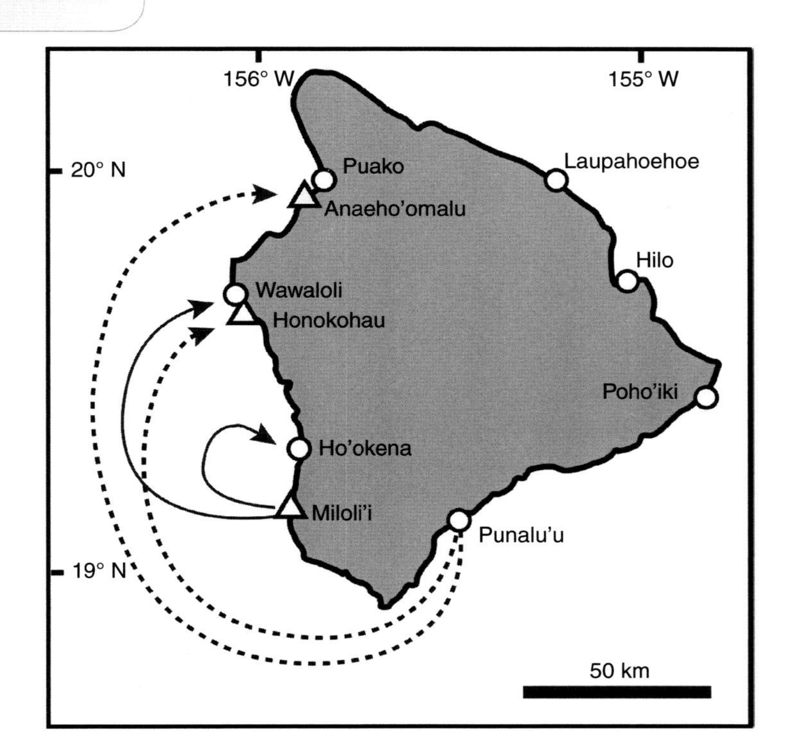


**Fig. 15.2** Dispersal of larvae of yellow tang between reef sites off the island of Hawaii. Triangles represent MPAs, and circles represent unprotected areas. Solid lines indicate the evidence of an MPA seeding unprotected sites. Reproduced with minor changes from Christie, M.R., Tissot, B.N., Albins, M.A., *et al.* (2010). Larval connectivity in an effective network of marine protected areas. *PLoS ONE*, 5, e15715. 

## Selection of MPAs

Given the various potential objectives for MPAs, the way they are selected, structured, and managed can vary greatly. Many MPAs have been created on an *ad hoc* basis as opportunities have arisen and with little regard for ecological criteria. Being an attractive location for diving and tourism has often strongly influenced site selection, contributing to a bias towards reef habitats in MPA coverage. Increasingly, however, governments and conservation agencies are attempting more systematic approaches, notably those based on protecting representative areas of habitat within a biogeographic framework. Typically, legislation does not provide adequate guidelines for selection, and it is usually up to the administering authority to develop criteria (Tisdell & Broadus, 1989).

Principles and criteria relating to the selection of MPAs are discussed by Salm *et al.* (2000). Fundamental to this process is that the overall purpose and goals of any MPA system are clearly defined. The objectives may, for instance, be to establish MPAs specifically for endangered species or as critical fisheries habitat or to protect the range of major habitat types of a region from disturbance. Nevertheless, major steps in the site selection process are likely to be: (1) collection of data (including from existing sources and dedicated field surveys), its assessment (i.e. adequacy, consistency, and quality), and its storage in a suitable database; (2) data analysis to examine the relationship between the distribution of the resources in question and threats to those resources (e.g. using mapping techniques, GIS, statistical methods); (3) data synthesis to identify potential candidate sites; and (4) the application of criteria to select specific sites for protection. Depending on the purpose of the MPA, selection criteria can be used to assess the eligibility of sites and to prioritize them, taking into account, for instance, national policy and available resources. Selection criteria may include social criteria (e.g. public acceptance, recreation, aesthetics, accessibility, research, and education); economic criteria (e.g. importance to particular commercially important species, importance to fisheries, impact on patterns of use, economic benefits, and tourism); regional criteria (i.e. the contribution of a site to a regional network); pragmatic criteria (e.g. urgency, size, degree of threat, availability, restorability); and ecological criteria.

A number of ecological criteria may be used to assess sites, depending on objectives (Salm *et al.*, 2000), notably:

- Biodiversity: the variety or richness of ecosystems, habitats, communities, and species.
- Naturalness: the lack of disturbance or degradation.
- Dependency: the degree to which a species depends on an area or the degree to which an ecosystem depends on ecological processes occurring in the area.
- Representativeness: the degree to which an area represents a habitat type, ecological process, biological community, physiographic feature, or other natural characteristic.
- Uniqueness: whether an area is 'one of a kind'.
- Integrity: the degree to which the area is a functional unit – an effective, self-sustaining ecological entity.
- Productivity: the degree to which productive processes within the area contribute benefits to species or to humans.
- Vulnerability: the area's susceptibility to degradation by natural events or the activities of people.

For protecting biodiversity, the prime strategy must be to establish no-take MPAs. To optimize the potential benefits of such MPAs we need to consider their siting, configuration, and coverage (Lubchenco *et al.*, 2003) – how they can make for effective networks.

## MPA Networks

It has been argued since the early 1960s that an important goal in marine conservation is the establishment of a global network of MPAs to protect and manage representative examples of the world's marine biodiversity. An MPA network can be defined as 'a collection of individual MPAs or reserves operating cooperatively and synergistically, at various spatial scales, and with a range of protection levels that are designed to meet objectives that a single reserve cannot achieve' (IUCN-WCPA, 2008). Effective networks of no-take MPAs share certain key features. Importantly, they need to be representative, replicated, of sufficient connectivity, and large enough to be self-sustaining (Ballantine & Langlois, 2008; McLeod *et al.*, 2009). Recommendations for the design of MPA networks are summarized in Table 15.3.

A primary aim should be to develop MPA networks that include representative examples of all the major habitats within biogeographic regions (e.g. Shears *et al.*, 2008). So far, however, of the world's 232 coastal ecoregions (see Chapter 1), only 20% have more than 10% MPA coverage, and 46% have less than 1% coverage (Watson *et al.*, 2014). Furthermore, few of these MPAs will be no-take, underlining the urgent need for governments to greatly increase the establishment of effective MPAs in their waters. A number of countries have undertaken comprehensive appraisals of the marine habitats within their national jurisdiction as a means to identify potential MPAs. Hiscock (2014), for instance, summarizes the approach used in Great Britain to identify coastal locations of nature conservation importance. Undertaking habitat mapping (see Chapter 10) to adequately characterize a region's biodiversity is, however, a potential constraint, particularly in the developing world.

**Table 15.3** General Recommendations for the Design of MPA Networks

| Factor | Recommendation |
|---|---|
| Size | At least 10–20 km in diameter to be large enough to protect the full range of habitat types and the ecological processes on which they depend and to accommodate self-seeding by short distance dispersers |
| Shape | Simple shapes, such as squares or rectangles, to minimize edge effects and maximize interior protected area |
| Representation | At least 20–40% of each habitat type |
| Replication | At least three examples of each habitat type |
| Spread | Replicates widely separated to reduce the chance they will all be affected by the same disturbance event |
| Critical areas | Protect areas that are biologically or ecologically critical, such as nursery grounds, spawning aggregations, and areas of high species diversity |
| Connectivity | Take biological patterns of connectivity into account to ensure MPA networks are mutually replenishing to facilitate recovery after disturbance. MPAs should be up to 15 km apart to allow for replenishment via larval dispersal, with smaller reserves closer together. Accommodate movement of adults of mobile species by including whole ecological units and a buffer around the core area of interest. Where this is not possible, protect larger rather than smaller areas. Take connectivity among habitat types into account (e.g. protect adjacent areas of coral reefs, seagrass beds, and mangroves). |
| Ecosystem function | Maintain healthy populations of key functional groups |
| Ecosystem-based management | Embed MPAs in broader management frameworks (e.g. integrated coastal zone management or an ecosystem approach to fishing) to address threats external to their boundaries. Address sources of pollution (e.g. nutrient enrichment). |
| Duration | Long-term (at least 20–40 years), preferably permanent |

Republished with permission of The Ecological Society of America, from: Designing marine protected area networks to address the impacts of climate change, *Frontiers in Ecology and the Environment*, McLeod, E., Salm, R., Green, A. & Almany, J., Vol. 7 (7) 2009; permission conveyed through Copyright Clearance Center, Inc. And from: Designing marine reserves for fisheries management, biodiversity conservation, and climate change adaptation, Green, A.L., Fernandes, L., Almany, G., *et al.*, *Coastal Management*, 2014 Taylor & Francis, reprinted by permission of the publisher Taylor & Francis Ltd.

There are good reasons for having replicate MPAs within a network. Replication enables possible differences in biological variables inside and outside MPAs to be tested statistically. At least three replicates of each habitat type are recommended. Replicates also provide some insurance against local catastrophic events and recruitment failure. They should be geographically spread to lessen this risk but without unduly affecting their connectivity: 'the demographic linking of local populations through the dispersal among them of individuals as larvae, juveniles, or adults' (Sale *et al.*, 2005).

MPAs of a no-take network need to be connected by transfer of individuals so as to be self-sustaining and to assist recovery from disturbance. Connectivity depends on a range of factors but in particular the oceanographic setting (especially as this influences the dispersal of eggs and larvae), size and configuration of the MPAs (affecting particularly spillover and export of propagules), and biological characteristics of the species concerned (for example, larval duration, home range, migratory behaviour, and ontogenetic shifts in habitat use). This makes designing an effective network far from straightforward, even if adequate information is available. For conservation of biodiversity, where protection of a wide range of species is important, then at least moderate-size (< 20 km diameter) MPAs are recommended. The larger the reserve, the greater the likelihood of populations being self-sustaining through larvae settling within their natal area, and more species with different home ranges can be accommodated. Small reserves need to be more closely spaced to ensure they are adequately connected. In general, reserves spaced up to about 15 km apart should provide for connectivity and spreading of risk in the event of localized disturbance (e.g. Green *et al.*, 2014). Fig. 15.3 illustrates how connectivity and the potential benefits inside and outside reserves could influence design options for MPA networks.

A key question is what total area should be included in no-take MPAs? Parties to the Convention on Biological Diversity have agreed that 'By 2020, at least … 10 per cent of coastal and marine areas, especially areas of particular importance for biodiversity and ecosystem services, are conserved through effectively and equitably managed, ecologically representative and well connected systems of protected areas and other effective area-based conservation measures, and integrated into the wider landscapes and seascapes' (CBD, 2011). An analysis by O'Leary *et al.* (2016) indicates, however, that to meet major environmental and socio-economic objectives, highly protected MPAs should cover at least 30% of the sea. But no-take areas so far comprise only about 1% of the global ocean and tend to be small (on average about 1.7 $km^2$) (Thomas *et al.*, 2014; Costello & Ballantine, 2015). Current data on the extent of marine habitats may well be inadequate to measure whether such conservation targets can actually be met. For coral reefs, Wabnitz *et al.* (2009) found huge discrepancies (of up to 1300%) between their estimates of coral reef area derived from high-resolution satellite imagery and existing published data, indicating that most countries need to undertake accurate inventories of their marine habitats to assess their progress in meeting CBD commitments.

A number of countries have developed programmes for establishing no-take MPA networks. From a comparative evaluation of the approaches used in New Zealand and New South Wales, Australia, Banks and Skilleter (2010) identified certain key elements for the successful implementation of marine reserve networks. To start with there must be appropriate leadership and commitment at the political level and by the agencies responsible for establishing MPAs. Countries need to have developed frameworks for marine conservation policy and within which there is specific MPA legislation for the primary purpose of protecting marine biodiversity. Programmes for gathering and analysing spatial information on natural and social features have to be suitably resourced, and there must be processes in place for stakeholder involvement and collaboration. Also, just a single lead agency should be mandated to oversee the process and implement MPA networks.

Given the diversity of approaches to spatial planning in conservation, an important development, particularly relevant to the design of protected area networks, has been the evolution of a methodology known as systematic conservation planning, a structured approach that aims to ensure that areas for biodiversity conservation are representative and protected for the long term. Systematic conservation planning has several key characteristics (Margules & Pressey, 2000):

- Requires clear choices about the features to be used as surrogates for overall biodiversity.
- Based on explicit goals, preferably quantitative targets.
- Recognizes the extent to which conservation goals have been met in existing reserves.
- Uses simple explicit methods for locating and designing new reserves to complement existing ones.
- Applies explicit criteria for implementing conservation action on the ground.
- Adopts explicit objectives and mechanisms for maintaining the conditions within reserves that are required to foster the persistence of key natural features.

A step-wise process to implement such a conservation plan is shown in Box 15.1.

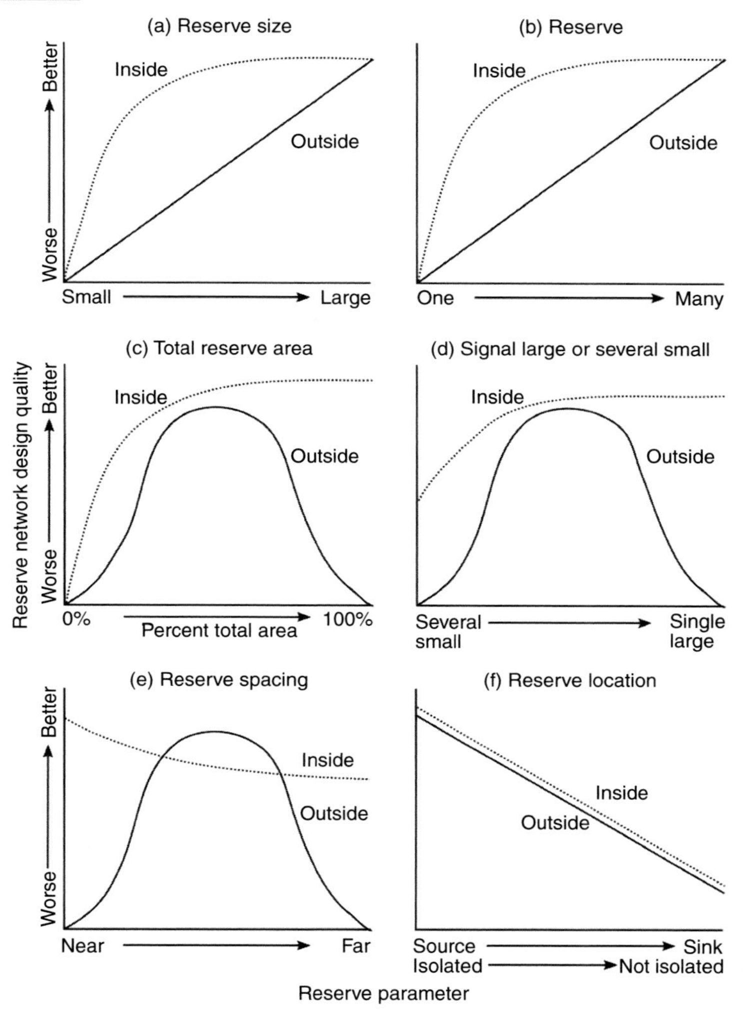


**Fig. 15.3** Connectivity and the design of MPA networks for biodiversity conservation, showing how the benefits inside and outside MPAs vary with different aspects of design. (a) Individual reserve size. Very small reserves provide minimal recruitment benefits either inside or outside. Larger reserves protect significantly larger populations. (b) Number of reserves. Assuming reserve sizes are the same, more is better. Recruitment subsidies and connectivity should increase with number of reserves. (c) Total reserve area (as percentage of total habitat). If all available habitat is protected, benefits would be maximized inside but none provided outside. Recommendations for the proportion of habitat inside generally range from 20% to 50%. (d) Single large versus several small. Which strategy maximizes the number of species inside reserves, having a single large or several small reserves (the so-called SLOSS debate)? Several medium-size reserves are likely to maximize recruitment subsidies because subsidies should increase as a function of the circumferences of all reserves. (e) Reserve spacing. Inside populations should have slightly higher persistence if reserves are close together because of increased recruitment subsidies. However, spacing will have a marked effect on recruitment subsidies outside: too close and this restricts the area in between, but if reserves are too far apart, they will not receive recruits from others. In general, the lower the effective dispersal, the closer reserves need to be to provide benefits outside. (f) Reserve location (source versus sink, isolated versus clustered). Larval sources, rather than sinks, make better reserves because they must be self-sustaining to persist and will provide above-average recruitment subsidies. Also, isolated locations are likely to be important because their biotas often have unique elements. Reproduced from Jones, G.P., Srinivasan, M. & Almany, G.R. (2007). Population connectivity and conservation of marine biodiversity. *Oceanography*, 20 (3), 101–11. doi.org/10.5670/oceanog.2007.33

## Box 15.1 A Framework for Systematic Conservation Planning

Although the process is shown as a sequence, some steps will be undertaken simultaneously, and there will be feedback from later to earlier stages.

1. **Scoping and Costing of the Planning Process**
   Deciding on the boundaries of the planning region, planning team, budget, required funds, and approach to each step in the process.
2. **Identifying and Involving Stakeholders**
   Involving, communicating with, and building capacity for stakeholders who will influence or be affected by conservation decisions and implementation of conservation action.
3. **Identifying the Context for Conservation Areas**
   Assessing the social, economic and political context for the planning process, including constraints on and opportunities for establishing conservation areas.
4. **Identifying Conservation Goals**
   Progressively refining the values of stakeholders from a broad vision statement to specific qualitative goals that shape the rest of the process.
5. **Collecting Socio-economic Data**
   Collecting and evaluating spatially explicit data on tenure, extractive uses, costs, threats, and existing management as a basis for planning decisions.
6. **Collecting Data on Biodiversity and Other Natural Features**
   Collecting and evaluating spatially explicit data on biodiversity pattern and process, ecosystem services, and previous disturbance to potential conservation areas.
7. **Setting Conservation Targets**
   Translating goals into quantitative targets that reflect the conservation requirements of biodiversity and other natural features.
8. **Reviewing Target Achievement in Existing Conservation Areas**
   Assessing, by remote data and/or field survey, the achievement of targets in different types of existing conservation areas.
9. **Selecting Additional Conservation Areas**
   With stakeholders, designing an expanded system of conservation areas that achieves targets whilst integrating commitments, exclusions, and preferences.
10. **Applying Conservation Actions to Selected Areas**
    Working through the technical and institutional tasks involved in applying effective conservation actions to areas identified in the conservation plan.
11. **Maintaining and Monitoring Established Conservation Areas**
    Applying and monitoring long-term management in established conservation areas to promote the persistence of the values for which they were identified.

Tools to facilitate systematic conservation planning include computer programs that enable large amounts of data to be analysed to generate cost-effective network designs. A commonly used such tool to assist in MPA selection is the software Marxan. This uses an optimization technique (simulated annealing) to minimize a combination of the cost of the MPA network and its boundary length whilst still achieving biodiversity objectives (Ball *et al.*, 2009). Marxan was used to assist with the rezoning of the Great Barrier Reef Marine Park (see Chapter 13). A further development, Marxan with Zones, is able to set priorities for MPAs with differing protection levels. Klein *et al.* (2010) used it to design a network of four types of MPAs with different fishing restrictions and biodiversity conservation targets along the coast of California. Their zoning configuration minimizes loss of fishing value without compromising conservation goals.

Obtaining adequate information on the extent of habitats can present significant challenges given the large areas that may be involved. Whilst a thorough, systematic approach to planning and implementing a network of representative MPAs may be the ideal, in reality there are often significant practical constraints. Carrying out detailed surveys and collating and analysing the required data are costly and time-consuming exercises if these data do not already exist and might delay management targets. In the face of such limitations, Hansen *et al.* (2011) found that an opportunistic approach based on community input, combined if possible with at least coarse biophysical data, could still deliver valuable MPAs. They advise that 'a lack of data suitable for systematic planning should not delay conservation decisions, as long as conservation action is taken with integrity, transparency, and an acknowledgement of uncertainties'. In this case, planning around basic ecosystem processes and ensuring adequate representation of basic habitat types can result in both biodiversity and fisheries gains at the regional scale (e.g. Wing & Jack, 2014).

## Community Involvement

Establishing MPAs depends not only on the gathering and appraising of biophysical information; there are also important socio-economic dimensions. The success of MPAs depends a great deal on social factors – how user groups may be affected by MPAs and the extent to which they are involved in setting up and managing them. Not properly taking account of the social aspects is likely to undermine the effectiveness of an MPA at the outset (Agardy *et al.*, 2011). It is to be expected that local communities will be encouraged to support the conservation objectives of protected areas if they are actively involved and have a vested interest in their success. This is a central tenet of the biosphere reserves programme (see below) and has been subsequently promoted in other international initiatives concerned with sustainable use, including the World Conservation Strategy and the Convention on Biological Diversity.

Community involvement in MPA programmes can take place at various stages and levels, from consultation and collaboration with regulatory authorities, identification and promotion of MPAs, continued traditional use of areas by local people, to community-based reserves (Wells & White, 1995). Alcala *et al.* (2006) compared MPAs in the Philippines that were co-managed by the local communities and local agencies with centrally managed MPAs where there was minimal involvement of local communities. Both groups of MPAs were similar in improving fisheries and biodiversity, but the community-based MPAs tended to be more successful in managing user conflicts. The main problems with centrally managed MPAs were that the fishers and local community were excluded from decision making and felt they did not benefit from MPAs. In many parts of the world, especially in developing nations, setting aside areas entirely free from human use may be unrealistic, and management may need to permit some sustainable use by local populations that is compatible as far as possible with conservation goals. This may be achievable within multiple-use MPAs that also include strictly protected core areas.

Potential MPAs may be in areas with a long history of traditional use by coastal communities, where marine environmental knowledge accumulated over generations can help in site selection and management. The continued welfare of affected people is an important aspect of MPA policy, as is a willingness to consider them as an integral part of any proposed new system of management. MPAs may provide opportunities to help preserve traditional fishing practices in a sustainable context (Johannes, 2002).

MPAs can attract large numbers of visitors and can have marked economic benefits to local communities in terms of tourism revenue and conservation status. The Leigh Marine Reserve in northern New Zealand has proved increasingly popular and attracts about 300 000 visitors per year, which is at least 10 times more than for nearby areas where fishing is permitted (Ballantine & Langlois, 2008). The danger is that such sites become victims of their own success and that through visitor numbers ecological values are compromised. Dive tourists respond positively to the greater abundance, variety, and size of fishes found in MPAs (Williams & Polunin, 2000). As one of the world's fastest-growing sports, diving is likely to put increasing pressure on MPAs that offer a rewarding diving experience, as at some heavily used coral reef sites (see Chapter 13).

Wolfenden *et al.* (1994) found, overall, a high level of public support for the establishment of MPAs in New Zealand, but respondents were less positive about having an MPA in their vicinity. In fact MPAs, in particular no-take reserves for biodiversity conservation, can generate strong opposition, with users concerned about possible impacts on fishing, livelihoods, and cultural traditions. Voyer *et al.* (2014) found that MPA opposition was often due to divergent views on objectives, with the agency's policy position focused on biodiversity protection whereas opponents focused on fisheries management, and furthermore perceived managers as being unreceptive to the local knowledge of fishers. A way to help overcome this and foster local support is to incorporate where possible both ecological and social objectives. Zoning may achieve this whilst still retaining no-take areas.

## Zoning

To cater for different objectives within an MPA requires a mix of management tools to regulate human uses. Although objectives need to be compatible, conflicts of interest can arise where ecological sensitivity of a particular area requires strict regulation of a potentially damaging impact. A form of spatial control that is commonly used to accommodate different uses is to assign subareas or zones for different functions. We saw an example of zoning in the case of the Great Barrier Reef Marine Park (see Chapter 13). Zoning may take many forms depending on the objectives of the MPA, its biota, and the type of pressures and uses to which the MPA is subject. Commonly employed is the concept of 'core' and 'buffer' areas (Laffoley, 1995; Salm *et al.*, 2000). The core represents a strictly protected area, typically for the conservation of representative environments and/or particularly vulnerable ecosystems, habitats, and species. Core areas may also be valuable as research or education sites where non-destructive research can be undertaken free from other human influences. A surrounding buffer area provides a zone where limitations on activities apply so as to safeguard the core from encroachment of damaging impacts and 'edge effects' – where bordering communities influence each other. Beyond this there may also be a transition area where, for instance, forms of sustainable development are pursued by the local population. But whether core areas can be adequately insulated from surrounding anthropogenic stresses may be questionable given the connectedness of marine systems. It may, for instance, be necessary to have a large buffer zone where there is strict enforcement of anti-pollution legislation. An important practical point is that zoning information and restrictions need to be readily understood and applied by users (Laffoley, 1995). These concerns highlight the vulnerability of marine resources to diffuse degradation from nutrients and toxins, sedimentation, and spread of pests, aspects that need to be managed at a broad spatial scale.

Vertical zoning may be considered, for instance, where pelagic fishing is allowed but where greater protection is sought for the benthic environment. Whilst some jurisdictions have introduced vertical zoning, it has significant drawbacks. It assumes that pelagic-benthic links are not important, and having different regulations at the same site may be hard to police (Dudley, 2008).

## Biosphere Reserves

The type of zoning described above, with core and buffer areas, is central to the concept of biosphere reserves, the first of which were designated in 1976 under UNESCO's Man and the Biosphere Programme. There are now more than 630, of which around 100 have marine or coastal components, although none is wholly marine. Biosphere reserves, as large multipurpose protected areas, are intended to fulfil three fundamental but integrated roles: conservation, research, and sustainable development (Batisse, 1990). The primary objective is the conservation of genetic resources and ecosystems on a world-wide basis, but each reserve should also be part of a network in an international programme of research and monitoring as well as being closely linked to the sustainable use of natural resources by the local populations. An undisturbed core area provides for strict protection of biota and possibly scientific monitoring. The surrounding buffer zone may be used for certain traditional human activities, tourism, education, and non-destructive research. And a transitional zone can provide for some forms of sustainable development by the local community as well as appropriate tourism and research activities. There is thus a transition from a central highly protected core to successive zones that permit increasing human activity (Fig. 15.4).

The biosphere reserve model can be well suited to application in the coastal environment where human use and ecosystem processes often conflict and integrated planning is crucial. Sustainable use of natural resources within buffer and transition zones is perhaps the most difficult of the roles to realize for biosphere reserves, especially for coastal reserves, which have the problem of being

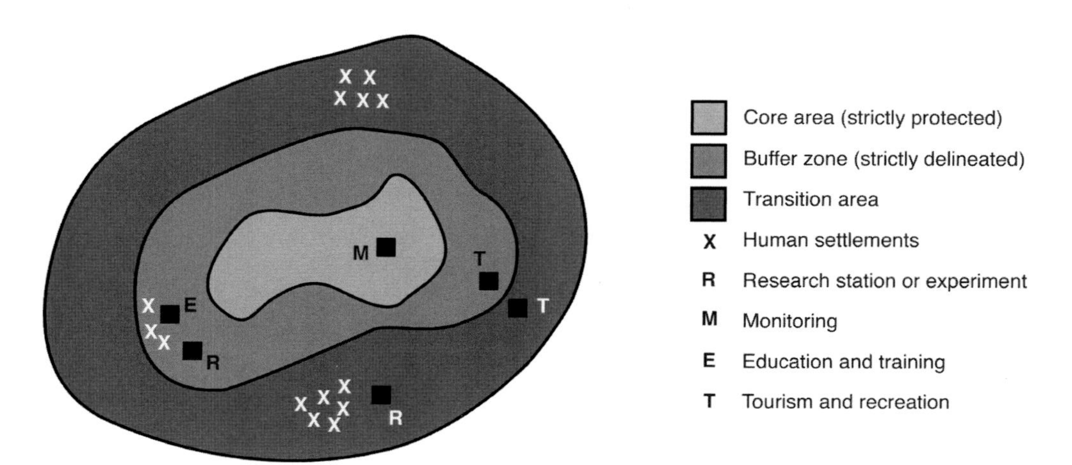


**Fig. 15.4** Schematic zonation of a biosphere reserve. From Batisse, M.,Development and implementation of the Biosphere Reserve concept and its applicability to coastal regions. *Environmental Conservation*, 17 (2), 111–16, 1990, reproduced with permission.

on an interface where scientific and administrative interests and responsibilities tend to be split. Nevertheless, such core areas should encompass both terrestrial and marine components so that the target area is a truly functional unit where the integrity of ecological processes can be secured as far as possible (Agardy, 1997). This offers a distinct advantage of being able to consider ecological processes that connect land and sea, such as catchments and water flow, diadromous species, and the flux of materials at the coastal margin.

Other international initiatives for establishing protected areas include the Ramsar Convention for the designation of wetland sites (see Chapter 11) and MARPOL, under which parties can identify special areas with a higher level of pollution protection (see Chapter 2). There are also protocols for designating MPAs under UNEP's Regional Seas Programme (see Chapter 16) and UNESCO's World Heritage Convention.

The World Heritage Convention came into force in 1975 for the designation of exceptional natural and/or cultural sites, including natural formations and areas of 'outstanding universal value' from the point of view of conservation. Only 5% of the more than 1050 World Heritage sites are recognized for their marine natural values, although some are very large (Box 15.2). The convention requires that sites have adequate conservation management systems in place, which are periodically monitored. Potential benefits of World Heritage listing include enhanced public awareness and funding. Criteria for listing have been biased towards terrestrial systems, but recent guidance on applying the criteria in the marine context should help achieve better representation of marine sites. Most of the marine World Heritage sites are tropical, and three-quarters of nearshore biogeographic provinces are not represented. Also, the convention targets sites that are under national jurisdiction; it currently has no mechanism for designating high-seas sites (Abdulla *et al.*, 2014).

## Effectiveness

An essential question is assessing the effectiveness of MPAs – to what extent are the objectives of an MPA being achieved? And on what time scales are results achieved? Studies have indicated that where there is sufficient information to measure performance of MPAs (the minority of MPAs), then at most a third might be judged as achieving management objectives (Pomeroy *et al.*, 2005). Common reasons for failure include insufficient financial and technical resources, lack of data for management decisions, lack of public support, inadequate commitment to enforcing management, and the impact of external factors such as pollution and overexploitation (Kelleher *et al.*, 1995). Several initiatives for assessing MPA effectiveness have now been developed (Fox *et al.*, 2014). The methodology of Pomeroy *et al.* (2005) identifies a set of biophysical, socio-economic, and governance indicators that can be selected and adapted according to the evaluation needs for the MPA. Objective evaluations can help pinpoint specific issues that are affecting goals being attained and improve performance through adaptive management.

Assessing the protection effects of MPAs requires, ideally, adequate information on the site before the MPA is established and monitoring once it is in place. Adequacy of data can often, however, be an issue, particularly in obtaining 'before' data. Where practicable, a suitable sampling programme needs to be instituted, such as a Before-After-Control-Impact design (Osenberg *et al.*, 2011; Chapter 2).

## Box 15.2 World Heritage Convention Sites

There are currently about 50 sites on the World Heritage list that have been designated primarily for their outstanding marine natural features (http://whc.unesco.org/en/marine-programme/). These include:

**Aldabra Atoll (designated in 1982, 350 $km^2$)**
One of the world's largest atolls, in the western Indian Ocean, relatively free of human impact, with important coral reefs, seagrass beds, mangroves, seabird, shorebird, and turtle populations.

**Banc d'Arguin (1989, 12 000 $km^2$)**
On the coast of Mauritania, west Africa, globally significant for its coastal wetlands and migratory waders.

**Galápagos Islands (1978, 140 665 $km^2$)**
An archipelago with complex oceanography in the eastern equatorial Pacific with a highly distinctive marine biota. Conspicuous species include endemic species of marine iguana, penguin, cormorant, fur seal, and sea lion (all threatened).

**Great Barrier Reef (1981, 348 700 $km^2$)**
The world's largest coral reef system (see Chapter 13).

**Papahãnaumokuãkea (2010, 362 075 $km^2$)**
A very large MPA encompassing the northwestern Hawaiian islands with a diversity of habitats including abyssal areas, seamounts, coral reefs, lagoons, and many important species including the Hawaiian monk seal (endangered), Laysan albatross, and black-footed albatross.

**Shark Bay, Western Australia (1991, 22 000 $km^2$)**
Renowned for its seagrass beds, dugongs, and stromatolites (see Chapter 11).

**Subantarctic Islands**
The World Heritage list includes several subantarctic islands notable (among other things) for their seabirds (see Chapter 8).

**Sundarbans (1987, 1400 $km^2$)**
The world's largest mangrove forests in the Bay of Bengal (see Chapter 12).

**Wadden Sea (2009, 11 430 $km^2$)**
The world's largest unbroken system of intertidal sediment flats encompassing a variety of depositional habitats including extensive eelgrass and saltmarsh areas and highly significant for the millions of migratory waders it supports.

## Very Large MPAs

Established in 1975 along Australia's Queensland coast, the Great Barrier Reef Marine Park covering 344 000 $km^2$ was for many years by far the largest MPA (see Chapter 13). But since 2000, other very large MPAs have been designated, mostly no-take MPAs established within national jurisdictions. There are now several larger than 100 000 $km^2$ and a few that exceed 1 million $km^2$, such as the Natural Park of the Coral Sea, a 1.3 million $km^2$ multi-use MPA established by New Caledonia in 2014, and the Pitcairn Marine Reserve (834 000 $km^2$), a no-take MPA (apart from small-scale traditional fishing by the local population).

Very large MPAs have obvious attractions. They can potentially encompass entire ecosystems, can contain extensive areas of deep-sea and open-ocean habitats, including mobile pelagic habitats such as eddies and upwelling systems, and benefit wide-ranging species. They have also been established in areas that are among the least impacted by human activity, supporting relatively intact ecosystems, and their size renders them more resilient to large-scale disturbances and reduces edge effects. They need to include extensive no-take areas to be of most use to biodiversity conservation, but they could be multiple-use MPAs and zoned to also accommodate, for instance, opportunities for indigenous groups to pursue traditional practices (Toonen *et al.*, 2013; Wilhelm *et al.*, 2014). Young *et al.* (2015) demonstrate the potential for large pelagic MPAs to protect foraging habitat for highly mobile pelagic predators, in this case three species of booby in the central Pacific.

A factor driving the establishment of very large MPAs is the pressure on states to meet international conservation obligations, in particular the CBD target of at least 10% of coastal and marine areas to be protected by 2020 (Leenhardt *et al.*, 2013). The designation of very large protected areas has certainly greatly increased the global MPA estate. They now account for more than 80% of the world's marine area under some form of protection, although not necessarily as no-take areas (Toonen *et al.*, 2013). Huge MPAs have been sited in remote locations where they are likely to face limited opposition, and there is concern that they may be established more for political than scientific reasons. Other concerns include the difficulties and cost of adequate surveillance and enforcement over vast ocean areas. Small island nations may not have the resources to manage such MPAs effectively and to prevent illegal fishing (Wilhelm *et al.*, 2014).

Particularly contentious has been the UK government's proclamation of the Chagos MPA in 2010, a no-take reserve of 640 000 $km^2$ in the middle of the Indian Ocean. The Chagos archipelago is regarded as highly significant for its marine biodiversity, containing, for example, 25–50% of the Indian Ocean's healthy coral reefs and important breeding sites for seabirds and turtles (Koldewey *et al.*, 2010; Sheppard *et al.*, 2012). Mauritius, however, disputes UK sovereignty of the islands and legality of the MPA. The UK expelled the native Chagossians in the late 1960s to early 1970s to allow for development of a US military base on Diego Garcia. But if allowed to return they would be prohibited by the MPA from supporting themselves from local marine resources. In 2015 the Permanent Court of Arbitration (an intergovernmental organization concerned with international dispute resolution) found the declaration of Chagos MPA to be illegal under UNCLOS and that the UK had failed to properly consider Mauritian rights in exercising sovereignty. The UK and Mauritius will need to renegotiate. The finding has implications for future large-scale MPAs, in particular the need to take account of possible third-party rights in affected areas (Appleby, 2015).

## High-Seas MPAs

Efforts at establishing MPAs have focused very largely on coastal areas, where indeed human impacts are generally most severe (see Chapter 10). The high seas (used here to include the international seabed, the Area) have received far less attention. Although the high seas cover about 60% of the global ocean, less than 1% of their area lies within MPAs (Thomas *et al.*, 2014). Yet as we have seen, open-ocean and deep-sea communities and habitats are increasingly being impacted by human activities, especially by fishing, and there is growing interest in designating MPAs to protect vulnerable deep-sea benthic habitats (see Chapter 14).

The potential role of MPAs specifically to protect pelagic environments has, on the other hand, largely been overlooked or discounted (Game *et al.*, 2009). Pelagic MPAs could, however, be used to protect the biodiversity of distinct habitats, such as fronts, upwellings, or eddies, which although not static features can be sufficiently predictable in space and time to be delineated. For example, the Pelagos Sanctuary (see Chapter 9), a largely high-seas MPA in the NW Mediterranean, contains the Ligurian gyre (Guidetti *et al.*, 2013). Also, closures could be applied seasonally if appropriate. But even MPAs that track a pelagic feature or a species' migration route are not unrealistic. Mobile fisheries closures are already in use with up-to-date information relayed to vessels. Also effective for highly mobile species may be MPAs that target critical habitat, such as spawning or nursery sites where species are stationary for a period of time. Such locations may constitute a very small proportion of a species' total range, but their protection could greatly reduce overall mortality (Game *et al.*, 2009; Breen *et al.*, 2015).

Designing any representative network of MPAs requires a suitable biogeographic framework, in particular a hierarchical classification ranging from large-scale realms and provinces to local ecoregions and habitats (see Chapter 1). However, this represents a considerably greater challenge in the high seas than for coastal regions given their vast extent and scant biological information and depends very largely on using surrogate biophysical variables (see Chapter 10). Spalding *et al.* (2012) have provided a broad-scale classification of pelagic biogeographic provinces of open-ocean waters (see Chapter 1), and for the benthic environment, major biogeographic provinces have been proposed for the deep-ocean floor (Watling *et al.*, 2013; Chapter 14). To add further structure to these broad bioregions for the seabed, Harris and Whiteway (2009) used multivariate analysis of global biophysical data (depth, seabed slope, sediment thickness, primary production, bottom water dissolved oxygen, and bottom temperature) to subdivide the deep ocean floor into 11 categories, or 'seascapes', as a basis for designing MPA networks in the high seas.

In planning a global network of reserves for the high seas, Roberts *et al.* (2006) adopted a minimum reserve size of 5° latitude × 5° longitude, which they regarded as the minimum viable area for a high-seas reserve. At the equator this is an area of about 314 000 km$^2$. Their network of reserves covers 40% of the global ocean and includes 29 candidate reserves that represent all ocean biogeographic zones.

Issues of physical and biological complexity and design are perceived as significant barriers to establishing high-seas MPAs. Other major challenges concern enforcement – notably the problem of illegal fishing – and governance (Game *et al.*, 2009; Hobday *et al.*, 2011). There is currently no single body with the global mandate to establish MPAs on the high seas. Various arrangements do exist, but the coverage is piecemeal and not consistent (Ardron *et al.*, 2008). Some regional bodies have designated MPAs, responding under Article 118 of UNCLOS that calls on states to cooperate in the conservation and management of living resources of the high seas. Six high-seas MPAs were designated in 2010 under the regional seas agreement for the NE Atlantic (the OSPAR Convention), although this was only made possible as the regional marine fisheries organization (North East Atlantic Fisheries Commission) had closed the areas to bottom fishing. However, only four of the regional seas agreements include areas beyond national jurisdiction, the RFMO coverage is incomplete, and measures adopted by regional bodies are only binding on the contracting parties (Gjerde & Rulska-Domino, 2012).

A new multilateral agreement under UNCLOS, comparable to the UN Fish Stocks Agreement (see Chapter 5), would provide the legal mandate for integrated ecosystem-based management of the world's oceans, including the establishment of representative MPA networks (Allsopp *et al.*, 2009; Gjerde & Rulska-Domino, 2012). White and Costello (2014) suggest that a new legal agreement could include making the entire high seas a no-take MPA, at least for highly mobile species, and that this would in fact be compatible with the objectives of UNCLOS in terms of resource management and ecosystem protection. Their bioeconomic model indicates that closing the high seas to fishing would benefit fisheries overall, by more than 30% in terms of fisheries yields. The high seas would act as a refuge for migratory stocks, and as these rebuilt, spillover would considerably enhance fishing in EEZs. (An analysis by Rogers *et al.* (2014) indicates that less than 1% of the global commercially important fish species are caught exclusively on the high seas.) The UN has started negotiating a global treaty for conservation of the high seas, including the use of MPAs (Chapter 18). In the meantime, despite their shortcomings, regional agreements still need to be pursued, as do other initiatives, such as those under the Convention on Biological Diversity for identifying 'ecologically or biologically significant marine areas' (EBSAs) in the open ocean and deep sea (see Chapter 14), and the obligation of parties to cooperate in the conservation of biodiversity in areas beyond national jurisdiction (Gjerde & Rulska-Domino, 2012; Ban *et al.*, 2014).

## Marine Spatial Planning

Protected areas that are appropriately implemented and managed can play a key role in the conservation of marine ecosystems. MPAs are not, however, sufficient on their own to meet broad-scale objectives for ecosystem management (Allison *et al.*, 1998; Boersma & Parrish, 1999; Mora & Sale, 2011). Marine environments are subject to many pressures, and their management and conservation call for a variety of approaches and instruments. Establishing and managing MPAs may be a futile exercise unless embedded within regimes that provide for integrated management of more diffuse stressors from the adjacent land and sea areas (Kelleher *et al.*, 1995). MPAs are subject to potentially damaging influences that are outside the area of management, such as upstream sources of pollutants, illegal fishing, introduced species, and climate change (Mora & Sale, 2011; Partelow *et al.*, 2015). Their effectiveness may be significantly curtailed by external stresses.

The use of MPAs needs to be integrated into wider, ecosystem-based approaches to management that seek to balance human uses and the objectives of environmental and ecosystem protection. One such approach is integrated coastal zone management, which many states have adopted (ICZM; Chapter 10). In this regard, Cicin-Sain and Belfiore (2005) provide a series of guiding principles to facilitate the management of MPAs within this broader context of governance. However, although originally envisioned as extending from costal watersheds to the edge of the continental shelf, most ICZM programmes have focused on a much narrower zone along the land-sea interface. A more comprehensive ecosystem-based management tool that has emerged more recently is marine spatial planning (MSP), defined as 'a public process of analysing and allocating the spatial and temporal distribution of human activities in marine areas to achieve ecological, economic, and social objectives that are usually specified through a political process' (Ehler & Douvere, 2009). Individual sectoral interests, such as transportation, mining, fisheries, aquaculture, and conservation, have tended to drive the use of ocean space, whereas MSP seeks instead to accommodate multiple uses, typically by means of zoning plans (Fig. 15.5). The origins of MSP date back to the management approach

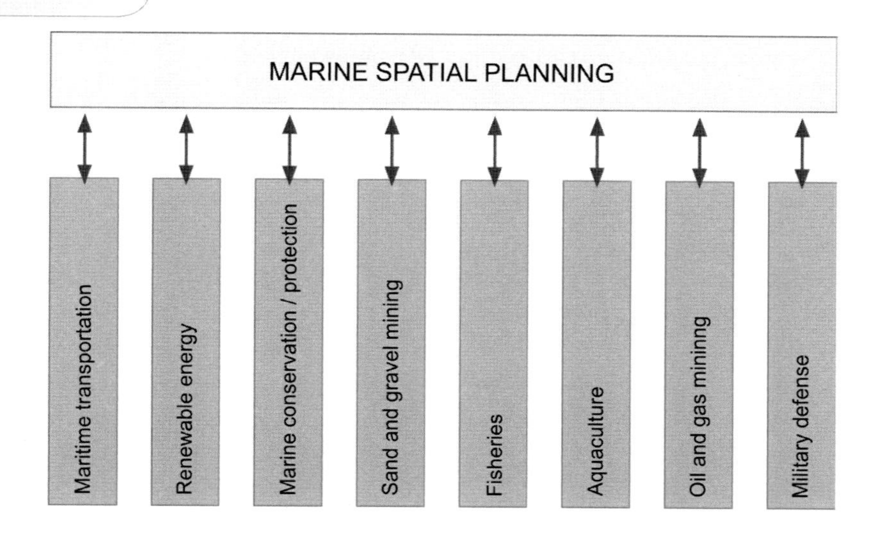


**Fig. 15.5** Marine spatial planning and single sector planning. Reproduced from Ehler, C. & Douvere, F. (2009). *Marine Spatial Planning: A Step-by-Step Approach Toward Ecosystem-Based Management*. Intergovernmental Oceanographic Commission and Man and the Biosphere Programme. IOC Manual and Guides No. 53, ICAM Dossier No. 6. Paris: UNESCO.

adopted in setting up the Great Barrier Reef Marine Park in 1975 and its subsequent rezoning (see Chapter 13) – multi-use whilst maintaining ecosystem health. Numerous countries have since developed MSP programmes. Foley *et al.* (2010) present core ecological principles for supporting the goals of ecosystem-based MSP, notably the need to maintain species and habitat diversity, key species, and connectivity. In their model, these ecological principles, together with social, economic, and governance principles, are used to designate the siting and use of ocean space and help deliver the overall goals of healthy ecosystems, delivery of ecosystem services, and sustainable uses (Fig. 15.6).

Marine protected areas have a variety of potential roles to play in marine conservation, although much research is still needed to establish how to use them most effectively, such as how best to design MPAs and configure networks, the use of no-take and multi-purpose MPAs, the contribution that reserves can make to adjacent areas in terms of spillover and larval export, and the potential role of flexible and adaptive management strategies. Also, MPAs range from many small coastal reserves to vast open-ocean areas. How are these best incorporated into more holistic approaches to managing use of marine space? In the next chapter we explore further the theme of large-scale management by examining regional-scale programmes and initiatives for marine environmental protection.

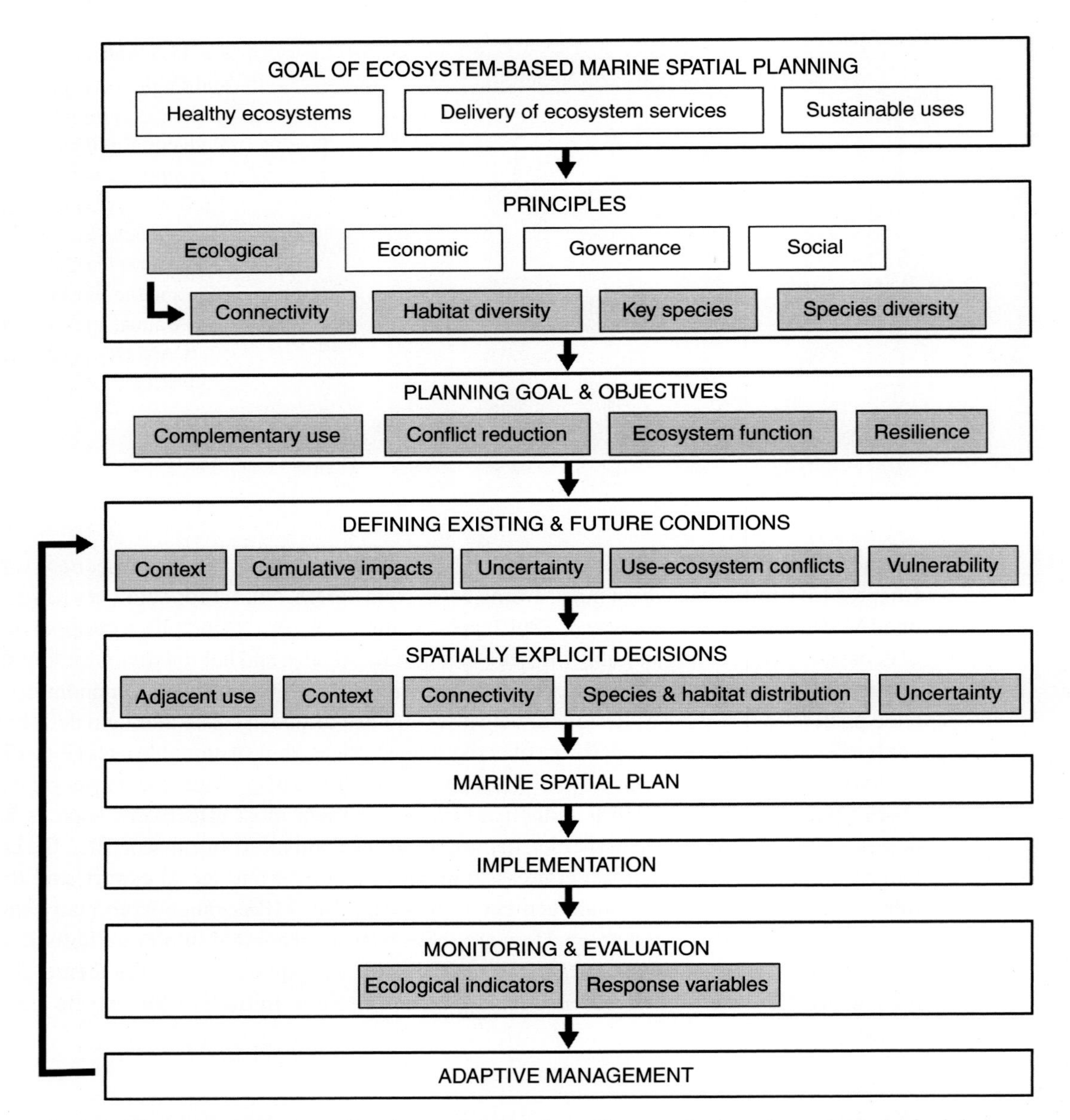


**Fig. 15.6** The main components of marine spatial planning with an emphasis on the ecological principles. Boxes that relate specifically to an ecosystem-based approach are shaded grey. Similar diagrams would be developed for the economic, governance, and social principles to provide a comprehensive marine spatial plan. Reprinted from *Marine Policy*, Vol. 34, Foley, M.M., Halpern, B.S., Micheli, F., et al., Guiding ecological principles for marine spatial planning, pages 955–66, copyright 2010, with permission from Elsevier.

## REFERENCES


Abdulla, A., Obura, D., Bertzky, B. & Shi, Y. (2014). Marine World Heritage: creating a globally more balanced and representative list. *Aquatic Conservation: Marine and Freshwater Ecosystems*, **24** (Suppl. 2), 59–74.

Adams, A.B. (ed.) (1962). *First World Conference on National Parks. Proceedings of a conference organized by the International Union for Conservation of Nature and Natural Resources*. Seattle, Washington, 30 June–7 July 1962. Washington, DC: National Park Service, US Department of the Interior.

Agardy, T.S. (1997). *Marine Protected Areas and Ocean Conservation*. Georgetown, TX: R.G. Landes & San Francisco, CA: Academic Press.

Agardy, T., Notarbartolo di Sciara, G. & Christie, P. (2011). Mind the gap: addressing the shortcomings of marine protected areas through large scale marine spatial planning. *Marine Policy*, **35**, 226–32.

Al-Abdulrazzak, D. & Trombulak, S.C. (2012). Classifying levels of protection in Marine Protected Areas. *Marine Policy*, **36**, 576–82.

Alcala, A.C., Russ, G.R. & Nillos P. (2006). Collaborative and community-based conservation of coral reefs, with reference to marine reserves in the Philippines. In *Coral Reef Conservation*, ed. I.M. Coté & J.D. Reynolds, pp. 392–418. Cambridge, UK: Cambridge University Press.

Allison, G.W., Lubchenco, J. & Carr, M.H. (1998). Marine reserves are necessary but not sufficient for marine conservation. *Ecological Applications*, **8** (Suppl. 1), S79–S92.

Allsopp, M., Page, R., Johnston, P. & Santillo, D. (2009). *State of the World's Oceans*. London: Springer.

Appleby, T. (2015). The Chagos marine protected arbitration—a battle of four losers? *Journal of Environmental Law*, **27**, 529–40.

Ardron, J., Gjerde, K., Pullen, S. & Tilot, V. (2008). Marine spatial planning in the high seas. *Marine Policy*, **32**, 832–9.

Babcock, R.C., Shears, N.T., Alcala, A.C., *et al.* (2010). Decadal trends in marine reserves reveal differential rates of change in direct and indirect effects. *Proceedings of the National Academy of Sciences*, **107**, 18256–61.

Ball, I.R., Possingham, H.P. & Watts, M.E. (2009). Marxan and relatives: software for spatial conservation prioritization. In *Spatial Conservation Prioritization: Quantitative Methods and Computational Tools*, ed. A. Moilanen, K.A. Wilson & H. Possingham, pp. 185–95. Oxford, UK: Oxford University Press.

Ballantine, B. (2014). Fifty years on: lessons from marine reserves in New Zealand and principles for a worldwide network. *Biological Conservation*, **176**, 297–307.

Ballantine, W.J. & Langlois, T.J. (2008). Marine reserves: the need for systems. *Hydrobiologia*, **606**, 35–44.

Ban, N.C., Bax, N.J., Gjerde, K.M., *et al.* (2014). Systematic conservation planning: a better recipe for managing the high seas for biodiversity conservation and sustainable use. *Conservation Letters*, **7**, 41–54.

Banks, S.A. & Skilleter, G.A. (2010). Implementing marine reserve networks: a comparison of approaches in New South Wales (Australia) and New Zealand. *Marine Policy*, **34**, 197–207.

Batisse, M. (1990). Development and implementation of the Biosphere Reserve concept and its applicability to coastal regions. *Environmental Conservation*, **17**, 111–16.

Boersma, P.D. & Parrish, J.K. (1999). Limiting abuse: marine protected areas, a limited solution. *Ecological Economics*, **31**, 287–304.

Breen, P., Posen, P. & Righton, D. (2015). Temperate marine protected areas and highly mobile fish: a review. *Ocean & Coastal Management*, **105**, 75–83.

CBD (Secretariat of the Convention on Biological Diversity) (2003). *Interlinkages between Biological Diversity and Climate Change. Advice on the Integration of Biodiversity Considerations into the Implementation of the United Nations Framework Convention on Climate Change and Its Kyoto Protocol*. Montreal, Canada: SCBD. *CBD Technical Series, No. 10,* 154 pp.

CBD (2011). *Convention on Biological Diversity COP 10 Decision X/2: Strategic Plan for Biodiversity 2011–2020*. www.cbd.int/decision/cop/?id=12268.

Christie, M.R., Tissot, B.N., Albins, M.A., *et al.* (2010). Larval connectivity in an effective network of marine protected areas. *PLoS ONE*, **5**, e15715.

Cicin-Sain, B. & Belfiore, S. (2005). Linking marine protected areas to integrated coastal and ocean management: a review of theory and practice. *Ocean & Coastal Management*, **48**, 847–68.

Claudet, J. (ed.) (2011). *Marine Protected Areas: A Multidisciplinary Approach*. Cambridge, UK: Cambridge University Press.

Costello, M.J. & Ballantine, B. (2015). Biodiversity conservation should focus on no-take Marine Reserves: 94% of Marine Protected Areas allow fishing. *Trends in Ecology & Evolution*, **30**, 507–9.

Day J., Dudley N., Hockings M., *et al.* (2012). *Guidelines for Applying the IUCN Protected Area Management Categories to Marine Protected Areas*. Gland, Switzerland: IUCN. 36 pp.

Dudley, N. (ed.) (2008). *Guidelines for Applying Protected Area Management Categories*. Gland, Switzerland: IUCN.

Edgar, G.J., Stuart-Smith, R.D., Willis, T.J., *et al.* (2014). Global conservation outcomes depend on marine protected areas with five key features. *Nature*, **506**, 216–20.

Ehler, C. & Douvere, F. (2009). *Marine Spatial Planning: A Step-by-Step Approach Toward Ecosystem-Based Management*. Intergovernmental Oceanographic Commission and Man and the Biosphere Programme. IOC Manual and Guides No. 53, ICAM Dossier No. 6. Paris: UNESCO.

Foley, M.M., Halpern, B.S., Micheli, F., *et al.* (2010). Guiding ecological principles for marine spatial planning. *Marine Policy*, **34**, 955–66.

Fox, H.E., Holtzman, J.L., Haisfield, K.M., *et al.* (2014). How are our MPAs doing? Challenges in assessing global patterns in marine protected area performance. *Coastal Management*, **42**, 207–26.

Game, E.T., Grantham, H.S., Hobday, A.J., *et al.* (2009). Pelagic protected areas: the missing dimension in ocean conservation. *Trends in Ecology and Evolution*, **24**, 360–9.

Gjerde, K.M. & Rulska-Domino, A. (2012). Marine protected areas beyond national jurisdiction: some practical perspectives for moving ahead. *The International Journal of Marine and Coastal Law*, **27**, 351–73.

Green, A.L., Fernandes, L., Almany, G., *et al.* (2014). Designing marine reserves for fisheries management, biodiversity conservation, and climate change adaptation. *Coastal Management*, **42**, 143–59.

Guidetti, P., Notarbartolo di Sciara, G. & Agardy, T. (2013). Integrating pelagic and coastal MPAs into large-scale ecosystem-wide management. *Aquatic Conservation: Marine and Freshwater Ecosystems*, **23**, 179–82.

Hansen, G.J.A., Ban, N.C., Jones, M.L., *et al.* (2011). Hindsight in marine protected area selection: a comparison of ecological representation arising from opportunistic and systematic approaches. *Biological Conservation*, **144**, 1866–75.

Harris, P.T. & Whiteway, T. (2009). High seas marine protected areas: benthic environmental conservation priorities from a GIS analysis of global ocean biophysical data. *Ocean and Coastal Management*, **52**, 22–38.

Harrison, H.B., Williamson, D.H., Evans, R.D., *et al.* (2012). Larval export from marine reserves and the recruitment benefit for fish and fisheries. *Current Biology*, **22**, 1023–8.

Hiscock, K. (2014). *Marine Biodiversity Conservation: A Practical Approach*. Abingdon, UK: Routledge.

Hobday, A.J., Game, E.T., Grantham, H.S. & Richardson, A.J. (2011). Conserving the largest habitat on Earth: protected areas in the pelagic ocean. In *Marine Protected Areas: a Multidisciplinary Approach*, ed. J. Claudet, pp.347–72. Cambridge, UK: Cambridge University Press.

IUCN-WCPA (2008). *Establishing Marine Protected Area Networks—Making It Happen*. Washington, DC: IUCN-WCPA, National Oceanic and Atmospheric Administration and The Nature Conservancy. 118 pp.

Jack, L. & Wing, S.R. (2010). Maintenance of old-growth size structure and fecundity of the red rock lobster *Jasus edwardsii* among marine protected areas in Fiordland, New Zealand. *Marine Ecology Progress Series*, **404**, 161–72.

Johannes, R.E. (2002). The renaissance of community-based marine resource management in Oceania. *Annual Review of Ecology and Systematics*, **33**, 317–40.

Jones, G.P., Srinivasan, M. & Almany, G.R. (2007). Population connectivity and conservation of marine biodiversity. *Oceanography*, **20**, 101–11.

Jones, P.J.S. (1994). A review and analysis of the objectives of marine nature reserves. *Ocean & Coastal Management*, **24**, 149–78.

Kelleher, G. (1999). *Guidelines for Marine Protected Areas.* Gland, Switzerland and Cambridge, UK: IUCN.

Kelleher, G., Bleakley, C. & Wells, S. (eds.) (1995). *A Global Representative System of Marine Protected Areas* (Volumes I–IV). Washington, DC: The World Bank.

Kelleher, G. & Kenchington, R. (1992). *Guidelines for Establishing Marine Protected Areas: A Marine Conservation and Development Report*. Gland, Switzerland: IUCN.

Klein, C.J., Steinback, C., Watts, M., Scholz, A.J. & Possingham, H.P. (2010). Spatial marine zoning for fisheries and conservation. *Frontiers in Ecology and the Environment*, **8**, 349–53.

Koldewey, H.J., Curnick, D., Harding, S., Harrison, L.R. & Gollock, M. (2010). Potential benefits to fisheries and biodiversity of the Chagos Archipelago/British Indian Ocean Territory as a no-take marine reserve. *Marine Pollution Bulletin*, **60**, 1906–15.

Laffoley, D. (1995). Techniques for managing marine protected areas: zoning. In *Marine Protected Areas: Principles and Techniques for Management*, ed. S. Gubbay, pp. 103–18. London: Chapman & Hall.

Leenhardt, P., Cazalet, B., Salvat, B., Claudet, J. & Feral, F. (2013). The rise of large-scale marine protected areas: conservation or geopolitics? *Ocean & Coastal Management*, **85**, 112–18.

Leleu, K., Remy-Zephir, B., Grace, R. & Costello, M.J. (2012). Mapping habitats in a marine reserve showed how a 30-year trophic cascade altered ecosystem structure. *Biological Conservation*, **155**, 193–201.

Lester, S.E., Halpern, B.S., Grorud-Colvert, K., *et al.* (2009). Biological effects within no-take marine reserves: a global synthesis. *Marine Ecology Progress Series*, **384**, 33–46.

Lubchenco, J. & Grorud-Colvert, K. (2015). Making waves: the science and politics of ocean protection. *Science*, **350**, 382–4.

Lubchenco, J., Palumbi, S.R., Gaines, S.D. & Andelman, S. (2003). Plugging a hole in the ocean: the emerging science of marine reserves. *Ecological Applications*, **13** (Suppl.), S3–S7.

Margules, C.R. & Pressey, R.L. (2000). Systematic conservation planning. *Nature*, **405**, 243–53.

McCauley, D.J., Pinsky, M.L., Palumbi, S.R., *et al.* (2015). Marine defaunation: animal loss in the global ocean. *Science*, **347**, 247–53.

McLeod, E., Salm, R., Green, A. & Almany, J. (2009). Designing marine protected area networks to address the impacts of climate change. *Frontiers in Ecology and the Environment*, **7**, 362–70.

Mora, C. & Sale, P.F. (2011). Ongoing global biodiversity loss and the need to move beyond protected areas: a review of the technical and practical shortcomings of protected areas on land and sea. *Marine Ecology Progress Series*, **434**, 251–66.

O'Leary, B.C., Winther-Janson, M., Bainbridge, J.M., *et al.* (2016). Effective coverage targets for ocean protection. *Conservation Letters*, **9**, 398–404.

Osenberg, C.W., Shima, J.S., Miller, S.L. & Stier, A.C. (2011). Assessing effects of marine protected areas: confounding in space and possible solutions. In *Marine Protected Areas: A Multidisciplinary Approach*, ed. J. Claudet, pp.143–67. Cambridge, UK: Cambridge University Press.

Partelow, S., von Wehrden, H. & Horn, O. (2015). Pollution exposure on marine protected areas: a global assessment. *Marine Pollution Bulletin*, **100**, 352–8.

Pomeroy, R.S., Watson, L.M., Parks, J.E. & Cid, G.A. (2005). How is your MPA doing? A methodology for evaluating the management effectiveness of marine protected areas. *Ocean & Coastal Management*, **48**, 485–502.

Pressey, R.L. & Bottrill, M.C. (2008). Opportunism, threats, and the evolution of systematic conservation planning. *Conservation Biology*, **22**, 1340–5.

Roberts, C.M., Mason, L., Hawkins, J.P., *et al.* (2006). *Roadmap to Recovery: A Global Network of Marine Reserves*. Amsterdam: Greenpeace International. 58 pp.

Roff, J. & Zacharias, M. (2011). *Marine Conservation Ecology*. London: Earthscan.

Rogers, A.D., Sumaila, U.R., Hussain, S.S. & Baulcomb, C. (2014). *The High Seas and Us: Understanding the Value of High-Seas Ecosystems*. Oxford, UK: Global Ocean Commission.

Sale, P.F., Cowen, R.K., Danilowicz, B.S., *et al.* (2005). Critical science gaps impede use of no-take fishery reserves. *Trends in Ecology and Evolution*, **20**, 74–80.

Salm, R.V., Clark, J. & Siirila, E. (2000). *Marine and Coastal Protected Areas: A Guide for Planners and Managers*, 3rd edn. Washington, DC: IUCN.

Shears, N.T. & Babcock, R.C. (2003). Continuing trophic cascade effects after 25 years of no-take marine reserve protection. *Marine Ecology Progress Series*, **246**, 1–16.

Shears, N.T., Smith, F., Babcock, R.C., Duffy, C.A.J. & Villouta, E. (2008). Evaluation of biogeographic classification schemes for conservation planning: application to New Zealand's coastal marine environment. *Conservation Biology*, **22**, 467–81.

Sheppard, C.R.C., Ateweberhan, M., Bowen, B.W., *et al.* (2012). Reefs and islands of the Chagos Archipelago, Indian Ocean: why it is the world's largest no-take marine protected area. *Aquatic Conservation: Marine Freshwater Ecosystems*, **22**, 232–61.

Sobel, J. & Dahlgren, C. (2004). *Marine Reserves: A Guide to Science, Design, and Use*. Washington, DC: Island Press.

Spalding, M.D., Agostini, V.N., Rice, J. & Grant, S.M. (2012). Pelagic provinces of the world: a biogeographic classification of the world's surface pelagic waters. *Ocean & Coastal Management*, **60**, 19–30.

Taylor, B.M. & McIlwain, J.L. (2010). Beyond abundance and biomass: effects of marine protected areas on the demography of a highly exploited reef fish. *Marine Ecology Progress Series*, **411**, 243–58.

Thomas, H.L., MacSharry, B., Morgan, L., *et al.* (2014). Evaluating official marine protected area coverage for Aichi Target 11: appraising the data and methods that define our progress. *Aquatic Conservation: Marine and Freshwater Ecosystems*, **24** (Suppl. 2), 8–23.

Tisdell, C. & Broadus, J.M. (1989). Policy issues related to the establishment and management of marine reserves. *Coastal Management*, **17**, 37–53.

Toonen, R.J., Wilhelm, T.A., Maxwell, S.M., *et al.* (2013). One size does not fit all: the emerging frontier in large-scale marine conservation. *Marine Pollution Bulletin*, **77**, 7–10.

Voyer, M., Gladstone, W. & Goodall, H. (2014). Understanding marine park opposition: the relationship between social impacts, environmental knowledge and motivation to fish. *Aquatic Conservation: Marine and Freshwater Ecosystems*, **24**, 441–62.

Wabnitz, C.C.C., Andréfouët, S. & Muller-Karger, F.E. (2009). Measuring progress toward global marine conservation targets. *Frontiers in Ecology and the Environment*, **8**, 124–9.

Watling, L., Guinotte, J., Clark, M.R. & Smith, C.R. (2013). A proposed biogeography of the deep ocean floor. *Progress in Oceanography*, **111**, 91–112.

Watson, J.E.M., Dudley, N., Segan, D.B. & Hockings, M. (2014). The performance and potential of protected areas. *Nature*, **515**, 67–73.

Wells, S. & White, A.T. (1995). Involving the community. In *Marine Protected Areas: Principles and Techniques for Management*, ed. S. Gubbay, pp. 61–84. London: Chapman & Hall.

White, C. & Costello, C. (2014). Close the high seas to fishing? *PLOS Biology*, **12**, e1001826.

Wilhelm, T.A., Sheppard, C.R.C., Sheppard, A.L.S., *et al.* (2014). Large marine protected areas – advantages and challenges of going big. *Aquatic Conservation: Marine and Freshwater Ecosystems*, **24** (Suppl. 2), 24–30.

Williams, I.D. & Polunin, N.V.C. (2000). Differences between protected and unprotected reefs of the western Caribbean in attributes preferred by dive tourists. *Environmental Conservation*, **27**, 382–91.

Wing, S.R. & Jack, L. (2013). Marine reserve networks conserve biodiversity by stabilizing communities and maintaining food web structure. *Ecosphere*, **4**, article 135.

Wing, S.R. & Jack, L. (2014). Fiordland: the ecological basis for ecosystem management. *New Zealand Journal of Marine and Freshwater Research*, **48**, 577–93.

Wolfenden, J., Cram, F. & Kirkwood, B. (1994). Marine reserves in New Zealand: a survey of community reactions. *Ocean & Coastal Management*, **25**, 31–51.

Young, H.S., Maxwell, S.M., Conners, M.G. & Shaffer, S.A. (2015). Pelagic marine protected areas protect foraging habitat for multiple breeding seabirds in the central Pacific. *Biological Conservation*, **181**, 226–35.

# 16 The Mediterranean Sea

In previous chapters we looked at marine conservation mainly with regard to particular species and habitats. Whilst this reflects where much of the effort in marine conservation has been directed, we also need to consider more holistic approaches that take account of the range of species and habitats within regions and the unifying factors of oceanography and biogeography. The need for regional approaches to management and conservation was endorsed under UNCLOS. Article 197 calls for states to co-operate on a global and, as appropriate, on a regional basis for the protection and preservation of the marine environment. Articles 122–3 apply in particular to states bordering an enclosed or semi-enclosed sea 'to coordinate the management, conservation, exploration and exploitation of the living resources of the sea'. An important regional approach in this regard is the Regional Seas Programme of the United Nations Environment Programme (UNEP).

## Regional Seas Programme

The Regional Seas Programme has its origins in the 1972 UN Conference on the Human Environment (see Chapter 3) and was launched in 1974. The programme provides a mechanism for neighbouring coastal nations to protect their shared marine environment. Given the pressing need to combat pollution of coastal waters, the programme has focused on coastal areas, especially enclosed or semi-enclosed seas. The programme involves more than 140 countries and covers 18 regions, 13 of them established under UNEP and the other five partner programmes established independently (Table 16.1).

Central to each programme is an action plan that sets out an environmental protection strategy tailored to the issues and priorities of the region in question. An action plan typically covers:

- Environmental assessment (evaluating the environmental problems of the region, their causes, magnitude, and impacts);
- Environmental management (including proposed management of coastal habitats, pollution control, and response);
- Environmental legislation (usually provided by a legally binding convention plus associated protocols dealing with specific issues); and
- Institutional and financial arrangements.

The programmes differ widely, however, in their level of activity and effectiveness. Some, such as the Mediterranean programme, have been very active in terms of environmental assessment,

**Table 16.1** Regional Seas Programmes

| Programme | Convention | Year adopted | Year entered into force |
|---|---|---|---|
| Mediterranean | Barcelona | 1976/1995 | 1978/2004 |
| ROPME Sea Area[1] | Kuwait | 1978 | 1979 |
| Western and Central Africa | Abidjan | 1981 | 1984 |
| South-East Pacific | Lima | 1981 | 1986 |
| Red Sea and Gulf of Aden | Jeddah | 1982 | 1985 |
| Wider Caribbean | Cartagena | 1983 | 1986 |
| Eastern Africa | Nairobi | 1985 | 1996 |
| South Pacific | Noumea | 1986 | 1990 |
| Black Sea | Bucharest | 1992 | 1994 |
| North-East Pacific | Antigua | 2002 | Action plan in force |
| East Asian Seas | None | 1984 (revised 1993) | Action plan in force |
| Northwest Pacific | None | 1994 | Action plan in force |
| South Asian Seas | None | 1995 | Action plan in force |
| Baltic Sea* | Helsinki | 1974/1992 | 1980/2000 |
| North-East Atlantic* | Oslo/Paris/OSPAR | 1974/1978/1992 | 1998 |
| Antarctic* | Antarctic Treaty/CCAMLR[2] | 1959/1980 | 1961/1982 |
| Caspian Sea* | Tehran | 2003 | 2006 |
| Arctic/PAME[3]* | None (but Arctic Council 1996) | | |

[1] Regional Organization for the Protection of the Marine Environment (ROPME) Sea Area is the Persian Gulf.

[2] Convention on the Conservation of Antarctic Marine Living Resources

[3] Protection of the Arctic Marine Environment

* = partner programmes

management, and legislation, whereas others, like the Asian seas programmes, have stagnated, significant factors being the level of political dispute, commitment, funding, and support from non-governmental organizations (Van Dyke, 2013). There is a pressing need for the programme as a whole to be revitalized.

The traditional focus of the Regional Seas Programme has been pollution control, but few of the programmes have adopted protocols to address land-based sources of pollution despite their acknowledged importance. There is also a wider need in the Regional Seas Programme for more (adequately supported) ecosystem-based approaches that integrate environmental protection and resource management, as espoused by the various international forums on sustainable development (see Chapter 3). This has been recognized and taken up by some of the programmes (Van Dyke, 2013; Johnson *et al.*, 2014).

The Mediterranean Sea was the first area to come under UNEP's Regional Seas Programme and has served as a model for programme development in other regions. Initially, the Mediterranean programme mainly concerned water quality and pollution, but it was subsequently broadened to encompass the management and conservation of living resources in a more comprehensive, integrated approach.

## The Mediterranean Sea

The Mediterranean region, often called the cradle of Western civilization, could also be considered a cradle of marine science, evidenced by pioneering studies in the nineteenth century and the early establishment of marine stations, notably the Stazione Zoologica at Naples dating from 1874. The Mediterranean Sea is also where a regional approach to marine environmental management was initiated. It has the reputation of being severely degraded by human impact, especially as a result of pollution, coastal development, and the heavy exploitation of its living marine resources. Much of the Mediterranean coast is densely populated, and the popularity of the region as a tourist destination continues to grow. Although a vital element in the Mediterranean economy, tourism adds enormous environmental pressures. The ecology of the eastern Mediterranean has, in addition, been altered by major engineering projects, notably the building of the Suez Canal and the damming of the Nile.

## Physical and Biological Characteristics

The Mediterranean Sea, almost landlocked by Europe, Africa, and Asia, is separated from the North Atlantic by the Strait of Gibraltar (14 km wide at its narrowest point) and from the Black Sea by the Sea of Marmara and the narrow straits of the Dardanelles and Bosporus. For more than a century it has also been connected to the Red Sea by the Suez Canal. At about 2.5 million km$^2$, it is the world's largest enclosed sea. It is divided by the Strait of Sicily (400 m deep) into western (0.85 million km$^2$) and eastern (1.65 million km$^2$) basins and includes several subsidiary seas (Fig. 16.1). The coastline is about 46 000 km long – with roughly similar amounts of rocky and sedimentary coast – and with islands accounting for 19 000 km (42%) (Benoit & Comeau, 2005). The Mediterranean Sea has mean and maximum water depths of about 1500 m and 5000 m, with only about 20% of the sea within shelf depths (< 200 m) (Miller, 1983).

The climate, with a prevalence of warm sunny days and dry winds, produces a strong evaporative loss from the sea that is about twice the volume of the inputs from rainfall and runoff. This deficit

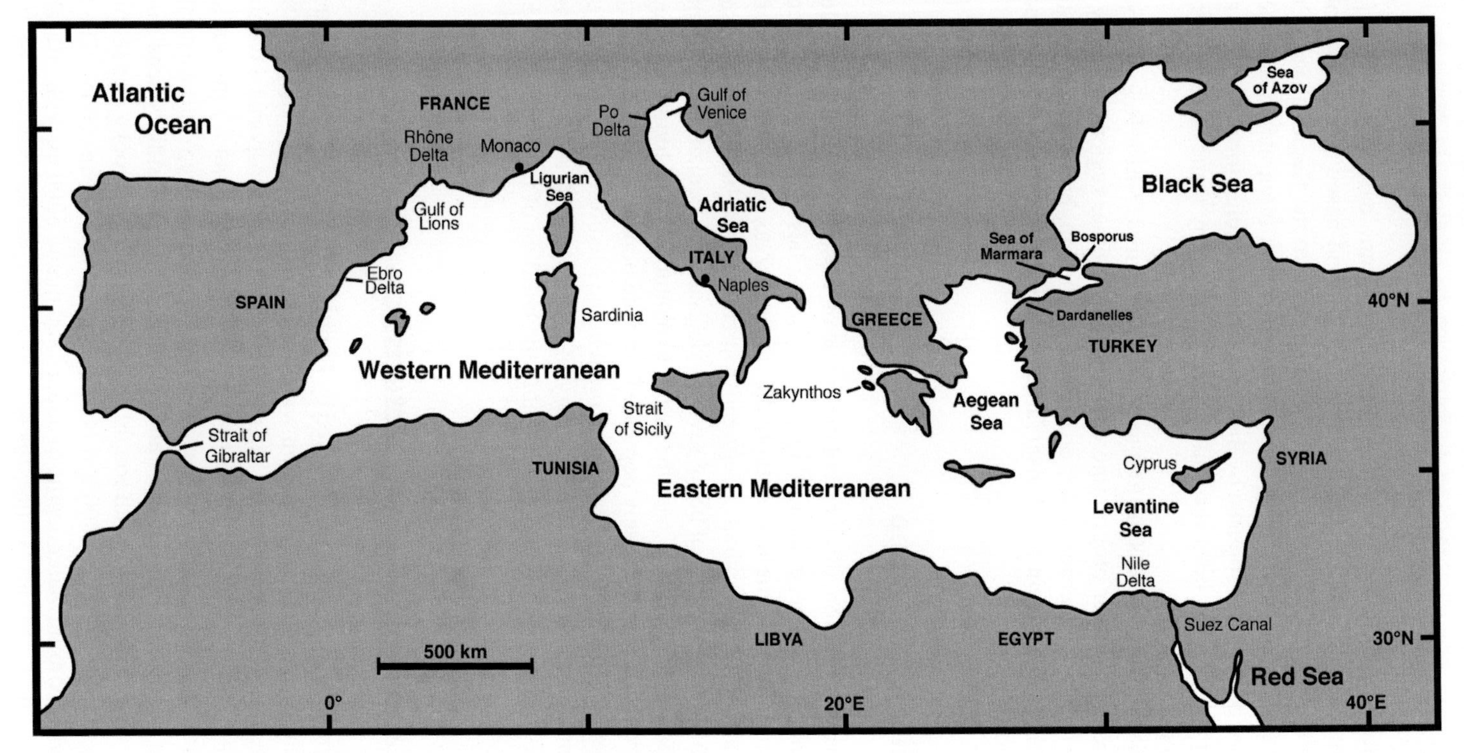


**Fig. 16.1** The Mediterranean Sea, showing places referred to in the text.

is largely compensated by a net inflow of surface Atlantic water through the Strait of Gibraltar. Surface circulation is dominated by this eastward flow and the large eddies it generates. With the high evaporation, surface salinity gradually increases from about 36 near Gibraltar to 38–39 in the eastern Mediterranean. The sinking of this denser, higher salinity water produces a return subsurface current that eventually spills over the sill at Gibraltar to become a North Atlantic water mass at about 1000 m depth. The residence time of water in the Mediterranean is some 80–100 years (Turley, 1999). Whilst this deeper outflow is nutrient rich, the Atlantic inflow is of surface water impoverished in nutrients. An important consequence of this pattern of circulation, together with the generally weak vertical circulation and narrow shelves, is that it maintains oligotrophic conditions (Miller, 1983). Total primary productivity estimates for major regions of the Mediterranean basin range from only about 60 to 100 g C $m^{-2}$ $y^{-1}$, with a general trend of decreasing productivity from west to east (Uitz *et al.*, 2012). Areas of high primary productivity do, however, occur in coastal areas following seasonal mixing, off large rivers, and as a result of anthropogenic nutrient inputs.

The Mediterranean Sea has a diverse biota, attributable to a complex geological history and a mix of biogeographic elements, although its fauna has most in common with that of the Atlantic (Ben-Tuvia, 1983; Bianchi & Morri, 2000). Species richness generally declines from north-west to south-east in line with the productivity gradient, with the most depleted biota occurring in the Levantine Basin. So far, about 17 000 marine species are known from the Mediterranean, but many groups, notably among the microbiota, are still poorly known (Coll *et al.*, 2010). On average, about 20% of Mediterranean marine species are endemic, but for some groups, such as sponges, up to half the species are endemic (Fredj *et al.*, 1992; Coll *et al.*, 2010). The deep-sea fauna of the Mediterranean is still poorly known but may be no less diverse than that of the Atlantic (Coll *et al.*, 2010).

## Selected Habitats and Species

The Mediterranean Sea is a marine biodiversity hotspot, but one that is subject to a wide range of human impacts (Coll *et al.*, 2010). Here we look at some of the region's more characteristic habitats and species and conservation issues that they raise.

### Lagoons and Wetlands

The Mediterranean coast includes numerous coastal lagoons and wetlands, with extensive systems, for instance, in the Gulf of Lions, the Gulf of Venice, and the area of the Nile delta. The total area of lagoons is estimated at 7000 $km^2$, nearly half of which is accounted for by a series of large lagoons along the Egyptian coast (Ben-Tuvia, 1983). Mediterranean lagoons are significant for their biodiversity, such as for their macrophyte and fish species. They have traditionally supported important fisheries, with an average catch of around 90 kg $ha^{-1}$ $y^{-1}$. Their high productivity also makes them attractive for aquaculture, but such developments may compromise their role as refuges, nursery areas, and feeding grounds (Pérez-Ruzafa *et al.*, 2011). Many of these lagoons and wetlands are Ramsar sites. Among the most important wetland sites in the western Mediterranean is the Camargue on the Rhône Delta, an area that includes a mosaic of lagoons, channels, saltings, and dune formations. This is an important site for large numbers of migratory birds and one of the few regular breeding sites in the Mediterranean for greater flamingo (*Phoenicopterus ruber*) ($<$ 22 000 pairs), which forage in the hypersaline salt pans and brackish lagoons (Bechet & Johnson, 2008).

## Seagrass Meadows

Four seagrass species occur in the Mediterranean: *Posidonia oceanica*, an endemic species; *Cymodocea nodosa*, a warm-water species of the eastern Atlantic; and *Zostera marina* and *Z. noltei*, both widespread northern temperate species. *P. oceanica* is the largest (< 75 cm leaf length) and most widespread seagrass in the Mediterranean (Fig. 16.2). It occurs on sandy and rocky bottoms from low water to depths of 30–40 m, depending on water clarity, where it can form extensive meadows. These support a high diversity of associated species, act as spawning or nursery grounds for several species of fish, and represent an important carbon sink in the Mediterranean for climate change mitigation. In many areas, however, *Posidonia* populations have been receding at an alarming rate. The total known area of *P. oceanica* meadows is now estimated at 12 247 km$^2$, with a third of the meadow area lost in the past 50 years (Telesca *et al.*, 2015). Losses are attributable to various physical impacts (such as coastal works, trawling, anchoring and boating, turbidity, and erosion) and eutrophication (Fig. 16.3). *P. oceanica* is one of the slowest-growing and longest-lived plants known. There is evidence that it can form enormous clones (up to several km) with life spans of thousands to tens of thousands of years (Arnaud-Haond *et al.*, 2012). Recolonization and development of meadows will likely take centuries at least (Marbà *et al.*, 2014; Pérez-Lloréns *et al.*, 2014).

**Fig. 16.2** Meadow of the seagrass *Posidonia oceanica*. Reproduced with no changes from Arnaud-Haond, S., Duarte, C.M., Diaz-Almela, E., et al. (2012). Implications of extreme life span in clonal organisms: millenary clones in meadows of the threatened seagrass Posidonia oceanica. *PLoS ONE*, 7, e30454. Photograph: M. San Félix. Licensed under a Creative Commons Attribution 4.0 International License: https://creativecommons.org/licenses/by/4.0/.

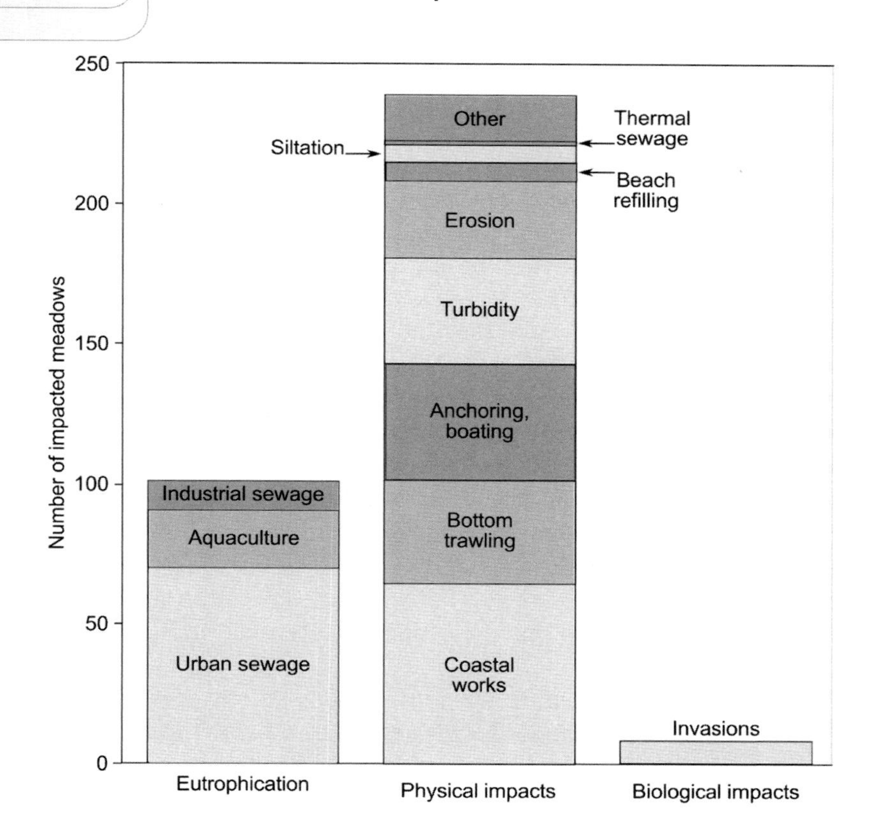


**Fig. 16.3** Pressures identified as causing the decline of *Posidonia oceanica* meadows. Reprinted from *Biological Conservation*, Vol. 176, Marbà, N., Díaz-Almela, E. & Duarte, C.M., Mediterranean seagrass (Posidonia oceanica) loss between 1842 and 2009, pages 183–90, copyright 2014, with permission from Elsevier.

## Hard-Bottom Communities

A notable and widespread feature of the Mediterranean benthos is coralligenous habitat, typically at water depths of between 20 and 120 m. This results from the accumulation of calcareous encrusting algae growing in low light conditions below the limit of the larger seaweeds. Among the main algal builders are *Mesophyllum alternans* in shallow water and *Lithophyllum stictaeforme, L. cabiochiae*, and *Neogoniolithon mamillosum* in deeper water. The habitat supports a highly diverse biota, with the number of species recorded from coralligenous communities totalling at least 315 macroalgae, 1240 invertebrates, and 110 fishes. Protection measures need to address several sources of disturbance that may degrade and damage coralligenous communities, in particular wastewater discharges and trawling (Ballesteros, 2006).

The red coral *Corallium rubrum*, an octocoral endemic to the Mediterranean and neighbouring Atlantic coasts, is a distinctive species of the coralligenous habitat and of deeper rocky outcrops, occurring mainly at water depths of 30–200 m. Colonies can reach 50 cm in height, may live for more than a century, and have average growth rates of only about 0.25 mm $y^{-1}$ in base diameter (Tsounis *et al.*, 2010; Bramanti *et al.*, 2014). Red coral is much prized for jewellery and has been harvested for millennia. But as a slow-growing, long-lived species with relatively low rates of natural mortality and recruitment, it is vulnerable to overexploitation. The major commercial stocks, which are mainly in the western Mediterranean, have been heavily overfished, with a sharp decline evident since the 1980s. Annual landings from the four major source countries (Italy, Spain, Tunisia, and France) declined from 97 t in 1976 to 12 t in 1992, but with an increase to 26 t by 2006 (Bruckner, 2009). The most accessible populations, at around 10–50 m depth, are now characterized by small colonies. These, however, are now protected by a ban on the exploitation of red coral at depths less than 50 m (a recommendation of the General Fisheries Commission for the Mediterranean in 2011), so now only the deeper-water populations (typically at 60–130 m) can be legally harvested (Cannas *et al.*, 2016). Red coral populations show strong genetic structuring due to the low dispersal capability of the larvae and hydrological factors, which helps explain the very low rates of recovery of overexploited banks (Santangelo & Abbiati, 2001). Barriers to gene flow are evident not only among areas but also between shallow and deep populations in the same area. This means that the recovery of shallow, overexploited sites depends on their strict protection, as the deep areas cannot be regarded as reproductive refugia (Cannas *et al.*, 2016).

As for red coral, the collection and use of sponges in the Mediterranean date back at least to ancient civilizations (Pronzato & Manconi, 2008). Four species have been exploited as bath sponges: *Spongia officinalis, S. zimocca, S. lamella*, and *Hippospongia communis. S. officinalis* and *S. zimocca* occur mainly on hard substrata from the shallow subtidal to 30–40 m, *S. lamella* has been recorded from depths of 30–50 m to over 100 m on sediment bottoms but also on rocky cliffs in shallow water, whilst. *H. communis* tends to occur in seagrass beds in shallow waters to depths of 15 m. Commercial sponge fishing, which has taken place mainly in the Eastern basin, intensified from the late nineteenth century with the development of diving suits and dredging methods. Exploitation peaked at 250 t in 1985 (Voultsiadou *et al.*, 2011). But depletion of stocks and disease outbreaks have brought many sponge populations to near extinction and effectively curtailed commercial harvesting (Bell *et al.*, 2015).

The large limpet *Patella ferruginea* is considered the most threatened endemic species in the Mediterranean. Once widely distributed on western Mediterranean rocky shores, it has declined dramatically as a result of human exploitation and loss and degradation of habitat and is now confined to just a few sites (Espinosa *et al.*, 2014). Individuals can reach 10 cm in length, and it is the large ones that are targeted for food, as bait for fishing, and by shell collectors. Large individuals are usually females as the limpets are protandrous hermaphrodites – they start off as males and later change into females. A short annual spawning period, a larval stage of only up to 10 days, and slow growth rate also contribute to their vulnerability to overexploitation. Coppa *et al.* (2016) found the small population of *P. ferruginea* in a marine protected area in Sardinia to be declining rapidly, mainly due to poaching. Population viability analysis (see Chapter 4) indicated that the population faces extinction within a decade given the lack of enforcement.

## Marine Turtles

Suitable nesting beaches for turtles were once widespread through the Mediterranean, but many such sites have been destroyed by urban and tourist development. In addition, Mediterranean turtle populations are affected by incidental catch in fisheries, intentional killing, pollution, and boat strikes (Casale & Margaritoulis, 2010; Coll *et al.*, 2010; Lewison *et al.*, 2014). Two species regularly nest in the eastern Mediterranean – loggerhead and green turtles. For loggerheads, the average number of nests for the entire Mediterranean is estimated at more than 7200, with the main nesting beaches in Greece, Turkey, Cyprus, and Libya. Green turtle nesting is mainly confined to a few beaches in Turkey, Cyprus, and Syria, with the average number of nests for the Mediterranean at more than 1500 per year. Green turtles mainly frequent the Levantine Basin, but loggerhead turtles occur more or less throughout the Mediterranean, with Atlantic individuals often abundant in the western Mediterranean (Casale & Margaritoulis, 2010). The Mediterranean populations of both species are, however, genetically distinct, indicating that they are not sustained by immigrants from the Atlantic, thus underlining the importance of regional conservation efforts (Broderick *et al.*, 2002). The most important nesting site for loggerheads in the Mediterranean is Laganas Bay on the

Greek island of Zakynthos, with an average number of 1200 nests per year (Katselidis *et al*., 2013). Venizelos and Corbett (2005) illustrate the struggle from 1980 of trying to establish protection for this rookery in the face of powerful tourism interests and bureaucratic difficulties. Overall, there appears to be inadequate implementation and enforcement of legislation and a need for greater protection of nesting beaches (Casale & Margaritoulis, 2010).

## Seabirds

Only 15 seabird species breed in the Mediterranean, and all have small populations, reflecting the sea's oligotrophic nature (Coll *et al*., 2010). Three of the species breed only in the Mediterranean: the Balearic shearwater (*Puffinus mauretanicus*), Yelkouan shearwater (*P. yelkouan*), and Audouin's gull (*Larus audouinii*). The two shearwaters are threatened species, due in particular to predation by introduced mammals (notably rats and cats), losses through fisheries bycatch, and their underlying vulnerability as *K*-selected species (www.iucnredlist.org). BirdLife International has developed action plans for certain Mediterranean birds, including Audouin's gull and Balearic shearwater, which identify the key conservation measures for restoring the populations (Gallo-Orsi, 2003).

## Marine Mammals

Eight cetacean species have resident subpopulations in the Mediterranean Sea. The data for three of them are inadequate to assess extinction risk (i.e. data deficient), but the Mediterranean populations of the other five species are threatened (Table 16.2).

The only pinniped that lives in the Mediterranean is the Mediterranean monk seal. It once occurred throughout the sea but is now endangered, with the total Mediterranean population

**Table 16.2** Cetacea with Mediterranean Subpopulations and Their IUCN Red List Category
EN: endangered, VU: vulnerable, DD: data deficient

| Species | Common name | Red List category |
|---|---|---|
| *Delphinus delphis* | Short-beaked common dolphin | EN |
| *Physeter macrocephalus* | Sperm whale | EN |
| *Tursiops truncatus* | Common bottlenose dolphin | VU |
| *Balaenoptera physalus* | Fin whale | VU |
| *Stenella coeruleoalba* | Striped dolphin | VU |
| *Ziphius cavirostris* | Cuvier's beaked whale | DD |
| *Globicephala melas* | Long-finned pilot whale | DD |
| *Grampus griseus* | Risso's dolphin | DD |

Extracted from www.iucnredlist.org (accessed 24 June 2016).

numbering about 350–450 individuals, most of them in the Aegean area. The species is now largely restricted to remote cliffed coasts and small islands and uses inaccessible caves for hauling out and breeding. Major reasons for its decline appear to be human disturbance of breeding sites and reduction of reproductive rate, deliberate persecution, and interaction with fisheries. Monk seals feed principally on fish and octopus and may also have suffered from the commercial overexploitation of prey species. Fishers regard the seals as competitors, and deliberate killing of them has been common. Seals also become entangled in fishing gear and drown. The small population is now highly fragmented, reducing its genetic diversity and further endangering the species (IUCN, 2012). The seal is legally protected, and a few reserves have been established, but further efforts are needed to minimize fishing interactions, halt the deliberate persecution, and increase public awareness. Numbers have continued to decline, and the species' future looks precarious.

## Human Impacts

The Mediterranean coastal region currently has a resident population of about 150 million but receives in addition some 250 million tourists each year. Over the next decade, permanent inhabitants and tourists are projected to increase by about 25 and 50 million respectively (Benoit & Comeau, 2005). Littoralization of the population around the Mediterranean has been increasing, and human population pressure is greatly impacting the coastal environment, particularly from mass tourism with the growth of resorts and associated infrastructure. Forty percent of the coastline had already been built on by 2000, and with the current rate of development the proportion is likely to reach 50% by 2025 (Benoit & Comeau, 2005). Other anthropogenic pressures that are markedly impacting the Mediterranean marine environment come from exploitation of its marine resources, inputs of pollutants, alteration of river flows, and the introduction of alien species (Laubier, 2005).

### Fisheries

Fishing and seafood have been integral to the Mediterranean way of life since earliest times. Major fisheries today include those for small pelagic species such as sardine (*Sardina pilchardus*), anchovy (*Engraulis encrasicolus*), and round sardinella (*Sardinella aurita*) taken by purse seine and pelagic trawl. The most important large pelagic species are Atlantic bluefin tuna (*Thunnus thynnus*) and swordfish (*Xiphias gladius*), taken by longlines and seines but also by illegal driftnetting. The demersal fisheries of the Mediterranean target numerous species, mainly by trawl, including mullets (*Mullus barbatus barbatus* and *M. surmuletus*), hake (*Merluccius merluccius*), blue and red shrimp (*Aristeus antennatus*), and striped venus clam (*Chamelea gallina*) (Papaconstantinou & Farrugio, 2000; Lleonart & Maynou, 2003).

Since 2000, total fisheries catches in the Mediterranean have fluctuated around 900 000 t, with many stocks fully exploited or heavily overexploited. Stocks in the eastern Mediterranean tend to be in the worst state, with more than half assessed as overexploited or collapsed (Tsikliras *et al.*, 2015). Furthermore, many of the fisheries are based on juvenile fish that are only in their first or second year of life. Colloca *et al.* (2013) show that if the size at which these fish are caught were to be increased towards the size at which unexploited fish cohorts achieve their maximum biomass or optimal length (see Chapter 5), this would produce far higher biomass for the exploited stocks as well as higher economic yield and help restore ecosystem structure.

Intensive trawling on the generally narrow continental shelf of the Mediterranean has not only severely depleted demersal stocks but also damaged or destroyed nursery grounds, notably seagrass beds. The General Fisheries Commission for the Mediterranean has prohibited bottom trawling deeper than 1000 m to protect vulnerable benthic habitats. Land-based sources of pollution and the expansion of tourism also pose fisheries management problems.

Fisheries management in the Mediterranean faces a number of other issues, such as the large number and diversity of local small-scale fisheries, the wide range of target species, and the lack of reliable official statistics. Quotas have not in general been applied, and reducing overall effort has been the main priority (Papaconstantinou & Farrugio, 2000). Lleonart and Maynou (2003) outline measures that would help improve the assessment and management of Mediterranean fisheries, particularly the need for the nations concerned to develop co-ordinated management programmes. These would involve the regional fisheries management organizations responsible for management and conservation of fisheries stocks in the Mediterranean: the General Fisheries Commission for the Mediterranean and the International Commission for the Conservation of Atlantic Tunas (ICCAT).

One of the most pressing issues in Mediterranean fisheries concerns the management of Atlantic bluefin tuna, a threatened species whose spawning stock biomass has declined dramatically since the mid-1970s. The species illustrates several of the problems of heavily overexploited species (see Chapters 5 and 6). It is a common-property resource of the high seas, and its catch quotas are set by ICCAT. These, however, have been far higher than those recommended by the commission's scientists. In addition, the illegal, unreported, and unregulated catch in recent years may be as much as twice the total reported catch. Overfishing is also exacerbated by government subsidies (Sumaila & Huang, 2012). Starting in 2007, ICCAT introduced a 15-year recovery plan for bluefin tuna in the East Atlantic and Mediterranean yet still allowed fishing to continue (MacKenzie *et al.*, 2009); indeed, with evidence that stocks could be starting to rebuild, ICCAT increased the total allowable catch from 2014. Although Atlantic bluefin tuna are highly migratory, they congregate to spawn. As the eastern stock spawns mainly in the Mediterranean Sea in summer, an MPA may be an effective conservation measure (Sumaila & Huang, 2012).

Aquaculture production in the Mediterranean has grown rapidly in recent decades. By 2009 marine aquaculture production was more than 355 000 t, almost entirely accounted for by gilthead seabream (*Sparus aurata*), European seabass (*Dicentrarchus labrax*), and the Mediterranean mussel (*Mytilus galloprovincialis*) (Rosa *et al.*, 2014). The amount of nitrogen released to the Mediterranean Sea from fish farming is comparable to that from terrestrial farming activities, and potentially contributing to eutrophication (see below) (Karydis & Kitsiou, 2012).

## Pollution

At least since the 1960s, serious concerns have been expressed about the state of the Mediterranean marine environment as the impact of pollution became only too apparent, most obviously next to large urban areas. Land-based sources account for the great majority of pollutant inputs to the Mediterranean, and most of this is borne by rivers, especially those along the highly populated northern coasts such as the Rhône, Po, and Ebro and, on the southern coast, the Nile. Land-based pollution derives chiefly from agricultural runoff, domestic sewage, and industrial discharges. Pollutants include organic matter, nutrients, heavy metals, petroleum hydrocarbons, persistent organic pollutants, and debris (Civili, 2004). In some respects the pollution status of the Mediterranean appears to have improved somewhat over recent decades, although more extensive data are needed to examine these trends (UNEP/MAP, 2009).

Sewage pollution has long been a problem in many coastal areas of the Mediterranean. It is estimated that 30% of sewage of the coastal cities is discharged into the Mediterranean, often untreated, and that 21–24% of the coastal cities of more than 10 000 inhabitants do not have a wastewater treatment plant, and treatment facilities that do exist often cannot cope with the massive influx of tourists in summer (UNEP/MAP/WHO, 2000).

Nutrient enrichment, particularly from wastewater inputs and agricultural runoff, induces eutrophication (see Chapter 2) in many areas of the Mediterranean, particularly areas along the northern coasts subject to high nutrient input and limited circulation and exchange. Among the basins most impacted by large-scale eutrophication are the Gulf of Lions, the northern Aegean Sea, and the Adriatic Sea (Karydis & Kitsiou, 2012). The northern Adriatic illustrates the importance of the interplay between weather, hydrological conditions, and nutrient inputs in affecting the ecological impact. The 1970s to early 1990s was a period of increasing eutrophication, usually occurring each summer just after the peak of the tourist season, such that oxygen depletion of the bottom water led to localized extinction of benthic populations. Stachowitsch (1984) describes anoxic conditions resulting in the mass mortality of a sponge- and brittlestar-dominated benthos in the Gulf of Trieste, where the impacted area covered several hundred square kilometres. But from the early 1990s to 2012, severe hypoxic events have occurred less frequently. During this period there has been less precipitation, reducing the discharge of the Po River, the main river flowing into the northern Adriatic, as well as other runoff, and a shift in circulation has meant a stronger inflow of high-salinity, oligotrophic water from the central Adriatic (Djakovac *et al.*, 2012).

Oil is a widespread pollutant in the Mediterranean. The sea is a major route for oil tankers, especially those from North Africa and the Middle East, and has numerous oil refineries on its coast. By the 1970s, the Mediterranean Sea had become notorious for its oil pollution (Hinrichsen, 1998). The input of petroleum hydrocarbons to the Mediterranean by the late 1980s was estimated at 635 000 tonnes per year, equivalent then to about 20% of the global input to the marine environment into only 1% of the of world's sea area. In common with the global pattern, about half was from land-based discharges and half from marine transportation (UNEP, 1989). Oil pollution in the Mediterranean appears to have declined over recent decades as a result of economic factors, better operational procedures, and international agreements. Under the MARPOL convention (see Chapter 2), the Mediterranean Sea is a special area where deliberate discharges of oily wastes (> 15 ppm) from vessels are prohibited. Nevertheless, illicit discharges are still common. Satellite surveillance of the Mediterranean during the years 1999 to 2002 detected 7000 oil spills, chronic pollution that is likely to harm marine wildlife, in particular seabirds (Camphuysen, 2007).

Gómez-Gutiérrez *et al.* (2007) assessed the contamination of Mediterranean sediments by persistent organic pollutants (POPs), in particular geographical and temporal trends in the levels of PCBs, DDTs, and the fungicide hexachlorobenzene over the period 1971 to 2005. Contamination tends to be localized and associated with land-based sources with most hotspots along the northern coastline. In general, concentrations have been declining over time, more so for DDTs than for PCBs. This probably reflects increasing restrictions on the use of POPs. The great majority of Mediterranean countries have ratified the Stockholm Convention, the international treaty concerned with the reduction and elimination of POPs (see Chapter 2).

Anthropogenic inputs of heavy metals into the Mediterranean also result in coastal hotspots, but concentrations are now generally low and declining as a result of tighter controls. High mercury levels in biota and sediments have been a particular concern, although these can be largely the result of natural geochemical sources in the Mediterranean region (EEA, 1999).

A mounting problem in the Mediterranean is the accumulation of persistent debris, especially plastics, in surface waters, on the seabed, and along the shore. The average plastic concentration in Mediterranean surface waters has been estimated at 423 g $km^{-2}$ (predominantly microplastics), giving a total load of surface plastics of around 1000 t (Cózar *et al.*, 2015). This density of floating plastic debris is comparable to the concentrations measured in the accumulation zones of the subtropical ocean gyres (see Chapter 2). As an enclosed sea with a densely populated coast and major tourism and shipping activities, the Mediterranean is susceptible to particularly high plastic concentrations, a problem exacerbated by its circulation, where the main outflow is the deep subsurface current. Even larger amounts of debris can accumulate on the seabed, for example up to 400 kg $km^{-2}$ of litter (predominantly plastic) on the western continental slope (Pham *et al.*, 2014). And high densities of persistent litter, mainly plastic items from recreational sources, accumulate on Mediterranean beaches (e.g. Gabrielides *et al.*, 1991). Plastic debris is a potential threat to a wide range of marine animals through ingestion or entanglement (see Chapter 2). Among the concerns in the Mediterranean is the high incidence (< 94%) of plastics being ingested by seabirds, especially by the endemic and threatened shearwaters (Cory's (*Calonectris diomedea*), Yelkouan, and Balearic shearwaters) (Codina-García *et al.*, 2013). High densities of anthropogenic debris along the main shipping lane to and from the Strait of Gibraltar suggest that dumping at sea is still occurring despite MARPOL Annex V (see Chapter 2) (Suaria & Aliani, 2014). Marine litter is a focus area of the European Union's Marine Strategy Framework Directive adopted in 2008. This should help as member states are required to take measures so that 'properties and quantities of marine litter do not cause harm to the coastal and marine environment'.

## Aswan High Dam

Riverine input to the Mediterranean has been substantially modified by dam construction and abstraction of water for irrigation. The flow of the Nile in particular has been drastically affected by construction of the Aswan High Dam, which became fully operational in 1965, preventing the annual Nile flood from discharging into the Mediterranean. The flood waters contained high nutrient concentrations, triggering massive diatom blooms off the delta coast, which in turn supported an important sardinella fishery. With completion of the dam, the autumn flood was cut by about 90%, nutrient concentrations fell, the phytoplankton blooms disappeared, and the annual fish catch fell from around 20 000 t in the early 1960s to less than half that by the mid-1970s. The fishery has since recovered, with landings now more than three times the pre-dam levels. Changes in access to fishing grounds may partly explain the fall and rise in fish catch (Biswas & Tortajada, 2012). However, the rapid recovery of the fishery coincides with large increases in fertilizer use and sewage discharge in Egypt, and these anthropogenic nutrient sources may have more than compensated for the loss of the Nile's naturally fertile flood waters (Oczkowski *et al.*, 2009).

## Alien Species

The biodiversity of the Mediterranean Sea has been greatly impacted by introduced species. Around 700–900 multicellular non-indigenous species have now been recorded in the Mediterranean, mostly benthic invertebrates (in particular molluscs, crustaceans, and polychaetes), fishes, and macroalgae (Zenetos *et al.*, 2012; Galil *et al.*, 2014) (Table 16.3). Establishing what is an alien species is not, however, straightforward. Without a detailed knowledge of the native biota, it may for example be difficult to determine whether rare species with a plausible biogeography are introductions, such as eastern Atlantic species that are found in the Mediterranean.

About half the alien species in the Mediterranean have come from the Red Sea via the Suez Canal, so-called Lessepsian species (after the French diplomat de Lesseps who promoted the canal). The 160-km sea-level canal runs from the Gulf of Suez on the Red Sea to Port Said, east of the Nile delta. Its opening in 1869 provided a corridor between the Atlantic-Mediterranean and Indo-Pacific biogeographical regions. The Mediterranean and Red Sea biotas had been separated since the late

**Table 16.3** Major Groups of Multicellular Organisms Contributing Alien Species to the Marine Biota of the Mediterranean Sea

| Group of organisms | % |
|---|---|
| Molluscs | 23 |
| Crustaceans | 17 |
| Polychaetes | 14 |
| Macrophytes | 14 |
| Fishes | 14 |
| Cnidarians | 6 |
| Bryozoans | 3 |
| Ascidiaceans | 2 |
| Other | 6 |

With permission from Zenetos, A., Gofas, S., Morri, C., *et al.* (2012). Alien species in the Mediterranean Sea by 2012. A contribution to the application of European Union's Marine Strategy Framework Directive (MSFD). Part 2. Introduction trends and pathways. *Mediterranean Marine Science*, 13, 328–52.

Pliocene when the Isthmus of Suez finally emerged. A major obstacle to potential migrants has been the lakes that the canal traverses, which range from hypersaline to brackish. This salinity range has, however, gradually reduced. The canal has been widened and deepened, and with the damming of the Nile and the control of flooding there is no longer the dramatic seasonal drop in salinity at the northern end of the canal. These changes have all lessened the barriers to migration and contributed to an accelerating rate of invasion (Rilov & Galil, 2009).

Lessepsian species are largely confined to the Levantine Sea, an invasion that has profoundly impacted this south-eastern part of the Mediterranean. These include 90 fish species, of which some two-thirds have established themselves in the eastern Mediterranean, making this the area of the world with the most fish invasions. Several of these species are now among the predominant fishes of the eastern Mediterranean and are targeted by fisheries. With the altered fish communities there is evidence of trophic competition and food-web restructuring (Edelist *et al.*, 2013; Fanelli *et al.*, 2015).

Movement of species through the Suez Canal has been almost entirely one way – very few Mediterranean species have invaded the Red Sea – mainly because the current flow in the canal is predominantly northwards and the low species richness of the eastern Mediterranean makes it vulnerable to invasion (Rilov & Galil, 2009). Continuing expansion of the Suez Canal and plans for a new channel mean greatly increased opportunities for bioinvasion (Galil *et al.*, 2015). Re-establishing a salinity barrier, such as a hypersaline section between locks, has been suggested as a way of minimizing the passage of further alien species.

The Suez Canal has been the main route for alien species entering the Mediterranean, but other important vectors have been vessels (~25% of species) and aquaculture (~10%) (UNEP/MAP, 2009;

Galil *et al.*, 2014). The aquarium trade, however, appears implicated in the most notorious introductions – green seaweeds of the genus *Caulerpa. C. taxifolia* is native to the Indian Ocean, but a hardy strain that had been selected for aquarium use was discovered in 1984 on the Monaco coast below the aquarium of the Musée Océanographique – seemingly the result of an accidental release. In the early 1990s another non-indigenous species of *Caulerpa* was discovered in the Mediterranean, *C. cylindracea*, initially at sites in Libya and southern Italy. This species, native to south-western Australia, may have arrived via shipping and the aquarium trade (Klein & Verlaque, 2008; Davidson *et al.*, 2015). Both these species have spread rapidly, particularly along the northern coasts, colonizing a variety of substrata. They readily propagate vegetatively from fragments, overgrow competitors, and secrete allelopathic substances that deter herbivores (Papini *et al.*, 2013). The *Caulerpa* invasions have been especially harmful to the native macroalgae, reducing their species richness and abundance, and have also degraded seagrass beds (Rilov & Galil, 2009; Davidson *et al.*, 2015). The devastating impact of exotic seaweeds in the Mediterranean has prompted other countries to ban the aquarium trade in *Caulerpa* species, but a high level of enforcement and public education is needed for such a measure to be effective (Diaz *et al.*, 2012).

## Mediterranean Action Plan

As a means to identify and address common environmental problems, Mediterranean countries adopted the Mediterranean Action Plan (MAP) in 1975 (Manos, 1991). A major component of the plan has been the establishment of a co-ordinated programme of pollution monitoring and research. Mediterranean countries have as a result undertaken major studies to determine the sources and quantities of pollutants. Legal treaties for the protection of the Mediterranean environment are another key component of the plan and comprise the Barcelona Convention and a series of protocols covering pollution, protected areas, and coastal zone management (UNEP, 2011) (Box 16.1). The 21 riparian countries of the Mediterranean plus the European Union are contracting parties to the convention and its protocols. The land-based sources protocol is the principal means under the convention of combating pollution, given that most pollutants entering the Mediterranean come from the land. The protocol lists certain substance groups that, on the basis of characteristics such as toxicity, persistence, and bioaccumulation, parties undertake to eliminate as pollutants. The initial focus of MAP was primarily pollution control, but following the 1992 UN Conference on Environment and Development its scope was widened to be more consistent with Agenda 21 and the aims of sustainable development (see Chapter 3). To that end a revised MAP and convention were adopted in 1995.

Frantzi and Lovett (2008) examined the overall environmental effectiveness of MAP but came to no obvious conclusion – whether or not the environmental state of the Mediterranean Sea has improved as a direct result of the regime. A complex of other factors and policies have also been operating, and despite the large environmental assessment programme of MAP, they concluded that the environmental information attained is not adequate for measuring the effectiveness of the regime. However, in spite of some strong political differences, the Mediterranean countries have come together under MAP to examine common environmental issues, and this can only have helped in mitigating deterioration of the sea. Nevertheless, it is the disposal of wastewater, often untreated, and other land-based sources of pollution that have long constituted one of the major problems facing the Mediterranean marine environment (Massoud *et al.*, 2003). Particularly beneficial in this regard, therefore, are programmes at the national level that aim to improve the water quality of effluents.

Within the framework of the Barcelona Convention, Mediterranean countries have adopted action plans for particular taxa and habitats: for the monk seal, marine turtles, cetaceans, birds,

### Box 16.1 Mediterranean Action Plan

The legal framework of the Mediterranean Action Plan (MAP) comprises the Barcelona Convention (adopted 1976) and seven protocols that concern:

- Pollution by dumping from ships and aircraft
- Pollution from ships and emergency situations
- Pollution from land-based sources and activities
- Specially protected areas of biological diversity
- Pollution from exploration and exploitation of the continental shelf
- Pollution by hazardous wastes and their disposal
- Integrated coastal zone management

Amended versions of the MAP[1] and the Barcelona Convention[2] were adopted in 1995 to broaden the scope from the initial focus on marine pollution control to sustainable management and socio-economic development.

MAP also includes a programme to attend to socio-economic aspects of environmental protection, in particular the Blue Plan launched in 1979.

MAP includes several other agreements and initiatives as well as linking with UN bodies and NGOs involved in environmental protection of the Mediterranean Sea.

[1] The Action Plan for the Protection of the Marine Environment and the Sustainable Development of the Coastal Areas of the Mediterranean (MAP Phase II)

[2] Convention for the Protection of the Marine Environment and the Coastal Region of the Mediterranean

marine vegetation, coralligenous formations, and seamounts, canyons, and chemosynthetic habitats. The plans, which are not legally binding, call on contracting parties to collaborate in the conservation of the species and habitats in question.

## Marine Protected Areas

Numerous marine and coastal protected areas have been established by Mediterranean countries. Included are important wetlands, lagoons, seagrass meadows, red coral and coralligenous formations, and sites valuable for their seabirds, monk seals, or turtles. Several of the sites are internationally recognized, for example, as Ramsar sites (see Chapter 11), biosphere reserves, or World Heritage sites (see Chapter 15). Gabrié *et al.* (2012) identified 677 MPAs in the Mediterranean Sea, with a total surface area of 114 600 $km^2$, which amounts to 4.6% of the sea's area. But if the large Pelagos Sanctuary (see Chapter 9) in the Ligurian Sea is excluded, only 1% of the Mediterranean Sea is within MPAs. Areas that are in some way protected have a variety of designations, such as natural marine reserves, marine parks, national parks, fishing reserves, and cantonnements, and the protective measures can differ greatly. Less than 0.1% of Mediterranean Sea is included in strict protection and/or no-take areas. Whilst MPA coverage in the Mediterranean has increased significantly over recent years, it is still far short of the target of 10% agreed by parties to the Convention on Biological Diversity (implemented through the Barcelona Convention Protocol concerning specially protected areas) (see Chapter 15). The MPA network also needs to be more representative of the sea's ecoregions. Currently, the great majority of the MPAs are coastal and in the northern part of the sea.

In terms of maritime boundaries, most Mediterranean states have claimed only a territorial sea (< 12 nm). As a result, much of the sea lies beyond national jurisdiction. The Pelagos Sanctuary for marine mammals illustrates that large MPAs in the high seas can at least be established in the Mediterranean (Notarbartolo di Sciara *et al.*, 2008). To assist in the conservation of the Mediterranean's high seas biodiversity, a proposed network of representative MPAs has been endorsed by parties to the Barcelona Convention. The network, based on eight ecoregions (a modification of those proposed by Spalding *et al.* (2007), see Chapter 1), comprises 11 areas, each containing a number of potential MPAs. The criteria for these areas are those for ecologically or biologically significant marine areas, as defined by the Convention on Biological Diversity (see Chapter 14). In order to advance this MPA planning and Mediterranean conservation, Portman *et al.* (2013) recommend a multi-step process suited to this complex geopolitical region (Fig. 16.4), a process based on the broad stages of systematic conservation planning (see Chapter 15, Box 15.1).

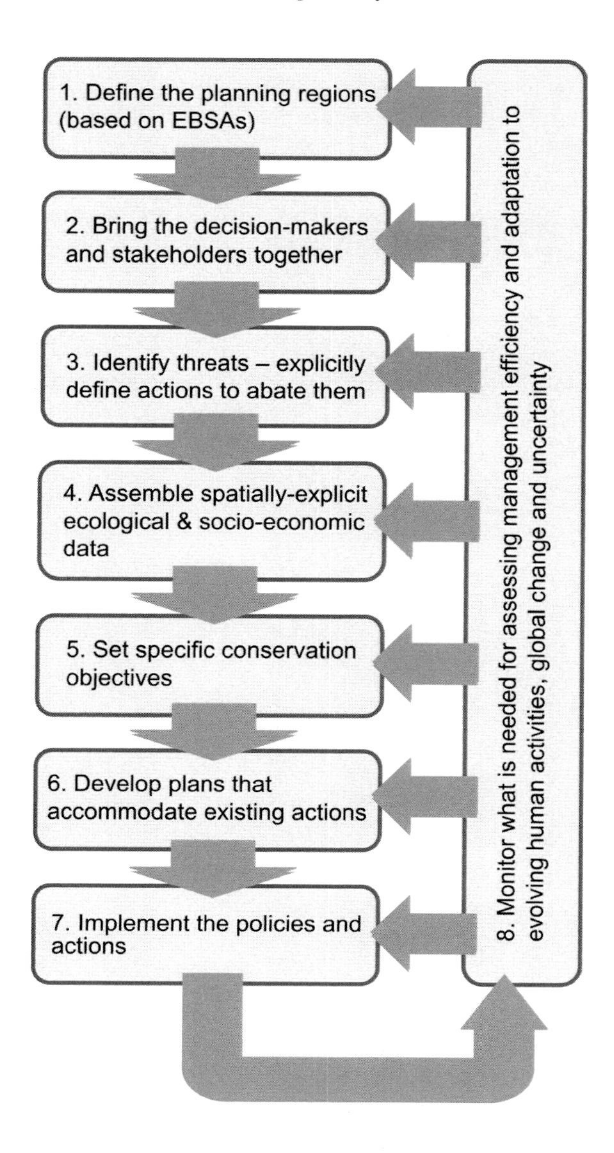


**Fig. 16.4** A process for marine spatial planning applicable to conservation of the Mediterranean. Reprinted from *Marine Policy*, Vol. 42, Portman, M.E., Notarbartolo-di-Sciara, G., Agardy, T., et al., He who hesitates is lost: why conservation in the Mediterranean Sea is necessary and possible now, pages 270–9, copyright 2013, with permission from Elsevier.

A necessary step in systematic conservation planning, but one that is rarely or adequately included in designing MPA networks, is the economic cost, most obviously the opportunity cost of commercial and recreational activities that are relinquished. In complex, populous regions with wide socio-economic differences, such as the Mediterranean, opportunity costs may have a major bearing on the protected area coverage that is feasible. Mazor *et al.* (2014) incorporated the opportunity cost of fishing and aquaculture in setting protected area targets in the Mediterranean. They found that for less than 10% of the sea's area, their conservation targets could be achieved whilst incurring opportunity costs of less than 1%.

Several studies in the Mediterranean Sea in recent years have applied principles of systematic conservation planning to MPA designation. These illustrate, however, that data on the spatial distribution of ecological features are often lacking or of uneven availability and quality (data being generally better in the European western sector) and that within the region there are numerous schemes for classifying marine habitats. This emphasizes the urgent need in the Mediterranean – as indeed elsewhere – for a common integrated framework for marine conservation planning, including coordinated approaches to marine mapping, centralized databases, and data sharing (Levin *et al.*, 2014).

The Mediterranean region illustrates the important role of economic and political stability in facilitating such trans-boundary collaboration and effecting large-scale marine conservation initiatives. Beyond its territorial seas, the Mediterranean is a complex of maritime zones and disputed areas. There is gathering interest among the riparian states to declare EEZs. If they all did so, international waters in the Mediterranean Sea would cease to exist. Katsanevakis *et al.* (2015) argue that such a shift in jurisdiction towards national policies and responsibilities would provide an important opportunity for the designation of MPAs in EEZs and the development of Mediterranean-wide conservation of marine biodiversity.

The Regional Seas Programme is based on geopolitical boundaries, although several of the regions are recognized biogeographic provinces or ecoregions (see Chapter 1). In the next chapter we look at a region that contrasts strongly with the Mediterranean Sea in terms of its oceanography, human impacts, and the development of marine environmental protection and where a more ecosystem-oriented approach to management has been emphasized.

## REFERENCES

Arnaud-Haond, S., Duarte, C.M., Diaz-Almela, E., *et al.* (2012) Implications of extreme life span in clonal organisms: millenary clones in meadows of the threatened seagrass *Posidonia oceanica*. *PLoS ONE*, **7**, e30454.

Ballesteros, E. (2006). Mediterranean coralligenous assemblages: a synthesis of present knowledge. *Oceanography and Marine Biology: An Annual Review*, **44**, 123–95.

Bechet, A. & Johnson, A.R. (2008). Anthropogenic and environmental determinants of Greater Flamingo *Phoenicopterus roseus* breeding numbers and productivity in the Camargue (Rhone delta, southern France). *Ibis*, **150**, 69–79.

Bell, J.J., McGrath, E., Biggerstaff, A., *et al.* (2015). Global conservation status of sponges. *Conservation Biology*, **29**, 42–53.

Ben-Tuvia, A. (1983). The Mediterranean Sea, B. Biological aspects. In *Estuaries and Enclosed Seas*, ed. B.H. Ketchum, pp. 239–51. Amsterdam: Elsevier.

Benoit, G. & Comeau, A. (2005). *A Sustainable Future for the Mediterranean: The Blue Plan's Environment and Development Outlook*. Oxford, UK: Taylor and Francis.

Bianchi, C.N. & Morri, C. (2000). Marine biodiversity of the Mediterranean Sea: situation, problems and prospects for future research. *Marine Pollution Bulletin*, **40**, 367–76.

Biswas, A.K. & Tortajada, C. (2012). Impacts of the High Aswan Dam. In *Impacts of Large Dams: A Global Assessment*, ed. C. Tortajada, D. Altinbilek & A.K. Biswas, pp. 379–95. Berlin: Springer-Verlag.

Bramanti, L., Vielmini, I., Rossi, S., *et al.* (2014). Demographic parameters of two populations of red coral (*Corallium rubrum* L. 1758) in the North Western Mediterranean. *Marine Biology*, **161**, 1015–26.

Broderick, A.C., Glen, F., Godley, B.J. & Hays, G.C. (2002). Estimating the number of green and loggerhead turtles nesting annually in the Mediterranean. *Oryx*, **36**, 227–35.

Bruckner, A.A. (2009). Rate and extent of decline in *Corallium* (pink and red coral) populations: existing data meet the requirements for a CITES Appendix II listing. *Marine Ecology Progress Series*, **397**, 319–32.

Camphuysen, C.J. (2007). *Chronic Oil Pollution in Europe. Status report*. Brussels: International Fund for Animal Welfare, 85 pp.

Cannas, R., Sacco, F., Cau, A., *et al.* (2016). Genetic monitoring of deep-water exploited banks of the precious Sardinia coral *Corallium rubrum* (L., 1758): useful data for a sustainable management. *Aquatic Conservation: Marine and Freshwater Ecosystems*, **26**, 236–50.

Casale, P. & Margaritoulis, D. (eds.). (2010). *Sea Turtles in the Mediterranean: Distribution, Threats and Conservation Priorities*. Gland, Switzerland: IUCN. 294 pp.

Civili, F.S. (2004). The pollution of the Mediterranean: present state and prospects. In *Environmental Challenges in the Mediterranean 2000–2050*, ed. A. Marquina, pp. 363–76. Dordrecht: Kluwer Academic Publishers.

Codina-García, M., Militão, T., Moreno, J. & González-Solís, J. (2013). Plastic debris in Mediterranean seabirds. *Marine Pollution Bulletin*, **77**, 220–6.

Coll, M., Piroddi, C., Steenbeek, J., *et al.* (2010) The biodiversity of the Mediterranean Sea: estimates, patterns, and threats. *PLoS ONE*, **5**, e11842.

Colloca, F., Cardinale, M., Maynou, F., *et al.* (2013). Rebuilding Mediterranean fisheries: a new paradigm for ecological sustainability. *Fish and Fisheries*, **14**, 89–109.

Coppa, S., De Lucia, G.A., Massaro, G., *et al.* (2016). Is the establishment of MPAs enough to preserve endangered intertidal species? The case of *Patella ferruginea* in Mal di Ventre Island (W Sardinia, Italy). *Aquatic Conservation: Marine and Freshwater Ecosystems*, **26**, 623–38.

Cózar, A., Sanz-Martín, M., Martí, E., *et al.* (2015). Plastic accumulation in the Mediterranean Sea. *PLoS ONE*, **10**, e0121762.

Davidson, A.D., Campbell, M.L., Hewitt, C.L. & Schaffelke, B. (2015). Assessing the impacts of nonindigenous marine macroalgae: an update of current knowledge. *Botanica Marina*, **58**, 55–79.

Diaz, S., Smith, J.R., Zaleski, S.F. & Murray, S.N. (2012). Effectiveness of the California state ban on the sale of *Caulerpa* species in aquarium retail stores in Southern California. *Environmental Management*, **50**, 89–96.

Djakovac, T., Degobbis, D., Supić, N. & Precali, R. (2012). Marked reduction of eutrophication pressure in the northeastern Adriatic in the period 2000–2009. *Estuarine, Coastal and Shelf Science*, **115**, 25–32.

Edelist, D., Rilov, G., Golani, D., Carlton, J.T.& Spanier, E. (2013). Restructuring the sea: profound shifts in the world's most invaded marine ecosystem. *Diversity and Distributions*, **19**, 69–77.

EEA (1999). *State and Pressures of the Marine and Coastal Mediterranean Environment*. Copenhagen: European Environment Agency.

Espinosa, F., Rivera-Ingraham, G.A., Maestre, M., *et al.* (2014). Updated global distribution of the threatened marine limpet *Patella ferruginea* (Gastropoda: Patellidae): an example of biodiversity loss in the Mediterranean. *Oryx*, **48**, 266–75.

Fanelli, E., Azzurro, E., Bariche, M., Cartes, J.E. & Maynou, F. (2015). Depicting the novel Eastern Mediterranean food web: a stable isotopes study following Lessepsian fish invasion. *Biological Invasions*, **17**, 2163–78.

Frantzi, S. & Lovett, J.C. (2008). Is science the driving force in the operation of environmental regimes? A case study of the Mediterranean Action Plan. *Ocean & Coastal Management*, **51**, 229–45.

Fredj, G., Bellan-Santini, D. & Meinardi, M. (1992). État des connaissances sur la faune marine méditeranéenne. *Bulletin de l'Institut Océanographique, Monaco*, no. Spécial, **9**, 133–45.

Gabrié, C., Lagabrielle, E., Bissery C., *et al.* (2012). *The Status of Marine Protected Areas in the Mediterranean Sea*. The network of managers of Marine Protected Areas in the Mediterranean (MedPAN) & the Regional Activity Centre for Specially Protected Areas (RAC/SPA). MedPAN Collection. 254 pp.

Gabrielides, G.P., Golik, A., Loizides, L., Marino, M.G., Bingel, F. & Torregrossa, M.V. (1991). Man-made garbage pollution on the Mediterranean coastline. *Marine Pollution Bulletin*, **23**, 437–41.

Galil, B.S., Boero, F., Campbell, M.L., *et al.* (2015). 'Double trouble': the expansion of the Suez Canal and marine bioinvasions in the Mediterranean Sea. *Biological Invasions*, **17**, 973–6.

Galil, B.S., Marchini, A., Occhipinti-Ambrogi, A., *et al.* (2014). International arrivals: widespread bioinvasions in European Seas. *Ethology Ecology & Evolution*, **26**, 152–71.

Gallo-Orsi, U. (2003). Species Action Plans for the conservation of seabirds in the Mediterranean Sea: Audouin's gull, Balearic shearwater and Mediterranean shag. *Scientia Marina*, **67** (Suppl. 2), 47–55.

Gómez-Gutiérrez, A., Garnacho, E., Bayona, J.M. & Albaigés, J. (2007). Assessment of the Mediterranean sediments contamination by persistent organic pollutants. *Environmental Pollution*, **148**, 396–408.

Hinrichsen, D. (1998). *Coastal Waters of the World: Trends, Threats, and Strategies*. Washington, DC: Island Press.

IUCN (2012). *Marine Mammals and Sea Turtles of the Mediterranean and Black Seas*. Gland, Switzerland and Malaga, Spain: IUCN. 32 pp.

Johnson, D.E., Martinez, C., Vestergaard, O., *et al.* (2014). Building the regional perspective: platforms for success. *Aquatic Conservation: Marine and Freshwater Ecosystems*, **24** (Suppl. 2), 75–93.

Karydis, M. & Kitsiou, D. (2012). Eutrophication and environmental policy in the Mediterranean Sea: a review. *Environmental Monitoring and Assessment*, **184**, 4931–84.

Katsanevakis, S., Levin, N., Coll, M., *et al.* (2015). Marine conservation challenges in an era of economic crisis and geopolitical instability: the case of the Mediterranean Sea. *Marine Policy*, **51**, 31–9.

Katselidis, K.A., Schofield, G., Stamou, G., Dimopoulos, P. & Pantis, J.D. (2013). Evidence-based management to regulate the impact of tourism at a key marine turtle rookery on Zakynthos Island, Greece. *Oryx*, **47**, 584–94.

Klein, J. & Verlaque, M. (2008). The *Caulerpa racemosa* invasion: a critical review. *Marine Pollution Bulletin*, **56**, 205–25.

Laubier, L. (2005). Mediterranean Sea and humans: improving a conflictual partnership. In *The Mediterranean Sea (Handbook of Environmental Chemistry Volume 5 K)*, ed. A. Saliot, pp. 3–27. Berlin Heidelberg: Springer-Verlag.

Levin, N., Coll, M., Fraschetti, S., *et al.* (2014). Biodiversity data requirements for systematic conservation planning in the Mediterranean Sea. *Marine Ecology Progress Series*, **508**, 261–81.

Lewison, R.L., Crowder, L.B., Wallace, B.P., *et al.* (2014). Global patterns of marine mammal, seabird, and sea turtle bycatch reveal taxa-specific and cumulative megafauna hotspots. *Proceedings of the National Academy of Sciences*, **111**, 5271–6.

Lleonart, J. & Maynou, F. (2003). Fish stock assessments in the Mediterranean: state of the art. *Scientia Marina*, **67** (Suppl. 1), 37–49.

MacKenzie, B.R., Mosegaard, H. & Rosenberg, A.A. (2009). Impending collapse of bluefin tuna in the northeast Atlantic and Mediterranean. *Conservation Letters*, **2**, 25–34.

Manos, A. (1991). An international programme for the protection of a semi-enclosed sea – the Mediterranean Action Plan. *Marine Pollution Bulletin*, **23**, 489–96.

Marbà, N., Díaz-Almela, E. & Duarte, C.M. (2014). Mediterranean seagrass (*Posidonia oceanica*) loss between 1842 and 2009. *Biological Conservation*, **176**, 183–90.

Massoud, M.A., Scrimshaw, M.D. & Lester, J.N. (2003). Qualitative assessment of the effectiveness of the Mediterranean action plan: wastewater management in the Mediterranean region. *Ocean & Coastal Management*, **46**, 875–99.

Mazor, T., Giakoumi, S., Kark, S. & Possingham, H.P. (2014). Large-scale conservation planning in a multinational marine environment: cost matters. *Ecological Applications*, **24**, 1115–30.

Miller, A.R. (1983). The Mediterranean Sea, A. Physical aspects. In *Estuaries and Enclosed Seas*, ed. B.H. Ketchum, pp. 219–38. Amsterdam: Elsevier.

Notarbartolo di Sciara, G., Agardy, T., Hyrenbach, D., Scovazzi, T. & van Klaveren, P. (2008). The Pelagos Sanctuary for Mediterranean marine mammals. *Aquatic Conservation: Marine and Freshwater Ecosystems*, **18**, 367–91.

Oczkowski, A.J., Nixon, S.W., Granger, S.L., El-Sayed, A.-F. M. & McKinney, R.A. (2009). Anthropogenic enhancement of Egypt's Mediterranean fishery. *Proceedings of the National Academy of Sciences*, **106**, 1364–7.

Papaconstantinou, C. & Farrugio, H. (2000). Fisheries in the Mediterranean. *Mediterranean Marine Science*, **1**, 5–18.

Papini, A., Mosti, S. & Santosuosso, U. (2013). Tracking the origin of the invading *Caulerpa* (Caulerpales, Chlorophyta) with Geographic Profiling, a criminological technique for a killer alga. *Biological Invasions*, **15**, 1613–21.

Pérez-Lloréns, J.L., Vergara, J.J., Olivé, I., *et al.* (2014). Autochthonous seagrasses. In *The Mediterranean Sea: its History and Present Challenges*, ed. S. Goffredo & Z. Dubinsky, pp. 137–58. Dordrecht, Netherlands: Springer.

Pérez-Ruzafa, A., Marcos, C. & Pérez-Ruzafa, I.M. (2011). Mediterranean coastal lagoons in an ecosystem and aquatic resources management context. *Physics and Chemistry of the Earth*, **36**, 160–6.

Pham, C.K., Ramirez-Llodra, E., Alt, C.H.S., *et al.* (2014). Marine litter distribution and density in European seas, from the shelves to deep basins. *PLoS ONE*, **9**, e95839.

Portman, M.E., Notarbartolo-di-Sciara, G., Agardy, T., *et al.* (2013). He who hesitates is lost: why conservation in the Mediterranean Sea is necessary and possible now. *Marine Policy*, **42**, 270–9.

Pronzato, R, & Manconi, R. (2008). Mediterranean commercial sponges: over 5000 years of natural history and cultural heritage. *Marine Ecology*, **29**, 146–66.

Rilov, G. & Galil, B. (2009). Marine bioinvasions in the Mediterranean Sea – history, distribution and ecology. In *Biological Invasions in Marine Ecosystems: Ecological, Management, and Geographic Perspectives*, ed. G. Rilov & J.A. Crooks, pp. 549–75. Berlin: Springer-Verlag.

Rosa, R., Marques, A. & Nunes, M.L. (2014). Mediterranean aquaculture in a changing climate. In *The Mediterranean Sea: its History and Present Challenges*, ed. S. Goffredo & Z. Dubinsky, pp. 605–16. Dordrecht, Netherlands: Springer.

Santangelo, G. & Abbiati, M. (2001). Red coral: conservation and management of an over-exploited Mediterranean species. *Aquatic Conservation: Marine and Freshwater Ecosystems*, **11**, 253–9.

Spalding, M.D., Fox, H.E., Allen, G.R., *et al.* (2007). Marine ecoregions of the world: a bioregionalization of coastal and shelf areas. *BioScience*, **57**, 573–83.

Stachowitsch, M. (1984). Mass mortality in the Gulf of Trieste: the course of community destruction. *Marine Ecology: Pubblicazione, della Stazione Zoologica di Napoli I*, **5**, 243–64.

Suaria, G. & Aliani, S. (2014). Floating debris in the Mediterranean Sea. *Marine Pollution Bulletin*, **86**, 494–504.

Sumaila, U.R. & Huang, L. (2012). Managing bluefin tuna in the Mediterranean Sea. *Marine Policy*, **36**, 502–11.

Telesca, L., Belluscio, A., Criscoli, A., *et al.* (2015). Seagrass meadows (*Posidonia oceanica*) distribution and trajectories of change. *Scientific Reports*, **5**, article 12505.

Tsikliras, A.C., Dinouli, A., Tsiros, V.-Z. & Tsalkou, E. (2015). The Mediterranean and Black Sea fisheries at risk from overexploitation. *PLoS ONE*, **10**, e0121188.

Tsounis, G., Rossi, S., Grigg, R., *et al.* (2010). The exploitation and conservation of precious corals. *Oceanography and Marine Biology: An Annual Review*, **48**, 161–212.

Turley, C.M. (1999). The changing Mediterranean Sea – a sensitive ecosystem? *Progress in Oceanography*, **44**, 387–400.

Uitz, J., Stramski, D., Gentili, B., D'Ortenzio, F. & Claustre, H. (2012). Estimates of phytoplankton class-specific and total primary production in the Mediterranean Sea from satellite ocean color observations. *Global Biogeochemical Cycles*, **26**, GB2024.

UNEP (1989). State of the Mediterranean Marine Environment. *MAP Technical Reports Series*, 28. Athens: United Nations Environment Programme.

UNEP (2011). *Convention for the Protection of the Marine Environment and the Coastal Region of the Mediterranean and its Protocols*. Athens: United Nations Environment Programme.

UNEP/MAP (2009). *Plan Bleu: State of the Environment and Development in the Mediterranean*. Athens: UNEP/MAP.

UNEP/MAP/WHO (2000). Municipal Wastewater Treatment Plants in Mediterranean Coastal Cities. *MAP Technical Report Series*, 128, 63 pp. Athens: UNEP/MAP.

Van Dyke, J.M. (2013). Whither the UNEP Regional Seas Programmes? In *Regions, Institutions, and Law of the Sea: Studies in Ocean Governance*, ed. H.N. Scheiber & J.-H. Paik, pp. 89–110. Leiden, Netherlands: Nijhoff.

Venizelos, L. & Corbett, K. (2005). Zakynthos sea turtle odyssey – a political ball game. *Marine Turtle Newsletter*, **108**, 10–12.

Voultsiadou, E., Dailianis, T., Antoniadou, C., *et al.* (2011). Aegean bath sponges: historical data and current status. *Reviews in Fisheries Science*, **19**, 34–51.

Zenetos, A., Gofas, S., Morri, C., *et al.* (2012). Alien species in the Mediterranean Sea by 2012. A contribution to the application of European Union's Marine Strategy Framework Directive (MSFD). Part 2. Introduction trends and pathways. *Mediterranean Marine Science*, **13**, 328–52.

# 17 The Southern Ocean

An important shift in the approach to regional marine conservation and management over recent decades has been a greater focus on areas that are meaningful in terms of ecological function. This has included a move towards large-scale ecosystem-based management of marine resources and the concept of large marine ecosystems.

## Large Marine Ecosystems

Large marine ecosystems (LMEs) have been defined as relatively large regions, generally more than 200 000 $km^2$, adjacent to landmasses and characterized by distinct bathymetry, hydrography, productivity, and trophically linked populations (Sherman *et al.*, 2005). The concept has been used to identify 66 LMEs (Fig. 17.1). Most are open shelf regions, whilst others are defined by coastal currents, such as the California, Humboldt, Canary, Benguela, Agulhas, and Kuroshio Currents. On the whole they correspond well with biogeographic realms and provinces (Watson *et al.*, 2003; Spalding *et al.*, 2012). Several LMEs that are semi-enclosed seas, such as the Black, Baltic, Mediterranean, and Caribbean, overlap with areas in the Regional Seas Programme (see Chapter 16). The LME concept was developed initially to provide a framework for fisheries management – LMEs account for the great majority of the world's fish catch – but has emerged as a basis for implementing ecosystem-based management for coastal oceans in general. As such, the LME approach has been taken up by numerous governments and agencies and has attracted funding from the Global Environment Facility to aid development and implementation of LME-oriented projects.

Widespread support for ecosystem-based management (EBM) is evidenced by an increasing number of agreements, governments, and agencies that have embraced the concept. But what exactly EBM encompasses and how it is put into practice are the subject of considerable debate. There are various definitions of EBM, but all recognize the basic criteria of sustainability, ecological health, and the inclusion of humans in an ecosystem (Arkema *et al.*, 2006). A definition would be 'an integrated approach to management that considers the entire ecosystem, including humans. The goal of ecosystem-based management is to maintain an ecosystem in a healthy, productive and resilient condition so that it can provide the services humans want and need. Ecosystem-based management differs from current approaches that usually focus on a single species, sector, activity or concern; it considers the cumulative impacts of different sectors' (McLeod & Leslie, 2009). Importantly, then, EBM adopts a holistic rather than a sectoral approach to management and conservation with a focus on ensuring the long-term integrity and potential of ecosystems to deliver ecosystem services, recognizing in this respect the interactions between the social and ecological domains. The management regime for the Southern Ocean provides an interesting example of a move towards EBM.

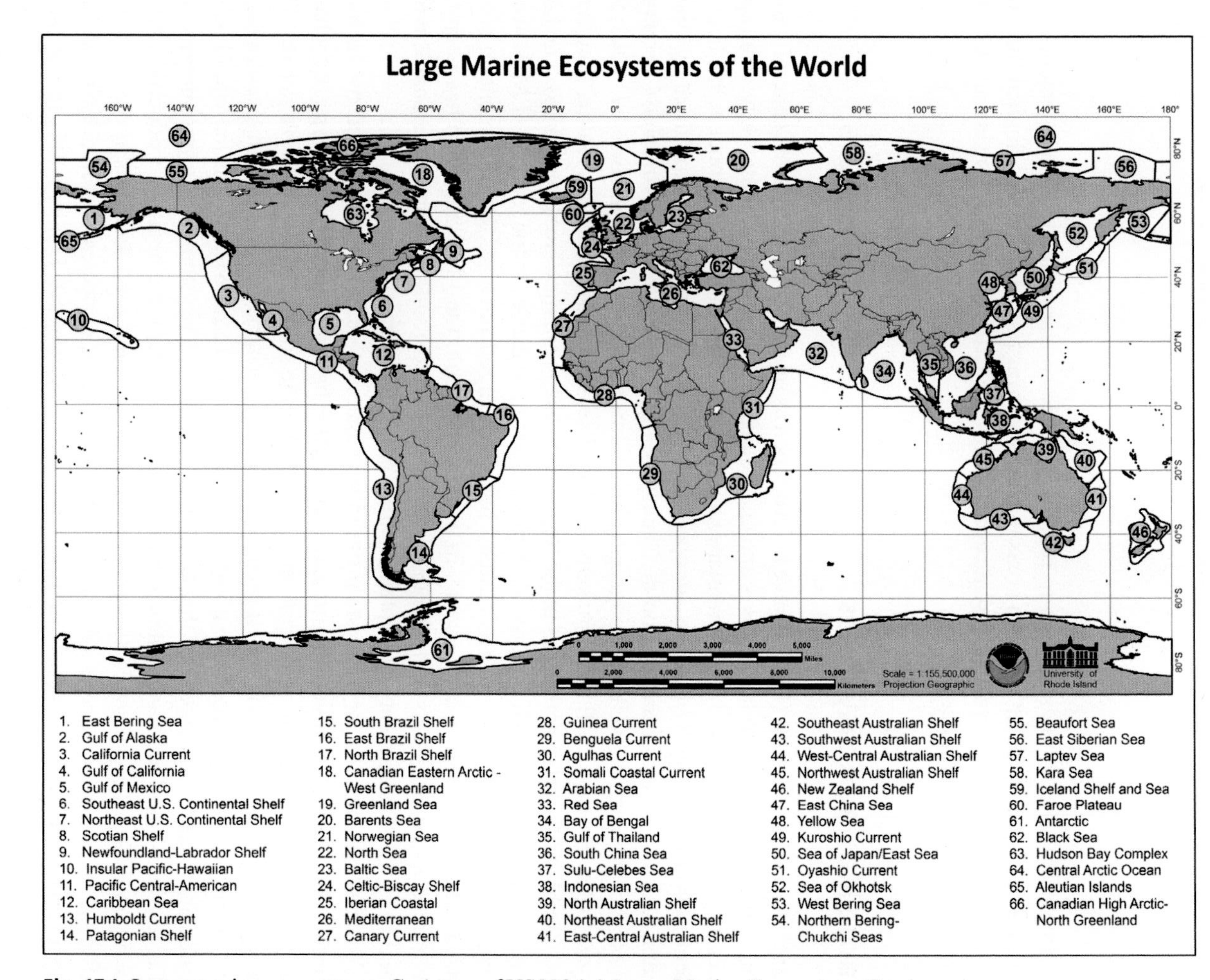


**Fig. 17.1** Large marine ecosystems. Courtesy of US NOAA Large Marine Ecosystems Program.

## The Southern Ocean

Polar regions have long been a magnet for explorers, hunters, scientists, and now, increasingly, tourists. Many have been drawn to these uttermost areas, their challenges, stark beauty, and seemingly inexhaustible stocks of large vertebrates. Polar marine environments raise important issues concerning the role and management of key species in ecosystems. The Antarctic region is unusual in a number of respects that have major implications for the area's marine life and its conservation. Remote from major centres of human population, its waters are among the least contaminated. But by contrast, the historical exploitation of its living resources represents a massive disruption of an ecosystem. Antarctica has no indigenous population nor recognized sovereignty, a unique political status that has helped enable an international, ecosystem-based approach to conservation.

The Southern or Antarctic Ocean is defined as the waters between the Antarctic continent (~ 65–75° S) and the Antarctic Polar Front at 50–60° S (Fig. 17.2). It encompasses some 30 million $km^2$, or 8% of the world's ocean area. Water masses spreading north from the Antarctic make major inputs to the deep circulation of the global ocean (see Chapter 1). On the other hand, the Antarctic Polar Front and Circumpolar Current tend to isolate the Southern Ocean, contributing to its distinctive biogeography.

A huge expanse of sea ice surrounds the continent of Antarctica. A zone of 'fast ice' is semi-permanently attached to the land or to the ice shelves at the edge of glaciers. Beyond this is the seasonal pack-ice zone, which varies enormously in extent from a summer minimum of about 3 million $km^2$ to 20 million $km^2$ in winter. Pack ice provides habitat for a wide range of organisms. Bacteria and diatoms living in the ice and on its undersurface support a diversity of grazers. Among the important zooplankton grazers is the Antarctic krill, a key species in the Southern Ocean ecosystem dependent on sea ice as foraging and nursery habitat. Increased production associated with the sea ice is exploited by seabirds and marine mammals, and the ice provides a platform for penguins and seals (Brierley & Thomas, 2002).

The intertidal zone is severely limited by ice cover and scour, and areas that are temporarily ice-free mainly support only a few species of lichens, algae, and diatoms. But beyond the reach of ice action is a diverse subtidal benthos. The Antarctic continental shelf is generally less than 100 km in width, the main exceptions being in the Weddell and Ross Seas. It is also unusually deep, depressed by the weight of the continental ice sheet, with an average water depth of 450–500 m but in places exceeding 1000 m (Brandt *et al.*, 2012). Beyond the shelf are deep basins (4000–5000 m).

The Antarctic marine biota is among the world's oldest and most isolated. The break-up of Gondwana allowed the Antarctic Circumpolar Current to develop some 25–30 million years ago and in turn its associated fronts, an oceanographic regime that has hampered north-south dispersal (Dayton *et al.*, 1994; Chown *et al.*, 2015). These factors have contributed to high levels of endemism and invertebrate species richness and to the distinctive elements of the biota. High species richness of the benthos may also be due to the size of the Southern Ocean, the stable conditions prevailing on the deep shelf, and the importance in some areas of large sessile organisms such as sponges that increase habitat complexity and the diversity of the associated fauna (Gray, 2001).

By contrast, the vertebrate fauna, although highly characteristic, is far less diverse. Among the fish, 222 demersal species are known from the Antarctic shelf and upper slope. Almost half of these are Antarctic icefishes or notothenioids, which dominate too in terms of abundance and biomass (both > 90%). Such a level of dominance by one group of fishes is unique (Eastman, 2005). Endemicity of the Antarctic fish fauna is also exceptionally high at nearly 90% of the demersal species (Eastman, 2005).

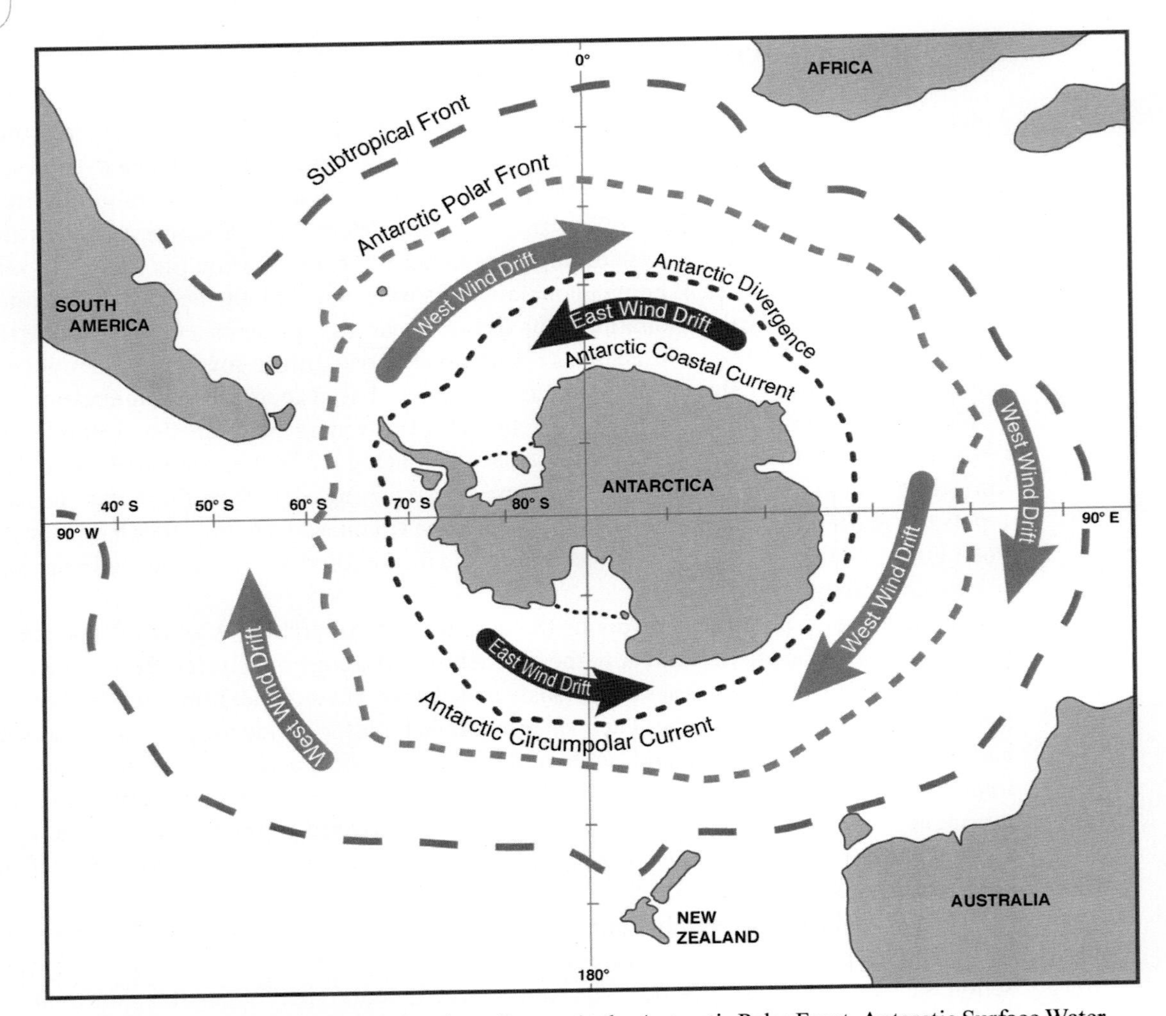


**Fig. 17.2** Surface circulation of the Southern Ocean. At the Antarctic Polar Front, Antarctic Surface Water (~2 to -1.8°C) sinks below warmer (~ 4°C) Subantarctic Surface Water. Antarctic Surface Water originates from the upwelling of Circumpolar Deep Water at the Antarctic Divergence. The divergence is caused by the interaction of opposing surface currents. North of the divergence is the Antarctic Circumpolar Current, being moved in an easterly direction by the West Wind Drift and dominating the surface circulation of the Southern Ocean, whilst between the divergence and the continent is the smaller Antarctic Coastal Current driven by the East Wind Drift. Where Antarctic Surface Water downwells at the Antarctic Polar Front it forms Antarctic Intermediate Water, a flow at intermediate depths (typically 700–1500 m). Surface water spreading southward from the divergence increases in density as it becomes colder and more saline (taking up brine released from the freezing of seawater). It sinks at the edge of the continent to form Antarctic Bottom Water, the main input to abyssal circulation of the world ocean. North of the Antarctic Polar Front is the Subtropical Front, where Subantarctic Surface Water meets Subtropical Surface Water. The subantarctic region lies between these two fronts. Place names referred to in the text are shown in Fig. 17.10.

High levels of endemism of at least 50% for benthic invertebrate classes have also been recorded. Some taxa, however, are notably absent or poorly represented in the Southern Ocean, like the decapod crustaceans, whilst others such as pycnogonids, amphipods, echinoderms, and many of the large suspension feeding groups are particularly diverse (Clarke & Johnston, 2003; Brandt *et al.*, 2012).

## Pelagic Food Web

Overall, primary production of the Southern Ocean is relatively low. For much of the year photosynthesis is constrained by day length and ice cover or the availability of micronutrients such as iron. Average rates of primary productivity for the Southern Ocean are estimated at only 55–60 g C $m^{-2}$ $y^{-1}$, but with the shelf areas being the most productive at about twice these rates (Arrigo *et al.*, 2008). Phytoplankton growth in the Southern Ocean can be very patchy and seasonal, and dense blooms, predominantly of diatoms, can suddenly develop (Dayton *et al.*, 1994; Thomas *et al.*, 2008).

Principal herbivores are copepods, salps, and krill (Euphausiacea). The Antarctic krill (*Euphausia superba*) occupies a pivotal role in the Southern Ocean as the dominant herbivore (Fig. 17.3). This is a large, long-lived species of krill, reaching a length of about 6 cm and with a lifespan of 5–8 years (Thomas *et al.*, 2008). It occurs mainly south of the Antarctic Divergence but also around South Georgia and feeds primarily on diatom blooms. Huge, dense swarms of krill amounting to several thousand tonnes are not uncommon. The total biomass of Antarctic krill is estimated to be of the order of 117–379 million t (Atkinson *et al.*, 2009), comparable to the global biomass of humans (see Chapter 2). Many Southern Ocean species depend on krill, notably various seals, whales, birds, fish, and squid. Current populations of predators are estimated to consume 128–470 million t of krill each year, with the major share taken by crabeater seals and minke whales (Atkinson *et al.*, 2009; Hewitt & Lipsky, 2009).

**Fig. 17.3** Antarctic krill (*Euphausia superba*), a key species in the Southern Ocean food web. Photograph: Getty Images/Gerald and Buff Corsi/Visuals Unlimited, Inc. (A black and white version of this figure will appear in some formats. For the colour version, please refer to the plate section.)

## Role of Endotherms

A marked feature of the Southern Ocean is the importance at the top of the food web of endotherms, especially the seals and penguins, which are permanent residents of the region, and the large whales, which migrate to Antarctic waters in summer (Table 17.1). Being able to maintain a high metabolic rate at low ambient temperatures is clearly advantageous, and the strong seasonal build-up of prey provides rich feeding grounds. The Southern Ocean probably supports more than half the world's marine mammal biomass (Boyd, 2002).

**Table 17.1** Major Species of Penguins, Seals, and Whales of the Antarctic and Their IUCN Red List Conservation Status

CR: critically endangered, EN: endangered, VU: vulnerable, NT: near threatened, LC: least concern, DD: data deficient

| | | |
|---|---|---|
| Penguins | | |
| Emperor penguin | *Aptenodytes forsteri* | NT |
| King penguin | *Aptenodytes patagonicus* | LC |
| Adélie penguin | *Pygoscelis adeliae* | NT |
| Chinstrap penguin | *Pygoscelis antarcticus* | LC |
| Gentoo penguin | *Pygoscelis papua* | NT |
| Macaroni penguin | *Eudyptes chrysolophus* | VU |
| Seals | | |
| Antarctic fur seal | *Arctocephalus gazella* | LC |
| Southern elephant seal | *Mirounga leonina* | LC |
| Weddell seal | *Leptonychotes weddellii* | LC |
| Ross seal | *Ommatophoca rossii* | LC |
| Crabeater seal | *Lobodon carcinophagus* | LC |
| Leopard seal | *Hydrurga leptonyx* | LC |
| Whales | | |
| Southern right whale | *Eubalaena australis* | LC |
| Humpback whale | *Megaptera novaeangliae* | LC |
| Common minke whale | *Balaenoptera acutorostrata* | LC |
| Antarctic minke whale | *Balaenoptera bonaerensis* | DD |
| Sei whale | *Balaenoptera borealis* | EN |
| Fin whale | *Balaenoptera physalus* | EN |
| Antarctic blue whale | *Balaenoptera musculus intermedia* | CR |
| Pygmy blue whale | *Balaenoptera musculus brevicauda* | DD |
| Sperm whale | *Physeter macrocephalus* | VU |
| Killer whale | *Orcinus orca* | DD |

Extracted from www.iucnredlist.org (accessed 24 June 2016).

Six species of seal occur in these waters: the Antarctic fur seal (Otariidae) and the southern elephant, Weddell, Ross, crabeater, and leopard seals (Phocidae). Weddell, Ross, crabeater, and leopard seals are adapted for living in the pack-ice or fast-ice zones, whereas fur seals and elephant seals have a more northerly distribution beyond the pack-ice zone and use land rather than ice during pupping (Siniff, 1991). By far the most abundant seal is the crabeater seal, with a likely population of around 5–10 million individuals and which feeds almost exclusively on krill (Bengtson, 2009). The other five pinnipeds have total populations of between 0.2 and 1 million. Antarctic fur seals also feed largely on krill, whereas southern elephant, Weddell, and Ross seals mainly take squid and fish. Leopard seals consume a wide range of prey, in particular krill, young crabeater seals, and penguins.

Seven species of baleen whales occur in Antarctic waters: the southern right, humpback, common minke, Antarctic minke, sei, fin, and blue (with two subspecies). The humpback and rorquals all feed chiefly on krill. There are also two large species of toothed whales, the sperm whale (although only adult males migrate this far south) and the killer whale, as well as several smaller odontocetes that appear to be comparatively rare and are poorly known.

**Fig. 17.4** Adélie penguins are considered vulnerable to effects of climate change, including loss of sea ice and a decline in krill abundance. Parent birds at Cape Bird, Ross Island, returning to sea to forage for their chicks. Photograph: Kim Westerskov. (A black and white version of this figure will appear in some formats. For the colour version, please refer to the plate section.)

About 40 seabird species breed in the region. Most prominent of the flying species are the albatrosses and petrels and the skuas, gulls, and terns. But the dominant group, accounting for about 90% of the bird biomass, are the penguins (Fig. 17.4). Six species are characteristically Antarctic: two large species, the emperor and king penguins, and four smaller species, Adélie, chinstrap, gentoo, and macaroni penguins. King, Adélie, chinstrap, and macaroni penguins have breeding populations of between 2 and 8 million pairs, and emperor and gentoo penguins some 0.24 and 0.39 million breeding pairs, respectively (Garcia Borboroglu & Boersma, 2013). Emperor and king penguins are deep divers that feed on fish. The four smaller species feed mainly on krill and in this regard are very significant in the pelagic food web. Macaroni penguins in particular, as the leading consumer among the seabirds, have been estimated to remove annually more than 9 million t of krill from the Southern Ocean (Brooke, 2004).

Pelagic food-web structure of the Southern Ocean has in the past been considered to be relatively simple, with few key groups, principally diatoms, krill, and marine mammals. More recent studies have shown the system to be far more regionalized, dynamic, and interlinked than previously realized and not radically different from other ocean areas, particularly in the role played by microplankton and microbial processes (Knox, 2007; Thomas *et al.*, 2008) (see Chapter 1). Nevertheless, at higher trophic levels the Southern Ocean food web can in general be portrayed as one with relatively few major components and with much of the energy flow through krill and a relatively small number of vertebrate predators (Fig. 17.5). Exploitation of top consumers has been by far the greatest human impact on the Southern Ocean, significantly disrupting ecosystem structure and with major ramifications for management and conservation.

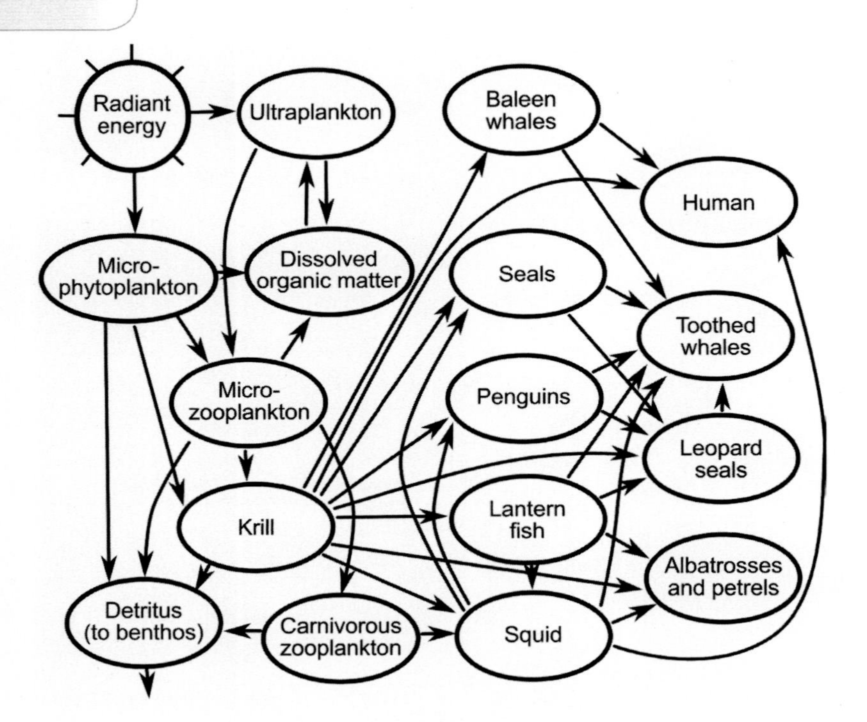


**Fig. 17.5** Simplified pelagic food web for the Southern Ocean. Redrawn from OUP Material: *The Biology of Polar Regions* by Thomas et al. (2008), Fig. 6.10(b), p. 177. By permission of Oxford University Press, www.oup.com.

# Human Impacts

## Exploitation of Living Resources

Historically, exploitation of the Southern Ocean's living resources covers four main phases, targeting in turn seals, whales, finfish, and krill (Sage, 1985; Miller, 1991).

### Sealing

The sealers' main target was initially the Antarctic fur seal, which has its major breeding population on South Georgia, with smaller numbers mainly on other islands of the Atlantic sector (Bonner, 1984). The first sealing expedition arrived on South Georgia in 1778, but by the early 1820s at least 1.2 million skins had been taken, and the species was nearly extinct there. Over the next few years the populations on the other main islands met a similar fate, and apart from a brief revival in the 1870s, sealing in this region was no longer worthwhile. Fur seals have since made a good recovery with the total population now likely to be several million. Sealers also took southern elephant seals for the oil in their thick blubber. Whilst exploitation was less severe than for fur seals, many island populations were decimated. However, stocks recovered more quickly than those of fur seals, and from 1910 to 1964 an elephant sealing industry was operating on South Georgia but this time restricted to a quota of 6000 adult males each season (Bonner, 1984). The other four species – Weddell, Ross, crabeater, and leopard seals – inhabit the sea-ice region and being far less accessible have suffered only small-scale exploitation (Siniff, 1991). None of these pinnipeds is now directly threatened by human activities.

### Whaling

The Southern Ocean saw the development of industrial whaling on an unprecedented scale, which hastened the first international attempts to regulate whaling (see Chapter 9). Modern whaling in these waters dates from 1904 when the first whaling station was established on South Georgia (Bonner, 1984) (Fig. 17.6). Over the next 20 years, shore-based whaling flourished here and at

**Fig. 17.6** The establishment of whaling stations on South Georgia and the South Shetland Islands heralded a period of intense whaling in the Southern Ocean from the early 1900s. The whaling station at Grytviken, South Georgia, about 1914–16. Photograph: Scott Polar Research Institute, University of Cambridge, with permission.

other islands in the South Atlantic sector, so that over the period 1910–25 the Southern Ocean was already accounting for half the world's annual whale catch (Knox, 2007). These shore-based operations were under the administration of the Falkland Island Dependencies, which exerted some control over the whaling. But such restrictions ended in the 1920s with the introduction of factory ships with stern slipways, enabling whaling to be carried out entirely at sea beyond any territorial restrictions. Catches then rose dramatically, peaking at 46 000 whales in the 1936–7 season. Post-war, annual catches were of the order of 30 000 whales up to the mid-1960s but thereafter fell rapidly as stocks dwindled and international regulatory measures became more restrictive (Knox, 2007). Over the past century more than 2 million whales have been taken from Antarctic waters. Stocks were probably already doomed by the 1930s given the highly *K*-selected characteristics of the species, the efficiency of factory ships, and the demands of an already over-capitalized industry (Bonner, 1984).

Whale catches in the Southern Ocean are a striking example of serial depletion (Fig. 17.7). Humpback whales were initially taken as they occurred close to the shore-based stations. But with the introduction of factory ships giving unrestricted access to open ocean, whalers targeted the largest, most valuable species, the blue and fin whales. With the collapse of these stocks, the industry turned to species previously largely ignored: sei and, eventually, minke whales. With the decline of baleen whales, sperm whales have also been taken, though in smaller numbers, mainly from the mid-1940s to the late 1970s. Species have been impacted very differently by whaling. Most severely affected have been the humpback and blue whale, which are now at only a few percent of their pre-exploitation levels. Fin whales have been reduced to about a fifth and sei and sperm whale numbers halved, but minke whales are probably as abundant now as they were initially (Berkman, 1992).

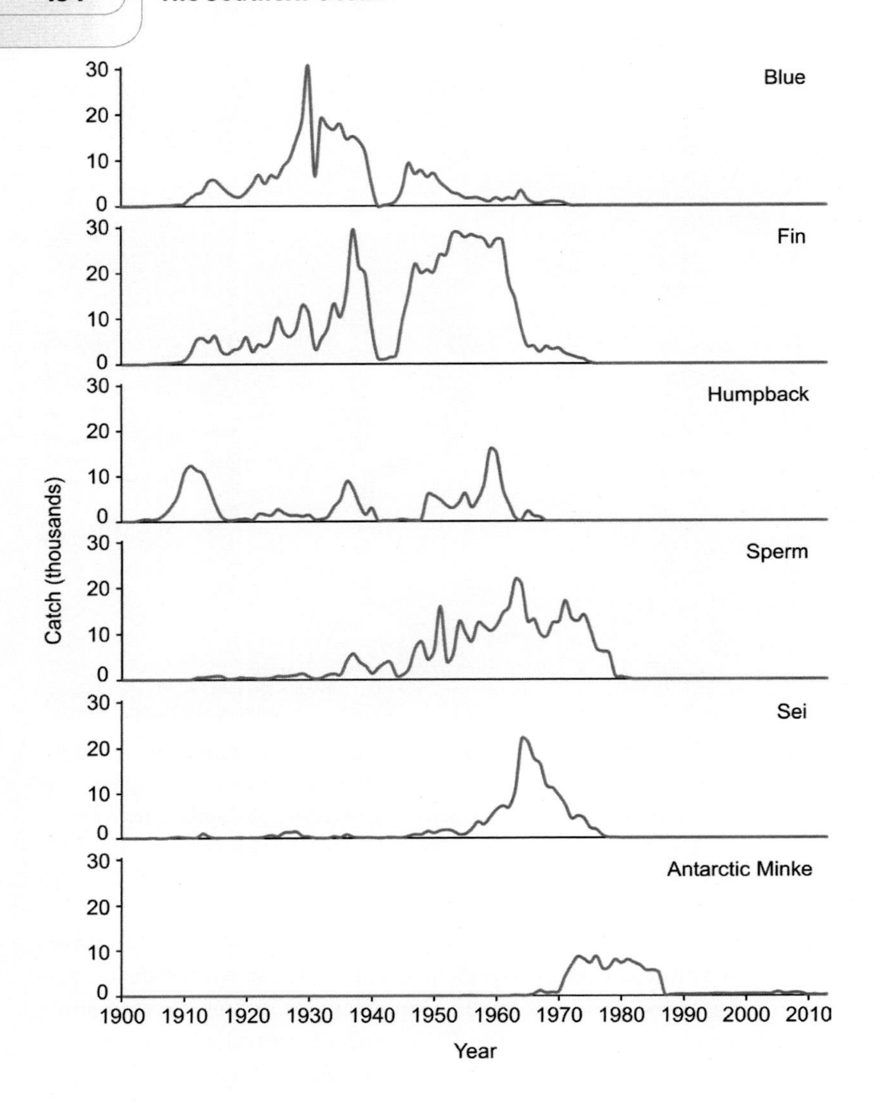


**Fig. 17.7** Annual catches of blue, fin, humpback, sperm, sei, and Antarctic minke whales in the Southern Hemisphere. Data from C. Allison, IWC, with permission.

## Finfish

Interest in the development of Southern Ocean fisheries dates from the early twentieth century when the South Georgia whaling station was established. But intensive commercial fishing in these waters started only in the late 1960s, as traditional grounds became depleted and appropriation of EEZs encouraged nations with distant-water fleets to seek new grounds. Several countries have been involved in Antarctic fisheries, but particularly Russia. The main areas exploited have been in the vicinity of the Kerguelen Islands and the Scotia Sea. Total catches of nearly 400 000 t were taken during the 1970s, mainly of the marbled rockcod (*Notothenia rossii*), a large nototheniid, and the mackerel icefish (*Champsocephalus gunnari*), but catches declined in the 1980s to less than 100 000 t, and by the early 1990s these stocks had effectively collapsed (Miller, 2000).

By this time a new fishery had developed, for the Patagonian toothfish (*Dissostichus eleginoides*), a large, long-lived nototheniid; individuals can live for 50 years and reach more than 2 m in length (Collins *et al.*, 2010). The species lives at shelf to slope depths with genetically distinct populations in Antarctic to southern temperate waters. The fishery, mainly by demersal longline, grew rapidly from the 1980s to a peak in reported landings of 40 000 t in 1995. The total allowable catch (TAC) is now about 25 000 t, but illegal, unreported, or unregulated fishing has been an ongoing problem and likely to significantly increase the total catch. A further issue has been the bycatch of seabirds, mainly of white-chinned petrels (*Procellaria aequinoctialis*) (Barbraud *et al.*, 2008). However, various mitigation measures (see Chapter 8) such as streamer lines, night setting, and seasonal closures have now greatly reduced this incidental mortality.

Patagonian toothfish typify the problem of fisheries that rapidly develop before there is adequate knowledge of the species' biology and ecology to guide management. This is a particular problem in the case of deep-sea demersal species with life-history traits that render them highly vulnerable to overexploitation (see Chapter 6). A second species of toothfish, the Antarctic toothfish (*Dissostichus mawsoni*), which occurs south of the Antarctic Polar Front, is the target of an emerging fishery. Pinkerton and Bradford-Grieve (2014) used a model of the Ross Sea food web to examine the potential impact on trophic groups of fishing for Antarctic toothfish in this region. Their analysis indicates that toothfish abundance has a strong top-down effect on medium-sized demersal fish, but as this group has only low trophic importance in the system, its release from predator pressure would be unlikely to trigger a trophic cascade. The model suggests that toothfish are only a small part of the diet of its main predators, Weddell seals and killer whales, and that fishing does not significantly impact the diet of these marine mammals. The longline fishery for Antarctic toothfish in the Ross Sea has been certified by the Marine Stewardship Council (see Chapter 5).

## Krill

Antarctic krill was initially seen as a potentially huge source of protein for human consumption but is now taken mainly for aquaculture feeds and pharmaceuticals. Commercial fishing for krill began in the early 1970s, carried out mainly by the former Soviet Union and Japan, and within a decade annual catches had reached more than 500 000 t. Catches fell with the break-up of the Soviet Union and the withdrawal of its subsidized fleet. Several other countries subsequently entered the fishery, and since the mid-1990s catches have been around 100 000–200 000 t (Miller & Agnew, 2000; Nicol *et al.*, 2012) (Fig. 17.8). The fishery operates chiefly in the Atlantic sector from late summer to mid-winter, traditionally using large pelagic trawls, but other methods are being developed including continuous pumping. The fishery for Antarctic krill is potentially the largest in the world, but the cost of fishing in these remote waters and difficulties in processing krill have held back the fishery (Nicol *et al.*, 2012). Nevertheless, given the size of the resource and the pressure on traditional stocks, the fishery is sure to grow.

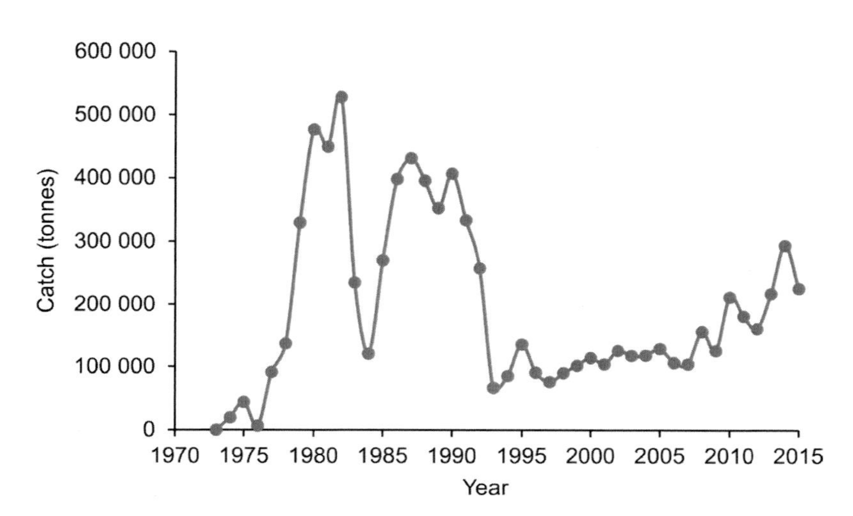


**Fig. 17.8** Reported total catches of krill from the Southern Ocean 1973–2015. Catches have been almost entirely from the Atlantic sector (FAO Area 48), the main exception being those from the Indian Ocean sector (Area 58) in the early 1980s (< 155 000 t). From CCAMLR. (2016). *Statistical Bulletin*, 28 (database version), Hobart: CCAMLR. www.ccamlr.org/en/fisheries/krill-fisheries.

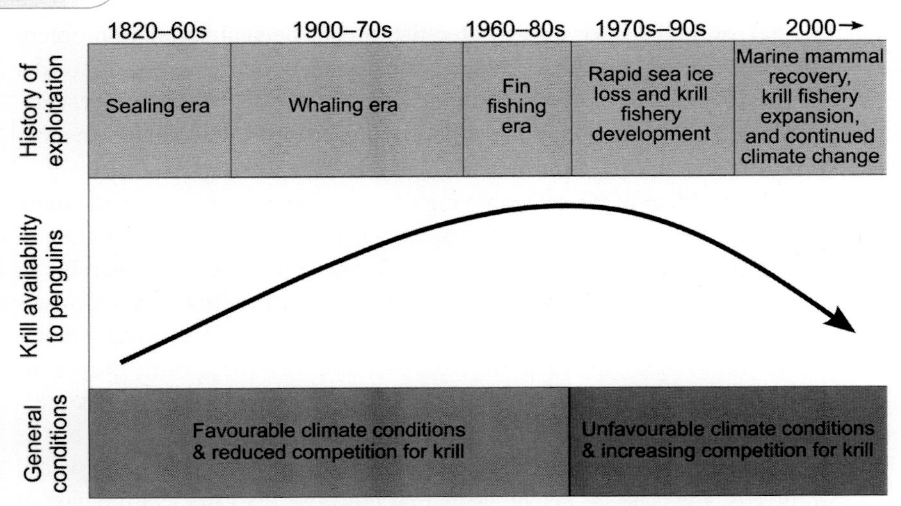


**Fig. 17.9** Ecosystem perturbations in the Scotia Sea and the effect of krill availability on populations of Adélie and chinstrap penguins. The serial depletion of Antarctic fur seals, large baleen whales, and finfish (ice fishes and notothenioids), and hence reduced competition for krill, was associated with large increases in Adélie and chinstrap penguin populations throughout the Scotia Sea region. But since the late twentieth century, the availability of krill to penguins has declined, a trend that is forecast to continue. This change in availability is due to a combination of reduced krill productivity from climate-change effects, increased competition for krill from recovering marine mammal populations, and an expanding krill fishery. Reproduced with permission from Trivelpiece, W.Z., Hinke, J.T., Miller, A.K., et al. (2011). Variability in krill biomass links harvesting and climate warming to penguin population changes in Antarctica. *Proceedings of the National Academy of Sciences*, 108, 7625–8.

The overexploitation of krill-dependent predators has undoubtedly caused a significant restructuring of the Southern Ocean ecosystem. The removal of large baleen whales in particular is believed to have resulted in a 'surplus' of krill for other consumers. There is evidence that krill-eating seals and penguins increased in abundance in the mid-twentieth century after whaling had peaked and that such increases may be directly related to greater availability of krill (Laws, 1977; Mori & Butterworth, 2006). However, any krill surplus was curtailed by a marked decline in primary productivity in the last quarter of the twentieth century. Various factors, including those associated with climate change, are likely to affect krill productivity. Trivelpiece *et al.* (2011) argue that the marked decline in krill abundance over recent decades has been driven by ocean warming and reductions in sea ice and that these factors combined with increased competition for krill from recovering whale and fur seal populations have been responsible for the severe decline in Adélie and chinstrap penguin populations in the Scotia Sea region (Fig. 17.9). There is the possibility too that krill productivity is linked to whaling. Pre-exploitation populations of baleen whales would have returned large amounts of iron, a limiting nutrient, from their krill prey back into surface waters for use by phytoplankton (Nicol *et al.*, 2010). If maintaining productivity of the Southern Ocean depends on healthy populations of baleen whales for this feedback, then their recovery is likely to take far longer than expected (Surma *et al.*, 2014).

## Other Human Impacts

### Visitors

Until recently, most visitors to the Antarctic were scientists and support staff. There are about 80 seasonal or year-round research stations in Antarctica, most of them coastal, with total personnel peaking at more than 4000, a quarter of them at the US McMurdo Station on Ross Island in the Ross

Sea (www.comnap.aq). There has sometimes been scant regard for the environmental impact of these operations, resulting in significant localized pollution, including from sewage, hydrocarbons, metals, and debris (Lenihan *et al.*, 1990; Stark *et al.*, 2014). The situation has, however, improved given the environmental protection provisions of the Madrid Protocol (see below), which requires that wastes (except sewage) be removed from Antarctica.

Most visitors to the Antarctic are now tourists, the great majority as ship-borne passengers. Tour ships have regularly visited the Southern Ocean since the mid-1960s, mostly to the Antarctic Peninsula region during summer (November to March). Tourist numbers were initially quite modest, up to 1000 visitors per year, but have risen sharply since the 1980s to about 37 000 in recent seasons (www.iaato.org). Tourism brings environmental risks. Tourist ships could spill oil or release ballast water containing exotic cold-water species, and since many cruises visit the same sites, there are concerns about the disruptive effects of repeated visits on penguin rookeries and other wildlife (Tin *et al.*, 2009; Lynch *et al.*, 2010). In 1991, major tour operators formed the International Association of Antarctica Tour Operators (IAATO), one of its aims being to promote environmentally responsible private-sector travel to the Antarctic. As yet, there is limited information on the effects of tourism in the Antarctic. Tourism ventures must also operate in accordance with the Madrid Protocol. But given the increasing scale and potential environmental impact of tourism, there is an urgent need to establish long-term programmes of research and monitoring and for parties to the Antarctic Treaty System to develop a comprehensive management framework for tourism (Verbitsky, 2013).

## Pollution and Global Climate Change

Concentrations of contaminants in water and biota in the Southern Ocean are generally far lower than those measured in the Northern Hemisphere (Knox, 2007). To a large extent this may be explained by the circumpolar circulation constraining the input of contaminants from lower latitudes. Nevertheless, contaminants such as pesticides and other organochlorines are detectable at low concentrations throughout the region, indicating input from atmospheric deposition. The most obvious inputs of marine pollutants have been fuel and oil from ship groundings and wastes from scientific bases. However, such impacts have been localized, and detrimental environmental effects appear to have been limited. Plastics and other persistent debris are a potential hazard to wildlife, affecting particularly seabirds and seals through ingestion and entanglement, and may become more significant with the growth of fishing and tourism. Antarctic waters are designated as a special area under Annexes I, II, and V of MARPOL, which relate particularly to the disposal from vessels of oil, noxious liquid substances, and plastics (see Chapter 2). Similar provisions, although with more restrictive discharge standards, are contained in Annex IV of the Madrid Protocol (see below).

Potentially of far greater significance to the Southern Ocean are the effects of global climate change, notably from ocean warming and acidification (see Chapter 2). There is still much uncertainty about the ultimate effects of climate change on the Southern Ocean. Considerable regional variation in responses is likely, but projections indicate overall a southward shift of ocean fronts, warming and freshening of surface waters, increased stratification (reducing upward mixing of nutrients into the euphotic zone), and a decrease in the extent of sea ice (Constable *et al.*, 2014). The impact of ocean acidification is expected to be greater in polar waters than at lower latitudes since the solubility of carbon dioxide increases as water temperature falls and calcification becomes energetically more demanding (Fabry *et al.*, 2009). There is evidence that ocean acidification is already affecting calcifying organisms in the Southern Ocean. Pteropods, pelagic gastropods that make their shells mainly from aragonite, can dominate the plankton biomass in parts of the Southern Ocean. Bednaršek *et al.* (2012) recovered live pteropods showing severe levels of shell dissolution from surface waters of the Scotia Sea, waters that were under-saturated with respect to aragonite, at least partly due to oceanic absorption of anthropogenic $CO_2$. But also of concern are possible physiological effects of reduced pH on non-calcifying organisms, such as Antarctic krill (Flores *et al.*, 2012). Climate change is expected to affect marine mammals and birds of the Southern Ocean mainly as a consequence of fronts shifting position and changes to the extent and duration of sea ice, as these have implications for the location of feeding grounds and the energetic cost of trips between foraging and breeding sites (Constable *et al.*, 2014). For species that depend on sea ice, such as the emperor penguin, climate warming will likely bring about major population declines (Jenouvrier *et al.*, 2014).

The springtime depletion of stratospheric ozone over Antarctica, the so-called ozone hole, and resulting enhanced flux of UV radiation may also have consequences for the Southern Ocean ecosystem, including a reduction in primary production and damaging effects on organisms living in surface waters (Arrigo *et al.*, 2003). Ozone depletion also has effects on atmospheric and oceanic circulation, such as enhancing the Antarctic Circumpolar Current, and these may have a greater impact on marine ecosystems than the increase in UV (Robinson & Erickson, 2015). However, there is now evidence that the ozone layer is beginning to heal (see Chapter 2).

Climate-change processes will have various direct and indirect effects on the major habitats – sea-ice, pelagic, and benthic – with almost the entire Southern Ocean ecosystem projected to be significantly impacted by one or more of these environmental stressors (Gutt *et al.*, 2015). As projections develop, it will be important to establish the extent to which areas expected to be most affected overlap with areas of high conservation importance and to incorporate climate-change effects into ecosystem-based management.

## Management and Conservation

Measures to manage and conserve marine life of the Southern Ocean are covered mainly by international whaling agreements and the Antarctic Treaty System. Attempts at managing whale stocks of the region date from the 1930s (see Chapter 9), including the establishment in 1938 of a baleen whale sanctuary in the Pacific sector south of 40° S. Whaling had not been carried out in this area – it was not thought to have large whale stocks – so the sanctuary met with little opposition. But as stocks declined elsewhere, pressure mounted for the area to be opened up, and the sanctuary lasted only until 1955 (Holt, 2000). Internationally agreed-upon whale quotas were introduced in 1944, but being based on the oil production of species rather than their different protection needs, the approach failed to check the serial overexploitation. In 1946, the International Whaling Commission (IWC) was set up to regulate catch limits. It successively reduced quotas, introduced new management methods, and eventually brought in protection for humpback and blue whales in the 1960s and later for fin and sei whale stocks. In the mid-1980s, the IWC brought in a moratorium for all commercial whaling, though subject to subsequent reassessment. But by this stage most Southern Ocean whale populations had already crashed. In 1994, the IWC adopted the Southern Ocean Sanctuary as an area in which commercial whaling is prohibited (Fig. 9.9). However, there is still whaling in the Southern Ocean as Japan has continued to take minke whales – up to 440 each year since 1987 – supposedly for scientific purposes (see Chapter 9).

### Antarctic Treaty System

Seven states have made territorial claims in the Antarctic, mostly pie-shaped sectors that extend to 50° or 60° S. The legitimacy of these claims has long been disputed by the international community and by the 1950s had become a source of international discord. The situation was partly defused by the Antarctic Treaty of 1959 under which the 12 nations then active in Antarctica agreed to hold all territorial claims in abeyance. Other major provisions of the treaty are to ensure that Antarctica is used for peaceful purposes and to promote freedom of scientific investigation and international collaboration in such research. The treaty, now with more than 50 signatories (www.ats.aq), applies to all land and ice shelves south of 60° S (Fig. 17.10) but not, under Article VI, the high seas (see below). The treaty has no expiry date; it was declared to remain in force for a minimum of 30 years, after which any of the contracting parties could seek a review of its operation, but none has done so.

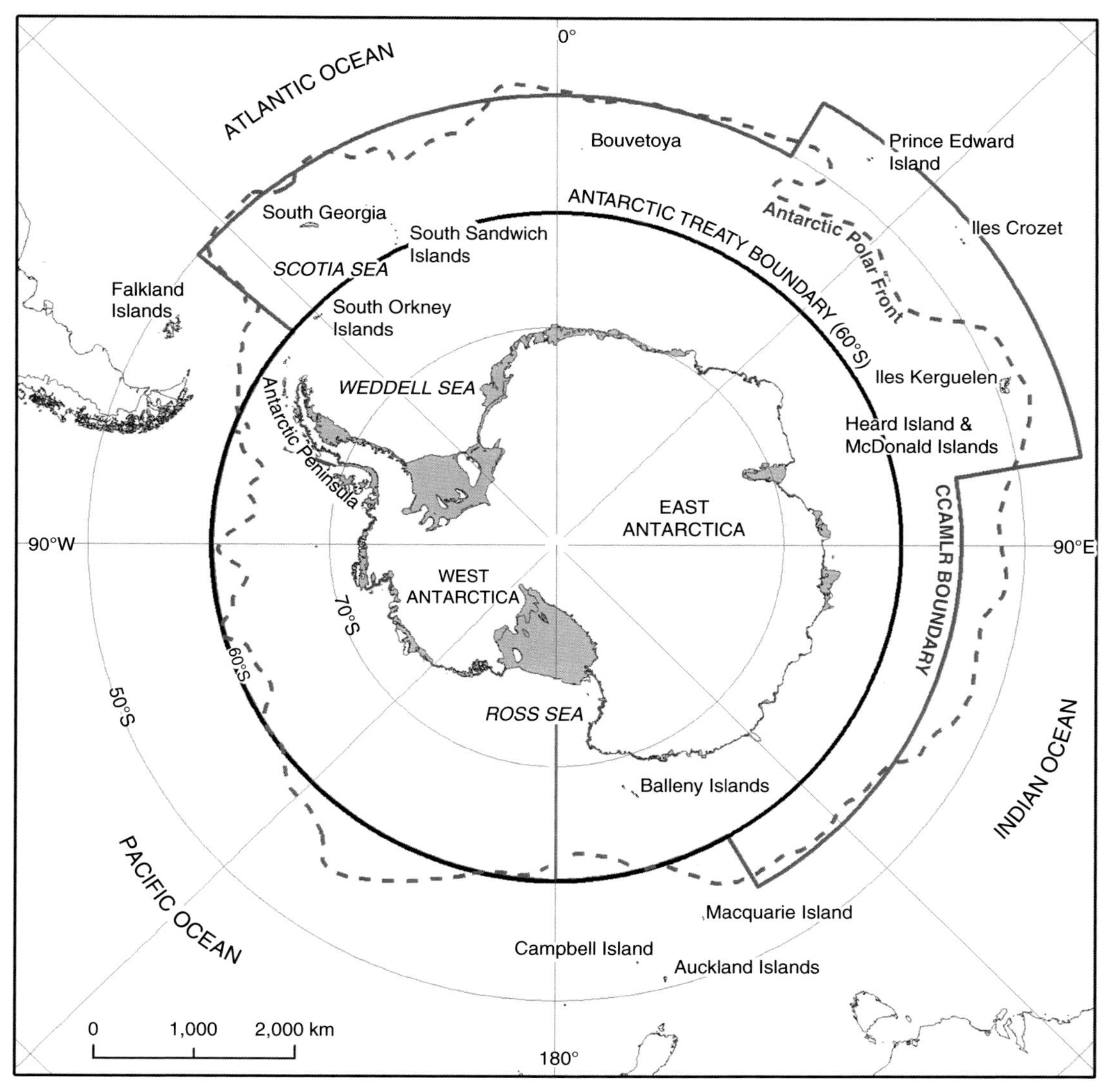


**Fig. 17.10** Antarctica and the Southern Ocean showing the boundary for the Antarctic Treaty at 60° S and for CCAMLR, which approximates to the position of the Antarctic Polar Front. Reproduced with permission from Grant, S.M., Convey, P., Hughes, K.A., Phillips, R.A. & Trathan, P.N. (2012). Conservation and management of Antarctic ecosystems. In *Antarctic Ecosystems: An Extreme Environment in a Changing World*, ed. A.D. Rogers, N.M. Johnston, E.J. Murphy & A. Clarke, pp. 492–525. Chichester, UK: Wiley-Blackwell. © 2012 by Blackwell Publishing Ltd.

Whilst the Antarctic Treaty itself does not specifically address environmental management and conservation, it enjoins the parties to consider and recommend measures for the preservation and conservation of living resources in Antarctica (Article IX), and such measures are included in a series of subsequent related agreements. The founding treaty and its related agreements are known collectively as the Antarctic Treaty System (ATS) (Box 17.1).

## Box 17.1 Main Components of the Antarctic Treaty System

The dates in brackets are the years of signing and of entry into force.

### The Antarctic Treaty (1959, 1961)

Related agreements:
Agreed Measures for the Conservation of Antarctic Fauna and Flora (1964, 1982)
Convention for the Conservation of Antarctic Seals (1972, 1978)
Convention on the Conservation of Antarctic Marine Living Resources (1980, 1982)
Protocol on Environmental Protection to the Antarctic Treaty (1991, 1998) (the Madrid Protocol) Includes:

- Annex I: Environmental Impact Assessment (1991, 1998)
- Annex II: Conservation of Antarctic Fauna and Flora (1991, 1998)
- Annex III: Waste Disposal and Waste Management (1991, 1998)
- Annex IV: Prevention of Marine Pollution (1991, 1998)
- Annex V: Area Protection and Management (1991, 2002)
- Annex VI: Liability Arising from Environmental Emergencies (2005, not yet in force)

The first such agreement was the Agreed Measures for the Conservation of Antarctic Fauna and Flora (1964), adopted for the protection of native mammals, birds, and plants, to prevent uncontrolled introduction of non-indigenous organisms, and to provide for the designation of specially protected areas and species, including fur and Ross seals. As with the treaty itself, the agreed measures apply only to land south of 60° S. To protect seals in the sea or on floating ice, further measures were adopted in 1972 with the Convention for the Conservation of Antarctic Seals (CCAS). Almost no commercial sealing has taken place in the Antarctic Treaty Area since the almost total extermination of fur seals in the South Shetland Islands by the mid-1820s. Nevertheless, CCAS provides a means to regulate sealing in the unlikely event that it should resume. It also includes provisions for closed seasons, sealing zones, and three seal reserves (Bonner, 1984).

Exploitation of marine mammals of the high seas in the Antarctic Treaty area is thus largely regulated by the CCAS and IWC (an exception being the smaller cetaceans). There are, however, other major living resources of potential commercial interest. In particular, the krill resource could rival all existing world fisheries. Large-scale extraction of krill would have major repercussions for dependent predators, notably among the seabirds, seals, and whales. To address such concerns, Antarctic Treaty nations concluded in 1980 the Convention on the Conservation of Antarctic Marine Living Resources (see below).

Discussions have also taken place among Antarctic Treaty nations on the possibility of mineral exploitation and, in the event, of the need for adequate environmental protection. The Convention on the Regulation of Antarctic Mineral Resource Activities (CRAMRA) was adopted in 1988 but remains unratified because of concerns about the efficacy of the environmental safeguards and whether the agreement would encourage rather than control mining. But as a result, parties recognized the need for a more comprehensive approach to environmental protection in the region, and the Protocol on Environmental Protection to the Antarctic Treaty – the Madrid Protocol – was signed in 1991. This aims to provide for the comprehensive protection of the Antarctic environment and dependent and associated ecosystems and to designate it as a natural reserve. Further to CRAMRA, the protocol prohibits any activity relating to mineral resources, other than scientific research, at least for 50 years, after which any of the contracting parties can seek a review. However, several countries have stepped up their research into Antarctic mineral resources, which is seen as a preparation for prospecting and eventual exploitation (Sutherland *et al.*, 2014). Other major provisions of the protocol concern environmental impact assessment for all activities likely to entail significant adverse effects, conservation of native fauna and flora (strengthening the agreed measures), waste disposal and management (including the banning of certain substances from the area), prevention of marine pollution (based on MARPOL), and management of protected areas (rationalizing previous categories) (Box 17.1).

Given that claims of sovereignty in the Antarctic are not recognized internationally, Antarctic circumpolar waters may be regarded as high seas that extend right up to the ice shelves and the continent's shoreline. Also, by this argument, the underlying seabed and its mineral resources should come under the jurisdiction of the International Seabed Authority (see Chapter 14), a view obviously opposed by claimant states.

## Convention on the Conservation of Antarctic Marine Living Resources

The Convention on the Conservation of Antarctic Marine Living Resources (CCAMLR) entered into force in 1982, with 30 states now party to the convention. A commission is responsible for implementing the convention. CCAMLR applies to a larger area than the Antarctic Treaty as its northern limit approximates to the Antarctic Polar Front (Fig. 17.10). CCAMLR is of particular significance to marine conservation as it was the first international convention to take an ecosystem approach to high-seas fisheries management. In this way it pioneered principles of the large marine ecosystem concept, and (unusually) it was concluded before the species it sought to protect had come under heavy commercial pressure (Bowman *et al.*, 2010). The convention's objective is 'the conservation of Antarctic marine living resources', where conservation includes rational use (Article II) and where Antarctic marine living resources means 'the populations of fin fish, molluscs, crustaceans and all other species of living organisms, including birds, found south of the Antarctic Convergence [Antarctic Polar Front]' (Article I). CCAMLR excludes seals and whales, which are covered by the CCAS and the International Convention for the Regulation of Whaling, respectively (see Chapter 9), although it cooperates closely with these agreements.

The key provisions (Article II 3) underpinning CCAMLR's ecosystem approach to exploitation of stocks require that any harvesting and associated activities be conducted in accordance with certain principles of conservation (Box 17.2). Essentially, the aims are to prevent exploited populations being reduced below their level of maximum sustainable yield and maintain stability of recruitment of target and associated species, to restore depleted populations, and to ensure that impacts of fishing are reversible within a reasonable period. In the case of krill, exploitation must not jeopardize its predator populations of baleen whales, seals, and penguins nor impede the recovery of such populations.

### Box 17.2 Principles of Conservation Set Out in the Convention on the Conservation of Antarctic Marine Living Resources (Article II 3)

Any harvesting and associated activities shall be conducted in accordance with the following principles of conservation:

(a) prevention of decrease in the size of any harvested population to levels below those which ensure its stable recruitment. For this purpose its size should not be allowed to fall below a level close to that which ensures the greatest net annual increment;
(b) maintenance of the ecological relationships between harvested, dependent and related populations of Antarctic marine living resources and the restoration of depleted populations to the levels defined in (a) above; and
(c) prevention of changes or minimization of the risk of changes in the marine ecosystem which are not potentially reversible over two or three decades, taking into account the state of available knowledge of the direct and indirect impact of harvesting, the effect of the introduction of alien species, the effects of associated activities on the marine ecosystem and of the effects of environmental changes, with the aim of making possible the sustained conservation of Antarctic marine living resources.



At least in its early years, the commission of the CCAMLR took a reasonably conventional approach to fisheries management with a focus on TACs of single species for finfish stocks, but an approach that was generally unsuccessful, being confounded, for example, by inadequate information on life-history characteristics of target species and problems of dealing with uncertainty and implementation of decisions (Constable *et al.*, 2000). But from the mid-1980s, the commission began to examine more critically how to achieve its ecosystem management objectives and to develop a more precautionary approach to fisheries. In particular, criteria were adopted in 1994 for setting catch limits for krill and rules established to calculate yield as a proportion of the estimated pre-exploitation biomass. The catch limit would not necessarily stay at the same level for the whole period but be revised as improved information arises.

Krill fishing currently occurs only in the SW Atlantic sector. Whilst the TAC has been calculated at 5.6 million t, the commission has limited the catch to within a 620 000 t 'trigger' level, equivalent to about 1% of the unexploited krill biomass of this region (www.ccamlr.org). This is a catch level that should not be exceeded until there is sufficient understanding of the scale of predator-prey interactions to enable small-scale management units to be triggered. There is otherwise the danger of fishing concentrating on one small area and adversely impacting dependent species (Constable, 2011). A similar approach to precautionary catch limits has been developed and applied to finfish species.

The commission has adopted numerous conservation measures (in accordance with Article IX of the convention). These come under four major categories: compliance (including catch documentation, satellite-linked vessel monitoring systems, port inspection of vessels), general fishery matters (gear restrictions, measures to reduce incidental mortality of seabirds and marine mammals), fishery regulations (TACs, fishing seasons, closed areas), and protected areas (www.ccamlr.org).

A problem for the commission, undermining its precautionary approach, has been the level of illegal, unregulated, and unreported (IUU) fishing in the convention area. Particularly serious has been illegal fishing for toothfish since the 1990s (Collins *et al.*, 2010). By its nature, the scale of the IUU fishing is hard to quantify, but estimates indicate for the 1996–7 season the IUU catch of toothfish catch peaked at around 35 000 t, four times the quota authorized by CCAMLR. As a result of various measures, the IUU catch appears to have been reduced to less than 5000 t over recent years (Österblom & Bodin, 2012). To combat IUU fishing, the commission has, for instance, introduced mandatory use of satellite-linked vessel monitoring devices on toothfish vessels and a catch documentation scheme requiring CCAMLR members to certify that toothfish have been taken in accordance with CCAMLR conservation measures and to make it harder for illegally caught fish to be sold (Collins *et al.*, 2010). But there are loopholes that illegal operators exploit. Most pirate fishing is by vessels registered in non-CCAMLR nations, especially flag-of-convenience countries that flout international law (see Chapter 5). Also, deliberately or not, toothfish is traded under a variety of names (some of them used for other fish), allowing it to slip more readily through regulatory mechanisms.

## Marine Protected Areas

The Antarctic Treaty System includes various provisions for marine protected areas (Grant, 2005). The treaty area itself is designated as 'a natural reserve, devoted to peace and science' (Madrid Protocol, Article 2), and the larger CCAMLR area can be considered an IUCN Category VI protected area, i.e. one that includes sustainable use of natural resources (Table 15.1). More specifically, the ATS provides for the establishment of Antarctic Specially Protected Areas (ASPAs) and Antarctic Specially Managed Areas (ASMAs) under Annex V of the Madrid Protocol. Many of these areas are important for their penguin and seal colonies, but so far only a few include a substantial marine component, 'a significant lost opportunity for the development of Antarctic environmental protection' (Scott, 2013).

The CCAMLR commission has particular potential for protecting marine areas. It has provision to close areas to fishing to protect vulnerable marine ecosystems (see Chapter 14), areas with an abundance of indicator taxa such as sponges, tunicates, bryozoans, and scleractinian and stylasterid corals (Jones & Lockhart, 2011). The commission is also developing a framework for setting up a network of MPAs to protect representative examples of marine ecosystems in the convention area. In 2009 it established a 94 000 km$^2$ no-take MPA south of the South Orkney Islands, an area that includes a frontal system between the Weddell and Scotia Seas, foraging areas for Adélie penguins, and shelf and seamount benthic habitats; and in 2016 agreement was eventually reached to establish a Ross Sea MPA, comprising a protected area of 1.55 million km$^2$ of which 72% is a no-take zone, although with protections to expire after 35 years. Other MPA proposals examined by the CCAMLR commission have failed to be adopted, mainly because of disagreement between members over jurisdictional issues and the extent to which fishing should be regulated within the MPA (Scott, 2013, 2014). Having adequate biogeographic information on which to base an MPA network for the region has been a shortcoming, but this deficiency is being addressed, such as through the Census of Antarctic Marine Life (De Broyer *et al.*, 2014), a Census of Marine Life programme (see Chapter 1).

Certain changes to the ATS could facilitate the designation of MPAs in the Southern Ocean (Scott, 2013). Extending the remit of the Madrid Protocol to the Antarctic Polar Front would allow ASPAs and ASMAs to be sited within the whole CCAMLR area, and greater cooperation between the CCAMLR commission and the parties to the Madrid Protocol would permit the designation of joint MPAs.

## Alternative Regimes

Alternative regimes for the conservation of Antarctica and the Southern Ocean have been proposed in recent decades. These include options for an Antarctic world park, both within and outside the framework of the ATS. Another proposal is that the region be declared the Common Heritage of Mankind (CHM), a concept that is already understood from the principles in UNCLOS (see Chapter 3). Under a CHM regime, the area would not be subject to appropriation; it would be administered by the international community for its own benefit, any benefits of natural resource exploitation would be shared internationally, the continent would be used solely for peaceful purposes, and freedom of scientific research would be guaranteed (Joyner, 1989). Even though the ATS is exclusive, this is in major respects what it already achieves, the main points of difference being in the decision-making and distribution of benefits from resource exploitation. Claimant states would, however, be unwilling to relinquish their territorial sectors, and moving to replace the ATS might precipitate further attempts at appropriation.

The management regime that has developed for the Southern Ocean incorporates many of the key features of ecosystem-based management (Gaichas *et al.*, 2014). Nevertheless, the region faces major challenges, particularly pressure on its living resources, increasing human presence, and global climate change, all of which have substantial implications for marine ecosystem structure and function (Chown *et al.*, 2012; Grant *et al.*, 2012). Much will depend on ongoing research to guide management and conservation, maintaining a suitably precautionary approach, and international cooperation and resolve to protect this unique region.

# REFERENCES


Arkema, K.K., Abramson, S.C. & Dewsbury, B.M. (2006). Marine ecosystem-based management: from characterization to implementation. *Frontiers in Ecology and the Environment*, **4**, 525–32.

Arrigo, K.R., Lubin, D., van Dijken, G.L., Holm-Hansen, O. & Morrow, E. (2003). Impact of a deep ozone hole on Southern Ocean primary production. *Journal of Geophysical Research*, **108** (C5), article 3154.

Arrigo, K.R., van Dijken, G.L. & Bushinsky, S. (2008). Primary production in the Southern Ocean, 1997–2006. *Journal of Geophysical Research*, **113** (C8), article C08004.

Atkinson, A., Siegel, V., Pakhomov, E.A., Jessopp, M.J. & Loeb, V. (2009). A re-appraisal of the total biomass and annual production of Antarctic krill. *Deep-Sea Research I*, **56**, 727–40.

Barbraud, C., Marteau, C., Ridoux, V., Delord, K. & Weimerskirch, H. (2008). Demographic response of a population of white-chinned petrels *Procellaria aequinoctialis* to climate and longline fishery bycatch. *Journal of Applied Ecology*, **45**, 1460–7.

Bednaršek, N., Tarling, G.A., Bakker, D.C.E., *et al.* (2012). Extensive dissolution of live pteropods in the Southern Ocean. *Nature Geoscience*, **5**, 881–5.

Bengtson, J.L. (2009). Crabeater seal *Lobodon carcinophaga*. In *Encyclopedia of Marine Mammals*, 2nd edn, ed. W.F. Perrin, B. Wursig & J.G.M. Thewissen, pp. 290–2. San Diego, CA: Academic Press.

Berkman, P.A. (1992). The Antarctic marine ecosystem and humankind. *Reviews in Aquatic Sciences*, **6**, 295–333.

Bonner, W.N. (1984). Conservation and the Antarctic. In *Antarctic Ecology* Vol. 2, ed. R.M. Laws, pp. 821–50. London: Academic Press.

Bowman, M., Davies, P. & Redgwell, C. (2010). *Lyster's International Wildlife Law*, 2nd edn. Cambridge, UK: Cambridge University Press.

Boyd, I.L. (2002). Antarctic marine mammals. In *Encyclopedia of Marine Mammals*, ed. W.F. Perrin, B. Würsig & J.G.M Thewissen, pp. 30–6. San Diego, CA: Academic Press.

Brandt, A., De Broyer, C., Ebbe, B., *et al.* (2012). Southern Ocean deep benthic biodiversity. In *Antarctic Ecosystems: An Extreme Environment in a Changing World*, ed. A.D. Rogers, N.M. Johnston, E.J. Murphy & A. Clarke, pp. 291–334. Chichester, UK: Wiley-Blackwell.

Brierley, A.S. & Thomas, D.N. (2002). Ecology of Southern Ocean pack ice. *Advances in Marine Biology*, **43**, 171–276.

Brooke, M. de L. (2004). The food consumption of the world's seabirds. *Proceedings of the Royal Society of London B*, **271** (Suppl. 4), S246–8.

Chown, S.L., Clarke, A., Fraser, C.I., *et al.* (2015). The changing form of Antarctic biodiversity. *Nature*, **522**, 431–8.

Chown, S.L., Lee, J.E., Hughes, K.A., *et al.* (2012). Challenges to the future conservation of the Antarctic. *Science*, **337**, 158–9.

Clarke, A. & Johnston, N.M. (2003). Antarctic marine benthic diversity. *Oceanography and Marine Biology: An Annual Review*, **41**, 47–114.

Collins, M.A., Brickle, P., Brown, J. & Belchier, M. (2010). The Patagonian toothfish: biology, ecology and fishery. *Advances in Marine Biology*, **58**, 227–300.

Constable, A.J. (2011). Lessons from CCAMLR on the implementation of the ecosystem approach to managing fisheries. *Fish and Fisheries*, **12**, 138–51.

Constable, A.J., de la Mare, W.K., Agnew, D.J., Everson, I. & Miller, D. (2000). Managing fisheries to conserve the Antarctic marine ecosystem: practical implementation of the Convention on the Conservation of Antarctic Marine Living Resources (CCAMLR). *ICES Journal of Marine Science*, **57**, 778–91.

Constable, A.J., Melbourne-Thomas, J., Corney, S.P., *et al.* (2014). Climate change and Southern Ocean ecosystems I: how changes in physical habitats directly affect marine biota. *Global Change Biology*, **20**, 3004–25.

Dayton, P.K., Mordida, B.J. & Bacon, F. (1994). Polar marine communities. *American Zoologist*, **34**, 90–9.

De Broyer C., Koubbi P., Griffiths H.J., *et al.* (eds.). (2014). *Biogeographic Atlas of the Southern Ocean*. Cambridge, UK: Scientific Committee on Antarctic Research.

Eastman, J.T. (2005). The nature of the diversity of Antarctic fishes. *Polar Biology*, **28**, 93–107.

Fabry, V.J., McClintock, J.B., Mathis, J.T. & Grebmeier, J.M. (2009). Ocean acidification at high latitudes: the bellwether. *Oceanography*, **22**, 160–71.

Flores, H., Atkinson, A., Kawaguchi, S., *et al.* (2012). Impact of climate change on Antarctic krill. *Marine Ecology Progress Series*, **458**, 1–19.

Gaichas, S., Reiss, C. & Koen-Alonso, M. (2014). Ecosystem-based management in high latitude ecosystems. In *The Sea, Vol. 16, Marine Ecosystem-Based Management*, ed. M.J. Fogarty & J.J. McCarthy, pp. 277–324. Cambridge, MA: Harvard University Press.

Garcia Borboroglu, P. & Boersma, P.D. (ed.). (2013). *Penguins: Natural History and Conservation*. Seattle, WA: University of Washington Press.

Grant, S.M. (2005). The applicability of international conservation instruments to the establishment of marine protected areas in Antarctica. *Ocean & Coastal Management*, **48**, 782–812.

Grant, S.M., Convey, P., Hughes, K.A., Phillips, R.A. & Trathan, P.N. (2012). Conservation and management of Antarctic ecosystems. In *Antarctic Ecosystems: an Extreme Environment in a Changing World*, ed. A.D. Rogers, N.M. Johnston, E.J. Murphy & A. Clarke, pp. 492–525. Chichester, UK: Wiley-Blackwell.

Gray, J.S. (2001). Antarctic marine benthic biodiversity in a world-wide latitudinal context. *Polar Biology*, **24**, 633–41.

Gutt, J., Bertler, N., Bracegirdle, T.J., *et al.* (2015). The Southern Ocean ecosystem under multiple climate change stresses – an integrated circumpolar assessment. *Global Change Biology*, **21**, 1434–53.

Hewitt, R. & Lipsky, J.D. (2009). Krill and other plankton. In *Encyclopedia of Marine Mammals*, 2nd edn, ed. W.F. Perrin, B. Wursig & J.G.M. Thewissen, pp. 657–64. San Diego, CA: Academic Press.

Holt, S. (2000). Whales and whaling. In *Seas at the Millennium: An Environmental Evaluation*, vol. III, ed. C. Sheppard, pp. 73–88. Oxford, UK: Elsevier Science Ltd.

Jenouvrier, S., Holland, M., Stroeve, J., *et al.* (2014). Projected continent-wide declines of the emperor penguin under climate change. *Nature Climate Change*, **4**, 715–18.

Jones, C.D. & Lockhart, S.J. (2011). Detecting Vulnerable Marine Ecosystems in the Southern Ocean using research trawls and underwater imagery. *Marine Policy*, **35**, 732–6.

Joyner, C.C. (1989). The evolving Antarctic legal regime. *American Journal of International Law*, **83**, 605–26.

Knox, G.A. (2007). *Biology of the Southern Ocean*, 2nd edn. Boca Raton, FL: CRC Press.

Laws, R.M. (1977). Seals and whales of the Southern Ocean. *Philosophical Transactions of the Royal Society of London B*, **279**, 81–96.

Lenihan, H.S., Oliver, J.S., Oakden, J.M. & Stephenson, M.D. (1990). Intense and localized benthic marine pollution around McMurdo Station, Antarctica. *Marine Pollution Bulletin*, **21**, 422–30.

Lynch, H.J., Crosbie, K., Fagan, W.F. & Naveen, R. (2010). Spatial patterns of tour ship traffic in the Antarctic Peninsula region. *Antarctic Science*, **22**, 123–30.

McLeod, K.L. & Leslie, H.M. (2009). Why ecosystem-based management? In *Ecosystem-Based Management for the Oceans*, ed. K. McLeod & H. Leslie, pp. 3–12. Washington, DC: Island Press.

Miller, D.G.M. (1991). Exploitation of Antarctic marine living resources: a brief history and a possible approach to managing the krill fishery. *South African Journal of Marine Science*, **10**, 321–39.

Miller, D.G.M. (2000). The Southern Ocean: a global view. *Ocean Yearbook*, **14**, 468–513.

Miller, D. & Agnew, D. (2000). Management of krill fisheries in the Southern Ocean. In *Krill: Biology, Ecology and Fisheries*, ed. I. Everson, pp. 300–37. Oxford, UK: Blackwell Science.

Mori, M. & Butterworth, D. S. (2006). A first step towards modelling the krill–predator dynamics of the Antarctic ecosystem. *CCAMLR Science*, **13**, 217–77.

Nicol, S., Bowie, A., Jarman, S., *et al.* (2010). Southern Ocean iron fertilization by baleen whales and Antarctic krill. *Fish and Fisheries*, **11**, 203–9.

Nicol, S., Foster, J. & Kawaguchi, S. (2012). The fishery for Antarctic krill – recent developments. *Fish and Fisheries*, **13**, 30–40.

Österblom, H. & Bodin, Ö (2012). Global cooperation among diverse organizations to reduce illegal fishing in the Southern Ocean. *Conservation Biology*, **26**, 638–48.

Pinkerton, M.H. & Bradford-Grieve, J.M. (2014). Characterizing foodweb structure to identify potential ecosystem effects of fishing in the Ross Sea, Antarctica. *ICES Journal of Marine Science*, **71**, 1542–53.

Robinson, S.A. & Erickson, D.J. III (2015). Not just about sunburn – the ozone hole's profound effect on climate has significant implications for Southern Hemisphere ecosystems. *Global Change Biology*, **21**, 515–27.

Sage, B. (1985). Conservation and exploitation. In *Key Environments – Antarctica*, ed. W.N. Bonner & D.W.H. Walton, pp. 351–69. Oxford, UK: Pergamon Press.

Scott, K.N. (2013). Marine protected areas in the Southern Ocean. In *The Law of the Sea and the Polar Regions: Interactions Between Global and Regional Regimes*, ed. E.J. Molenaar, A.G. Oude Elferink & D.R. Rothwell, pp. 113–37. Leiden, Netherlands: Koninklijke Brill.

Scott, K.N. (2014). Protecting the last ocean: the proposed Ross Sea MPA. Prospects and Progress. In *Jurisdiction and Control at Sea: Some Environmental and Security Issues*, ed. G. Andreone, pp. 79–90. Naples, Italy: Giannini Editore.

Sherman, K., Sissenwine, M., Christensen, V., *et al.* (2005). A global movement toward an ecosystem approach to management of marine resources. *Marine Ecology Progress Series*, **300**, 275–9.

Siniff, D.B. (1991). An overview of the ecology of Antarctic seals. *American Zoologist*, **31**, 143–9.

Spalding, M.D., Agostini, V.N., Rice, J. & Grant, S.M. (2012). Pelagic provinces of the world: a biogeographic classification of the world's surface pelagic waters. *Ocean & Coastal Management*, **60**, 19–30.

Stark, J.S., Kim, S.L. & Oliver, J.S. (2014). Anthropogenic disturbance and biodiversity of marine benthic communities in Antarctica: a regional comparison. *PLoS ONE*, **9**, e98802.

Surma, S., Pakhomov, E.A. & Pitcher, T.J. (2014). Effects of whaling on the structure of the Southern Ocean food web: insights on the 'krill surplus' from ecosystem modelling. *PLoS ONE*, **9**, e114978.

Sutherland, W.J., Aveling, R., Brooks, T.M., *et al.* (2014). A horizon scan of global conservation issues for 2014. *Trends in Ecology & Evolution*, **29**, 15–22.

Thomas, D.N., Fogg, G.E., Convey, P., *et al.* (2008). *The Biology of Polar Regions*. Oxford, UK: Oxford University Press.

Tin, T., Fleming, Z.L., Hughes, K.A., *et al.* (2009). Impacts of local human activities on the Antarctic environment. *Antarctic Science*, **21**, 3–33.

Trivelpiece, W.Z., Hinke, J.T., Miller, A.K., *et al.* (2011). Variability in krill biomass links harvesting and climate warming to penguin population changes in Antarctica. *Proceedings of the National Academy of Sciences*, **108**, 7625–8.

Verbitsky, J. (2013). Antarctic tourism management and regulation: the need for change. *Polar Record*, **49**, 278–85.

Watson, R., Pauly, D., Christensen, V., *et al.* (2003). Mapping fisheries onto marine ecosystems for regional, oceanic and global integrations. In *Large Marine Ecosystems of the World: Trends in Exploitation, Protection, and Research*, ed. G. Hempel & K. Sherman, pp. 375–95. Amsterdam: Elsevier.

# 18 Epilogue

To conclude, let us briefly review why we should care about the marine environment, why marine conservation matters, what the key issues are, and what can we do about them. (References to the literature are mainly limited to the closing discussion of underlying socio-economic factors where we cover areas only briefly touched on in earlier chapters.)

## Why Should We Care about the Marine Environment?

The overall aim of marine conservation is to sustain the structure and function of marine ecosystems. Why that is important has various facets (Chapter 3). Human welfare ultimately depends on healthily functioning ecosystems – on maintaining biodiversity, food-web structure, and ecosystem processes. Seafood is important, or in many countries essential, to the nutritional needs of millions of people, and we use marine organisms for a host of other reasons, including as a source of pharmaceutical compounds, for biomedical research, and for educational purposes. Marine ecosystems also provide a range of indispensible services such as nutrient cycling, maintaining water quality, protecting coastlines, and supporting recreational opportunities. We have shifted the baseline too far for us now to regain a pre-human state of the marine environment. Nevertheless, we must as far as possible safeguard the sustainability of natural resources and the quality of services that ecosystems can provide. There are also ethical, aesthetic, and scientific arguments for marine conservation that further illustrate why we should care about the marine environment (Chapter 3).

## Why Does Marine Conservation Matter?

Many human activities threaten the integrity of marine ecosystems and must be addressed by marine conservation (Chapter 2).

Overexploitation of living resources represents the greatest direct impact on marine ecosystems. When all sources are considered, capture fisheries remove from the world's oceans each year a biomass of more than 100 million tonnes. And about 90% of assessed fish stocks are now fully or overexploited.

Marine aquaculture has grown dramatically in recent decades but in some aspects detrimentally, such as using fishmeal to farm high-trophic-level species. Aquaculture has an important part to play in providing food for humans, but the focus has to be on sustainable practices.

Coastal habitat is also being lost to urban and industrial developments. This loss represents a major impact on the biodiversity and function of these systems, such as their role in coastal protection, in carbon sequestration, and as nursery areas.

A wide range of pollutants are harmful to marine ecosystems including organic wastes, oil, heavy metals, persistent organic pollutants, and debris. Nutrient enrichment is a serious problem impacting coastal waters in many parts of the world as a consequence of discharges and runoff from heavily populated and intensively farmed areas. Subsequent eutrophication and oxygen depletion can result in impoverished pollution-tolerant communities or in some cases dead zones. Coastal as well as oceanic waters are extensively polluted by plastic debris. Conspicuously impacted are turtles, seabirds, and marine mammals that ingest or become entangled in macroplastic items. Possibly more serious are microplastics, which are accumulating in the oceans in vast quantities. These non-nutritive particles can enter food webs via filter feeders and deposit feeders and may carry harmful adsorbed chemicals.

We also continue to alter the composition of marine communities by introducing non-native species, especially to enclosed waters through ballast water discharges. Probably the great majority of harbours support exotic species, in some cases as dominant organisms.

Compounding these impacts is global climate change. Ocean warming has fundamental implications for the structure and functioning of marine ecosystems given, for instance, its effects on species' distributions, nutrient renewal to surface waters, and primary productivity. The predicted rise in mean sea level will particularly impact intertidal and low-lying coastal habitats. Also, the rising level of atmospheric carbon dioxide and associated decrease in seawater pH render calcifying organisms especially vulnerable. Major areas of the tropics are projected to become unsuitable as coral reef habitat under the combined assault of ocean warming and acidification.

The stress exerted on marine ecosystems by human activities is often a combination of factors, such as overfishing and pollution. Anthropogenic stress may also exacerbate natural events, as in the case of coral bleaching induced by El Niño conditions but made worse by global warming. Such interactions, including possible synergistic or antagonistic effects, indicate the potential difficulties of disentangling and forecasting the effects of human pressure on marine systems and the importance of maintaining a precautionary approach to environmental protection.

Reducing the impact of pollutants on marine ecosystems has been most readily achieved in the case of substances such as oil from vessels and TBT anti-fouling paint. Land-based activities are, however, the main source of marine pollutants, but global action on this front continues to be impeded by socio-economic pressures and the scale and complexity of the problem. Some progress to regulate such sources has been made through regional agreements, an approach that needs to be more widely developed. Improved wastewater treatment and more efficient agricultural practices to make better use of fertilizers could significantly reduce nutrient inputs to coastal waters. And in the case of plastics, efforts to minimize waste, recycle, and ban non-essential uses must be more actively pursued.

With regard to invasive species, important developments are the Ballast Water Management Convention entering into force in 2017 and having IMO-approved ballast water treatment methods in place.

Of critical importance is the Paris Agreement adopted in 2015 by parties to the UN Framework Convention on Climate Change. This entered into force in November 2016 and pledges governments to limit the global average temperature increase to well below 2°C above pre-industrial levels and 'to pursue efforts' to limit this increase to 1.5°C. This requires greenhouse gas emissions to be reduced dramatically and as rapidly as possible.

Conservation practices can help address these problems and are actively doing so. Nonetheless, there are some key generic issues that loom large, including the main sources of stress to marine ecosystems, the methodology of marine conservation, and the socio-economic context.

## Key Marine Conservation Issues

### Global Fisheries

Management of the world's fisheries is a challenge that has wide-ranging implications for marine conservation (Chapter 5). Fisheries have major impacts on marine ecosystems, including reducing the average body size and age of fished populations, diminishing the role of higher-trophic-level species, restructuring communities (sometimes resulting in regime shifts), and reducing the heterogeneity of benthic habitats and their associated biodiversity. Target species are often those with 'slow' life-history traits that are particularly vulnerable to overexploitation, as are many bycatch species, from deep-sea corals to marine mammals.

Essential tools needed for sustainable fisheries and for protecting marine ecosystems are set out in the Code of Conduct for Responsible Fisheries and other fisheries management instruments that augment provisions of the UN Convention on the Law of the Sea (UNCLOS). The code, however, is voluntary, and compliance is low. Political inaction or interference, subsidies, and other economic factors commonly impede such initiatives.

Global management of fisheries is urgently needed, with international legislation to address the problem of overexploitation of high-seas resources. In 2016, UN members began negotiations on developing an international legally binding instrument under UNCLOS on the conservation and sustainable use of marine biodiversity of areas beyond national jurisdiction (subsequent to UNGA Resolution 69/292). Negotiations are to include the use of area-based management tools including marine protected areas, environmental impact assessments, and benefit sharing. This process will likely be protracted, and the development of regional initiatives also needs to continue, including improving the efficacy of regional fisheries management organizations or preferably the creation of regional ocean management organizations with a broader remit to provide for precautionary ecosystem-based management (Global Ocean Commission, 2014).

### Integrated Management

Management and conservation in the marine realm have tended to be fragmented, involving a range of sectoral interests and pursued at a range of scales. This fragmentation is very evident in the coastal zone where stakeholders are accountable to agencies having a diversity of separate terrestrial and marine responsibilities and where administrative structures rarely align with ecological processes across the land-sea interface. The institutional barriers can be considerable. Nevertheless, an essential task for conservation is to explore and further develop frameworks to support integrated management, given the importance of connections that exist between and within systems. Developments in this area, notably ecosystem-based management (Chapter 5), integrated coastal zone management (Chapter 10), and marine spatial planning (Chapter 15), are contributing to integrated management. The large marine ecosystem approach provides a basis for implementing ecosystem-based management for relatively large ocean areas, as in the case of the Southern Ocean (Chapter 17).

### Marine Protected Area Networks

Marine protected areas (MPAs) are often viewed as the principal tool of marine conservation – a way to limit human impact over a defined area with the goal of attaining a more natural ecosystem structure and function (Chapter 15). This approach usually means restricting resource exploitation to protect vulnerable species and habitats and associated biodiversity. There is abundant evidence that MPAs allow target species' populations to recover and regain their normal role, providing that fishing is prohibited or severely curtailed. However, only about 1% of the global ocean area is highly protected, far short of the proportion considered necessary to significantly assist biodiversity conservation and fisheries management. As a start, governments must make every effort to meet the Convention on Biological Diversity target of at least 10% of the total marine area in protected areas by 2020. However, there is a danger of undue emphasis on total areas that are designated as protected. To contribute effectively to conservation goals, MPAs need not only to be well-enforced no-take reserves but also to be within suitably designed networks in order to be sustainable and to best represent the biodiversity of a region. MPA networks should also be embedded in integrated management schemes. Even so, MPAs are still vulnerable to pervasive impacts, such as from pollution, illegal fishing, introduced species, and global climate change. Nevertheless, by supporting more intact ecosystems, protected areas may have a greater capacity to withstand external stressors, recover more rapidly from disturbances, and be better able to sustain source populations for colonizing degraded areas elsewhere.

## Socio-economic Drivers

The state of the global environment – and the need for conservation action – depends ultimately on human population pressure and socio-economic practices. This pressure can be expressed in terms of environmental footprint accounting – whether humanity's demands on resource appropriation and waste generation each year are within the regenerative capacity of the biosphere (Kitzes *et al.*, 2008; Hoekstra & Widemann, 2014). Footprints per capita vary greatly, with the richest industrialized nations having footprints an order of magnitude larger than the poorest. On the other hand, it is usually argued that poor communities, in order to survive, are driven to overexploit resources, in turn resulting in worse impoverishment. Inequality, however, is often the ultimate driver of such unsustainable practices, with wealthier players exploiting the socio-economic divide (Glavovic, 2008). Overall, humanity is now demanding 1.6 times more capacity than the biosphere can supply each year (Global Footprint Network, 2016), resulting in inexorable degradation of ecosystems. To regain sustainability requires a reduction in human birth rates and major changes in food production, consumption and energy use, and levels of recycling (Hoekstra & Widemann, 2014; Worldwatch Institute, 2016).

The human population, world economy, and environmental footprints have grown rapidly since the mid-twentieth century. Traditional economic models have focussed on growth, with little or no regard for resource sustainability, and have ignored environmental externalities. Converting a mangrove forest to aquaculture ponds directly benefits a shrimp farmer, but there are third-party costs such as for shoreline protection and carbon sequestration, and the value of ecosystem services can be expressed in monetary terms (Costanza *et al.*, 2014). Schemes such as 'blue carbon' initiatives (to account for services provided by coastal wetlands) are gaining ground but need to be fully integrated into economic planning.

An economic approach that properly accounts for ecosystem health and resource sustainability is called for, essentially an amalgamation of ecology and economics. The concept of sustainable development assumes that resources are used more efficiently so that economic growth does not incur greater environmental impact. But whether that decoupling actually occurs needs to be carefully examined. Using consumption of materials as an indicator, Wiedmann *et al.* (2015) found that pressure on natural resources does not necessarily diminish as nations become more affluent. It can appear otherwise because rich nations increasingly make use of raw materials and processing in developing countries, the cost of which is not accounted for in domestic material consumption, the metric traditionally used.

An analysis by Bos *et al.* (2015) indicates that marine conservation is massively underfunded, especially in the tropics. They make several recommendations, including the need for financial strategies, development of financial mechanisms, integration of financial and conservation planning, and addressing environmental externalities.

## What Should We Do?

It is difficult to be optimistic about marine conservation in the face of our past and projected impacts on marine ecosystems and the social, political, and economic factors that obstruct remedial action. On the other hand, we understand the issues and already have appropriate tools – at least for proximate action – and further instruments are being developed to facilitate the conservation and sustainable use of the world ocean.

Principles of ocean governance for sustainable use were delineated some time ago (e.g. Costanza *et al.*, 1998; IWCO, 1998). And in the case of fishing and its attendant impacts, the Code of Conduct for Responsible Fisheries sets out what is required to achieve sustainable exploitation of living aquatic resources with due regard for ecosystem health and biodiversity. Overexploitation and other harmful effects of fishing are problems that could in principle be addressed within a reasonable timeframe, with sustainable capture fisheries supplemented by appropriate (mainly low-trophic-level) aquaculture. We have legislation targeting pollutants and evidence that it can be effective. The main difficulty is nutrient enrichment, although that too could be greatly mitigated by the smarter use of fertilizer and better wastewater treatment. Climate change as a result of anthropogenic greenhouse gas emissions remains the chief stumbling block to alleviating human pressure on marine ecosystems. Even under a stringent mitigation scenario we are locked in to decades of a warmer, more acidic ocean. This makes attending to those stressors that can be lessened in the shorter term, and promoting ecosystem resilience, all the more imperative.

The 1982 UN Convention on the Law of the Sea recognized that 'the problems of ocean space are closely interrelated and need to be considered as a whole' (Preamble), a holistic approach that remains paramount. Nevertheless, our use of ocean space has since moved on, particularly the exploitation of high-seas resources, and we need additional new instruments to promote integrated management and conservation of the world ocean (notably the UN agreement currently under negotiation). And we still lack a suitable international forum, a UN Ocean Assembly, for making and overseeing an integrated ocean policy (Borgese, 1998).

But we also need society at large to value and care about the global ocean. To help effect change requires greater awareness of the vital importance of the marine realm. The well-being of humanity depends on our wise use of ocean space, a message that we should appreciate from childhood onwards. Basic ecology is perhaps the most important subject that can be taught in schools. It should become clear to young minds why protecting the natural world matters and that crucial to our survival is the capacity of intact ecosystems to provide sustainable resources and services. We also, however, derive much more from the natural world, from its scientific, cultural, and aesthetic dimensions. As Wilson (1992) reminds us, 'the loss of life's diversity endangers not just the body but the spirit'. Society can perhaps be most readily engaged to care about the living world by its sheer diversity and beauty, its constant appeal to our sense of wonder. Accessibility makes this harder to realize for the marine environment. Nevertheless, marine conservation affects us all in some way.

More intractable are the fundamental issues of human population pressure combined with an economic order that essentially disregards environmental costs. The world population is currently estimated at about 7.4 billion. Although human birth rates worldwide are slowly declining, the population is still projected to exceed 11 billion by 2100, an increase accounted for very largely by developing countries. To achieve further declines in fertility requires greater investment in social and reproductive health programmes and increased empowerment of women (UN, Department of Economic and Social Affairs, Population Division, 2015; Worldwatch Institute, 2016). An easing of population pressure has to be combined with economic activity that recognizes environmental realities. To discount the ecological price of economic activity is fundamentally misleading. There are mechanisms to address the illusion that such accounting creates, such as economic incentives to achieve ecological goals (Costanza, 2000). For healthy ecosystems worldwide to support sustainable use requires developed nations to accept greatly reduced environmental footprints and a more equitable demand on ecosystems across all humanity – a shrink-and-share path (Kitzes *et al.*, 2008). Maintaining the integrity of marine ecosystems – marine conservation – is crucial to this process.

## REFERENCES

Borgese, E.M. (1998). *The Oceanic Circle: Governing the Seas as a Global Resource*. Tokyo: United Nations University Press.

Bos, M., Pressey, R.L. & Stoeckl, N. (2015). Marine conservation finance: the need for and scope of an emerging field. *Ocean & Coastal Management*, **114**, 116–28.

Costanza, R. (2000). The ecological, economic, and social importance of the oceans. In *Seas at the Millennium: an Environmental Evaluation*, ed. C. Sheppard, pp. 393–403. Amsterdam; New York: Pergamon.

Costanza, R., Andrade, F., Antunes, P., *et al.* (1998). Principles for sustainable governance of the oceans. *Science*, **281**, 198–9.

Costanza, R., de Groot, R., Sutton, P., *et al.* (2014). Changes in the global value of ecosystem services. *Global Environmental Change*, **26**, 152–8.

Glavovic, B. (2008). Poverty and inequity at sea: challenges for ecological economics. In *Ecological Economics of the Oceans and Coasts*, ed. M. Patterson & B. Glavovic, pp. 244–65. Cheltenham, UK: Edward Elgar.

Global Footprint Network (2016). World Footprint. www.footprintnetwork.org/en/index.php/GFN/page/world_footprint/ (accessed 17 June 2016).

Global Ocean Commission (2014). *From Decline to Recovery – A Rescue Package for the Global Ocean*. Oxford, UK: Global Ocean Commission.

Hoekstra, A.Y. & Widemann, T.O. (2014). Humanity's unsustainable environmental footprint. *Science*, **344**, 1114–7.

IWCO (1998). *The Ocean ... Our Future*. The Report of the Independent World Commission on the Oceans. Cambridge, UK: Cambridge University Press.

Kitzes, J., Wackernagel, M., Loh. J., *et al.* (2008). Shrink and share: humanity's present and future Ecological Footprint. *Philosophical Transactions of the Royal Society B*, **363**, 467–75.

UN, Department of Economic and Social Affairs, Population Division (2015). *World Population Prospects: The 2015 Revision, Key Findings and Advance Tables*. Working Paper No. ESA/P/WP.241. New York: United Nations.

Wiedmann, T.O., Schandl, H., Lenzen, M., *et al.* (2015). The material footprint of nations. *Proceedings of the National Academy of Sciences*, **112**, 6271–6.

Wilson, E.O. (1992). *The Diversity of Life*. Cambridge, MA: Harvard University Press.

Worldwatch Institute (2016). Nine Population Strategies to Stop Short of 9 Billion. www.worldwatch.org/nine-population-strategies-stop-short-9-billion (accessed 20 June 2016).

# APPENDIX 1
# Abbreviations

| | |
|---|---|
| **APF** | Antarctic Polar Front |
| **ASMA** | Antarctic Specially Managed Area |
| **ASPA** | Antarctic Specially Protected Area |
| **ATS** | Antarctic Treaty System |
| **BACI** | Before-After-Control-Impact |
| **BTF** | back-to-the-future |
| **BWU** | blue whale unit |
| **CBD** | Convention on Biological Diversity |
| **CHM** | Common Heritage of Mankind |
| **CCAMLR** | Convention on the Conservation of Antarctic Marine Living Resources |
| **CCAS** | Convention for the Conservation of Antarctic Seals |
| **CFC** | chlorofluorocarbon |
| **CITES** | Convention on International Trade in Endangered Species of Wild Fauna and Flora |
| **CPUE** | catch per unit effort |
| **CRAMRA** | Convention on the Regulation of Antarctic Mineral Resource Activities |
| **CWA** | Clean Water Act |
| **DDE** | dichlorodiphenyldichloroethylene |
| **DDT** | dichlorodiphenyltrichloroethane |
| **DNA** | deoxyribonucleic acid |
| **EAF** | ecosystem approach to fisheries |
| **EBFM** | ecosystem-based fishery management |
| **EBSA** | ecologically or biologically significant marine area |
| **EEZ** | exclusive economic zone |
| **EIA** | environmental impact assessment |
| **ENSO** | El Niño-Southern Oscillation |
| **EPA** | Environmental Protection Agency [US] |
| **ESU** | evolutionary significant unit |
| **EUNIS** | European Union Nature Information System |
| **FAO** | Food and Agriculture Organization of the United Nations |
| **FOC** | flag of convenience |
| **GBR** | Great Barrier Reef |
| **GBRMPA** | Great Barrier Reef Marine Park Authority |
| **GEF** | Global Environment Facility |
| **GESAMP** | Joint Group of Experts on the Scientific Aspects of Marine Environmental Protection |
| **GPA** | Global Programme of Action for the Protection of the Marine Environment from Land-based Activities |
| **IAATO** | International Association of Antarctica Tour Operators |
| **IAEA** | International Atomic Energy Agency |
| **IAIA** | International Association for Impact Assessment |
| **IBA** | Important bird area |
| **ICCAT** | International Commission for the Conservation of Atlantic Tunas |
| **ICES** | International Council for the Exploration of the Sea |
| **ICRW** | International Convention for the Regulation of Whaling |

| | |
|---|---|
| **ICZM** | integrated coastal zone management |
| **IMO** | International Maritime Organization |
| **IMTA** | integrated multi-trophic aquaculture |
| **IOC** | Intergovernmental Oceanographic Commission |
| **IPCC** | Intergovernmental Panel on Climate Change |
| **ITQ** | individual transferable quota |
| **IUCN** | International Union for Conservation of Nature |
| **IUU** | illegal, unregulated, and unreported [fishing] |
| **IWC** | International Whaling Commission |
| **JARPA** | Japanese Whale Research Program under Special Permit in the Antarctic |
| **L** | litre |
| **LME** | large marine ecosystem |
| **MAP** | Mediterranean Action Plan |
| **MARPOL** | International Convention for the Prevention of Pollution from Ships |
| **mg** | milligram |
| **MLS** | minimum legal size |
| **MMPA** | Marine Mammal Protection Act [US] |
| **MNCR** | Marine Nature Conservation Review [Great Britain] |
| **MPA** | marine protected area |
| **MSP** | marine spatial planning |
| **MSY** | maximum sustainable yield |
| **mtDNA** | mitochondrial DNA |
| **nm** | nautical mile [1.85 km] |
| **NAO** | North Atlantic Oscillation |
| **NEP** | National Estuary Program [US] |
| **NGO** | non-governmental organization |
| **NMP** | New Management Procedure |
| **OSP** | optimum sustainable population |
| **OSPAR** | From the former Oslo and Paris Conventions (and Commissions), which became the OSPAR Convention (and Commission) (for the Protection of the Marine Environment of the North-East Atlantic) |
| **PAH** | polycyclic aromatic hydrocarbon |
| **PBDE** | polybrominated diphenyl ether |
| **PBq** | petabecquerel |
| **PBR** | potential biological removal |
| **PCB** | polychlorinated biphenyl |
| **PFC** | perfluorinated compound |
| **POP** | persistent organic pollutant |
| **ppm** | parts per million |
| **REACH** | Registration, Evaluation, Authorisation and Restriction of Chemicals |
| **RFMO** | Regional Fisheries Management Organization |
| **RMP** | Revised Management Procedure |
| **scuba** | self-contained underwater breathing apparatus |
| **SDM** | species distribution model |
| **SLED** | sea lion exclusion device |
| **sp.** | species [singular] |
| **spp.** | species [plural] |
| **SSC** | Species Survival Commission [IUCN] |
| **SST** | sea surface temperature |
| **t** | tonne [metric ton] |
| **TBT** | tributyltin |
| **UN** | United Nations |

| | |
|---|---|
| **UNCED** | United Nations Conference on Environment and Development |
| **UNCLOS** | United Nations Convention on the Law of the Sea |
| **UNEP** | United Nations Environment Programme |
| **UNESCO** | United Nations Educational, Scientific and Cultural Organization |
| **UNFCCC** | United Nations Framework Convention on Climate Change |
| **UNGA** | United Nations General Assembly |
| **UV** | ultraviolet |
| **VME** | vulnerable marine ecosystem |
| **WCPA** | World Commission on Protected Areas |
| **WHO** | World Health Organization |
| **WMO** | World Meteorological Organization |
| **WHSRN** | Western Hemisphere Shorebird Reserve Network |
| **WSSD** | World Summit on Sustainable Development |
| **WTO** | World Trade Organization |
| **WWF** | World Wide Fund for Nature |

# APPENDIX 2
# Species List

Scientific and common names of organisms. Where a species does not have a recognized common name, a general common name is given in parentheses.
Nomenclature generally follows the World Register of Marine Species.

| SEAWEEDS | |
|---|---|
| *Caulerpa cylindracea* | (green alga) |
| *Caulerpa racemosa* | (green alga) |
| *Caulerpa taxifolia* | (green alga) |
| *Chaetomorpha* spp. | (green algae) |
| *Cladophora* spp. | (filamentous green algae) |
| *Codium fragile* | branched velvet weed |
| *Ecklonia radiata* | common kelp |
| *Eisenia arborea* | (kelp) |
| *Eucheuma* spp. | (red algae) |
| *Eualaria fistulosa* | dragon kelp |
| *Gigartina australis* | Turkish towel alga |
| *Kappaphycus alvarezii* | Elkhorn sea moss |
| *Laminaria farlowii* | (kelp) |
| *Lithophyllum cabiochiae* | (coralline alga) |
| *Lithophyllum stictaeforme* | (coralline alga) |
| *Macrocystis pyrifera* | bladder kelp, giant kelp |
| *Mazzaella laminarioides* | (red alga) |
| *Mesophyllum alternans* | (coralline alga) |
| *Neogoniolithon mamillosum* | (coralline alga) |
| *Nereocystis luetkeana* | ribbon kelp, bull kelp |
| *Pterygophora californica* | stalked kelp |
| *Saccharina japonica* | kombu |
| *Sargassum muticum* | Japanese wireweed |
| *Scytothalia dorycarpa* | (brown alga) |
| *Ulva* spp. | sea lettuce, green laver |
| *Undaria pinnatifida* | wakame |
| *Vanvoorstia bennettiana* | Bennett's seaweed |

| FLOWERING PLANTS | |
|---|---|
| *Avicennia officinalis* | Indian mangrove |
| *Cymodocea nodosa* | slender seagrass |
| *Distichlis palmeri* | nipa grass |
| *Distichlis spicata* | seashore saltgrass |
| *Halodule wrightii* | shoalgrass |
| *Phragmites australis* | common reed |
| *Posidonia oceanica* | Neptune grass |
| *Rhizophora mangle* | red mangrove |

| | |
|---|---|
| *Sonneratia apetala* | (mangrove apple) |
| *Spartina alterniflora* | smooth cordgrass |
| *Spartina anglica* | common cordgrass, English cordgrass |
| *Spartina maritima* | small cordgrass |
| *Spartina patens* | saltmeadow cordgrass |
| *Thalassia testudinum* | turtlegrass |
| *Typha angustifolia* | lesser bulrush |
| *Typha latifolia* | common bulrush, broadleaf cattail |
| *Zostera japonica* | dwarf eelgrass, Japanese eelgrass |
| *Zostera marina* | eelgrass |
| *Zostera noltei* | dwarf eelgrass |
| SPONGES | |
| *Hippospongia communis* | honeycomb bath sponge |
| *Spongia lamella* | elephant ear |
| *Spongia officinalis* | bath sponge |
| *Spongia zimocca* | leather sponge |
| CNIDARIANS | |
| *Acropora cervicornis* | staghorm coral |
| *Acropora palmata* | elkhorn coral |
| *Corallium rubrum* | red coral |
| *Edwardsia ivelli* | Ivell's sea anemone |
| *Enallopsammia profunda* | (scleractinian coral) |
| *Goniocorella dumosa* | (scleractinian coral) |
| *Lophelia pertusa* | (scleractinian coral) |
| *Madrepora oculata* | (scleractinian coral) |
| *Millepora boschmai* | (fire coral) |
| *Orbicella annularis* | boulder star coral |
| *Siderastrea glynni* | (starlet coral) |
| *Solenosmilia variabilis* | (scleractinian coral) |
| CTENOPHORES | |
| *Mnemiopsis leidyi* | warty comb jelly, sea walnut |
| POLYCHAETES | |
| *Arenicola marina* | lugworm |
| *Capitella capitata* | gallery worm |
| *Glycera dibranchiata* | bloodworm |
| *Hediste diversicolor* | ragworm |
| MOLLUSCS | |
| *Cerastoderma edule* | common cockle |
| *Chamelea gallina* | striped venus clam |
| *Charonia tritonis* | giant triton |
| *Concholepas concholepas* | loco |

| | |
|---|---|
| *Conus* spp. | cone snails |
| *Conus gloriamaris* | glory-of-the-seas cone snail |
| *Crassostrea gigas* | Pacific oyster |
| *Crassostrea virginica* | eastern oyster |
| *Crepidula fornicata* | slipper limpet |
| *Cymbiola rossiniana* | (volute) |
| *Cypraecassis rufa* | bullmouth helmet, red helmet shell |
| *Fissurella* spp. | keyhole limpets |
| *Haliotis sorenseni* | white abalone |
| *Littoraria flammea* | (periwinkle) |
| *Lobatus gigas* | queen conch |
| *Lottia alveus* | Atlantic eelgrass limpet |
| *Lottia edmitchelli* | (limpet) |
| *Macoma balthica* | Baltic tellin |
| *Mya arenaria* | soft-shell clam |
| *Mytilus edulis* | blue mussel |
| *Mytilus galloprovincialis* | Mediterranean mussel |
| *Nucella lapillus* | dogwhelk |
| *Ommastrephes bartramii* | flying squid |
| *Patella ferruginea* | ribbed Mediterranean limpet |
| *Perumytilus purpuratus* | purple mussel |
| *Tectus niloticus* | trochus |
| *Tentaoculus balantiophaga* | (limpet) |
| *Tridacna derasa* | southern giant clam |
| *Tridacna gigas* | giant clam |
| *Urosalpinx cinerea* | American oyster drill |
| CRUSTACEANS | |
| *Aristeus antennatus* | blue and red shrimp |
| *Austrominius modestus* | Australasian barnacle |
| *Carcinus aestuarii* | European green shore-crab |
| *Carcinus maenas* | European green shore-crab |
| *Corophium volutator* | (amphipod) |
| *Euphausia superba* | Antarctic krill |
| *Jasus edwardsii* | southern spiny lobster |
| *Jehlius cirratus* | picacho barnacle |
| *Notochthamalus scabrosus* | (barnacle) |
| *Palinurus elephas* | European spiny lobster |
| BRYOZOANS | |
| *Electra tenella* | (bryozoan) |
| ECHINODERMS | |
| *Acanthaster planci* | crown-of-thorns starfish |
| *Centrostephanus rodgersii* | black sea urchin |
| *Diadema antillarum* | long-spined sea urchin |
| *Echinometra mathaei* | rock-boring urchin |
| *Evechinus chloroticus* | kina |

| TUNICATES | |
|---|---|
| *Ciona intestinalis* | sea vase |

| FISHES | |
|---|---|
| *Acanthocybium solandri* | wahoo |
| *Acipenser baerii* | Siberian sturgeon |
| *Acipenser brevirostrum* | shortnose sturgeon |
| *Acipenser gueldenstaedtii* | Russian sturgeon |
| *Acipenser persicus* | Persian sturgeon |
| *Acipenser ruthenus* | sterlet |
| *Acipenser stellatus* | stellate sturgeon |
| *Acipenser sturio* | Atlantic sturgeon |
| *Alopias* spp. | thresher sharks |
| *Alosa fallax* | twaite shad |
| *Ammodytes dubius* | northern sand lance |
| *Ammodytes marinus* | lesser sandeel |
| *Anampses viridis* | green wrasse |
| *Anguilla* spp. | freshwater eels |
| *Anoplopoma fimbria* | sablefish |
| *Aphanopus carbo* | black scabbardfish |
| *Azurina eupalama* | Galápagos damselfish |
| *Beryx splendens* | alfonsino |
| *Brevoortia* spp. | menhaden |
| *Brevoortia tyrannus* | Atlantic menhaden |
| *Brosme brosme* | tusk |
| *Carcharhinus falciformis* | silky shark |
| *Carcharhinus leucas* | bull shark |
| *Carcharhinus obscurus* | dusky shark |
| *Carcharias taurus* | grey nurse shark, sand tiger shark |
| *Carcharodon carcharias* | great white shark |
| *Cetorhinus maximus* | basking shark |
| *Champsocephalus gunnari* | mackerel icefish |
| *Chanos chanos* | milkfish |
| *Cheilinus undulatus* | humphead wrasse |
| *Chrysoblephus laticeps* | Roman seabream |
| *Chrysophrys auratus* | snapper |
| *Clupea harengus* | Atlantic herring |
| *Cololabis* spp. | sauries |
| *Coryphaena* spp. | dolphinfishes |
| *Coryphaenoides rupestris* | roundnose grenadier |
| *Decapterus* spp. | scads |
| *Dipturus batis* | flapper skate |
| *Dissostichus eleginoides* | Patagonian toothfish |
| *Dissostichus mawsoni* | Antarctic toothfish |
| *Engraulis encrasicolus* | European anchovy |
| *Engraulis japonicus* | Japanese anchovy |
| *Engraulis ringens* | Peruvian anchovy, anchoveta |
| *Epigonus telescopus* | black cardinalfish |
| *Epinephelus striatus* | Nassau grouper |
| *Fundulus heteroclitus heteroclitus* | mummichog |

| | |
|---|---|
| *Gadus morhua* | Atlantic cod |
| *Galeocerdo cuvier* | tiger shark |
| *Galeorhinus galeus* | tope, school shark |
| *Glyptocephalus cynoglossus* | witch flounder |
| *Haemulon sciurus* | bluestriped grunt |
| *Hippocampus kelloggi* | great seahorse |
| *Hippocampus kuda* | spotted seahorse |
| *Hippocampus spinosissimus* | hedgehog seahorse |
| *Hippocampus trimaculatus* | three-spot seahorse |
| *Hippoglossus hippoglossus* | Atlantic halibut |
| *Hoplostethus atlanticus* | orange roughy |
| *Huso huso* | beluga |
| *Isurus oxyrinchus* | shortfin mako |
| *Katsuwonus pelamis* | skipjack tuna |
| *Lamna nasus* | porbeagle |
| *Latimeria chalumnae* | coelacanth |
| *Latimeria menadoensis* | Sulawesi coelacanth |
| *Lethrinus harak* | thumbprint emperor |
| *Macrourus berglax* | roughhead grenadier |
| *Macruronus novaezelandiae* | hoki, blue grenadier |
| *Mallotus villosus* | capelin |
| *Manta* spp. | manta rays |
| *Melannogrammus aeglefinus* | haddock |
| *Merlangius merlangus* | whiting |
| *Merluccius capensis* | shallow-water Cape hake |
| *Merluccius merluccius* | European hake |
| *Merluccius paradoxus* | deep-water Cape hake |
| *Micromesistius poutassou* | blue whiting |
| *Mugil cephalus* | flathead grey mullet |
| *Mullus barbatus barbatus* | red mullet |
| *Mullus surmuletus* | surmullet |
| *Mustelus antarcticus* | gummy shark |
| *Mustelus lenticulatus* | rig |
| *Notothenia rossii* | marbled rockcod |
| *Oncorhynchus* spp. | Pacific salmon |
| *Osmerus eperlanus* | smelt |
| *Parapercis colias* | blue cod |
| *Petromyzon marinus* | sea lamprey |
| *Platichthys flesus* | European flounder |
| *Pomatomus saltatrix* | bluefish |
| *Prionace glauca* | blue shark |
| *Pseudopentaceros richardsoni* | southern boarfish, pelagic armourhead |
| *Pterois miles* | common lionfish |
| *Pterois volitans* | red lionfish |
| *Reinhardtius hippoglossoides* | Greenland halibut |
| *Rhincodon typus* | whale shark |
| *Rhinoptera bonasus* | cownose ray |
| *Ruvettus pretiosus* | oilfish |
| *Salmo salar* | Atlantic salmon |
| *Sarda sarda* | Atlantic bonito |
| *Sardina pilchardus* | European pilchard, sardine |
| *Sardinella* spp. | sardinellas |

| | |
|---|---|
| *Sardinella aurita* | round sardinella |
| *Sardinella longiceps* | Indian oil sardine |
| *Scarus guacamaia* | rainbow parrotfish |
| *Scomber japonicus* | chub mackerel |
| *Scomber scombrus* | Atlantic mackerel |
| *Scyliorhinus canicula* | small-spotted catshark |
| *Sebastes melanops* | black rockfish |
| *Sebastes viviparus* | Norway redfish |
| *Sparus aurata* | gilthead seabream |
| *Sphyrna* spp. | hammerhead sharks |
| *Sprattus sprattus* | European sprat |
| *Squalus acanthias* | spiny dogfish |
| *Theragra chalcogramma* | Alaska pollock |
| *Thunnus alalunga* | albacore |
| *Thunnus albacares* | yellowfin tuna |
| *Thunnus maccoyii* | southern bluefin tuna |
| *Thunnus obesus* | bigeye tuna |
| *Thunnus orientalis* | Pacific bluefin tuna |
| *Thunnus thynnus* | Atlantic bluefin tuna |
| *Totoaba macdonaldi* | totoaba |
| *Trachurus mediterraneus* | Mediterranean horse mackerel |
| *Trichiurus lepturus* | largehead hairtail |
| *Xiphias gladius* | swordfish |
| *Zebrasoma flavescens* | yellow tang |
| REPTILES | |
| *Amblyrhynchus cristatus* | marine iguana |
| *Caretta caretta* | loggerhead turtle |
| *Chelonia mydas* | green turtle |
| *Crocodylus porosus* | saltwater crocodile |
| *Dermochelys coriacea* | leatherback turtle |
| *Enhydrina schistosa* | beaked sea snake |
| *Eretmochelys imbricata* | hawksbill turtle |
| *Hydrophis ornatus* | ornate sea snake |
| *Laticauda laticaudata* | brown-lipped sea krait |
| *Lepidochelys kempii* | Kemp's ridley |
| *Lepidochelys olivacea* | olive ridley |
| *Natator depressa* | flatback turtle |
| *Pelamis platura* | yellow-bellied sea snake |
| *Pseudolaticauda semifasciata* | Chinese sea snake |
| BIRDS | |
| *Alca torda* | razorbill |
| *Anas penelope* | wigeon |
| *Anous stolidus* | brown noddy |
| *Aptenodytes forsteri* | emperor penguin |
| *Aptenodytes patagonicus* | king penguin |
| *Arenaria interpres* | turnstone |
| *Brachyramphus brevirostris* | Kittlitz's murrelet |
| *Branta bernicla* | brent goose |

| | |
|---|---|
| *Bulweria bifax* | small St Helena petrel |
| *Calidris alba* | sanderling |
| *Calidris alpina* | dunlin |
| *Calidris canutus* | knot |
| *Calidris pusilla* | semipalmated sandpiper |
| *Calidris tenuirostris* | great knot |
| *Calonectris diomedea* | Cory's shearwater |
| *Camptorhynchus labradorius* | Labrador duck |
| *Cygnus olor* | mute swan |
| *Daption capense* | Cape petrel |
| *Diomedea dabbenena* | Tristan albatross |
| *Diomedea exulans* | wandering albatross |
| *Diomedea exulans amsterdamensis* | Amsterdam albatross |
| *Diomedea sanfordi* | northern royal albatross |
| *Eudyptes chrysocome* | southern rockhopper penguin |
| *Eudyptes chrysolophus* | macaroni penguin |
| *Fratercula arctica* | Atlantic puffin |
| *Fregata andrewsi* | Christmas frigatebird |
| *Fregetta maoriana* | New Zealand storm-petrel |
| *Fulmarus glacialis* | northern fulmar |
| *Fulmarus glacialoides* | southern fulmar |
| *Haematopus meadewaldoi* | Canary Islands oystercatcher |
| *Haematopus ostralegus* | Eurasian oystercatcher |
| *Larus argentatus* | herring gull |
| *Larus audouinii* | Audouin's gull |
| *Larus dominicanus* | kelp gull |
| *Larus marinus* | great black-backed gull |
| *Larus ridibundus* | black-headed gull |
| *Limosa lapponica* | bar-tailed godwit |
| *Megadyptes antipodes* | yellow-eyed penguin |
| *Mergus australis* | Auckland Islands merganser |
| *Morus bassanus* | northern gannet |
| *Oceanodroma macrodactyla* | Guadalupe storm-petrel |
| *Onychoprion fuscatus* | sooty tern |
| *Pagophila eburnea* | ivory gull |
| *Pelecanoides urinatrix* | common diving-petrel |
| *Pelecanus occidentalis* | brown pelican |
| *Phaethon rubricauda* | red-tailed tropicbird |
| *Phalacrocorax aristotelis* | European shag |
| *Phalacrocorax atriceps* | imperial shag |
| *Phalacrocorax bougainvillii* | Guanay cormorant |
| *Phalacrocorax carbo* | great cormorant |
| *Phalacrocorax harrisi* | Galápagos cormorant |
| *Phalacrocorax onslowi* | Chatham shag |
| *Phalacrocorax perspicillatus* | spectacled cormorant |
| *Phoebastria albatrus* | short-tailed albatross |
| *Phoebastria immutabilis* | Laysan albatross |
| *Phoebastria irrorata* | waved albatross |
| *Phoebastria nigripes* | black-footed albatross |
| *Phoenicopterus ruber* | greater flamingo |
| *Pinguinus impennis* | great auk |
| *Procellaria aequinoctialis* | white-chinned petrel |

| | |
|---|---|
| *Pseudobulweria aterrima* | Mascarene petrel |
| *Pseudobulweria becki* | Beck's petrel |
| *Pseudobulweria macgillivrayi* | Fiji petrel |
| *Pterodroma axillaris* | Chatham petrel |
| *Pterodroma cahow* | Bermuda petrel |
| *Pterodroma caribbaea* | Jamaica petrel |
| *Pterodroma cookii* | Cook's petrel |
| *Pterodroma hasitata* | black-capped petrel |
| *Pterodroma magentae* | magenta petrel |
| *Pterodroma phaeopygia* | Galápagos petrel |
| *Pterodroma rupinarum* | large St Helena petrel |
| *Puffinus auricularis* | Townsend's shearwater |
| *Puffinus griseus* | sooty shearwater |
| *Puffinus mauretanicus* | Balearic shearwater |
| *Puffinus puffinus* | Manx shearwater |
| *Puffinus tenuirostris* | short-tailed shearwater |
| *Puffinus yelkouan* | Yelkouan shearwater |
| *Pygoscelis adeliae* | Adélie penguin |
| *Pygoscelis antarcticus* | chinstrap penguin |
| *Pygoscelis papua* | gentoo penguin |
| *Spheniscus demersus* | African penguin |
| *Spheniscus magellanicus* | Magellanic penguin |
| *Spheniscus mendiculus* | Galápagos penguin |
| *Stercorarius parasiticus* | parasitic jaeger, Arctic skua |
| *Stercorarius skua* | great skua |
| *Sterna bernsteini* | Chinese crested tern |
| *Sterna paradisaea* | Arctic tern |
| *Sula dactylatra* | masked booby |
| *Sula sula* | red-footed booby |
| *Sula variegata* | Peruvian booby |
| *Synthliboramphus antiquus* | ancient murrelet |
| *Tadorna tadorna* | common shelduck |
| *Thalassarche cauta* | shy albatross |
| *Thalassarche eremita* | Chatham albatross |
| *Thalassarche melanophris* | black-browed albatross |
| *Tringa totanus* | redshank |
| *Uria aalge* | common guillemot, common murre |
| *Uria lomvia* | thick-billed murre |
| MAMMALS | |
| *Arctocephalus australis forsteri* | New Zealand fur seal |
| *Arctocephalus galapagoensis* | Galápagos fur seal |
| *Arctocephalus gazella* | Antarctic fur seal |
| *Arctocephalus pusillus* | Cape fur seal |
| *Balaena mysticetus* | bowhead whale |
| *Balaenoptera acutorostrata* | common minke whale |
| *Balaenoptera bonaerensis* | Antarctic minke whale |
| *Balaenoptera borealis* | sei whale |
| *Balaenoptera edeni* | Bryde's whale |
| *Balaenoptera musculus* | blue whale |
| *Balaenoptera musculus intermedia* | Antarctic blue whale |

| | |
|---|---|
| *Balaenoptera musculus brevicauda* | pygmy blue whale |
| *Balaenoptera physalus* | fin whale |
| *Callorhinus ursinus* | northern fur seal |
| *Cephalorhynchus hectori* | Hector's dolphin |
| *Cephalorhynchus hectori maui* | Maui's dolphin |
| *Cystophora cristata* | hooded seal |
| *Delphinapterus leucas* | beluga, white whale |
| *Delphinus capensis* | long-beaked common dolphin |
| *Delphinus delphis* | short-beaked common dolphin |
| *Dugong dugon* | dugong |
| *Enhydra lutris* | sea otter |
| *Erignathus barbatus* | bearded seal |
| *Eschrichtius robustus* | gray whale |
| *Eubalaena australis* | southern right whale |
| *Eubalaena glacialis* | North Atlantic right whale |
| *Eubalaena japonica* | North Pacific right whale |
| *Eumetopias jubatus* | Steller sea lion |
| *Globicephala macrorhynchus* | short-finned pilot whale |
| *Globicephala melas* | long-finned pilot whale |
| *Hydrodamalis gigas* | Steller's sea cow |
| *Hydrurga leptonyx* | leopard seal |
| *Hyperoodon ampullatus* | northern bottlenose whale |
| *Kogia sima* | dwarf sperm whale |
| *Lagenorhynchus obliquidens* | Pacific white-sided dolphin |
| *Lagenorhynchus obscurus* | dusky dolphin |
| *Leptonychotes weddellii* | Weddell seal |
| *Lissodelphis borealis* | northern right whale dolphin |
| *Lobodon carcinophagus* | crabeater seal |
| *Lutra felina* | marine otter |
| *Megaptera novaeangliae* | humpback whale |
| *Mesoplodon densirostris* | Blainville's beaked whale |
| *Mesoplodon europaeus* | Gervais' Beaked whale |
| *Mirounga angustirostris* | northern elephant seal |
| *Mirounga leonina* | southern elephant seal |
| *Monachus monachus* | Mediterranean monk seal |
| *Monachus schauinslandi* | Hawaiian monk seal |
| *Monachus tropicalis* | Caribbean monk seal |
| *Monodon monoceros* | narwhal |
| *Neophoca cinerea* | Australian sea lion |
| *Neophocaena asiaeorientalis* | finless porpoise |
| *Neophocaena phocaenoides* | Indo-Pacific finless porpoise |
| *Neovison macrodon* | sea mink |
| *Odobenus rosmarus* | walrus |
| *Ommatophoca rossii* | Ross seal |
| *Orcinus orca* | killer whale |
| *Pagophilus groenlandicus* | harp seal |
| *Phoca vitulina* | harbour seal |
| *Phocarctos hookeri* | New Zealand sea lion |
| *Phocoena phocoena* | harbour porpoise |
| *Phocoena spinipinnis* | Burmeister's porpoise |
| *Phocoena sinus* | vaquita |
| *Phocoenoides dalli* | Dall's porpoise |

| | |
|---|---|
| *Physeter macrocephalus* | sperm whale |
| *Pontoporia blainvillei* | fransicana |
| *Pusa caspica* | Caspian seal |
| *Pusa hispida* | ringed seal |
| *Sousa chinensis* | Indo-Pacific humpbacked dolphin |
| *Stenella attenuata* | pantropical spotted dolphin |
| *Stenella coeruleoalba* | striped dolphin |
| *Stenella longirostris* | spinner dolphin |
| *Trichechus manatus* | West Indian manatee |
| *Trichechus manatus latirostris* | Florida manatee |
| *Trichechus senegalensis* | West African manatee |
| *Tursiops truncatus* | common bottlenose dolphin |
| *Ursus maritimus* | polar bear |
| *Zalophus californianus* | California sea lion |
| *Zalophus japonicus* | Japanese sea lion |
| *Zalophus wollebaeki* | Galápagos sea lion |
| *Ziphius cavirostris* | Cuvier's beaked whale |

# APPENDIX 3
# Glossary

**Abyssal** Relating to water depths of 3000–6000 m.

**Actiniaria** Order of Anthozoa comprising the sea anemones. Solitary polypoid animals that lack a hard exoskeleton.

**Actinopteri** Class of vertebrates that includes the sturgeons and teleosts.

**Ahermatypic** Non-reef-building (of scleractinian corals).

**Algae** Photosynthetic eukaryotes ranging from unicellular microalgae to large non-vascular organisms (macroalgae or seaweeds). Whilst 'algae' is a convenient term, it encompasses organisms with very different evolutionary histories and biology. Singular, alga.

**Allee effect** Decrease in per capita population growth rate at low population size, often due to the lower probability of mate encounters.

**Allelopathic substance** A chemical produced by an organism that has an inhibitory effect on other organisms.

**Amphipoda** Mostly small peracarid crustaceans with a laterally compressed body and ventral brood pouch; sandhoppers.

**Anadromous** Of fish that ascend rivers to spawn.

**Anaerobic** Living without free oxygen.

**Anchialine ecosystem** Special type of estuary occupying subterranean crevices and caverns on limestone and lava-flow shores.

**Annelida** Phylum of segmented worms that includes polychaetes, oligochaetes, and leeches.

**Anomura** Group of decapod crustaceans that includes the squat lobsters, king crabs, and hermit crabs.

**Anoxia** Depletion of dissolved oxygen ($< 1$ mg $L^{-1}$).

**Anthozoa** Class of Cnidaria that includes sea anemones and corals.

**Anthropocene** Recent period of geological time, at least since the industrial era, characterized by the global influence of human activities on ecosystems.

**Antipatharia** Order of Anthozoa comprising the black corals, with erect branching colonies with a dark-coloured axial skeleton, mainly in deep water.

**Aquaculture** Farming of aquatic organisms. Marine aquaculture (mariculture) mainly involves cultivation of seaweeds, molluscs, crustaceans, and fishes.

**Aragonite** Form of calcium carbonate important as skeletal material, such as in certain scleractinian corals, molluscs, and bryozoans.

**Area** Seabed beyond the limits of national jurisdiction.

**Arribada** Synchronized mass nesting of turtles, as in olive and Kemp's ridleys; Spanish for 'arrival'.

**Arrow worms** See Chaetognatha.

**Arthropoda** Phylum of invertebrates that have an exoskeleton, segmented body, and jointed appendages. Includes insects, horseshoe crabs, sea spiders, and crustaceans.

**Artificial reef** Structure placed or constructed on the seabed, usually to provide reef-like habitat to enhance local biodiversity and particularly to attract fish. May be purpose-built (e.g. from concrete modules) or utilize disused structures that are sunk on site (e.g. scuttled vessels).

**Ascidiacea** Class of tunicates, solitary or colonial, that are sessile suspension feeders; sea squirts.

**Asteroidea** Class of echinoderms with five (or sometimes many) arms; mostly predators of benthic invertebrates; sea stars or starfish.

**Atoll** Ring-shaped coral reef or string of low-lying coral islands enclosing a lagoon.

**Autotrophs** Organisms that, as primary producers, synthesize organic compounds using light (photosynthesis) or a chemical reaction (chemosynthesis) as their energy source (*cf.* heterotrophs).

**Bacteria** Prokaryotic microorganisms important as decomposers of organic matter, as primary producers (using photosynthesis or chemosynthesis), and as pathogens.

**Baleen** Bristly plates of keratin used by baleen whales for filter feeding.

**Barnacle** See Cirripedia.

**Bathyal** Relating to water depths of about 200–3000 m; continental slope depths.

**Benthos** Organisms that live on or in the seabed.

**Bioaccumulation** Build-up in an organism's tissues of substances that are not readily excreted (e.g. heavy metals, organochlorines).

**Biodeposition** Deposition of material on the seabed by benthic organisms, in particular faeces and pseudofaeces.

**Biodiversity (biological diversity)** Variability among living organisms, including diversity within species, between species, and ecosystems.

**Biogenic** Produced by living organisms, e.g. biogenic habitats such as maerl beds, mussel banks, bryozoan reefs.

**Biogeography** Study of the geographical distribution of organisms.

**Bioluminescence** Light produced by living organisms.

**Biomass** Weight of living matter for a given area or volume.

**Biome** Marine biome can be used to refer to the entire marine environment (as opposed to the terrestrial biome) or to a group of biogeographic provinces with common oceanographic processes.

**Biosphere** Entire space inhabited by living organisms.

**Biotope** A space with particular environmental conditions that is sufficiently uniform to support a characteristic assemblage of organisms; often used more or less synonymously with habitat.

**Bivalvia** Class of molluscs characterized by a shell of two valves and mostly suspension feeders; includes mussels, oysters, and clams.

**Black corals** See Antipatharia.

**Bleaching** Loss of zooxanthellae from corals causing them to turn white.

**Boreal** Of northern cold-temperate/subarctic latitudes.

**Bottom-up control** Refers to food webs where basal resources, such as nutrient supply, have a strong control on community structure (*cf.* top-down control).

**Brachiopoda** Phylum of benthic invertebrates characterized by a bivalve shell and a ciliated lophophore for suspension feeding; lamp shells.

**Brachyura** Group of decapod crustaceans comprising the true crabs.

**Brackish** Of water that has a salinity between that of freshwater and seawater.

**Brittle stars** See Ophiuroidea.

**Brown algae** See Phaeophyceae.

**Bryozoa** Phylum of benthic invertebrates that form colonies, often calcareous, of suspension-feeding, lophophorate individuals; moss animals.

**Bycatch** Incidental take by a fishery of non-target organisms or, more broadly, catch that is unused or unmanaged.

**Calcareous** Containing or consisting of calcium carbonate.

**Calcite** Form of calcium carbonate used by many organisms as skeletal material.

**Caridea** Shrimps and prawns.

**Carnivora** Order of mammals that includes the pinnipeds (sea lions, seals, walrus), otters, and polar bear.

**Carrageenans** Polysaccharides found in red seaweeds that are widely used in food and other industries.

**Cartilaginous fishes** See Chondrichthyes.

**Catadromous** Of fish that move from freshwater to the sea to spawn.

**Catch per unit effort** Catch of a fishery divided by the amount of effort used to harvest the catch (e.g. number of vessels, duration of fishing). Changes in CPUE are an indication of the level of exploitation.

**Cephalopoda** Class of molluscs that includes the squids and octopuses.

**Cetacea** Group of mammals comprising the Mysticeti (baleen whales) and Odontoceti (toothed whales, dolphins, porpoises).

**Chaetognatha** Phylum of planktonic predators; arrow worms.

**Chelicerata** Group of arthropods containing the horseshoe crabs (Merostomata), sea spiders (Pycnogonida), and mites (Acari).

**Chemosynthesis** Primary production by microorganisms that use the oxidation of inorganic substrates (often hydrogen sulphide) as an energy source (*cf.* photosynthesis).

**Chitons** See Polyplacophora.

**Chlorophyta** Phylum of plants that includes the green seaweeds. Green on account of the chlorophyll pigments they contain.

**Chondrichthyes** Cartilaginous fishes that include the sharks, rays, and chimaeras.

**Chordata** Phylum that includes the tunicates and vertebrates.

**Cirripedia** Group of crustaceans, mostly sessile suspension feeders enclosed in calcareous plates; barnacles.

**Class** Taxonomic group between phylum and order: related orders comprise a class, and related classes comprise a phylum.

**Clupeidae** Family of fishes that includes the herrings, sardines, pilchards, and sprats.

**Cnidaria** Phylum that includes the sea anemones, corals, hydrozoans, and jellyfishes and characterized by tentacles with stinging cells and a polyp and/or medusa body form.

**Coastal waters** Sea areas significantly influenced by their proximity to land; in general, marine and estuarine waters seaward of the high-tide line to the shelf break.

**Coastal zone** Zone of interface between terrestrial and marine environments encompassing areas of hinterland, shore, and adjacent sea, with landward and seaward limits taken as the respective reach of significant marine and terrestrial influences.

**Coccolithophores** Unicellular algae covered in calcareous plates (coccoliths).

**Comb jellies** See Ctenophora.

**Commercial extinction** When a stock has been reduced to such a level that exploiting it is no longer profitable.

**Community** Populations of species that occur together and interact (e.g. through competition and trophic relationships).

**Connectivity** Degree of exchange among subpopulations of a species (e.g. between marine protected areas).

**Contamination** Anthropogenic introduction of substances or energy into the environment at concentrations above natural levels but without producing significant adverse effects (*cf.* pollution).

**Continental margin** Continental shelf and slope comprising the submerged edge of a continent, consisting of continental crust and distinct from seaward oceanic crust.

**Continental shelf** Gently sloping seabed extending from the coastline to the shelf break, typically to a water depth of 100–200 m.

**Continental slope** Sloping seabed extending from the edge of the continental shelf, or shelf break, seaward to about 3000 m water depth and the abyssal plain.

**Copepoda** Group of crustaceans particularly important in the zooplankton as consumers of phytoplankton.

**Coralligenous formation** A biogenic substratum produced by the accumulation of calcareous encrusting algae growing in low light conditions.

**Coralline algae** Red algae that have a calcareous skeleton; includes maerl.

**Crinoidea** Class of echinoderms with long arms and tube feet used for suspension feeding; feather stars, sea lilies.

**Crustacea** Subphylum of arthropods that includes barnacles, copepods, amphipods, krill, shrimps, crabs, and lobsters.

**Cryptic species** Species that are morphologically indistinguishable.

**Ctenophora** Phylum of gelatinous, planktonic predators; comb jellies.

**Cyanobacteria** Group of photosynthetic bacteria; blue-green bacteria.

**Decapoda** Order of crustaceans that includes the shrimps, prawns, hermit crabs, true crabs, and lobsters.

**Deep scattering layer** Aggregations of mesopelagic fish and large zooplankton that form distinct sonar-reflecting layers.

**Demersal** Of fish that are bottom-living.

**Deposit feeder** Animal that feeds on particulate organic matter in sediment, such as benthic microalgae, bacteria, plankton detritus, macrophyte debris.

**Detritus** Particulate material from the decomposition of organisms.

**Diadromous** Of fish that migrate between the sea and freshwater to complete their life cycle.

**Diatom** Unicellular alga with a siliceous test, important in pelagic and benthic habitats.

**Dinoflagellate** Planktonic, flagellated protist; heterotrophic and/or photosynthetic.

**Echinodermata** Phylum of invertebrates characterized by pentamerous symmetry, calcareous plates and spines, and a water vascular system with tube feet. Includes sea stars, brittle stars, sea urchins, sea cucumbers, and feather stars.

**Echinoidea** Class of echinoderms with a globular to flattened spiny test; sea urchins, heart urchins, sand dollars.

**Ecological extinction** Point at which a species that is decreasing in abundance ceases to play a significant role in its ecosystem (e.g. as a key herbivore or predator).

**Ecoregions** Small-scale biogeographical units that are areas of relatively homogeneous species composition, clearly distinct from adjacent systems.

**Ecosystem** Community of organisms plus its abiotic environment that together comprise an interacting system in terms of the flow of materials and energy.

**Ecosystem-based management** An integrated approach to management that considers the entire ecosystem, including humans.

**Ecosystem engineer** Organism that modifies or constructs habitat and thereby influences associated species. Ecosystem engineers include species that rework seabed sediments and reef-builders.

**Ecosystem services** Benefits humans derive from ecosystems (e.g. seafood, coastal protection, recreation).

**Ectotherm** Animal whose body temperature is mainly governed by the temperature of its environment; 'cold-blooded' (*cf.* endotherm).

**El Niño-Southern Oscillation** A large-scale ocean-atmosphere interaction involving the movement of warm surface water in the tropical Pacific Ocean. The build-up every few years of warm water in the eastern tropical Pacific (El Niño) suppresses upwelling along this coast and has global effects on oceanographic conditions.

**Endemic** Having a geographical distribution restricted to a particular region.

**Endotherm** Animal that can maintain a constant body temperature that is essentially independent of the temperature of its environment; 'warm-blooded' (*cf.* ectotherm).

**Epibenthos/epifauna** Organisms/animals that live on the surface of the seabed.

**Epipelagic** Upper zone of the pelagic environment, from the ocean surface to a depth of about 200 m; includes the euphotic zone.

**Epiphyte** Alga living on the surface of a seaweed or plant.

**Equilibrium species** Species characteristic of mature communities; *K*-selected species.

**Estuary** Typically an inlet of the sea markedly affected by freshwater inflow; usually a river mouth.

**Euhaline** Fully saline water, with a salinity of 30–40.

**Eukaryotes** Organisms with cells that have discrete nuclei and organelles; includes protists, plants, and animals (*cf.* prokaryotes).

**Euphausiacea** Planktonic, filter-feeding crustaceans up to several cm long; krill.

**Euphotic zone** Uppermost zone of the water column that receives enough light for net photosynthesis; up to about 200 m depth in clear ocean water but much less in more turbid coastal water.

**Euryhaline** Of organisms tolerant of a wide range of salinity (*cf.* stenohaline).

**Eustatic** Relating to global sea level and the change in sea level due to a change in the volume of seawater (as opposed to local changes in sea level due to tectonic movements).

**Eutrophic** Of waters that are well supplied with nutrients.

**Eutrophication** Nutrient enrichment resulting in excessive primary production, build-up of organic detritus, and a depletion in dissolved oxygen concentration.

**Exclusive economic zone (EEZ)** Zone up to 200 nm from the coast over which a coastal state can claim sovereign rights, including use of its natural resources.

**Extinction** Disappearance of a species (or other taxon). Usually applied when the extinction is global but also used for local extinction (extirpation) and ecological extinction.

**Extirpation** Local extinction of a species.

**Family** Taxonomic rank between genus and order: related genera comprise a family and related families comprise an order.

**Fjord** Deep, narrow inlet formed as a result of glacial erosion with estuarine circulation.

**Flag of convenience** Registering a vessel in a country that is not that of the vessel's owner. Flag states tend to have little regard for marine conservation.

**Flagship species** Species used to raise awareness, public support, and funding for conservation programmes.

**Focal species** Species advocated, for ecological or social reasons, for the management and conservation of natural environments; includes indicator, keystone, and flagship species.

**Food web** Trophic network of a community.

**Foraminifera** Amoeboid protists, benthic or planktonic, with an agglutinated or a calcareous (often chambered) test.

**Front** Boundary between two distinct water masses that typically differ in terms of their physico-chemical properties (e.g. temperature, salinity) and biology.

**Gastropoda** Class of molluscs that includes limpets, abalones, snails, sea slugs.

**Gene** A unit of hereditary information composed of DNA on a chromosome that codes for a specific characteristic of an individual.

**Gene flow** Exchange of genetic material between populations, e.g. in the connectivity of marine protected areas.

**Genera** Plural of genus.

**Genome** The total genetic information of an organism.

**Genus** Taxonomic group between species and family: related species comprise a genus and related genera (plural of genus) comprise a family; plural, genera.

**Ghost fishing** Fishing that continues to occur by gear that is lost or abandoned at sea.

**Global warming** Rise in the Earth's temperature attributable to increased levels of greenhouse gases in the atmosphere as a result of human activities.

**Gondwana** The ancient supercontinent that broke up to give the present-day South America, Africa, Arabia, India, Australia, New Zealand, and Antarctica.

**Green algae** See Chlorophyta.

**Greenhouse gases** Gases such as carbon dioxide and methane that trap infrared radiation, thereby warming the Earth's surface and lower atmosphere.

**Gross primary production** Total primary production not taking account of the loss by respiration

**Gyre** Major surface flow within an ocean basin. Gyres move clockwise in the Northern Hemisphere and anticlockwise in the Southern Hemisphere.

**Habitat** Place where an organism normally lives, particularly in relation to the suite of environmental factors that characterize that locality (e.g. a rocky shore habitat).

**Hadal** Relating to water depths of 6000–11 000 m (mainly in ocean trenches).

**Halogenated** Of compounds that contain halogen atoms (e.g. chlorine, bromine).

**Head-starting** Rearing hatchling turtles in captivity to a size that may increase their survival rate when released.

**Heavy metal** Metallic elements of relatively high density but typically applied to those that are potentially toxic (e.g. mercury, lead, cadmium).

**Hermatypic** Reef-building.

**Heterotrophs** Organisms that depend on complex organic substances for their nutrition, i.e. animals and many bacteria (*cf.* autotrophs).

**Heterozygous** Having two different alternative forms of a gene (alleles) at a particular locus of a chromosome pair.

**High seas** Ocean areas beyond national jurisdiction; international waters.

**Holothuroidea** Class of echinoderms, usually elongate, soft bodied, and with mouth surrounded by feeding tentacles for deposit or suspension feeding; sea cucumbers.

**Hydroids** Hydrozoans consisting typically of a branched colony of polyps attached to a substratum and forming small bushy or feathery growths; sea firs.

**Hydrothermal vents** Volcanically active sites, usually in the deep sea, where geothermally heated water rich in leachates (particularly sulphides) vents from the seafloor.

**Hydrozoa** Class of Cnidaria that includes the hydroids, milleporid corals (fire corals), stylasterid corals, and siphonophores.

**Hypersaline** Water of high salinity (above 40).

**Hypoxia** Low concentration of dissolved oxygen ($< 2$ mg $L^{-1}$).

**Indicator species** Species whose presence denotes the composition or condition of a particular habitat, community or ecosystem.

**Infauna** Benthic animals that live within seabed sediments.

**Intertidal zone** Zone between the levels of the highest and lowest tides; shore.

**Introduced species** Species introduced to a location outside its natural geographical range. Anthropogenic introductions may be accidental (e.g. in ballast water) or intentional (e.g. for fisheries). Other terms used: alien species, exotic species, non-indigenous species.

**Invasive species** Introduced species that has an adverse ecological impact.

**Invertebrates** Animals that lack a vertebral column, i.e. all animals except those of the subphylum Vertebrata.

**Isomers** Compounds with the same molecular formula but with different atomic arrangements and hence different properties, e.g. polychlorinated biphenyls.

**Isopoda** Small to medium-sized peracarid crustaceans, often dorso-ventrally flattened, with a brood pouch, and mostly benthic; sea slaters.

**Isotherm** Line that connects points of equal temperature.

***K*-selected species** Species characteristic of predictable environments with life-history traits that render them vulnerable to disturbance and exploitation (e.g. large body size, long life span, slow growth, delayed maturity); equilibrium species (*cf.* *r*-selected species).

**Kelp** Large brown seaweeds, mostly of the order Laminariales.

**Keystone species** Species that have a disproportionate influence on community structure by their activity; applied usually to trophic activity (e.g. a key predator).

**Krill** See Euphausiacea.

**Lagoon** Shallow body of saline water separated from the adjacent sea by a sand or shingle barrier; also the body of water enclosed by an atoll.

**Lessepsian species** Alien species in the Mediterranean Sea that have come from the Red Sea via the Suez Canal.

**Lichen** Mutualistic association between a fungus and an alga or a cyanobacterium.

**Lipophilic** Having an affinity for lipids (e.g. organochlorines that accumulate in fatty tissues).

**Littoralization** The increasing proportion of the human population living in coastal areas.

**Longline** Fishing line with baited hooks (up to thousands) attached at regular intervals and set for pelagic or demersal target species.

**Lophophore** Structure of ciliated tentacles used for suspension feeding in brachiopods and bryozoans.

**Macroalgae** See Seaweeds.

**Macrobenthos/macrofauna** Macroscopic benthic organisms/animals; those that would be retained on a 0.5 mm mesh.

**Macrophytes** Aquatic vascular plants (notably seagrasses, mangroves, and saltmarsh plants) and macroalgae.

**Maerl** Unattached branching coralline algae that form calcareous gravelly deposits known as rhodolith or maerl beds.

**Mangrove** Coastal wetland community of tropical to subtropical latitudes dominated by trees and shrubs.

**Mantis shrimps (Stomatopoda)** Order of predatory crustaceans with raptorial claws.

**Mariculture** Marine aquaculture.

**Marine protected area** Defined geographical area where human activities are restricted, at least to some extent, primarily so as to help conserve biodiversity and/or for fisheries management.

**Maximum sustainable yield** Maximum catch that can be taken from a stock whilst still maintaining its productive capacity.

**Megafauna** Large animals; used in benthic ecology for those visible in bottom photographs and taken by trawls.

**Meiofauna** Small benthic invertebrates and large protists (notably foraminiferans) that pass through a 0.5 mm mesh but are retained on a mesh of around 30–60 μm (*cf.* macrofauna).

**Mesopelagic zone** Pelagic zone between 200 and 1000 m water depth.

**Metapopulation** Group of populations of the same species that are spatially separate but interact to some extent.

**Methane hydrate** Ice-like solid containing methane formed at low temperature and high pressure, especially in continental slope sediments.

**Microalgae** Unicellular algae in pelagic and benthic environments, such as diatoms and photosynthetic dinoflagellates.

**Microbenthos** Microscopic benthic organisms; benthic protists and bacteria.

**Mid-ocean ridge** Submarine mountain range between two tectonic plates that are moving apart and where new seafloor is being formed; sites of hydrothermal vents.

**Minimum viable population** The smallest isolated population that has a particular probability of remaining extant.

**Mitochondrial DNA** DNA located in the mitochondria (organelles in the cells of eukaryotes) and maternally inherited.

**Molluscs** Phylum that includes the limpets, bivalves, gastropods, and cephalopods.

**Muro-ami** Destructive method of fishing used on coral reefs where swimmers move in a line across a reef, pounding the reef with stone weights to drive fish into a net.

**Mysids** Order of peracarid crustaceans; opossum shrimps.

**Mysticeti** Baleen whales.

**Nematoda** Phylum of free-living and parasitic unsegmented worms; roundworms. They usually dominate the meiofauna.

**Nemertea** Phylum of unsegmented worms with an eversible proboscis for catching prey; ribbon worms.

**Neogastropoda** Order of gastropods with a shell characterized by a well-developed canal for the incurrent siphon; carnivores and scavengers. Includes whelks and cone snails.

**Neritic** Relating to coastal waters; waters overlying the continental shelf (*cf.* oceanic).

**Niche** Place where an organism lives and its role in that habitat; for example, parrotfishes of coral reef habitat may have various roles such as bioeroder, scraping herbivore, or grazer.

**Net primary production** Total primary production minus that used in respiration.

**No-take MPA** Marine protected area where no extractive activities are allowed.

**Ocean acidification** Decline in seawater pH due to increased carbon dioxide concentration in the atmosphere and its uptake by the oceans.

**Ocean trench** Narrow, elongate depression in the seafloor formed at a subduction zone, typically at water depths greater than 6000 m.

**Oceanic** Relating to waters seaward of the shelf break.

**Octocorallia** Subclass of Anthozoa that includes the soft corals, gorgonians, sea pens, and blue coral.

**Octopuses** Order of Cephalopoda comprising carnivorous molluscs lacking a shell and with eight arms.

**Odontoceti** Toothed whales, dolphins, and porpoises.

**Oligochaetes** Group of annelid worms with small macrofaunal and meiofaunal representatives (but best known as earthworms).

**Oligotrophic** Of waters with low nutrient concentrations and thus low productivity.

**Ontogeny** The life history of an individual (*cf.* phylogeny).

**Ophiuroidea** Class of echinoderms with a distinct central disc and long slender arms; brittle stars, basket stars.

**Opportunistic species** See *r*-selected species.

**Organochlorines (chlorinated hydrocarbons)** Organic compounds containing chlorine, mostly synthetic and not readily degraded. Includes organochlorine pesticides (e.g. DDT) and polychlorinated biphenyls.

**Organometal** Organic compound with carbon to metal links, e.g. tributyltin (TBT).

**Osteichthyes** Bony fishes including teleosts.

**Ostracods** Class of small crustaceans with a bivalved carapace enclosing the body; seed shrimps.

**Otoliths** Ear bones of vertebrates. Annual growth increments of the otoliths of bony fishes are used for ageing.

**Overfishing** Exploitation that exceeds a species' regenerative capacity.

**Oviparous** Of species that have internal fertilization, but where the eggs then develop and hatch externally, as in marine turtles and birds (but sometimes applied more widely to include species that spawn) (*cf.* ovoviviparous).

**Ovoviviparous** Of species that have internal fertilization and where the eggs, nourished by the egg yolk rather than directly by the mother, develop and hatch internally, as in many sharks, and most sea snakes (*cf.* oviparous).

**Ozone** Form of oxygen with three oxygen atoms in the molecule ($O_3$). The ozone layer of the stratosphere shields the Earth from the sun's UV radiation.

**Oxygen minimum zone** Zone below the thermocline, typically at several hundred metres water depth, where bacterial degradation of accumulated particulate organic matter results in oxygen depletion.

**Pelagic** Living in the water column (*cf.* benthic).

**Penaeidae** Family of shrimps or prawns that are important in wild-caught fisheries and aquaculture.

**Peracarida** Large group of small to medium-size crustaceans that brood their young; includes the amphipods and isopods.

**Persistent organic pollutants (POPs)** Organic compounds, mostly synthetic and halogenated, resistant to breakdown, which tend to bioaccumulate and are potentially toxic; includes organochlorines.

**pH** Measure of acidity/alkalinity of an aqueous solution from 0 (acid), 7 (neutral), to 14 (alkaline) (as the hydrogen ion concentration on a logarithmic scale). Seawater normally has a pH of about 8.

**Phaeophyceae** Class of eukaryotes comprising the brown algae or brown seaweeds, including kelps. Brown on account of an accessory photosynthetic pigment they contain.

**Phylogeny** The evolutionary history of a taxon (*cf.* ontogeny).

**Phytoplankton** Microscopic planktonic organisms that photosynthesize, including diatoms, dinoflagellates, and cyanobacteria.

**Photosynthesis** Primary production that uses sunlight energy to produce carbohydrates from carbon dioxide and water in the presence of chlorophyll (and releases oxygen) (*cf.* chemosynthesis).

**Phylum** Taxonomic rank between kingdom and class and a way of categorizing organisms that share the same basic body plan (e.g. Cnidaria, Mollusca, Arthropoda, Echinodermata); plural, phyla.

**Pinnipedia** Group of mammals comprising the sea lions, seals, and walrus.

**Plankton** Mostly small to microscopic organisms living in the water column that drift passively or have only weak swimming ability relative to water currents, including bacteria, phytoplankton, and zooplankton.

**Platyhelminthes** Phylum of worms that includes free-living flatworms and parasitic species (flukes and tapeworms).

**Pneumatophore** Aerial root of mangroves for gas exchange that emerges vertically above the sediment surface.

**Polar** Areas within the Arctic and Antarctic regions (north and south of 60°).

**Pollution** Anthropogenic introduction of substances or energy into the environment resulting in harm to an ecosystem, including its living resources (*cf.* contamination).

**Polychaeta** Large, diverse class of mostly benthic annelid worms; bristleworms.

**Polychlorinated biphenyls (PCBs)** Synthetic organochlorine compounds, very stable and usually toxic, used in a wide range of industries.

**Polygyny** Reproductive behaviour where a male mates with several females during a breeding season (e.g. in fur seals and sea lions).

**Polyplacophora** Class of molluscs characterized by a shell of eight plates; mostly grazers on intertidal and shallow subtidal rocky substrata; chitons.

**Polyploid organisms** Organisms with extra sets of chromosomes (e.g. as induced in some aquaculture species).

**Population viability analysis** Methods for computing the probability of extinction within a certain timeframe based on a species' population parameters and the various processes and forces impacting them.

**Porifera** Phylum comprising the sponges. Body essentially a system of canals and chambers for suspension feeding and skeletal elements.

**Precautionary principle** The precept that where an activity might cause significant environmental harm, preventative measures should not be delayed because of scientific uncertainty.

**Primary production** Synthesis of organic compounds by living organisms using carbon dioxide as the carbon source and sunlight (photosynthesis by phytoplankton and macrophytes) or a chemical reaction (chemosynthesis by e.g. sulphur bacteria) as the energy source.

**Primary productivity** Rate of primary production.

**Prokaryotes** Unicellular organisms that lack discrete nuclei and organelles; includes the bacteria (*cf.* eukaryotes).

**Protandrous** Relating to a hermaphrodite organism that produces male gametes and later switches to producing female gametes.

**Protists** Unicellular eukaryotes, e.g. ciliates, foraminiferans, diatoms, dinoflagellates.

**Provinces** Biogeographical units that are large areas defined by the presence of distinct biotas that have at least some cohesion over evolutionary time frames.

**Pseudofaeces** Particulate material collected by suspension feeders as possible food but rejected before ingestion; applied mainly to bivalves.

**Pteropods** Group of pelagic gastropods.

**Pycnogonida** Group of chelicerate arthropods characterized by small body, eight long legs, and suctorial proboscis; sea spiders.

***r*-selected species** Species with characteristics enabling them to exploit unpredictable environments and rapidly colonize newly available habitat (e.g. small body size, rapid reproduction, high dispersal); opportunistic species (*cf.* *K*-selected species).

**Radionuclide (radioisotope)** Unstable isotope that undergoes radioactive decay.

**Realms** Biogeographical units that are very large regions of coastal, benthic, or pelagic ocean across which biotas are internally coherent at higher taxonomic levels.

**Recruitment** Addition of a new cohort to a population, in particular those juveniles that survive and are subsequently detected as a new influx.

**Red algae** See Rhodophyta.

**Red List** List complied by the IUCN of the conservation status of species. Uses certain criteria to assess extinction risk, which includes categories for threatened species.

**Reef** Area of seabed of raised relief that is rocky or biogenic.

**Regime shift** A major, relatively abrupt shift in ecosystem structure and function to a new persistent regime.

**Remineralization** Breakdown of organic matter to inorganic forms, with release of nutrients into the sea.

**Remipedia** Class of (probably ancient) predatory Crustacea found in anchialine caves, characterized by numerous trunk segments with swimming appendages.

**Remote sensing** Obtaining environmental data by using remote sensors on aircraft or satellites (e.g. sea surface temperature, chlorophyll).

**Remotely operated vehicle (ROV)** Unoccupied, underwater vehicle equipped with cameras and sampling devices that is operated and manoeuvred from a ship via a cable.

**Reptiles** Class of vertebrates that includes turtles, snakes, and crocodiles.

**Rhodophyta** Phylum of plants comprising the red algae or red seaweeds. Red on account of accessory photosynthetic pigments they contain; includes coralline algae.

**Ribbon worms** See Nemertea.

**Rookery** Breeding or nesting site of colonial seabirds, pinnipeds, and turtles.

**Rorquals** Baleen whales of the family Balaenopteridae; includes the fin, sei, blue, and minke whales.

**Roundworms** See Nematoda.

**Runoff** Freshwater that drains off land into coastal waters; may carry pollutants (e.g. nutrients, pesticides).

**Salinity** Measure of the concentration of dissolved salts in seawater. Open ocean seawater has a salinity of about 35 (salinity has no units), equivalent to about 35 g of dissolved salts per kg of water.

**Salps** Pelagic tunicates, solitary or colonial; often important in the zooplankton.

**Saltmarsh** Coastal wetland vegetated by herbs, grasses, or low shrubs, occurring mainly at sheltered upper shore sites at temperate latitudes.

**Scleractinia** Order of Anthozoa comprising the true or stony corals. The main contributors to tropical coral reefs but also important in cold-water biogenic habitats.

**Scombroids** Fishes belonging to the suborder that includes the tunas, mackerels, and bonitos.

**Scyphozoa** Class of cnidarians comprising the jellyfish.

**Sea anemones** See Actiniaria.

**Seabirds** Birds that normally feed at sea and breed on offshore islands or at coastal sites.

**Seagrasses** Flowering plants of four families, most with blade-like leaves, that root in shallow subtidal to intertidal environments.

**Seamount** Discrete undersea mountain (usually volcanic) that rises steeply from the deep seafloor with a height of at least 100 m but not reaching the sea surface.

**Sea star** See Asteroidea.

**Sea urchin** See Echinoidea.

**Seaweeds** Macroscopic marine algae (macroalgae) including green (Chlorophyta), brown (Phaeophyceae), and red (Rhodophyta) algae.

**Serpulid polychaetes** Family of sessile suspension-feeding polychaetes that build calcareous tubes.

**Sessile** Attached to a substratum.

**Shelf break** Seaward edge of the continental shelf where it gives way to the steeper gradient of the continental slope.

**Shellfish** Invertebrates with an exoskeleton that are taken as seafood, particularly various molluscs and crustaceans.

**Shifting baseline** Concept that succeeding generations have an altered perception of what constitutes a healthy ecosystem, each in turn accepting a slightly more degraded system as normal.

**Shore** Intertidal zone.

**Sill** Submarine ridge separating two deeper water bodies, such as the sill at the mouth of a fjord.

**Siphonophores** Group of planktonic, colonial hydrozoans. Most float at the sea surface and have long tentacles with stinging cells to capture prey (e.g. Portuguese man-of-war).

**Sirenia** Order of mammals that includes the dugong and manatees.

**Slope** Continental slope.

**Sonar (echo-sounding)** Measuring water depth from the time interval between the emission of an acoustic pulse and the return of the reflected sound waves (recorded by a transducer on the vessel's hull). Multibeam echo-sounding, for seabed mapping, uses hundreds of beams projected sidewards enabling a wide swath of seabed to be surveyed along each track line.

**Spawning** Reproduction where eggs and sperm are released into the surrounding water for external fertilization, as in most marine invertebrates and bony fishes.

**Species** A group of individuals that can interbreed in nature and produce fertile offspring. (Such a definition, suitable for most organisms of marine conservation interest, is not universally applicable.) Species is the taxon below genus and regarded as the basic unit of biological classification. Subspecies may, however, be recognized (e.g. of the blue whale). Species have a two-part scientific name: the first part denoting the genus and the second part denoting the species within that genus.

**Species diversity** Measure of the number of species in a sample or community (richness) and their relative abundance (evenness) but often used just to denote species richness.

**Species richness** Number of species in a collection, community, or particular area.

**Sponges** See Porifera.

**Squids** Order of pelagic carnivorous cephalopods with eight arms and two tentacles.

**Stenohaline** Of organisms tolerant of a narrow range of salinity (*cf.* euryhaline).

**Stochasticity** In ecology, stochasticity is used for unpredictable fluctuations in environmental conditions. This becomes increasingly important as population size diminishes.

**Stock** A more or less discrete subpopulation of a fisheries species.

**Stony corals** See Scleractinia.

**Stromatolite** Laminated structure accreted as a result of a benthic microbial community trapping and binding sediment.

**Stylasteridae** Family of hydrozoan corals.

**Subduction zone** Area where two tectonic plates converge and one plate descends beneath the other. Ocean trenches are typically the result of subduction.

**Submarine canyon** Canyon incised in the continental slope acting as a conduit for sediment to be transported down the slope.

**Submersible** Crewed underwater vehicle deployed from a surface support vessel (*cf.* remotely operated vehicle).

**Suspension feeder** Animal that feeds on particulate organic matter in the water column, such as phytoplankton, bacteria, organic detritus.

**Symbiosis** Any close relationship between two organisms of different species where one or both partners benefit from the association. Includes mutualism where both benefit (e.g. zooxanthellae and corals) and parasitism where one benefits whereas the other (host) is harmed.

**Tanaidacea** Group of small peracarid crustaceans of benthic habitats with a pair of thoracic appendages bearing pincers (i.e. chelate); high deep-sea diversity.

**Taxon** A taxonomic group (e.g. phylum, family, species); plural, taxa.

**Tectonic plates** Plates composed of crust and uppermost mantle that together comprise the Earth's outer shell.

**Teleosts** Predominant group of bony fish.

**Temperate** Zone between the tropics and the polar circles.

**Territorial sea** Zone of up to 12 nm adjacent to the coast that the coastal state can claim as sovereign territory.

**Thermocline** Layer in the water column where the temperature changes sharply with depth.

**Threatened species** Species in the IUCN Red List categorized as critically endangered, endangered, or vulnerable.

**Tides** Periodic rise and fall of the sea surface resulting from gravitational attraction of the moon and sun on the Earth.

**Top-down control** Refers to food webs where abundance of high trophic level animals have a strong control on community structure (*cf.* bottom-up control).

**Trade winds** Persistent winds blowing from the subtropics towards the equator, from the north-east in the Northern Hemisphere, and from the south-east in the Southern Hemisphere.

**Tragedy of the commons** Use of a common resource where individuals gain a benefit at the expense of the group as a whole (e.g. open access fisheries and discharge of pollutants).

**Trophic** Relating to nutrition.

**Trophic cascade** The propagation of strong interactions between three or more trophic levels, typically where loss of a predator enables a herbivorous prey species to increase in abundance, which in turn suppresses the abundance of a primary producer.

**Trophic level** Position an organism occupies in a food web relative to its source of nutrition, between low trophic level organisms (e.g. primary producers) to high trophic level organisms (e.g. predators).

**Tropical** Zone between the tropics of Cancer and Capricorn (~23° 30' N and S).

**Tsunami** Wave generated by a submarine earthquake. Its increased amplitude on reaching shallow water can be highly destructive.

**Tube feet** Tubular extensions of the water vascular system of echinoderms used for locomotion and feeding.

**Tunicata** Subphylum of chordates that includes the benthic ascidians and pelagic salps; may be solitary or colonial.

**Ultraviolet** Solar radiation beyond the violet end of the visible light spectrum (wavelengths of 10–400 nanometres).

**Upwelling** Upward movement of deep (usually cold, nutrient-rich) water to the surface. In coastal upwelling, water pushed away from the coast as a wind-driven surface current is replaced by upwelled water.

**Vertebrates** Subphylum of the Chordata that includes fishes, reptiles, birds, and mammals.

**Waders** Birds that typically feed on the shore and migrate to inland breeding sites; shorebirds.

**Water mass** Water body distinguishable by its physico-chemical characteristics (e.g. temperature, salinity, nutrient concentrations).

**Wetland** Low-lying area, submerged or periodically inundated by water, where rooted plants are typically conspicuous.

**Zonation** Occurrence of species in distinguishable bands related to environmental gradients (e.g. gradients related to water depth, tidal emersion, a pollutant).

**Zooplankton** Animal plankton (*cf.* phytoplankton).

**Zooxanthellae** Single-celled algae that live as symbionts in coral polyps (and certain other invertebrates). They provide nutrition to the polyps through photosynthesis and enhance calcification of the coral skeleton.

# INDEX

Page numbers in *italics* refer to figures and tables

Printed in the United States
By Bookmasters